SCHAUM'S OUTLINE OF

THEORY AND PROBLEMS

OF

STRENGTH OF MATERIALS

Fourth Edition

•

WILLIAM A. NASH, Ph.D.
Professor of Civil Engineering
University of Massachusetts

•

SCHAUM'S OUTLINE SERIES

McGRAW-HILL

New York San Francisco Washington, D.C. Auckland Bogotá Caracas Lisbon London Madrid Mexico City Milan Montreal New Delhi San Juan Singapore Sydney Tokyo Toronto

This book is dedicated by the author to his parents, William A. Nash and Rose Nash, for their years of patient guidance toward his career.

WILLIAM A. NASH is Professor of Civil Engineering at the University of Massachusetts, Amherst. He received his B.S. and M.S. from the Illinois Institute of Technology and his Ph.D. from the University of Michigan. He served as Structural Research Engineer at the David Taylor Research Center of the Navy Department in Washington, D.C., and was a faculty member at the University of Florida for 13 years prior to his present affiliation. He has had extensive consulting experience with the U.S. Air Force, the U.S. Navy Department, Lockheed Aerospace Corp., and the General Electric Co. His special areas of interest are structural dynamics and structural stability.

Schaum's Outline of Theory and Problems of
STRENGTH OF MATERIALS

6 7 8 9 10 11 12 13 14 15 16 17 18 19 20 VFM VFM 0 9 8 7 6 5 4

ISBN 0-07-046617-3

Sponsoring Editor: Barbara Gilson
Production Supervisor: Pamela Pelton
Editing Supervisor: Maureen B. Walker
Project Supervision: Keyword Publishing Services Ltd

Library of Congress Cataloging-in-Publication Data

Nash, William A.
Schaum's outline of theory and problems of strength of materials / William A. Nash. -- 4th ed.
p. cm. -- (Schaum's outline series)
Includes index.
ISBN 0-07-046617-3
1. Strength of materials--Problems, exercises, etc. 2. Strength of materials--Outlines, syllabi, etc. I. Title.
TA407.4.N37 1998
620.1'12--dc21 98-28410
CIP

McGraw-Hill

A Division of The ***McGraw-Hill*** *Companies*

Preface

This Fourth Edition of *Schaum's Outline of Theory and Problems of Strength of Materials* adheres to the basic plan of the third edition but has several distinctive features.

1. Problem solutions are given in both SI (metric) and USCS units.
2. About fourteen computer programs are offered in either FORTRAN or BASIC for those types of problems that otherwise involve long, tedious computation. For example, beam stresses and deflections are readily determined by the programs given. All of these programs may be utilized on most PC systems with only modest changes in input format.
3. The presentation passes from elementary to more complex cases for a variety of structural elements subject to practical conditions of loading and support. Generalized treatments, such as elastic energy approaches, as well as plastic analysis and design are treated in detail.

The author is much indebted to Kathleen Derwin for preparation of most of the computer programs as well as careful checking of some of the new problems.

WILLIAM A. NASH

Contents

Chapter 1

Tension and Compression

INTERNAL EFFECTS OF FORCES

In this book we shall be concerned with what might be called the *internal effects* of forces acting on a body. The bodies themselves will no longer be considered to be perfectly rigid as was assumed in statics; instead, the calculation of the deformations of various bodies under a variety of loads will be one of our primary concerns in the study of strength of materials.

Axially Loaded Bar

The simplest case to consider at the start is that of an initially straight metal bar of constant cross section, loaded at its ends by a pair of oppositely directed collinear forces coinciding with the longitudinal axis of the bar and acting through the centroid of each cross section. For static equilibrium the magnitudes of the forces must be equal. If the forces are directed away from the bar, the bar is said to be in *tension*; if they are directed toward the bar, a state of *compression* exists. These two conditions are illustrated in Fig. 1-1.

Under the action of this pair of applied forces, internal resisting forces are set up within the bar and their characteristics may be studied by imagining a plane to be passed through the bar anywhere along its length and oriented perpendicular to the longitudinal axis of the bar. Such a plane is designated as *a-a* in Fig. 1-2(*a*). If for purposes of analysis the portion of the bar to the right of this plane is considered to be removed, as in Fig. 1-2(*b*), then it must be replaced by whatever effect it exerts upon the left portion. By this technique of introducing a cutting plane, the originally internal forces now become external with respect to the remaining portion of the body. For equilibrium of the portion to the left this "effect" must be a horizontal force of magnitude P. However, this force P acting normal to the cross-section *a-a* is actually the resultant of distributed forces acting over this cross section in a direction normal to it.

At this point it is necessary to make some assumption regarding the manner of variation of these distributed forces, and since the applied force P acts through the centroid it is commonly assumed that they are uniform across the cross section.

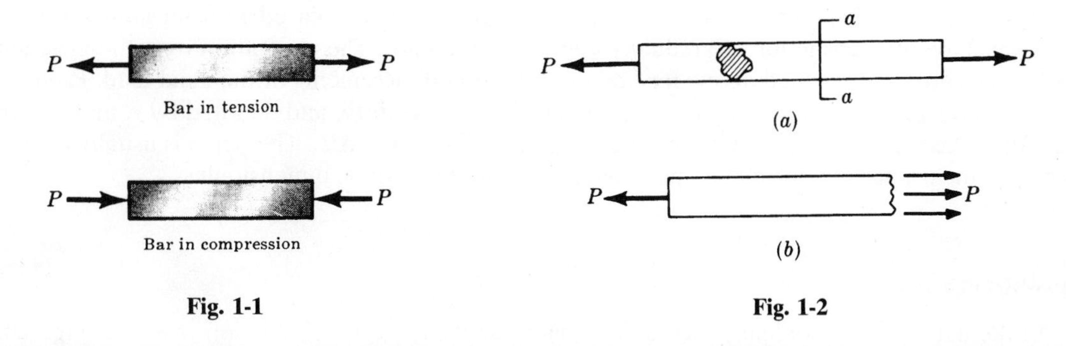

Fig. 1-1

Fig. 1-2

Normal Stress

Instead of speaking of the internal force acting on some small element of area, it is better for comparative purposes to treat the normal force acting over a *unit* area of the cross section. The intensity of normal force per unit area is termed the normal *stress* and is expressed in units of force per unit area, e.g., lb/in^2 or N/m^2. If the forces applied to the ends of the bar are such that the bar is in tension, then *tensile stresses* are set up in the bar; if the bar is in compression we have *compressive stresses*. It is essential that the line of action of the applied end forces pass through the centroid of each cross section of the bar.

Test Specimens

The axial loading shown in Fig. 1-2(a) occurs frequently in structural and machine design problems. To simulate this loading in the laboratory, a test specimen is held in the grips of either an electrically driven gear-type testing machine or a hydraulic machine. Both of these machines are commonly used in materials testing laboratories for applying axial tension.

In an effort to standardize materials testing techniques the American Society for Testing Materials (ASTM) has issued specifications that are in common use. Only two of these will be mentioned here, one for metal plates thicker than $\frac{3}{16}$ in (4.76 mm) and appearing as in Fig. 1-3, the other for metals over 1.5 in (38 mm) thick and having the appearance shown in Fig. 1-4. As may be seen from these figures, the central portion of the specimen is somewhat smaller than the end regions so that failure will not take place in the gripped portion. The rounded fillets shown are provided so that no stress concentrations will arise at the transition between the two lateral dimensions. The standard gage length over which elongations are measured is 8 in (203 mm) for the specimen shown in Fig. 1-3 and 2 in (57 mm) for that shown in Fig. 1-4.

The elongations are measured by either mechanical or optical extensometers or by cementing an electric resistance-type strain gage to the surface of the material. This resistance strain gage consists of a number of very fine wires oriented in the axial direction of the bar. As the bar elongates, the electrical resistance of the wires changes and this change of resistance is detected on a Wheatstone bridge and interpreted as elongation.

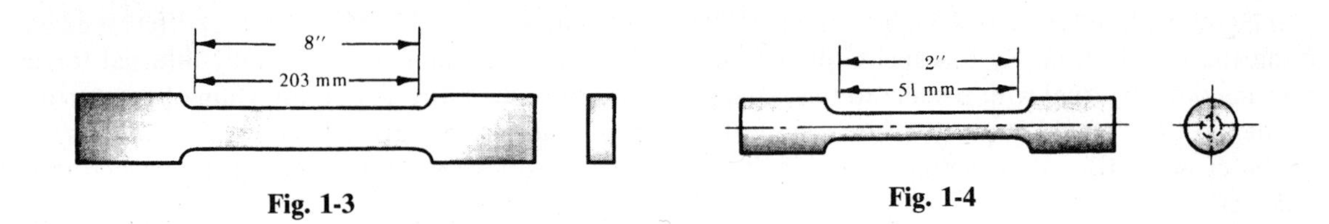

Fig. 1-3 Fig. 1-4

Normal Strain

Let us suppose that one of these tension specimens has been placed in a tension-compression testing machine and tensile forces gradually applied to the ends. The elongation over the gage length may be measured as indicated above for any predetermined increments of the axial load. From these values the elongation per unit length, which is termed *normal strain* and denoted by ϵ, may be found by dividing the total elongation Δ by the gage length L, that is, $\epsilon = \Delta/L$. The strain is usually expressed in units of inches per inch or meters per meter and consequently is dimensionless.

Stress-Strain Curve

As the axial load is gradually increased in increments, the total elongation over the gage length is measured at each increment of load and this is continued until fracture of the specimen takes place. Knowing the original cross-sectional area of the test specimen the *normal stress*, denoted by σ, may be obtained for any value of the axial load by the use of the relation

$$\sigma = \frac{P}{A}$$

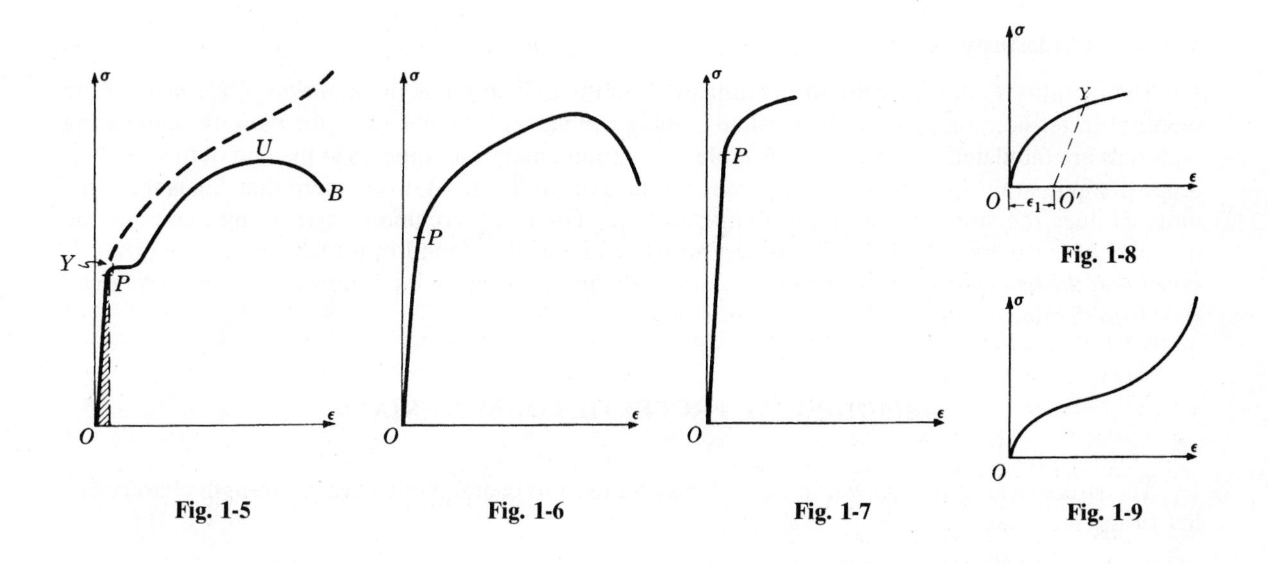

Fig. 1-5 Fig. 1-6 Fig. 1-7 Fig. 1-8 Fig. 1-9

where P denotes the axial load in pounds or Newtons and A the original cross-sectional area. Having obtained numerous pairs of values of normal stress σ and normal strain ϵ, the experimental data may be plotted with these quantities considered as ordinate and abscissa, respectively. This is the *stress-strain curve* or *diagram* of the material for this type of loading. Stress-strain diagrams assume widely differing forms for various materials. Figure 1-5 is the stress-strain diagram for a medium-carbon structural steel, Fig. 1-6 is for an alloy steel, and Fig. 1-7 is for hard steels and certain nonferrous alloys. For nonferrous alloys and cast iron the diagram has the form indicated in Fig. 1-8, while for rubber the plot of Fig. 1-9 is typical.

Ductile and Brittle Materials

Metallic engineering materials are commonly classed as either *ductile* or *brittle* materials. A *ductile material* is one having a relatively large tensile strain up to the point of rupture (for example, structural steel or aluminum) whereas a *brittle material* has a relatively small strain up to this same point. An arbitrary strain of 0.05 in/in (or mm/mm) is frequently taken as the dividing line between these two classes of materials. Cast iron and concrete are examples of brittle materials.

Hooke's Law

For any material having a stress-strain curve of the form shown in Fig. 1-5, 1-6, or 1-7, it is evident that the relation between stress and strain is linear for comparatively small values of the strain. This linear relation between elongation and the axial force causing it (since these quantities respectively differ from the strain or the stress only by a constant factor) was first noticed by Sir Robert Hooke in 1678 and is called *Hooke's law*. To describe this initial linear range of action of the material we may consequently write

$$\sigma = E\epsilon$$

where E denotes the slope of the straight-line portion OP of each of the curves in Figs. 1-5, 1-6, and 1-7.

Modulus of Elasticity

The quantity E, i.e., the ratio of the unit stress to the unit strain, is the *modulus of elasticity* of the material in tension, or, as it is often called, *Young's modulus*.* Values of E for various engineering materials are tabulated in handbooks. A table for common materials appears at the end of this chapter. Since the unit strain ϵ is a pure number (being a ratio of two lengths) it is evident that E has the same units as does the stress, for example lb/in^2, or N/m^2. For many common engineering materials the modulus of elasticity in compression is very nearly equal to that found in tension. *It is to be carefully noted that the behavior of materials under load as discussed in this book is restricted (unless otherwise stated) to the linear region of the stress-strain curve.*

MECHANICAL PROPERTIES OF MATERIALS

The stress-strain curve shown in Fig. 1-5 may be used to characterize several strength characteristics of the material. They are:

Proportional Limit

The ordinate of the point P is known as the *proportional limit*, i.e., the maximum stress that may be developed during a simple tension test such that the stress is a linear function of strain. For a material having the stress-strain curve shown in Fig. 1-8 there is no proportional limit.

Elastic Limit

The ordinate of a point almost coincident with P is known as the *elastic limit*, i.e., the maximum stress that may be developed during a simple tension test such that there is no permanent or residual deformation when the load is entirely removed. For many materials the numerical values of the elastic limit and the proportional limit are almost identical and the terms are sometimes used synonymously. In those cases where the distinction between the two values is evident the elastic limit is almost always greater than the proportional limit.

Elastic and Plastic Ranges

That region of the stress-strain curve extending from the origin to the proportional limit is called the *elastic range*; that region of the stress-strain curve extending from the proportional limit to the point of rupture is called the *plastic range*.

Yield Point

The ordinate of the point Y in Fig. 1-5, denoted by σ_{yp}, at which there is an increase in strain with no increase in stress is known as the *yield point* of the material. After loading has progressed to the point Y, yielding is said to take place. Some materials exhibit two points on the stress-strain curve at which there is an increase of strain without an increase of stress. These are called *upper* and *lower yield points*.

---

*Thomas Young was an English physicist, born in 1773, who worked in a number of areas such as mechanics, light, and heat. Before Young, historians had been unable to decipher stone tablets cut or painted in the characters (hieroglyphics) employed by Egyptians several thousand years B.C. Young, a master of eleven languages, was the first to successfully decipher any of the characters based upon study of the famous Rosetta stone found in 1799. His work, followed by that of Champollion in France, led to complete decipherment of the ancient language.

Ultimate Strength or Tensile Strength

The ordinate of the point U in Fig. 1-5, the maximum ordinate to the curve, is known either as the *ultimate strength* or the *tensile strength* of the material.

Breaking Strength

The ordinate of the point B in Fig. 1-5 is called the *breaking strength* of the material.

Modulus of Resilience

The work done on a unit volume of material, as a simple tensile force is gradually increased from zero to such a value that the proportional limit of the material is reached, is defined as the *modulus of resilience*. This may be calculated as the area under the stress-strain curve from the origin up to the proportional limit and is represented as the shaded area in Fig. 1-5. The units of this quantity are $\text{in} \cdot \text{lb/in}^3$, or $\text{N} \cdot \text{m/m}^3$ in the SI system. Thus, resilience of a material is its ability to absorb energy in the elastic range.

Modulus of Toughness

The work done on a unit volume of material as a simple tensile force is gradually increased from zero to the value causing rupture is defined as the *modulus of toughness*. This may be calculated as the entire area under the stress-strain curve from the origin to rupture. Toughness of a material is its ability to absorb energy in the plastic range of the material.

Percentage Reduction in Area

The decrease in cross-sectional area from the original area upon fracture divided by the *original* area and multiplied by 100 is termed *percentage reduction in area*. It is to be noted that when tensile forces act upon a bar, the cross-sectional area decreases, but calculations for the normal stress are usually made upon the basis of the original area. This is the case for the curve shown in Fig. 1-5. As the strains become increasingly larger it is more important to consider the instantaneous values of the cross-sectional area (which are decreasing), and if this is done the *true* stess-strain curve is obtained. Such a curve has the appearance shown by the dashed line in Fig. 1-5.

Percentage Elongation

The increase in length (of the gage length) after fracture divided by the initial length and multiplied by 100 is the *percentage elongation*. Both the percentage reduction in area and the percentage elongation are considered to be measures of the *ductility* of a material.

Working Stress

The above-mentioned strength characteristics may be used to select a *working stress*. Frequently such a stress is determined merely by dividing either the stress at yield or the ultimate stress by a number termed the *safety factor*. Selection of the safety factor is based upon the designer's judgment and experience. Specific safety factors are sometimes specified in design codes.

Strain Hardening

If a ductile material can be stressed considerably beyond the yield point without failure, it is said to *strain-harden*. This is true of many structural metals.

The nonlinear stress-strain curve of a brittle material, shown in Fig. 1-8, characterizes several other strength measures that cannot be introduced if the stress-strain curve has a linear region. They are:

Yield Strength

The ordinate to the stress-strain curve such that the material has a predetermined permanent deformation or "set" when the load is removed is called the *yield strength* of the material. The permanent set is often taken to be either 0.002 or 0.0035 in per in or mm per mm. These values are of course arbitrary. In Fig. 1-8 a set ϵ_1 is denoted on the strain axis and the line $O'Y$ is drawn parallel to the initial tangent to the curve. The ordinate of Y represents the yield strength of the material, sometimes called the *proof stress*.

Tangent Modulus

The rate of change of stress with respect to strain is known as the *tangent modulus* of the material. It is essentially an instantaneous modulus given by $E_t = d\sigma/d\epsilon$.

Coefficient of Linear Expansion

This is defined as the change of length per unit length of a straight bar subject to a temperature change of one degree and is usually denoted by α. The value of this coefficient is independent of the unit of length but does depend upon the temperature scale used. For example, from Table 1-1 at the end of this chapter the coefficient for steel is 6.5×10^{-6}/°F but 12×10^{-6}/°C. Temperature changes in a structure give rise to internal stresses, just as do applied loads.

Poisson's Ratio

When a bar is subject to a simple tensile loading there is an increase in length of the bar in the direction of the load, but a decrease in the lateral dimensions perpendicular to the load. The ratio of the strain in the lateral direction to that in the axial direction is defined as *Poisson's ratio*. It is denoted in this book by the Greek letter μ. For most metals it lies in the range 0.25 to 0.35. For cork, μ is very nearly zero. One new and unique material, so far of interest only in laboratory investigations, actually has a *negative* value of Poisson's ratio; i.e., if stretched in one direction it *expands* in every other direction. See Problems 1.19 through 1.24.

General Form of Hooke's Law

The simple form of Hooke's law has been given for axial tension when the loading is entirely along one straight line, i.e., uniaxial. Only the deformation in the direction of the load was considered and it was given by

$$\epsilon = \frac{\sigma}{E}$$

In the more general case an element of material is subject to three mutually perpendicular normal stresses σ_x, σ_y, σ_z, which are accompanied by the strains ϵ_x, ϵ_y, ϵ_z, respectively. By superposing the strain components arising from lateral contraction due to Poisson's effect upon the direct strains we obtain the general statement of Hooke's law:

$$\epsilon_x = \frac{1}{E}[\sigma_x - \mu(\sigma_y + \sigma_z)] \qquad \epsilon_y = \frac{1}{E}[\sigma_y - \mu(\sigma_x + \sigma_z)] \qquad \epsilon_z = \frac{1}{E}[\sigma_z - \mu(\sigma_z + \sigma_y)]$$

See Problems 1.20 and 1.23.

Specific Strength

This quantity is defined as the ratio of the ultimate (or tensile) strength to specific weight, i.e., weight per unit volume. Thus, in the USCS system, we have

$$\frac{\text{lb}}{\text{in}^2} \Big/ \frac{\text{lb}}{\text{in}^3} = \text{in}$$

and, in the SI system, we have

$$\frac{\text{N}}{\text{m}^2} \Big/ \frac{\text{N}}{\text{m}^3} = \text{m}$$

so that in either system specific strength has units of *length*. This parameter is useful for comparisons of material efficiencies. See Problem 1.25.

Specific Modulus

This quantity is defined as the ratio of the Young's modulus to specific weight. Substitution of units indicates that specific modulus has physical units of length in either the USCS or SI systems. See Problem 1.25.

DYNAMIC EFFECTS

In determination of mechanical properties of a material through a tension or compression test, the rate at which loading is applied sometimes has a significant influence upon the results. In general, ductile materials exhibit the greatest sensitivity to variations in loading rate, whereas the effect of testing speed on brittle materials, such as cast iron, has been found to be negligible. In the case of mild steel, a ductile material, it has been found that the yield point may be increased as much as 170 percent by extremely rapid application of axial force. It is of interest to note, however, that for this case the total elongation remains unchanged from that found for slower loadings.

CLASSIFICATION OF MATERIALS

Up to now, this entire discussion has been based upon the assumptions that two characteristics prevail in the material. They are that we have

A *homogeneous material*, one with the same elastic properties (E, μ) at all points in the body

An *isotropic material*, one having the same elastic properties in all directions at any one point of the body.

Not all materials are isotropic. If a material does not possess any kind of elastic symmetry it is called *anisotropic*, or sometimes *aeolotropic*. Instead of having two independent elastic constants (E, μ) as an isotropic material does, such a substance has 21 elastic constants. If the material has three mutually perpendicular planes of elastic symmetry it is said to be *orthotropic*. The number of independent constants is nine in this case. Modern filamentary reinforced *composite materials*, such as shown in Fig. 1-10, are excellent examples of anisotropic substances.

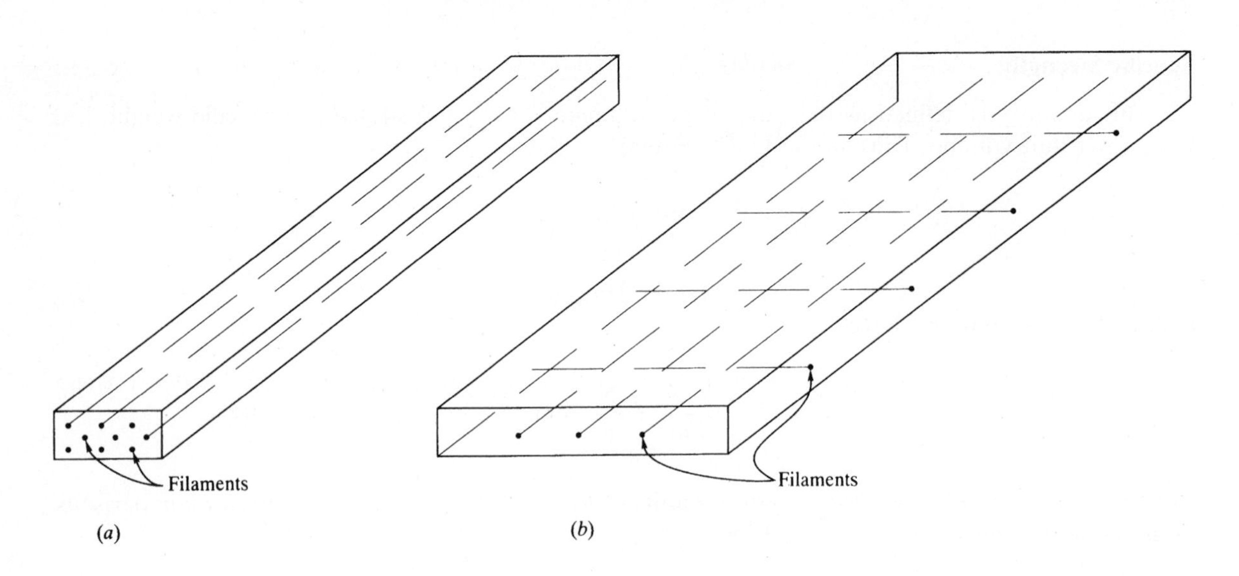

Fig. 1-10 (*a*) Epoxy bar reinforced by fine filaments in one direction; (*b*) epoxy plate reinforced by fine filaments in two directions.

ELASTIC VERSUS PLASTIC ANALYSIS

Stresses and deformations in the plastic range of action of a material are frequently permitted in certain structures. Some building codes allow particular structural members to undergo plastic deformation, and certain components of aircraft and missile structures are deliberately designed to act in the plastic range so as to achieve weight savings. Furthermore, many metal-forming processes involve plastic action of the material. For small plastic strains of low- and medium-carbon structural steels the stress-strain curve of Fig. 1-11 is usually idealized by two straight lines, one with a slope of E, representing the elastic range, the other with zero slope representing the plastic range. This plot, shown in Fig. 1-11, represents a so-called *elastic, perfectly plastic material*. It takes no account of still larger plastic strains occurring in the strain-hardening region shown as the right portion of the stress-strain curve of Fig. 1-5. See Problem 1.26.

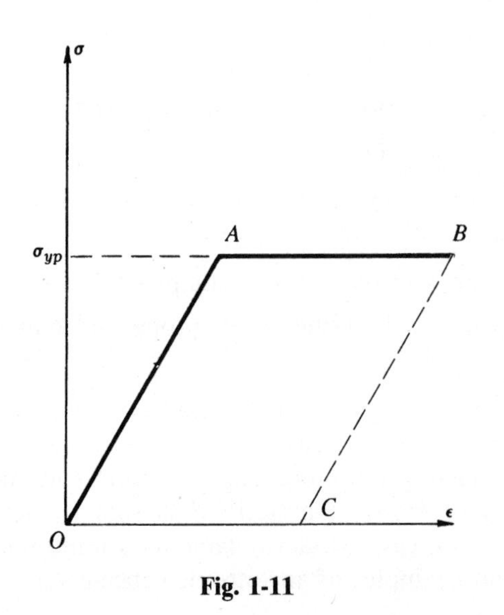

Fig. 1-11

If the load increases so as to bring about the strain corresponding to point B in Fig. 1-11, and then the load is removed, unloading takes place along the line BC so that complete removal of the load leaves a permanent "set" or elongation corresponding to the strain OC.

Solved Problems

1.1. In Fig. 1-12, determine the total elongation of an initially straight bar of length L, cross-sectional area A, and modulus of elasticity E if a tensile load P acts on the ends of the bar.

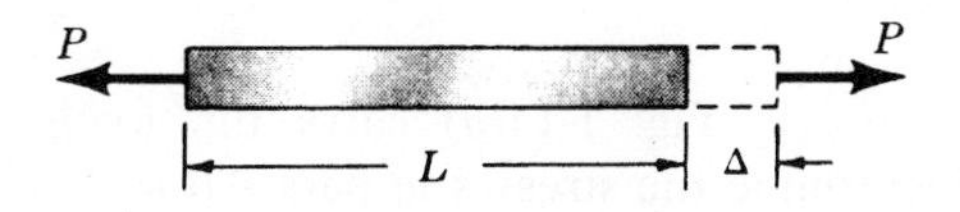

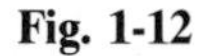

Fig. 1-12

The unit stress in the direction of the force P is merely the load divided by the cross-sectional area, that is, $\sigma = P/A$. Also the unit strain ϵ is given by the total elongation Δ divided by the original length, i.e., $\epsilon = \Delta/L$. By definition the modulus of elasticity E is the ratio of σ to ϵ, that is,

$$E = \frac{\sigma}{\epsilon} = \frac{P/A}{\Delta/L} = \frac{PL}{A\Delta} \quad \text{or} \quad \Delta = \frac{PL}{AE}$$

Note that Δ has the units of length, perhaps inches or meters.

1.2. A steel bar of cross section 500 mm^2 is acted upon by the forces shown in Fig. 1-13(a). Determine the total elongation of the bar. For steel, consider $E = 200$ GPa.

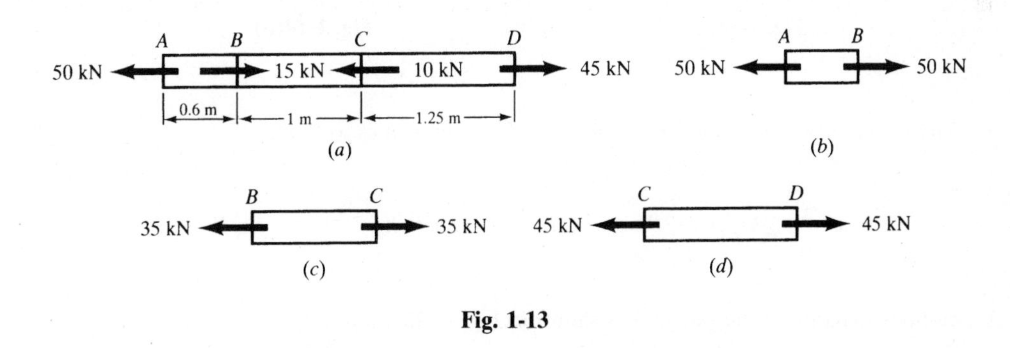

Fig. 1-13

The entire bar is in equilibrium, and hence all portions of it are also. The portion between A and B has a resultant force of 50 kN acting over every cross section and a free-body diagram of this 0.6-m length appears as in Fig. 1-13(b). The force at the right end of this segment must be 50 kN to maintain equilibrium with the applied load at A. The elongation of this portion is, from Problem 1.1:

$$\Delta_1 = \frac{(50{,}000\ \text{N})(0.6\ \text{m})}{(500 \times 10^{-6}\ \text{m}^2)(200 \times 10^9\ \text{N/m}^2)} = 0.0003\ \text{m}$$

The force acting in the segment between B and C is found by considering the algebraic sum of the forces to the left of any section between B and C, i.e., a resultant force of 35 kN acts to the left, so that a tensile force exists. The free-body diagram of the segment between B and C is shown in Fig. 1-13(c) and the elongation of it is

$$\Delta_2 = \frac{(35{,}000\ \text{N})(1\ \text{m})}{(500 \times 10^{-6}\ \text{m}^2)(200 \times 10^9\ \text{N/m}^2)} = 0.00035\ \text{m}$$

Similarly, the force acting over any cross section between C and D must be 45 kN to maintain equilibrium with the applied load at D. The elongation of CD is

$$\Delta_3 = \frac{(45{,}000\ \text{N})(1.25\ \text{m})}{(500 \times 10^{-6}\ \text{m}^2)(200 \times 10^9\ \text{N/m}^2)} = 0.00056\ \text{m}$$

The total elongation is

$$\Delta = \Delta_1 + \Delta_2 + \Delta_3 = 0.00121\ \text{m} \quad \text{or} \quad 1.21\ \text{mm}$$

1.3. The pinned members shown in Fig. 1-14(a) carry the loads P and $2P$. All bars have cross-sectional area A_1. Determine the stresses in bars AB and AF.

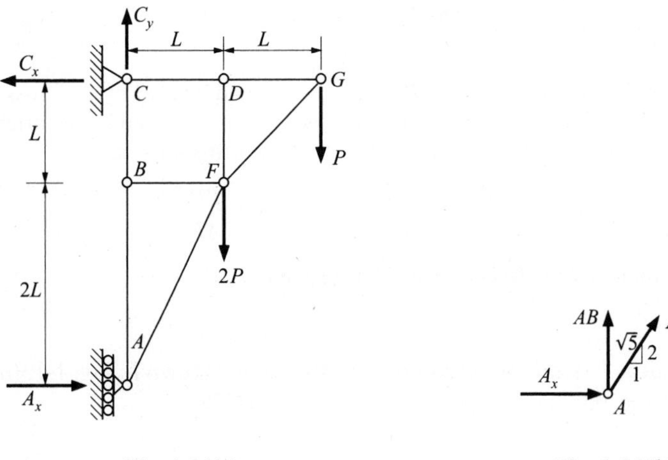

Fig. 1-14(a)

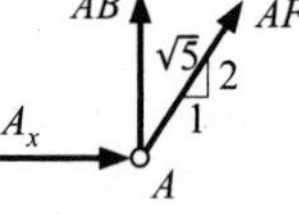

Fig. 1-14(b)

The reactions are indicated by C_x, C_y, and A_x. From statics we have

$$\Sigma M_c = -(2PL) - P(2L) + A_x(3L) = 0; \qquad A_x = \frac{4}{3}P$$

A free-body diagram of the pin at A is shown in Fig. 1.14(b). From statics:

$$\Sigma F_x = \frac{4P}{3} + \frac{1}{\sqrt{5}}(AF) = 0; \qquad AF = -\frac{4P\sqrt{5}}{3}$$

$$\Sigma F_y = (AB) + \frac{2}{\sqrt{5}}(AF) = 0; \qquad AB = -\frac{8}{3}P$$

The bar stresses are

$$\sigma_{AF} = -\frac{4P\sqrt{5}}{3A}; \qquad \sigma_{AB} = -\frac{8P}{3A}$$

1.4. A component of a power generator consists of a torus supported by six tie rods from an overhead central point as shown in Fig. 1-15. The weight of the torus is 2000 N per meter of circumferential length. The point of attachment A is 1.25 m above the plane of the torus. The radius of the middle line of the torus is 0.5 m. Each tie rod has a cross-sectional area of 25 mm^2. Determine the vertical displacement of the torus due to its own weight.

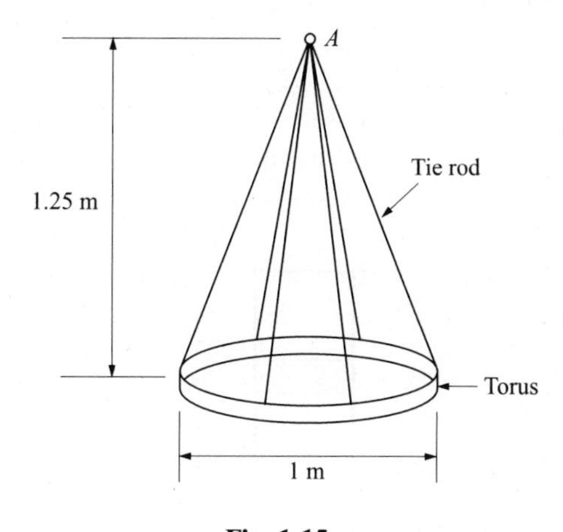

Fig. 1-15

A free-body diagram of the torus appears in Fig. 1-16 where T denotes the tensile force in each rod. Summing forces vertically:

$$6T\left(\frac{1.25}{1.34}\right) - \left(2000\,\frac{\text{N}}{\text{m}}\right)2\pi(0.5\text{ m}) = 0$$

$$T = 1120\text{ N}$$

Let us examine the deformation of a typical tie rod, such as AB. Figure 1-17 shows how AB elongates an amount BB' given by

$$\Delta = BB' = \frac{TL}{AE} = \frac{(1120\text{ N})(1.34\text{ m})}{(25\text{ mm}^2)\left(\dfrac{\text{m}}{10^3\text{ mm}}\right)^2\left(200\times10^9\,\dfrac{\text{N}}{\text{m}^2}\right)} = 0.0003\text{ m} \quad \text{or} \quad 0.3\text{ mm}$$

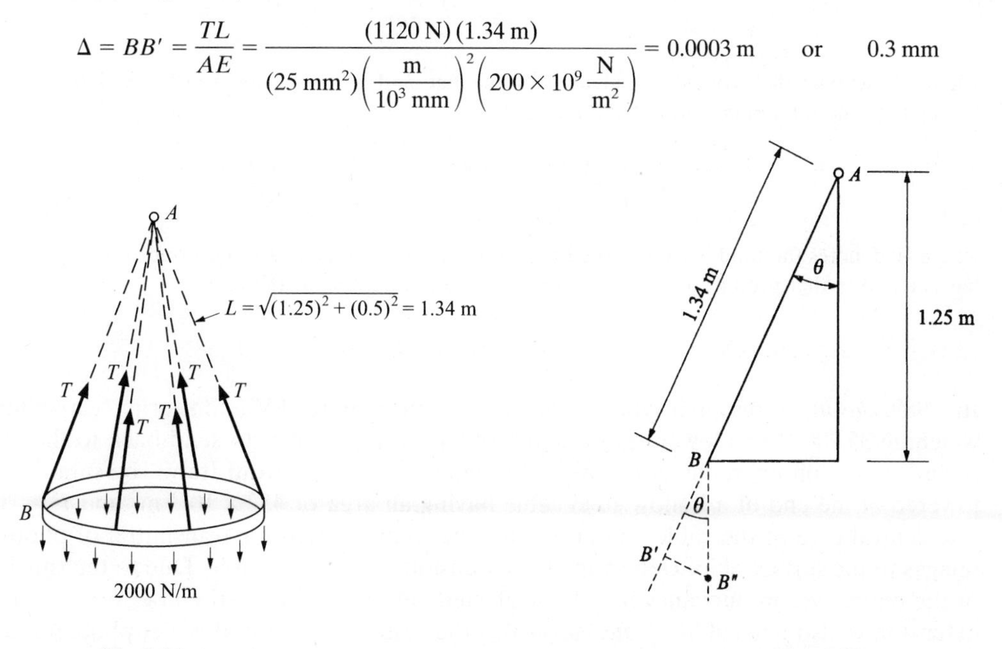

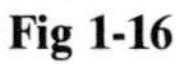

Fig 1-16

Fig. 1-17

Since B is on the torus, it (B) must move to B'' which is vertical below B. From Fig. 1-16 we have

$$BB'' = (0.3)\frac{1}{\cos\theta} = \frac{0.3}{\left(\dfrac{1.25}{1.34}\right)} = 0.32 \text{ mm}$$

which is the vertical displacement of the rigid torus.

1.5. In Fig. 1-18, determine the total increase of length of a bar of constant cross section hanging vertically and subject to its own weight as the only load. The bar is initially straight.

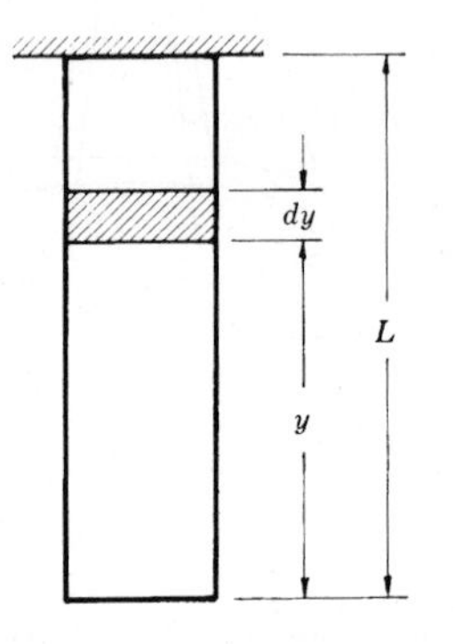

Fig. 1-18

The normal stress (tensile) over any horizontal cross section is caused by the weight of the material below that section. The elongation of the element of thickness dy shown is

$$d\Delta = \frac{Ay\gamma}{AE}dy$$

where A denotes the cross-sectional area of the bar and γ its specific weight (weight/unit volume). Integrating, the total elongation of the bar is

$$\Delta = \int_0^L \frac{Ay\gamma\,dy}{AE} = \frac{A\gamma}{AE}\frac{L^2}{2} = \frac{(A\gamma L)L}{2AE} = \frac{WL}{2AE}$$

where W denotes the total weight of the bar. Note that the total elongation produced by the weight of the bar is equal to that produced by a load of half its weight applied at the end.

1.6. In 1989, *Jason*, a research-type submersible with remote TV monitoring capabilities and weighing 35,200 N was lowered to a depth of 646 m in an effort to send back to the attending surface vessel photographs of a sunken Roman ship offshore from Italy. The submersible was lowered at the end of a hollow steel cable having an area of $452 \times 10^{-6}\text{ m}^2$ and $E = 200$ GPa. The central core of the cable contained the fiber-optic system for transmittal of photographic images to the surface ship. Determine the extension of the steel cable. Due to the small volume of the entire system buoyancy may be neglected, and the effect of the fiber optic cable on the extension is also negligible. (*Note: Jason* was the system that took the first photographs of the sunken *Titanic* in 1986.)

The total cable extension is the sum of the extensions due to (*a*) the weight of *Jason*, and (*b*) the weight of the steel cable. From Problem 1.1, we have for (*a*)

$$\Delta_1 = \frac{PL}{AE} = \frac{(35{,}200\ \text{N})(646\ \text{m})}{(452 \times 10^{-6}\ \text{m}^2)(200 \times 10^9\ \text{N/m}^2)} = 0.252\ \text{m}$$

and from Problem 1.5, we have for (*b*)

$$\Delta_2 = \frac{WL}{2AE}$$

where W is the weight of the cable. W may be found as the volume of the cable

$$(452 \times 10^{-6}\ \text{m}^2)(646\ \text{m}) = 0.292\ \text{m}^3$$

which must be multiplied by the weight of steel per unit volume which, from Table 1-1 at the end of the chapter is 77 kN/m^3. Thus, the cable weight is

$$W = (0.292\ \text{m}^3)(77\ \text{kN/m}^3) = 22{,}484\ \text{N}$$

so that the elongation due to the weight of the cable is

$$\Delta_2 = \frac{(22{,}484\ \text{N})(646\ \text{m})}{2(452 \times 10^{-6}\ \text{m}^2)(200 \times 10^9\ \text{N/m}^2)} = 0.080\ \text{m}$$

The total elongation is the sum of the effects,

$$\Delta = \Delta_1 + \Delta_2 = 0.252 + 0.080 = 0.332\ \text{m}$$

1.7. Two prismatic bars are rigidly fastened together and support a vertical load of 10,000 lb, as shown in Fig. 1-19. The upper bar is steel having specific weight 0.283 lb/in^3, length 35 ft, and cross-sectional area 10 in^2. The lower bar is brass having specific weight 0.300 lb/in^3, length 20 ft, and cross-sectional area 8 in^2. For steel $E = 30 \times 10^6$ lb/in^2, for brass $E = 13 \times 10^6$ lb/in^2. Determine the maximum stress in each material.

The maximum stress in the brass bar occurs just below the junction at section *B*-*B*. There, the vertical normal stress is caused by the combined effect of the load of 10,000 lb together with the weight of the entire brass bar below *B*-*B*.

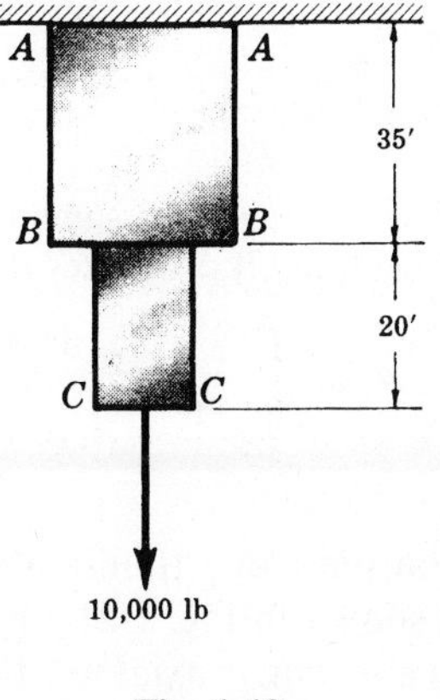

Fig. 1-19

The weight of the brass bar is $W_b = (20 \times 12)(8)(0.300) = 576$ lb.

The stress at this section is

$$\sigma = \frac{P}{A} = \frac{10{,}000 + 576}{8} = 1320 \text{ lb/in}^2$$

The maximum stress in the steel bar occurs at section A-A, the point of suspension, because there the entire weight of the steel and brass bars gives rise to normal stress, whereas at any lower section only a portion of the weight of the steel would be effective in causing stress.

The weight of the steel bar is $W_s = (35 \times 12)(10)(0.283) = 1185$ lb.

The stress across section A-A is

$$\sigma = \frac{P}{A} = \frac{10{,}000 + 576 + 1185}{10} = 1180 \text{ lb/in}^2$$

1.8. A solid truncated conical bar of circular cross section tapers uniformly from a diameter d at its small end to D at the large end. The length of the bar is L. Determine the elongation due to an axial force P applied at each end. See Fig. 1-20.

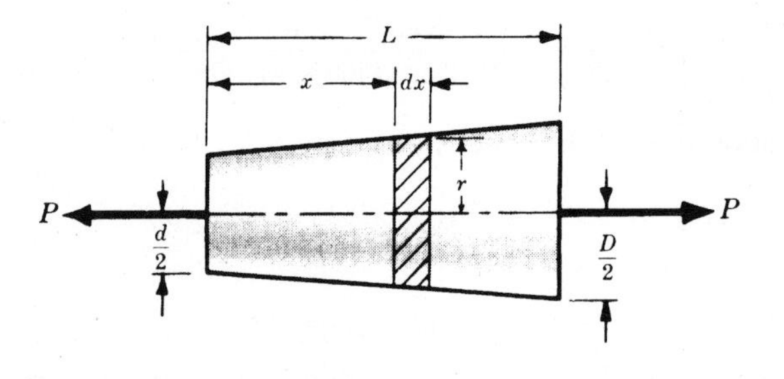

Fig. 1-20

The coordinate x describes the distance from the small end of a disc-like element of thickness dx. The radius of this small element is readily found by similar triangles:

$$r = \frac{d}{2} + \frac{x}{L}\left(\frac{D-d}{2}\right)$$

The elongation of this disc-like element may be found by applying the formula for extension due to axial loading, $\Delta = PL/AE$. For the element, this expression becomes

$$d\Delta = \frac{P\,dx}{\pi\left[\frac{d}{2} + \frac{x}{L}\left(\frac{D-d}{2}\right)\right]^2 E}$$

The extension of the entire bar is obtained by summing the elongations of all such elements over the bar. This is of course done by integrating. If Δ denotes the elongation of the entire bar,

$$\Delta = \int_0^L d\Delta = \int_0^L \frac{4P\,dx}{\pi[d + (x/L)(D-d)]^2 E} = \frac{4PL}{\pi DdE}$$

1.9. Two solid circular cross-section bars, one titanium and the other steel, each in the form of a truncated cone, are joined as shown in Fig. 1-21(a) and attached to a rigid vertical wall at the left. The system is subject to a concentric axial tensile force of 500 kN at the right end, together with an axisymmetric ring-type load applied at the junction of the bars as shown and having a horizontal resultant of 1000 kN. Determine the change of length of the system. For titanium, $E = 110$ GPa, and for steel, $E = 200$ GPa.

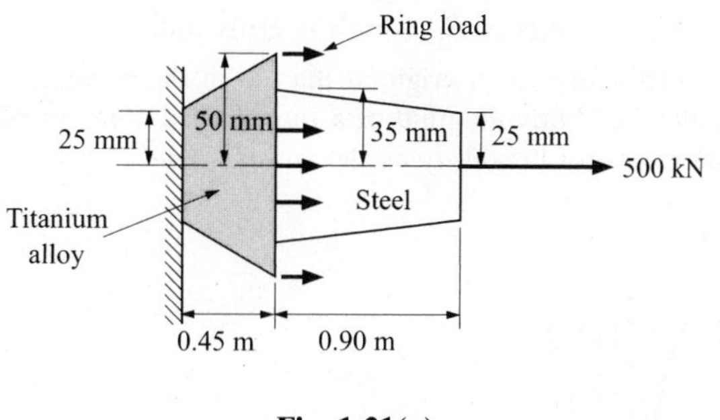

Fig. 1-21(*a*)

A free-body diagram of the system appears as shown in Fig. 1-21(*b*)

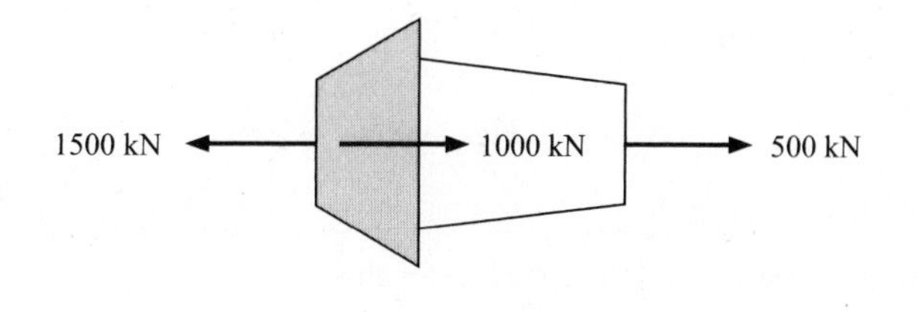

Fig. 1-21(*b*)

and a free-body diagram of each bar is shown in Fig. 1-21(*c*).

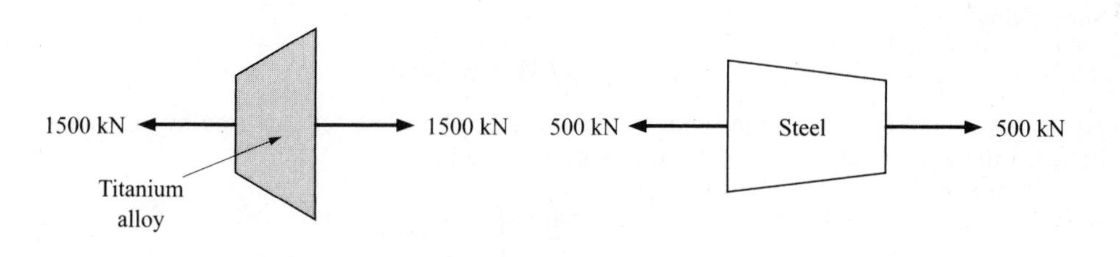

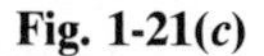

Fig. 1-21(*c*)

We may now apply the result of Problem 1.8 to each bar and obtain

$$\Delta_{Ti} = \frac{4(1{,}500{,}000\ \text{N})\,(0.45\ \text{m})}{\pi(0.10\ \text{m})\,(0.05\ \text{m})\,(110 \times 10^9\ \text{N/m}^2)} = 0.00156\ \text{m}$$

$$\Delta_{ST} = \frac{4(500{,}000\ \text{N})\,(0.90\ \text{m})}{\pi(0.07\ \text{m})\,(0.05\ \text{m})\,(200 \times 10^9\ \text{N/m}^2)} = 0.00082$$

Using superposition,

$$\Delta = \Delta_{ST} + \Delta_{Ti} = 0.00238\ \text{m} \qquad \text{or} \qquad 2.38\ \text{mm}$$

1.10. A large-scale pumping system to lift water consists of a pump of weight W in a circular cylindrical housing (with vertical axis) suspended from an axisymmetric thick-walled tube of variable radial thickness [see Fig. 1-22(a)]. Find the variation in outer radius R along the height so that the normal (vertical) stress in the tube is constant. The specific weight of the tube material is γ and the inner radius is R_i, which is constant.

We introduce the coordinate y, with origin at the top of the pump and extending positive upward as shown. Let us consider the free-body diagram of a ring-shaped element of the tube located a distance y above the top of the pump and of height dy as shown in Fig. 1-22(b).

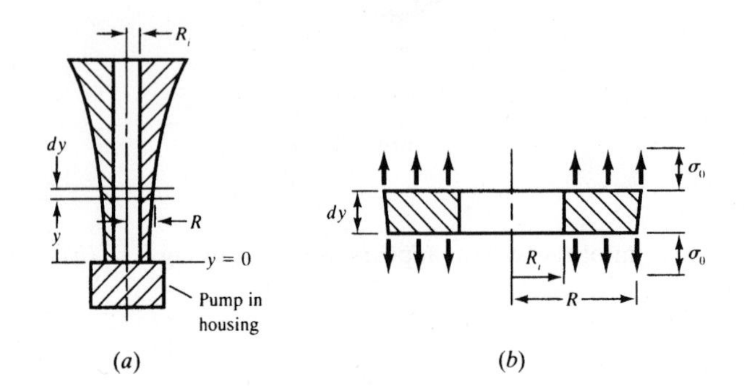

Fig. 1-22

The cross-sectional area of the lower surface of this ring is

$$A = \pi(R^2 - R_i^2) \tag{1}$$

and the area of the upper surface is $(A + dA)$. The weight of the material in the ring is $\gamma A\, dy$. For vertical equilibrium we have

$$\sigma_0(A + dA) - \sigma_0(A) - \gamma A(dy) = 0 \tag{2}$$

Simplifying:

$$\sigma_0(dA) = \gamma A(dy) \tag{3}$$

At the lower end ($y = 0$) of the tube, we denote the tube cross-section area by A_0. Integrating Eq. (3) between the lower end ($y = 0$) and the elevation y, we have

$$\int_{A_0}^{A} \frac{dA}{A} = \int_{y=0}^{y} \frac{\gamma}{\sigma_0}(dy) \tag{4}$$

Thus:

$$\ln\frac{A}{A_0} = \frac{\gamma}{\sigma_0}(y) \tag{5}$$

$$A = A_0 e^{\gamma y/\sigma_0} \tag{6}$$

At $y = 0$, we have for vertical equilibrium

$$\sigma_0 = \frac{W}{A_0} \tag{7}$$

so from (1), (6), and (7) we have the radius at any elevation y as

$$R^2 = R_i^2 + \frac{W}{\pi\sigma_0} e^{\gamma y/\sigma_0} \tag{8}$$

1.11. The pin-connected framework shown in Fig. 1-23(*a*) consists of two identical upper rods AB and AC, two shorter, lower rods BD and DC, together with a rigid horizontal brace BC. All bars have cross-sectional area A and modulus of elasticity E. Determine the vertical displacement of point D due to the action of the vertical load P applied at D as well as the distributed load q per unit length.

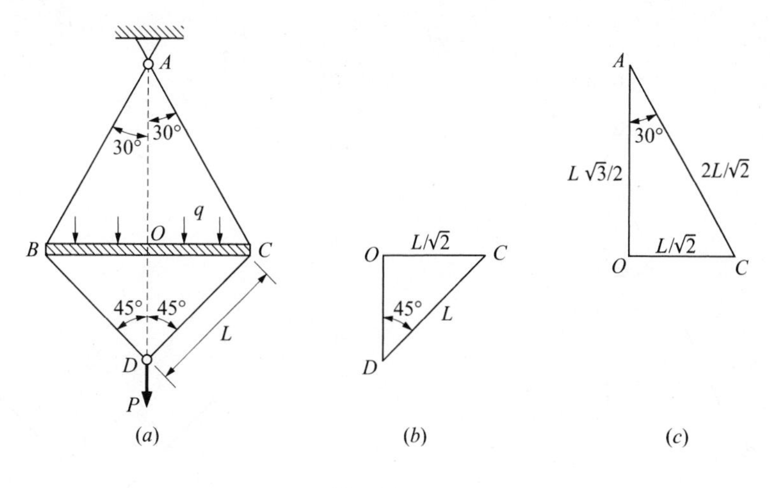

Fig. 1-23

Let us consider a horizontal cutting plane passed through the system slightly above BC. The free-body diagram is shown in Fig. 1-24 where F_2 represents the force in each of the bars AB and AC. From statics:

$$\Sigma F_v = -P - q\left(\frac{L}{\sqrt{2}}\right)(2) + 2F_2 \sin 60^\circ = 0$$

$$F_2 = \frac{P + \left(\frac{2}{\sqrt{2}}\right)qL}{\sqrt{3}} \tag{1}$$

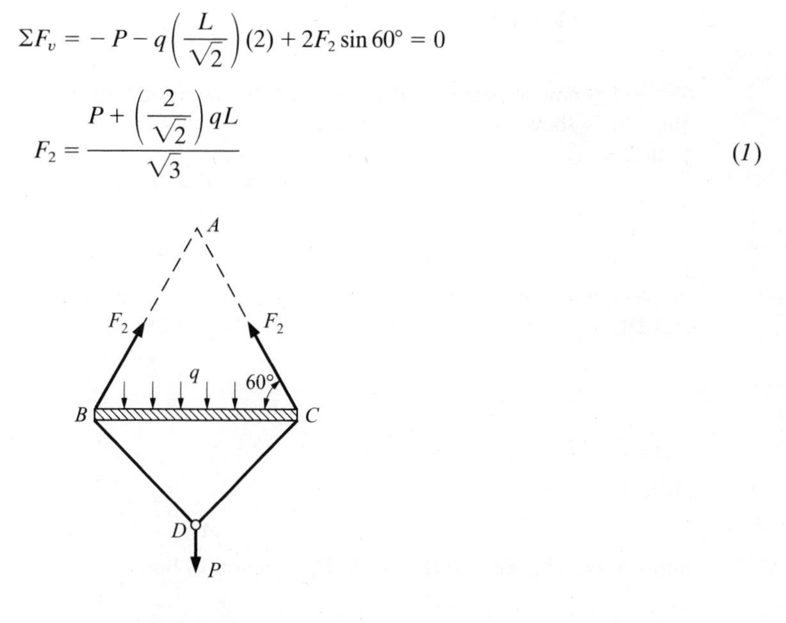

Fig. 1-24

To determine the dropping of bar BC we consider the deformation of bar AB, as shown in Fig. 1-25. The increase of length of AB is given by

$$\Delta_{AB} = \frac{F_2\left(\frac{2L}{\sqrt{2}}\right)}{AE}$$

and the vertical projection of this is

$$B'B'' = \frac{F_2\left(\dfrac{2L}{\sqrt{2}}\right)}{AE\cos 30^\circ}$$

Substituting F_2 from (*1*), this is

$$BB'' = \frac{4PL}{3\sqrt{2}AE} + \frac{4qL^2}{3AE} \tag{2}$$

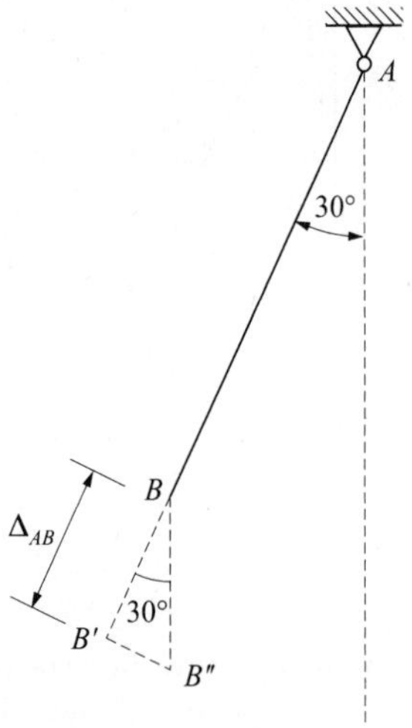

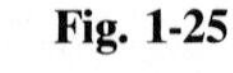

Fig. 1-25

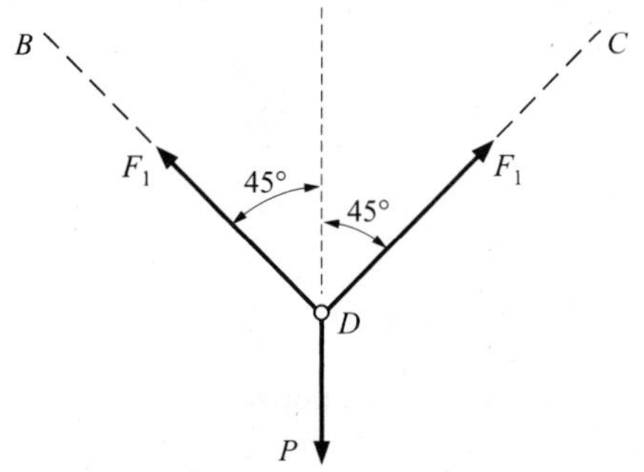

Fig. 1-26

Let us now consider another horizontal plane passed through the system just below BC. The free-body diagram is shown in Fig. 1-26 where F_1 represents the force in each of the bars BD and DC. From statics:

$$\Sigma F_v = -P + 2F_1\cos 45^\circ = 0$$

$$F_1 = \frac{P\sqrt{2}}{2} \tag{3}$$

We must now determine the lowering of point D due to the action of the load P acting on bars BD and DC (see Fig. 1-27). The increase of length of BD is

$$\frac{F_1 L}{AE}$$

and the vertical projection of this is

$$\frac{F_1 L}{AE\cos 45^\circ}$$

Substituting (*3*), we find the vertical projection to be

$$\frac{PL}{AE} \tag{4}$$

The actual drop of point D is the sum of (*2*) and (*4*):

$$\Delta_D = \frac{4PL}{3\sqrt{2}AE} + \frac{PL}{AE} + \frac{4qL^2}{3AE}$$

$$= 1.942\frac{PL}{AE} + 1.333\frac{qL^2}{AE} \tag{5}$$

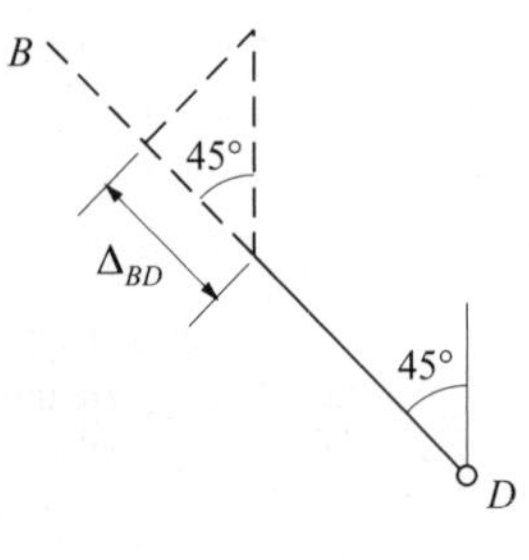

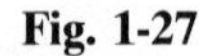

Fig. 1-27

1.12. Consider the system of two pinned end bars AB and CB (which is vertical) subject to the single horizontal force P applied at the pin B (see Fig. 1-28). Bar AB has area A_1, length L_1, and Young's modulus E_1. The corresponding quantities for bar CB are A_2, L_2, and E_2. Determine the horizontal and vertical components of displacement of pin B.

The free-body diagram of the pin is shown in Fig. 1-29(a) where F_1 and F_2 denote the forces bars AB and CB, respectively, exert on that pin. Each of these bar forces has been assumed to be positive in the direction shown; i.e., each bar is assumed to be in tension. Should the equilibrium equations indicate a negative value for either of these bar forces, that would signify that we have assumed the direction incorrectly and that the bar is in compression. Figures 1-29(b) and 1-29(c) indicate the effects that the pin at B exerts on bars AB and CB, respectively. These are of course equal and opposite to the values shown in Fig. 1-29(a).

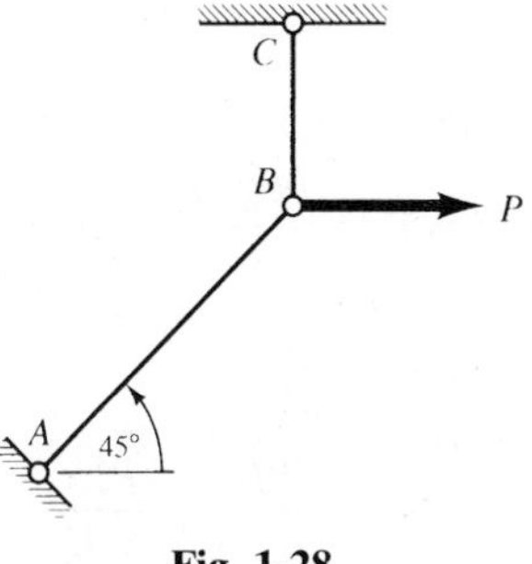

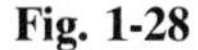
Fig. 1-28

For equilibrium of the pin at B, we have

$$\Sigma F_x = P - F_1 \cos 45° = 0 \tag{1}$$

$$\Sigma F_y = F_2 - F_1 \sin 45° = 0 \tag{2}$$

Solving,

$$F_1 = P\sqrt{2} \qquad F_2 = P \tag{3}$$

which indicates tension in each bar. Let us think of temporarily unlocking the bars at B by removing pin B. Bar AB then stretches an amount BB' and bar CB stretches an amount BB'', as shown in Fig. 1-30. These extensions are found from Problem 1.1 to be

$$BB' = \frac{F_1 L_1}{A_1 E_1} = \frac{P\sqrt{2}L_1}{A_1 E_1} \tag{4}$$

$$BB'' = \frac{F_2 L_2}{A_2 E_2} = \frac{PL_2}{A_2 E_2} \tag{5}$$

However, the final position of the pin must be the same after the pin is considered to be reintroduced, so the bar AB must undergo a rigid-body rotation about pin A and bar CB must rotate about pin C. The point B' on AB (extended) must move along a circular arc with center at A, but for the very small deformations that we consider this arc may be replaced by a dotted straight line $B'B'''$ perpendicular to AB'. Likewise point B'' on CB (extended) must move along the horizontal dotted line $B''B'''$ as rotation takes place about pin C. The intersection of these two dotted lines at B''' must be the true, final position of the pin B.

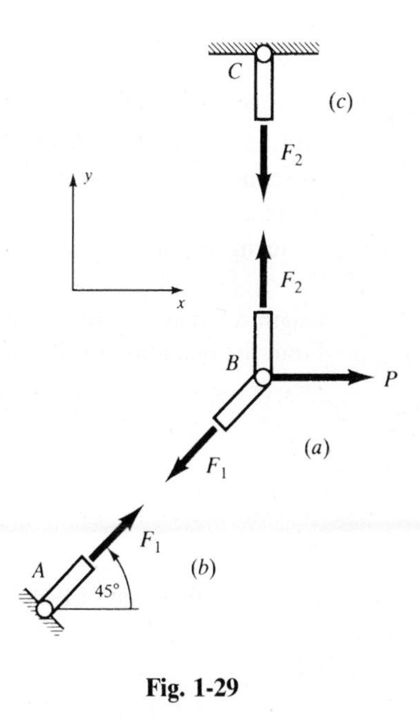

Fig. 1-29

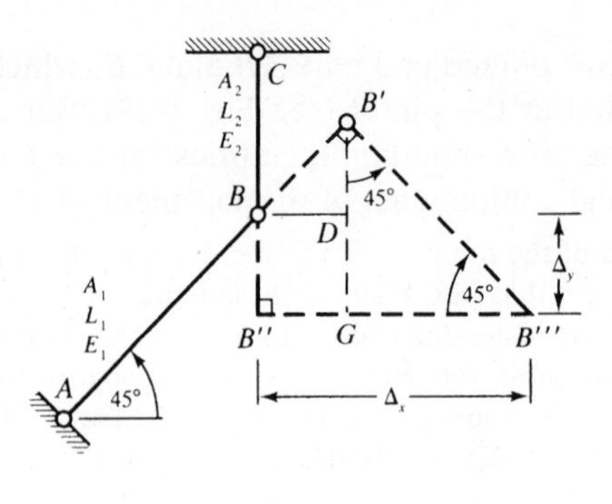

Fig. 1-30

From the geometry of Fig. 1-30 we have

$$BD = BB' \cos 45° = \frac{\sqrt{2}PL_1}{A_1E_1} \cdot \frac{1}{\sqrt{2}} = \frac{PL_1}{A_1E_1} \tag{6}$$

$$B'D = BB' \sin 45° = \frac{PL_1}{A_1E_1} \tag{7}$$

$$B'G = B'D + DG = \frac{PL_1}{A_1E_1} + \frac{PL_2}{A_2E_2} \tag{8}$$

$$GB''' = B'G \qquad (45° \text{ triangle}) \tag{9}$$

$$B''B''' = BD + GB''' = \frac{PL_1}{A_1E_1} + \left(\frac{PL_1}{A_1E_1} + \frac{PL_2}{A_2E_2}\right) \tag{10}$$

$$= \frac{2PL_1}{A_1E_1} + \frac{PL_2}{A_2E_2} = \Delta_x \tag{11}$$

Finally, from Fig. 1-30 the vertical displacement of B is

$$\Delta_y = BB'' = \frac{PL_2}{A_2E_2} \tag{12}$$

1.13. In 1989 a new fiber-optic cable capable of handling 40,000 telephone calls simultaneously was laid under the Pacific Ocean from California to Japan, a distance of 13,300 km. The cable was unreeled from shipboard at a mean temperature of 22°C and dropped to the ocean floor having a mean temperature of 5°C. The coefficient of linear expansion of the cable is 75×10^{-6}/°C. Determine the length of cable that must be carried on the ship to span the 13,300 km.

The length of cable that must be carried on board ship consists of the 13,300 km plus an unknown length ΔL that will allow for contraction to a final length of 13,300 km when resting on the ocean floor. From the definition of the coefficient of thermal expansion (Chap. 1), we have

$$\Delta L = \alpha L(\Delta T)$$

$$\Delta L = (75 \times 10^{-6}/°\text{C})\,[13{,}300 \text{ km} + \underline{\Delta L}]\,(22 - 5)°\text{C} \tag{a}$$

Solving, we find

$$\Delta L = 16.96 \text{ km}$$

The percent change of length is thus

$$\frac{(16.96)(100)}{13{,}300 + 16.96} = 0.13\%$$

so that the underlined term in Eq. (*a*) is of minor consequence. Thus, the required length of cable at shipboard temperature is approximately 13,317 km.

1.14. An elastic bar of variable cross section is loaded by axial tension or compression at its ends as shown in Fig. 1-31. The variation of cross-sectional dimension may be known either analytically or numerically along the dimension in the axial direction. Write a FORTRAN program for change of length of the bar for the cases of (*a*) a bar of solid circular cross section and (*b*) a flat slab of constant thickness t as shown in Figs. 1-31(*b*) and 1-31(*c*), respectively. The contour of the bar is described by the equation $y = Ae^{Bx}$, where x is the axial coordinate.

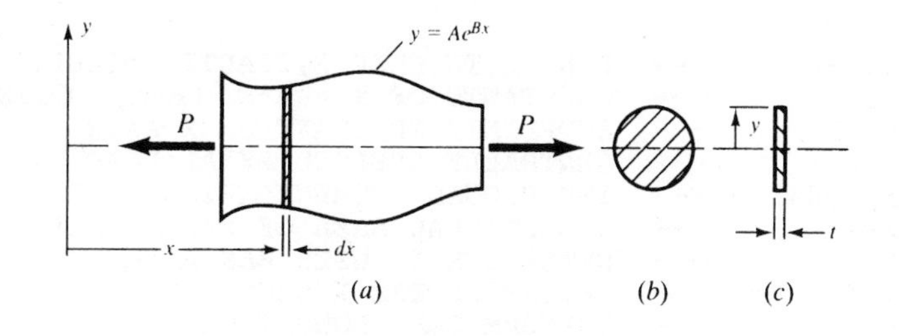

Fig. 1-31

The equation derived in Problem 1.1 may be applied to each subsegment of length dx as shown in Fig. 1-31(*a*). The cross-sectional area of each such subsegment is taken to be constant and we then apply the relation

$$\Delta = \frac{PL}{AE}$$

to this segment, where the length of the segment is dx and A is the cross-sectional area of the segment. Clearly A may be found if the equation $y = y(x)$ for the cross section is known, or, alternatively, measurements may be made at a number of stations along the length of the bar and the area found numerically at each such station.

This approach is represented by the following FORTRAN program which is self-prompting. Tensile loadings are regarded as positive and compressives as negative.

Note that in the equation describing the shape of the bar, $y = Ae^{Bx}$, e represents the base of natural logs, and A and B are parameters of the contour. Note in particular that this A is *not* cross-sectional area.

```
00010*****************************************************************************
00020                        PROGRAM SLBTEN2(INPUT,OUTPUT)
00030*****************************************************************************
00040*
00050*              AUTHOR: KATHLEEN DERWIN
00060*              DATE  : FEBRUARY 5, 1989
00070*
00080*     BRIEF DESCRIPTION:
00090*         THIS PROGRAM DETERMINES THE CHANGE OF LENGTH OF A BAR DUE
00100*     TO AXIAL TENSION OR COMPRESSION. THE BAR MAY BE A CONSTANT
00110*     THICKNESS, VARIABLE WIDTH RECTANGULAR SLAB, OR A SOLID CIRCULAR
00120*     ROD WITH VARIABLE DIAMETER. IN EITHER CASE THE SHAFT IS CENTRALLY
```

```
00130*     LOADED BY AN AXIAL FORCE.
00140*        THE VARYING WIDTH (OF THE SLAB) OR DIAMETER (OF THE ROD) MAY
00150*     BE DESCRIBED EITHER ANALYTICALLY AS  Y = A*E ^(B*X)  WHERE X IS THE
00160*     GEOMETRIC AXIS OF THE BAR, OR NUMERICALLY USING THE MAGNITUDE OF
00170*     Y AT EACH END OF  N SEGMENTS, MEANING N+1 VALUES.
00180*
00190*     INPUT:
00200*        THE USER IS PROMPTED FOR THE TOTAL BAR LENGTH, THE ELASTIC
00210*     MODULUS, AND THE AXIAL LOAD. THE USER IS THEN ASKED IF THE
00220*     BAR IS BOUNDED BY A KNOWN FUNCTION, AS WELL AS THE SHAPE OF ITS
00230*     X-SECTION. FOR THE CASE OF THE SLAB, THE UNIFORM THICKNESS IS
00240*     ALSO ASKED FOR... IF THE FUNCTION IS KNOWN, THE CONSTANTS ARE
00250*     THEN PROMPTED AND THE ENDPOINTS OF THE BAR ON THE X-AXIS INPUTTED;
00260*     ALTERNATELY, THE NUMBER OF SEGMENTS AND MEASURED HEIGHTS/DIAMETERS
00270*     MUST BE ENTERED.
00280*
00290*     OUTPUT:
00300*        THE TOTAL ELONGATION OF THE BAR IS DETERMINED AND PRINTED.
00310*
00320*     VARIABLES:
00330*        L,T,EM       ---   LENGTH,THICKNESS,ELASTIC MODULUS OF BAR
00340*        A,B          ---   CONSTANTS OF Y = A*E ^(B*X)  GOVERNING BAR BOUNDA
00350*        X0,XN        ---   ENDPOINTS OF SHAFT ON X-AXIS
00360*        P            ---   CENTRALLY APPLIED AXIAL LOAD
00370*        AA(100)      ---   INDIVIDUAL SEGMENT HEIGHTS/DIAMETERS
00380*        AREA         ---   X-SECTIONAL AREA OF EACH SMALL INCREMENT
00390*        ANS          ---   DETERMINE IF USER HAS A KNOWN FUNCTION
00400*        TYPE         ---   DETERMINE BAR X-SECTION
00410*        DELTA        ---   UNIFORM BAR ELONGATION
00420*        LEN          ---   LENGTH OF INCREMENTAL ELEMENT
00430*
00440***********************************************************************
00450***********************************************************************
00460*                           MAIN PROGRAM
00470***********************************************************************
00480***********************************************************************
00490*
00500*     VARIABLE DECLARATION
00510*
00520      REAL I,T,L,EM,A,B,X0,XN,P,DELTA,AA(100),AREA,LEN
00530      INTEGER ANS,TYPE,NUM,J
00540*
00550*         USER INPUT PROMPTS
00560*
00570      PRINT*,'ENTER THE TOTAL LENGTH OF THE BAR (IN M OR INCHES):'
00580      READ*,L
00590      PRINT*,'ENTER THE ELASTIC MODULUS (IN PASCALS OR PSI) :'
00600      READ*,EM
00610      PRINT*,'ENTER THE UNIFORM AXIAL LOAD (IN NEWTONS OR LBS) :'
00620      READ*,P
00630      PRINT*,'PLEASE DENOTE THE BAR X-SECTIONAL SHAPE:'
00640      PRINT*,'ENTER  1--SLAB  ;   2--CIRCULAR ROD'
00650      READ*,TYPE
00660*
00670*         IF A SLAB, PROMPT FOR ITS THICKNESS
00680*
00690      IF (TYPE.EQ.1) THEN
00700         PRINT*,'ENTER THE THICKNESS OF THE SLAB (IN M OR INCHES):'
00710         READ*,T
00720      ENDIF
00730      PRINT*,'DO YOU KNOW THE FUNCTION DESCRIBING THE BAR?'
00740      PRINT*,'ENTER  1--YES  ;   2--NO'
00750      READ*,ANS
00760*
00770*         IF ANS EQUALS ONE, THE USER KNOWS FUNCTION. PROMPT
00780*         FOR CONSTANTS AND ENDPOINTS.
00790*
00800      IF (ANS.EQ.1) THEN
```

```
00810        PRINT*,'F(X) = A*E ^(B*X)'
00820        PRINT*,'ENTER A,B:'
00830        READ*,A,B
00840        PRINT*,'ENTER THE X-COORDINATE FOR BOTH ENDS OF THE BAR:'
00850        PRINT*,'(IN M OR INCHES):'
00860        READ*,X0,XN
00870*
00880     AREA = 0
00890     L=XN-X0
00900     LEN=L/50
00910     DO 20 I = X0,XN,LEN
00920        Y1=(A*(2.71828**(B*I)))*2
00930        Y2=(A*(2.71828**(B*(I + LEN))))*2
00940        Y=(Y1+Y2)/2
00950        IF(TYPE.EQ.1) THEN
00960           AREA=1/(Y*T) + AREA
00970        ELSE
00980           AREA=4/(3.14159*(Y**2)) + AREA
00990        ENDIF
01000 20  CONTINUE
01010*
01020*          IF ANS EQUALS TWO, THE USER DOES NOT KNOW FUNCTION.
01030*          PROMPT FOR NUMBER OF SEGMENTS AND MEASURED HEIGHTS/DIAMETERS.
01040*
01050     ELSE
01060        PRINT*,'ENTER THE NUMBER OF SECTIONS TO BE CALCULATED:'
01070        READ*,NUM
01080        IF(TYPE.EQ.1) THEN
01090           PRINT*,'ENTER THE HEIGHTS OF THE ENDS FOR SECTIONS 1 TO N:'
01100           PRINT*,'(IN M OR INCHES):'
01110        ELSE
01120           PRINT*,'ENTER THE DIAMETERS OF THE ENDS FOR SECTIONS 1 TO N:'
01130           PRINT*,'(IN M OR INCHES):'
01140        ENDIF
01150*
01160*          INPUT MEASURED HEIGHTS/DIAMETERS
01170*
01180        DO 30 J=1,NUM+1
01190           READ*,AA(J)
01200 30     CONTINUE
01210*
01220        AREA = 0
01230        LEN = L/NUM
01240        DO 40 J = 1,NUM+1
01250           Y=(AA(J)+AA(J+1))/2
01260           IF(TYPE.EQ.1) THEN
01270              AREA = 1/(Y*T) + AREA
01280           ELSE
01290              AREA = 4/(3.14159*(Y**2)) + AREA
01300           ENDIF
01310 40     CONTINUE
01320     ENDIF
01330*
01340*          DETERMINING THE ELONGATION OF THE LOADED BAR
01350*
01360     DELTA=(P*LEN*AREA)/EM
01370*
01380     PRINT 50,DELTA
01390*
01400 50  FORMAT(2X,'THE DEFORMATION OF THE BAR IS:',F8.5,' (M OR IN.)')
01410*
01420     STOP
01430     END
```

1.15. A bar of variable solid circular cross section is bounded by the curve $y = 8e^{-0.01x}$ and extends from $x = 0$ to $x = 180$ in. It is subject to an axial tensile load of 100,000 lb as shown in Fig. 1-32. The material is steel, for which $E = 30 \times 10^6$ lb/in^2. Use the FORTRAN program of Problem 1.14 to determine the elongation of the bar.

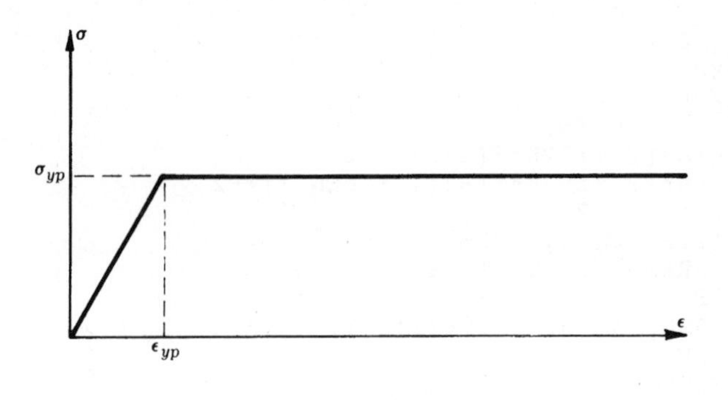

Fig. 1-32

Since the contour is bounded by the curve of the form $y = Ae^{Bx}$, we have $A = 8$ and $B = -0.01$. The bar extends from $x = 0$ to $x = 180$ in and entry of these data into the program of Problem 1.14 leads to an axial elongation of 0.03176 in.

```
run
 ENTER THE TOTAL LENGTH OF THE BAR (IN M OR INCHES):
? 180
 ENTER THE ELASTIC MODULUS (IN PASCALS OR PSI) :
? 30E+6
 ENTER THE UNIFORM AXIAL LOAD (IN NEWTONS OR LBS) :
? 100000
 PLEASE DENOTE THE BAR X-SECTIONAL SHAPE:
 ENTER  1--SLAB  :   2--CIRCULAR ROD
? 2
 DO YOU KNOW THE FUNCTION DESCRIBING THE BAR?
 ENTER  1--YES  ;   2--NO
? 1
 F(X) = A*E^(B*X)
 ENTER A,B:
? 8,-0.01
 ENTER THE X-COORDINATE FOR BOTH ENDS OF THE BAR:
 (IN M OR INCHES):
? 0,180
  THE DEFORMATION OF THE BAR IS:  .03176 (M OR IN)

SRU       0.804 UNTS.
```

1.16. A flat slab of variable depth is bounded by the curve $y = 0.25e^{0.025x}$ and extends from $x = 4$ m to $x = 10$ m as shown in Fig. 1-33. The slab is 10 mm thick and is subject to an axial tensile force of 385 kN. Use the FORTRAN program of Problem 1.14 to determine the elongation of the slab. Take $E = 200$ GPa.

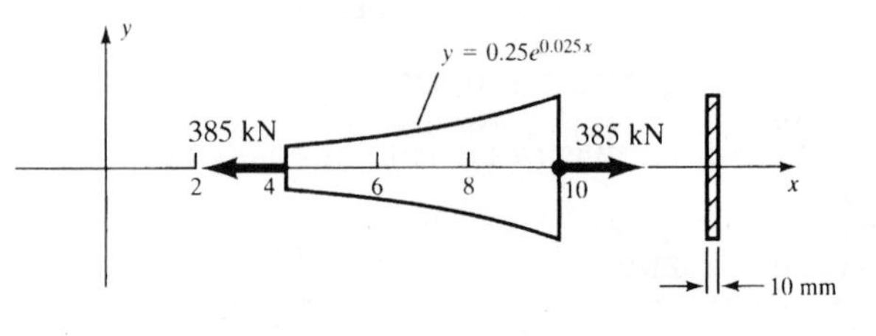

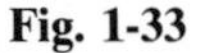

Fig. 1-33

To enter the program of Problem 1.14, we must set $A = 0.25$ and $B = 0.025$. The input data then appear as

```
run
 ENTER THE TOTAL LENGTH OF THE BAR (IN M OR INCHES):
? 6
 ENTER THE ELASTIC MODULUS (IN PASCALS OR PSI) :
? 200E+9
 ENTER THE UNIFORM AXIAL LOAD (IN NEWTONS OR LBS) :
? 385000
 PLEASE DENOTE THE BAR X-SECTIONAL SHAPE:
 ENTER  1--SLAB  :   2--CIRCULAR ROD
? 1
 ENTER THE THICKNESS OF THE SLAB (IN M OR INCHES):
? 0.01
 DO YOU KNOW THE FUNCTION DESCRIBING THE BAR?
 ENTER  1--YES  ;   2--NO
? 1
 F(X) = A*E^(B*X)
 ENTER A,B:
? 0.25,0.025
 ENTER THE X-COORDINATE FOR BOTH ENDS OF THE BAR:
 (IN M OR INCHES):
? 4,10
  THE DEFORMATION OF THE BAR IS:  .00198 (M OR IN)
```

The elongation of the bar is thus 0.00198 m or 1.98 mm.

1.17. Consider two thin rods or wires as shown in Fig. 1-34(a), which are pinned at A, B, and C and are initially horizontal and of length L when no load is applied. The weight of each wire is negligible. A force Q is then applied (gradually) at the point B. Determine the magnitude of Q so as to produce a prescribed vertical deflection δ of the point B.

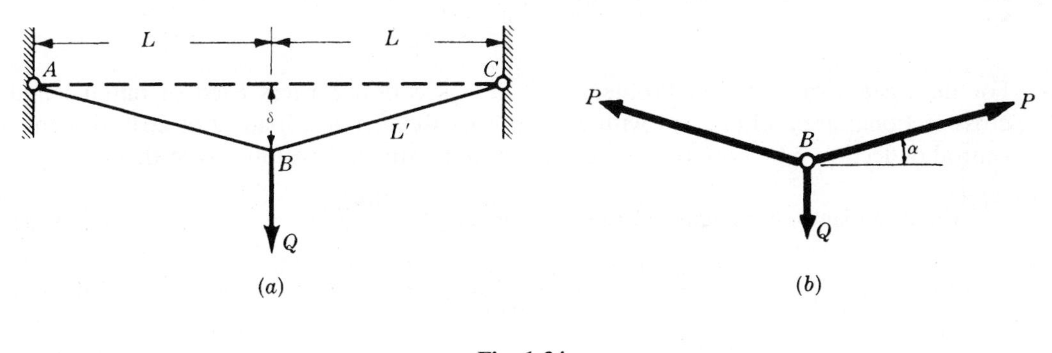

Fig. 1-34

This is an extremely interesting example of a system in which the elongations of all the individual members satisfy Hooke's law and yet for geometric reasons deflection is *not* proportional to force.

Each bar obeys the relation $\Delta = PL/AE$ where P is the axial force in each bar and Δ the axial elongation. Initially each bar is of length L and after the entire load Q has been applied the length is L'. Thus

$$L' - L = \frac{PL}{AE} \tag{1}$$

The free-body diagram of the pin at B is shown in Fig. 1-34(b). From statics,

$$\Sigma F_v = 2P\sin\alpha - Q = 0 \qquad \text{or} \qquad Q = 2P\left(\frac{\delta}{L'}\right)$$

Using (*1*),
$$Q = 2\frac{(L'-L)AE}{L}\frac{\delta}{L'} = \frac{2\delta AE}{L}\left(1-\frac{L}{L'}\right) \tag{2}$$

But
$$(L')^2 = L^2 + \delta^2 \tag{3}$$

Consequently
$$Q = \frac{2\delta AE}{L}\left(1-\frac{L}{\sqrt{L^2+\delta^2}}\right) \tag{4}$$

Also, from the binomial theorem we have

$$\sqrt{L^2+\delta^2} = L\left(1+\frac{\delta^2}{L^2}\right)^{1/2} = L\left(1+\frac{1}{2}\frac{\delta^2}{L^2}+\cdots\right) \tag{5}$$

and thus
$$1-\frac{L}{L\left(1+\dfrac{1}{2}\dfrac{\delta^2}{L^2}\right)} \approx 1-\left(1-\frac{1}{2}\frac{\delta^2}{L^2}\right) = \frac{1}{2}\frac{\delta^2}{L^2} \tag{6}$$

From this we have the approximate relation between force and displacement,

$$Q \approx \frac{2AE\delta}{L}\frac{\delta^2}{2L^2} = \frac{AE\delta^3}{L^3} \tag{7}$$

which corresponds to (*4*).

Thus the displacement δ is *not* proportional to the force Q even though Hooke's law holds for each bar individually. It is to be noted that Q becomes more nearly proportional to δ as δ becomes larger, assuming that Hooke's law still holds for the elongations of the bars. In this example superposition does *not* hold. The characteristic of this system is that the action of the external forces is *appreciably* affected by the small deformations which take place. In this event the stresses and displacements are not linear functions of the applied loads and superposition does not apply.

Summary: A material must follow Hooke's law if superposition is to apply. But this requirement alone is not sufficient. We must see whether or not the action of the applied loads is affected by small deformations of the structure. If the effect is substantial, superposition does not hold.

1.18. For the system discussed in Problem 1.17, let us consider wires each of initial length 5 ft, cross-sectional area 0.1 in^2, and with $E = 30 \times 16^6\ \text{lb/in}^2$. For a load Q of 20 lb determine the central deflection δ by both the exact and the approximate relations given there.

The exact expression relating force and deflection is $Q = \dfrac{2\delta AE}{L}\left(1-\dfrac{L}{\sqrt{L^2+\delta^2}}\right)$. Substituting the given numerical values, $20 = \dfrac{2\delta(0.1)(30\times10^6)}{(60)}\left(1-\dfrac{60}{\sqrt{(60)^2+\delta^2}}\right)$. Solving by trial and error we find $\delta = 1.131$ in.

The approximate relation between force and deflection is $Q \approx \dfrac{AE\delta^3}{L^3}$. Substituting,

$$20 \approx \frac{(0.1)(30\times10^6)\delta^3}{(60)^3} \qquad \text{from which} \qquad \delta \approx 1.129 \text{ in}$$

1.19. A square steel bar 50 mm on a side and 1 m long is subject to an axial tensile force of 250 kN. Determine the decrease in the lateral dimension due to this load. Consider $E = 200$ GPa and $\mu = 0.3$.

The loading is axial, hence the stress in the direction of the load is given by

$$\sigma = \frac{P}{A} = \frac{(250 \times 10^3\ \text{N})}{(0.05\ \text{m})(0.05\ \text{m})} = 100\ \text{MPa}$$

The simple form of Hooke's law for uniaxial loading states that $E = \sigma/\epsilon$. The strain ϵ in the direction of the load is thus $(100 \times 10^6)/(200 \times 10^9) = 5 \times 10^{-4}$.

The ratio of the lateral strain to the axial strain is denoted as Poisson's ratio, i.e.,

$$\mu = \frac{\text{lateral strain}}{\text{axial strain}}$$

The axial strain has been found to be 5×10^{-4}. Consequently, the lateral strain is μ times that value, or $(0.3)(5 \times 10^{-4}) = 1.5 \times 10^{-4}$. Since the lateral strain is 1.5×10^{-4}, the change in a 50 mm length is 7.5×10^{-3} mm, which represents the decrease in the lateral dimension of the bar.

It is to be noted that the definition of Poisson's ratio of two strains presumes that only a single uniaxial load acts on the member.

1.20. Consider a state of stress of an element such that a stress σ_x is exerted in one direction, lateral contraction is free to occur in a second (z) direction, but is completely restrained in the third (y) direction. Find the ratio of the stress in the x-direction to the strain in that direction. Also, find the ratio of the strain in the z-direction to that in the x-direction.

Let us examine the general statement of Hooke's law discussed earlier. If in those equations we set $\sigma_z = 0$, $\epsilon_y = 0$ so as to satisfy the conditions of the problem, then Hooke's law becomes

$$\epsilon_x = \frac{1}{E}[\sigma_x - \mu(\sigma_y + 0)] \qquad (a)$$

$$\epsilon_y = \frac{1}{E}[\sigma_y - \mu(\sigma_x + 0)] = 0 \qquad (b)$$

$$\epsilon_z = \frac{1}{E}[0 - \mu(\sigma_x + \sigma_y)] \qquad (c)$$

From (b),

$$\sigma_y = \mu\sigma_x$$

Consequently, from (a)

$$\epsilon_x = \frac{1}{E}(\sigma_x - \mu^2\sigma_x) = \frac{1-\mu^2}{E}\sigma_x$$

Solving this equation for σ_x as a function of ϵ_x and substituting in (c), we have

$$\epsilon_z = -\frac{\mu}{E}(\sigma_x + \mu\sigma_x) = -\frac{\mu(1+\mu)}{E}\frac{\epsilon_x E}{1-\mu^2} = -\frac{\mu\epsilon_x}{1-\mu}$$

We may now form the ratios

$$\frac{\sigma_x}{\epsilon_x} = \frac{E}{1-\mu^2} \quad \text{and} \quad -\frac{\epsilon_z}{\epsilon_x} = \frac{\mu}{1-\mu}$$

The first quantity, $E/(1-\mu^2)$, is usually denoted as the *effective modulus of elasticity* and is useful in the theory of thin plates and shells. The second ratio, $\mu/(1-\mu)$, is called the *effective value of Poisson's ratio.*

1.21. Consider an elemental block subject to uniaxial tension (see Fig. 1-35). Derive approximate expressions for the change of volume per unit volume due to this loading.

The strain in the direction of the forces may be denoted by ϵ_x. The strains in the other two orthogonal directions are then each $-\mu\epsilon_x$. Consequently, if the initial dimensions of the element are dx, dy, and dz then the final dimensions are

$$(1+\epsilon_x)\,dx \qquad (1-\mu\epsilon_x)\,dy \qquad (1-\mu\epsilon_x)\,dz$$

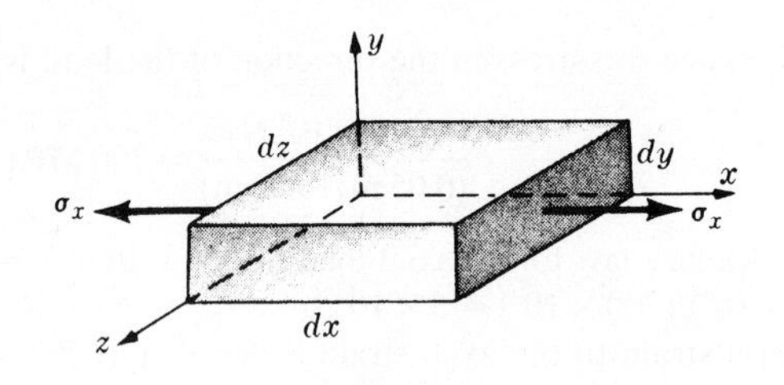

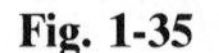
Fig. 1-35

and the volume after deformation is

$$\begin{aligned} V' &= [(1+\epsilon_x)\,dx]\,[(1-\mu\epsilon_x)\,dy]\,[(1-\mu\epsilon_x)\,dz] \\ &= (1+\epsilon_x)(1-2\mu\epsilon_x)\,dx\,dy\,dz \\ &= (1-2\mu\epsilon_x+\epsilon_x)\,dx\,dy\,dz \end{aligned}$$

since the deformations are so small that the *squares* and *products* of strains may be neglected.

Since the initial volume was $dx\,dy\,dz$, the change of volume per unit volume is

$$\frac{\Delta V}{V} = (1-2\mu)\epsilon_x$$

Hence, for a tensile force the volume increases slightly, for a compressive force it decreases.

Also, the cross-sectional area of the element in a plane normal to the direction of the applied force is given approximately by $A = (1-\mu\epsilon_x)^2\,dy\,dz = (1-2\mu\epsilon_x)\,dy\,dz$.

1.22. A square bar of aluminum 50 mm on a side and 250 mm long is loaded by axial tensile forces at the ends. Experimentally, it is found that the strain in the direction of the load is 0.001. Determine the volume of the bar when the load is acting. Consider $\mu = 0.33$.

From Problem 1.21 the change of volume per unit volume is given by

$$\frac{\Delta V}{V} = \epsilon(1-2\mu) = 0.001(1-0.66) = 0.00034$$

Consequently, the change of volume of the entire bar is given by

$$\Delta V = (50)(50)(250)(0.00034) = 212.5\text{ mm}^3$$

The original volume of the bar in the unstrained state is 6.25×10^5 mm^3. Since a tensile force increases the volume, the final volume under load is 6.252125×10^5 mm^3. Measurements made with the aid of lasers do permit determination of the final volume under load to the indicated accuracy of seven significant figures. Ordinary methods of measurement do not of course lead to such accuracy.

1.23. The general three-dimensional form of Hooke's law in which strain components are expressed as functions of stress components has already been presented. Occasionally it is necessary to express the stress components as functions of the strain components. Derive these expressions.

Given the previous expressions

$$\epsilon_x = \frac{1}{E}[\sigma_x - \mu(\sigma_y+\sigma_z)] \tag{1}$$

$$\epsilon_y = \frac{1}{E}[\sigma_y - \mu(\sigma_x+\sigma_z)] \tag{2}$$

$$\epsilon_z = \frac{1}{E}[\sigma_z - \mu(\sigma_x+\sigma_y)] \tag{3}$$

let us introduce the notation

$$e = \epsilon_x + \epsilon_y + \epsilon_z \tag{4}$$

$$\theta = \sigma_x + \sigma_y + \sigma_z \tag{5}$$

With this notation, (*1*), (*2*), and (*3*) may be readily solved by determinants for the unknowns σ_x, σ_y, σ_z to yield

$$\sigma_x = \frac{\mu E}{(1+\mu)(1-2\mu)} e + \frac{E}{1+\mu}\epsilon_x \tag{6}$$

$$\sigma_y = \frac{\mu E}{(1+\mu)(1-2\mu)} e + \frac{E}{1+\mu}\epsilon_y \tag{7}$$

$$\sigma_z = \frac{\mu E}{(1+\mu)(1-2\mu)} e + \frac{E}{1+\mu}\epsilon_z \tag{8}$$

These are the desired expressions.

Further information may also be obtained from (*1*) through (*5*). If (*1*), (*2*), and (*3*) are added and the symbols e and θ introduced, we have

$$e = \frac{1}{E}(1-2\mu)\theta \tag{9}$$

For the special case of a solid subjected to uniform hydrostatic pressure p, $\sigma_x = \sigma_y = \sigma_z = -p$. Hence

$$e = \frac{-3(1-2\mu)p}{E} \quad \text{or} \quad \frac{p}{e} = -\frac{E}{3(1-2\mu)} \tag{10}$$

The quantity $E/3(1-2\mu)$ is often denoted by K and is called the *bulk modulus* or *modulus of volume expansion* of the material. Physically, the bulk modulus K is a measure of the resistance of a material to change of volume without change of shape or form.

We see that the final volume of an element having sides dx, dy, dz prior to loading and subject to strains ϵ_x, ϵ_y, ϵ_z is $(1+\epsilon_x)\,dx\,(1+\epsilon_y)\,dy\,(1+\epsilon_z)\,dz = (1+\epsilon_x+\epsilon_y+\epsilon_z)\,dx\,dy\,dz$.

Thus the ratio of the increase in volume to the original volume is given approximately by

$$e = \epsilon_x + \epsilon_y + \epsilon_z$$

This change of volume per unit volume, e, is defined as the *dilatation*.

1.24. A steel cube is subject to a hydrostatic pressure of 1.5 MPa. Because of this pressure the volume decreases to give a dilatation of -10^{-5}. The Young's modulus of the material is 200 GPa. Determine Poisson's ratio of the material and also the bulk modulus.

From Problem 1.23 for hydrostatic loading the dilatation e is given by Eq. (*10*)

$$e = \frac{-3(1-2\mu)p}{E}$$

Substituting the given numerical values, we have

$$-10^{-5} = \frac{-3(1-2\mu)(1.5\times 10^6\ \text{N/m}^2)}{200\times 10^9\ \text{N/m}^2}$$

from which $\mu = 0.278$. Also from Problem 1.23 the bulk modulus is

$$K = \frac{E}{3(1-2\mu)}$$

which becomes

$$K = \frac{200\times 10^9\ \text{N/m}^2}{3(1-0.556)} = 150\ \text{MPa}$$

1.25. Determine the specific strength and also the specific modulus in the USCS system of (*a*) aluminum alloy, (*b*) titanium alloy, and (*c*) S-glass epoxy. Use materials properties given in Table 1-1.

By definition, specific strength is the ratio of the ultimate stress to the specific weight of the material and specific modulus is the ratio of Young's modulus to the specific weight.

(*a*) From aluminum alloy we have

$$\text{Specific strength} = \frac{80{,}000\ \text{lb/in}^2}{0.0984\ \text{lb/in}^3} = 813{,}000\ \text{in}$$

$$\text{Specific modulus} = \frac{12 \times 10^6\ \text{lb/in}^2}{0.0984\ \text{lb/in}^3} = 122 \times 10^6\ \text{in}$$

(*b*) For titanium alloy we have

$$\text{Specific strength} = \frac{140{,}000\ \text{lb/in}^2}{0.162\ \text{lb/in}^3} = 864{,}200\ \text{in}$$

$$\text{Specific modulus} = \frac{17 \times 10^6\ \text{lb/in}^2}{0.162\ \text{lb/in}^3} = 105 \times 10^6\ \text{in}$$

(*c*) For S-glass epoxy we have

$$\text{Specific strength} = \frac{275{,}000\ \text{lb/in}^2}{0.0766\ \text{lb/in}^3} = 3.6 \times 10^6\ \text{in}$$

$$\text{Specific modulus} = \frac{9.6 \times 10^6\ \text{lb/in}^2}{0.0766\ \text{lb/in}^3} = 125 \times 10^6\ \text{in}$$

Comparison of these specific strengths reveals that the composite material (S-glass epoxy) is much stronger on a unit weight basis than either of the metals, and it also has a slightly higher modulus, indicating greater rigidity than either of the metals.

1.26. Consider a low-carbon square steel bar 20 mm on a side and 1.7 m long having a material yield point of 275 MPa and $E = 200$ GPa. An applied axial load gradually builds up from zero to a value such that the elongation of the bar is 15 mm, after which the load is removed. Determine the permanent elongation of the bar after removal of the load. Assume elastic, perfectly plastic behavior as shown in Fig. 1-36.

Yield begins when the applied load reaches a value of

$$\begin{aligned} P &= \sigma_{yp}\ (\text{area}) \\ &= (275 \times 10^6\ \text{N/m}^2)(0.020\ \text{m})^2 \\ &= 110{,}000\ \text{N} \end{aligned}$$

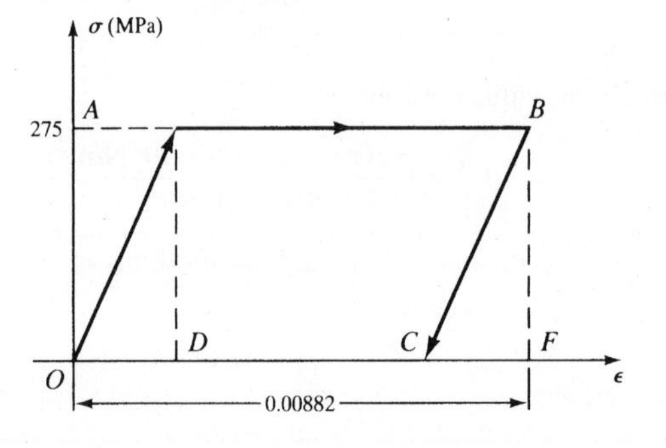

Fig. 1-36

which corresponds to point A of Fig. 1-36. Note that in that figure the ordinate is stress and the abscissa is strain. However, values on each of these axes differ only by constants from those on a force-elongation plot.

When the elongation is 15 mm, corresponding to point B in Fig. 1-36, unloading begins and the axial strain at the initiation of unloading is

$$\frac{15\ \text{mm}}{1700\ \text{mm}} = 0.00882$$

Unloading follows along line BC (parallel to AO) until the horizontal axis is reached, so that OC corresponds to the strain after complete removal of the load. We next find the strain CF—but this is readily found from using the similar triangles OAD and CBF to be

$$E = \frac{\sigma}{\epsilon}$$

$$\epsilon = \frac{275 \times 10^6\ \text{Pa}}{200 \times 10^9\ \text{Pa}} = 1.375 \times 10^{-3}$$

Thus, after load removal the residual strain is

$$OC = OF - CF$$

$$= 0.00882 - 0.00138 = 0.00744$$

The elongation of the 1.7-m long bar is consequently

$$(1.7\ \text{m})(0.00744) = 0.0126\ \text{m} \qquad \text{or} \qquad 12.6\ \text{mm}$$

Supplementary Problems

1.27. Forces acting in the articulated joints in the human vertebrae may lead to excessive stresses and eventual rupture of the spinal discs. Measurements of the adult disc indicate a surface area of approximately 1000 mm^2. Additional measurements during a lifting exercise indicate that a normal force of 708 N has been developed. Determine the normal stress in the disc. *Ans.* 708 kPa

1.28. Laboratory tests on human teeth indicate that the area effective during chewing is approximately 0.04 in^2 and that the tooth length is about 0.41 in. If the applied load in the vertical direction is 200 lb and the measured shortening is 0.0015 in, determine Young's modulus. *Ans.* 1.37×10^6 lb/in^2

1.29. A hollow right-circular cylinder is made of cast iron and has an outside diameter of 75 mm and an inside diameter of 60 mm. If the cylinder is loaded by an axial compressive force of 50 kN, determine the total shortening in a 600-mm length. Also determine the normal stress under this load. Take the modulus of elasticity to be 100 GPa and neglect any possibility of lateral buckling of the cylinder.
Ans. $\Delta = 0.188$ mm, $\sigma = 31.45$ MPa

1.30. A solid circular steel rod 6 mm in diameter and 500 mm long is rigidly fastened to the end of a square brass bar 25 mm on a side and 400 mm long, the geometric axes of the bars lying along the same line. An axial tensile force of 5 kN is applied at each of the extreme ends. Determine the total elongation of the assembly. For steel, $E = 200$ GPa and for brass $E = 90$ GPa. *Ans.* 0.477 mm

1.31. A high-performance jet aircraft cruises at three times the speed of sound at an altitude of 25,000 m. It has a long, slender titanium body reinforced by titanium ribs. The length of the aircraft is 30 m and the coefficient of thermal expansion of the titanium is 10×10^{-6}/°C. Determine the increase of overall length of the aircraft at cruise altitude over its length on the ground if the temperature while cruising is 500°C above ground temperature. (*Note:* This change of length is of importance since the designer must account for it because it changes the performance characteristics of the system.) *Ans.* 0.150 m

1.32. One of the most promising materials for use as a superconductor is composed of yttrium (a rare earth metal), barium, copper, and oxygen. This material acts as a superconductor (i.e., transmits electricity with essentially no resistance losses) at temperatures up to $-178°C$. If the temperature is then raised to 67°C, and the coefficient of thermal expansion is $11.0 \times 10^{-6}/°C$, determine the elongation of a 100-m long segment due to this temperature differential. *Ans.* 0.27 m

1.33 A solid circular cross-section bar in the form of a truncated cone is made of aluminum and has the dimensions shown in Fig. 1-37. The bar is loaded by an axial tensile force of 80,000 lb and $E = 10 \times 10^6$ lb/in². Find the elongation of the bar. *Ans.* 0.00874 in

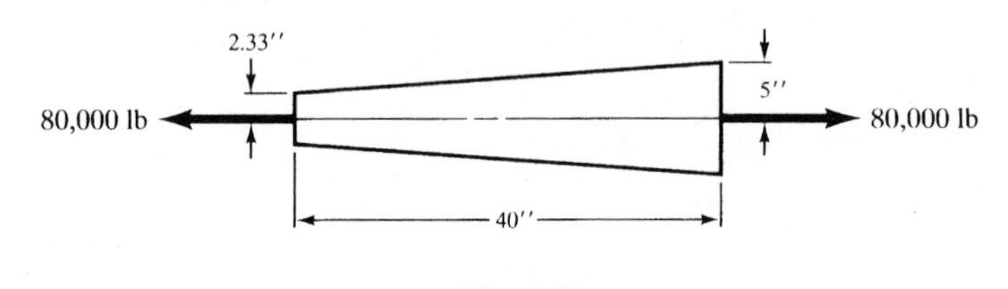

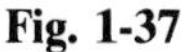

Fig. 1-37

1.34. A solid conical bar of circular cross section is suspended vertically, as shown in Fig. 1-38. The length of the bar is L, the diameter of the base is D, the modulus of elasticity is E, and the weight per unit volume is γ. Determine the elongation of the bar due to its own weight.

Ans. $\Delta = \dfrac{\gamma L^2}{6E}$

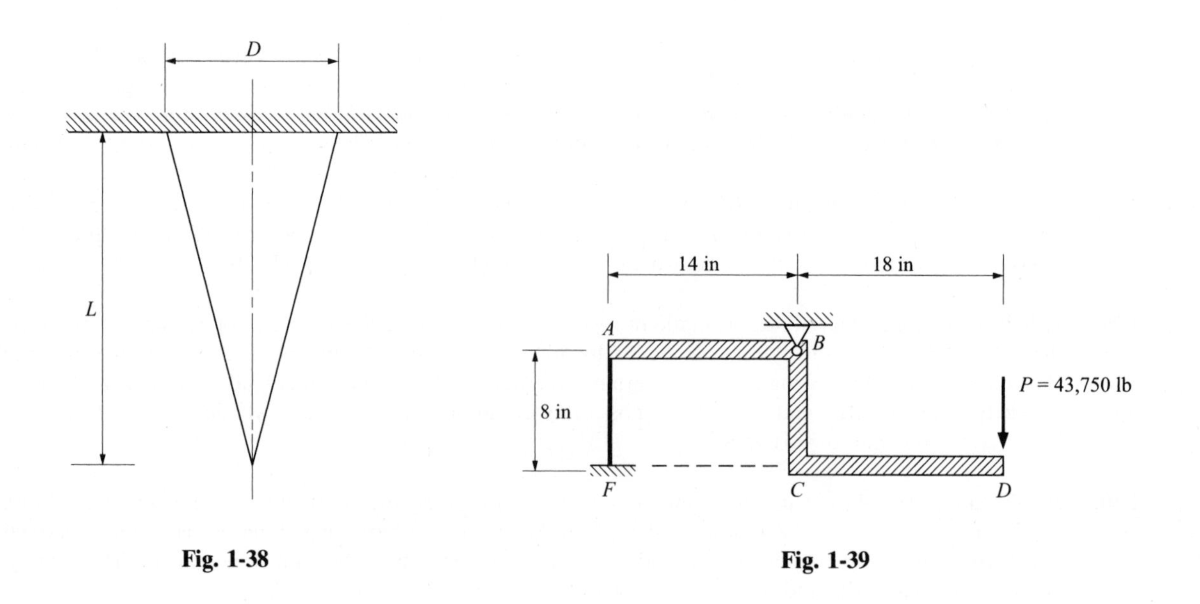

Fig. 1-38 **Fig. 1-39**

1.35. A Z-shaped rigid bar $ABCD$, shown in Fig. 1-39, is suspended by a pin at B, and loaded by a vertical force P. At A a steel tie rod AF connects the section to a firm ground support at F. Take $E = 30 \times 10^6$ lb/in². Determine the vertical deflection at D. *Ans.* 0.099 in

1.36. The rigid bar ABC is pinned at B and at A attached to a vertical steel bar AD which in turn is attached to a larger steel bar DF which is firmly attached to a rigid foundation. The geometry of the system is shown in Fig. 1-40. If a vertical force P of magnitude 40 kN is applied at C, determine the vertical displacement of point C. *Ans.* 9.17 mm

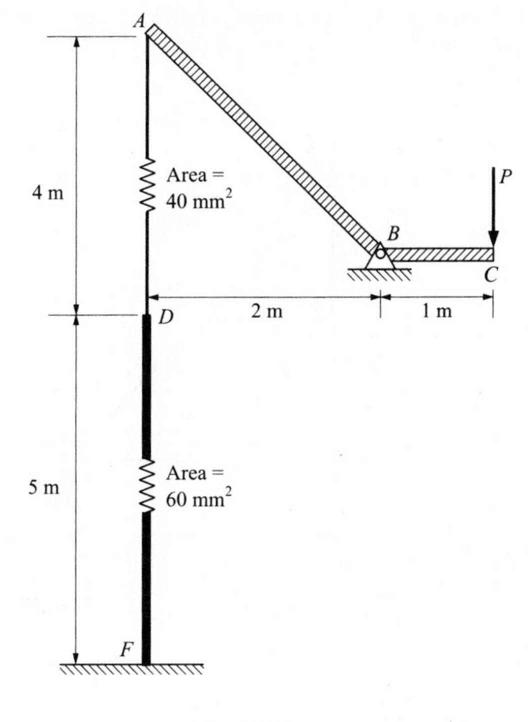

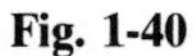
Fig. 1-40

1.37. A body having the form of a solid of revolution supports a load P as shown in Fig. 1-41. The radius of the upper base of the body is r_0 and the specific weight of the material is γ per unit volume. Determine how the radius should vary with the altitude in order that the compressive stress at all cross sections should be constant. The weight of the solid is not negligible. *Ans.* $r = r_0 e^{\gamma \pi r_0^2 y/2P}$

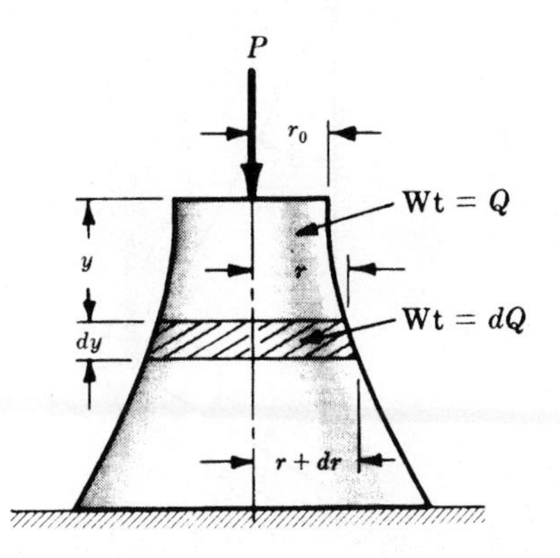

Fig. 1-41

1.38. In Problem 1.12 consider the force P to be 20,000 lb, $A_1 = 1.2\ \text{in}^2$, $L_1 = 5$ ft, $E_1 = 16 \times 10^6\ \text{lb/in}^2$, $A_2 = 1.5\ \text{in}^2$, $L_2 = 4$ ft, and $E_2 = 10 \times 10^6\ \text{lb/in}^2$. Find the horizontal and vertical components of displacement of pin B. *Ans.* $\Delta_x = 0.189$ in; $\Delta_y = 0.064$ in

1.39. In Fig. 1-42, AB, AC, BC, CD, and BD are pin-connected rods. Point B is attached to point E by a spring whose unstretched length is 1 m and whose spring constant is 4 kN/m. Neglecting the weight of all bars and the spring, determine the magnitude of the load W applied at D that makes CD horizontal. *Ans.* 583 N

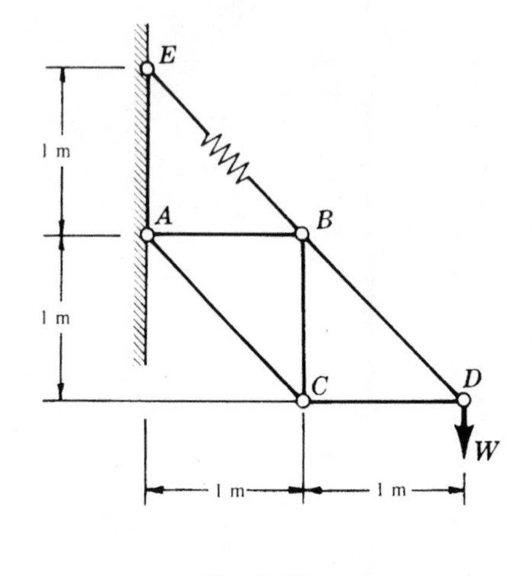

Fig. 1-42

1.40. The steel bars AB and BC are pinned at each end and support the load of 200 kN, as shown in Fig. 1-43. The material is structural steel, having a yield point of 200 MPa, and safety factors of 2 and 3.5 are satisfactory for tension and compression, respectively. Determine the size of each bar and also the horizontal and vertical components of displacement of point B. Take $E = 200$ GPa. Neglect any possibility of lateral buckling of bar BC.
Ans. Area $AB = 1732\ \text{mm}^2$, area $BC = 1750\ \text{mm}^2$, $\Delta_h = 0.37$ mm (to right), $\Delta_v = 1.78$ mm (downward)

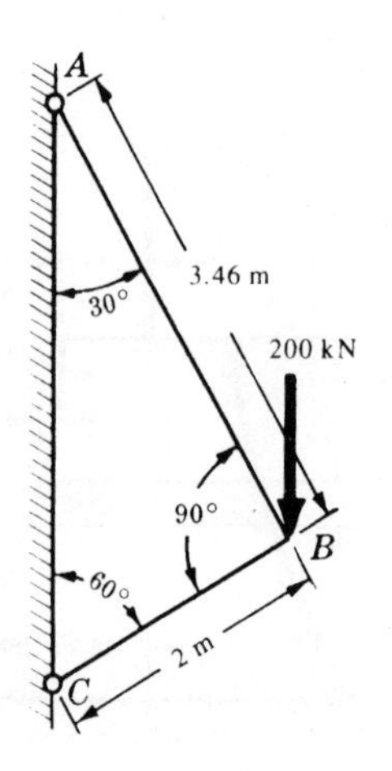

Fig. 1-43

1.41. The two bars AB and CB shown in Fig. 1-44 are pinned at each end and subject to a single vertical force P. The geometric and elastic constants of each are as indicated. Determine the horizontal and vertical components of displacement of pin B.

Ans. $\Delta_x = -\dfrac{PL_1}{\sqrt{3}A_1E_1} + \dfrac{PL_2}{\sqrt{3}A_2E_2}, \quad \Delta_y = \dfrac{PL_1}{3A_1E_1} + \dfrac{PL_2}{3A_2E_2}$

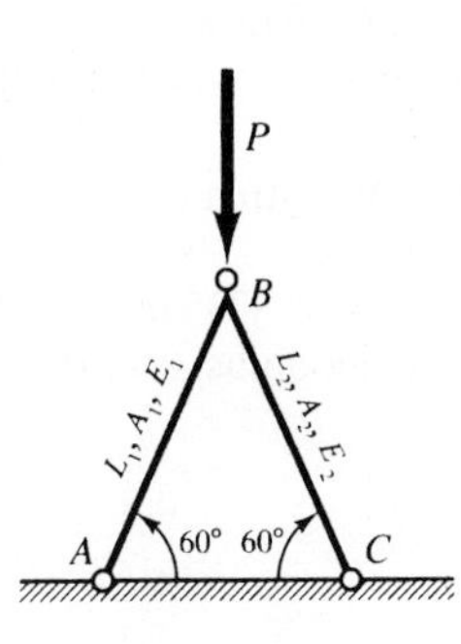

Fig. 1-44

1.42. In Problem 1-41, the bar AB is titanium, having an area of 1000 mm^2, length of 2.4 m, and $E_1 = 110$ GPa. Bar CB is steel having an area of 400 mm^2, length of 2.4 m, and $E_2 = 200$ GPa. What are the horizontal and vertical components of displacement of the pin B if $P = 600$ kN? *Ans.* $\Delta_x = 2.83$ mm, $\Delta_y = 10.4$ mm

1.43. A flat slab of variable width is bounded by the curve $y = 10e^{-0.25x}$ and extends from the origin to $x = 5$ in. It is subject to an axial tensile load of 20,000 lb and the material is steel for which $E = 30 \times 10^6$ lb/in^2. The slab thickness is 0.125 in. Use the FORTRAN program of Problem 1.14 to determine the elongation of the slab. *Ans.* 0.00275 in

1.44. A steel bar of solid circular cross section is bounded by the curve $y = 0.07e^{-0.05x}$ and extends from the origin to $x = 5$ m. It is subject to an axial tensile load of 1.5 MN and Young's modulus is 200 GPa. Use the FORTRAN program of Problem 1.14 to determine the elongation of the bar. *Ans.* 3.24 mm

1.45. Consider a state of stress of an element in which a stress σ_x is exerted in one direction and lateral contraction is completely restrained in each of the other two directions. Find the effective modulus of elasticity and also the effective value of Poisson's ratio.

Ans. eff. mod. $= \dfrac{E(1-\mu)}{(1-2\mu)(1+\mu)}$, eff. Poisson's ratio $= 0$

1.46. A block of aluminum alloy is 400 mm long and of rectangular cross section 25 by 30 mm. A compressive force $P = 60$ kN is applied in the direction of the 400-mm dimension and lateral contraction is completely restrained in each of the other two directions. Find the effective modulus of elasticity as well as the change of the 400-mm length. Take $E = 75$ GPa and Poisson's ratio to be 0.33.
Ans. eff. mod. = 114.5 GPa, change of length = −0.286 mm

1.47. Consider the state of stress in a bar subject to compression in the axial direction. Lateral expansion is restrained to half the amount it would ordinarily be if the lateral faces were load free. Find the effective modulus of elasticity.

Ans. $\dfrac{E(1-\mu)}{1-\mu-\mu^2}$

1.48. A bar of uniform cross section is subject to uniaxial tension and develops a strain in the direction of the force of 1/800. Calculate the change of volume per unit volume. Assume $\mu = 1/3$. *Ans.* 1/2400 (increase)

1.49. A square steel bar is 50 mm on a side and 250 mm long. It is loaded by an axial tensile force of 200 kN. If $E = 200$ GPa and $\mu = 0.3$, determine the change of volume per unit volume. *Ans.* 0.00016

1.50. Consider a low-carbon steel square steel bar 1 in on a side and 70 in long having a material yield point of 40,000 lb/in^2 and a Young's modulus of 30×10^6 lb/in^2. An axial tensile load gradually builds up from zero to a value such that the elongation of the bar is 0.6 in, after which the load is removed. Determine the permanent elongation of the bar. Assume that the material is elastic, perfectly plastic. *Ans.* 0.509 in

1.51. Determine, from Table 1-1, the specific strength and also the specific modulus of (*a*) nickel, and (*b*) boron epoxy composite. Use the SI system.
Ans. (*a*) nickel: specific strength = 3563 to 8736 m, specific modulus = 2.41×10^6 m; (*b*) boron epoxy: specific strength = 71.8×10^3 m, specific modulus = 11.0×10^6 m

Table 1-1. Properties of Common Engineering Materials at 68 °F (20 °C)

Material	Specific weight		Young's modulus		Ultimate stress		Coefficient of linear thermal expansion		Poisson's ratio
	lb/in^3	kN/m^3	lb/in^2	GPa	lb/in^2	kPa	10e–6/°F	10e–6/°C	
I. Metals in slab, bar, or block form									
Aluminum alloy	0.0984	27	10–12e6	70–79	45–80e3	310–550	13	23	0.33
Brass	0.307	84	14–16e6	96–110	43–85e3	300–590	11	20	0.34
Copper	0.322	87	16–18e6	112–120	33–55e3	230–380	9.5	17	0.33
Nickel	0.318	87	30e6	210	45–110e3	310–760	7.2	13	0.31
Steel	0.283	77	28–30e6	195–210	80–200e3	550–1400	6.5	12	0.30
Titanium alloy	0.162	44	15–17e6	105–120	130–140e3	900–970	4.5–5.5	8–10	0.33
II. Nonmetallics in slab, bar, or block form									
Concrete (composite)	0.0868	24	3.6e6	25	4000–6000	28–41	6	11	
Glass	0.0955	26	7–12e6	48–83	10,000	70	3–6	5–11	0.23
III. Materials in filamentary (whisker) form: [dia. <0.001 in (0.025 mm)]									
Aluminum oxide	0.141	38	100–350e6	690–2410	2–4e6	13,800–27,600			
Barium carbide	0.090	25	65e6	450	1e6	6900			
Glass			50e6	345	1–3e6	7000–20,000			
Graphite	0.081	22	142e6	980	3e6	20,000			
IV. Composite materials (unidirectionally reinforced in direction of loading)									
Boron epoxy	0.071	19	31e6	210	198,000	1365	2.5	4.5	
S-glass-reinforced epoxy	0.0766	21	9.6e6	66.2	275,000	1900			
V. Others									
Graphite-reinforced epoxy	0.054	15	15e6	104	190,000	1310			
Kevlar–49 epoxy*	0.050	13.7	12.5e6	86	220,000	1520			

*Tradename of E. I. duPont Co.

Chapter 2

Statically Indeterminate Force Systems Tension and Compression

DEFINITION OF A DETERMINATE FORCE SYSTEM

If the values of all the external forces which act on a body can be determined by the equations of static equilibrium alone, then the force system is *statically determinate*. The problems in Chap. 1 were all of this type.

Example 1

The bar shown in Fig. 2-1 is loaded by the force P. The reactions are R_1, R_2, and R_3. The system is statically determinate because there are three equations of static equilibrium available for the system and these are sufficient to determine the three unknowns.

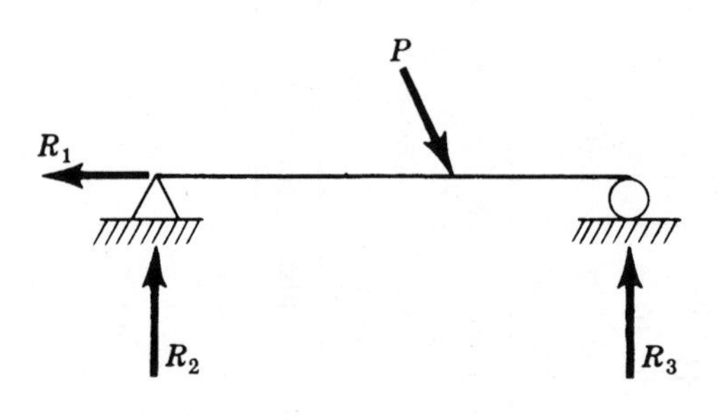

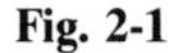

Fig. 2-1

Example 2

The truss $ABCD$ shown in Fig. 2-2 is loaded by the forces P_1 and P_2. The reactions are R_1, R_2, and R_3. Again, since there are three equations of static equilibrium available, all three unknown reactions may be determined and consequently the external force system is statically determinate.

The above two illustrations refer only to external reactions and the force systems may be defined as statically determinate *externally*.

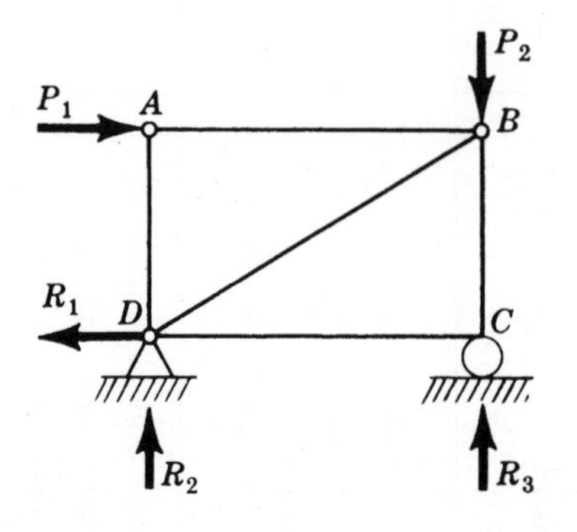

Fig. 2-2

DEFINITION OF AN INDETERMINATE FORCE SYSTEM

In many cases the forces acting on a body cannot be determined by the equations of statics alone because there are more unknown forces than there are equations of equilibrium. In such a case the force system is said to be *statically indeterminate.*

Example 3

The bar shown in Fig. 2-3 is loaded by the force P. The reactions are R_1, R_2, R_3, and R_4. The force system is statically indeterminate because there are four unknown reactions but only three equations of static equilibrium. Such a force system is said to be *indeterminate to the first degree.*

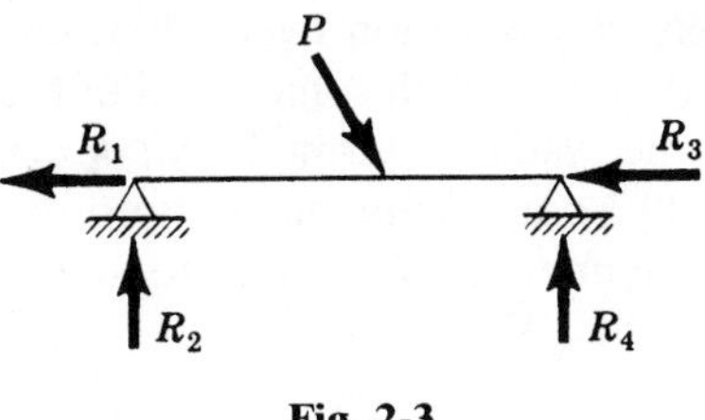

Fig. 2-3

Example 4

The bar shown in Fig. 2-4 is statically indeterminate to the second degree because there are five unknown reactions R_1, R_2, R_3, R_4, and M_1 but only three equations of static equilibrium. Consequently the values of all reactions cannot be determined by use of statics equations alone.

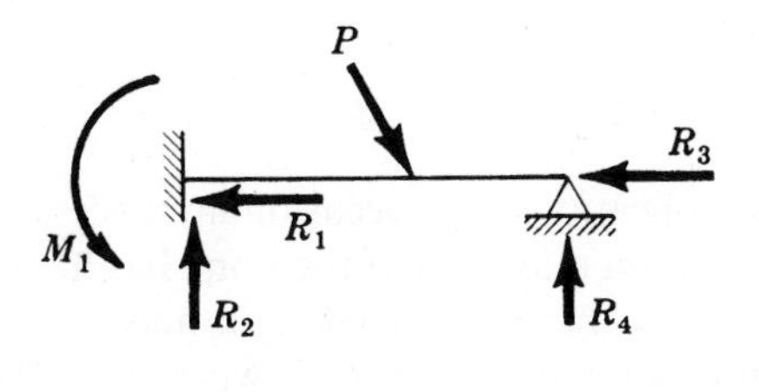

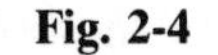

Fig. 2-4

METHOD OF ELASTIC ANALYSIS

The approach that we will consider here is called the *deformation method* because it considers the deformations in the system. Briefly, the procedure to be followed in analyzing an indeterminate system is first to write all equations of static equilibrium that pertain to the system and then *supplement* these equations with additional equations based upon the deformations of the structure. Enough equations involving deformations must be written so that the total number of equations from both statics and deformations is equal to the number of unknown forces involved. See Problems 2.1 through 2.12.

ANALYSIS FOR ULTIMATE STRENGTH (LIMIT DESIGN)

We consider that the stress-strain curve for the material is of the form indicated in Fig. 2-5, i.e., one characterizing an extremely ductile material such as structural steel. Such idealized elastoplastic behavior is a good representation of low-carbon steel. This representation assumes that the material is incapable of developing stresses greater than the yield point.

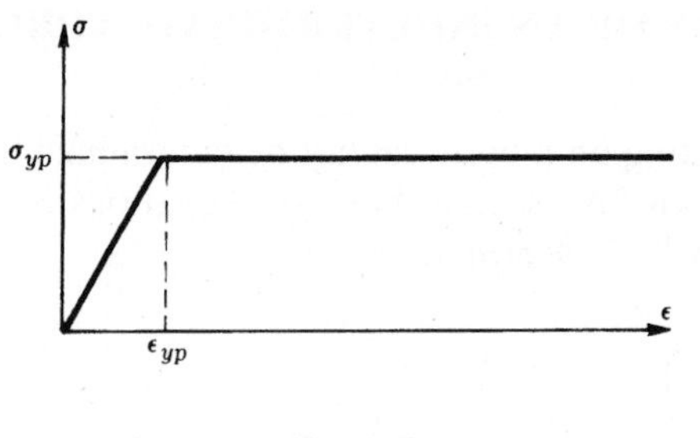

Fig. 2-5

In a statically indeterminate system any inelastic action changes the conditions of constraint. Under these altered conditions the loading that the system can carry usually increases over that predicted on the basis of completely elastic action everywhere in the system. Design of a statically indeterminate structure for that load under which some or all of the regions of the structure reach the yield point and cause "collapse" of the system is termed *limit design*. The *ultimate load* corresponding to such design is of course divided by some factor of safety to determine a *working load*. The term "limit design," when used in this manner, applies only to statically indeterminate structures. For applications, see Problems 2.13 through 2.17.

Solved Problems

Elastic Analysis

In Problems 2.1 through 2.12 it is assumed that the system is acting within the linear elastic range of action of the material.

2.1. In medical (orthopedic) applications it is occasionally necessary to lengthen a main bone of a human leg or arm. This situation may arise if the bone has healed in a wrong configuration after some accident, or alternatively the improper length may be due to a birth defect. One way to accomplish this lengthening is for the surgeon to weaken the bone through the introduction of one or two cuts near the outer surface of the bone, then attach the mechanical system shown in Fig. 2-6 to the exterior of the leg. This system consists of a pair of metallic rings which encircle the leg, with the rings being connected by a pair of parallel brass rods which are threaded at each end. The distance between the rings can be varied over the months of treatment by turning the nut at each end of each rod. Typically, the bone has a cross-sectional area of $1.2\ \text{in}^2$, a modulus of elasticity of $4.6 \times 10^6\ \text{lb/in}^2$, and a length of 8 in. The two brass rods have a total cross-sectional area of $0.05\ \text{in}^2$, a modulus of $13.5 \times 10^6\ \text{lb/in}^2$, and 32 threads per inch. If the nut at the end of the bar is turned $\frac{1}{8}$ of a revolution to stretch the bone, determine the axial stress arising in the bone.

Let us consider a section to be passed through the bone and perpendicular to the axial dimension of the bone. The free-body diagram of the system is shown in Fig. 2-7 where P_{bone} represents the axial force in the bone and P_{rod} is the axial force in each brass bar. For equilibrium:

$$P_{\text{bone}} = P_{\text{rod}} \tag{1}$$

From deformations of the system, we realize that the extension of the bone plus the shortening of each rod is equal to the displacement of the nut along the bar. This latter quantity is $\frac{1}{8}(\frac{1}{32}\ \text{in})$. Thus, we have

$$\frac{P_{\text{bone}}(8\ \text{in})}{(1.2\ \text{in}^2)(4.6 \times 10^6\ \text{lb/in}^2)} + \frac{P_{\text{rod}}(8\ \text{in})}{(0.05\ \text{in}^2)(13.5 \times 10^6\ \text{lb/in}^2)} = (\tfrac{1}{8})(\tfrac{1}{32}\ \text{in}) \tag{2}$$

Solving (*1*) and (*2*) we find

$$P_{\text{bone}} = 588 \text{ lb}$$

$$\sigma_{\text{bone}} = \frac{588 \text{ lb}}{1.2 \text{ in}^2} = 490 \text{ lb/in}^2$$

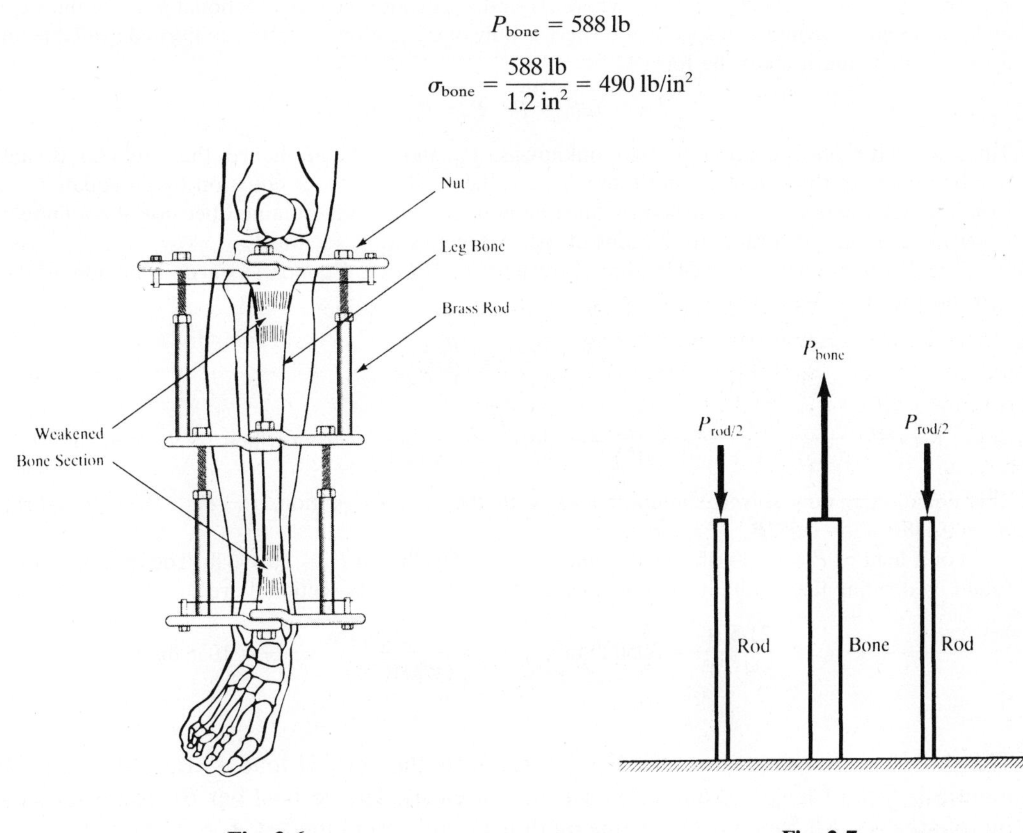

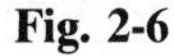
Fig. 2-6

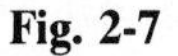
Fig. 2-7

2.2. Consider a steel tube surrounding a solid aluminum cylinder, the assembly being compressed between infinitely rigid cover plates by centrally applied forces as shown in Fig. 2-8(*a*). The aluminum cylinder is 3 in in diameter and the outside diameter of the steel tube is 3.5 in. If $P = 48{,}000$ lb, find the stress in the steel and also in the aluminum. For steel, $E = 30 \times 10^6$ lb/in² and for aluminum $E = 12 \times 10^6$ lb/in².

Let us pass a horizontal plane through the assembly at any elevation except in the immediate vicinity of the cover plates and then remove one portion or the other, say the upper portion. In that event the portion that we have removed must be replaced by the effect it exerted upon the remaining portion and that effect consists of vertical normal stresses distributed over the two materials. The free-body diagram of the portion of the assembly below this cutting plane is shown in Fig. 2-8(*b*) where σ_{st} and σ_{al} denote the normal stresses existing in the steel and aluminum respectively.

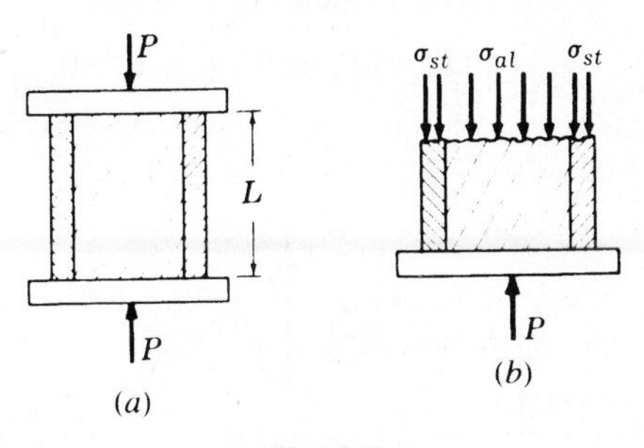

Fig. 2-8

Let us denote the resultant force carried by the steel by P_{st} (lb) and that carried by the aluminum by P_{al}. Then $P_{st} = A_{st}\sigma_{st}$ and $P_{al} = A_{al}\sigma_{al}$ where A_{st} and A_{al} denote the cross-sectional areas of the steel tube and the aluminum cylinder, respectively. There is only one equation of static equilibrium available for such a force system and it takes the form

$$\Sigma F_v = P - P_{st} - P_{al} = 0$$

Thus, we have one equation in two unknowns, P_{st} and P_{al}, and hence the problem is statically indeterminate. In that event we must supplement the available statics equation by an equation derived from the deformations of the structure. Such an equation is readily obtained because the infinitely rigid cover plates force the axial deformations of the two metals to be identical.

The deformation due to axial loading is given by $\Delta = PL/AE$. Equating axial deformations of the steel and the aluminum we have

$$\frac{P_{st}L}{A_{st}E_{st}} = \frac{P_{al}L}{A_{al}E_{al}}$$

or

$$\frac{P_{st}L}{(\pi/4)\,[(3.5)^2 - (3)^2]\,(30 \times 10^6)} = \frac{P_{al}L}{(\pi/4)\,(3)^2\,(12 \times 10^6)} \qquad \text{from which} \qquad P_{st} = 1.23P_{al}$$

This equation is now solved simultaneously with the statics equation, $P - P_{st} - P_{al} = 0$, and we find $P_{al} = 0.448P$, $P_{st} = 0.552P$.

For a load of $P = 48{,}000$ lb this becomes $P_{al} = 21{,}504$ lb and $P_{st} = 26{,}496$ lb. The desired stresses are found by dividing the resultant force in each material by its cross-sectional area:

$$\sigma_{al} = \frac{21{,}504}{(\pi/4)\,(3)^2} = 3050 \text{ lb/in}^2 \qquad \sigma_{st} = \frac{26{,}496}{(\pi/4)\,[(3.5)^2 - (3)^2]} = 1038 \text{ lb/in}^2$$

2.3. The three-bar assembly shown in Fig. 2-9 supports the vertical load P. Bars AB and BD are identical, each of length L and cross-sectional area A_1. The vertical bar BC is also of length L but of area A_2. All bars have the same modulus E and are pinned at A, B, C, and D. Determine the axial force in each of the bars.

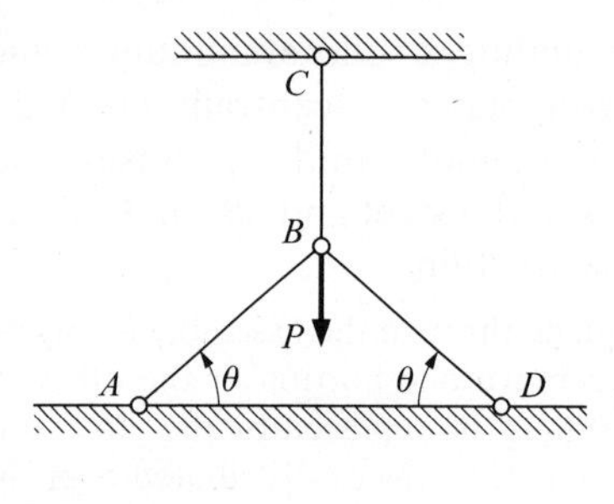

Fig. 2-9

First, we draw a free-body diagram of the pin at B. The forces in each of the bars are represented by P_1 and P_2 as shown in Fig. 2-10. For vertical equilibrium we find:

$$\Sigma F_v = 2P_1 \sin\theta + P_2 - P = 0 \qquad (1)$$

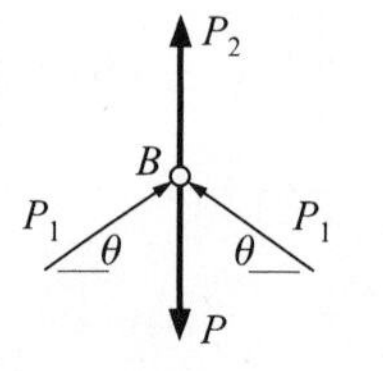

Fig. 2-10

We assume, temporarily, that the pin at B is removed. Next we examine deformations. Under the action of the axial force P_2 the vertical bar extends downward an amount

$$\Delta_1 = \frac{P_2 L}{A_2 E} \tag{2}$$

so that the lower end (originally at B) moves to B' as shown in Fig. 2-11.

Fig. 2-11 Fig. 2-12

The compressive force in AB causes it to shorten an amount Δ shown as BB'' in Fig. 2-12. The bar AB then rotates about A as a rigid body so that B'' moves to B''' directly below point C. From Fig. 2-12 the vertical component of Δ is

$$BB''' = \frac{P_1 L}{A_1 E \sin\theta}$$

Next, we consider the pin to be reinserted in the system. The points B' and B''' must coincide so that

$$\frac{P_2 L}{A_2 E} = \frac{P_1 L}{A_1 E \sin\theta} \tag{3}$$

Substituting Eq. (3) in Eq. (1) we find

$$P_1 = \frac{P\sin\theta}{2\sin^2\theta + \alpha}$$

$$P_2 = \frac{P\alpha}{2\sin^2\theta + \alpha}$$

where $\alpha = A_2/A_1$.

2.4. Consider the two identical bars AB and AC, each 0.5 m long, each with area A and $E = 200$ GPa. They are pinned at A, B, and C. Bar DF has area $2A$ and $E = 200$ GPa. Bar DF is accidentally made 0.8 mm too short to extend between A and D. Points A and F must be brought together mechanically to form a frame consisting of the two isosceles triangles shown in Fig. 2-13. Find

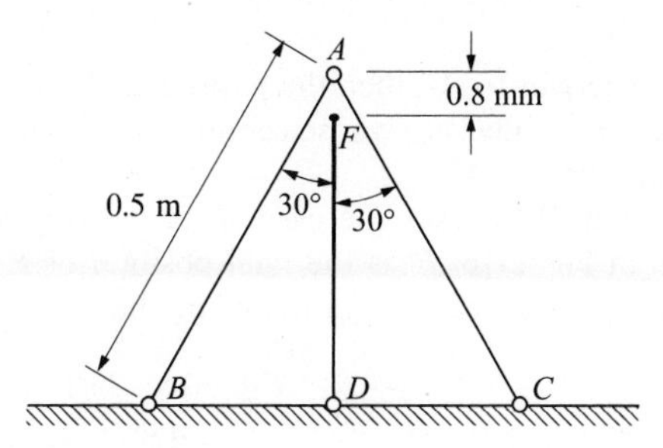

Fig. 2-13

the initial stresses in the bars prior to application of any external loading. The system of bars lies on a frictionless horizontal plane.

It is evident that point A must be forced downward (creating compression in AB and AC) and the end F of the vertical bar must be pulled upward to meet the (lowered) point A. The meeting point of A and F is *not* necessarily midway between the initial locations of A and F. After these two points have met, they are joined by a pin. At this stage there are no external applied loads on the three-bar system. However, there are locked-in stresses in each of the bars.

We may find these initial stresses by designating compressive forces in AB and AC by P_2 and the tensile force in FD by P_1 (Newtons). After these bars have been jointed by a pin, the free-body diagram of that pin appears as shown in Fig. 2-14.

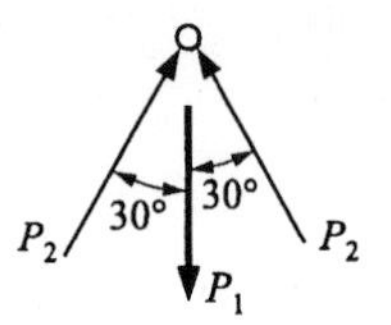

Fig. 2-14

For equilibrium of the pin:

$$2P_2 \cos 30° - P_1 = 0; \qquad \text{or} \qquad P_1 = P_2\sqrt{3} \tag{1}$$

As point A is mechanically forced downward, each of the bars AB and AC shortens an amount

$$\Delta_1 = \frac{P_2(500)}{AE}$$

in the direction of the respective bar. With the pin at A removed, the deformed configuration appears as shown in Fig. 2-15. The vertical component of Δ_1 is given by

$$\frac{P_2(500)}{AE \cos 30°}$$

The deformation of the inclined bars may be visualized (see Fig. 2-15) by realizing that the compressed

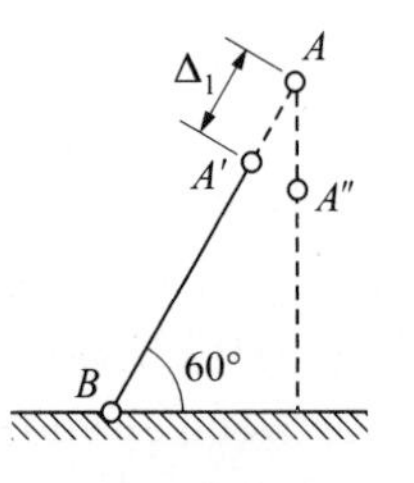

Fig. 2-15

bar AB first shortens as A moves to A', then the entire bar AB rotates as a rigid body about B so that A' moves to A'' actually along a circular arc whose center is at B, but for small angles of rotation the arc may be replaced by the straight line $A'A''$.

The tensile force in bar DF causes the point F in the originally stress-free bar to move vertically upward to F', as shown in Fig. 2-16. F' is the final position of F after the pin has been inserted at the junction of all three bars. The vertical elongation of the bar is

$$\Delta_2 = \frac{P_1(500 \cos 30°)}{2AE} \tag{2}$$

where $2A$ is the cross-sectional area of bar DF.

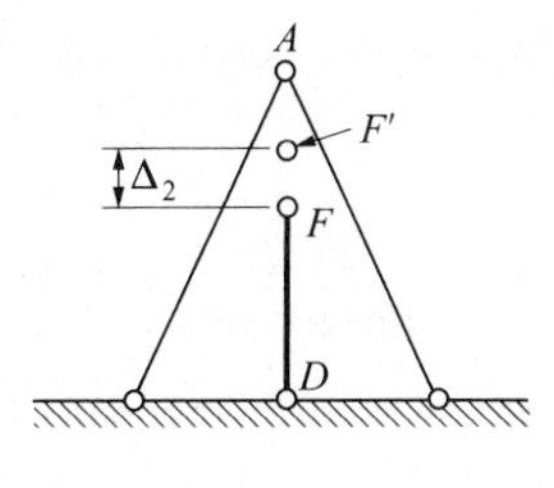

Fig. 2-16

Thus, to close the gap of 0.8 mm between the bars, we must have

$$\frac{P_2(500)}{AE\cos 30°}+\frac{P_1(500\cos 30°)}{2AE}=0.8\text{ mm} \tag{3}$$

Substituting Eq. (*1*) in Eq. (*3*) we find

$$577.4\frac{P_2}{AE}+\frac{(216.5)(P_2\sqrt{3})}{AE}=0.8$$

But $E = 200$ GPa, so solving the above equations for normal stresses in the bars we find

$$\sigma_2=\frac{P_2}{A}=168\text{ MPa}$$

$$\sigma_1=\frac{P_1}{2A}=145.5\text{ MPa}$$

2.5. The composite bar shown in Fig. 2-17(*a*) is rigidly attached to the two supports. The left portion of the bar is copper, of uniform cross-sectional area 12 in^2 and length 12 in. The right portion is aluminum, of uniform cross-sectional area 3 in^2 and length 8 in. At a temperature of 80°F the entire assembly is stress free. The temperature of the structure drops and during this process the right support yields 0.001 in in the direction of the contracting metal. Determine the minimum temperature to which the assembly may be subjected in order that the stress in the aluminum does not exceed 24,000 lb/in^2. For copper $E = 16\times10^6$ lb/in^2, $\alpha = 9.3\times10^{-6}$/°F and for aluminum $E = 10\times10^6$ lb/in^2, $\alpha = 12.8\times10^{-6}$/°F.

It is perhaps simplest to consider that the bar is cut just to the left of the supporting wall at the right and is then free to contract due to the temperature drop ΔT. The total shortening of the composite bar is given by

$$(9.3\times10^{-6})(12)\Delta T+(12.8\times10^{-6})(8)\Delta T$$

according to the definition of the coefficient of linear expansion. It is to be noted that the shape of the cross section has no influence upon the change in length of the bar due to a temperature change.

Even though the bar has contracted this amount, it is still stress free. However, this is not the complete analysis because the reaction of the wall at the right has been neglected by cutting the bar there. Consequently, we must represent the action of the wall by an axial force P applied to the bar, as shown

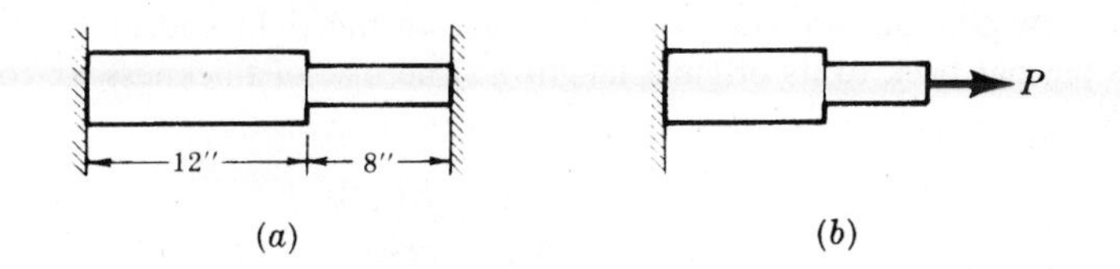

Fig. 2-17

in Fig. 2-17(*b*). For equilibrium, the resultant force acting over any cross section of either the copper or the aluminum must be equal to P. The application of the force P stretches the composite bar by an amount

$$\frac{P(12)}{12(16 \times 10^6)} + \frac{P(8)}{3(10 \times 10^6)}$$

If the right support were unyielding, we would equate the last expression to the expression giving the total shortening due to the temperature drop. Actually the right support yields 0.001 in and consequently we may write

$$\frac{P(12)}{12(16 \times 10^6)} + \frac{P(8)}{3(10 \times 10^6)} = (9.3 \times 10^{-6})(12)\Delta T + (12.8 \times 10^{-6})(8)\Delta T - 0.001$$

The stress in the aluminum is not to exceed 24,000 lb/in^2, and since it is given by the formula $\sigma = P/A$, the maximum force P becomes $P = A\sigma = 3(24{,}000) = 72{,}000$ lb. Substituting this value of P in the above equation relating deformations, we find $\Delta T = 115°$F. Therefore the temperature may drop 115°F from the original 80°F. The final temperature would be $-35°$F.

2.6. A bar (see Fig. 2-18) in the shape of a solid, truncated cone of circular cross section is situated between two rigid supports which constrain the bar from any change of axial length. The temperature of the entire bar is then raised ΔT. Assume that the cross sections perpendicular to the longitudinal axis of symmetry remain plane and neglect localized end effects due to the end supports. Determine the normal stress at any point in the bar.

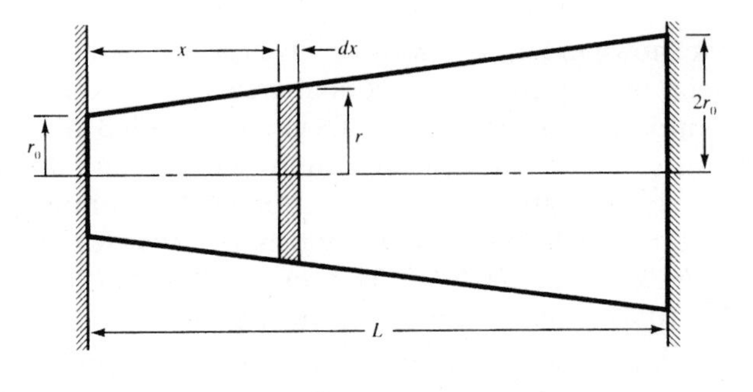

Fig. 2-18

Let us introduce the coordinate system shown in Fig. 2-18 where x denotes the distance of a thin disc from the left end of the bar, and dx is the thickness of the disc in the direction of the x-axis. The radius of this disc is found from geometry to be

$$r = r_0 + \frac{r_0 x}{L}$$

If the support at the right end of the bar is considered to be temporarily removed, the entire bar will expand in length an amount $\alpha(L)(\Delta T)$, where α is the coefficient of thermal expansion of the material.

We may now consider an axial force N to act on the right end of the bar, as shown in Fig. 2-19, to compress the bar back to its original length L. The disc of thickness dx compresses an amount (see Problem 1.1)

$$\frac{N\,dx}{AE} = \frac{N(dx)}{\pi r^2(E)}$$

because of this axial force N (which, for equilibrium, must be constant over any cross section of the bar).

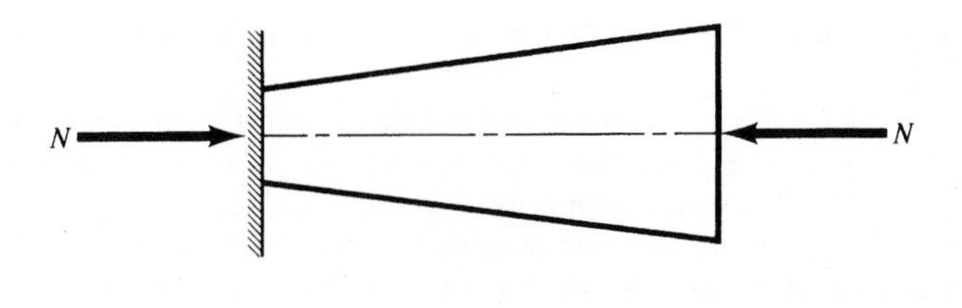

Fig. 2-19

The total compression of the bar due to N is found by summing the changes of length of all discs from $x = 0$ to $x = L$:

$$\int_{x=0}^{x=L} \frac{N(dx)}{\pi r^2 E} = \frac{NL^2}{E\pi r_0^2} \int_0^L \frac{dx}{(L+x)^2}$$

Integrating,

$$\int_0^L \frac{dx}{(L+x)^2} = \frac{1}{2L}$$

and setting the bar extension due to heating equal to bar compression due to the axial force N, we find

$$\alpha(L)(\Delta T) = \frac{NL^2}{E\pi r_0^2}\left(\frac{1}{2L}\right)$$

$$N = 2\alpha(\Delta T)E\pi r_0^2$$

The axial normal stress is now found by dividing the force N by the cross-sectional area at any station x,

$$\sigma = \frac{N}{\pi r^2} = \frac{2\alpha(\Delta T)E}{(1 + x/L)^2}$$

2.7. A hollow steel cylinder surrounds a solid copper cylinder and the assembly is subject to an axial loading of 50,000 lb as shown in Fig. 2-20(a). The cross-sectional area of the steel is 3 in^2, while that of the copper is 10 in^2. Both cylinders are the same length before the load is applied. Determine the temperature rise of the entire system required to place all of the load on the copper cylinder. The cover plate at the top of the assembly is rigid. For copper $E = 16 \times 10^6$ lb/in^2, $\alpha = 9.3 \times 10^{-6}$/°F, while for steel $E = 30 \times 10^6$ lb/in^2, $\alpha = 6.5 \times 10^{-6}$/°F.

One method of analyzing this problem is to assume that the load as well as the upper cover plate are removed and that the system is allowed to freely expand vertically because of a temperature rise ΔT. In

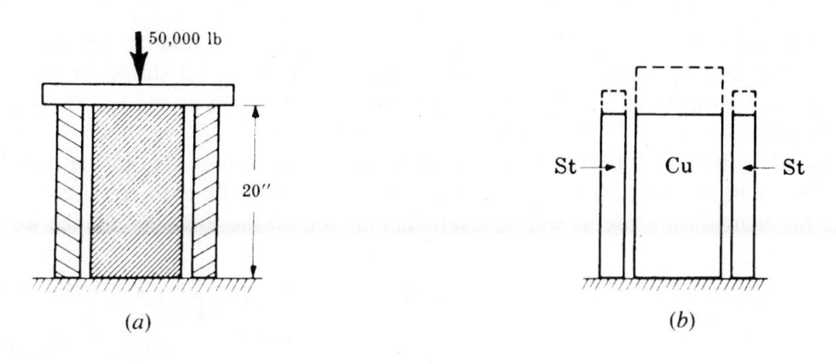

Fig. 2-20

that event the upper ends of the cylinders assume the positions shown by the dashed lines in Fig. 2-20(b).

The copper cylinder naturally expands upward more than the steel one because the coefficient of linear expansion of copper is greater than that of steel. The upward expansion of the steel cylinder is $(6.5 \times 10^{-6})(20)\Delta T$, while that of the copper is $(9.3 \times 10^{-6})(20)\Delta T$.

This is not of course the true situation because the load of 50,000 lb has not as yet been considered. If all of this axial load is carried by the copper then only the copper will be compressed and the compression of the copper is given by

$$\Delta_{cu} = \frac{PL}{AE} = \frac{50{,}000(20)}{10(16 \times 10^6)}$$

The condition of the problem states that the temperature rise ΔT is just sufficient so that all of the load is carried by the copper. Thus, the expanded length of the copper indicated by the dashed lines in the above sketch will be decreased by the action of the force. The net expansion of the copper is the expansion caused by the rise of temperature minus the compression due to the load. The change of length of the steel is due only to the temperature rise. Consequently we may write

$$(9.3 \times 10^{-6})(20)\Delta T - \frac{50{,}000(20)}{10(16 \times 10^6)} = (6.5 \times 10^{-6})(20)\Delta T \qquad \text{or} \qquad \Delta T = 111°F$$

2.8. The rigid bar AD is pinned at A and attached to the bars BC and ED, as shown in Fig. 2-21(a). The entire system is initially stress free and the weights of all bars are negligible. The temperature of bar BC is lowered 25°C and that of the bar ED is raised 25°C. Neglecting any possibility of lateral buckling, find the normal stresses in bars BC and ED. For BC, which is brass, assume $E = 90$ GPa, $\alpha = 20 \times 10^{-6}$/°C, and for ED, which is steel, take $E = 200$ GPa and $\alpha = 12 \times 10^{-6}$/°C. The cross-sectional area of BC is 500 mm² and of ED is 250 mm².

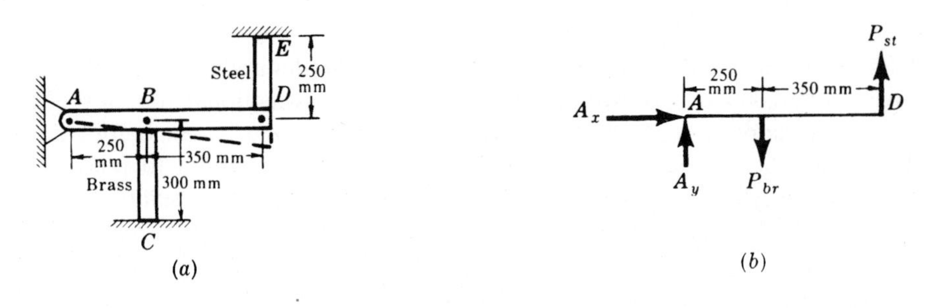

Fig. 2-20

Let us denote the forces on AD by P_{st} and P_{br} acting in the assumed directions shown in the free-body diagram, Fig. 2-21(b). Since AD rotates as a rigid body about A (as shown by the dashed line) we have $\Delta_{br}/250 = \Delta_{st}/350$ where Δ_{br} and Δ_{st} denote the axial compression of BC and the axial elongation of DE, respectively.

The total change of length of BC is composed of a shortening due to the temperature drop as well as a lengthening due to the axial force P_{br}. The total change of length of DE is composed of a lengthening due to the temperature rise as well as a lengthening due to the force P_{st}. Hence we have

$$\left(\frac{25}{60}\right)\left[(12 \times 10^{-6})(250)(25) + \frac{P_{st}(250)}{(250)(200 \times 10^9 \times 10^{-6})}\right] = -(20 \times 10^{-6})(300)(25) + \frac{P_{br}(300)}{(500)(90 \times 10^9 \times 10^{-6})}$$

or

$$6.66P_{br} - 2.08P_{st} = 153.0 \times 10^3$$

From statics,

$$\Sigma M_A = 250P_{br} - 600P_{st} = 0$$

Solving these equations simultaneously, $P_{st} = 10.99$ kN and $P_{br} = 26.3$ kN. Using $\sigma = P/A$ for each bar, we obtain $\sigma_{st} = 43.9$ MPa and $\sigma_{br} = 52.6$ kN.

2.9. Consider the statically indeterminate pin-connected framework shown in Fig. 2-22(*a*). Before the load P is applied the entire system is stress free. Find the axial force in each bar caused by the vertical load P. The two outer bars are identical and have cross-sectional area A_i, while the middle bar has area A_v. All bars have the same modulus of elasticity, E.

The free-body diagram of the pin at A appears as in Fig. 2-22(*b*) where F_1 and F_2 denote axial forces (lb) in the vertical and inclined bars. From statics we have

$$\Sigma F_v = F_1 + 2F_2 \cos\theta - P = 0$$

This is the only statics equation available since we have made use of symmetry in stating that the forces in the inclined bars are equal. Since it contains two unknowns, F_1 and F_2, the force system is statically

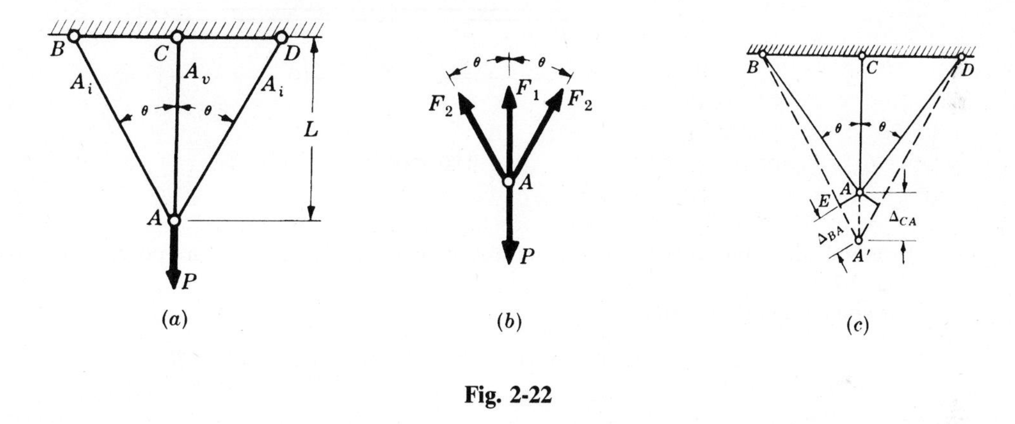

Fig. 2-22

indeterminate. Hence we must examine the deformations of the system to obtain another equation. Under the action of the load P the bars assume the positions shown by the dashed lines in Fig. 2-22(*c*).

Because the deformations of the system are *small*, the basic geometry is essentially unchanged and the angle $BA'A$ may be taken to be θ. AEA' is a right triangle and AE, which is actually an arc having a radius equal in length to the length of the inclined bars, is perpendicular to BA'. The elongation of the vertical bar is thus represented by AA' and that of the inclined bars by EA'. From this small triangle we have the relation

$$\Delta_{BA} = \Delta_{CA} \cos\theta$$

where Δ_{BA} and Δ_{CA} denote elongations of the inclined and vertical bars, respectively.

Since these bars are subject to axial loading their elongations are given by $\Delta = PL/AE$. From that expression we have

$$\Delta_{BA} = \frac{F_2(L/\cos\theta)}{A_i E} \quad \text{and} \quad \Delta_{CA} = \frac{F_1 L}{A_v E}$$

Substituting these in the above equation relating Δ_{BA} and Δ_{CA} we have

$$\frac{F_2 L}{A_i E \cos\theta} = \frac{F_1 L}{A_v E}\cos\theta \quad \text{or} \quad F_2 = F_1 \frac{A_i}{A_v}\cos^2\theta$$

Substituting this in the statics equation we find $F_1 + 2F_1(A_i/A_v)\cos^3\theta = P$, or

$$F_1 = \frac{P}{1 + 2(A_i/A_v)\cos^3\theta} \quad \text{and} \quad F_2 = \frac{P\cos^2\theta}{(A_v/A_i) + 2\cos^3\theta} \tag{1}$$

2.10. Two initially horizontal rigid bars AC and DG are pinned at A and G and are also connected by elastic vertical bars BD and CF, each of rigidity AE, as shown in Fig. 2-23. The temperature of bar BD is then raised by an amount ΔT. Determine the force in the two vertical bars.

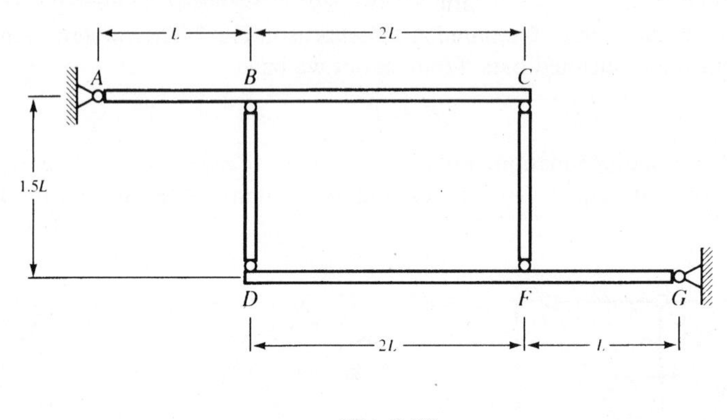

Fig. 2-23

Free-body diagrams of the components, assuming all unknown forces are positive, in tension appear as in Fig. 2-24.

For equilibrium of bar DG, we have

$$+\uparrow\Sigma M_G = -F_2(L) - F_1(3L) = 0 \qquad \therefore F_2 + 3F_1 = 0 \tag{1}$$

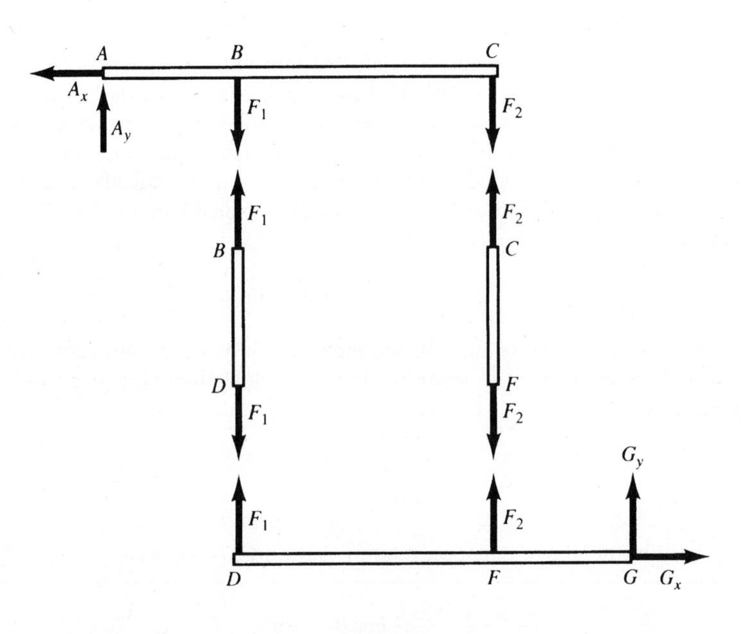

Fig. 2-24

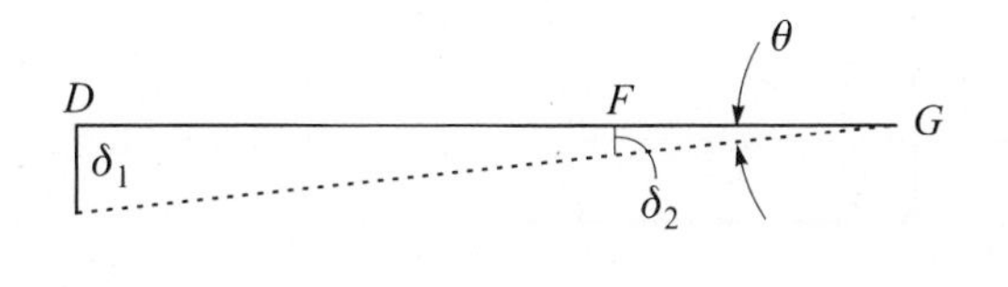

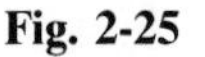

Fig. 2-25

We must now examine deformations of the system. To simplify this analysis, it is permissible to assume that the upper bar AC remains horizontal and that all distortion is due to rigid-body rotation of the lower bar DG about G. This leads to the deformed position of DG as shown by the dotted line in Fig. 2-25. The changes of length of the vertical bars are indicated by Δ_1 and Δ_2 in that figure. From geometry, for a small angle of rotation, we have

$$\theta = \frac{\delta_1}{3L} = \frac{\delta_2}{L}$$

from which

$$\delta_1 = 3\delta_2 \tag{2}$$

The increase in length of bar BD is due partially to the force F_1 it carries and partially to the increase in temperature. It is

$$\delta_1 = \frac{F_1(1.5L)}{AE} + \alpha(\Delta T)(1.5L) \tag{3}$$

For bar CF, the increase of length is due only to the force F_2 in it, so we have

$$\delta_2 = \frac{F_2(1.5L)}{AE} \tag{4}$$

Solving Eqs. (*1*), (*2*), (*3*), and (*4*) simultaneously, we have

$$F_1 = -\frac{\alpha(\Delta T)AE}{10}$$

$$F_2 = \frac{3\alpha(\Delta T)AE}{10}$$

The negative sign accompanying the bar force F_1 indicates that bar BD is in compression, whereas bar CF is in tension.

2.11. A two-dimensional framework consists of two bars AB and BH forming a 30° triangle with pins at A, B, and H together with a horizontal bar GD, as shown in Fig. 2-26. Because of a manufacturing error, the bar GD is slightly short of the length $2L$. All bars have axial rigidity AE. Determine the axial force in bar GD when the gap Δ is closed by mechanical action.

First, let us examine the forces acting at point B. In particular, we apply a horizontal force F at the node B, and a free-body diagram of that node is shown in Fig. 2-27. For horizontal equilibrium, we have

$$\Sigma F_x = F - F_{BA}\cos 30^\circ = 0$$

from which

$$F_{BA} = \frac{F}{\cos 30^\circ} \tag{1}$$

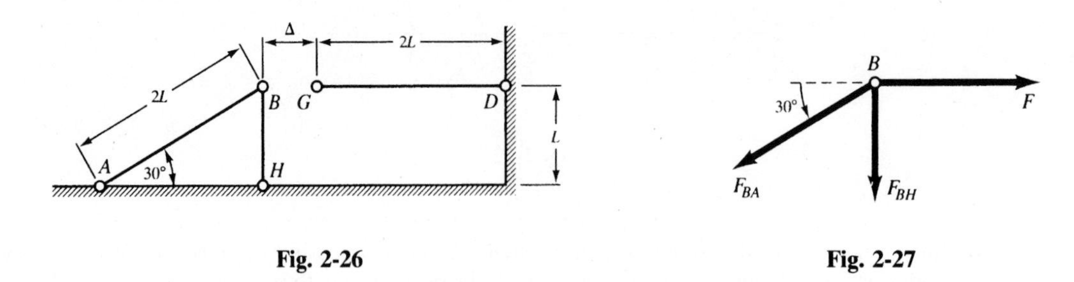

Fig. 2-26

Fig. 2-27

Next, let us examine displacements at the node B. Since we have just found that bar AB is in tension, it will lengthen an amount Δ_{BA}, as shown in Fig. 2-28, where

$$\Delta_{BA} = \frac{F_{BA}(2L)}{AE} \tag{2}$$

The bar AB will then rotate as a rigid body about point A through the circular arc from B'' to B', which for small deformations we approximate as a straight line from B'' to B'. The horizontal projections of BB'' and $B''B'$ are denoted by Δ_3 and Δ_2, respectively. From geometry we have

$$\Delta_2 = \Delta_1 \sin 30° = \Delta_{BA}(\tan 30°)(\sin 30°) \tag{3}$$

$$\Delta_3 = \Delta_{BA} \cos 30° \tag{4}$$

The bar GD is subject to an equal and opposite force F, as shown in Fig. 2-29, and it elongates an amount

$$\frac{F(2L)}{AE} \tag{5}$$

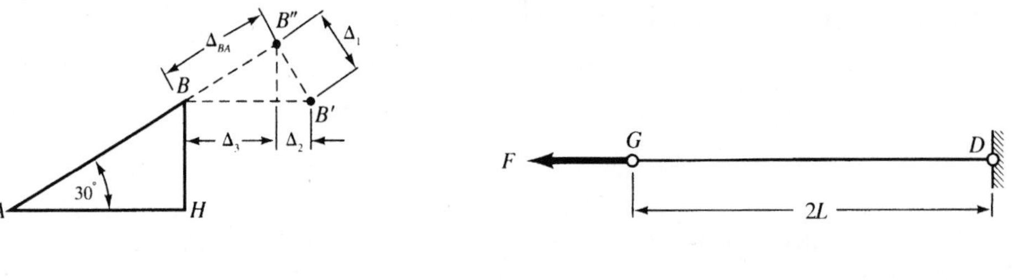

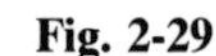

Fig. 2-28

Fig. 2-29

Thus, to bring points B and G together and close the gap, we have

$$\Delta_2 + \Delta_3 + \frac{F(2L)}{AE} = \Delta \tag{6}$$

From Eqs. (*1*) through (*6*), we have the required force in bar GD to close the gap Δ:

$$F = \frac{AE\Delta}{2L(1 + \tan^2 30°)} = \frac{2AE\Delta}{5L}$$

2.12. The rigid horizontal bar ABC is supported by vertical elastic posts and restrained against horizontal movement at A as shown in Fig. 2-30. A vertical load P acts at C. The extensional rigidity of each post is indicated in the figure and each is of length L. Find the axial force in each of the three posts.

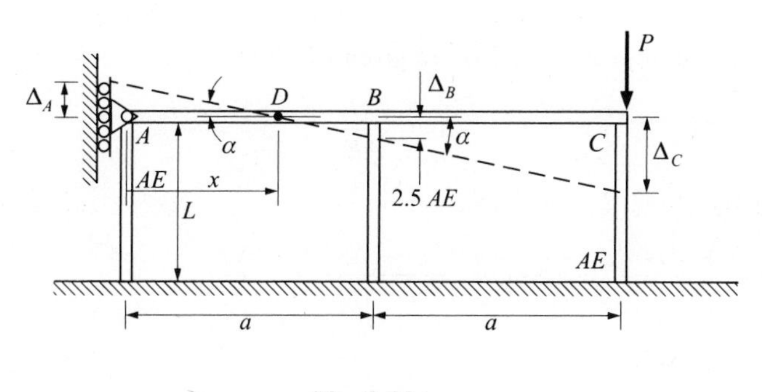

Fig. 2-30

Due to the load P the originally horizontal bar deforms to the configuration indicated by the dotted line in Fig. 2-31. That is, it rotates as a rigid body about some point D (whose location is unknown) through the angle α.

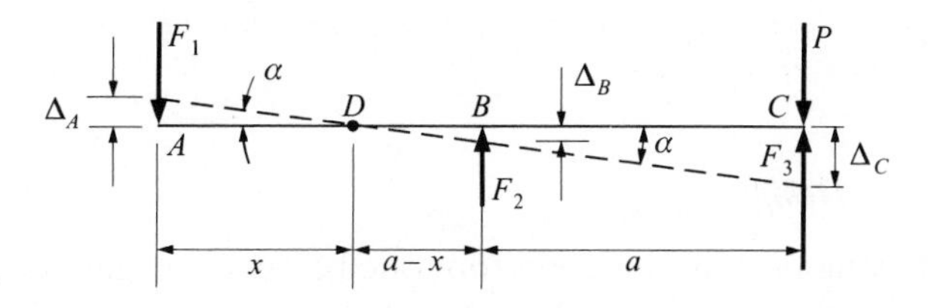

Fig. 2-31

Figure 2-31 shows a free-body diagram of ABC where the forces exerted on ABC by the posts are represented by F_1, F_2, and F_3. The change of length of each post is indicated by Δ in the figure. From the geometry of the deformed system we have

$$\alpha = \frac{\Delta_A}{x} = \frac{(F_1 L/AE)}{x} = \frac{(F_2 L/2.5AE)}{a - x} \tag{1}$$

For this parallel force system there are two equations of static equilibrium. For the first equation we set

$$\Sigma M_C = F_1(2a) - F_2(a) = 0$$

from which

$$F_1 = \frac{F_2}{2} \tag{2}$$

If we now substitute (2) in (1), we find

$$\frac{(F_2 L/2AE)}{x} = \frac{(F_2 L/2.5AE)}{a - x}$$

from which

$$x = \left(\frac{5}{9}\right) a \tag{3}$$

For the second statics equation we write

$$\Sigma M_B = -F_3 a + Pa - F_1 a = 0$$

Thus

$$F_3 = P - F_1 \tag{4}$$

The changes of length of posts C and A are given by

$$\Delta_C = \frac{F_3 L}{AE}; \qquad \Delta_A = \frac{F_1 L}{AE} \tag{5}$$

From the geometry of Fig. 2-31 we have

$$\frac{\Delta_C}{a + (a - x)} = \frac{\Delta_A}{x}$$

and from Eq. (3):

$$\Delta_C = \left(\frac{13}{5}\right) \Delta_A \tag{6}$$

Substituting (5) in (4), we find

$$F_3 = (13/9) F_1 \tag{7}$$

and from (2) and (7) we have

$$F_1 = \left(\frac{9}{22}\right) P; \qquad F_2 = \left(\frac{9}{11}\right) P \tag{8}$$

Ultimate Strength (Limit Design)

In each of the following problems the elastoplastic behavior of the material is assumed to follow the idealized stress-strain curve of Fig. 2-32.

The ultimate load, or limit load, determined in each of the following problems is the maximum possible load that can be applied to each system provided the stress-strain curve is of the type indicated and the material has infinite ductility, i.e., the flat region of the curve extends indefinitely to the right.

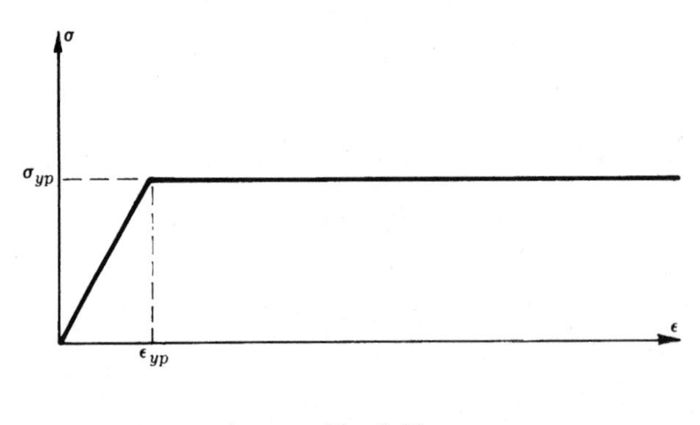

Fig. 2-32

2.13. Consider the system composed of three vertical bars as indicated in Fig. 2-33(*a*). The outer bars of length L are equally spaced from the central bar and a load P is applied to the rigid horizontal member. Using limit design, determine the ultimate load P. The values of A and E are identical in all three bars.

Let us analyze the action as the load P increases from an initial value of zero, i.e., as it is slowly applied. For equilibrium we have

$$2P_1 + P_2 = P \tag{1}$$

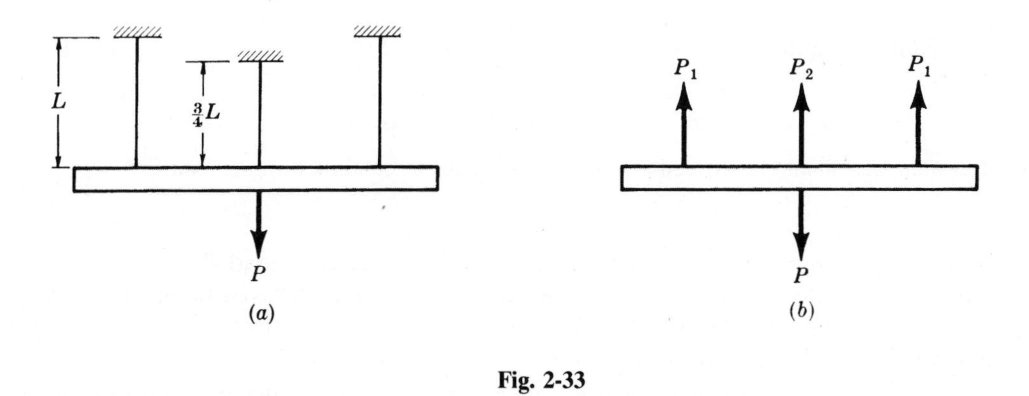

Fig. 2-33

where P_1 represents the force in each of the outer bars and P_2 is the force in the inner bar [see Fig. 2-33(*b*)]. Since the horizontal member is rigid, the vertical elongation of each of the outer bars must equal that of the central bar. Thus

$$\frac{P_1 L}{AE} = \frac{P_2(3L/4)}{AE} \tag{2}$$

or

$$P_1 = \tfrac{3}{4}P_2 \tag{3}$$

Substituting this value in (*1*) we find

$$P_2 = \tfrac{2}{5}P \qquad P_1 = \tfrac{3}{10}P \tag{4}$$

The system thus begins to yield when $P_2 = \sigma_{yp}A$. Thus

$$P_{yp} = \tfrac{5}{2}\sigma_{yp}A$$

From the time of yielding of the central bar, the system deforms as if supported by only the two outside bars (which still act elastically) together with a constant force $\sigma_{yp}A$ supplied by the central bar. The value of P increases until yielding begins in each of the outer bars, i.e., when $P_1 = \sigma_{yp}A$. The ultimate load is thus

$$P_u = 2P_1 + P_2 = 2\sigma_{yp}A + \sigma_{yp}A = 3\sigma_{yp}A$$

It is to be noted that the deformation equation (*2*) is not employed to determine the ultimate load.

2.14. Reconsider Problem 2.9 for the case of three bars of equal cross-sectional area. Determine the ultimate load-carrying capability of the system.

For $A_i = A_v = A$ the force in the vertical bar exceeds that in either inclined bar as indicated by (*1*) of Problem 2.9. Thus, as P increases, the central vertical bar is the first to enter the inelastic range of action and its stiffness (effective value of AE) decreases. Any additional increase in the load P will cause no further increase in F_1 which will remain at the limit value $F_1^* = \sigma_{yp}A$. The central bar can now be replaced by a constant upward vertical force F_1^* and the system is now reduced to a statically determinate system consisting of the two outer bars subject to an applied load $P - F_1^*$. The load P can now be increased until the outer bars also develop the yield stress. It is not necessary to consider deformations of the system; we need look only at the equilibrium relation

$$P = F_1^* + 2F_2 \cos\theta \tag{1}$$

As the load P increases still more, the outer bars also reach the yield point and the force in each of them becomes

$$F_2^* = \sigma_{yp}A \tag{2}$$

The ultimate load thus corresponds to the situation when $F_1^* = F_2^* = \sigma_{yp} A$ and this load is found from (*1*) as

$$P_u = \sigma_{yp} A(1 + 2\cos\theta) \tag{3}$$

This *limit load* should be divided by some safety factor to obtain a *working load.*

2.15. Suppose the three-bar system of Problem 2.9 is to withstand a load $P = 200$ kN. Compare the bar weights required if the design is based upon (*a*) the peak stress just reaching the yield point, and (*b*) ultimate load analysis. Assume that all bars are of identical cross section, that $\theta = 45°$, and take the yield point of the material to be 250 MPa.

(*a*) According to the elastic theory of Problem 2.9, the force in the vertical bar becomes

$$F_1 = \frac{2P}{2 + \sqrt{2}} = 117 \text{ kN}$$

If the stress in that bar is equal to the yield point, we have a required cross-sectional area of $F_1 = A_1 \sigma_{yp}$. Hence

$$117 \times 10^3 = A_1(250) \qquad \text{or} \qquad A_1 = 468 \text{ mm}^2$$

(*b*) If the ultimate load analysis of Problem 2.14 is employed, the stresses in all three bars are equal to the yield point and from (*3*) of Problem 2.14 we find a cross-sectional area of

$$200 \times 10^3 = 250A_2[1 + 2(0.707)] \qquad \text{or} \qquad A_2 = 331 \text{ mm}^2$$

Ultimate load analysis thus implies a 29 percent saving in cross-sectional area and the same weight saving.

2.16. The frame shown in Fig. 2-34 consists of three pinned end bars AD, BD, and CD. The bars are of identical material and cross section, and the ultimate load-carrying capacity of each is 30 kN. Determine the ultimate vertical load P_u that may be applied to the system at point D.

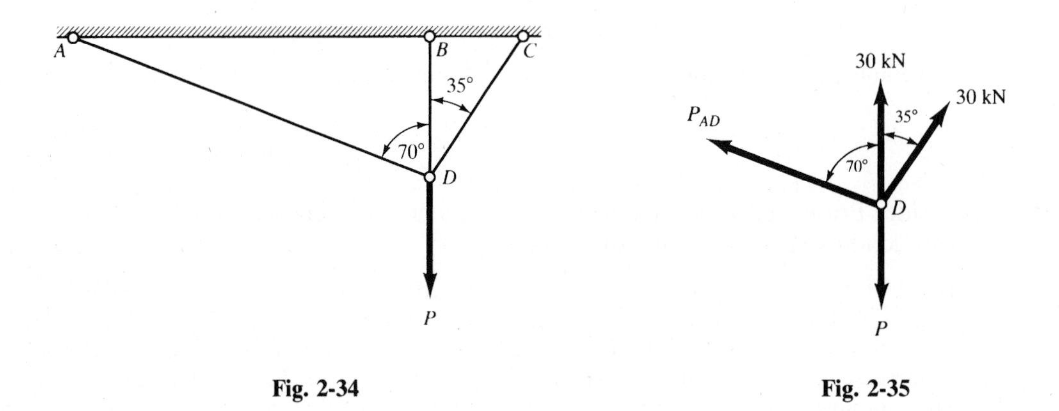

Fig. 2-34 **Fig. 2-35**

Let us assume that bars BD and CD have reached yield. Examination of a free-body diagram for the node D as shown in Fig. 2-35 leads to

$$\Sigma F_x = 30 \sin 35° - P_{AD} \sin 70° = 0$$

$$P_{AD} = 18.3 \text{ kN}$$

Thus, bar AD does not yield since the bar force for equilibrium is less than the 30 kN required for yield. Summing vertically for equilibrium we have

$$\Sigma F_y = -P_u + 18.3\cos 70° + 30 + 30\cos 35° = 0$$

$$P_u = 60.9 \text{ kN}$$

2.17. A system composed of a rigid horizontal member AB supported by four bars is indicated in Fig. 2-36(a). The bars have identical cross sections and are made of the same material. Determine the ultimate load P that may be applied to the system.

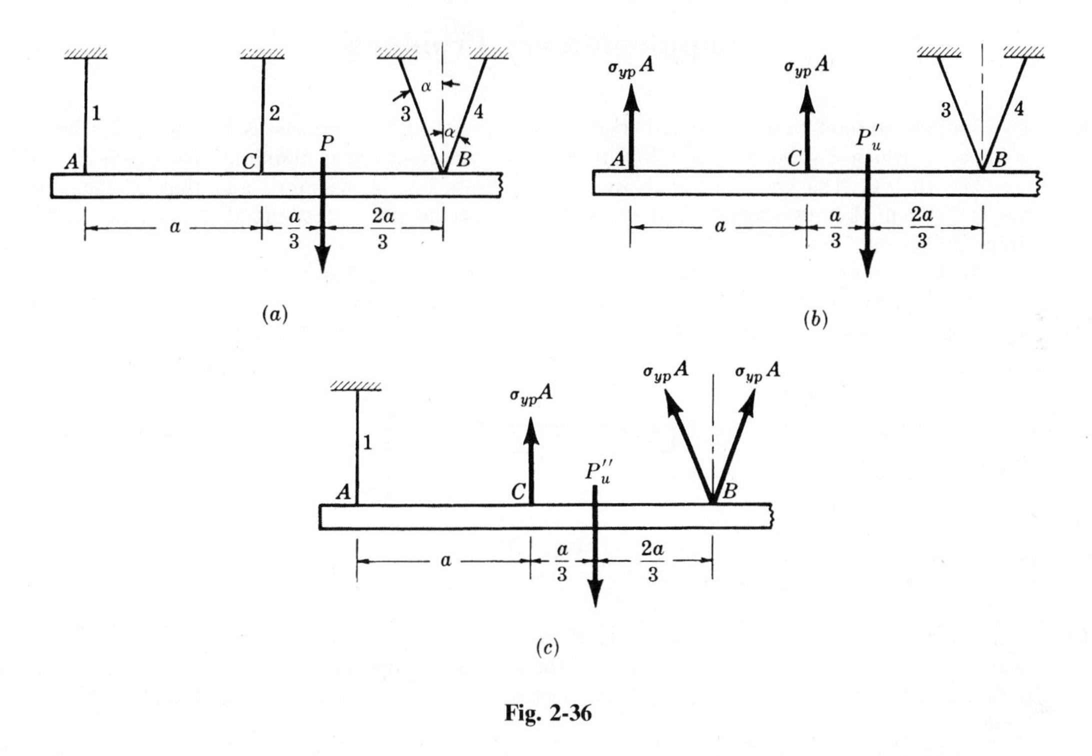

Fig. 2-36

Since the member AB is rigid, it is evident that, upon application of a sufficiently large load P, AB may rotate as a rigid body about either point A or point B. (The ultimate load implies plastic deformation in bar 2; hence it is not necessary to consider rotation about C.) It is necessary to determine the ultimate loads corresponding to these two possibilities and then to select the smaller.

Let us first assume that yielding first begins in bars 1 and 2, in which case their effect can be represented by the two constant forces $\sigma_{yp}A$ as indicated in Fig. 2-36(b). The bars 3 and 4 are still in the elastic range of action and the forces in them are unknown. However, it is not necessary to determine the forces since the ultimate load P'_u may be determined by summing moments about point B:

$$P'_u\left(\frac{2a}{3}\right) - \sigma_{yp}A(a) - \sigma_{yp}A(2a) = 0$$

Solving,

$$P'_u = 4.5\sigma_{yp}A$$

Next, let us consider that yielding begins in bars 2, 3, and 4 as indicated in Fig. 2-36(c). Bar 1 is still in the elastic range of action. Taking moments about point A:

$$(\sigma_{yp}A\cos\alpha)4a + \sigma_{yp}Aa - P''_u\frac{4a}{3} = 0$$

Solving,

$$P''_u = \tfrac{3}{4}\sigma_{yp}A(1 + 4\cos\alpha)$$

It is evident from inspection of P'_u and P''_u that, for all values of the angle α, the value of P''_u is the smaller of the two and thus P''_u represents the ultimate load. When the applied load reaches this value, the system is essentially converted into a mechanism and the rigid bar rotates about point A. Even in this condition bar 1 is not working to its full capacity.

Supplementary Problems

2.18. Two initially straight bars are joined together and attached to supports as in Fig. 2-37. The left bar is brass for which $E = 90$ GPa, $\alpha = 20 \times 10^{-6}$/°C, and the right bar is aluminum for which $E = 70$ GPa, $\alpha = 25 \times 10^{-6}$/°C. The cross-sectional area of the brass bar is 500 mm², and that of the aluminum bar is 750 mm². Let us suppose that the system is initially stress free and that the temperature then drops 20°C.
(*a*) If the supports are unyielding, find the normal stress in each bar.
(*b*) If the right support yields 0.1 mm, find the normal stress in each bar. The weight of the bars is negligible. *Ans.* (*a*) $\sigma_{br} = 41$ MPa, $\sigma_{al} = 27.33$ MPa; (*b*) $\sigma_{br} = 28.4$ MPa, $\sigma_{al} = 19$ MPa

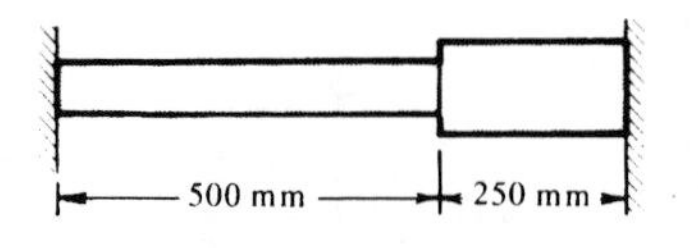

Fig. 2-37

2.19 The framework shown in Fig. 2-38 consists of bars AD, AC, BC, and BD pinned at A, B, C, and D, and also a fifth bar CD. The system is loaded by the equal and opposite forces P. All bars are of identical material and cross section. Determine the decrease of the distance between A and B due to these loads.

Ans. $\dfrac{PL\sqrt{2}}{AE}(\sqrt{2}+1)$

2.20. Refer to the framework shown in Fig. 2-38. Now, instead of the two loads P, the temperature of the entire system is raised by an amount ΔT. Determine the change of distance between A and B in terms of the geometry of the system and the coefficient of thermal expansion α of the material. *Ans.* $L\sqrt{2}\alpha(\Delta T)$

2.21. Refer to Problem 2.6. If the conical bar has a diameter at its small end of 100 mm, a length of 1 m, and is of steel having $E = 200$ GPa and a coefficient of thermal expansion of 12×10^{-6}/°C, determine the maximum axial stress in the bar due to a temperature drop of 20°C. *Ans.* 96 MPa

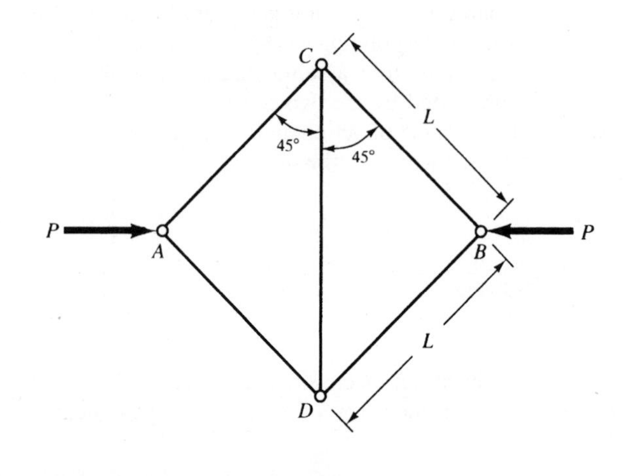

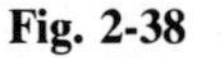

Fig. 2-38

2.22. A compound bar is composed of a strip of copper between two cold-rolled steel plates. The ends of the assembly are covered with infinitely rigid cover plates and an axial tensile load P is applied to the bar by means of a force acting on each rigid plate as shown in Fig. 2-39. The width of all bars is 4 in, the steel plates are each $\frac{1}{4}$ in thick and the copper is $\frac{3}{4}$ in thick. Determine the maximum load P that may be applied. The ultimate strength of the steel is 80,000 lb/in^2 and that of the copper is 30,000 lb/in^2. A safety factor of 3 based upon the ultimate strength of each material is satisfactory. For steel $E = 30 \times 10^6$ lb/in^2 and for copper $E = 13 \times 10^6$ lb/in^2. *Ans.* $P = 76{,}200$ lb

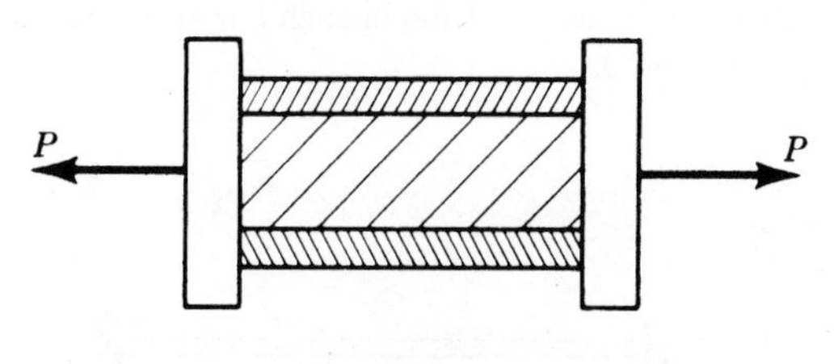

Fig. 2-39

2.23. An aluminum right-circular cylinder surrounds a steel cylinder as shown in Fig. 2-40. The axial compressive load of 200 kN is applied through the infinitely rigid cover plate shown. If the aluminum cylinder is originally 0.25 mm longer than the steel before any load is applied, find the normal stress in each when the temperature has dropped 20 K and the entire load is acting. For steel take $E = 200$ GPa; $\alpha = 12 \times 10^{-6}/°C$, and for aluminum assume $E = 70$ GPa, $\alpha = 25 \times 10^{-6}/°C$.
Ans. $\sigma_{st} = 9$ MPa, $\sigma_{al} = 15.5$ MPa

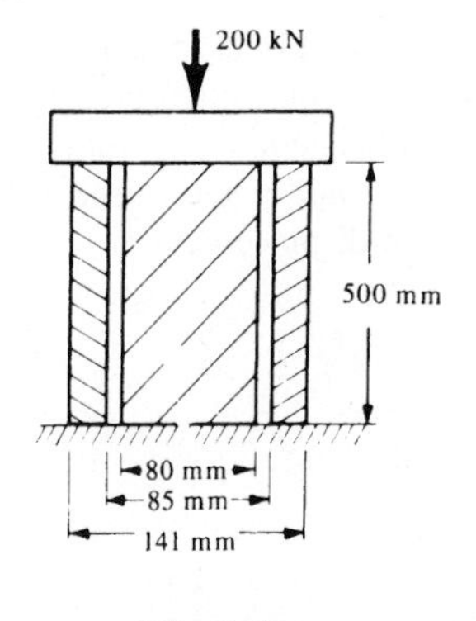

Fig. 2-40

2.24. The rigid horizontal bar AB is supported by three vertical wires as shown in Fig. 2-41 and carries a load of 24,000 lb. The weight of AB is negligible and the system is stress free before the 24,000-lb load is applied. After the load is applied, the temperature of all three wires is raised by 25°F. Find the stress in each wire

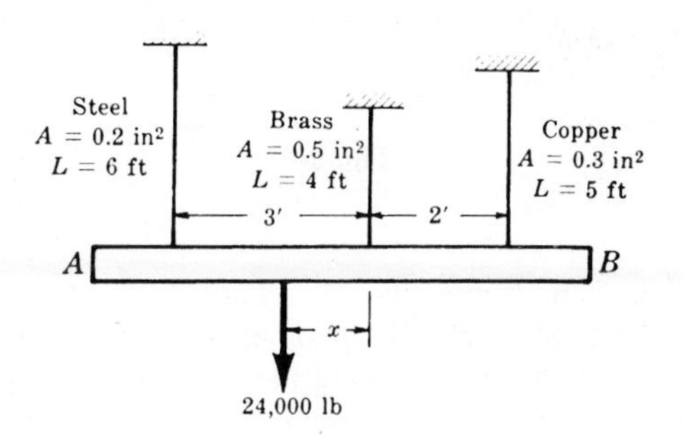

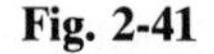
Fig. 2-41

as well as the location of the applied load in order that AB remains horizontal. For the steel wire take $E = 30 \times 10^6$ lb/in^2, $\alpha = 6.5 \times 10^{-6}$/°F, for the brass wire $E = 14 \times 10^6$ lb/in^2, $\alpha = 10.4 \times 10^{-6}$/°F, and for copper $E = 17 \times 10^6$ lb/in^2, $\alpha = 9.3 \times 10^{-6}$/°F. Neglect any possibility of lateral buckling of any of the wires. *Ans.* $\sigma_{st} = 32{,}300$ lb/in^2, $\sigma_{br} = 22{,}400$ lb/in^2, $\sigma_{cu} = 21{,}400$ lb/in^2, $x = 0.273$ ft

2.25. A system consists of two rigid end-plates, tied together by three horizontal bars as shown in Fig. 2-42. Through a fabrication error, the central bar, ②, is $0.0005L$ too short. All bars are of identical cross section and of steel having $E = 210$ GPa. Find the stress in each bar after the system has mechanically been pulled together so that the gap Δ is closed.

Ans $\sigma_1 = -35$ MPa

$\sigma_2 = 70$ MPa

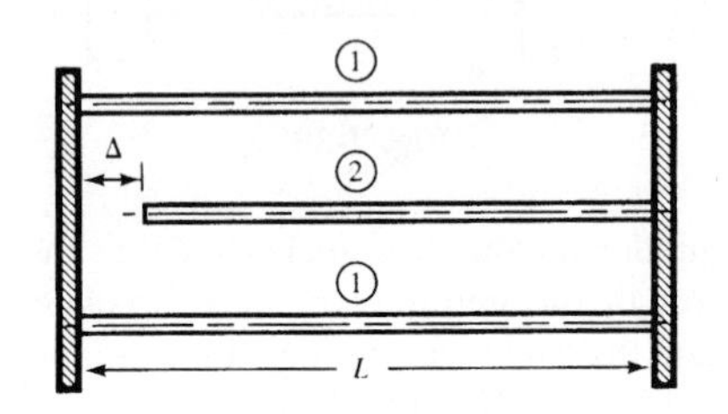

Fig. 2-42

2.26. A structural system consists of three joined bars of different materials and geometries, as shown in Fig. 2-43. Bar ① is aluminum alloy, bar ② is cold rolled brass, and bar ③ is tempered alloy steel. Properties and dimensions of all three are shown in the figure. Initially, the entire system is free of stresses, but then the right support is moved 3 mm to the right whereas the left support remains fixed in space. Determine the stress in each bar due to this 3 mm displacement.

Ans. $\sigma_1 = 223$ MPa

$\sigma_2 = 178$ MPa

$\sigma_3 = 446$ MPa

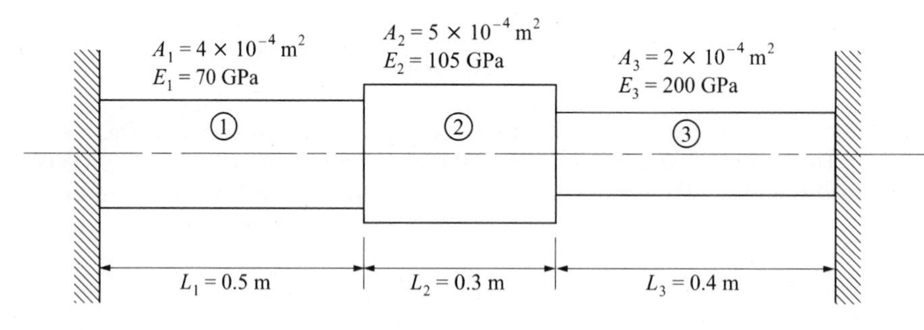

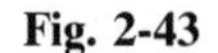

Fig. 2-43

2.27. The bar AC is absolutely rigid and is pinned at A and attached to bars DB and CE as shown in Fig. 2-44. The weight of AC is 50 kN and the weights of the other two bars are negligible. Consider the temperature of both bars DB and CE to be raised 35°C. Find the resulting normal stresses in these two bars. DB is copper for which $E = 90$ GPa, $\alpha = 18 \times 10^{-6}$/°C, and the cross-sectional area is 1000 mm^2, while CE is steel for which $E = 200$ GPa, $\alpha = 12 \times 10^{-6}$/°C, and the cross section is 500 mm^2. Neglect any possibility of lateral buckling of the bars. *Ans.* $\sigma_{st} = 72$ MPa, $\sigma_{cu} = -21.7$ MPa

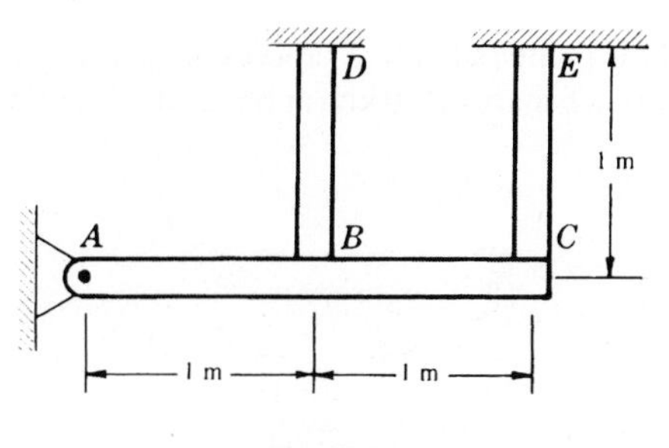

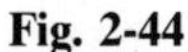
Fig. 2-44

2.28. The three bars shown in Fig. 2-45 support the vertical load of 5000 lb. The bars are all stress free and joined by the pin at A before the load is applied. The load is put on gradually and simultaneously the temperature of all three bars decreases by 15°F. Calculate the stress in each bar. The outer bars are each brass and of cross-sectional area 0.4 in^2. The central bar is steel and of area 0.3 in^2. For brass $E = 13 \times 10^6$ lb/in^2 and $\alpha = 10.4 \times 10^{-6}$/°F and for steel $E = 30 \times 10^6$ lb/in^2 and $\alpha = 6.3 \times 10^{-6}$/°F.
Ans. $\sigma_{br} = 3550$ lb/in^2, $\sigma_{st} = 10{,}000$ lb/in^2

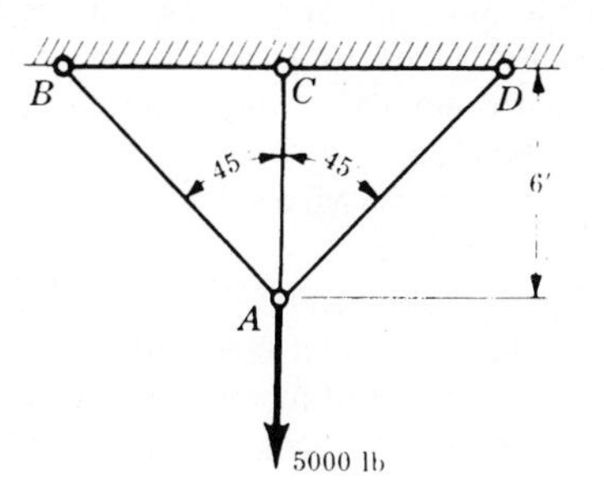

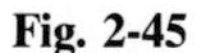
Fig. 2-45

2.29. A framework consists of three pinned bars AD, BD, and CD as shown in Fig. 2-46. The load $F = 8$ kN acts vertically at D. The cross-sectional areas of bars ① and ③ are each $200\sqrt{5}$ mm^2, the area of bar ② is 400 mm^2, $L = 3$ m, the elastic moduli are $E_1 = 200$ GPa, $E_2 = 80$ GPa, and $E_3 = 100$ GPa. Determine the horizontal and vertical components of displacement of point D as well as the axial force in bar ②.
Ans. -0.136 mm, -0.204 mm, 2.182 kN

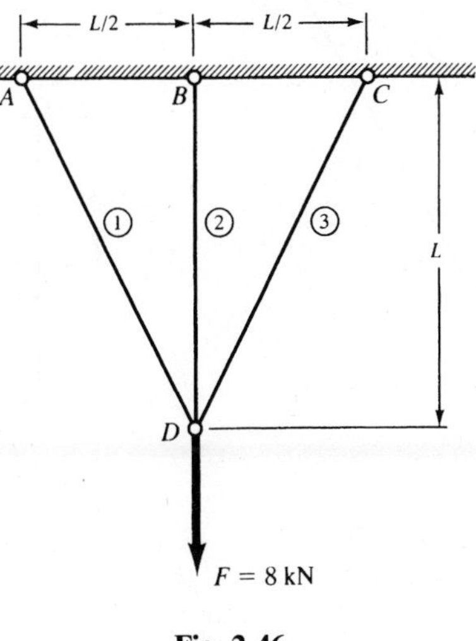

Fig. 2-46

2.30. The rigid bar AD in Fig. 2-47 is pinned at A and supported by a steel rod at D together with a linear spring at B. The bar carries a vertical load of 30 kN applied at C. Determine the vertical displacement of point D. *Ans.* 0.8 mm

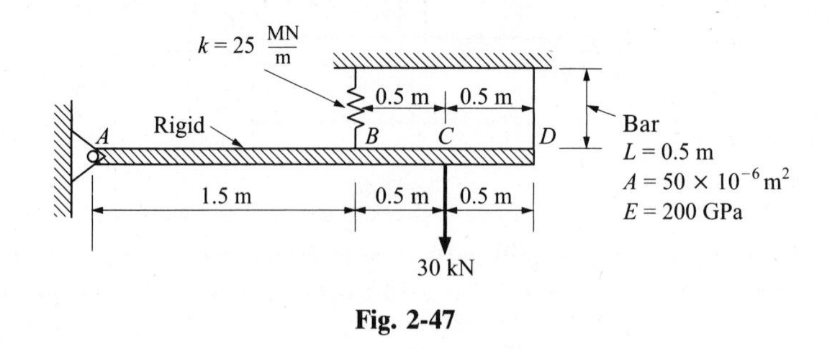

Fig. 2-47

2.31. The curved rigid bar ADB is joined to the two elastic bars OA and OB as shown in Fig. 2-48. For additional strength, it is desired to join bar OC to ADB at the midpoint D. However, through a manufacturing error OC is fabricated 1.8 mm too short. Determine the initial stresses in these three bars when point C is mechanically forced to D and these two points pinned together. The area of each outer bar is three times that of the central bar, and for all bars E = 200 GPa.
Ans. Outer bars 43.6 MPa, central bar 75.5 MPa

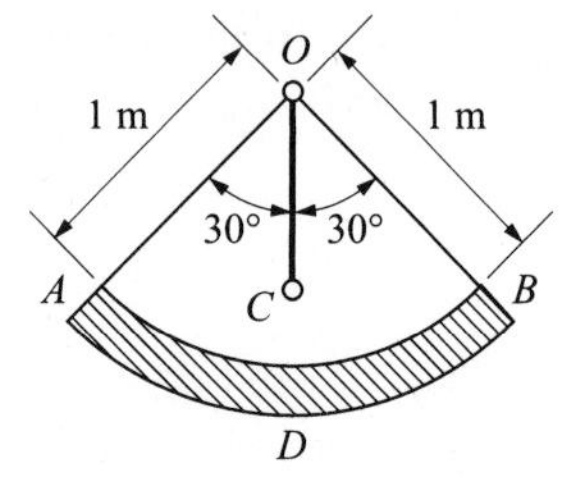

Fig. 2-48

2.32. The five-bar assembly of Fig. 2-49 was found to be slightly defective, i.e., points A and C which ought to have coincided failed to coincide by a distance Δ. After these points had been forced to coincide, the joint at that point was pinned. Determine the forces existing in each bar. All bars have the same cross-sectional area.

Ans. $F_1 = F_2 = F_3 = \left(\dfrac{\sqrt{3}}{2 + 3\sqrt{3}}\right)\dfrac{\Delta AE}{L}$ $\quad F_4 = F_5 = -\left(\dfrac{1}{2 + 3\sqrt{3}}\right)\dfrac{\Delta AE}{L}$

2.33. The rigid bar AB is supported by the four rods shown in Fig. 2-50. The rods are each circular in cross section and of 50 mm diameter. They have a yield point of 300 MPa. Using limit design determine the maximum weight of the bar AB. Assume that the weight is uniformly distributed along the length. *Ans.* 1.38 MN

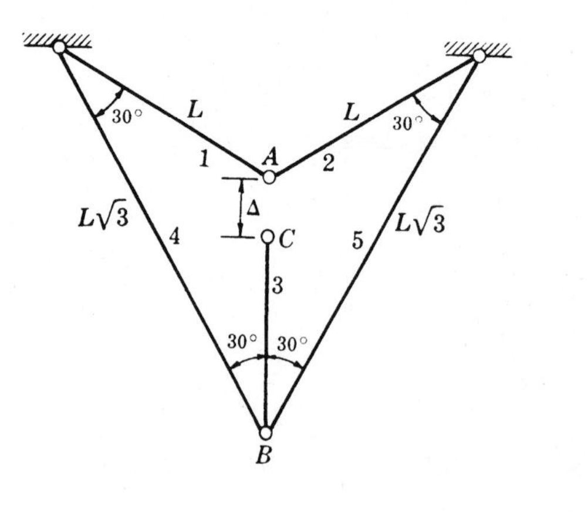

Fig. 2-49

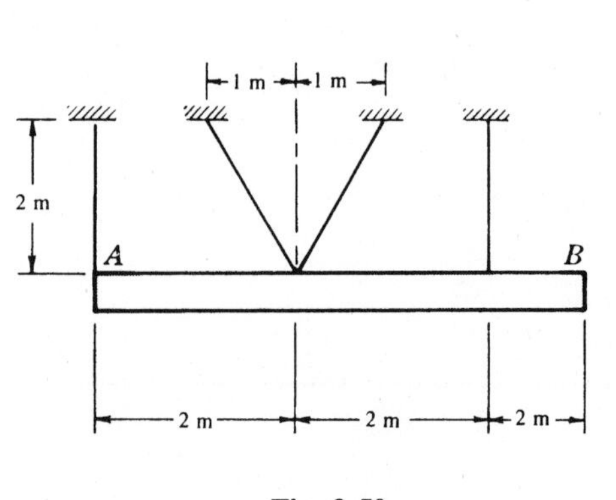

Fig. 2-50

Chapter 3

Thin-Walled Pressure Vessels

In Chaps. 1 and 2 we examined various cases involving uniform normal stresses acting in bars. Another application of uniformly distributed normal stresses occurs in the approximate analysis of thin-walled pressure vessels, such as cylindrical, spherical, conical, or toroidal shells subject to internal or external pressure from a gas or a liquid. In this chapter we will treat only thin shells of revolution and restrict ourselves to axisymmetric deformations of these shells.

NATURE OF STRESSES

The shell of revolution shown in Fig. 3-1 is formed by rotating a plane curve (the meridian) about an axis lying in the plane of the curve. The radius of curvature of the meridian is denoted by r_1 and this of course varies along the length of the meridian. This radius of curvature is defined by two lines perpendicular to the shell and passing through points B and C of Fig. 3-1. Another parameter, r_2, denotes the radius of curvature of the shell surface in a direction perpendicular to the meridian. This radius of curvature is defined by perpendiculars to the shell through points A and B of Fig. 3-1. The center of curvature corresponding to r_2 must lie on the axis of symmetry of the shell although the center for r_1 in general does not lie there. An internal pressure p acting normal to the curved surface of the shell gives rise to *meridional stresses* σ_ϕ and *hoop stresses* σ_θ as indicated in the figure. These stresses are orthogonal to one another and act in the plane of the shell wall.

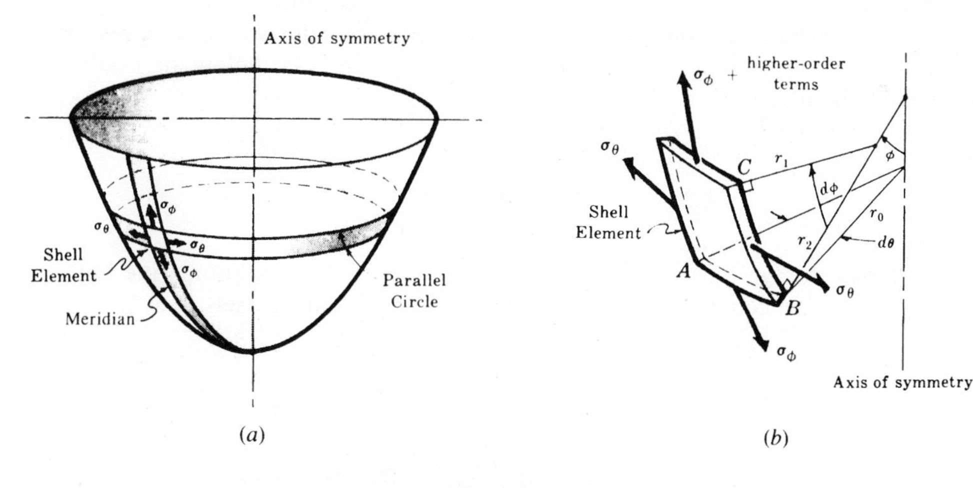

Fig. 3-1

In Problem 3.15 it is shown that

$$\frac{\sigma_\phi}{r_1} + \frac{\sigma_\theta}{r_2} = \frac{p}{h}$$

where h denotes the shell thickness. A second equation may be obtained by consideration of the vertical equilibrium of the entire shell above some convenient parallel circle, as indicated in Problem 3.15. The derivation of the above equation assumes that the stresses σ_ϕ and σ_θ are uniformly distributed over the wall thickness.

Applications of this analysis to cylindrical shells are to be found in Problems 3.1 through 3.6; to spherical shells in Problems 3.7 through 3.11, and 3.16, 3.17; to conical shells in Problem 3.14; and to toroidal shells in Problem 3.18.

LIMITATIONS

The ratio of the wall thickness to either radius of curvature should not exceed approximately 0.10. Also there must be no discontinuities in the structure. The simplified treatment presented here does not permit consideration of reinforcing rings on a cylindrical shell as shown in Fig. 3-2, nor does it give an accurate indication of the stresses and deformations in the vicinity of end closure plates on cylindrical pressure vessels. Even so, the treatment is satisfactory in many design problems.

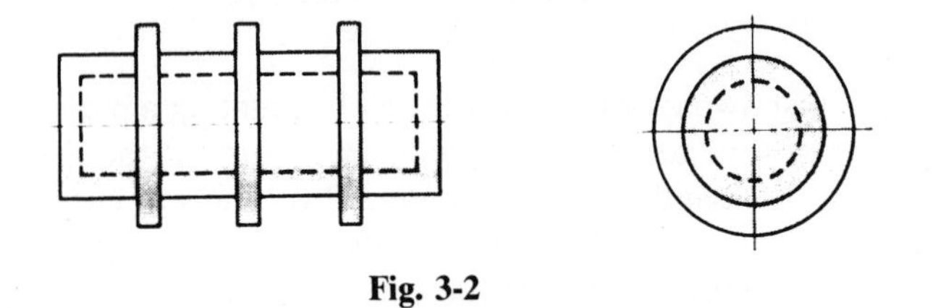

Fig. 3-2

The problems which follow are concerned with stresses arising from a uniform *internal* pressure acting on a thin shell of revolution. The formulas for the various stresses will be correct if the sense of the pressure is reversed, i.e., if external pressure acts on the container. However, it is to be noted that an additional consideration, beyond the scope of this book, must then be taken into account. Not only must the stress distribution be investigated but another study of an entirely different nature must be carried out to determine the load at which the shell will *buckle* due to the compression. A buckling or instability failure may take place even though the peak stress is far below the maximum allowable working stress of the material.

APPLICATIONS

Liquid and gas storage tanks and containers, water pipes, boilers, submarine hulls, and certain airplane components are common examples of thin-walled pressure vessels.

Solved Problems

3.1. Consider a thin-walled cylinder closed at both ends by cover plates and subject to a uniform internal pressure p. The wall thickness is h and the inner radius r. Neglecting the restraining effects of the end-plates, calculate the longitudinal (meridional) and circumferential (hoop) normal stresses existing in the walls due to this loading.

To determine the circumferential stress σ_c let us consider a section of the cylinder of length L to be removed from the vessel. The free-body diagram of half of this section appears as in Fig. 3-3(a). Note that the body has been cut in such a way that the originally *internal* effect (σ_c) now appears as an *external* force to this free body. Figure 3-3(b) shows the forces acting on a cross section.

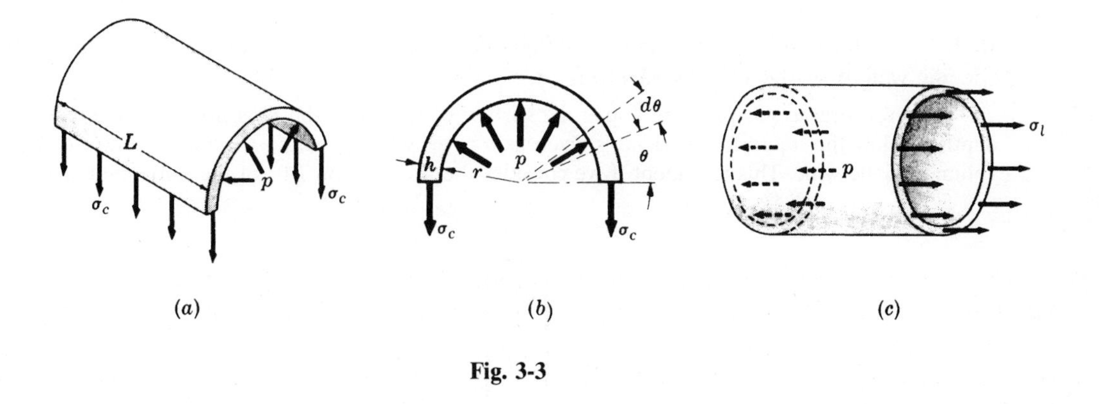

Fig. 3-3

The horizontal components of the radial pressures cancel one another by virtue of symmetry about the vertical centerline. In the vertical direction we have the equilibrium equation

$$\Sigma F_v = -2\sigma_c hL + \int_0^{\pi} pr(d\theta)(\sin\theta)L = 0$$

Integrating,

$$2\sigma_c hL = -prL[\cos\theta]_0^{\pi} \qquad \text{or} \qquad \sigma_c = \frac{pr}{h}$$

Note that the resultant vertical force due to the pressure p could have been obtained by multiplying the pressure by the horizontal *projected area* upon which the pressure acts.

To determine the longitudinal stress σ_l consider a section to be passed through the cylinder normal to its geometric axis. The free-body diagram of the remaining portion of the cylinder is shown in Fig. 3-3(*c*). For equilibrium

$$\Sigma F_h = -p\pi r^2 + 2\pi rh\sigma_l = 0 \qquad \text{or} \qquad \sigma_l = \frac{pr}{2h}$$

Consequently, the circumferential stress is twice the longitudinal stress. These rather simple expressions for stresses are not accurate in the immediate vicinity of the end closure plates.

3.2. The Space Simulator at the Jet Propulsion Laboratory in Pasadena, California, consists of a 27-ft-diameter cylindrical vessel which is 85 ft high. It is made of cold-rolled stainless steel having a proportional limit of 165,000 lb/in^2. The minimum operating pressure of the chamber is 10^{-6} torr, where 1 torr = 1/760 of a standard atmosphere, which in turn is approximately 14.7 lb/in^2. Determine the required wall thickness so that a working stress based upon the proportional limit together with a safety factor of 2.5 will not be exceeded. This solution will neglect the possibility of buckling due to the external pressure, and also the effects of certain hard-load points in the Simulator to which the test specimens are attached.

From Problem 3.1 the significant stress is the circumferential stress, given by $\sigma_c = pr/h$. The pressure to be used for design is essentially the atmospheric pressure acting on the outside of the shell, which is satisfactorily represented as 14.7 lb/in^2 since the internal pressure of 10^{-6} torr is negligible compared to 14.7 lb/in^2. We thus have

$$\frac{165{,}000}{2.5} = \frac{14.7(13.5)(12)}{h} \qquad \text{or} \qquad h = 0.036 \text{ in}$$

3.3. A vertical axis circular cylindrical wine storage tank, fabricated from stainless steel, has total height of 25 ft, a radius of 5 ft, and is filled to a depth of 20 ft with wine. An inert gas occupies the 5-ft height H_0 above the liquid-free surface and is pressurized to a value of p_0 of 12 lb/in^2.

If the working stress in the steel is 28,000 lb/in^2, determine the required wall thickness. The specific weight of the wine is 62.4 lb/ft^3.

If there were no gas pressure above the surface of the wine, the pressure (in any direction) at any depth y below the liquid-free surface is given as $p = \gamma y$, where γ is the specific weight (weight per unit volume) of the wine. This is evident if we consider the pressure on 1 ft^2 of the horizontal cross section a

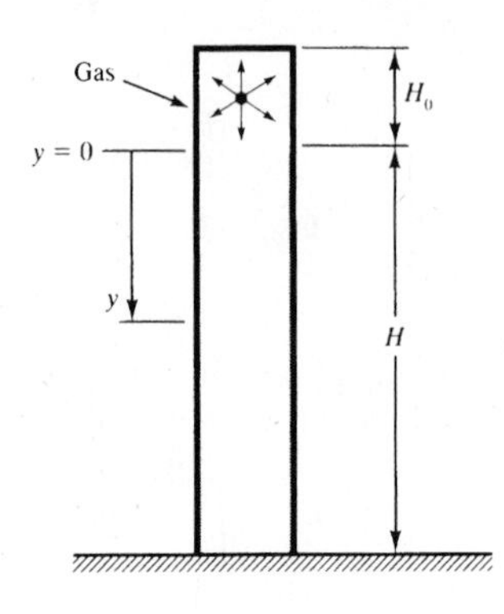

Fig. 3-4

distance y below the liquid surface to be given by the weight of the column of wine above that section divided by the 1-ft^2 area. The total pressure at the base ($y = H$) is thus $(p_0 + \gamma H)$ so that from Problem 3.1 the circumferential stress is

$$\sigma_c = \frac{(p_0 + \gamma H)R}{h} \tag{1}$$

where t is tank wall thickness.

The liquid has zero viscosity, and hence it can exert no tangential shearing stresses on the inside of the tank wall. For vertical equilibrium the upward thrust of the gas pressure p_0 must be balanced by longitudinal stresses σ_l distributed uniformly around the tank wall at the tank bottom as shown in Fig. 3-5. Thus

$$\Sigma F_y = \sigma_L(2\pi R)h - p_0 \pi R^2 = 0$$

$$\therefore \sigma_L = \frac{p_0 R}{2h} \qquad \text{(independent of } y\text{)} \tag{2}$$

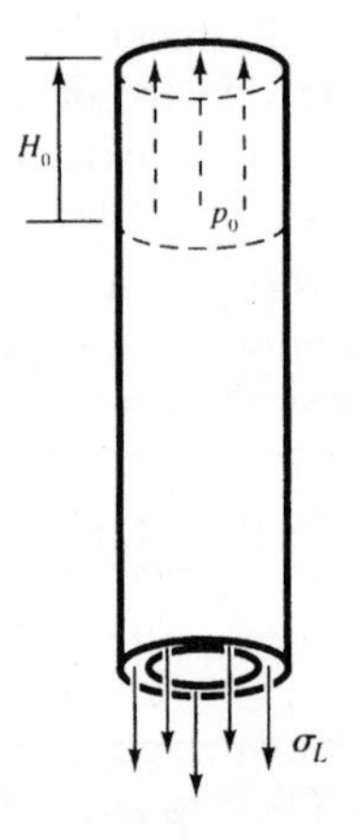

Fig. 3-5

The circumferential stress (*1*) is clearly larger than the longitudinal stress (*2*) and thus controls design. We have from (*1*)

$$\frac{\left[12\ \text{lb/in}^2 + (62.4\ \text{lb/ft}^3)\left(\frac{\text{ft}^3}{1728\ \text{in}^3}\right)(240\ \text{in})\right](60\ \text{in})}{h} = 18{,}000\ \text{lb/in}^2$$

from which the thickness is found to be $h = 0.055$ in.

3.4. A vertical axis circular cylindrical liquid storage tank of cross-sectional area A is filled to a depth of 15 m with a liquid whose specific weight (weight per unit volume) γ varies according to the law $\gamma = \gamma_{H_2O}(1 + 0.018z)$, where z is depth below the free surface of the liquid as shown in Fig. 3-6(*a*). The tank is 4 m in radius and is made of steel having a yield point of 240 MPa. The specific weight of water γ_{H_2O} is 9810 N/m^3. If a safety factor of 2 is applied, determine the required tank wall thickness.

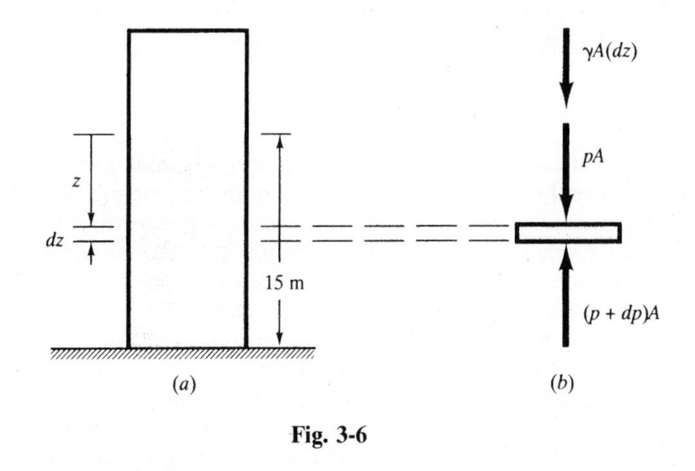

Fig. 3-6

Let us draw a free-body diagram of a thin layer of liquid situated at a distance z below the liquid free surface and of depth dz as shown in Fig. 3-6(*b*). The pressure at the top of the layer is p and at the bottom of the layer is $(p + dp)$. The weight of the layer of liquid is $\gamma A(dz)$ where it must be noted that γ in this problem is a function of z; that is, $\gamma = \gamma(z)$. Note that it is incorrect to use the equation $p = \gamma z$ from Problem 3.3 since its derivation assumes that γ is constant in the liquid whereas here γ varies with depth.

For vertical equilibrium of the element:

$$\Sigma F = (p + dp)A - pA - \gamma A\, dz = 0$$

from which

$$dp = \gamma_{H_2O}(1 + 0.018z)\, dz$$

Integrating:

$$p = \gamma_{H_2O}\left[z + 0.018\frac{z^2}{2}\right] + C$$

To find the constant of integration C, we note that at the liquid-free surface $z = 0$, $p = 0$. Thus, $C = 0$. Thus, the pressure at the tank bottom ($z = 15$ m) is

$$p_{max} = \gamma_{H_2O}[15 + 0.009(15)^2]$$

Since γ_{H_2O} is 9810 N/m^3, the peak pressure is

$$p_{max} = [9810\ \text{N/m}^3][15\ \text{m} + 0.009(15\ \text{m})^2] = 167{,}000\ \text{N/m}^2$$

From Problem 3.1, this is the significant pressure that controls design, so

$$\sigma_{max} = \frac{(p_{max})r}{h}$$

$$\frac{240 \times 10^6}{2} \text{N/m}^2 = \frac{(167{,}000 \text{ N/m}^2)(4 \text{ m})}{h}$$

from which the required tank wall thickness is

$$h = 0.0056 \text{ m} \quad \text{or} \quad 5.6 \text{ mm}$$

3.5. Calculate the increase in the radius of the cylinder considered in Problem 3.1 due to the internal pressure p.

Let us consider the longitudinal and circumferential loadings separately. Due to radial pressure p *only*, the circumferential stress is given by $\sigma_c = pr/h$, and because $\sigma = E\epsilon$ the circumferential strain is given by $\epsilon_c = pr/Eh$.

It is to be noted that ϵ_c is a unit strain. The length over which it acts is the circumference of the cylinder which is $2\pi r$. Hence the total elongation of the circumference is

$$\Delta = \epsilon_c(2\pi r) = \frac{2\pi pr^2}{Eh}$$

The final length of the circumference is thus $2\pi r + 2\pi pr^2/Eh$. Dividing this circumference by 2π we find the radius of the deformed cylinder to be $r + pr^2/Eh$, so that the increase in radius is pr^2/Eh.

Due to the axial pressure p *only*, longitudinal stresses $\sigma_l = pr/2h$ are set up. These longitudinal stresses give rise to longitudinal strains $\epsilon_l = pr/2Eh$. As in Chap. 1 an extension in the direction of loading, which is the longitudinal direction here, is accompanied by a decrease in the dimension perpendicular to the load. Thus here the circumferential dimension decreases. The ratio of the strain in the lateral direction to that in the direction of loading was defined in Chap. 1 to be Poisson's ratio, denoted by μ. Consequently the above strain ϵ_l induces a circumferential strain equal to $-\mu\epsilon_l$ and if this strain is denoted ϵ_c' we have $\epsilon_c' = -\mu pr/2Eh$, which tends to decrease the radius of the cylinder as shown by the negative sign.

In a manner exactly analogous to the treatment of the increase of radius due to radial loading only, the decrease of radius corresponding to the strain ϵ_c' is given by $\mu pr^2/2Eh$. The resultant increase of radius due to the internal pressure p is thus

$$\Delta r = \frac{pr^2}{Eh} - \frac{\mu pr^2}{2Eh} = \frac{pr^2}{Eh}\left(1 - \frac{\mu}{2}\right)$$

3.6. A thin-walled cylinder with rigid end closures is fabricated by welding long rectangular plates around a cylindrical form so that the completed pressure vessel has the form shown in Fig. 3-7. The angle that the helix makes with a generator of the shell is 35° at all points. The mean radius of the cylinder is 20 in, the wall thickness is $h = 0.5$ in, and the internal pressure is 400 lb/in^2. Neglect the localized effects at each end due to the end closure plates and determine the normal and shearing stresses acting on the helical weld in the curved plane of the cylinder wall.

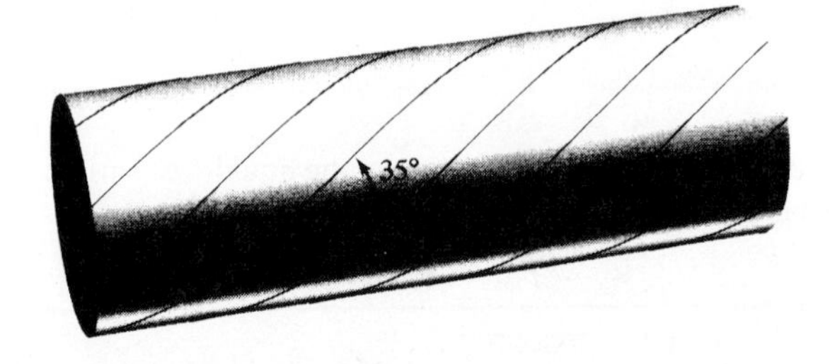

Fig. 3-7

From Problem 3.1 the circumferential and longitudinal stresses in the cylinder are

$$\sigma_c = \frac{pr}{h} = \frac{(400\ \text{lb/in}^2)(20\ \text{in})}{0.5\ \text{in}} = 16{,}000\ \text{lb/in}^2$$

$$\sigma_L = \frac{pr}{2h} = 8000\ \text{lb/in}^2$$

Let us consider a small triangular element to be removed from the cylinder wall, with the element being bounded on its hypotenuse by the weld and along the other two sides by a generator together with a circumference of the shell. The stresses found above (shown by solid vectors) act on the perpendicular sides as shown in Fig. 3-8, and on the inclined side of the element (coinciding with the helical weld) we have the unknown normal stress σ and shearing stress τ. The length of the hypotenuse of the element is taken to be ds, in which case the side along a generator has the length $ds \cos 35°$ and the length in the circumferential direction is $ds \sin 35°$.

It is convenient to introduce n- and t-axes perpendicular to and along the helical weld. The n and t components of the applied stresses are shown in Fig. 3-8 by dotted vectors. For equilibrium in the

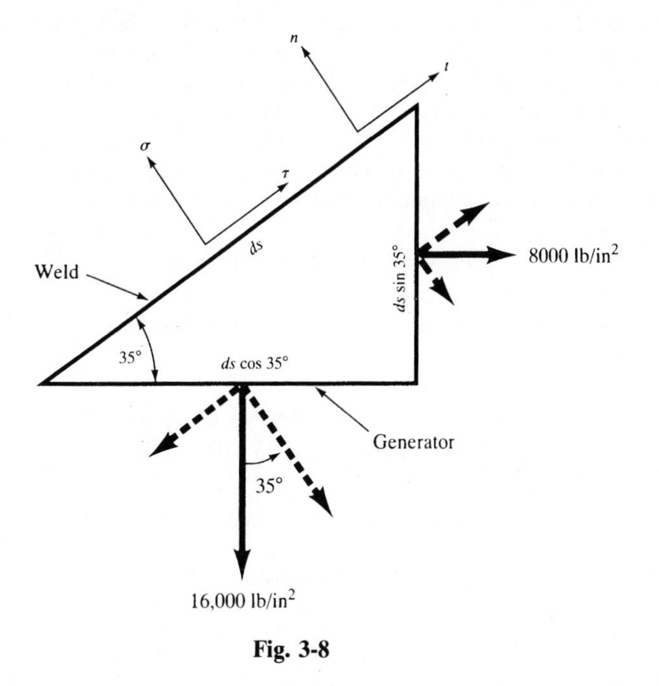

Fig. 3-8

n-direction, we have

$$\Sigma F_n = \sigma(ds)(h) - 8000(ds)(\sin 35°)(h)(\sin 35°) - 16{,}000(ds)(\cos 35°)(h)(\cos 35°) = 0$$

$$\therefore \sigma = 8000 \sin^2 35° + 16{,}000 \cos^2 35° = 13{,}370\ \text{lb/in}^2$$

Similarly, in the tangential direction (i.e., in the direction along the helix), we have

$$\Sigma F_t = \tau(ds)(h) + 8000(ds)(\sin 35°)(\cos 35°)(h) - 16{,}000(ds)(\cos 35°)(h)(\sin 35°) = 0$$

$$\therefore \tau = (8000)(\sin 35°)(\cos 35°) = 3760\ \text{lb/in}^2$$

3.7. Consider a closed thin-walled spherical shell subject to a uniform internal pressure p. The inside radius of the shell is r and its wall thickness is h. Derive an expression for the tensile stress existing in the wall.

For a free-body diagram, let us consider exactly half of the entire sphere. This body is acted upon by the applied internal pressure p as well as the forces that the other half of the sphere, which has been

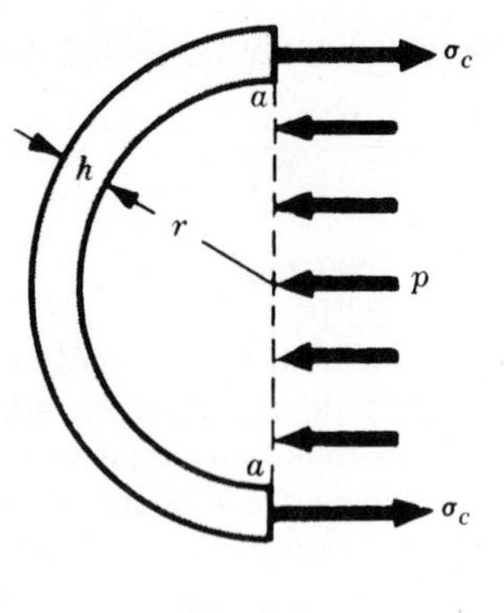

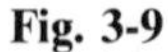

Fig. 3-9

removed, exerts upon the half under consideration. Because of the symmetry of loading and deformation, these forces may be represented by circumferential tensile stresses σ_c as shown in Fig. 3-9.

This free-body diagram represents the forces acting on the hemisphere, the diagram showing only a projection of the hemisphere on a vertical plane. Actually the pressure p acts over the entire inside surface of the hemisphere and in a direction perpendicular to the surface at every point. However, as mentioned in Problem 3.1, it is permissible to consider the force exerted by this same pressure p upon the *projection* of this area which in this case is the vertical circular area denoted by *a-a*. This is possible because the hemisphere is symmetric about the horizontal axis and the vertical components of the pressure annul one another. Only the horizontal components produce the tensile stress σ_c. For equilibrium we have

$$\Sigma F_h = \sigma_c 2\pi r h - p\pi r^2 = 0 \qquad \text{or} \qquad \sigma_c = \frac{pr}{2h}$$

From symmetry this circumferential stress is the same in all directions at any point in the wall of the sphere.

3.8. A 20-m-diameter spherical tank is to be used to store gas. The shell plating is 10 mm thick and the working stress of the material is 125 MPa. What is the maximum permissible gas pressure p?

From Problem 3.7 the tensile stress in all directions is uniform and given by $\sigma_c = pr/2h$. Substituting:

$$125 \times 10^6\ \text{N/m}^2 = \frac{p(10\ \text{m})}{2(0.010\ \text{m})}$$

$$p = 0.25\ \text{MPa}$$

3.9. The undersea research vehicle *Alvin* has a spherical pressure hull 1 m in radius and shell thickness of 30 mm. The pressure hull is steel having a yield point of 700 MPa. Determine the depth of submergence that would set up the yield point stress in the spherical shell. Consider sea water to have a specific weight of 10.07 kN/m^3.

From Problem 3.7 the compressive stress due to the external hydrostatic pressure is given by $\sigma_c = pr/2h$. The hydrostatic pressure corresponding to yield is thus

$$700 \times 10^6\ \text{N/m}^2 = \frac{p(1\ \text{m})}{2(0.03\ \text{m})} \qquad \text{or} \qquad p = 42\ \text{MPa}$$

Since, as in Problem 3.3, we have $p = \gamma h$, where γ is the specific weight of the sea water, we have

$$42 \times 10^6\ \text{N/m}^2 = (10.07 \times 10^3\ \text{N/m}^3)(h) \qquad \text{or} \qquad h = 4170\ \text{m}$$

It should be noted that this neglects the possibility of buckling of the sphere due to hydrostatic pressure as well as effects of entrance ports on its strength. These factors, beyond the scope of this treatment, result in a true operating depth of 1650 m.

3.10. Find the increase of volume of a thin-walled spherical shell subject to a uniform internal pressure p.

From Problem 3.7 we know that the circumferential stress is constant through the shell thickness and is given by

$$\sigma_c = \frac{pr}{2h}$$

in all directions at any point in the shell. From the two-dimensional form of Hooke's law (see Chap. 1), we have the circumferential strain as

$$\epsilon_c = \frac{1}{E}[\sigma_c - \mu\sigma_c] = \frac{pr}{2Eh}[1-\mu]$$

This strain is the change of length per unit length of the circumference of the sphere, so the increase of length of the circumference is

$$(2\pi r)\cdot\frac{pr}{2Eh}[1-\mu]$$

The radius of the spherical shell subject to internal pressure p is now found by dividing the circumference of the pressurized shell by the factor 2π. Thus the final radius is

$$\left[2\pi r + (2\pi r)\cdot\frac{pr}{2Eh}(1-\mu)\right]\Big/2\pi \qquad (1)$$

or

$$\left[r + \frac{pr^2}{2Eh}(1-\mu)\right] \qquad (2)$$

and the volume of the pressurized sphere is

$$\frac{4}{3}\pi\left[r + \frac{pr^2}{2Eh}(1-\mu)\right]^3 \qquad (3)$$

The desired increase of volume due to pressurization is found by subtracting from (*3*) the initial volume:

$$\Delta V = \frac{4\pi}{3}\left[r + \frac{pr^2}{2Eh}(1-\mu)\right]^3 - \frac{4}{3}\pi r^3$$

Expanding and dropping terms involving powers of (p/E), which is ordinarily of the order of 1/1000, we see that the increase of volume due to pressurization is

$$\Delta V = \frac{2\pi p r^4}{Eh}(1-\mu)$$

3.11. A thin-walled titanium alloy spherical shell has a 1-m inside diameter and is 7 mm thick. It is completely filled with an unpressurized, incompressible liquid. Through a small hole an additional 1000 cm^3 of the same liquid is pumped into the shell, thus increasing the shell radius. Find the pressure after the additional liquid has been introduced and the hole closed. For this titanium allow $E = 114$ GPa and the tensile yield point of the material to be 830 MPa.

The initial volume of the spherical shell is

$$V = \frac{4}{3}\pi r^3 = \frac{\pi}{6}d^3 \qquad (d = \text{diameter})$$

$$= \frac{\pi}{6}(1\ \text{m})^3 = 0.5236\ \text{m}^3$$

The volume of liquid pumped in is

$$1000\ \text{cm}^3\left(\frac{\text{m}}{100\ \text{cm}}\right)^3 = \frac{1}{10^3}\ \text{m}^3$$

so that the final volume of the incompressible liquid is

$$0.5236\ \text{m}^3 + 0.001\ \text{m}^3 = 0.5246\ \text{m}^3$$

which is equal to the volume of the expanded shell. The relation between pressure and volume change was found in Problem 3.10 to be

$$\Delta V = \frac{2\pi p r^4}{Eh}(1-\mu)$$

Substituting,

$$0.001\ \text{m}^3 = \frac{(2\pi)p(0.5\ \text{m})^4(0.67)}{(114\times 10^9\ \text{N/m}^2)(0.007\ \text{m})}$$

Solving,

$$p = 3.03\ \text{MPa}$$

It is well to check the normal stress in the titanium shell due to this pressure. From Problem 3.7 we have

$$\sigma = \frac{pr}{2h}$$

$$= \frac{(3.03\times 10^6\ \text{N/m}^2)(0.05\ \text{m})}{2(0.007\ \text{m})} = 109\ \text{MPa}$$

which is well below the yield point of the material.

3.12. Consider a laminated pressure vessel composed of two thin coaxial cylinders as shown in Fig. 3-10. In the state prior to assembly there is a slight "interference" between these shells, i.e., the inner one is too large to slide into the outer one. The outer cylinder is heated, placed on the inner, and allowed to cool, thus providing a "shrink fit." If both cylinders are steel and the mean diameter of the assembly is 100 mm, find the tangential stresses in each shell arising from the shrinking if the initial interference (of diameters) is 0.25 mm. The thickness of the inner shell is 2.5 mm, and that of the outer shell 2 mm. Take $E = 200$ GPa.

There is evidently an interfacial pressure p acting between the adjacent faces of the two shells. It is to be noted that there are no external applied loads. The pressure p may be considered to increase the diameter of the outer shell and decrease the diameter of the inner so that the inner shell may fit inside the outer. The radial expansion of a cylinder due to a radial pressure p was found in Problem 3.5 to be pr^2/Eh. No longitudinal forces are acting in this problem. The increase in radius of the outer shell due to p, plus

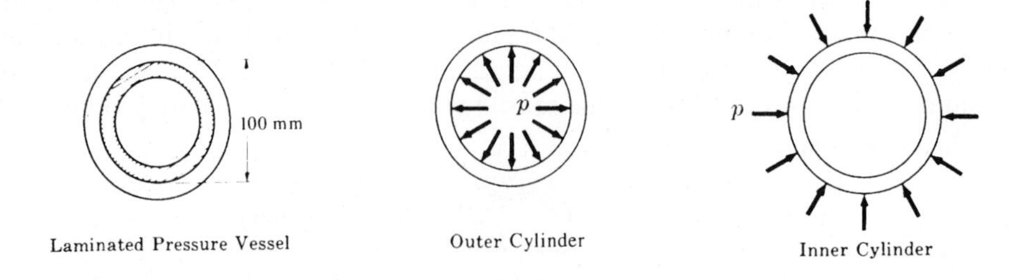

Fig. 3-10

the decrease in radius of the inner one due to p, must equal the initial interference between radii, or 0.25/2 mm. Thus we have

$$\frac{p(0.05\ \text{m})^2}{(200\times10^9\ \text{N/m}^2)(0.0025\ \text{m})}+\frac{p(0.05\ \text{m})^2}{(200\times10^9\ \text{N/m}^2)(0.002\ \text{m})}=\frac{0.125}{1000}\ \text{m}$$

$$p = 11.1\ \text{MPa}$$

This pressure, illustrated in the above figures, acts between the cylinders after the outer one has been shrunk onto the inner one. In the inner cylinder this pressure p gives rise to a stress

$$\sigma_c=\frac{pr}{h}=\frac{(11.1\times10^6\ \text{N/m}^2)(0.05\ \text{m})}{(0.0025\ \text{m})}=-222\ \text{MPa}$$

In the outer cylinder the circumferential stress due to the pressure p is

$$\sigma_c=\frac{pr}{h}=\frac{(11.1\times10^6\ \text{N/m}^2)(0.05\ \text{m})}{(0.002\ \text{m})}=277\ \text{MPa}$$

If, for example, the laminated shell is subject to a uniform internal pressure, these shrink-fit stresses would merely be added algebraically to the stresses found by the use of the simple formulas given in Problem 3.1.

3.13. The thin steel cylinder just fits over the inner copper cylinder as shown in Fig. 3-11. Find the tangential stresses in each shell due to a temperature rise of 60 °F. Do not consider the effects introduced by the accompanying longitudinal expansion. This arrangement is sometimes used for storing corrosive fluids. Take

$$E_{st}=30\times10^6\ \text{lb/in}^2 \qquad \alpha_{st}=6.5\times10^{-6}/^\circ\text{F}$$

$$E_{cu}=13\times10^6\ \text{lb/in}^2 \qquad \alpha_{cu}=9.3\times10^{-6}/^\circ\text{F}$$

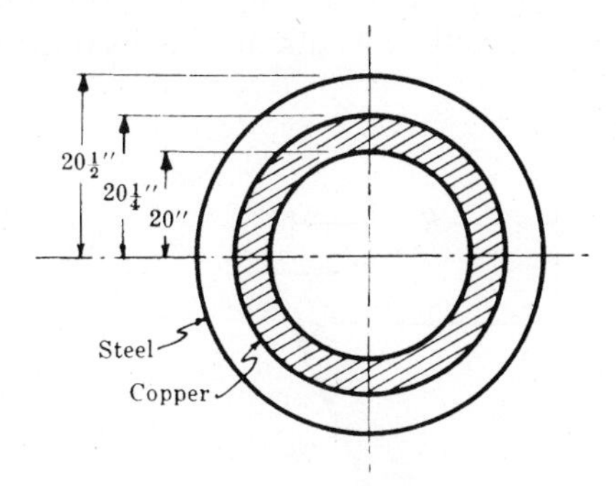

Fig. 3-11

The simplest approach is to first consider the two shells to be separated from one another so that they are no longer in contact.

Due to the temperature rise of 60 °F the circumference of the steel shell increases by an amount $2\pi(20.375)(60)(6.5\times10^{-6}) = 0.0498$ in. Also, the circumference of the copper shell increases an amount $2\pi(20.125)(60)(9.3\times10^{-6}) = 0.0705$ in. Thus the interference between the radii, i.e., the difference in radii, of the two shells (due to the heating) is $(0.0705 - 0.0498)/2\pi = 0.00345$ in. Again, there are no external loads acting on either cylinder.

However, from the statement of the problem the adjacent surfaces of the two shells are obviously in contact after the temperature rise. Hence there must be an interfacial pressure p between the two surfaces, i.e., a pressure tending to increase the radius of the steel shell and decrease the radius of the copper shell

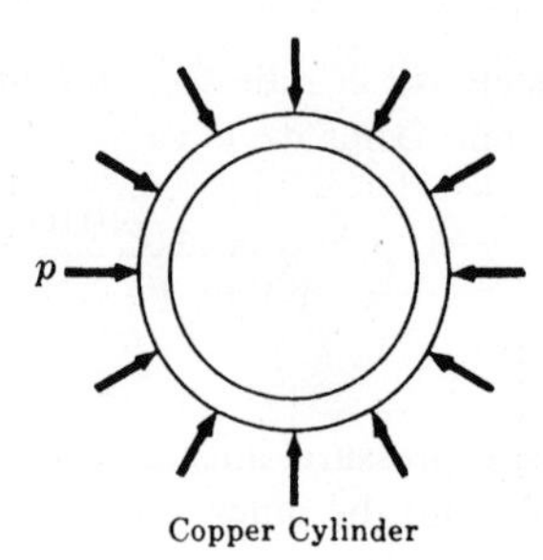

Fig. 3-12

so that the copper shell may fit inside the steel one. Such a pressure is shown in the free-body diagrams of Fig. 3-12.

In Problem 3.5 the change of radius of a cylinder due to a uniform radial pressure p (with no longitudinal forces acting) was found to be pr^2/Eh. Consequently the increase of radius of the steel shell due to p, added to the decrease of radius of the copper one due to p, must equal the interference; thus

$$\frac{p(20.375)^2}{(30 \times 10^6)(0.25)} + \frac{p(20.125)^2}{(13 \times 10^6)(0.25)} = 0.00345 \quad \text{or} \quad p = 19.2 \text{ lb/in}^2$$

This interfacial pressure creates the required continuity at the common surface of the two shells when they are in contact. Using the formula for the tangential stress, $\sigma_c = pr/h$, we find the tangential stresses in the steel and copper shells to be, respectively,

$$\sigma_{st} = \frac{19.2(20.375)}{0.25} = 1560 \text{ lb/in}^2 \quad \text{and} \quad \sigma_{cu} = -\frac{19.2(20.125)}{0.25} = -1550 \text{ lb/in}^2$$

3.14. Consider a thin-walled conical shell containing a liquid whose weight per unit volume is γ [see Fig. 3-13(a)]. The shell is supported around its upper rim and filled with liquid to a depth H. Determine the stresses in the shell walls due to this loading. The geometric axis of the shell is vertical.

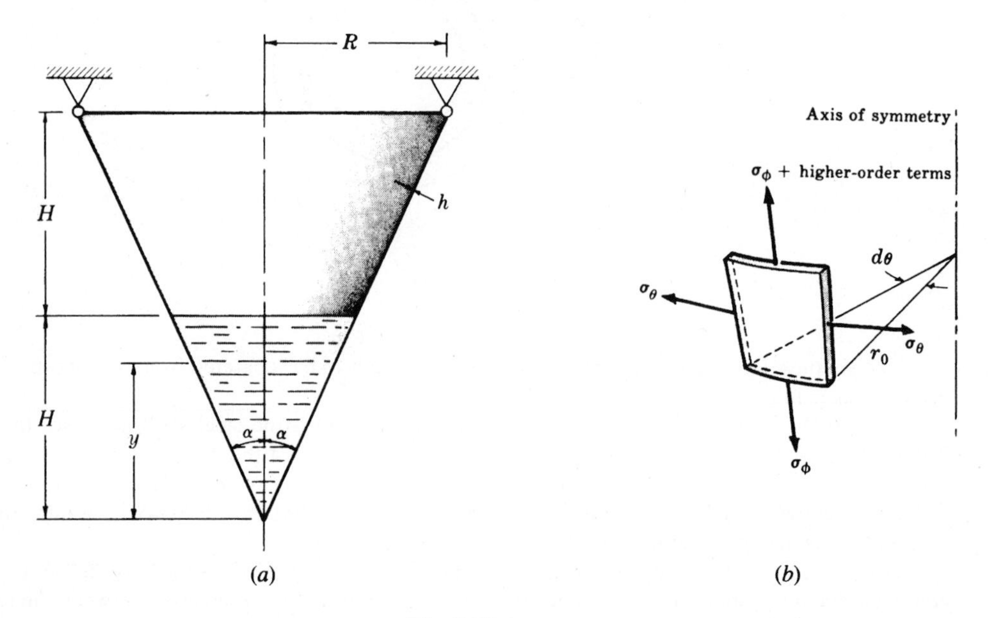

Fig. 3-13

The state of stress in this shell is obviously axisymmetric. It is assumed that the shell thickness h is small compared to H and R. The stresses may be determined by consideration of the equilibrium of a shell element bounded by two closely adjacent parallel circles whose planes are normal to the vertical axis of symmetry of the cone and by two closely adjacent generators of the cone. Such an element, together with the vectors representing the stresses σ_θ in the horizontal direction and σ_ϕ in the direction of a generator, is indicated in Fig. 3-13(b). The quantity σ_θ is called *hoop stress* and σ_ϕ is termed the *meridional stress*.

In the diagram θ represents the angular coordinate measured in a horizontal plane which is normal to the vertical axis of symmetry of the shell. The radius of the cone there is r_0, which is of course a function of the location of the element with respect to its position along the axis of symmetry. Another coordinate useful for defining the geometry of the cone is r_2, which corresponds to the radius of curvature of the shell surface in a direction perpendicular to the generator. This is best illustrated by examining a section of the cone formed by passing a vertical plane through the shell axis as indicated in Fig. 3-14(a) below. It is evident that $r_0 = r_2 \cos\alpha$.

From geometry we have

$$r_0 = y \tan\alpha \qquad \text{and so} \qquad r_2 = \frac{y \tan\alpha}{\cos\alpha}$$

The hoop *stresses* in Fig. 3-13(b) may be visualized more clearly by looking along the axis of symmetry, as shown in Fig. 3-14(b). It is evident that each of the hoop *forces* vectors $\sigma_\theta(dy/\cos\alpha)h$ makes an angle $d\theta/2$ with the tangent to the element. The resultant of these hoop forces is $2\sigma_\theta h(dy/\cos\alpha)\sin(d\theta/2)$ or, since $d\theta/2$ is small, $\sigma_\theta h(dy/\cos\alpha)\,d\theta$ acting in a horizontal plane and directed toward the geometric axis of the shell. From Fig. 3-14(a) we see that this resultant must be multiplied by $\cos\alpha$ to determine the component of this force acting in a direction normal to the shell surface. Also, it is evident that the meridional forces corresponding to Fig. 3-14(a) cancel one another. The liquid exerts a normal pressure

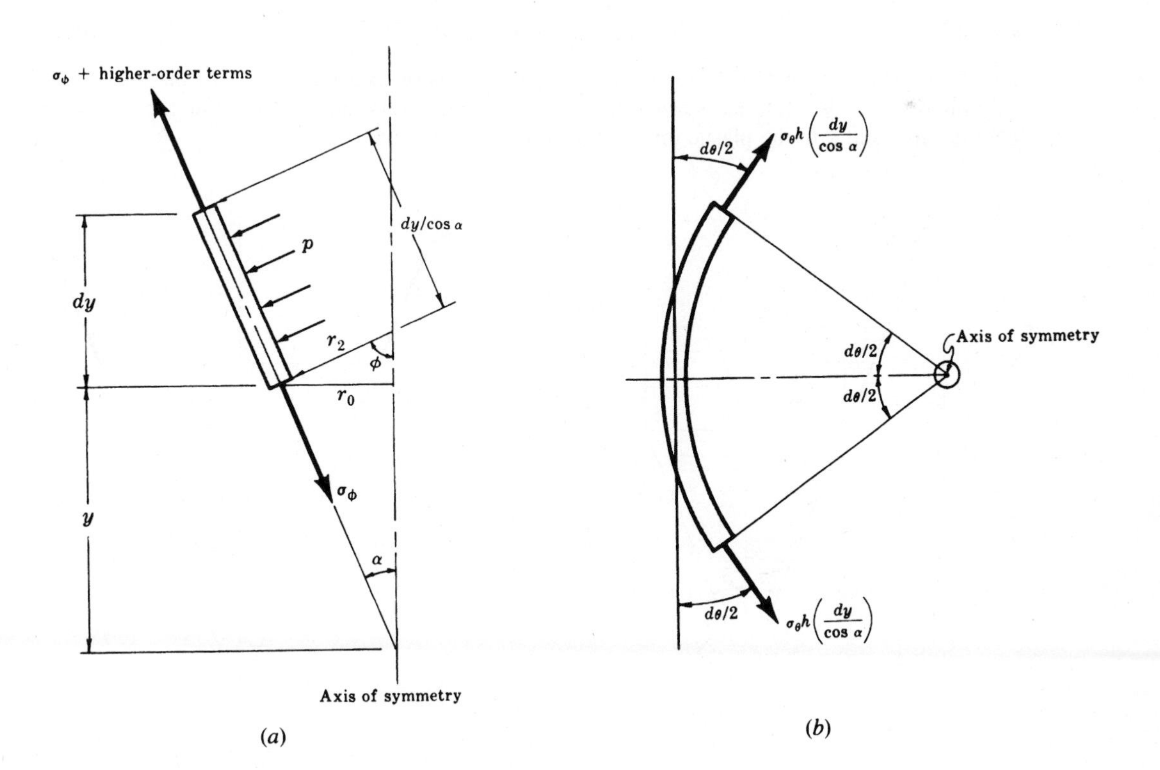

Fig. 3-14

p as indicated in the figure and it acts over an area $(r_0\,d\theta)\,(dy/\cos\alpha)$. Thus, for equilibrium of the element in a direction normal to the surface we have

$$\sigma_\theta h\left(\frac{dy}{\cos\alpha}\right)(d\theta)\cos\alpha - pr_0(d\theta)\frac{dy}{\cos\alpha} = 0 \tag{1}$$

or

$$\sigma_\theta = \frac{pr_0}{h\cos\alpha} = \frac{py\tan\alpha}{h\cos\alpha} = \frac{pr_2}{h} \tag{2}$$

This expression holds anywhere in the conical shell. In the lower half, $0<y<H$, we have $p = \gamma(H-y)$, so

$$\sigma_\theta = \frac{\gamma(H-y)y\tan\alpha}{h\cos\alpha} \qquad \text{for} \qquad 0<y<H \tag{3}$$

In the upper half, $H<y<2H$, $p = 0$, so $\sigma_\theta = 0$ in that region.

The other stress σ_ϕ may be found by considering the vertical equilibrium of the conical shell. For $0<y<H$ the weight of the liquid in the conical region *abo* plus that in the cylindrical region *abcd* is held in equilibrium by the forces corresponding to σ_ϕ and we have from Fig. 3-15(a)

$$\sigma_\phi h 2\pi y\tan\alpha\cos\alpha - \gamma[\tfrac{1}{3}\pi(y\tan\alpha)^2 y + (H-y)\pi(y\tan\alpha)^2] = 0 \tag{4}$$

or

$$\sigma_\phi = \frac{\gamma\tan\alpha}{h\cos\alpha}\left(\frac{Hy}{2} - \frac{y^2}{3}\right) \qquad \text{for} \qquad 0<y<H \tag{5}$$

Similarly, for $H<y<2H$, the weight of all the liquid is held in equilibrium by the forces corresponding to σ_ϕ so that from Fig. 3-15(b)

$$\sigma_\phi h(2\pi y)(\tan\alpha)\cos\alpha - \gamma\tfrac{1}{3}\pi r_0^2 H = 0 \tag{6}$$

Since $r_0 = H\tan\alpha$ we get

$$\sigma_\phi = \frac{\gamma H^3\tan\alpha}{6hy\cos\alpha} \qquad \text{for} \qquad H<y<2H \tag{7}$$

It is to be observed that the stresses associated with these axisymmetric deformations are statically determinate; i.e., it was not necessary to use any deformation relations to determine the stresses. Thus the relations are valid into the plastic range of action.

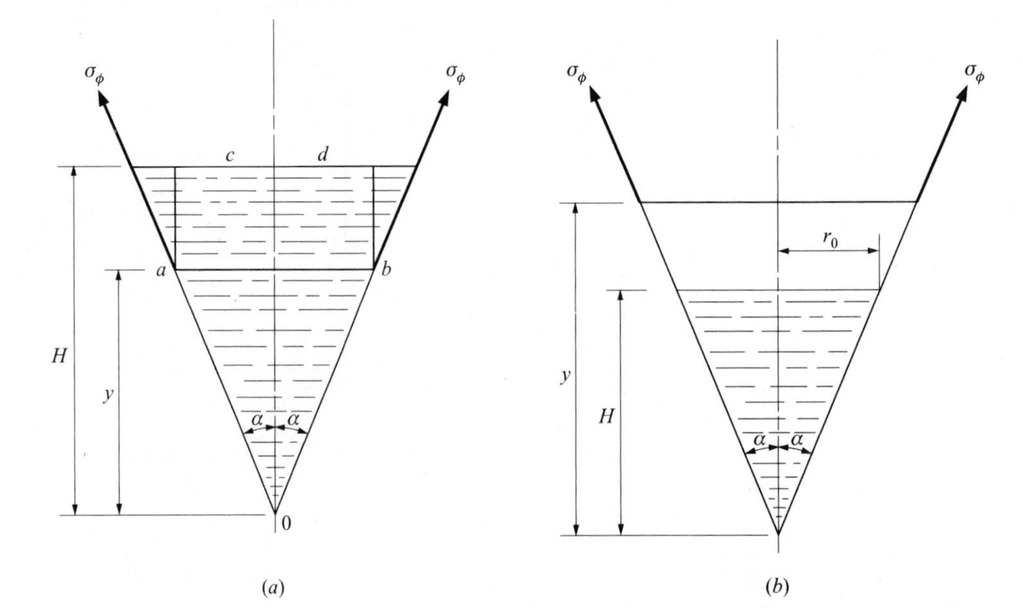

Fig. 3-15

3.15. Determine the hoop stresses and meridional stresses in a thin shell of revolution subject to an internal pressure p.

This problem is readily solved as a generalization of Problem 3.14. The stresses may be determined by consideration of the equilibrium of a shell element bounded by two closely adjacent parallel circles whose planes are normal to the vertical axis of a symmetry of the shell and by two closely adjacent generators, or meridians, of the shell (see Fig. 3-1). This element is analogous to that shown in Fig. 3-13(b) of Problem 3.14, except that the vertical sides are curved rather than straight.

The hoop stresses σ_θ and the meridional stresses σ_ϕ thus appear as shown in Fig. 3-16. We now require two radii of curvature to describe this element. We use r_1 to denote the radius of curvature of the meridian and r_2 to denote the radius of curvature of the shell surface in a direction perpendicular to the meridian. The center of curvature corresponding to r_2 must lie on the axis of symmetry although the center for r_1 does not (in general). Figure 3-17(a) shows the hoop forces as seen by looking along the axis of symmetry and, analogous to Problem 3.14, they have a horizontal component $2\sigma_\theta h r_1\, d\phi(d\theta/2)$ directed toward the shell axis. This is multiplied by $\sin\phi$ to obtain the component normal to the shell element. The meridional forces appear as in Fig. 3-17(b) and they have a component normal to the shell given by $\sigma_\phi h r_0\, d\theta\, d\phi$. The

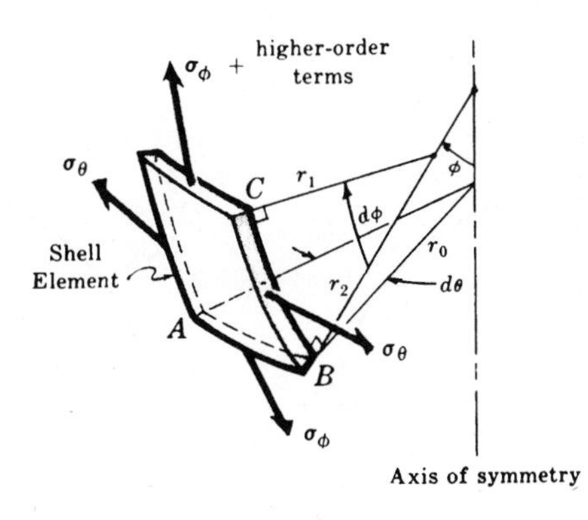

Fig. 3-16

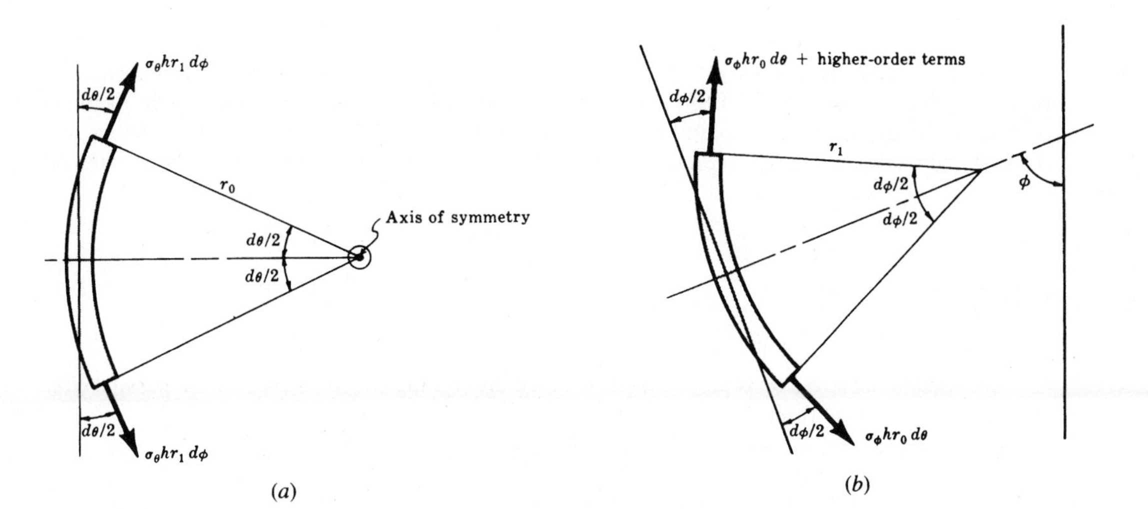

Fig. 3-17

pressure p acts over an area $(r_0\,d\theta)(r_1\,d\phi)$ so that the equation of equilibrium in the normal direction becomes

$$\sigma_\theta h r_1\,d\theta\,d\phi \sin\phi + \sigma_\phi h r_0\,d\theta\,d\phi - p r_0\,d\theta\,r_1\,d\phi = 0$$

or, since $r_0 = r_2 \sin\phi$, we get

$$\frac{\sigma_\phi}{r_1} + \frac{\sigma_\theta}{r_2} = \frac{p}{h} \tag{1}$$

This fundamental equation applies to axisymmetric deformations of all thin shells of revolution. A second equation is obtained as in Problem 3.14 by consideration of the vertical equilibrium of the entire shell above some convenient parallel circle. Again, these equations are valid into the plastic range of action.

3.16. Consider a constant-thickness thin-walled spherical dome of radius r loaded only by its own weight q per unit of surface area. The dome is supported by frictionless rollers around its lower boundary as shown in Fig. 3-18(a). Determine meridional and hoop stresses at all points in the system.

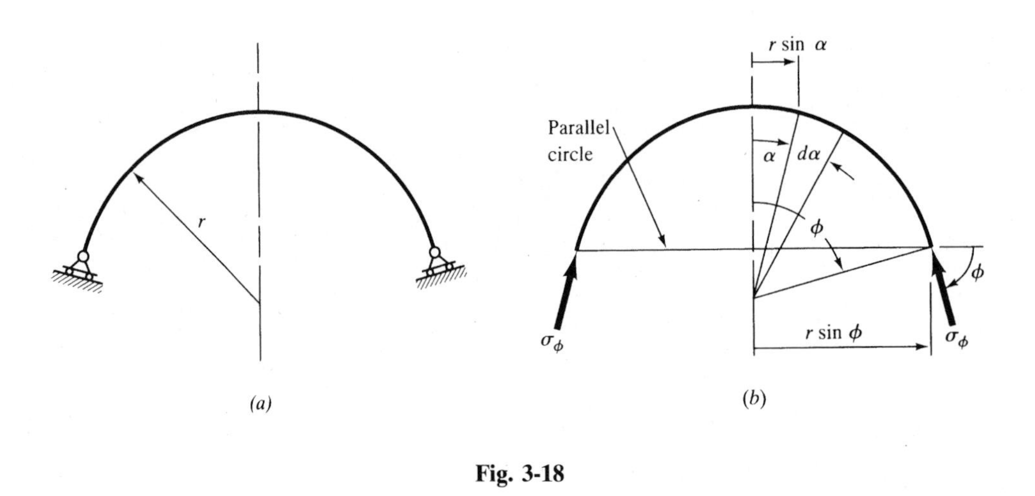

Fig. 3-18

Let us consider the vertical equilibrium of a portion of the dome above some parallel circle defined by the angle ϕ shown in Fig. 3-18(b). The variable angle α is introduced and the weight of the central portion of the dome above the parallel circle is found by considering a ring-shaped element of radius $(r\sin\alpha)$ and meridional length $(r\,d\alpha)$. The weight of the portion of the dome above the parallel circle is

$$\int_{\alpha=0}^{\alpha=\phi} q[2\pi(r\sin\alpha)]\,(r\,d\alpha)$$

which becomes

$$2\pi r^2 q(1-\cos\phi)$$

The meridional stress σ_ϕ is uniformly distributed around the circumference of the parallel circle and has an upward vertical resultant given by

$$2\pi(r\sin\phi)h\sigma_\phi(\sin\phi)$$

For vertical equilibrium of the dome above the parallel circle, we thus have

$$2\pi(r\sin\phi)h\sigma_\phi(\sin\phi) - 2\pi r^2 q(1-\cos\phi) = 0$$

or

$$\sigma_\phi = \frac{rq}{h(1+\cos\phi)} \qquad \text{(compression)} \tag{1}$$

This value, when introduced into Eq. (*1*) of Problem 3.15, leads to a hoop stress σ_θ given by

$$\sigma_\theta = \frac{rq}{h}\left[\frac{1}{1+\cos\phi} - \cos\phi\right] \tag{2}$$

3.17. The spherical dome of Problem 3.16 subtends an opening angle of 120°, has a wall thickness of 100 mm, and a radius of 50 m. It is constructed of concrete having a specific weight of 23.5 kN/m^3. Determine meridional and circumferential stresses at (*a*) the apex of the dome, and (*b*) the simply supported rim.

The meridional stress is given by Eq. (*1*) of Problem 3.16. In that equation q denotes weight per unit of surface area. Here, since the specific weight refers to a cube of concrete weighing 23.5 kN, the weight per unit surface area is found by considering the 100-mm thickness to be

$$q = (23{,}500\ \text{N/m}^3)\left(\frac{100}{1000}\right) = 2350\ \text{N/m}^2$$

The meridional stress at the apex, where $\phi = 0°$, is

$$\sigma_\phi = -\frac{(50\ \text{m})(2350\ \text{N/m}^2)}{(0.1\ \text{m})[1+\cos 0°]} = -587{,}500\ \text{N/m}^2 \qquad \text{or} \qquad -0.587\ \text{MPa}$$

and at the rim, where $\phi = 60°$, we have

$$\sigma_\phi = -\frac{(50\ \text{m})(2350\ \text{N/m}^2)}{(0.1\ \text{m})[1+\cos 60°]} = -786{,}000\ \text{N/m}^2 \qquad \text{or} \qquad -0.786\ \text{MPa}$$

The circumferential stress is given by Eq. (*2*) of Problem 3.16. At the apex this is

$$\sigma_\theta = \frac{(50\ \text{m})(2350\ \text{N/m}^2)}{(0.1\ \text{m})}\left[\frac{1}{1+\cos 0°} - \cos 0°\right] = -587{,}500\ \text{N/m}^2 \qquad \text{or} \qquad -0.588\ \text{MPa}$$

and at the rim, where $\phi = 60°$, it is

$$\sigma_\theta = \frac{(50\ \text{m})(2350\ \text{N/m}^2)}{(0.1\ \text{m})}\left[\frac{1}{1+\cos 60°} - \cos 60°\right] = 195{,}000\ \text{N/m}^2 \qquad \text{or} \qquad 0.195\ \text{MPa}$$

Thus the circumferential stress is tensile at the rim and compressive at the apex. From Eq. (*2*) of Problem 3.16, the circumferential stress is zero when

$$\frac{1}{1+\cos\phi_0} - \cos\phi_0 = 0$$

Solving by trial and error, we find $\phi_0 = 51.8°$.

3.18. Thin toroidal shells are sometimes employed as gas storage tanks in boosters for space vehicles. One design considered by the National Aeronautics and Space Administration for possible future use employs a torus of mean diameter $2b = 70$ ft with a cross-section diameter of $2R = 5$ ft as indicated in Fig. 3-19. The internal pressure p is 20 lb/in^2 and the shell material is 2219 T87 aluminum alloy, having a yield point of 50,000 lb/in^2 at room temperature. For this material the yield point increases at lower temperatures, reaching 120 percent of the above value at −300 °F. If a safety factor of 1.5 is employed, determine the required wall thickness.

First, we consider the vertical equilibrium of a ring-shaped portion of the toroidal shell above an

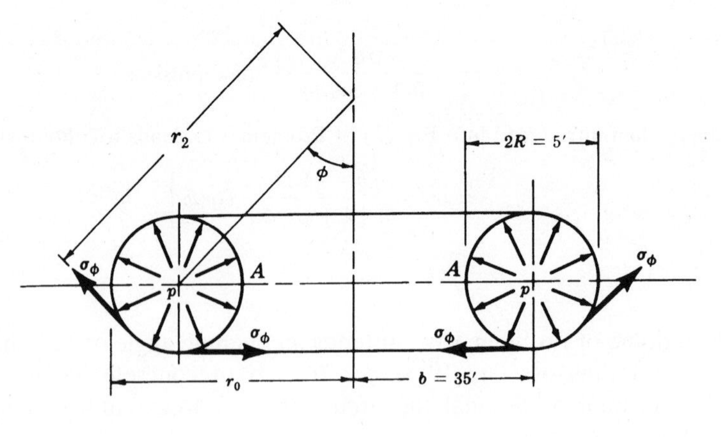

Fig. 3-19

arbitrary plane, as indicated by the angle ϕ. The meridional stress σ_ϕ is readily found by considering the pressure p to act on the horizontal projection of the curved area. Thus

$$2\pi r_0 \sigma_\phi h \sin\phi = \pi p(r_0^2 - b^2)$$

or since $\sin\phi = (r_0 - b)/R$

$$\sigma_\phi = \frac{pR(r_0 + b)}{2\pi r_0 h} \tag{1}$$

From (*1*) it is evident that the peak value of σ_ϕ occurs at the innermost points A where

$$(\sigma_\phi)_{max} = \frac{pR}{2h}\left(\frac{2b - R}{b - R}\right) \tag{2}$$

If $b = 0$, the torus reduces to a sphere and (*2*) coincides with the stresses in a sphere as found in Problem 3.7. For the given dimensions we have $R = 30$ in, $b = 420$ in, $p = 20$ lb/in^2, and (*2*) becomes

$$\frac{50{,}000}{1.5} = \frac{20(30)(840 - 30)}{2h(420 - 30)} \qquad \text{or} \qquad h = 0.0187 \text{ in} \tag{3}$$

If σ_ϕ as given by (*1*) is substituted into (*1*) of Problem 3.15 (which holds for axisymmetric deformation of any thin shell of revolution) we obtain, for $r_1 = R$ and $r_2 = (b + R\sin\phi)/\sin\phi$,

$$\sigma_\theta = \frac{pR}{2h} \tag{4}$$

at any point in the toroidal shell. Evidently the peak value of σ_ϕ as given by (*2*) exceeds the value of σ_θ and hence the maximum value of σ_ϕ controls the design. The required thickness is thus given by (*3*).

Supplementary Problems

3.19. One proposed design for an energy-efficient automobile involves an on-board tank storing hydrogen (in a special nonvolatile form) which would be released to a fuel cell. The tank is to be cylindrical, 0.4 m in diameter, made of type 302 stainless steel having a working stress in tension of 290 MPa, and closed by hemispherical end caps. The hydrogen would be pressurized to 15 MPa when the tank is initially filled. Determine the required wall thickness of the tank. *Ans.* $h = 5.2$ mm

3.20. A vertical cylindrical gasoline storage tank is 30 m in diameter and is filled to a depth of 15 m with gasoline whose specific gravity is 0.74. If the yield point of the shell plating is 250 MPa and a safety factor of 2.5 is adequate, calculate the required wall thickness at the bottom of the tank. *Ans.* $h = 16.7$ mm

3.21. The research deep submersible *Aluminaut* has a cylindrical pressure hull of outside diameter 8 ft and a wall thickness of 5.5 in. It is constructed of 7079-T6 aluminum alloy, having a yield point of 60,000 lb/in^2. Determine the circumferential stress in the cylindrical portion of the pressure hull when the vehicle is at its operating depth of 15,000 ft below the surface of the sea. Use the mean diameter of the shell in calculations, and consider sea water to weigh 64.0 lb/ft^3. *Ans.* 54,800 lb/in^2

3.22. Derive an expression for the increase of volume per unit volume of a thin-walled circular cylinder subjected to a uniform internal pressure p. The ends of the cylinder are closed by circular plates. Assume that the radial expansion is constant along the length.

Ans. $$\frac{\Delta V}{V} = \frac{pr}{Eh}\left(\frac{5}{2} - 2\mu\right)$$

3.23. Calculate the increase of volume per unit volume of a thin-walled steel circular cylinder closed at both ends and subjected to a uniform internal pressure of 0.5 MPa. The wall thickness is 1.5 mm, the radius 350 mm, and $\mu = \frac{1}{3}$. Consider $E = 200$ GPa. *Ans.* $\Delta V/V = 10^{-3}$

3.24. Consider a laminated cylinder consisting of a thin steel shell "shrunk" on an aluminum one. The thickness of each is 0.10 in and the mean diameter of the assembly is 4 in. The initial "interference" of the shells prior to assembly is 0.004 in measured on a diameter. Find the tangential stresses in each shell caused by this shrink fit. For aluminum $E = 10 \times 10^6$ lb/in^2 and for steel $E = 30 \times 10^6$ lb/in^2.
Ans. $\sigma_{st} = 7500$ lb/in^2, $\sigma_{al} = -7500$ lb/in^2

3.25. A spherical tank for storing gas under pressure is 25 m in diameter and is made of structural steel 15 mm thick. The yield point of the material is 250 MPa and a safety factor of 2.5 is adequate. Determine the maximum permissible internal pressure, assuming the welded seams between the various plates are as strong as the solid metal. Also, determine the permissible pressure if the seams are 75 percent as strong as the solid metal. *Ans.* $p = 0.24$ MPa, $p = 0.18$ MPa

3.26. A thin-walled spherical shell is subject to a temperature rise ΔT which is constant at all points in the shell as well as through the shell thickness. Find the increase of volume per unit volume of the shell. Let α denote the coefficient of thermal expansion of the material. *Ans.* $3\alpha(\Delta T)$

3.27. A liquid storage tank consists of a vertical axis circular cylindrical shell closed at its lower end by a hemispherical shell as shown in Fig. 3-20. The weight of the system is carried by a ring-like support at the top and the lower extremity is unsupported. A liquid of specific weight γ entirely fills the container. Determine the peak circumferential and meridional stress in the cylindrical region of the assembly, as well as the peak stresses in the hemispherical region.

Ans.

Cylinder: $\sigma_c = \dfrac{\gamma R}{h}(H - R) \qquad \sigma_L = \dfrac{\gamma R}{2h}\left(H - \dfrac{R}{3}\right)$

Hemisphere: $\dfrac{\gamma HR}{2h}$

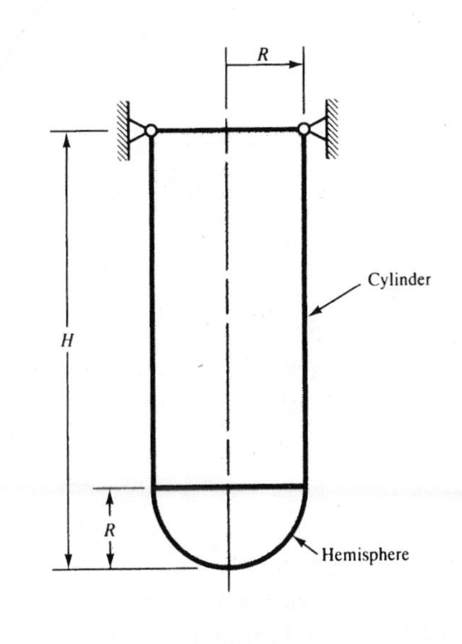

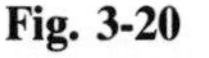
Fig. 3-20

3.28. Reexamine Problem 3.18 with all parameters as indicated there except that the shell material is now Ti-6Al-4V titanium alloy having a yield point of 126,000 lb/in^2 at room temperature. If a safety factor of 1.5 is used, determine the required wall thickness. *Ans.* 0.0074 in

Chapter 4

Direct Shear Stresses

DEFINITION OF SHEAR FORCE

If a plane is passed through a body, a force acting along this plane is called a *shear force* or *shearing force*. It will be denoted by F_s.

DEFINITION OF SHEAR STRESS

The shear force, divided by the area over which acts, is called the *shear stress* or *shearing stress*. It is denoted in this book by τ. Thus

$$\tau = \frac{F_s}{A} \tag{4.1}$$

COMPARISON OF SHEAR AND NORMAL STRESSES

Let us consider a bar cut by a plane *a-a* perpendicular to its axis, as shown in Fig. 4-1. A normal stress σ is perpendicular to this plane. This is the type of stress considered in Chaps. 1, 2, and 3.

A shear stress is one acting *along* the plane, as shown by the stress τ. Hence the distinction between normal stresses and shear stresses is one of *direction*.

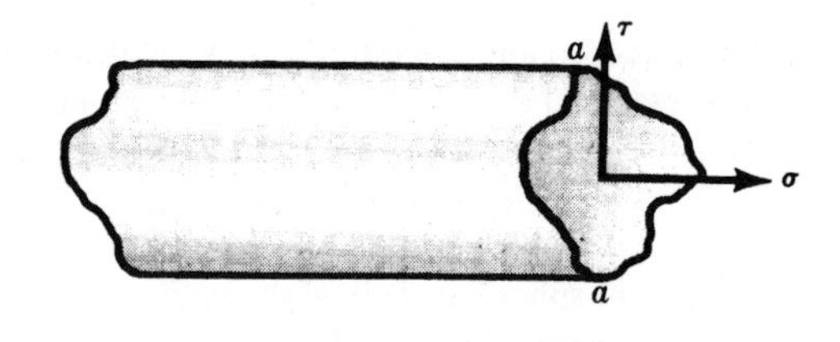

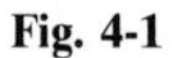

Fig. 4-1

ASSUMPTION

It is necessary to make some assumption regarding the manner of distribution of shear stresses, and for lack of any more precise knowledge it will be taken to be uniform in all problems discussed in this chapter. Thus the expression $\tau = F_s/A$ indicates an average shear stress over the area.

APPLICATIONS

Punching operations (Problem 4.2), wood test specimens (Problem 4.3), riveted joints (Problem 4.5), welded joints (Problem 4.6), and towing devices (Problem 4.10) are common examples of systems involving shear stresses.

DEFORMATIONS DUE TO SHEAR STRESSES

Let us consider the deformation of a plane rectangular element cut from a solid where the forces acting on the element are known to be shearing stresses τ in the directions shown in Fig. 4-2(*a*).

The faces of the element parallel to the plane of the paper are assumed to be load free. Since there are no normal stresses acting on the element, the lengths of the sides of the originally rectangular element will not change when the shearing stresses assume the value τ. However, there will be a distortion of the originally right *angles* of the element, and after this distortion due to the shearing stresses the element assumes the configuration shown by the dashed lines in Fig. 4-2(*b*).

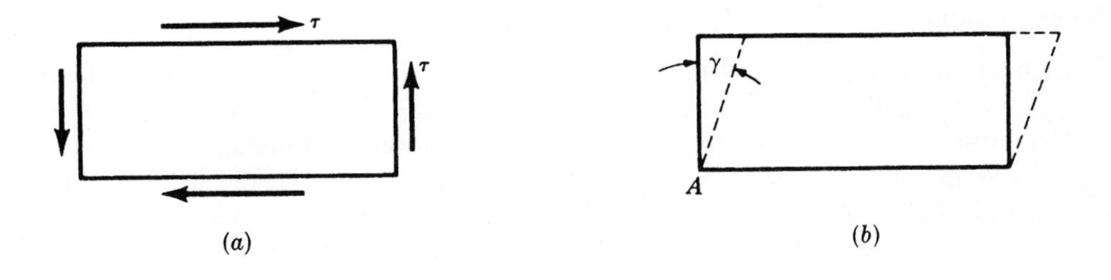

Fig. 4-2

SHEAR STRAIN

The change of angle at the corner of an originally rectangular element is defined as the *shear strain*. It must be expressed in radian measure and is usually denoted by γ.

MODULUS OF ELASTICITY IN SHEAR

The ratio of the shear stress τ to the shear strain γ is called the *modulus of elasticity in shear* and is usually denoted by G. Thus

$$G = \frac{\tau}{\gamma} \tag{4.2}$$

G is also known as the *modulus of rigidity*.

The units of G are the same as those of the shear stress, e.g., lb/in^2 or N/m^2, since the shear strain is dimensionless. The experimental determination of G and the region of linear action of τ and γ will be discussed in Chap. 5. Stress-strain diagrams for various materials may be drawn for shearing loads, just as they were drawn for normal loads in Chap. 1. They have the same general appearance as those sketched in Chap. 1 but the numerical values associated with the plots are of course different.

WELDED JOINTS

In addition to the traditional techniques of gas welding and electric arc welding, the past few decades have seen the emergence of two significant new methods, namely (*a*) electron beam welding and (*b*) laser beam welding.

Electron Beam Welding

In electron beam welding (EBW), coalescence of metals is achieved by having a focused beam of high-velocity electrons striking the surfaces to be joined. This beam of electrons carries a very high energy density that is capable of producing deep, narrow welds. Such welds can be produced much more quickly and with less distortion of the parent metals than with either gas or arc welding. Negative aspects of EBW are (i) surfaces to be joined must be very accurately aligned, and (ii) in certain situations EBW must be done in a partial vacuum. Also, safety precautions must be taken to protect personnel from the electron beam. (See Problem 4.12.)

Laser Beam Welding

In laser beam welding (LBW), joining of metals is carried out by having an optical energy source focused over a very small spot, such as the diameter of a circle ranging from 100 to 1000 μm (0.004 to 0.040 in). The term "laser" is an acronym for light amplification by stimulated emission of radiation. Energy densities of the order of 10^5 watts/cm^2 (6×10^6 watts/in^2) make the laser beam suitable for welding of metals. Laser beams can produce welds of high quality, but precautions must be taken to guard the operators of the laser, particularly with regard to damage to the human eye. One of the first successful applications involved laser welding of thermocouple gages in the Apollo lunar probe in the late 1960s. Types of systems in common use today include lasers of ruby, carbon dioxide, and various rare earth materials. Common commercial applications in the 1990s include sealing of batteries for digital watches and heart pacemakers, sealing of ink cartridges for fountain pens, joining telephone wires in circuits, and a host of other applications in aerospace, automotive, and electronic consumer items. (See Problem 4.13.)

Solved Problems

4.1. Consider the bolted joint shown in Fig. 4-3. The force P is 30 kN and the diameter of the bolt is 10 mm. Determine the average value of the shearing stress existing across either of the planes *a-a* or *b-b*.

Lacking any more precise information we can only assume that force P is equally divided between the sections *a-a* and *b-b*. Consequently a force of $\frac{1}{2}(30 \times 10^3) = 15 \times 10^3$ N acts across either of these planes over a cross-sectional area

$$\tfrac{1}{4}\pi(10)^2 = 78.6 \text{ mm}^2$$

Thus the average shearing stress across either plane is $\tau = \frac{1}{2}P/A = 15 \times 10^3/78.6 = 192$ MPa.

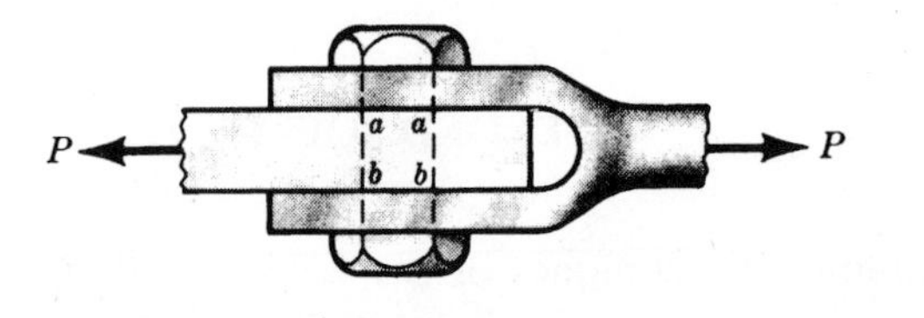

Fig. 4-3

4.2. Low-carbon structural steel has a shearing ultimate strength of approximately 45,000 lb/in^2. Determine the force P necessary to punch a 1-in-diameter hole through a plate of this steel $\frac{3}{8}$ in thick. If the modulus of elasticity in shear for this material is 12×10^6 lb/in^2, find the shear strain at the edge of this hole when the shear stress is 21,000 lb/in^2.

Let us assume uniform shearing on a cylindrical surface 1 in in diameter and $\frac{3}{8}$ in thick as shown in Fig. 4-4. For equilibrium the force P is $P = \tau A = \pi(1)\left(\frac{3}{8}\right)(45{,}000) = 53{,}100$ lb.

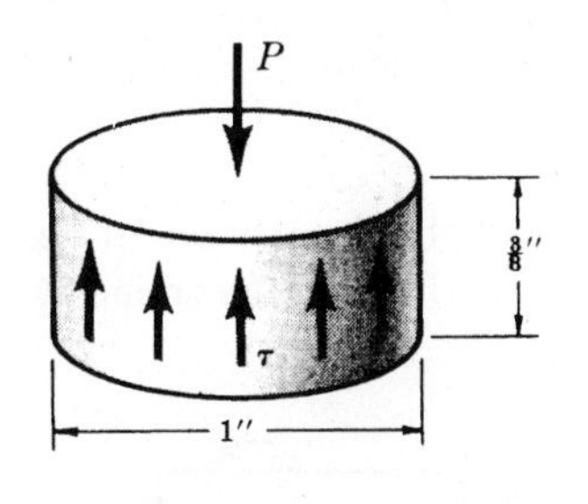

Fig. 4-4

To determine the shear strain γ when the shear stress τ is 21,000 lb/in^2, we employ the definition $G = \tau/\gamma$ to obtain $\gamma = \tau/G = 21{,}000/12{,}000{,}000 = 0.00175$ radian.

4.3. In the wood industries, inclined blocks of wood are sometimes used to determine the *compression-shear* strength of glued joints. Consider the pair of glued blocks A and B which are 1.5 in deep in a direction perpendicular to the plane of the paper. Determine the shearing ultimate strength of the glue if a vertical force of 9000 lb is required to cause rupture of the joint. It is to be noted that a good glue causes a large proportion of the failure to occur in the wood.

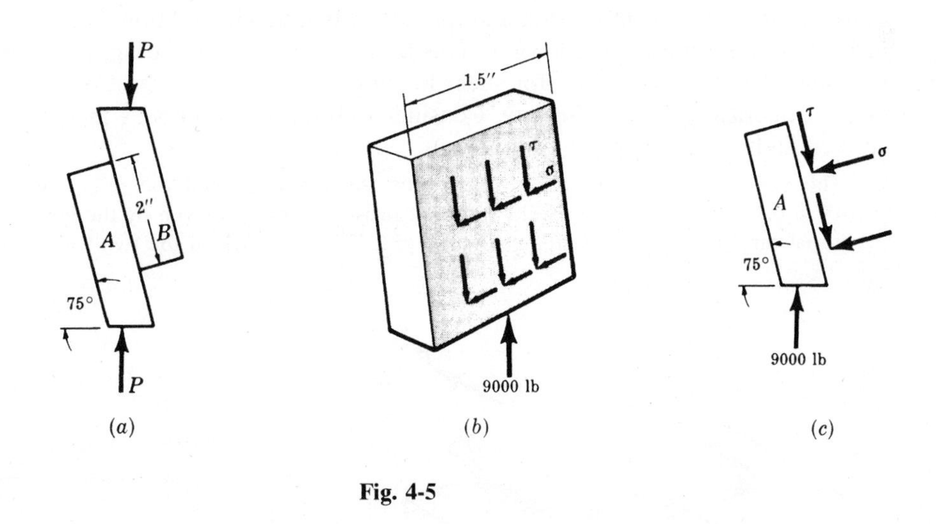

Fig. 4-5

Let us consider the equilibrium of the lower block, A. The reactions of the upper block B upon the lower one consist of both normal and shearing forces appearing as in the perspective and orthogonal views of Figs. 4-5(b) and 4-5(c).

Referring to Fig. 4-5(c) we see that for equilibrium in the horizontal direction

$$\Sigma F_h = \tau(2)(1.5)\cos 75^\circ - \sigma(2)(1.5)\cos 15^\circ = 0 \qquad \text{or} \qquad \sigma = 0.269\tau$$

For equilibrium in the vertical direction we have

$$\Sigma F_v = 9000 - \tau(2)(1.5)\sin 75^\circ - \sigma(2)(1.5)\sin 15^\circ = 0$$

Substituting $\sigma = 0.269\tau$ and solving, we find $\tau = 2900\ \text{lb/in}^2$.

4.4. The shearing stress in a piece of structural steel is 100 MPa. If the modulus of rigidity G is 85 GPa, find the shearing strain γ.

By definition, $G = \tau/\gamma$. Then the shearing strain $\gamma = \tau/G = (100 \times 10^6)/(85 \times 10^9) = 0.00117$ rad.

4.5. A single rivet is used to join two plates as shown in Fig. 4-6. If the diameter of the rivet is 20 mm and the load P is 30 kN, what is the average shearing stress developed in the rivet?

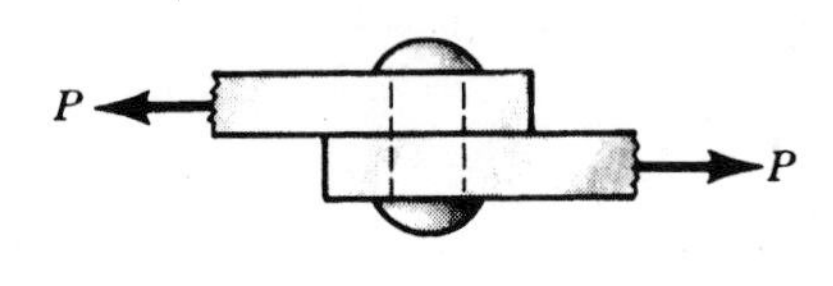

Fig. 4-6

Here the average shear stress in the rivet is P/A where A is the cross-sectional area of the rivet. However, rivet holes are usually 1.5 mm larger in diameter than the rivet and it is customary to assume that the rivet fills the hole completely. Hence the shearing stress is given by

$$\tau = \frac{30{,}000\ \text{N}}{(\pi/4)\,[0.0215\ \text{m}]^2} = 8.26 \times 10^7\ \text{N/m}^2 \qquad \text{or} \qquad 82.6\ \text{MPa}$$

4.6. One common type of weld for joining two plates is the *fillet weld*. This weld undergoes shear as well as tension or compression and frequently bending in addition. For the two plates shown in Fig. 4-7, determine the allowable tensile force P that may be applied using an allowable working stress of 11,300 lb/in² for shear loading as indicated by the Code for Fusion Welding of the American Welding Society. Consider only shearing stresses in the weld. The load is applied midway between the two welds.

The minimum dimension of the weld cross section is termed the *throat*, which in this case is $\frac{1}{2}\sin 45^\circ = 0.353$ in. The effective weld area that resists shearing is given by the length of the weld times the throat dimension, or weld area $= 7(0.353) = 2.47\ \text{in}^2$ for each of the two welds. Thus the allowable

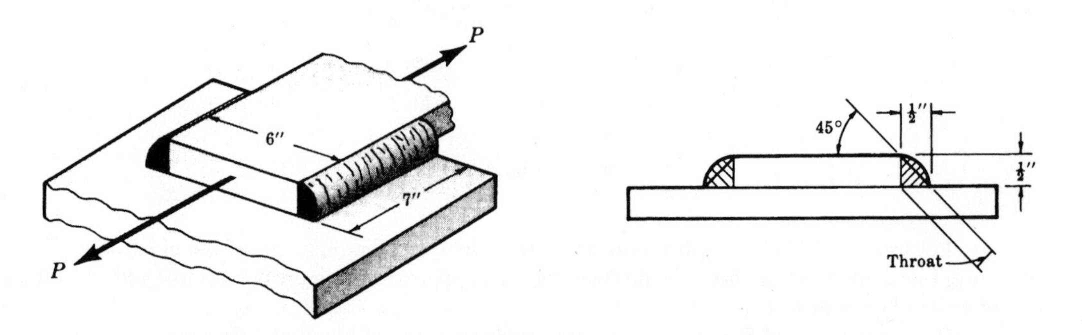

Fig. 4-7

tensile load P is given by the product of the working stress in shear times the area resisting shear, or $P = 11{,}300(2)(2.47) = 56{,}000$ lb.

4.7. Shafts and pulleys are usually fastened together by means of a key, as shown in Fig. 4-8(*a*). Consider a pulley subject to a turning moment T of 10,000 lb-in keyed by a $\frac{1}{2} \times \frac{1}{2} \times 3$ in key to the shaft. The shaft is 2 in in diameter. Determine the shear stress on a horizontal plane through the key.

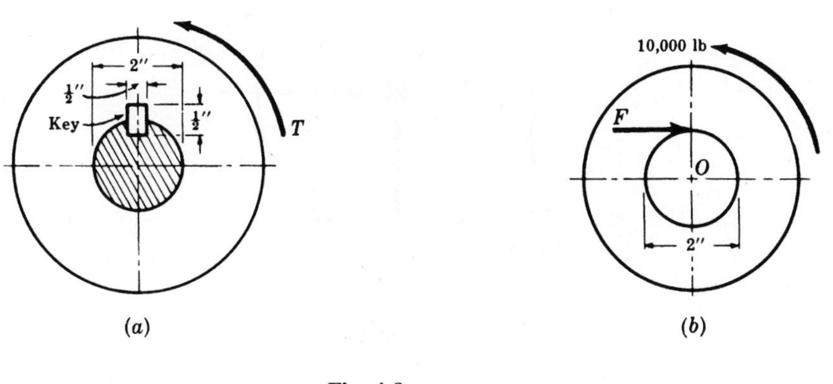

Fig. 4-8

Drawing a free-body diagram of the pulley alone, as shown in Fig. 4-8(*b*), we see that the applied turning moment of 10,000 lb-in must be resisted by a horizontal tangential force F exerted on the pulley by the key. For equilibrium of moments about the center of the pulley we have

$$\Sigma M_0 = 10{,}000 - F(1) = 0 \qquad \text{or} \qquad F = 10{,}000 \text{ lb}$$

It is to be noted that the shaft exerts additional forces, not shown, on the pulley. These act through the center O and do not enter the above moment equation. The resultant forces acting on the key appear as in Fig. 4-9(*a*). Actually the force F acting to the right is the resultant of distributed forces acting over the lower half of the left face. The other forces F shown likewise represent resultants of distributed force systems. The exact nature of the force distribution is not known.

The free-body diagram of the portion of the key below a horizontal plane *a-a* through its midsection is shown in Fig. 4-9(*b*). For equilibrium in the horizontal direction we have

$$\Sigma F_h = 10{,}000 - \tau(\tfrac{1}{2})(3) = 0 \qquad \text{or} \qquad \tau = 6670 \text{ lb/in}^2$$

This is the horizontal shear stress in the key.

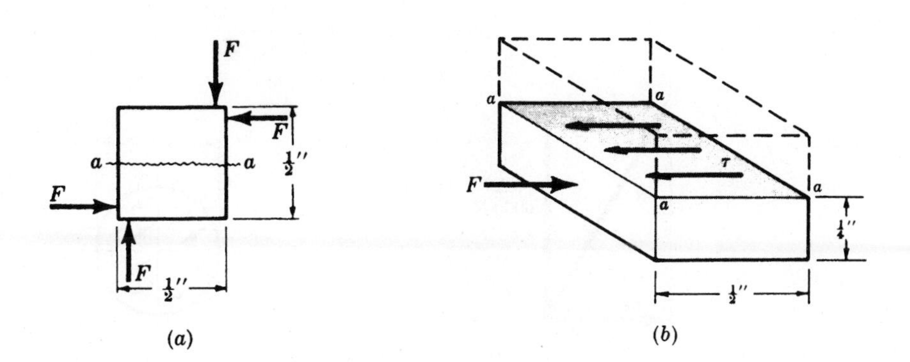

Fig. 4-9

4.8. A lifeboat on a seagoing cruise ship is supported at each end by a stranded steel cable passing over a pulley on a davit anchored to the top deck. The cable at each end carries a tension of 4000 N and the cable as well as the pulley are located in a vertical plane as shown in Fig. 4-10. The pulley may rotate freely about the horizontal circular axle indicated. Determine the diameter of this axle if the allowable transverse shearing stress is 50 MPa.

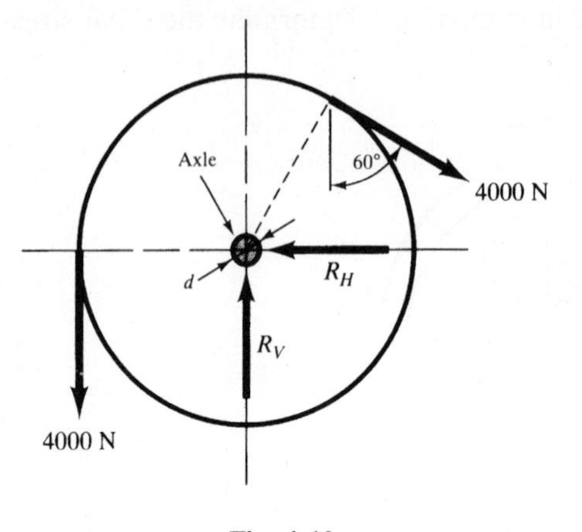

Fig. 4-10

The free-body diagram of the pulley shows not only the cable tensions but also the forces R_H and R_V exerted on the pulley by the circular axle. From statics we have

$$\Sigma F_H = -R_H + 4000 \sin 60° = 0$$

$$R_H = 3464 \text{ N} (\leftarrow\ )$$

$$\Sigma F_V = R_V - 4000 - 4000 \cos 60° = 0$$

$$R_V = 6000 \text{ N} (\uparrow)$$

The resultant of R_H and R_V is $R = \sqrt{(3464)^2 + (6000)^2} = 6930$ N oriented at an angle θ from the horizontal given by

$$\theta = \arctan \frac{6000 \text{ N}}{3464 \text{ N}}$$

$$= 60°$$

The force exerted by the pulley upon the axle is equal and opposite to that shown in Fig. 4-11. If we assume that the resultant force of 6930 N is uniformly distributed over the cross section of the axle, the transverse shearing stress has the appearance shown in Fig. 4-12. From Eq. (*4.1*), we have

$$50 \times 10^6 \text{ N/m}^2 = \frac{6930 \text{ N}}{(\pi/4)\, d^2}$$

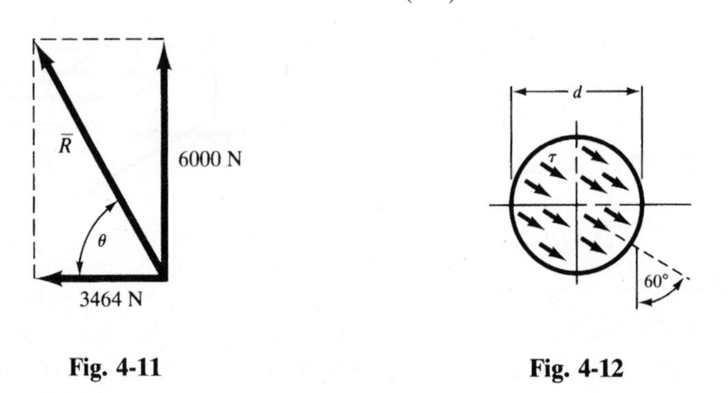

Fig. 4-11 **Fig. 4-12**

where d is the unknown axle diameter. Solving,

$$d = 13.3 \times 10^{-3}\ \text{m} \qquad \text{or} \qquad 13.3\ \text{mm}$$

4.9. A building that is 60 m tall has essentially the rectangular configuration shown in Fig. 4-13. Horizontal wind loads will act on the building exerting pressures on the vertical face that may be approximated as uniform within each of the three "layers" as shown. From empirical expressions for wind pressures at the midpoint of each of the three layers, we have a pressure of 781 N/m² on the lower layer, 1264 N/m² on the middle layer, and 1530 N/m² on the top layer. Determine the resisting shear that the foundation must develop to withstand this wind load.

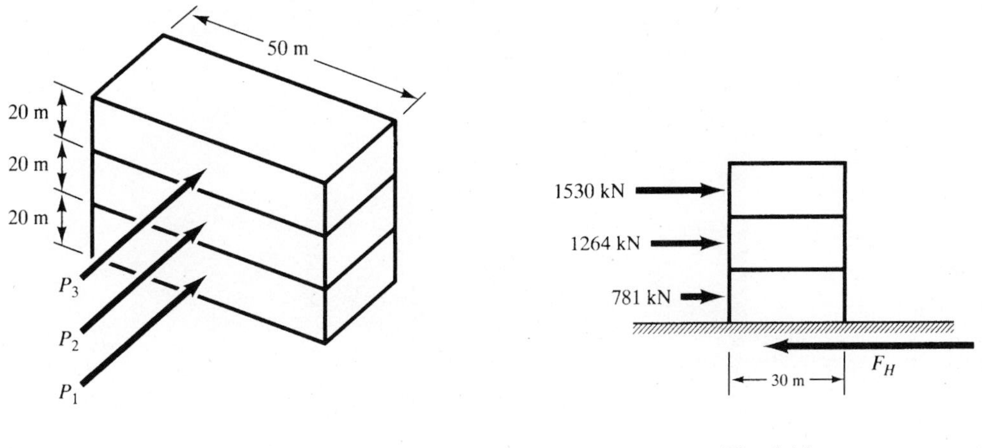

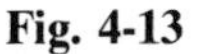
Fig. 4-13

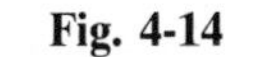
Fig. 4-14

The horizontal forces acting on these three layers are found to be

$$P_1 = (20\ \text{m})(50\ \text{m})(781\ \text{N/m}^2) = 781\ \text{kN}$$

$$P_2 = (20\ \text{m})(50\ \text{m})(1264\ \text{N/m}^2) = 1264\ \text{kN}$$

$$P_3 = (20\ \text{m})(50\ \text{m})(1530\ \text{N/m}^2) = 1530\ \text{kN}$$

These forces are taken to act at the midheight of each layer, so the free-body diagram of the building has the appearance of Fig. 4-14, where F_H denotes the horizontal shearing force exerted by the foundation upon the structure. From horizontal equilibrium, we have

$$\Sigma F_H = 1530 + 1264 + 781 - F_H = 0$$

$$F_H = 3575\ \text{kN}$$

If we assume that this horizontal reaction is uniformly distributed over the base of the structure, the horizontal shearing stress given by Eq. (*4.1*) is

$$\tau = \frac{3575\ \text{kN}}{(30\ \text{m})(50\ \text{m})} = 2.38\ \text{kN/m}^2$$

4.10. In the North Atlantic Ocean, large icebergs (often weighing more than 8000 MN) present a menace to ship navigation. A recently developed technique makes it possible to tow them to acceptable locations. The method involves the use of a remotely operated unmanned submersible vehicle which drills a hole in the iceberg about 30 m below the water surface and then inserts a cylindrical anchor in the hole as shown in Fig. 4-15. The anchor is a cylindrical steel tube of diameter 100 mm and it is secured to the hole in the iceberg by injecting gaseous carbon dioxide through small holes in the tube. This gas quickly freezes and fills the narrow annular space between the outside of the anchor and the inside of the hole in the ice. A connection from the exposed end of the anchor permits a cable to be run to the towing vessel. If the maximum allowable shear stress in the frozen carbon dioxide is 0.5 MPa, determine the minimum length of the cylindrical anchor so that it will not be pulled out from the iceberg under a towing force of 200 kN.

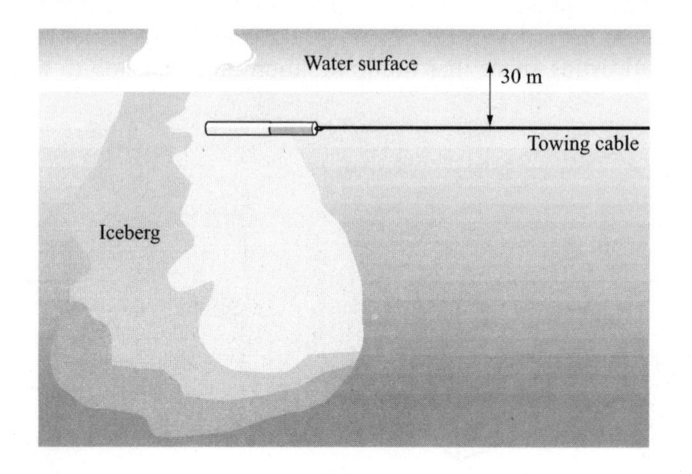

Fig. 4-15

A free-body diagram of the cylindrical tube (anchor) is shown in Fig. 4-16. There, T represents the towing force in the cable attached to the anchor and τ is the shearing stress in the frozen carbon dioxide. It is assumed that τ is uniform at all points along the length L of the anchor as well as around the circumference of the tube. If τ is 0.5 MPa for horizontal equilibrium, we have

$$\Sigma F_H = T - \pi D L \tau = 0$$

$$200{,}000 \text{ N} - \pi(0.1 \text{ m}) L (0.5 \times 10^6 \text{ N/m}^2) = 0$$

$$L = 1.27 \text{ m}$$

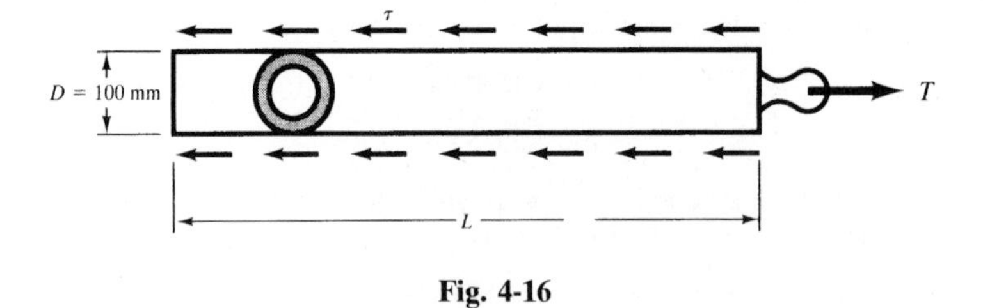

Fig. 4-16

4.11. It is occasionally desirable to design certai.. structural fasteners to be strong in tension yet somewhat weak in transverse shear. One example of this is to be found in contemporary design of four-engine wide-body aircraft. Each engine is attached to the main supporting frame inside the wing [see Fig. 4-17(a)] by aluminum alloy bolts that are adequately strong to support the dead weight of the engine plus additional loads occurring in flight. However, the alloying is such that each bolt can carry only moderate transverse shear in the unlikely event of a "wheels-up" emergency landing so that the engine will be torn free from the wing. If the ultimate transverse shear strength of each bolt is 120 MPa, the bolt diameter 20 mm, and four bolts secure the engine to the wing, determine the horizontal force that must act between the ground and the engine for separation of the engine from the wing to occur.

A free-body diagram of the engine together with the four bolts is shown in Fig. 4-17(*b*). There F_u represents ultimate shearing force in each bolt ($F_u = \tau_u A$), where τ_u represents the ultimate shear stress and A the cross-sectional area of each bolt. Also, F_g represents the force exerted by the ground on the bottom of the engine. Note that the underside of the aircraft fuselage is above the bottom of the engine. We have

$$F_u = \frac{\pi}{4}(0.020\ \text{m})^2(120 \times 10^6\ \text{N/m}^2) = 37.7\ \text{kN}$$

and for horizontal equilibrium (neglecting dynamic effects)

$$\Sigma F_H = F_g - 4F_u = 0$$

$$F_g = 4(37.7) = 151\ \text{kN}$$

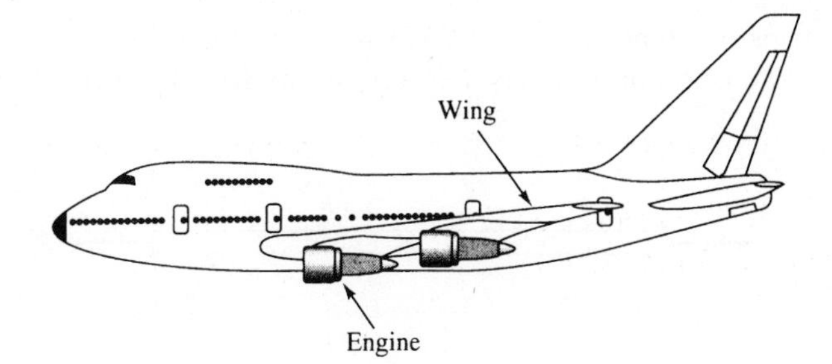

Fig. 4-17(*a*)

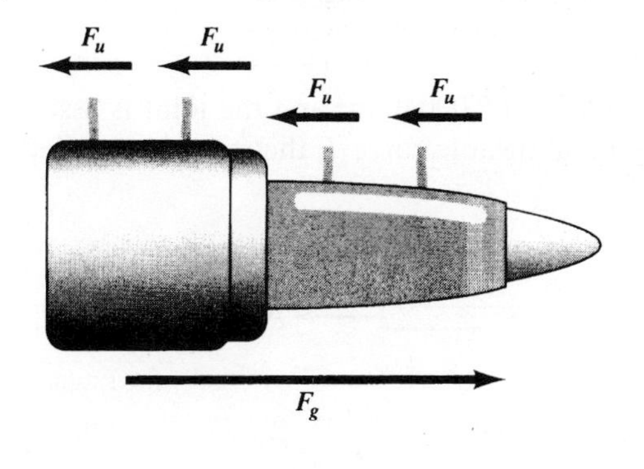

Fig. 4-17(*b*)

4.12. A power reactor has certain of its pressurized components (see Fig. 4-18) made of type 304 stainless steel, 2.5 in thick. Adjacent butt-welded sections are joined by electron beam welding in a partial vacuum using a 200 kW system. The ultimate strength of the parent steel is 160,000 lb/in². If the weld is assumed to be 100 percent efficient, determine the force that may be transmitted through each 14 in wide section. Also, determine the force if 80 percent efficiency is assumed.

For 100 percent effectiveness of the weld we determine the cross-sectional area of the 14 in by 2.5 in section to be (14 in) (2.5 in) = 35 in². The allowable load P is then given by

$$P = (35\ \text{in}^2)(160{,}000\ \text{lb/in}^2) = 5.6 \times 10^6\ \text{lb}$$

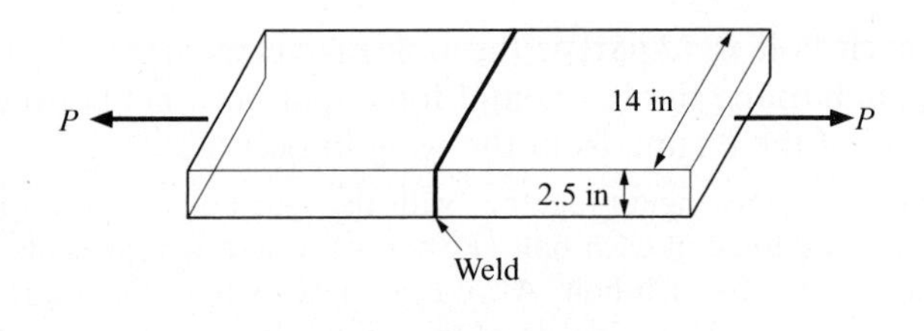

Fig. 4-18

For 80 percent effectiveness of the weld we have the allowable load

$$P = (5.6 \times 10^6 \text{ lb/in}^2)(0.80) = 4.48 \times 10^6 \text{ lb}$$

4.13. Two $\frac{1}{16}$ in thick strips of titanium alloy 1.75 in wide are joined by a 45° laser weld as shown in Fig. 4-19. A 100 kW carbon dioxide laser system is employed to form the joint. If the allowable shearing stress in the alloy is 65,000 lb/in² and the joint is assumed to be 100 percent efficient, determine the maximum allowable force P that may be applied.

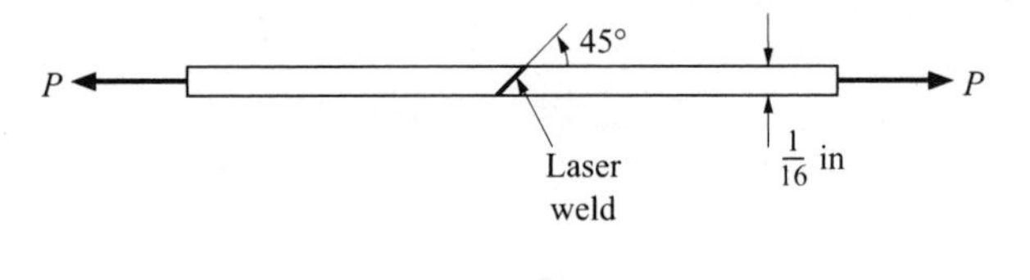

Fig. 4-19

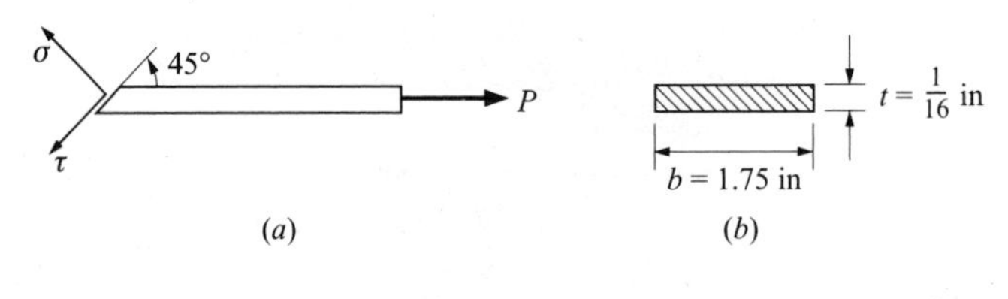

Fig. 4-20

A free-body diagram of the right strip has the form shown in Fig. 4-20. There, σ denotes normal stress in the weld on the 45° plane and τ the shearing stress. These are, of course, forces per unit area on the 45° plane and these must be multiplied by the area of the 45° plane which is $bt/\cos 45°$ where t denotes strip thickness and b the width. For horizontal equilibrium we have

$$\Sigma F_l = \tau\left(\frac{bt}{\cos 45°}\right) - P\cos 45° = 0$$

$$\tau = \frac{P\cos 45°}{bt}$$

$$65{,}000 \text{ lb/in}^2 = \frac{P(1/\sqrt{2})^2}{(1.75 \text{ in})(\frac{1}{16} \text{ in})} \qquad \text{or} \qquad P = 7110 \text{ lb}$$

Supplementary Problems

4.14. In Problem 4.1, if the maximum allowable working stress in shear is 14,000 lb/in^2, determine the required diameter of the bolt in order that this value is not exceeded. *Ans.* $d = 0.585$ in

4.15. A circular punch 20 mm in diameter is used to punch a hole through a steel plate 10 mm thick. If the force necessary to drive the punch through the metal is 250 kN, determine the maximum shearing stress developed in the material. *Ans.* $\tau = 400$ MPa

4.16. In structural practice, steel clip angles are commonly used to transfer loads from horizontal girders to vertical columns. If the reaction of the girder upon the angle is a downward force of 10,000 lb as shown in Fig. 4-21 and if two $\frac{7}{8}$-in-diameter rivets resist this force, find the average shearing stress in each of the rivets. As in Problem 4.5, assume that the rivet fills the hole, which is $\frac{1}{16}$ in larger in diameter than the rivet. *Ans.* 7200 lb/in^2

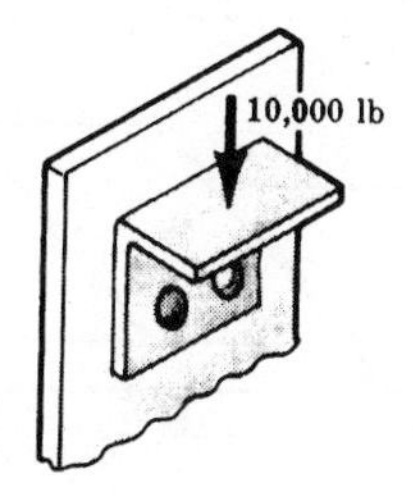

Fig. 4-21

4.17. A pulley is keyed (to prevent relative motion) to a 60-mm-diameter shaft. The unequal belt pulls, T_1 and T_2, on the two sides of the pulley give rise to a net turning moment of 120 N · m. The key is 10 mm by 15 mm in cross section and 75 mm long, as shown in Fig. 4-22. Determine the average shearing stress acting on a horizontal plane through the key. *Ans.* $\tau = 5.33$ MPa

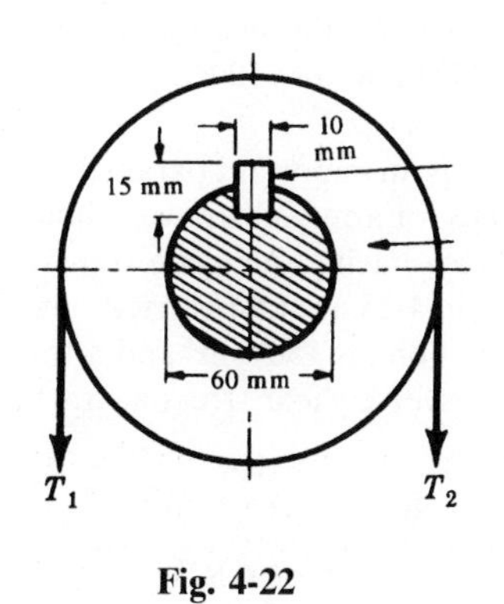

Fig. 4-22

4.18. Consider the balcony-type structure shown in Fig. 4-23. The horizontal balcony is loaded by a total load of 80 kN distributed in a radially symmetric fashion. The central support is a shaft 500 mm in diameter and the balcony is welded at both the upper and lower surfaces to this shaft by welds 10 mm on a side (or leg) as shown in the enlarged view at the right. Determine the average shearing stress existing between the shaft and the weld. *Ans.* 2.5 MPa

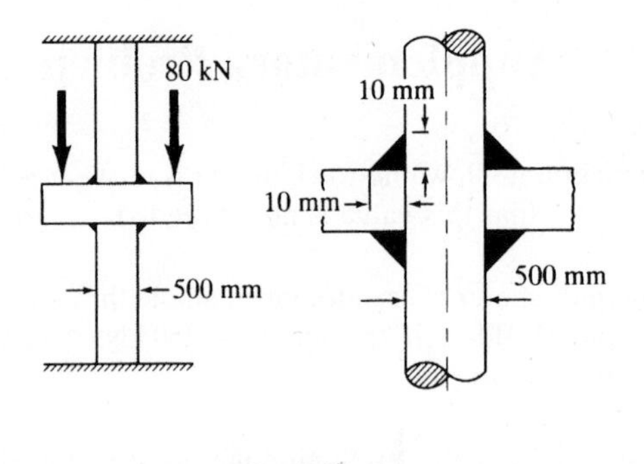

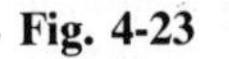
Fig. 4-23

4.19. Consider the two plates of equal thickness joined by two fillet welds as indicated in Fig. 4-24. Determine the maximum shearing stress in the welds. *Ans.* $\tau = 0.707P/ab$

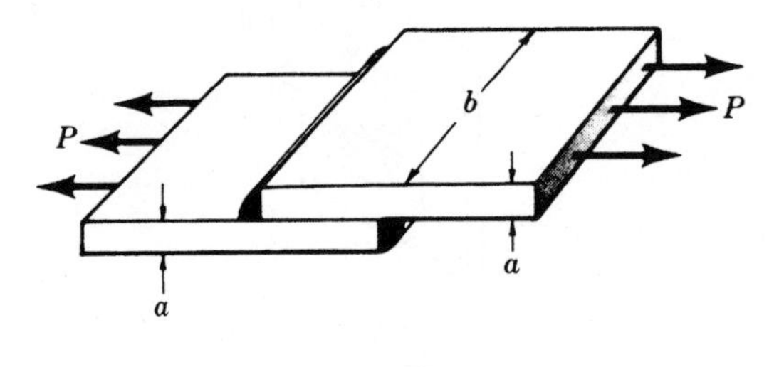

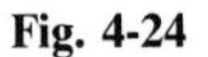
Fig. 4-24

4.20. A copper tube 55 mm in outside diameter and of wall thickness 5 mm fits loosely over a solid steel circular bar 40 mm in diameter. The two members are fastened together by two metal pins each 8 mm in diameter and passing transversely through both members, one pin being near each end of the assembly. At room temperature the assembly is stress free when the pins are in position. The temperature of the entire assembly is then raised 40°C. Calculate the average shear stress in the pins. For copper $E = 90\,\text{GPa}$, $\alpha = 18 \times 10^{-6}/°\text{C}$; for steel $E = 200\,\text{GPa}$, $\alpha = 12 \times 10^{-6}/°\text{C}$. *Ans.* $\tau = 132\,\text{MPa}$

4.21. The shear strength of human bone is an important parameter when implants must be employed to maintain the desired length of a fractured leg or arm. Substitute animal bone segments are sometimes employed but it is necessary to select a substance having the same transverse shear strength as human bone. For this purpose tests such as shown in Fig. 4-25 are first carried out on the substitute under consideration. If the cross-sectional area of the animal bone is $150\,\text{mm}^2$ and a transverse force $F = 600\,\text{N}$ is required to cause shear fracture, find the mean transverse shear stress at fracture. *Ans.* 2 MPa

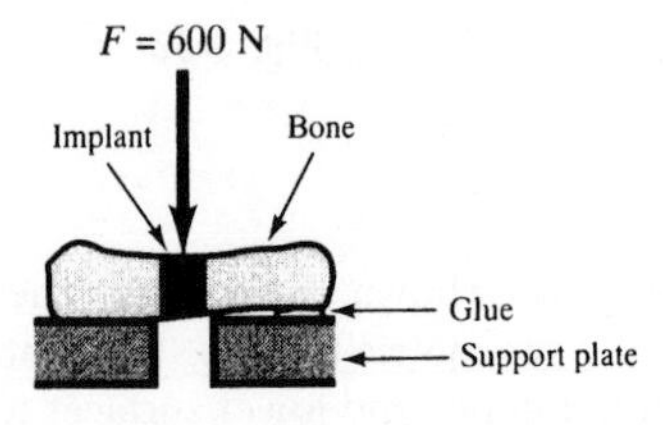

Fig. 4-25

4.22. In automotive as well as aircraft applications, two pieces of thin metal are often joined by a single lap shear joint, as shown in Fig. 4-26. Here, the metal has a thickness of 2.2 mm. The ultimate shearing strength of the epoxy adhesive joining the metals is 2.57×10^4 kPa, the shear modulus of the epoxy is 2.8 GPa, and the epoxy is effective over the 12.7×25.4-mm overlapping area. Determine the maximum axial load P the joint can carry. Neglect the slight bending effect that arises because the metal pieces are not in the same plane. *Ans.* 8290 N

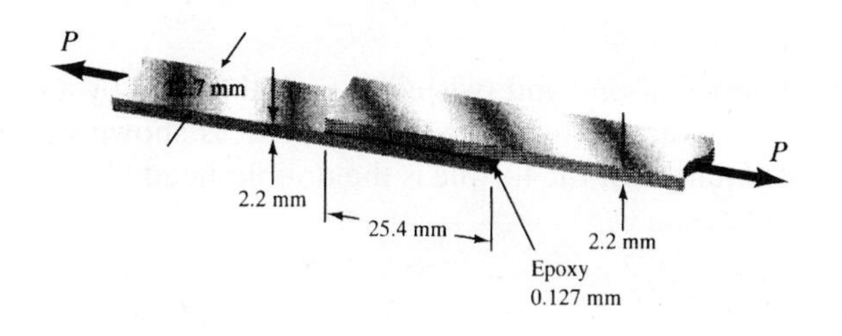

Fig. 4-26

4.23. If the shear modulus of the epoxy in Problem 4.22 is 2.8 GPa, determine the axial displacement of one piece of metal with respect to the other just prior to failure of the epoxy if the epoxy is 0.127 mm thick. *Ans.* 0.0017 mm

Chapter 5

Torsion

DEFINITION OF TORSION

Consider a bar rigidly clamped at one and twisted at the other end by a torque (twisting moment) $T = Fd$ applied in a plane perpendicular to the axis of the bar as shown in Fig. 5-1. Such a bar is in torsion. An alternative representation of the torque is the double-headed vector directed along the axis of the bar.

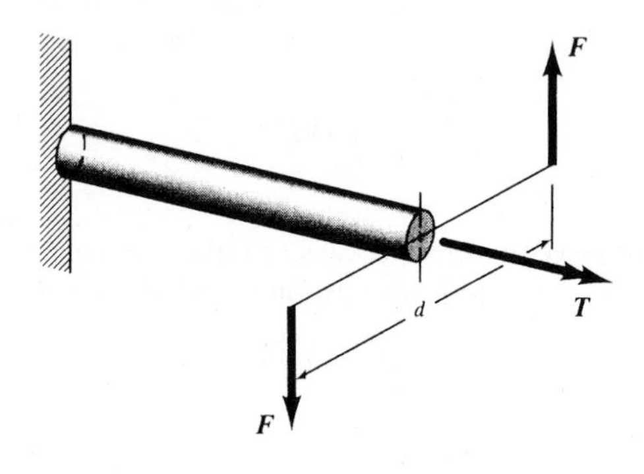

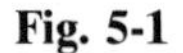

Fig. 5-1

TWISTING MOMENT

Occasionally a number of couples act along the length of a shaft. In that case it is convenient to introduce a new quantity, the *twisting moment*, which for any section along the bar is defined to be the algebraic sum of the moments of the applied couples that lie to one side of the section in question. The choice of side in any case is of course arbitrary.

POLAR MOMENT OF INERTIA

For a hollow circular shaft of outer diameter D_o with a concentric circular hole of diameter D_i the *polar moment of inertia* of the cross-sectional area, usually denoted by J, is given by

$$J = \frac{\pi}{32}(D_o^4 - D_i^4) \tag{5.1}$$

The polar moment of inertia for a solid shaft is obtained by setting $D_i = 0$. See Problem 5.1. This quantity J is a mathematical property of the geometry of the cross section which occurs in the study of the stresses set up in a circular shaft subject to torsion.

Occasionally it is convenient to rewrite the above equation in the form

$$J = \frac{\pi}{32}(D_o^2 + D_i^2)(D_o^2 - D_i^2)$$

$$= \frac{\pi}{32}(D_o^2 + D_i^2)(D_o + D_i)(D_o - D_i)$$

This last form is useful in numerical evaluation of J in those cases where the difference $(D_o - D_i)$ is small. See Problem 5.6.

TORSIONAL SHEARING STRESS

For either a solid or a hollow circular shaft subject to a twisting moment T the *torsional shearing stress* τ at a distance ρ from the center of the shaft is given by

$$\tau = \frac{T\rho}{J} \tag{5.2}$$

This expression is derived in Problem 5.2. For applications see Problems 5.4, 5.5, 5.9, 5.10, and 5.11. This stress distribution varies from zero at the center of the shaft (if it is solid) to a maximum at the outer fibers, as shown in Fig. 5-2. It is to be emphasized that no points of the bar are stressed beyond the proportional limit.

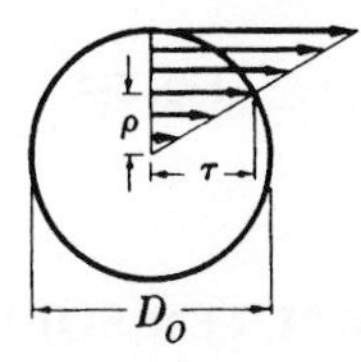

Fig. 5-2

SHEARING STRAIN

If a generator a-b is marked on the surface of the unloaded bar, then after the twisting moment T has been applied this line moves to a-b', as shown in Fig. 5-3. The angle γ, measured in radians, between the final and original positions of the generator is defined as the *shearing strain* at the surface of the bar. The same definition would hold at any interior point of the bar.

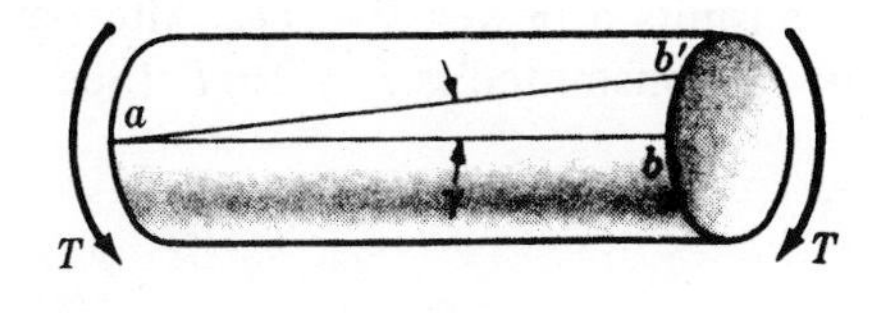

Fig. 5-3

MODULUS OF ELASTICITY IN SHEAR

The ratio of the shear stress τ to the shear strain γ is called the *modulus of elasticity in shear* and, as in Chap. 4, is given by

$$G = \frac{\tau}{\gamma} \tag{5.3}$$

Again the units of G are the same as those of shear stress, since the shear strain is dimensionless.

ANGLE OF TWIST

If a shaft of length L is subject to a constant twisting moment T along its length, then the angle θ through which one end of the bar will twist relative to the other is

$$\theta = \frac{TL}{GJ} \tag{5.4}$$

where J denotes the polar moment of inertia of the cross section. See Fig. 5-4. This equation is derived in Problem 5.3. For applications see Problems 5.5, 5.7, 5.8, 5.11, 5.12, and 5.13. This expression holds only for purely elastic action of the bar.

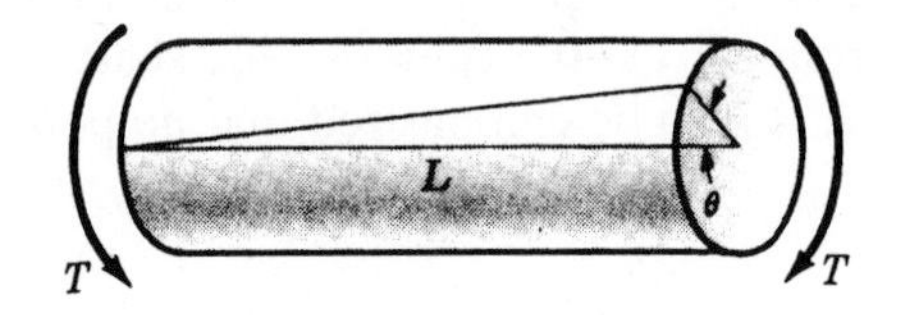

Fig. 5-4

COMPUTER SOLUTION

For a bar of circular cross section and variable diameter, the angle of twist θ is determined by dividing the bar into a number of segments along its length, such that in each segment the diameter may be taken to be constant. This procedure is well suited to computer implementation, and a FORTRAN program for implementing it is given in Problem 5.14. (See also Problem 5.15.)

POWER TRANSMISSION

A shaft rotating with constant angular velocity ω (radians per second) is being acted on by a twisting moment T and hence transmits a power $P = T\omega$. Alternatively, in terms of the number of revolutions per second f, the power transmitted is $P = 2\pi fT$. (See Problems 5.9, 5.10 and 5.11.)

PLASTIC TORSION OF CIRCULAR BARS

As the twisting moment acting on either a solid or hollow circular bar is increased, a value of the twisting moment is finally reached for which the extreme fibers of the bar have reached the yield point in shear of the material. This is the maximum possible elastic twisting moment that the bar can withstand and is denoted by T_e. A further increase in the value of the twisting moment puts the interior fibers at the yield point, with yielding progressing from the outer fibers inward. The limiting case occurs when all fibers are stressed to the yield point in shear and this represents the *fully plastic twisting moment*. It is denoted by T_p. Provided we do not consider stresses greater than the yield point in shear, this is the maximum possible twisting moment the bar can carry. For a solid circular bar subject to torsion it is shown in Problem 5.21 that $T_p = 4T_e/3$.

Solved Problems

5.1. Derive an expression for the polar moment of inertia of the cross-sectional area of a hollow circular shaft. What does this expression become for the special case of a solid circular shaft?

Let D_o denote the outside diameter of the shaft and D_i the inside diameter. Because of the circular symmetry involved, it is most convenient to adopt the polar coordinate system shown in Fig. 5-5.

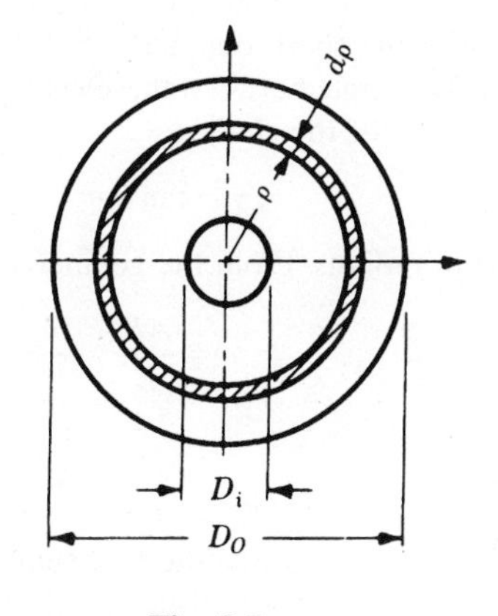

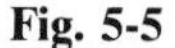

Fig. 5-5

By definition, the polar moment of inertia is given by the integral

$$J = \int_A \rho^2 \, da$$

where A indicates that the integral is to be evaluated over the entire cross-sectional area.

To evaluate this integral we select as an element of area a thin ring-shaped element of radius ρ and radial thickness $d\rho$ as shown. The area of the ring is $da = 2\pi\rho(d\rho)$. Thus

$$J = \int_{1/2D_i}^{1/2D_o} \rho^2(2\pi\rho)\, d\rho = \frac{\pi}{32}[D_o^4 - D_i^4]$$

The units of J are in^4 or m^4. For the special case of a solid circular shaft, the above becomes $J = \pi D^4/32$, where D denotes the diameter of the shaft.

5.2. Derive an expression relating the applied twisting moment acting on a shaft of circular cross section and the shearing stress at any point in the shaft.

In Fig. 5-6(a) the shaft is shown loaded by the two torques T and consequently is in static equilibrium. To determine the distribution of shearing stress in the shaft, let us cut the shaft by a plane passing through it in a direction perpendicular to the geometric axis of the bar.

The free-body diagram of the portion of the shaft to the left of this plane appears as in Fig. 5-6(b). Obviously a torque T must act over the cross section cut by the plane. This is true since the entire shaft

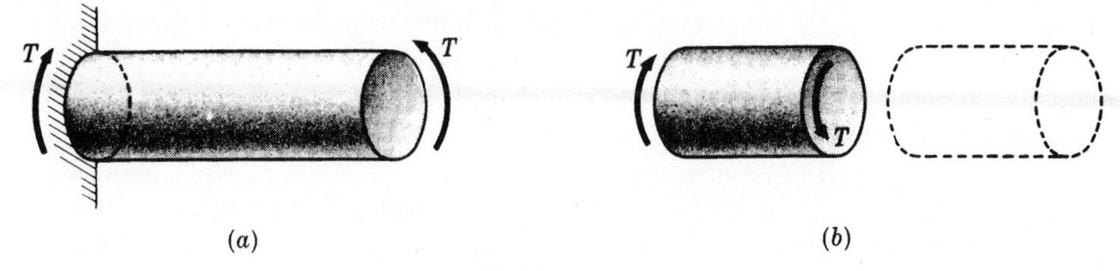

Fig. 5-6

is in equilibrium, and hence any portion of it also is. The torque T acting on the cut section represents the effect of the right portion of the shaft on the left portion. Since the right portion has been removed, it must be replaced by its effect on the left portion. This effect is represented by the torque T. This torque is of course a resultant of shearing stresses distributed over the cross section. It is now necessary to make certain assumptions in order to determine the nature of the variation of shear stress intensity over the cross section.

One fundamental assumption is that a plane section of the shaft normal to its axis before loads are applied remains plane and normal to the axis after loading. This may be verified experimentally for circular shafts, but this assumption is not valid for shafts of noncircular cross section.

A generator on the surface of the shaft, denoted by O_1A in Fig. 5-7, deforms into the configuration O_1B after torsion has occurred. The angle between these configurations is denoted by α. By definition, the shearing unit strain γ on the surface of the shaft is

$$\gamma = \tan\alpha \approx \alpha$$

where the angle α is measured in radians. From the geometry of the figure,

$$\alpha = \frac{AB}{L} = \frac{r\theta}{L}$$

Hence
$$\gamma = \frac{r\theta}{L}$$

But since a diameter of the shaft prior to loading is assumed to remain a diameter after torsion has occurred, the shearing unit strain at a general distance ρ from the center of the shaft may likewise be written $\gamma_\rho = \rho\theta/L$. Consequently the shearing strains of the longitudinal fibers vary linearly as the distances from the center of the shaft.

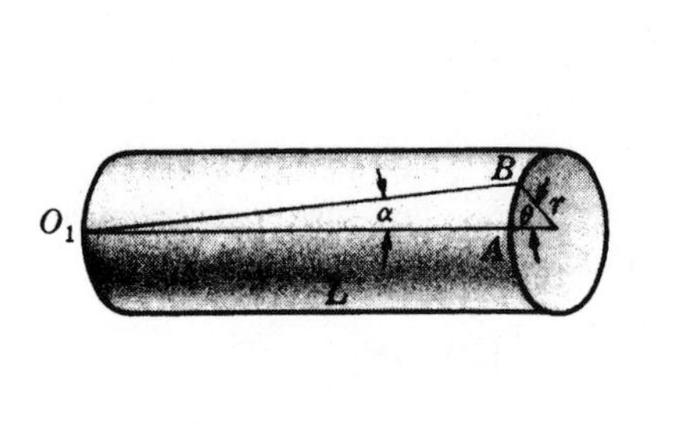

Fig. 5-7

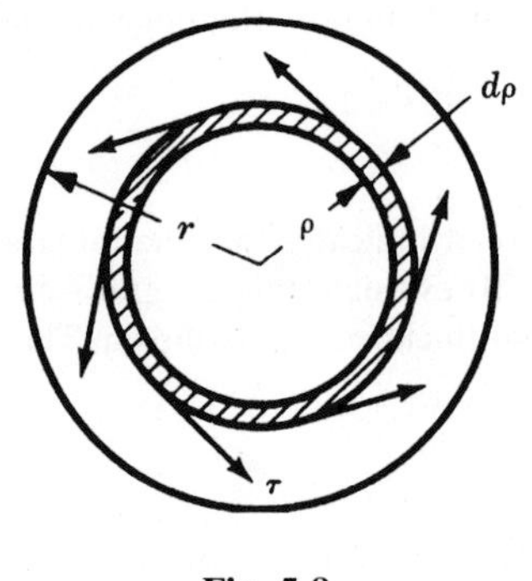

Fig. 5-8

If we assume that we are concerned only with the linear range of action of the material where the shearing stress is proportional to shearing strain, then it is evident that the shearing stresses of the longitudinal fibers vary linearly as the distances from the center of the shaft. Obviously the distribution of shearing stresses is symmetric around the geometric axis of the shaft. They have the appearance shown in Fig. 5-8. For equilibrium, the sum of the moments of these distributed shearing forces over the entire circular cross section is equal to the applied twisting moment. Also, the sum of the moments of these forces is exactly equal to the torque T shown in Fig. 5-6(b) above.

Thus we have

$$T = \int_0^r \tau\rho\, da$$

where da represents the area of the shaded ring-shaped element shown in Fig. 5-8. However, the shearing stresses vary as the distances from the geometric axis; hence

$$\frac{\tau_\rho}{\rho} = \frac{\tau_r}{r} = \text{constant}$$

where the subscripts on the shearing stress denote the distances of the element from the axis of the shaft.

Consequently we may write

$$T = \int_0^r \frac{\tau_\rho}{\rho}(\rho^2)\,da = \frac{\tau_\rho}{\rho}\int_0^r \rho^2\,da$$

since the ratio τ_ρ/ρ is a constant. However, the expression $\int_0^r \rho^2\,da$ is by definition (see Problem 5.1) the polar moment of inertia of the cross-sectional area. Values of this for solid and hollow circular shafts are derived in Problem 5.1. Hence the desired relationship is

$$T = \frac{\tau_\rho J}{\rho} \qquad \text{or} \qquad \tau_\rho = \frac{T\rho}{J}$$

It is to be emphasized that this expression holds *only* if no points of the bar are stressed beyond the proportional limit of the material.

5.3. Derive an expression for the angle of twist of a circular shaft as a function of the applied twisting moment. Assume that the entire shaft is acting within the elastic range of action of the material.

Let L denote the length of the shaft, J the polar moment of inertia of the cross section, T the applied twisting moment (assumed constant along the length of the bar), and G the modulus of elasticity in shear. The angle of twist in a length L is represented by θ in Fig. 5-9.

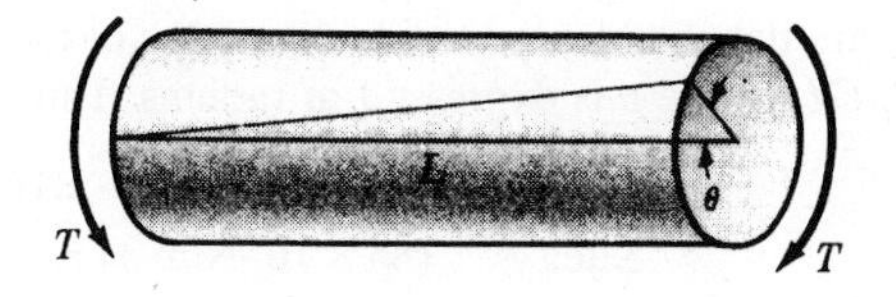

Fig. 5-9

From Problem 5.2 we have at the outer fibers where $\rho = r$:

$$\gamma_r = \frac{r\theta}{L} \qquad \text{and} \qquad \tau_r = \frac{Tr}{J}$$

By definition, the shearing modulus is given by $G = \dfrac{\tau}{\gamma} = \dfrac{Tr/J}{r\theta/L} = \dfrac{TL}{J\theta}$ from which $\theta = \dfrac{TL}{GJ}$. Note that θ is expressed in radians, i.e., it is dimensionless.

Occasionally the angle of twist in a unit length is useful. It is often denoted by ϕ and is given by $\phi = \theta/L = T/GJ$.

5.4. If a twisting moment of 10,000 lb · in is impressed upon a $1\frac{3}{4}$-in-diameter shaft, what is the maximum shearing stress developed? Also, what is the angle of twist in a 4-ft length of the shaft? The material is steel for which $G = 12 \times 10^6$ lb/in^2. Assume entirely elastic action.

From Problem 5.1 the polar moment of inertia of the cross-sectional area is

$$J = \frac{\pi}{32}(D_o)^4 = \frac{\pi}{32}\left(\frac{7}{4}\right)^4 = 0.92 \text{ in}^4$$

The torsional shearing stress τ at any distance ρ from the center of the shaft was shown in Problem 5.2 to be $\tau_\rho = T\rho/J$. The maximum shear stress is developed at the outer fibers and there at $\rho = \frac{7}{8}$ in

$$\tau_{\max} = \frac{10{,}000(\frac{7}{8})}{0.92} = 9500 \text{ lb/in}^2$$

Hence the shear stress varies linearly from zero at the center of the shaft to 9500 lb/in^2 at the outer fibers as shown in Fig. 5-10.

The angle of twist θ in a 4-ft length of the shaft is

$$\theta = \frac{TL}{GJ} = \frac{10{,}000(48 \text{ in})}{12 \times 10^6 (0.92)} = 0.0435 \text{ radian}$$

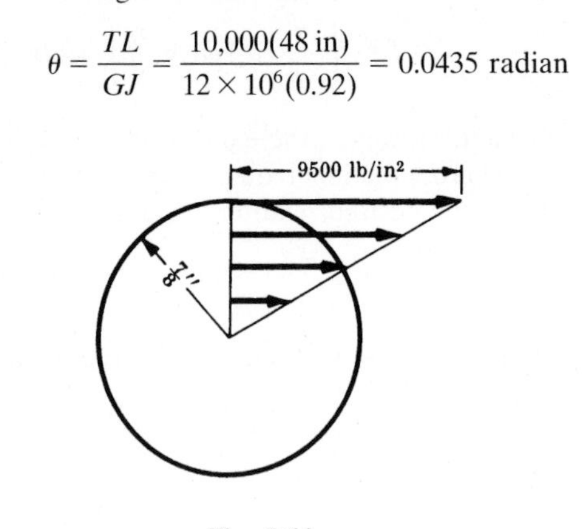

Fig. 5-10

5.5. A hollow steel shaft 3 m long must transmit a torque of 25 kN · m. The total angle of twist in this length is not to exceed 2.5° and the allowable shearing stress is 90 MPa. Determine the inside and outside diameter of the shaft if $G = 85$ GPa.

Let d_o and d_i designate the outside and inside diameters of the shaft, respectively. From Eq. (*5.4*) the angle of twist is $\theta = TL/GJ$, where θ is expressed in radians. Thus, in the 3-m length we have

$$2.5^\circ \left(\frac{\text{rad}}{57.3 \text{ deg}} \right) = \frac{(25{,}000 \text{ N} \cdot \text{m})(3 \text{ m})}{(85 \times 10^9 \text{ N/m}^2)(\pi/32)(d_o^4 - d_i^4)}$$

or

$$d_o^4 - d_i^4 = (206 \times 10^{-6}) \text{ m}^4$$

The maximum shearing stress occurs at the outer fibers where $\rho = d_o/2$. At these points from Eq. (*5.2*), we have

$$90 \times 10^6 \text{ N/m}^2 = \frac{(25{,}000 \text{ N} \cdot \text{m})(d_o/2)}{(\pi/32)(d_o^4 - d_i^4)}$$

or

$$d_o^4 - d_i^4 = (1414 d_o)(10^{-6}) \text{ m}^4$$

Comparison of the right-hand sides of these equations indicates that

$$206 \times 10^{-6} = 1414 d_o(10^{-6})$$

and thus $d = 0.145$ m or 145 mm. Substitution of this value into either of the equations then gives $d_i = 0.125$ m or 125 mm.

5.6. Let us consider a thin-walled tube subject to torsion. Derive an approximate expression for the allowable twisting moment if the working stress in shear is a given constant τ_w. Also, derive an approximate expression for the strength–weight ratio of such a tube. It is assumed the tube does not buckle, and the material is within the elastic range of action.

The polar moment of inertia of a hollow circular shaft of outer diameter D_o and inner diameter D_i is $J = (\pi/32)(D_o^4 - D_i^4)$. If R denotes the outer radius of the tube, then $D_o = 2R$, and further, if t denotes the wall thickness of the tube, then $D_i = 2R - 2t$.

The polar moment of inertia J may be written in the alternate form

$$J = \frac{\pi}{32}[(2R)^4 - (2R - 2t)^4] = \frac{\pi}{2}[R^4 - (R - t)^4] = \frac{\pi}{2}(4R^3 t - 6R^2 t^2 + 4Rt^3 - t^4)$$

$$= \frac{\pi}{2} R^4 \left[4\left(\frac{t}{R}\right) - 6\left(\frac{t}{R}\right)^2 + 4\left(\frac{t}{R}\right)^3 - \left(\frac{t}{R}\right)^4 \right]$$

Neglecting squares and higher powers of the ratio t/R, since we are considering a thin-walled tube, this becomes, approximately, $J = 2\pi R^3 t$.

The ordinary torsion formula is $T = \tau_w J/R$. For a thin-walled tube this becomes, for the allowable twisting moment, $T = 2\pi R^2 t\tau_w$.

The weight W of the tube is $W = \gamma LA$ where γ is the specific weight of the material, L the length of the tube, and A the cross-sectional area of the tube. The area is given by

$$A = \pi[R^2 - (R-t)^2] = \pi(2Rt - t^2) = \pi R^2\left[\frac{2t}{R} - \left(\frac{t}{R}\right)^2\right]$$

Again neglecting the square of the ratio t/R for a thin tube, this becomes $A = 2\pi Rt$.

The strength–weight ratio is defined to be T/W. This is given by

$$\frac{T}{W} = \frac{2\pi R^2 t\tau_w}{2\pi RtL\gamma} = \frac{R\tau_w}{L\gamma}$$

The ratio is of considerable importance in aircraft design.

5.7. A solid circular shaft has a slight taper extending uniformly from one end to the other. Denote the radius at the small end by a, that at the large end by b. Determine the error committed if the angle of twist for a given length is calculated using the mean radius of the shaft. The radius at the larger end is 1.2 times that at the smaller end.

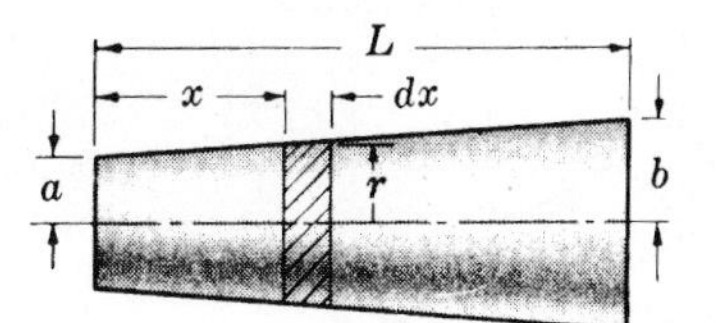

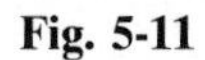

Fig. 5-11

Let us set up a coordinate system with the variable x denoting the distance from the small end of the shaft (see Fig. 5-11). The radius at a section at the distance x from the small end is

$$r = a + \frac{(b-a)x}{L}$$

where L is the length of the bar.

Provided the angle of taper is small, it is sufficient to consider the angle $d\theta$ through which the shaded element of length dx is twisted. This is obtained by applying the expression $\theta = TL/GJ$ to the element of length dx and radius $r = a + [(b-a)x/L]$. For such an element the polar moment of inertia is

$$J = \frac{\pi}{32}D^4 = \frac{\pi}{2}r^4 = \frac{\pi}{2}\left[a + \frac{(b-a)x}{L}\right]^4$$

Thus

$$d\theta = \frac{T\,dx}{G\dfrac{\pi}{2}\left[a + \dfrac{(b-a)x}{L}\right]^4}$$

The angle of twist in the length L is found by integrating the last equation. Thus

$$\theta = \frac{2T}{G\pi}\int_0^L \frac{dx}{\left[a + \dfrac{(b-a)x}{L}\right]^4} = \frac{2T}{G\pi}\left(-\frac{1}{3}\right)\left(\frac{L}{b-a}\right)\left[\frac{1}{\left[a + \dfrac{(b-a)x}{L}\right]^3}\right]_0^L = \frac{2TL}{3G\pi(b-a)}\left(-\frac{1}{b^3} + \frac{1}{a^3}\right)$$

If $b = 1.2a$, this becomes $\theta = 1.40433TL/G\pi a^4$. For a solid shaft of radius $1.1a$

$$\theta_1 = \frac{TL}{G\dfrac{\pi}{2}(1.1a)^4} = \frac{1.36602TL}{G\pi a^4}$$

Using these values of θ and θ_1, we find

$$\text{Percent error} = \frac{0.03831}{1.40433} \times 100 = 2.73\%$$

5.8. Consider two solid circular shafts connected by 2-in- and 10-in-pitch-diameter gears as in Fig. 5-12(a). The shafts are assumed to be supported by the bearings in such a manner that they undergo no bending. Find the angular rotation of D, the right end of one shaft, with respect to A, the left end of the other, caused by the torque of 2500 lb · in applied at D. The left shaft is steel for which $G = 12 \times 10^6$ lb/in^2 and the right is brass for which $G = 5 \times 10^6$ lb/in^2. Assume elastic action.

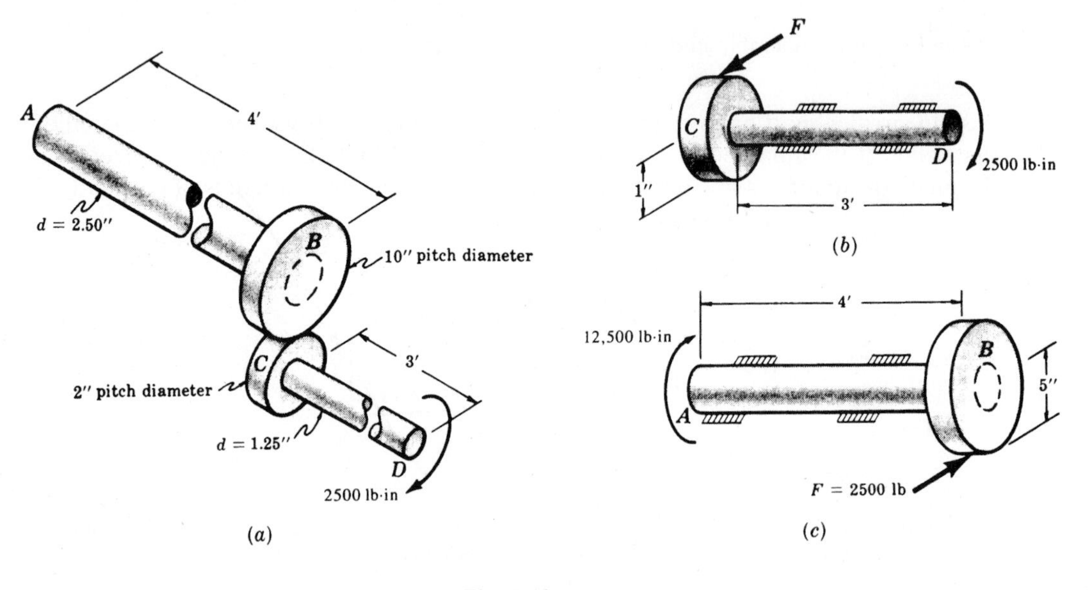

Fig. 5-12

A free-body diagram of the right shaft CD [Fig. 5-12(b)] reveals that a tangential force F must act on the smaller gear. For equilibrium, $F = 2500$ lb.

The angle of twist of the right shaft is

$$\theta_1 = \frac{TL}{GJ} = \frac{2500(36)}{5 \times 10^6 \dfrac{\pi}{32}(1.25)^4} = 0.0750 \text{ rad}$$

A free-body diagram of the left shaft AB is shown in Fig. 5-12(c). The force F is equal and opposite to that acting on the small gear C. This force F acts 5 in from the center line of the left shaft; hence it imparts a torque of 5(2500) = 12,500 lb · in to the shaft AB. Because of this torque there is a rotation of end B with respect to end A given by the angle θ_2, where

$$\theta_2 = \frac{12{,}500(48)}{12 \times 10^6(\pi/32)(2.5)^4} = 0.0130 \text{ rad}$$

It is to be carefully noted that this angle of rotation θ_2 induces a *rigid-body* rotation of the entire shaft CD because of the gears. In fact, the rotation of CD will be in the same ratio to that of AB as the ratio of the pitch diameters, or 5:1. Thus a rigid-body rotation of 5(0.0130) rad is imparted to shaft CD. Superposed on this rigid body movement of CD is the angular displacement of D with respect to C previously denoted by θ_1.

Hence the resultant angle of twist of D with respect to A is $\theta = 5(0.0130) + 0.075 = 0.140$ rad.

5.9. A solid circular shaft is required to transmit 200 kW while turning at 1.5 rev/s. The allowable shearing stress is 42 MPa. Find the required shaft diameter.

In the SI system the time rate of work (power) is expressed in N · m/s. By definition 1 N · m/s is 1 W. Power is thus given by $P = T\omega$, where T is twisting moment and ω is shaft angular velocity in radians/second. Or, alternatively, $P = 2\pi f T$, where f is revolutions per second or hertz. Thus we have

$$200{,}000 \text{ N}\cdot\text{m/s} = 2\pi(1.5 \text{ rev/s})T$$

$$T = 21{,}230 \text{ N}\cdot\text{m}$$

As in Problem 5.2, the outer fiber shearing stresses are maximum and given by

$$\tau = \frac{16T}{\pi d^3}$$

Thus,

$$42 \times 10^6 \text{ N/m}^2 = \frac{16(21{,}230 \text{ N}\cdot\text{m})}{\pi d^3}$$

Solving,

$$d = 138 \text{ mm}$$

5.10. It is required to transmit 70 hp from a turbine by a solid circular shaft turning at 200 r/min. If the allowable shearing stress is 7000 lb/in^2, determine the required shaft diameter.

In the USCS system the time rate of work (i.e., power) is expressed in lb · in/s. By definition 6600 lb · in/s is 1 hp. Power is thus given by $P = T\omega$, where T is the twisting moment and ω is shaft angular velocity in radians/second. Or, alternatively, $P = 2\pi f T$, where f is revolutions per second, usually termed hertz. Here, we have

$$70(6600 \text{ lb}\cdot\text{in/s}) = 2\pi\left(\frac{200 \text{ r}}{1 \text{ min}}\right)\left(\frac{1 \text{ min}}{60 \text{ s}}\right)T$$

from which $T = 22{,}070$ lb · in.

From Eq. (*5.2*), we have the peak shearing stresses at the outer fibers of the shaft as

$$\tau = \frac{T(d/2)}{J} = \frac{Td/2}{\pi d^4/32} = \frac{16T}{\pi d^3}$$

Thus

$$7000 \text{ lb/in}^2 = \frac{16{,}000 \text{ lb}\cdot\text{in}}{\pi d^3}$$

Solving, $d = 2.52$ in.

5.11. A solid circular shaft has a uniform diameter of 2 in and is 10 ft long. At its midpoint 65 hp is delivered to the shaft by means of a belt passing over a pulley. This power is used to drive two machines, one at the left end of the shaft consuming 25 hp and one at the right end consuming the remaining 40 hp. Determine the maximum shearing stress in the shaft and also the relative angle of twist between the two extreme ends of the shaft. The shaft turns at 200 r/min and the material is steel for which $G = 12 \times 10^6$ lb/in^2. Assume elastic action.

In the left half of the shaft we have 25 hp which corresponds to a torque T_1 given by

$$T_1 = \frac{63{,}000 \times \text{hp}}{n} = \frac{63{,}000(25)}{200} = 7880 \text{ lb}\cdot\text{in}$$

Similarly, in the right half we have 40 hp corresponding to a torque T_2 given by

$$T_2 = \frac{63{,}000(40)}{200} = 12{,}600 \text{ lb}\cdot\text{in}$$

The maximum shearing stress consequently occurs in the outer fibers in the right half and is given by the ordinary torsion formula:

$$\tau_\rho = \frac{T\rho}{J} \qquad \text{or} \qquad \tau = \frac{12{,}600(1)}{(\pi/32)\,(2)^4} = 8000\ \text{lb/in}^2$$

The angles of twist of the left and right ends relative to the center are, respectively,

$$\theta_1 = \frac{7880(60)}{12 \times 10^6(\pi/32)\,(2)^4} = 0.0250\ \text{rad} \qquad \text{and} \qquad \theta_2 = \frac{12{,}600(60)}{12 \times 10^6(\pi/32)\,(2)^4} = 0.0401\ \text{rad}$$

Since θ_1 and θ_2 are in the same direction, the relative angle of twist between the two ends of the shaft is $\theta = \theta_2 - \theta_1 = 0.015$ rad.

5.12. A circular cross-section bar is clamped at one end, free at the other, and loaded by a uniformly distributed twisting moment of magnitude t per unit length along its length [see Fig. 5-13(*a*)]. The torsional rigidity of the bar is GJ. Find the angle of twist of the free end of the bar.

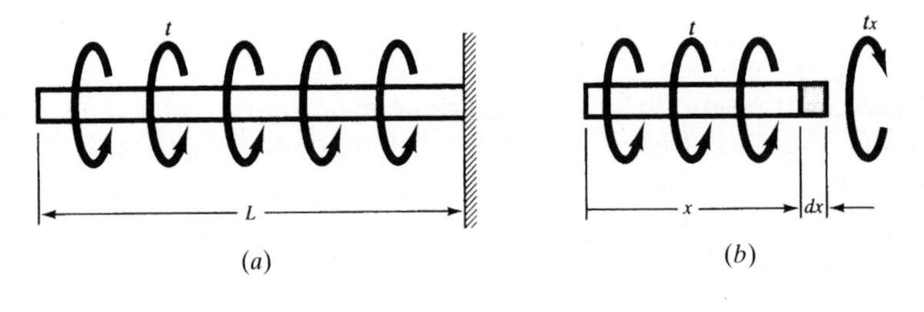

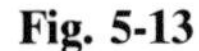
Fig. 5-13

The twisting moment per unit length is denoted by t, and the coordinate x having its origin at the left end is introduced. A free-body diagram of the portion of the bar between the left end and the section x is shown in Fig. 5-13(*b*). An element of length dx is shown in that figure and we wish to determine the angular rotation of the cylindrical element of length dx. For equilibrium of moments about the axis of the bar, a twisting moment tx must act at the right of the section shown. This twisting moment tx imparts to the element of length dx an angular rotation (from Problem 5.3)

$$d\theta = \frac{(tx)\,dx}{GJ}$$

The total rotation of the left end with respect to the right end is found by integration of all such elemental angles of twist to be

$$\theta = \int_{x=0}^{x=L} \frac{(tx)\,dx}{GJ} = \frac{tL^2}{2GJ}$$

5.13. A circular cross-section bar is clamped at one end, free at the other, and loaded by a twisting moment distributed parabolically along the length as shown in Fig. 5-14(*a*). The torsional rigidity of the bar is GJ and the moment intensity is T_0 at the clamped end. Find the angle of twist of the free end of the bar.

Let us introduce a coordinate x having origin at B and extending positive to the left. The equation of a parabola is of the general form

$$t_x = ax^2 + bx + c$$

and for the given loading we have the conditions (*a*) when $x = 0$, $t_x = 0$, (*b*) when $x = L$, $t_x = t_0$, and

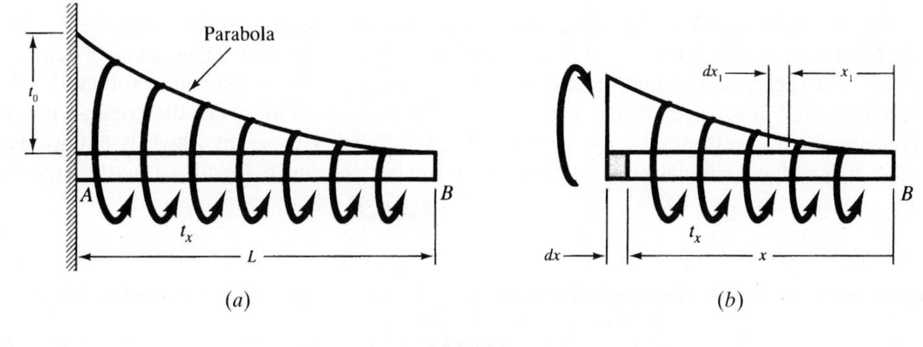

Fig. 5-14

(c) when $x = 0$, $dt_0/dx = 0$. From these conditions we find $a = t_0/L^2$, and $b = c = 0$. Thus, the loading intensity is described by the relation

$$t_x = \frac{t_0}{L^2}x^2$$

A free-body diagram of the portion of the bar between B and a section x is shown in Fig. 5-15(b). An element of length dx is also shown there and we seek to determine the angular rotation of that element. The moment acting on the element dx is found by equilibrium of twisting moments about the geometric axis of the bar to be equal to the sum of the distributed moments to the right of dx. This sum is found by introducing an auxiliary variable x_1 and we have

$$\int_{x_1=0}^{x_1=x} t_x\,dx_1 = \int_{x_1=0}^{x_1=x} \frac{t_0}{L^2}(x_1)^2\,dx_1 = \frac{t_0x^3}{3L^2}$$

From Problem 5.3, the angular rotation of the element dx is

$$d\theta = \frac{t_x\,dx}{GJ}$$

and the total angle of rotation between A and B is found by integration to be

$$\theta = \int_{x=0}^{x=L} d\theta = \int_{x=0}^{x=L} \frac{t_x\,dx}{GJ} = \int_{x=0}^{x=L} \frac{t_0x^2}{GJ}\,dx = \frac{t_0L^3}{3GJ}$$

5.14. An elastic bar of variable-diameter circular cross section is loaded in torsion at its ends as shown in Fig. 5-15. The variation of diameter may be known analytically, or through measurements at a number of locations along the axial direction. Write a FORTRAN program to give the angle of twist of one end of the bar with respect to the other.

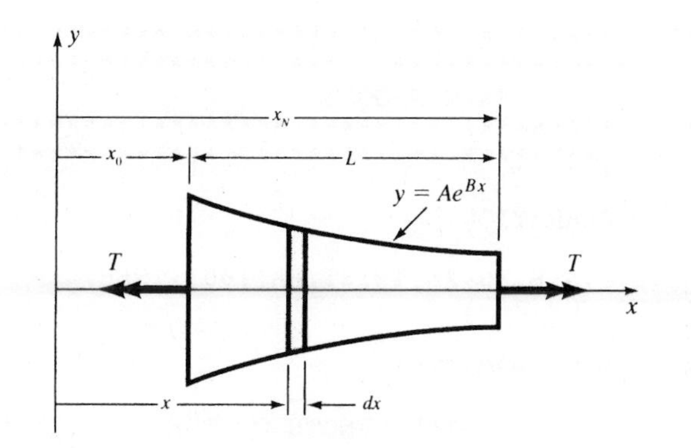

Fig. 5-15

Let us divide the bar of length L into a number of infinitesimal subsegments each of length dx, so that the cross section may be regarded as constant for each such element. Then, we may determine the angular rotation of each such element through use of the equation $\theta = TL/GJ$ from Problem 5.3. For the element of length dx, L is replaced by dx, and J is the polar moment of inertia of the cross section of the segment. This approach is represented by the following FORTRAN program which is applicable to any bar of arbitrarily varying circular cross section where the bar contour is described by the equation

$$y = Ae^{Bx}$$

```
00010*******************************************************************
00020                     PROGRAM TORSN2(INPUT,OUTPUT)
00030*******************************************************************
00040*
00050*          AUTHOR: KATHLEEN DERWIN
00060*          DATE  : FEBRUARY 5,1989
00070*
00080*    BRIEF DESCRIPTION:
00090*       THIS PROGRAM DETERMINES THE TOTAL ANGLE OF TWIST OF A CIRCULAR
00100*    ROD DUE TO TORSIONAL LOADING. CONSIDER THE ROD TO BE OF SOLID
00110*    CIRCULAR CROSS SECTION WITH A VARIABLE DIAMETER,  LOADED
00120*    BY A UNIFORM TORQUE.
00130*       THE VARYING  DIAMETER (OF THE ROD) MAY  BE DESCRIBED
00140*    EITHER ANALYTICALLY AS  Y = A*E   ^(B*X) ,  WHERE X IS THE
00150*    GEOMETRIC AXIS OF THE ROD, OR NUMERICALLY USING THE MAGNITUDE OF
00160*    Y AT EACH END OF  N SEGMENTS, MEANING N+1 VALUES.
00170*
00180*    INPUT:
00190*       THE USER IS PROMPTED FOR THE TOTAL SHAFT LENGTH, THE SHEAR
00200*    MODULUS, AND THE APPLIED TORQUE. THE USER IS THEN ASKED IF THE
00210*    ROD IS BOUNDED BY A KNOWN FUNCTION...IF THE FUNCTION IS KNOWN, THE
00220*    CONSTANTS AND THE ENDPOINTS OF THE ROD ON THE X-AXIS ARE INPUTTED;
00230*    ALTERNATELY, THE NUMBER OF SEGMENTS AND MEASURED  DIAMETERS
00240*    MUST BE ENTERED.
00250*
00260*    OUTPUT:
00270*       THE TOTAL ANGLE OF TWIST OF THE ROD IS DETERMINED AND PRINTED.
00280*
00290*    VARIABLES:
00300*       L,G           ---   LENGTH,SHEAR MODULUS OF ROD
00310*       A,B           ---   CONSTANTS OF Y=A*E   ^(B*X) GOVERNING ROD BOUNDAR
00320*       X0,XN         ---   ENDPOINTS OF SHAFT ON X-AXIS
00330*       T             ---   CENTRALLY APPLIED TORQUE
00340*       AA(100)       ---   INDIVIDUAL SEGMENT  DIAMETERS
00350*        INER         ---   POLAR MOMENT OF INERTIA OF EACH SMALL INCREMENT
00360*        ANS          ---   DETERMINE IF USER HAS A KNOWN FUNCTION
00370*      TWIST          ---   UNIFORM ANGLE OF TWIST
00380*        LEN          ---   LENGTH OF INCREMENTAL ELEMENT
00390*
00400************************************************************************
00410************************************************************************
00420*                            MAIN PROGRAM
00430************************************************************************
00440************************************************************************
00450*
00460*    VARIABLE DECLARATION
00470*
00480     REAL I,T,L,G,A,B,X0,XN,TWIST,AA(100),INER,LEN
00490     INTEGER ANS,NUM,J
00500*
00510*          USER INPUT PROMPTS
00520*
00530     PRINT*,'ENTER THE TOTAL LENGTH OF THE ROD (IN M OR INCHES):'
00540     READ*,L
```

```
00550      PRINT*,'ENTER THE SHEAR MODULUS (IN PASCALS OR PSI) :'
00560      READ*,G
00570      PRINT*,'ENTER THE UNIFORM TORQUE (IN N-M OR LB-IN) :'
00580      READ*,T
00590      PRINT*,'DO YOU KNOW THE FUNCTION DESCRIBING THE ROD?'
00600      PRINT*,'ENTER  1--YES  ;   2--NO'
00610      READ*,ANS
00620*
00630*          IF ANS EQUALS ONE, THE USER KNOWS FUNCTION. PROMPT
00640*          FOR CONSTANTS AND ENDPOINTS.
00650*
00660      INER  = 0
00670      IF (ANS.EQ.1) THEN
00680         PRINT*,'F(X) = A*E ^(B*X) '
00690         PRINT*,'ENTER A,B:'
00700         READ*,A,B
00710         PRINT*,'ENTER THE X-COORDINATE FOR BOTH ENDS OF THE ROD:'
00720         PRINT*,'(IN M OR INCHES):'
00730         READ*,X0,XN
00740*
00750      L=XN-X0
00760      LEN=L/50
00770      DO 20 I = X0,XN,LEN
00780         Y1=A*(2.71828**(B*I))
00790         Y2=A*(2.71828**( B*(I+LEN)))
00800         Y=(Y1+Y2)/2
00810         INER =(2./(3.14159*(Y**4)))+INER
00820 20   CONTINUE
00830*
00840*          IF ANS EQUALS TWO, THE USER DOES NOT KNOW FUNCTION.
00850*          PROMPT FOR NUMBER OF SEGMENTS AND MEASURED DIAMETERS.
00860*
00870      ELSE
00880         PRINT*,'ENTER THE NUMBER OF SECTIONS TO BE CALCULATED:'
00890         READ*,NUM
00900         PRINT*,'ENTER THE DIAMETERS OF THE ENDS FOR SECTIONS 1 TO N:'
00910         PRINT*,'(IN M OR INCHES):'
00920*
00930*          INPUT MEASURED  DIAMETERS
00940*
00950         DO 30 J=1,NUM+1
00960            READ*,AA(J)
00970 30      CONTINUE
00980*
00990         LEN = L/NUM
01000         DO 40 J = 1,NUM+1
01010            Y=(AA(J)+AA(J+1))/4
01020            INER =(2./(3.14159*(Y**4)))+INER
01030 40      CONTINUE
01040      ENDIF
01050*
01060      TWIST = (T*LEN*INER)/G
01070      TWIST = TWIST*180/3.14159
01080      PRINT 50,TWIST
01090*
01100 50   FORMAT(2X,'THE ANGLE OF TWIST IS:',F9.3,' DEGREES.')
01110*
01120      STOP
01130      END
```

5.15. A solid circular cross-section shaft (see Fig. 5-16) lies along the x-axis and has a contour described by the equation

$$y = 3e^{-0.05x}$$

The contour extends from $x = 0$ to $x = 25$ in. The shear modulus of the material is 12×10^6 lb/in^2 and the shaft is loaded by a twisting moment of 23,000 lb · in at each end. Use the FORTRAN program of Problem 5.14 to determine the angle of twist between the ends.

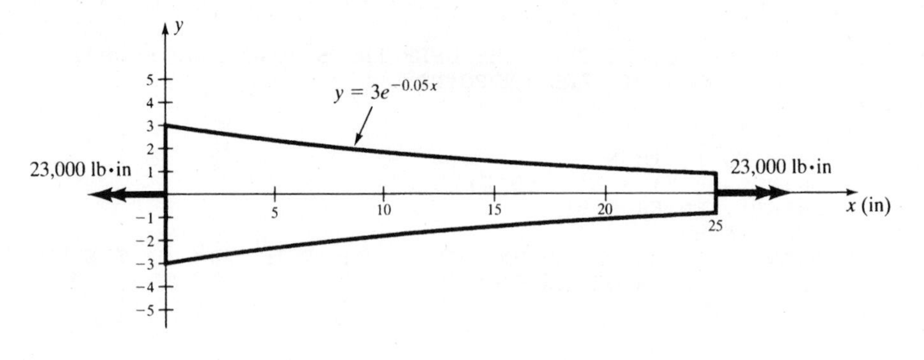

Fig. 5-16

Entering the above data into the program, we have the computer run:

```
 ENTER THE TOTAL LENGTH OF THE ROD (IN M OR INCHES):
? 25
 ENTER THE SHEAR MODULUS (IN PASCALS OR PSI) :
? 12E+6
 ENTER THE UNIFORM TORQUE (IN N-M OR LB-IN) :
? 23000
 DO YOU KNOW THE FUNCTION DESCRIBING THE ROD?
 ENTER  1--YES  ;   2--NO
? 1
 F(X) = A*E^(B*X)
 ENTER A,B:
? 3,-0.05
 ENTER THE X-COORDINATE FOR BOTH ENDS OF THE ROD:
 (IN M OR INCHES):
? 0,25
  THE ANGLE OF TWIST IS:      .703 DEGREES.
```

5.16. A circular cross-section bar is clamped at each end and loaded by the distributed twisting moments of magnitude t_1 per unit length of the bar in one direction in the left region AB and by the same intensity twisting moment but in the opposite direction in the right region BC (see Fig. 5-17). If $t_1 = 30$ N · m per meter of length, $L = 0.7$ m, and the maximum allowable shearing stress is 32 MPa, determine the required diameter of the bar.

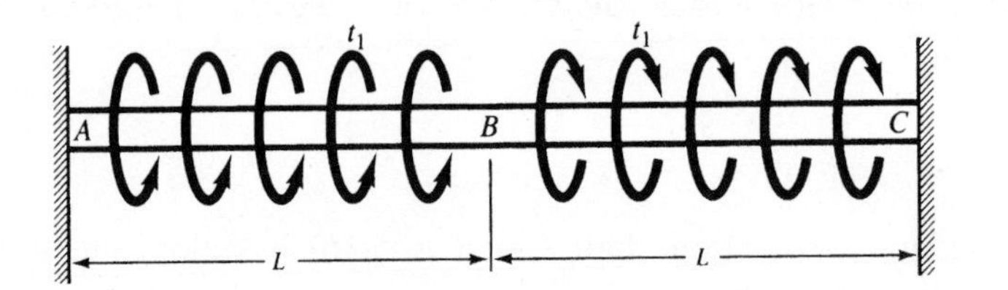

Fig. 5-17

Let us solve this problem by superposition of solutions of two subproblems. These problems are Fig. 5-18(a), labeled I, and Fig. 5-18(b), labeled II.

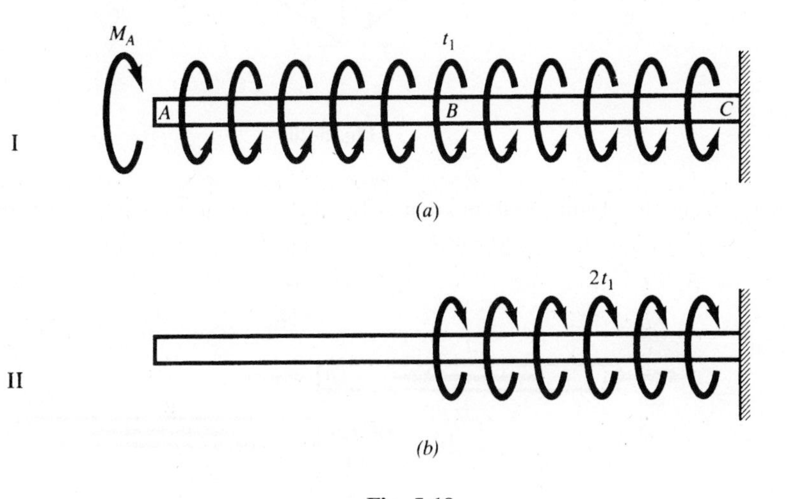

Fig. 5-18

Let us temporarily release the end A of the bar and determine the rotation of A due to an arbitrary end moment M_A plus the two distributed loadings I and II. Using the results of Problems 5.3 and 5.12, we find that the angular rotation at A is given by

$$\theta_A = \frac{t_1(2L)^2}{2GJ} - \frac{M_A(2L)}{GJ} - \frac{(2t_1)L^2}{2GJ}$$

However, since we know that end A is rigidly clamped, $\theta_A = 0$; solving we find

$$M_A = \frac{t_1 L}{2}$$

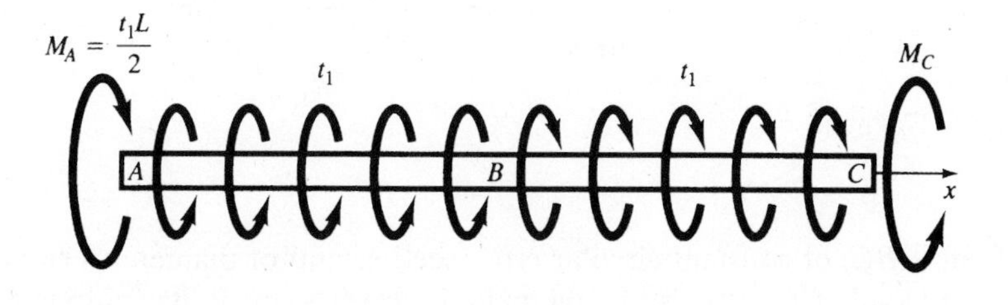

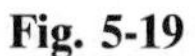

Fig. 5-19

Thus, the free-body diagram of the bar ABC appears as shown in Fig. 5-19.

From Fig. 5-19 the sum of the twisting moments about the x-axis is

$$\Sigma M_x = M_A - M_C + t_1 L - t_1 L = 0$$

which leads to

$$M_C = \frac{t_1 L}{2}$$

Thus, the variation of twisting moment along the length of the bar may be plotted as shown in Fig. 5-20.

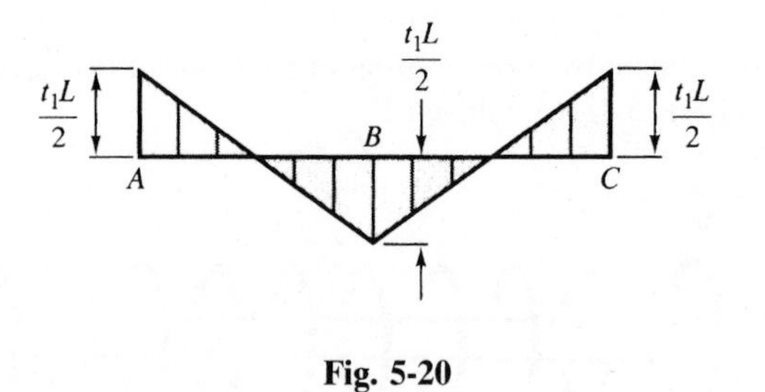

Fig. 5-20

Alternatively, using the vector representation of twisting moment, we see that the free-body diagrams of the left and right regions of ABC appear as shown in Fig. 5-21.

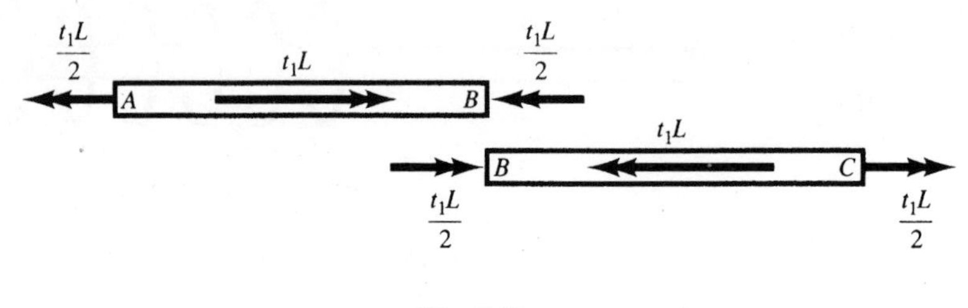

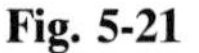

Fig. 5-21

The free-body diagram of AB indicates that there must be a twisting moment $t_1 L/2$ acting as shown at B. By Newton's law, there is an equal and opposite twisting moment acting at the left end of BC. Thus, there is a nonzero moment at the midpoint B, as indicated by Fig. 5-21. It can be shown that the angular rotation of the bar at B is zero.

From Fig. 5-21, the peak torque in the bar is $t_1 L/2$. The maximum shearing stress occurs at the outer fibers of ABC at the ends A and C as well as the midpoint B. The peak stress is, from Eq. (*5.2*):

$$\tau_{max} = \frac{T(d/2)}{\pi d^4/32}$$

$$32 \times 10^6\ \text{N/m}^2 = \frac{16\left[\dfrac{30\ \text{N}\cdot\text{m}}{1\ \text{m}}\cdot\dfrac{1}{2}\right](0.7\ \text{m})}{\pi d^3}$$

Solving, $d = 17.4$ mm.

5.17. A steel bar ABC, of constant circular cross section and of diameter 80 mm, is clamped at the left end A, loaded by a twisting moment of 6000 N · m at its midpoint B, and elastically restrained against twisting at the right end C (see Fig. 5-22). At end C the bar ABC is attached to vertical steel bars each of 16-mm diameter. The upper bar MN is attached to the end N of a horizontal diameter of the 80-mm bar ABC and the lower bar PQ is attached to the other end Q of this same horizontal diameter, as shown in Fig. 5-22(*a*). For all materials $E = 200$ GPa and $G = 80$ GPa. Determine the peak shearing stress in bar ABC as well as the tensile stress in bar MN.

Let us consider that bars MN and PQ are temporarily disconnected from the bar ABC. Then, from Problem 5.3 the angle of twist at B relative to A is

$$\theta = \frac{TL}{GJ} = \frac{(6000\ \text{N}\cdot\text{m})(0.75\ \text{m})}{(G)(\pi/32)(0.08\ \text{m})^4}$$

Since no additional twisting moments act between B and C, this same angle of twist due to the 6000-N · m loading exists at C, called θ_C.

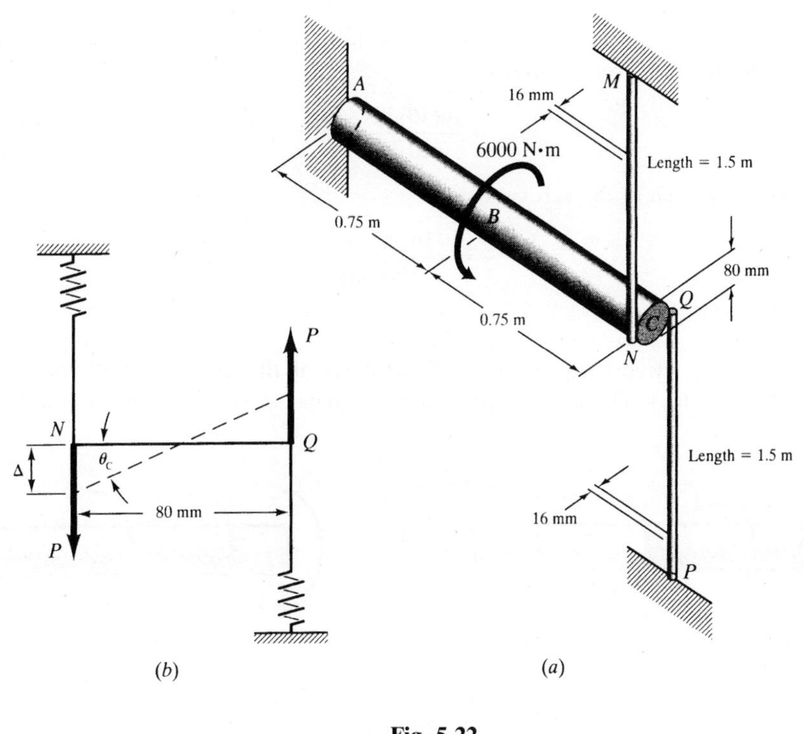

Fig. 5-22

From Fig. 5-22(b) the horizontal diameter NQ of bar ABC must rotate to some true, final position indicated by the dotted line. This is due to extension Δ of each of the vertical bars, which is accompanied by an axial force P in each bar. For a small angle of rotation θ, we have $\Delta = (0.040\text{ m})\theta_C$. The axial forces P constitute a couple of magnitude $P(0.08\text{ m}) = T_C$ which must act at the end C of bar ABC when the vertical bars are once again considered to be attached to the horizontal bar ABC. This couple must act in a sense opposite to the 6000-N·m load as shown in Fig. 5-22(a) since the elastic vertical bars tend to restrain angular rotation of the end C.

The elongation of each vertical bar may be found from Problem 1.1 to be

$$\Delta = \frac{PL}{AE} = \frac{P(1.5\text{ m})}{(\pi/4)(0.016\text{ m})^2 E} = \frac{(T_C/0.08)(1.5\text{ m})}{(\pi/4)(0.016\text{ m})^2 E}$$

The angular rotation of end C of bar ABC may now be determined by (a) considering the effect of the twisting moments of 6000 N·m and the end load T_C, and by (b) considering the angular rotation caused by the axial force P in the vertical bars. Thus, for the same rotation of end C we have

$$\frac{(6000\text{ N}\cdot\text{m})(0.75\text{ m})}{(G)(\pi/32)(0.08\text{ m})^4} - \frac{T_C(1.5\text{ m})}{(G)(\pi/32)(0.08)^4} = \frac{(T_C/0.8)(1.5\text{ m})}{(\pi/4)(0.016\text{ m})^2(0.04\text{ m})(E)}$$

Solving, $T_C = 1327\text{ N}\cdot\text{m}$ and $P = T_C/0.08 = 16{,}587\text{ N}$. The variation of twisting moment along ABC

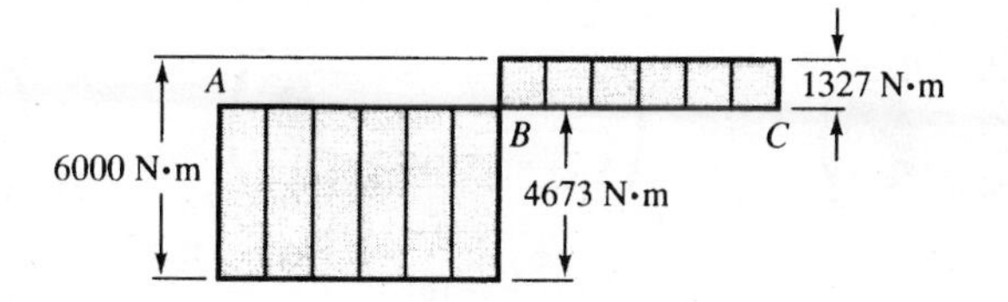

Fig. 5-23

appears as in Fig. 5-23 so that the peak torsional shearing stress occurs at the outer fibers at all points between A and B and is from Problem 5.2

$$\tau_{\max} = \frac{16(4673\ \text{N}\cdot\text{m})}{\pi(0.08\ \text{m})^3} = 46.5\ \text{MPa}$$

The axial stress in each of the vertical bars is

$$\sigma = \frac{P}{A} = \frac{16{,}587\ \text{N}}{\pi(0.008\ \text{m})^2} = 82.5\ \text{MPa}$$

5.18. Determine the reactive torques at the fixed ends of the circular shaft loaded by the couples shown in Fig. 5-24(a). The cross section of the bar is constant along the length. Assume elastic action.

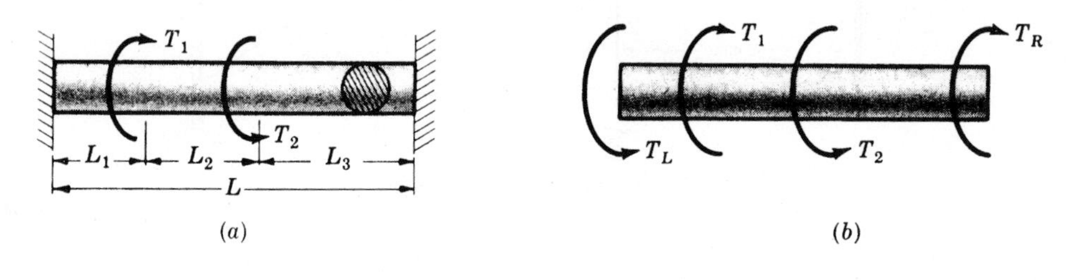

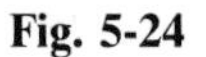

Fig. 5-24

Let us assume that the reactive torques T_L and T_R are positive in the directions shown in Fig. 5-24(b). From statics we have

$$T_L - T_1 + T_2 - T_R = 0 \tag{1}$$

This is the only equation of static equilibrium and it contains two unknowns. Hence this problem is statically indeterminate and it is necessary to augment this equation with another equation based on the deformations of the system.

The variation of torque with length along the bar may be represented by the plot shown in Fig. 5-25.

The free-body diagram of the left region of length L_1 appears as in Fig. 5-26(a).

Working from left to right along the shaft, the twisting moment in the central region of length L_2 is given by the algebraic sum of the torques to the left of this section, i.e., $T_1 - T_L$. The free-body diagram of this region appears as in Fig. 5-26(b).

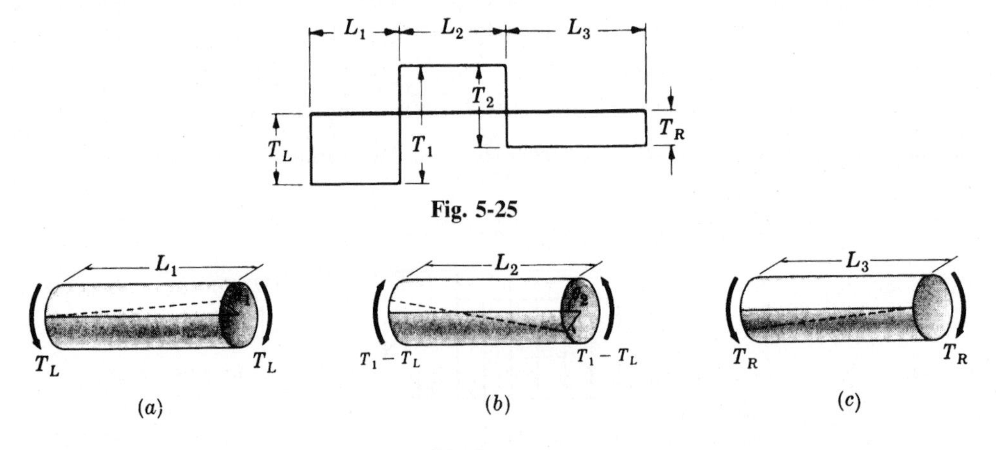

Fig. 5-25

Fig. 5-26

Finally, the free-body diagram of the right region of length L_3 appears as in Fig. 5-26(c).

Let θ_1 denote the angle of twist at the point of application of T_1, and θ_2 the angle at T_2. Then from a consideration of the regions of lengths L_1 and L_3 we immediately have

$$\theta_1 = \frac{T_L L_1}{GJ} \tag{2}$$

$$\theta_2 = \frac{T_R L_3}{GJ} \tag{3}$$

The original position of a generator on the surface of the shaft is shown by a solid line in Fig. 5-26, and the deformed position by a dashed line. Consideration of the central region of length L_2 reveals that the angle of twist of its right end with respect to its left end is $\theta_1 + \theta_2$. Hence, since the torque causing this deformation is $T_1 - T_L$, we have

$$\theta_1 + \theta_2 = \frac{(T_1 - T_L)L_2}{GJ} \tag{4}$$

Solving (*1*) through (*4*) simultaneously, we find

$$T_L = T_1\frac{L_2 + L_3}{L} - T_2\frac{L_3}{L} \qquad \text{and} \qquad T_R = -T_1\frac{L_1}{L} + T_2\frac{L_1 + L_2}{L}$$

It is of interest to examine the behavior of a generator on the surface of the shaft. Originally it was, of course, straight over the entire length L, but after application of T_1 and T_2 it has the appearance shown by the broken line in Fig. 5-27.

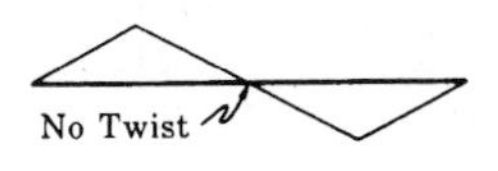

Fig. 5-27

5.19. Consider a composite shaft fabricated from a 2-in-diameter solid aluminum alloy, $G = 4 \times 10^6\ \text{lb/in}^2$, surrounded by a hollow steel circular shaft of outside diameter 2.5 in and inside diameter 2 in, $G = 12 \times 10^6\ \text{lb/in}^2$. The two metals are rigidly connected at their juncture. If the composite shaft is loaded by a twisting moment of 14,000 lb · in, calculate the shearing stress at the outer fibers of the steel and also at the extreme fibers of the aluminum. The action is elastic.

Let T_1 = torque carried by the aluminum shaft and T_2 = torque carried by the steel. For static equilibrium of moments about the geometric axis we have

$$T_1 + T_2 = T = 14{,}000$$

where T = external applied twisting moment. This is the only equation from statics available in this problem. Since it contains two unknowns, T_1 and T_2, it is necessary to supplement it with an additional equation coming from the deformations of the shaft. The structure is thus statically indeterminate.

Such an equation is easily found, since the two materials are rigidly joined; hence their angles of twist must be equal. In a length L of the shaft we have, using the formula $\theta = TL/GJ$,

$$\frac{T_1 L}{4 \times 10^6(\pi/32)(2)^4} = \frac{T_2 L}{12 \times 10^6(\pi/32)[(2.5)^4 - (2)^4]} \qquad \text{or} \qquad T_1 = 0.231T_2$$

This equation, together with the statics equation, may be solved simultaneously to yield

$$T_1 = 2600\ \text{lb} \cdot \text{in (carried by aluminum)} \qquad \text{and} \qquad T_2 = 11{,}400\ \text{lb} \cdot \text{in (carried by steel)}$$

The shearing stresses at the extreme fibers of the steel and of the aluminum are, respectively,

$$\tau_2 = \frac{11{,}400(1.25)}{(\pi/32)[(2.5)^4 - (2)^4]} = 6300\ \text{lb/in}^2 \qquad \text{and} \qquad \tau_1 = \frac{2600(1)}{(\pi/32)(2)^4} = 1650\ \text{lb/in}^2$$

5.20. A stepped shaft has the appearance shown in Fig. 5-28. The region AB is Al 2014-T6 alloy, having $G = 28$ GPa, and the region BC is steel, having $G = 84$ GPa. The aluminum portion is of solid circular cross section 45 mm in diameter, and the steel region is circular of 60-mm outside diameter and 30-mm inside diameter. Determine the peak shearing stress in each material as well as the angle of twist at B where a torsional load of 4000 N · m is applied. Ends A and C are rigidly clamped.

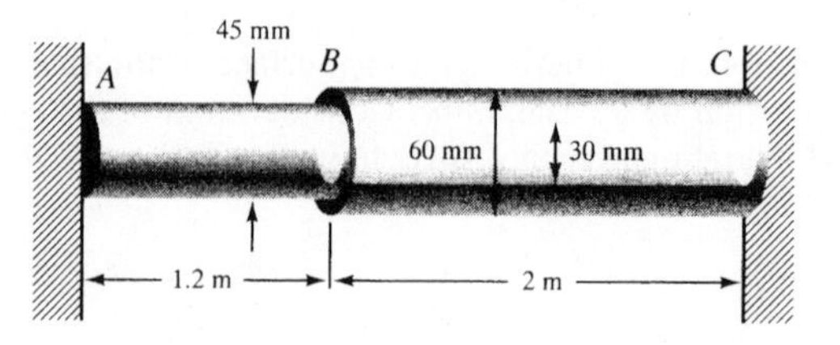

Fig. 5-28

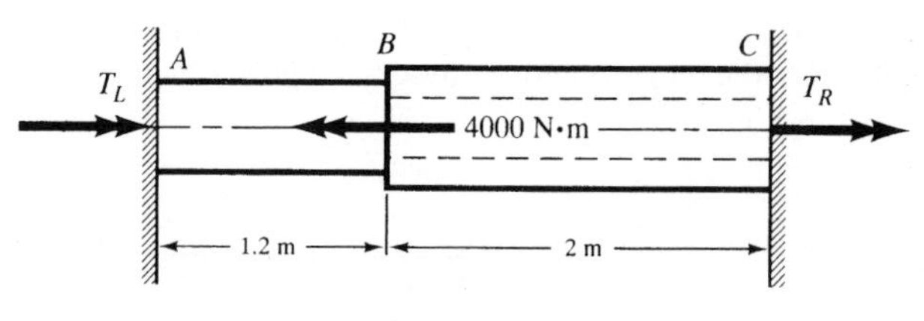

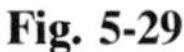

Fig. 5-29

The free-body diagram of the system is shown in Fig. 5-29.

The applied load of 4000 N · m as well as the unknown end reactive torques are indicated by the double-headed vectors above. There is only one equation of static equilibrium:

$$\Sigma M_x = T_L + T_R - 4000\ \text{N} \cdot \text{m} = 0$$

Since there are two unknowns T_L and T_R, another equation (based upon deformations) is required. This is set up by realizing that the angular rotation at B is the same if we determine it at the right end of AB or the left end of BC. Using Eq. (*5.4*), we thus have

$$\frac{T_L(1.2\ \text{m})}{(28 \times 10^9\ \text{N/m}^2)J_{\text{Al}}} = \frac{T_R(2.0\ \text{m})}{(84 \times 10^9\ \text{N/m}^2)J_{\text{ST}}} \tag{1}$$

The polar moment of inertia in AB is

$$J_{\text{Al}} = \frac{\pi(0.045\ \text{m})^4}{32} = 0.40 \times 10^{-6}\ \text{m}^4$$

and in BC it is

$$J_{\text{ST}} = \frac{\pi}{32}[(0.060\ \text{m})^4 - (0.030\ \text{m})^4] = 1.19 \times 10^{-6}\ \text{m}^4$$

Thus, from the above Eq. (*1*), we have

$$T_L = 0.187 T_R \tag{2}$$

Substituting this relation in Eq. (*1*), we find

$$T_L = 630\ \text{N} \cdot \text{m} \qquad \text{and} \qquad T_R = 3370\ \text{N} \cdot \text{m}$$

The outer fiber shearing stresses in AB are given by

$$\tau_{AB} = \frac{T\rho}{J} = \frac{(630\ \text{N}\cdot\text{m})\,(0.0225\ \text{m})}{0.40\times10^{-6}\ \text{m}^4} = 35.2\ \text{MPa}$$

and in BC by

$$\tau_{BC} = \frac{T\rho}{J} = \frac{(3370\ \text{N}\cdot\text{m})\,(0.030\ \text{m})}{1.19\times10^{-6}\ \text{m}^4} = 85.0\ \text{MPa}$$

The angle of twist at B, using parameters of the region AB, is

$$\theta_B = \frac{TL}{GJ} = \frac{(630\ \text{N}\cdot\text{m})\,(1.2\ \text{m})}{(28\times10^6\ \text{N/m}^2)\,(0.40\times10^{-6}\ \text{m}^4)} = 0.675\times10^{-3}\ \text{rad} \qquad \text{or} \qquad 0.039^\circ$$

5.21. Consider a bar of solid circular cross section subject to torsion. The material is considered to be elastic-perfectly plastic, i.e., the shear stress-strain diagram has the appearance indicated in Fig. 5-30(*a*). Determine the distance from the center at which plastic flow begins in terms of the twisting moment. Also determine the twisting moment for fully plastic action of the cross section.

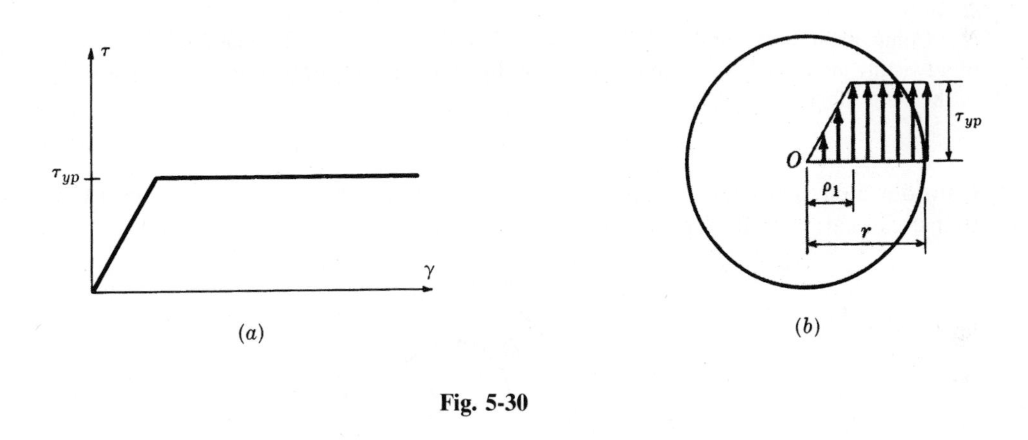

Fig. 5-30

Even though torsion of the bar has caused the outer portion to have yielded it is still realistic to assume that plane sections of the bar normal to its axis prior to loading remain plane after the torques have been applied, and further that a diameter in the section before deformation remains a diameter, or straight line, after deformation. Consequently the shearing strains of the longitudinal fibers vary linearly as the distances from the center of the bar.

Let us assume that plastic action begins at a distance ρ_1 from the center of the bar, so that the stress distribution appears as in Fig. 5-30(*b*). Thus, the shearing stresses vary linearly as the distance of the fiber from the center up to the point ρ_1 after which they are constant and equal to the yield point in shear.

From Fig. 5-30(*b*) we have for $\rho < \rho_1$:

$$\frac{\tau}{\rho} = \frac{\tau_{yp}}{\rho_1} \qquad \text{or} \qquad \tau = \left(\frac{\rho}{\rho_1}\right)\tau_{yp}$$

and for $\rho > \rho_1$: $\tau = \tau_{yp} =$ constant. Thus the twisting moment is

$$T = \int_0^r \tau\rho\, da \tag{1}$$

where da refers to the ring-shaped element shown in Fig. 5-8 of Problem 5.2. Using the above values of shearing stress in the inner elastic region and outer plastic region, we have

$$T = \int_0^{\rho_1} \left(\frac{\rho}{\rho_1}\right) \tau_{yp} \rho \, da + \int_{\rho_1}^{r} \tau_{yp} \rho \, da = \frac{\tau_{yp}}{\rho_1} \int_0^{\rho_1} \rho^2 \, da + \tau_{yp} \int_{\rho_1}^{r} \rho \, da$$

$$= \frac{\tau_{yp}}{\rho_1} \int_0^{\rho_1} \rho^2 2\pi\rho \, d\rho + \tau_{yp} \int_{\rho_1}^{r} \rho 2\pi\rho \, d\rho = \tau_{yp}\left(\frac{\pi}{2} - \frac{2\pi}{3}\right)\rho_1^3 + \frac{2\pi}{3}\tau_{yp} r^3$$

Solving for ρ_1,

$$\rho_1 = \left[4r^3 - \frac{6T}{\pi\tau_{yp}}\right]^{1/3} \tag{2}$$

as the distance from the center at which plastic flow begins. For fully plastic action, that is, $\tau = \tau_{yp}$ at all points of the cross section, we set $\rho_1 = 0$ to obtain the fully plastic twisting moment T_p:

$$T_p = \frac{2}{3}\pi r^3 \tau_{yp} = \frac{4}{3}\frac{J}{r}\tau_{yp} \tag{3}$$

But from Problem 5.2 if only the outer fibers of the bar are stressed to the yield point of the material and all interior fibers are in the elastic range of action we have the maximum possible elastic twisting moment T_e:

$$T_e = \frac{\tau_{yp}}{2}\pi r^3 \tag{4}$$

Comparison of (*3*) and (*4*) indicates that $T_p = 4T_e/3$, that is, fully plastic action permits application of a twisting moment $33\frac{1}{3}$ percent greater than the twisting moment that just causes plastic action to begin in the outer fibers.

5.22. Consider a circular shaft having a concentrically bored hole. Determine the twisting moment that it can carry for fully plastic action.

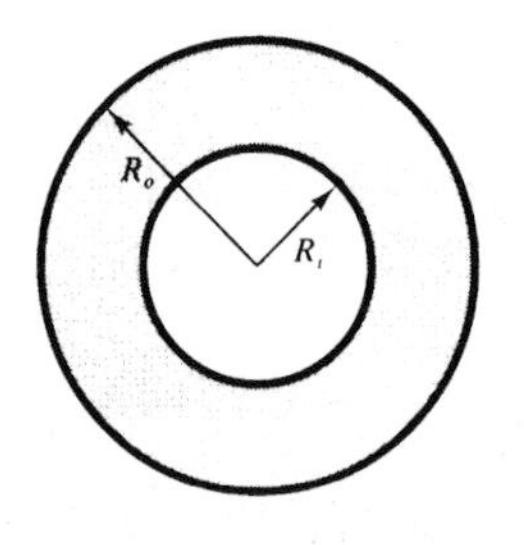

Fig. 5-31

As shown in Fig. 5-31, we denote the outer radius of the shaft by R_o and the inner radius by R_i. The yield point of the material in torsion is denoted by τ_{yp}. We return to Eq. (*1*) of Problem 5.21 and merely change the limits of integration. That is,

$$T = \int_{R_i}^{R_o} \tau_{yp} \rho \, da = \tau_{yp} \int_{Ri}^{R_o} \rho(2\pi\rho \, d\rho)$$

$$= \frac{2\pi}{3}\tau_{yp}[R_o^3 - R_i^3]$$

Note that if we express the fully plastic moment in Eq. (*3*) of Problem 5.21 in terms of J for the solid shaft it is not possible to obtain the correct fully plastic torsional loading for a hollow shaft merely by utilizing $(J_o - J_i)$ where these Js correspond to the outside and inside boundaries of the hollow shaft, respectively. It is necessary to determine the fully plastic load by returning to fundamentals and integrating as shown above.

Supplementary Problems

5.23. If a solid circular shaft of 1.25-in diameter is subject to a torque T of 2500 lb · in causing an angle of twist of 3.12° in a 5-ft length, determine the shear modulus of the material. *Ans.* $G = 11.5 \times 10^6$ lb/in^2

5.24. Determine the maximum shearing stress in a 4-in-diameter solid shaft carrying a torque of 228,000 lb · in. What is the angle of twist per unit length if the material is steel for which $G = 12 \times 10^6$ lb/in^2?
Ans. 18,100 lb/in^2, 0.000755 rad/in

5.25. A propeller shaft in a ship is 350 mm in diameter. The allowable working stress in shear is 50 MPa and the allowable angle of twist is 1° in 15 diameters of length. If $G = 85$ GPa, determine the maximum torque the shaft can transmit. *Ans.* 416 kN · m

5.26. Consider the same shaft described in Problem 5.25 but with a 175-mm axial hole bored throughout its length. The conditions on working stress and angle of twist remain as before. By what percentage is the torsional load-carrying capacity reduced? By what percentage is the weight of the shaft reduced?
Ans. 6 percent, 25 percent

5.27. A compound shaft is composed of a 24-in length of solid copper 4 in in diameter, joined to a 32-in length of solid steel 4.5 in in diameter. A torque of 120,000 lb · in is applied to each end of the shaft. Find the maximum shear stress in each material and the total angle of twist of the entire shaft. For copper $G = 6 \times 10^6$ lb/in^2, for steel $G = 12 \times 10^6$ lb/in^2.
Ans. in the copper, 9520 lb/in^2; in the steel, 6700 lb/in^2; $\theta = 0.027$ rad

5.28. In Fig. 5-32 the vertical shaft and pulley keyed to it may be considered to be weightless. The shaft rotates with a uniform angular velocity. The known belt pulls are indicated and the three pulleys are rigidly keyed to the shaft. If the working stress in shear is 50 MPa, determine the necessary diameter of a solid circular shaft. Neglect bending of the shaft because of the proximity of the bearings to the pulleys. *Ans.* 29 mm

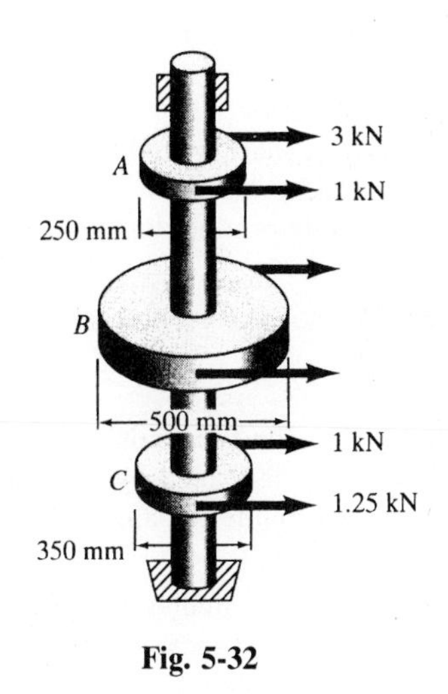

Fig. 5-32

5.29. Determine the reactive torques at the fixed ends of the circular shaft loaded by the three couples shown in Fig. 5-33. The cross section of the bar is constant along the length.
Ans. $T_L = 3600$ lb · in, $T_R = 13{,}600$ lb · in

5.30. A hollow steel shaft has an outside diameter of 4 in and an inside diameter of 3 in. Determine the maximum torque the shaft can transmit in fully plastic action if the yield point of the material in shear is 22,000 lb/in^2. *Ans.* 214,000 lb · in

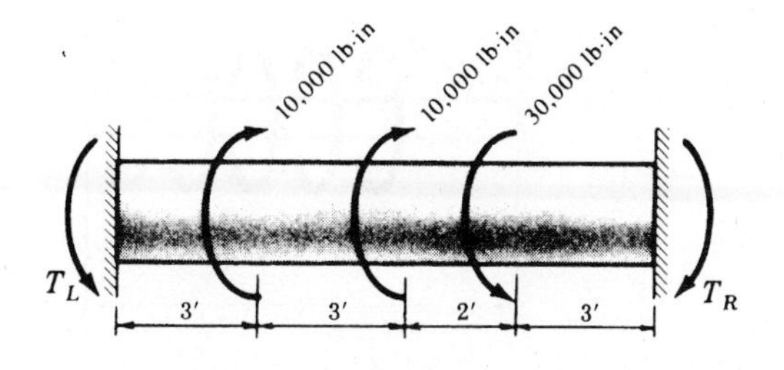

Fig. 5-33

5.31. A bar of circular cross section is clamped at its left end, free at the right, and loaded by a twisting moment t per unit length that is uniformly distributed along the middle third of the bar as shown in Fig. 5-34. Find the angle of twist of the free end of the bar.

Ans. $\frac{2}{9}\frac{tL^2}{GJ}$

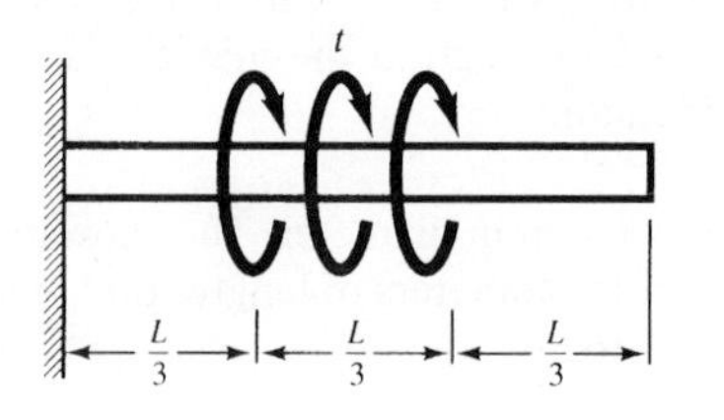

Fig. 5-34

5.32. It is desired to transmit 90 kW by means of a solid circular shaft rotating at 3.5 r/s. The allowable shearing stress is 45 MPa. Find the required shaft diameter. *Ans.* 77.4 mm

5.33. A hollow circular shaft whose outside diameter is three times its inner diameter transmits 110 hp at 120 r/min. If the maximum allowable shearing stress is 6500 lb/in^2, find the required outside diameter of the shaft. *Ans.* 3.58 in

5.34. A solid circular cross-section shaft lies along the x-axis and has a contour described by the equation

$$y = 0.074e^{-0.045x}$$

The shaft extends from $x = 0$ to $x = 3$ m. The shear modulus of the material is 83 GPa and the shaft is loaded by a twisting moment of 42,100 N · m at each end. Use the FORTRAN program of Problem 5.14 to determine the angle of twist between the ends of the bar. *Ans.* 2.518°

5.35. A solid circular cross-section shaft lies along the x-axis and has a contour described by the equation

$$y = 8e^{-0.01x}$$

The shaft extends from $x = 0$ to $x = 180$ in. The shear modulus of the material is 12×10^6 lb/in^2, and the shaft is loaded by a twisting moment of 65,000 lb · in. Use the FORTRAN program of Problem 5.14 to determine the angle of twist between the ends of the bar. *Ans.* 1.861°

5.36. A solid circular cross-section shaft is clamped at both ends and loaded by a twisting moment t per unit length as shown in Fig. 5-35. Determine the reactive twisting moments at each end of the bar.
Ans. $M_A = \frac{2}{9}tL$, $M_C = \frac{4}{9}tL$

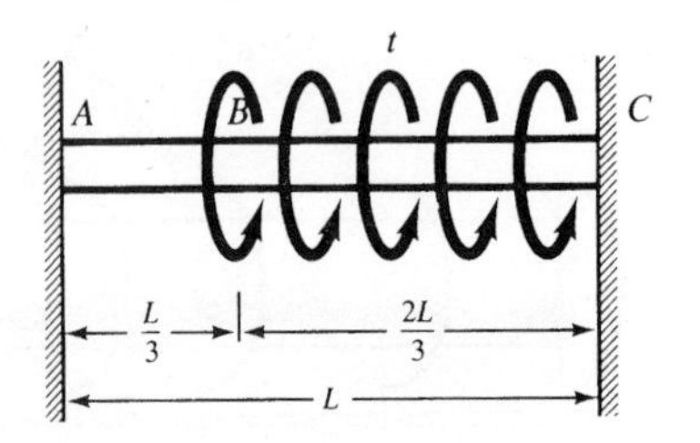

Fig. 5-35

5.37. A solid steel shaft of circular cross section has a length of 300 mm and is tapered from 50-mm diameter at the small end to 100-mm diameter at the large end, as shown in Fig. 5-36. The shaft is subject to a twisting moment of 1000 N · m applied at each end. For $G = 80$ GPa, determine the angle of twist between the ends and the peak shearing stress. *Ans.* 0.48°, 40.7 MPa

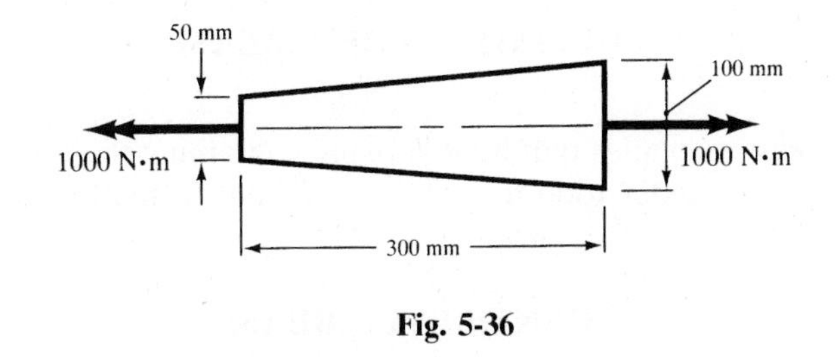

Fig. 5-36

5.38. A circular cross-section steel shaft is of diameter 50 mm over the left 150 mm of length and of diameter 100 mm over the right 150 mm, as shown in Fig. 5-37. Each end of the shaft is loaded by a twisting moment of 1000 N · m. If $G = 80$ GPa, determine the angle of twist between the ends of the shaft as well as the peak shearing stress. *Ans.* 1.09°, 40.7 MPa

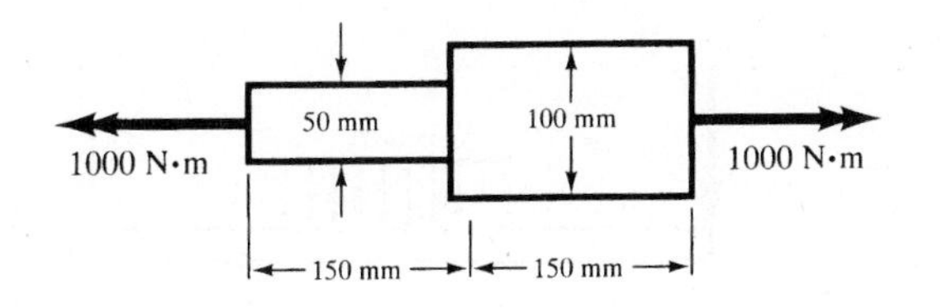

Fig. 5-37

Chapter 6

Shearing Force and Bending Moment

DEFINITION OF A BEAM

A bar subject to forces or couples that lie in a plane containing the longitudinal axis of the bar is called a *beam*. The forces are understood to act perpendicular to the longitudinal axis.

CANTILEVER BEAMS

If a beam is supported at only one end and in such a manner that the axis of the beam cannot rotate at that point, it is called a *cantilever beam*. This type of beam is illustrated in Fig. 6-1. The left end of the bar is free to deflect but the right end is rigidly clamped. The right end is usually said to be "restrained." The reaction of the supporting wall at the right upon the beam consists of a vertical force together with a couple acting in the plane of the applied loads shown.

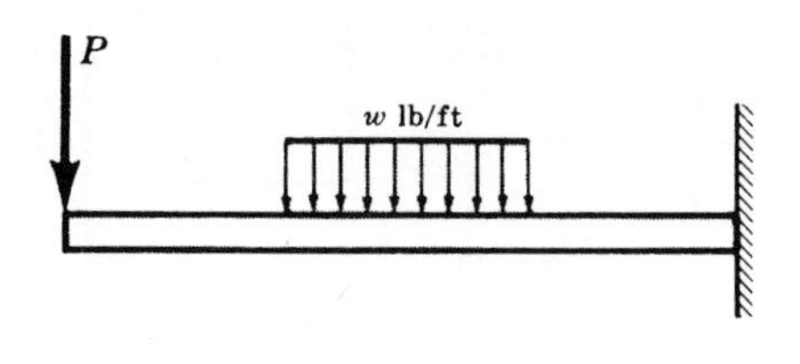

Fig. 6-1

SIMPLE BEAMS

A beam that is freely supported at both ends is called a *simple beam*. The term "freely supported" implies that the end supports are capable of exerting only forces upon the bar and are not capable of exerting any moments. Thus there is no restraint offered to the angular rotation of the ends of the bar at the supports as the bar deflects under the loads. Two simple beams are sketched in Fig. 6-2.

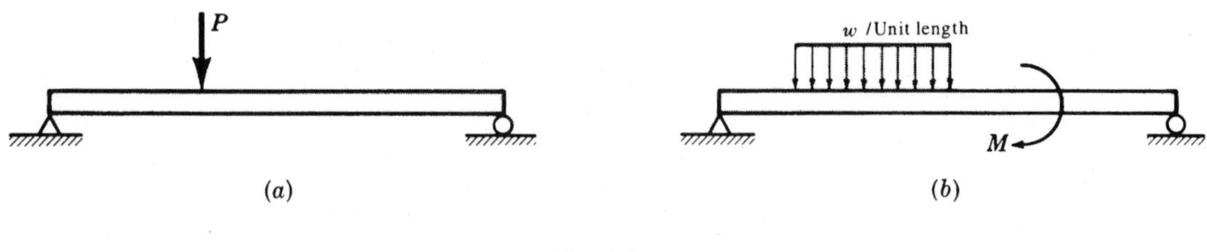

Fig. 6-2

It is to be observed that at least one of the supports must be capable of undergoing horizontal movement so that no force will exist in the direction of the axis of the beam. If neither end were free to move horizontally, then some axial force would arise in the beam as it deforms under load. Problems of this nature are not considered in this book.

The beam of Fig. 6-2(a) is said to be subject to a concentrated force; that of Fig. 6-2(b) is loaded by a uniformly distributed load as well as a couple.

OVERHANGING BEAMS

A beam freely supported at two points and having one or both ends extending beyond these supports is termed an *overhanging beam.* Two examples are given in Fig. 6-3.

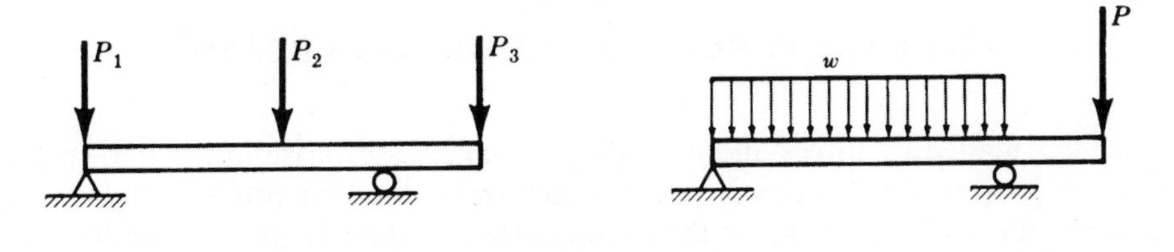

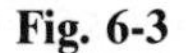

Fig. 6-3

STATICALLY DETERMINATE BEAMS

All the beams considered above, the cantilevers, simple beams, and overhanging beams, are ones in which the reactions of the supports may be determined by use of the equations of static equilibrium. The values of these reactions are independent of the deformations of the beam. Such beams are said to be *statically determinate*.

STATICALLY INDETERMINATE BEAMS

If the number of reactions exerted upon the beam exceeds the number of equations of static equilibrium, then the statics equations must be supplemented by equations based upon the deformations of the beam. In this case the beam is said to be *statically indeterminate*. Examples are shown in Fig. 6-4.

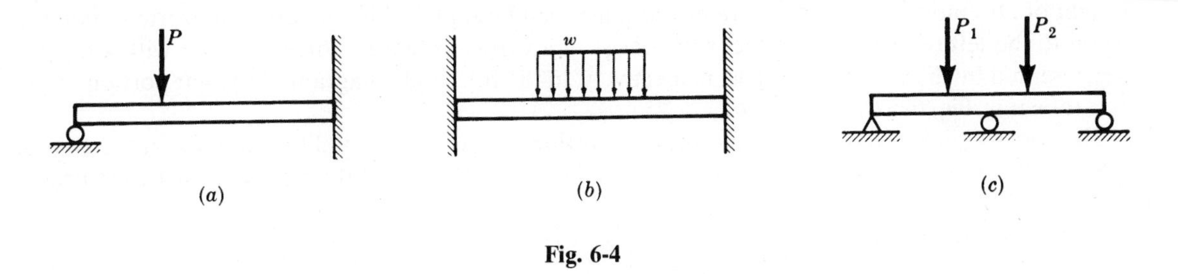

Fig. 6-4

TYPES OF LOADING

Loads commonly applied to a beam may consist of concentrated forces (applied at a point), uniformly distributed loads, in which case the magnitude is expressed as a certain number of pounds per foot or Newtons per meter of length of the beam, or uniformly varying loads. This last type of load is exemplified in Fig. 6-5.

A beam may also be loaded by an applied couple. The magnitude of the couple is usually expressed in lb · ft or N · m.

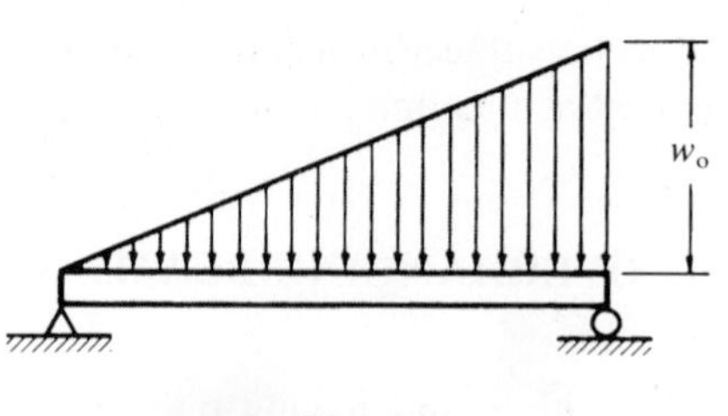

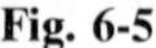

Fig. 6-5

INTERNAL FORCES AND MOMENTS IN BEAMS

When a beam is loaded by forces and couples, internal stresses arise in the bar. In general, both normal and shearing stresses will occur. In order to determine the magnitude of these stresses at any section of the beam, it is necessary to know the resultant force and moment acting at that section. These may be found by applying the equations of static equilibrium.

Example 1

Suppose several concentrated forces act on a simple beam as in Fig. 6-6(a).

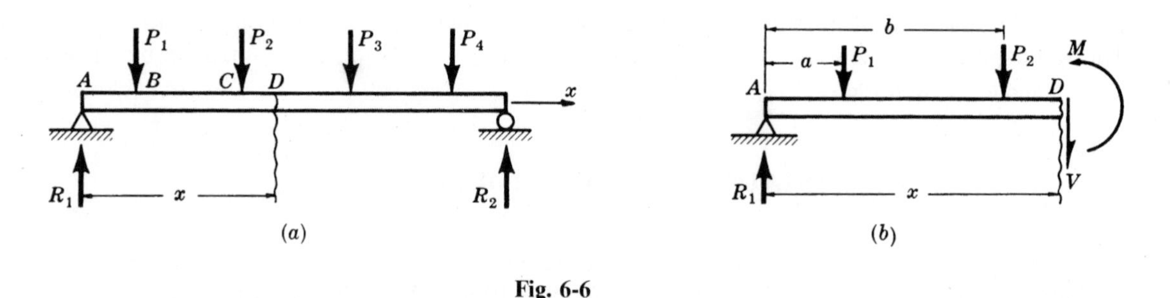

Fig. 6-6

It is desired to study the internal stresses across the section at D, located a distance x from the left end of the beam. To do this let us consider the beam to be cut at D and the portion of the beam to the right of D removed. The portion removed must then be replaced by the effect it exerted upon the portion to the left of D and this effect will consist of a vertical shearing force together with a couple, as represented by the vectors V and M, respectively, in the free-body diagram of the left portion of the beam shown in Fig. 6-6(b).

The force V and the couple M hold the left portion of the bar in equilibrium under the action of the forces R_1, P_1, P_2. The quantities V and M are taken to be positive if they have the senses indicated above.

RESISTING MOMENT

The couple M shown in Fig. 6-6(b) is called the *resisting moment* at section D. The magnitude of M may be found by use of a statics equation which states that the sum of the moments of all forces about an axis through D and perpendicular to the plane of the page is zero. Thus

$$\Sigma M_0 = M - R_1 x + P_1(x - a) + P_2(x - b) = 0 \qquad \text{or} \qquad M = R_1 x - P_1(x - a) - P_2(x - b)$$

Thus the resisting moment M is the moment at point D created by the moments of the reaction at A and the applied forces P_1 and P_2. The resisting moment M is the resultant couple due to stresses that are distributed over the vertical section at D. These stresses act in a horizontal direction and are tensile in certain portions of the cross section and compressive in others. Their nature will be discussed in detail in Chap. 8.

RESISTING SHEAR

The vertical force V shown in Fig. 6-6(b) is called the *resisting shear* at section D. For equilibrium of forces in the vertical direction,

$$\Sigma F_v = R_1 - P_1 - P_2 - V = 0 \qquad \text{or} \qquad V = R_1 - P_1 - P_2$$

This force V is actually the resultant of shearing stresses distributed over the vertical section at D. The nature of these stresses will be studied in Chap. 8.

BENDING MOMENT

The algebraic sum of the moments of the external forces to one side of the section D about an axis through D is called the *bending moment* at D. This is represented by

$$R_1 x - P_1(x - a) - P_2(x - b)$$

for the loading considered above. The quantity is considered in Problems 6.1 through 6.12. Thus the bending moment is opposite in direction to the resisting moment but is of the same magnitude. It is usually denoted by M also. Ordinarily the bending moment rather than the resisting moment is used in calculations because it can be represented directly in terms of the external loads.

SHEARING FORCE

The algebraic sum of all the vertical forces to one side, say the left side, of section D is called the *shearing force* at that section. This is represented by $R_1 - P_1 - P_2$ for the above loading. The shearing force is opposite in direction to the resisting shear but of the same magnitude. Usually it is denoted by V. It is ordinarily used in calculations, rather than the resisting shear. This quantity is considered in Problems 6.1 through 6.12.

SIGN CONVENTIONS

The customary sign conventions for shearing force and bending moment are represented in Fig. 6-7. Thus a force that tends to bend the beam so that it is concave upward is said to produce a positive bending moment. A force that tends to shear the left portion of the beam upward with respect to the right portion is said to produce a positive shearing force.

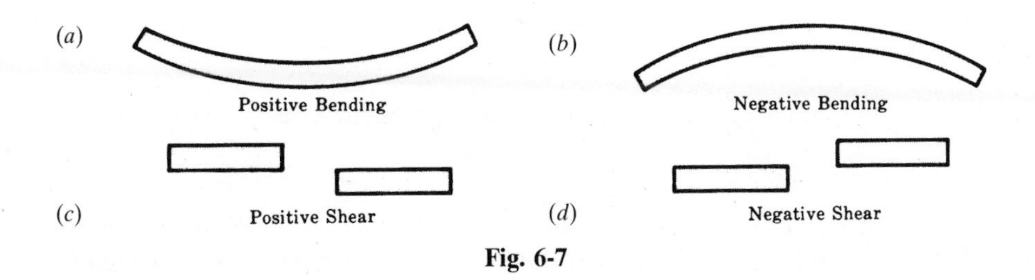

Fig. 6-7

An easier method for determining the algebraic sign of the bending moment at any section is to say that upward external forces produce positive bending moments, downward forces yield negative bending moments.

SHEAR AND MOMENT EQUATIONS

Usually it is convenient to introduce a coordinate system along the beam, with the origin at one end of the beam. It will be desirable to know the shearing force and bending moment at all sections along the beam and for this purpose two equations are written, one specifying the shearing force V as a function of the distance, say x, from one end of the beam, the other giving the bending moment M as a function of x.

SHEARING FORCE AND BENDING MOMENT DIAGRAMS

The plots of these equations for V and M are known as *shearing force* and *bending moment diagrams*, respectively. In these plots the abscissas (horizontals) indicate the position of the section along the beam and the ordinates (verticals) represent the values of the shearing force and bending moment, respectively. Thus these diagrams represent graphically the variation of shearing force and bending moment at any section along the length of the bar. From these plots it is quite easy to determine the maximum value of each of these quantities.

RELATIONS BETWEEN LOAD INTENSITY, SHEARING FORCE, AND BENDING MOMENT

A simple beam with a varying load indicated by $w(x)$ is sketched in Fig. 6-8. The coordinate system with origin at the left end A is established and distances to various sections in the beam are denoted by the variable x.

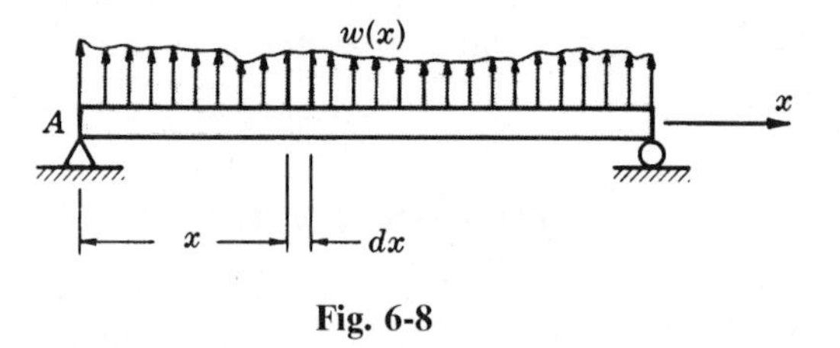

Fig. 6-8

For any value of x the relationship between the load $w(x)$ and the shearing force V is

$$w = \frac{dV}{dx}$$

and the relationship between shearing force and bending moment M is

$$V = \frac{dM}{dx}$$

These relations are derived in Problem 6.1. For applications see Problems 6.3 through 6.7.

SINGULARITY FUNCTIONS

For ease in treating problems involving concentrated forces and concentrated moments we introduce the function

$$f_n(x) = \langle x - a \rangle^n$$

where for $n > 0$ the quantity in pointed brackets is zero if $x < a$ and is the usual $(x - a)^n$ if $x > a$. This is the *singularity* or *half-range* function. Thus, if the argument is positive the pointed brackets behave just as ordinary parentheses. For applications see Problems 6.8 through 6.13.

COMPUTER IMPLEMENTATION

Determination of shearing forces and bending moments in a beam subject to a number of concentrated forces, moments, and distributed loadings is best carried out on a computer. A simple program suitable for PC implementation is given in Problem 6.13 and applications are given in Problems 6.14 and 6.15.

Solved Problems

6.1. Derive relationships between load intensity, shearing force and bending moment at any point in a beam.

Let us consider a beam subject to any type of transverse load of the general form shown in Fig. 6-9(a). Simple supports are illustrated but the following consideration holds for all types of beams. We will isolate from the beam the element of length dx shown and draw a free-body diagram of it. The shearing force V

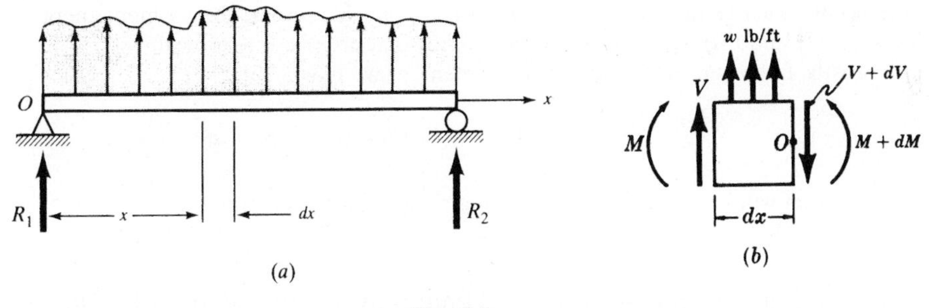

Fig. 6-9

acts on the left side of the element, and in passing through the distance dx the shearing force will in general change slightly to an amount $V + dV$. The bending moment M acts on the left side of the element and it changes to $M + dM$ on the right side. Since dx is extremely small, the applied load may be taken as uniform over the top of the beam and equal to w lb/ft. The free-body diagram of this element thus appears as in Fig. 6-9(b). For equilibrium of moments about O, we have

$$\Sigma M_0 = M - (M + dM) + V\,dx + w\,dx(dx/2) = 0 \qquad \text{or} \qquad dM = V\,dx + \tfrac{1}{2}w(dx)^2$$

Since the last term consists of the product of two differentials, it is negligible compared with the other forms involving only one differential. Hence

$$dM = V\,dx \qquad \text{or} \qquad V = \frac{dM}{dx}$$

Thus the shearing force is equal to the rate of change of the bending moment with respect to x.

This equation will prove to be of considerable value in drawing shearing force and bending moment diagrams for the more complicated types of loading. For example, from this equation it is evident that if the shearing force is positive at a certain section of the beam then the slope of the bending moment diagram is also positive at that point. Also, it demonstrates that an abrupt change in shear, corresponding to a concentrated force, is accompanied by an abrupt change in the slope of the bending moment diagram.

Further, at those points where the shear is zero, the slope of the bending moment diagram is zero. At these points where the tangent to the moment diagram is horizontal, the moment may have a maximum or minimum value. This follows from the usual calculus technique of obtaining maximum or minimum values of a function by equating the first derivative of the function to zero. Thus in Fig. 6-10 if the curves shown represent portions of a bending moment diagram then critical values may occur at points A and B.

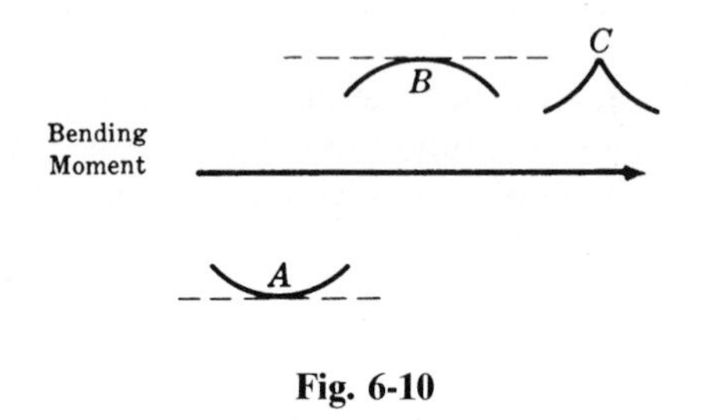

Fig. 6-10

To establish the direction of concavity at a point such as A or B, we may form the second derivative of M with respect to x, that is, d^2M/dx^2. If the value of this second derivative is positive, then the moment diagram is concave upward, as at A, and the moment assumes a minimum value. If the second derivative is negative the moment diagram is concave downward, as at B, and the moment assumes a maximum value.

However, it is to be carefully noted that the calculus method of obtaining critical values by use of the first derivative does not indicate possible maximum values at a cusp-like point in the moment diagram, if one occurs, such as that shown at C. If such a point is present, the moment there must be determined numerically and then compared to other values that are possibly critical.

Lastly, for vertical equilibrium of the element we have

$$w\,dx + V - (V + dV) = 0 \qquad \text{or} \qquad w = \frac{dV}{dx}$$

This relation will be of value in establishing shearing force diagrams.

6.2. For the cantilever beam subject to the uniformly distributed load of w N/m of length, as shown below in Fig. 6-11(a), write equations for the shearing force and bending moment at any point along the length of the bar. Also sketch the shearing force and bending moment diagrams.

It is not necessary to determine the reactions at the supporting wall. We shall choose the axis of the beam as the x-axis of a coordinate system with origin O at the left end of the bar. To determine the shearing force and bending moment at any section of the beam a distance x from the free end, we may replace the

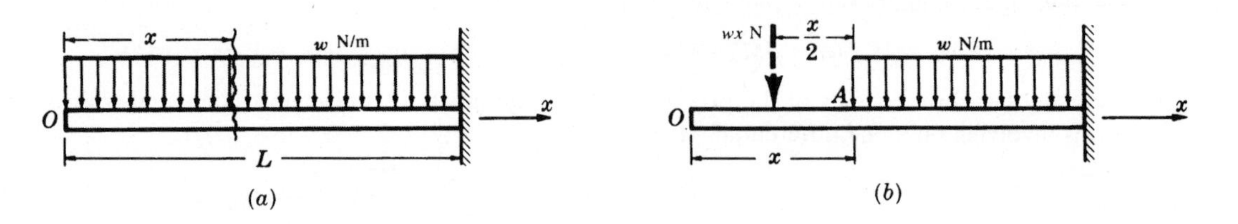

Fig. 6-11

portion of the distributed load to the left of this section by its resultant. As shown by the dashed vector in Fig. 6-11(b), the resultant is a downward force of wx N acting midway between O and the section x. Note that none of the load to the right of the section is included in calculating this resultant. Such a resultant force tends to shear the portion of the bar to the left of the section downward with respect to the portion to the right. By our sign convention this constitutes negative shear.

The shearing force at this section x is defined to be the sum of the forces to the left of the section. In this case, the sum is wx N acting downward; hence

$$V = -wx \text{ N}$$

This equation indicates that the shear is zero at $x = 0$ and when $x = L$ it is $-wL$. Since V is a first-degree function of x, the shearing force plots as a straight line connecting these values at the ends of the beam. It has the appearance shown in Fig. 6-12(a). The ordinate to this inclined line at any point represents the shearing force at that same point.

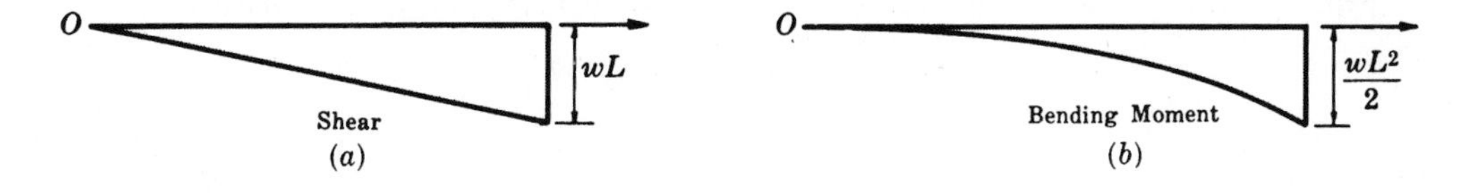

Fig. 6-12

The bending moment at this same section x is defined to be the sum of the moments of the forces to the left of this section about an axis through point A and perpendicular to the plane of the page. This sum of the moments is given by the moment of the resultant, wx N about an axis through A; it is

$$M = -wx\left(\frac{x}{2}\right) \text{N} \cdot \text{m}$$

The minus sign is necessary because downward loads indicate negative bending moments. By this equation the bending moment is zero at the left end of the bar and $-wL^2/2$ at the clamped end when $x = L$. The variation of bending moment is parabolic along the bar and may be plotted as in Fig. 6-12(b). The ordinate to this parabola at any point represents the bending moment at that same point.

It is to be noted that a downward uniform load as considered here leads to a bending moment diagram that is concave downward. This could be established by taking the second derivative of M with respect to x, the derivative in this particular case being $-w$. Since the second derivative is negative, the rules of calculus tell us that the curve must be concave downward.

6.3. Consider a simply supported beam 10 ft long and subject to a uniformly distributed vertical load of 120 lb per ft of length, as shown in Fig. 6.13(a). Draw shearing force and bending moment diagrams.

The total load on the beam is 1200 lb, and from symmetry each of the end reactions is 600 lb. We shall now consider any cross section of the beam at a distance x from the left end. The shearing force at this

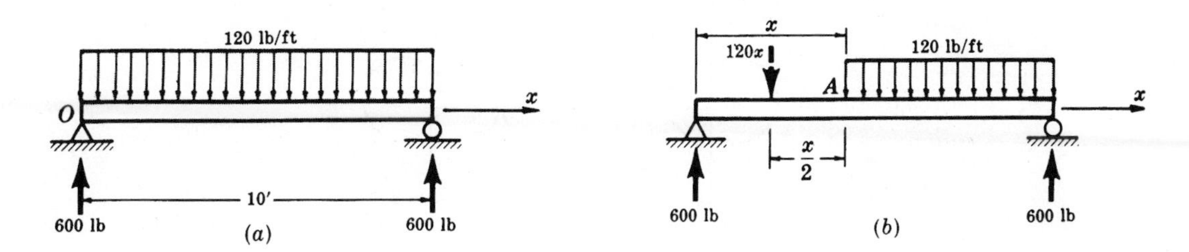

Fig. 6-13

section is given by the algebraic sum of the forces to the left of this section and these forces consist of the 600-lb reaction and the distributed load of 120 lb/ft extending over a length x ft. We may replace the portion of the distributed load to the left of the section at x by its resultant, which is $120x$ lb acting downward as shown by the dashed vector in Fig. 6-13(b). None of the load to the right of x is included in this resultant. The shearing force at x is then given by

$$V = 600 - 120x \text{ lb}$$

Since there are no concentrated loads acting on the beam, this equation is valid at all points along its length. Evidently the shearing force varies linearly from $V = 600$ lb at $x = 0$ to $V = 600 - 1200 = -600$ lb at $x = 10$ ft. The variation of shearing force along the length of the bar may then be represented by a straight line connecting these two end-point values. The shear diagram is shown in Fig. 6-14(a). The shear is zero at the center of the beam.

The bending moment at the section x is given by the algebraic sum of the moments of the 600-lb reaction and the distributed load of $120x$ lb about an axis through A perpendicular to the plane of the paper. Remembering that upward forces give positive bending moments, we have

$$M = 600x - 120x\left(\frac{x}{2}\right) \text{ lb} \cdot \text{ft}$$

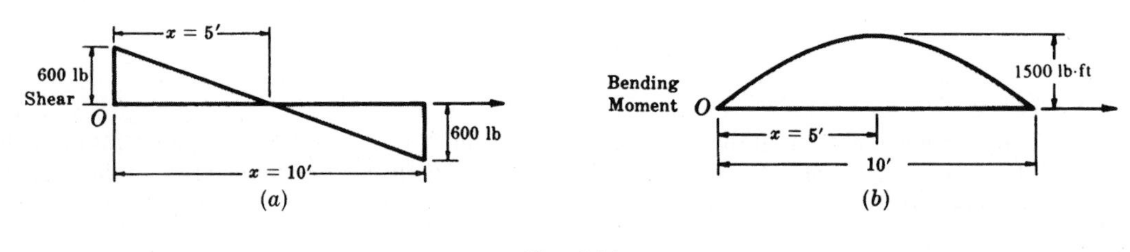

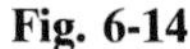
Fig. 6-14

Again, this equation holds along the entire length of the beam. It is to be noted that since the load is uniformly distributed the resultant indicated by the dashed vector acts at a distance $x/2$ from A, i.e., at the midpoint of the uniform load to the left of the section x where the bending moment is being calculated. From the above equation it is evident that the bending moment is represented by a parabola along the length of the beam. Since the bar is simply supported the moment is zero at either end and, because of the symmetry of loading, the bending moment must be a maximum at the center of the beam where $x = 5$ ft. The bending moment at that point is

$$M_{x=5} = 600(5) - 60(5)^2 = 1500 \text{ lb} \cdot \text{ft}$$

The parabolic variation of bending moment along the length of the bar may thus be represented by the ordinates to the bending moment diagram shown in Fig. 6-14(b).

6.4. The beam AD in Fig. 6-15 is supported between knife edges at B and C and subject to the end couples indicated. Draw the shearing force and bending moment diagrams.

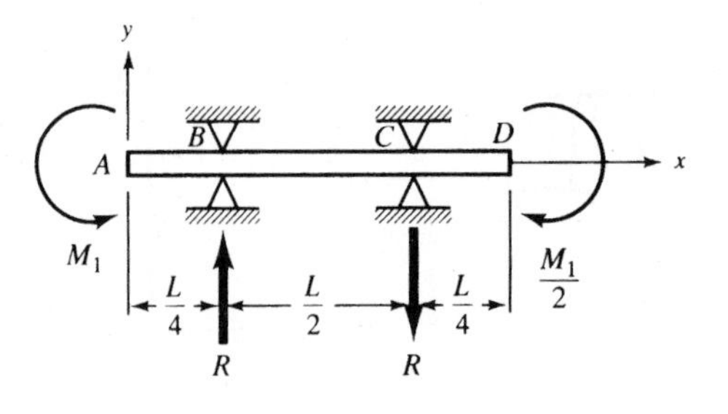

Fig. 6-15

The resultant of the end loadings is a couple

$$M_1 - \frac{M_1}{2} = \frac{M_1}{2}$$

which must be maintained in equilibrium by another couple of that magnitude but oppositely directed. This reactive couple arises from the vertical force reactions R at B and C. The moment of the couple corresponding to these forces must be $M_1/2$ for equilibrium, so we have

$$R \cdot \frac{L}{2} = \frac{M_1}{2}$$

$$R = \frac{M_1}{L}$$

For the coordinate system shown, the shearing force at any point a distance x to the right of A is given by the sum of all vertical forces to the left of x. Thus, for the three regions of the beam we have

$$V = 0 \qquad 0 < x < \frac{L}{4}$$

$$V = \frac{M_1}{L} \qquad \frac{L}{4} < x < \frac{3L}{4}$$

$$V = 0 \qquad \frac{3L}{4} < x < L$$

Analogously, the bending moment at the point x is given by the sum of the moments of all forces and couples to the left of x. Thus, we need the three equations

$$M = -M_1 \qquad 0 < x < \frac{L}{4}$$

$$M = -M_1 + \frac{M_1}{L}\left(x - \frac{L}{4}\right) \qquad \frac{L}{4} < x < \frac{3L}{4}$$

$$M = -M_1 + \frac{M_1}{L} \cdot \frac{L}{2} = -\frac{M_1}{2} \qquad \frac{3L}{4} < x < L$$

Plots of these equations appear in Figs. 6-16(*a*) and 6-16(*b*).

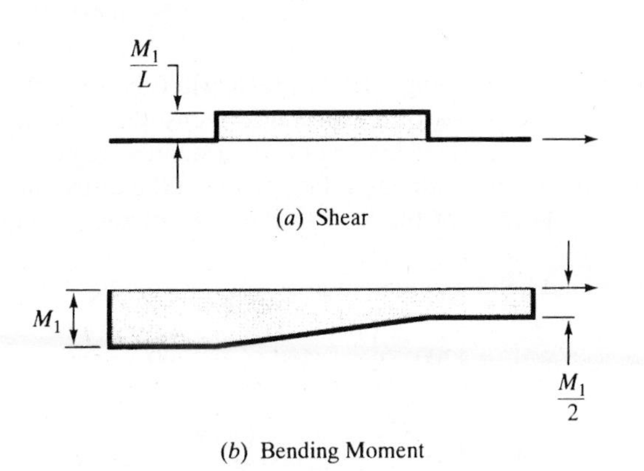

Fig. 6-16

6.5. The simply supported beam shown in Fig. 6-17(a) carries a vertical load that increases uniformly from zero at the left end to a maximum value of 600 lb/ft of length at the right end. Draw the shearing force and bending moment diagrams.

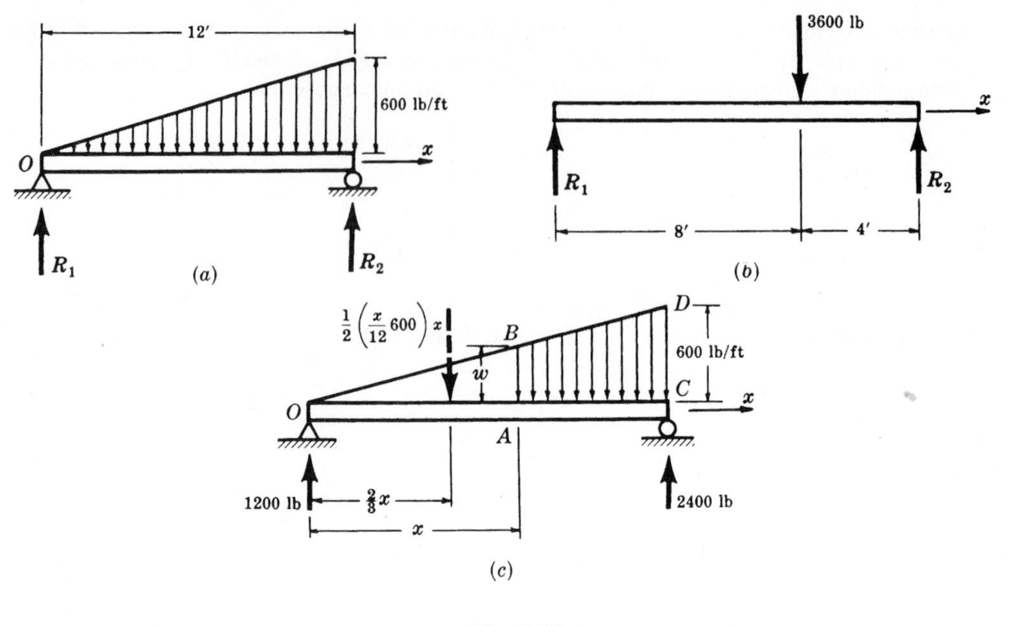

Fig. 6-17

For the purpose of determining the reactions R_1 and R_2 the entire distributed load may be replaced by its resultant which will act through the centroid of the triangular loading diagram. Since the load varies from 0 at the left end to 600 lb/ft at the right end, the average intensity is 300 lb/ft acting over a length of 12 ft. Hence the total load is 3600 lb applied 8 ft to the right of the left support. The free-body diagram to be used in determining the reactions is shown in Fig. 6-17(b). Applying the equations of static equilibrium to this bar, we find $R_1 = 1200$ lb and $R_2 = 2400$ lb.

However, this resultant cannot be used for the purpose of drawing shear and moment diagrams. We must consider the distributed load and determine the shear and moment at a section a distance x from the left end as shown in Fig. 6.17(c). At this section x the load intensity w may be found from the similar triangles OAB and OCD as follows:

$$\frac{w}{x} = \frac{600}{12} \qquad \text{or} \qquad w = \left(\frac{x}{12}\right) 600 \text{ lb/ft}$$

The average load intensity over the length x is $\frac{1}{2}(x/12)\,600$ lb/ft because the load is zero at the left end. The total load acting over the length x is the average intensity of loading multiplied by the length, or $\frac{1}{2}[(x/12)\,600]x$ lb. This acts through the centroid of the triangular region OAB shown, i.e., through a point located a distance $\frac{2}{3}x$ from O. The resultant of this portion of the distributed load is indicated by the dashed vector in Fig. 6-17(c). No portion of the load to the right of the section x is included in this resultant force.

The shearing force and bending moment at A are now readily found to be

$$V = 1200 - \frac{1}{2}\left(\frac{x}{12}600\right)x = 1200 - 25x^2$$

$$M = 1200x - \frac{1}{2}\left(\frac{x}{12}600\right)x\left(\frac{x}{3}\right) = 1200x - \frac{25}{3}x^3$$

These equations are true along the entire length of the beam. The shearing force thus plots as a parabola, having a value 1200 lb when $x = 0$ and -2400 lb when $x = 12$ ft. The bending moment is a third-degree polynomial. It vanishes at the ends and assumes a maximum value where the shear is zero. This is true because $V = dM/dx$, and hence the point of zero shear must be the point where the tangent to the moment diagram is horizontal. This point of zero shear may be found by setting $V = 0$:

$$0 = 1200 - 25x^2 \qquad \text{or} \qquad x = 6.94 \text{ ft}$$

The bending moment at this point is found by substitution in the general expression given above:

$$M_{x=6.94} = 1200(6.94) - \frac{25}{3}(6.94)^3 = 5520 \text{ lb} \cdot \text{ft}$$

The plots of the shear and moment equations appear in Fig. 6-18.

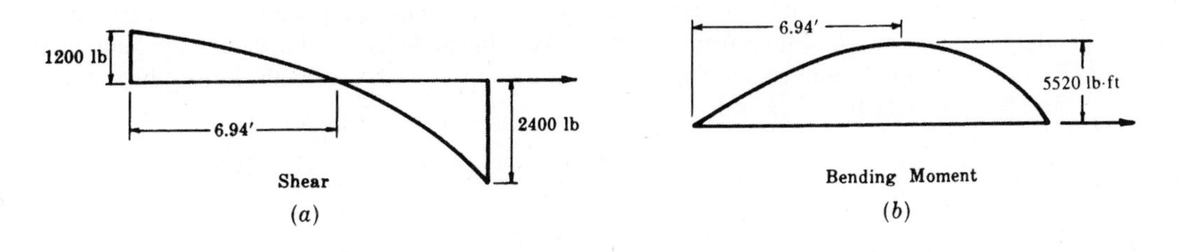

Fig. 6-18

6.6. The cantilever beam AC in Fig. 6-19 is loaded by the uniform load of 600 N/m over the length BC together with the couple of magnitude 4800 N · m at the tip C. Determine the shearing force and bending moment diagrams.

The reactions at A must consist of a vertical shearing force together with a moment to prevent angular rotation. To find these reactions, we write the statics equations

$$\Sigma F_y = R_A - (600 \text{ N/m})(2 \text{ m}) = 0 \tag{1}$$

$$\circlearrowright \Sigma M_A = M_A - 4800 \text{ N} \cdot \text{m} - (1200 \text{ N}) \cdot (3 \text{ m}) = 0 \tag{2}$$

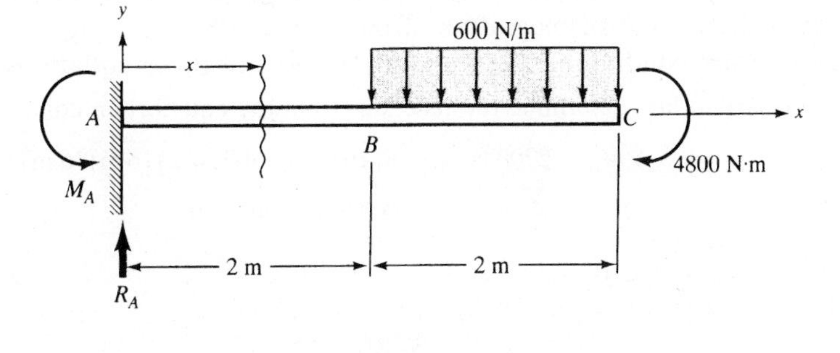

Fig. 6-19

Solving,

$$R_A = 1200 \text{ N} \qquad M_A = 8400 \text{ N} \cdot \text{m}$$

For the coordinate system shown, the shearing force at any point a distance x to the right of A is given by the sum of all forces to the left of x. Thus we must write the two equations

$$V = 1200 \text{ N} \qquad 0 < x < 2 \text{ m} \tag{3}$$

$$V = 1200 \text{ N} - 600(x - 2)N \qquad 2 < x < 4 \text{ m} \tag{4}$$

Likewise, the bending moment at this point x is given by the sum of the moments of all forces (and couples) to the left of x about point x. This is given by the two equations

$$M = -8400\ \text{N}\cdot\text{m} + 1200x \qquad 0<x<2\ \text{m} \tag{5}$$

$$M = -8400\ \text{N}\cdot\text{m} + 1200x - (600\ \text{N/m})\left[(x-2)\ \text{N}\frac{(x-2)\ \text{N}}{2}\right] \tag{6}$$

Plots of Eqs. (*3*) through (*6*) appear in Fig. 6-20(*a*) and 6-20(*b*), respectively. The nature of the concave region of the bending moment in BC is determined by taking the second derivative of the bending moment Eq. (*6*) in BC:

$$\frac{d^2M}{dx^2} = -600$$

Since this is negative for values of x in BC, the plot in BC of bending moment is concave downward. The bending moment in AB is seen from Eq. (*5*) to be a linear function of x; hence the bending moment in AB plots as a straight line connecting the end couple of $-8400\ \text{N}\cdot\text{m}$ with the bending moment at B of $-6000\ \text{N}\cdot\text{m}$ as determined from Eq. (*6*).

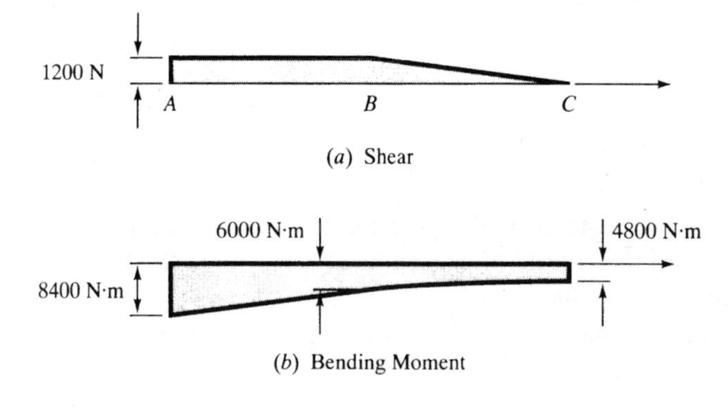

Fig. 6-20

6.7. The beam AC is simply supported at A and C and subject to the uniformly distributed load of 300 N/m plus the couple of magnitude $2700\ \text{N}\cdot\text{m}$ as shown in Fig. 6-21. Write equations for shearing force and bending moment and make plots of these equations.

It is necessary to first determine the reactions from the equilibrium equations

$$+\circlearrowright \Sigma M_A = 2700\ \text{N}\cdot\text{m} + R_C(6\ \text{m}) - (300\ \text{N/m})(6\ \text{m})(6\ \text{m}) = 0 \tag{1}$$

$$\Sigma F_y = R_A + R_C - (300\ \text{N/m})(6\ \text{m}) = 0 \tag{2}$$

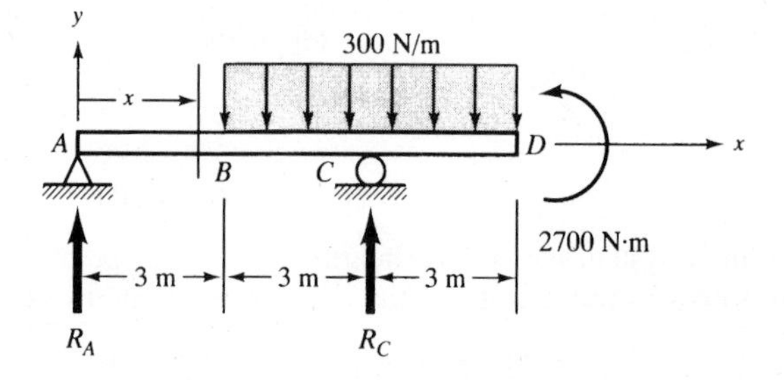

Fig. 6-21

Solving,

$$R_A = 450\ \text{N} \qquad R_C = 1350\ \text{N}$$

For the coordinate x as shown the shearing force a distance x from point A is described by the three relations

$$V = 450\ \text{N} \qquad 0 < x < 3\ \text{m} \tag{3}$$

$$V = [450 - 300(x-3)]\ \text{N} \qquad 3\ \text{m} < x < 6\ \text{m} \tag{4}$$

$$V = [450 - 300(x-3) + 1350]\ \text{N} \qquad 6\ \text{m} < x < 9\ \text{m} \tag{5}$$

Likewise the bending moment in each of these three regions of the beam is described by

$$M = (450x)\ \text{N}\cdot\text{m} \qquad 0 < x < 3\ \text{m} \tag{6}$$

$$M = \left[450x - 300(x-3)\left(\frac{x-3}{2}\right)\right]\ \text{N}\cdot\text{m} \qquad 3\ \text{m} < x < 6\ \text{m} \tag{7}$$

$$M = \left[450x - 300\frac{(x-3)^2}{2} + 1350(x-6)\right]\ \text{N}\cdot\text{m} \qquad 6\ \text{m} < x < 9\ \text{m} \tag{8}$$

Plots of these equations appear in Fig. 6-22. In regions BC and CD it is necessary to determine that the second derivative of the bending moment from Eq. (7) and Eq. (8) is negative in each of these regions, and that hence in each case the curvature of the bending moment plot is concave downward.

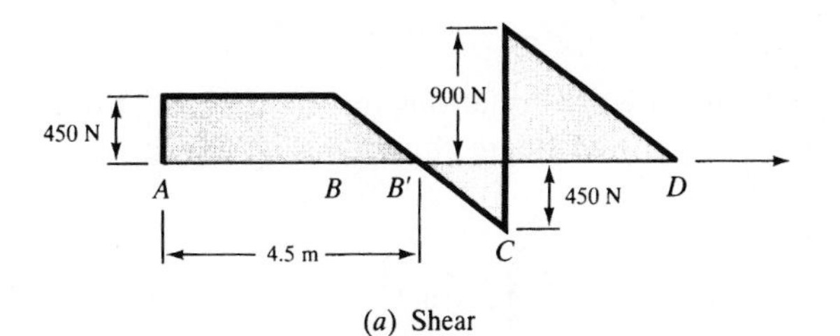

(a) Shear

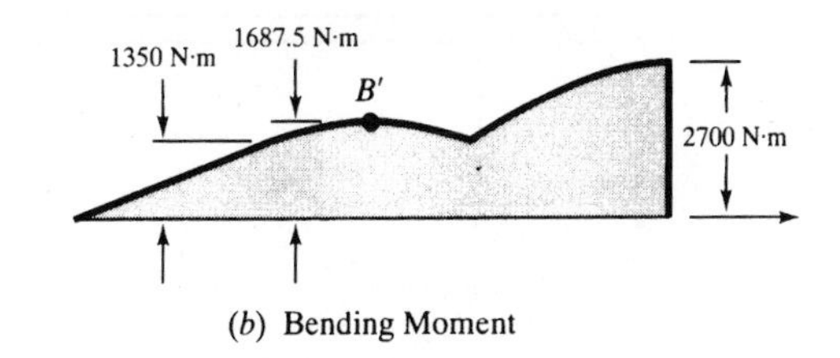

(b) Bending Moment

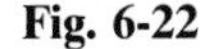

Fig. 6-22

Singularity Functions

The techniques discussed in the preceding problems are adequate if the loadings are continuously varying over the length of the beam. However, if concentrated forces or moments are present, a distinct pair of shearing force and bending moment equations must be written for each region between such concentrated forces or moments. Although this presents no fundamental difficulties, it usually leads to very cumbersome results. As we shall see in a later chapter, these results are particularly unwieldy to work with in dealing with deflections of beams.

At least some compactness of representation may be achieved by introduction of so-called *singularity* or *half-range* functions. Such functions were applied to beam analysis by Macauley in 1919 and this technique of analysis sometimes bears the name of Macauley's method, although the functions were actually used in the 19th century by A. Clebsch. Let us introduce, by definition, the pointed brackets $\langle x - a \rangle$ and define this quantity to be zero if $(x - a) < 0$, that is, $x < a$, and to be simply $(x - a)$ if $(x - a) > 0$, that is, $x > a$. That is, a half-range function is defined to have a value only when the argument is positive. When the argument is positive, the pointed brackets behave just as ordinary parentheses. The singularity function

$$f_n(x) = \langle x - a \rangle^n$$

obeys the integration law

$$\int_{-\infty}^{x} \langle y - a \rangle^n \, dy = \frac{\langle x - a \rangle^{n+1}}{n+1} \qquad \text{for} \qquad n \geq 0$$

The singularity function is very well suited for representation of shearing forces and bending moments in beams subject to loadings of the type discussed in Problems 6.4 through 6.7. This is clear since, say in Problem 6.4 for shearing force, the effect of a single concentrated load is not present (explicitly) in the equation for V for points along the beam to the left of that force, but it immediately appears in the equation for V when one considers values of x to the right of the point of application of the force.

The use of singularity functions for the representations of shearing force and bending moment makes it possible to describe each of these quantities by a single equation along the entire length of the beam, no matter how complex the loading may be. Most important, the singularity function approach leads to simple computer implementation.

6.8. Use singularity functions to write equations for the shearing force and bending moment at any position in the simply supported beam shown in Fig. 6-23.

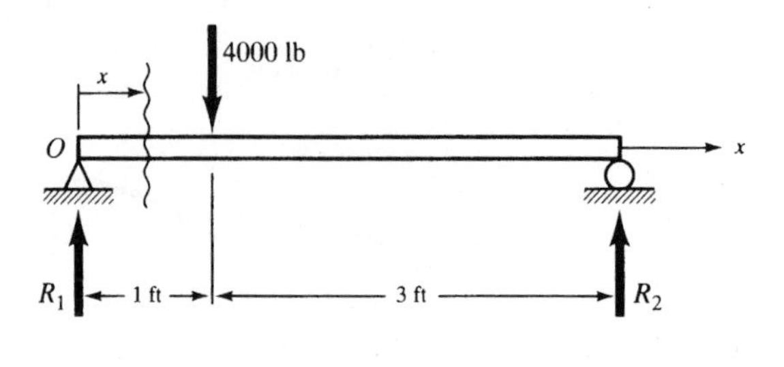

Fig. 6-23

From statics the reactions are easily found to be

$$R_1 = 3000 \text{ lb} \qquad R_2 = 1000 \text{ lb}$$

For the coordinate system shown, with origin at O, we may write

$$V = 3000\langle x \rangle^{\circ} - 4000\langle x - 1 \rangle^{\circ} \text{ lb}$$

which indicates that $V = 3000$ lb if $x < 1$ ft and $V = 3000 - 4000 = -1000$ lb if $x > 1$ ft.

Similarly,

$$M = 3000\langle x \rangle^1 - 4000\langle x - 1 \rangle^1 \text{ lb} \cdot \text{ft} \tag{2}$$

which tells us that $M = 3000x$ lb · ft if $x < 1$ ft and $M = 3000x - 4000\{x - 1\}$ lb · ft if $x > 1$ ft.

The relations (*1*) and (*2*) hold for all values of x provided we remember the definition of singularity functions. Use of these equations leads to the shearing force and bending moment diagrams shown in Fig. 6-24.

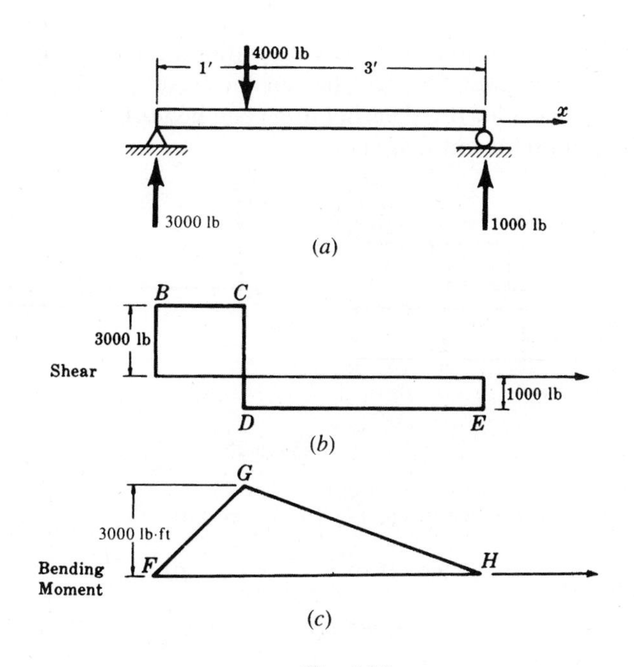

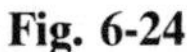
Fig. 6-24

6.9. Consider a cantilever beam loaded only by the couple of 200 lb · ft applied as shown in Fig. 6-25(*a*). Using singularity functions, write equations for the shearing force and bending moment at any position in the beam and plot the shear and moment diagrams.

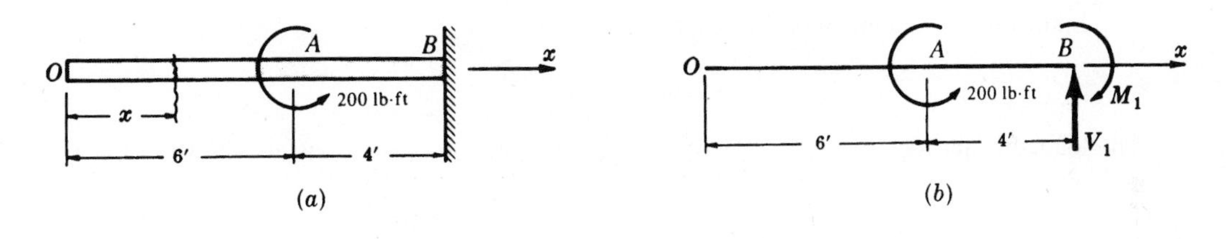

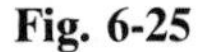
Fig. 6-25

A free-body diagram is shown in Fig. 6-25(*b*), where V_1 and M_1 denote the reactions of the supporting wall. From statics these are found to be $V_1 = 0$, $M_1 = 200$ lb · ft.

We introduce the coordinate system shown in which case the shearing force everywhere is

$$V = 0 \qquad (1)$$

In writing the expression for bending moment, working from left to right it is clear that there is no bending moment to the left of point A. At A the applied load of 200 lb · ft tends to bend the portion AB into a curvature that is concave downward, which according to our sign convention is negative bending. Thus the bending moment anywhere in the beam is

$$M = -200\langle x - 6\rangle^0 \text{ lb} \cdot \text{ft} \qquad (2)$$

Plots of (*1*) and (*2*) appear in Fig. 6-26.

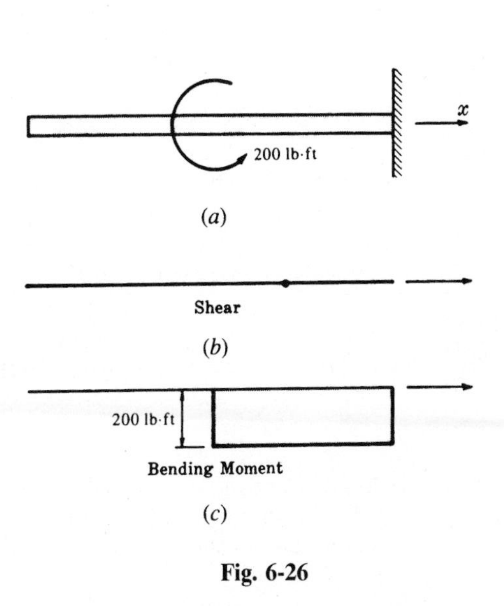

Fig. 6-26

6.10. Consider a cantilever beam loaded by a concentrated force at the free end together with a uniform load distributed over the right half of the beam [see Fig. 6-27(*a*)]. Using singularity functions, write equations for the shearing force and bending moment at any point in the beam and plot the shear and moment diagrams.

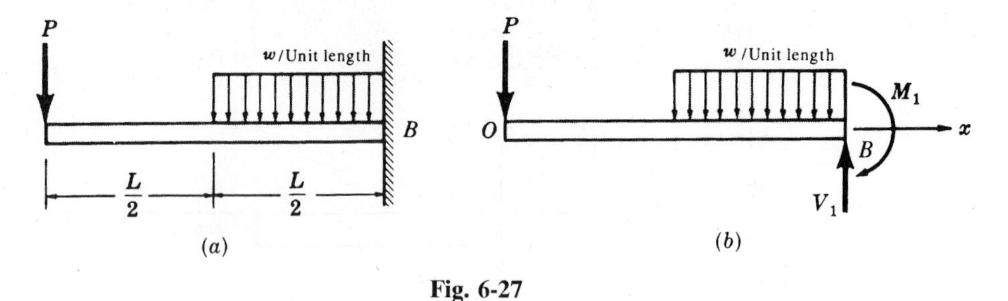

Fig. 6-27

A free-body diagram is shown in Fig. 6-27(*b*). From statics the wall reactions are found to be

$$V_1 = P + \frac{wL}{2} \qquad M_1 = PL + \frac{wL^2}{8}$$

although for the case of a cantilever it is not necessary to find these prior to writing shearing force and bending moment equations.

With the coordinate system shown, with origin at O, the effect of the concentrated force P as well as the distributed load is to produce negative shear according to our shearing force sign convention. Thus we may write

$$V = -P\langle x\rangle^0 - w\left\langle x - \frac{L}{2}\right\rangle^1 \tag{1}$$

which indicates shearing force at any position x if one remembers the definition of the bracketed term.

Likewise, the bending moment at any position x is

$$M = -P\langle x\rangle^1 - \frac{w}{2}\left\langle x - \frac{L}{2}\right\rangle^2 \tag{2}$$

The loaded beam together with plots of the shear and moment equations are shown in Fig. 6-28.

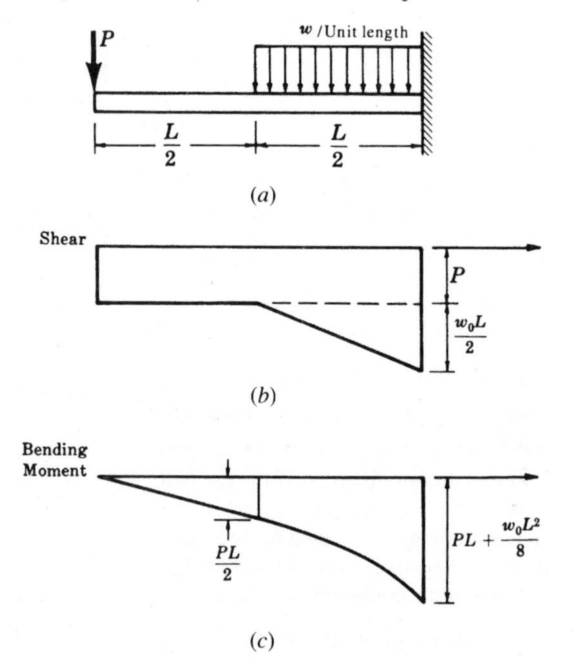

Fig. 6-28

6.11. In Fig. 6-29(a) a simply supported beam is loaded by the couple of 1 kN · m. Using singularity functions, write equations for the shearing force and bending moment at any point in the beam and plot the shear and moment diagrams.

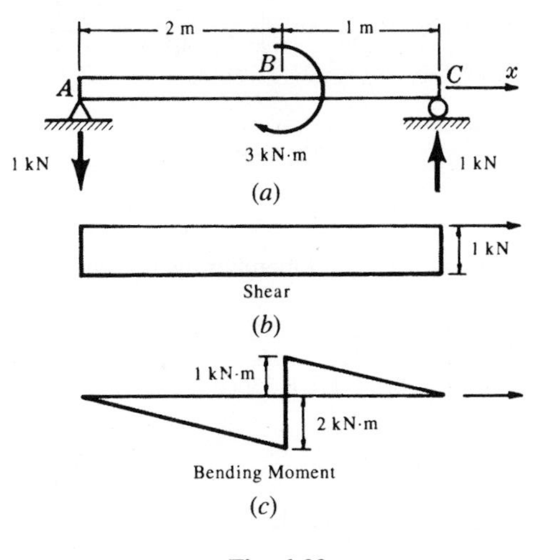

Fig. 6-29

The beam is loaded by one couple, and the only possible manner in which equilibrium may be created is for the reactions R at the supports A and C to constitute another couple. Thus, these reactions appear as in Fig. 6-29(b). For equilibrium,

$$\Sigma M_A = 3R - 3 = 0 \qquad \text{from which} \qquad R = 1\ \text{kN}$$

Thus the two forces R shown constitute the reactions necessary for equilibrium.

Inspection of the problem reveals that between A and B the shearing force is negative (according to our sign convention shown in Fig. 6-7) and also the bending moment is negative from the same figure. Just as soon as we consider points on the beam to the right of B, that couple of 3 kN · m tends to produce bending which is concave upward, and thus positive from Fig. 6-7. Therefore the expressions for V and M are

$$V = -(1)\langle x\rangle^0 \qquad \text{kN}$$

$$M = -(1)\langle x\rangle^1 + 3\langle x-2\rangle^0 \qquad \text{kN}\cdot\text{m}$$

Shear and moment diagrams are plotted in Figs. 6-29(b) and 6-29(c). From these it is evident that when a couple acts on a bar the bending moment diagram exhibits an abrupt jump or discontinuity at the point where the couple is applied.

6.12. The overhanging beam AE is subject to uniform normal loadings in the regions AB and DE, together with a couple acting at the midpoint C as shown in Fig. 6-30. Using singularity functions, write equations for the shearing force and bending moment at any point in the beam and plot the shear and moment diagram.

To first determine the reactions, we have from statics

$$+\circlearrowright\ \Sigma M_B = (300\ \text{lb/ft})(1\ \text{ft})(0.5\ \text{ft}) + 150\ \text{lb}\cdot\text{ft} + R_D(3\ \text{ft}) - (300\ \text{lb/ft})(1\ \text{ft})(3.5\ \text{ft}) = 0 \qquad (1)$$

$$\Sigma F_y = -300\ \text{lb} + R_B + R_D - 300\ \text{lb} = 0 \qquad (2)$$

Solving,

$$R_D = 250\ \text{lb} \qquad \text{and} \qquad R_B = 350\ \text{lb} \qquad (3)$$

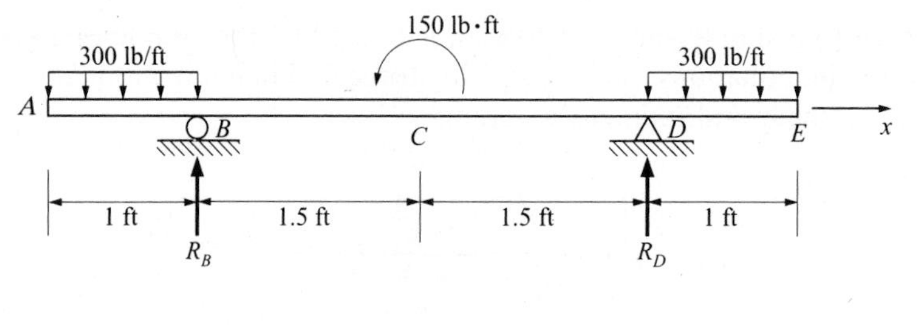

Fig. 6-30

For the coordinate system shown and remembering the definition of the singularity function, we may write

$$V = \overset{①}{-300\langle x\rangle^1} + \overset{②}{300\langle x-1\rangle^1} + \overset{③}{350\langle x-1\rangle^1} + \overset{④}{250\langle x-4\rangle^1} - \overset{⑤}{300\langle x-4\rangle^1} \tag{4}$$

$$M = \overset{⑥}{-300\langle x\rangle^1 \frac{\langle x\rangle^1}{2}} + \overset{⑦}{300\langle x-1\rangle^1 \frac{\langle x-1\rangle^1}{2}} + \overset{⑧}{350\langle x-1\rangle^1} - \overset{⑨}{150\langle x-2.5\rangle^0}$$

$$+ \overset{⑩}{250\langle x-4\rangle^1} - \overset{⑪}{300\langle x-4\rangle^1 \frac{\langle x-4\rangle^1}{2}} \tag{5}$$

Equations (*4*) and (*5*) each contain quantities designated by the numerals circled above the terms. Terms may be interpreted as follows for shearing force V:

I. The shearing force V acting in region OB of Fig. 6-30 is, for any value of the coordinate x in AB, simply the sum of all applied downward normal forces to the left of x, i.e., $300x$, which is term ①. Such forces tend to produce the type of displacement shown in Fig. 6-7(*d*), hence we must prefix the load $300\langle x\rangle$ by a negative sign.

II. Continuing, the first term ① in Eq. (*4*) holds for all values of x ranging from $x = 0$ to $x = 5$ ft. That is, the singularity functions are defined as being zero if the quantity in brackets $\langle\,\rangle$ is negative, but there is no way to specify an upper bound on the coordinate x shown in term ①. Consequently, we must annul the downward 300 lb/ft load to the right of point B and this may be accomplished by adding an upward (positive) uniform load to the right of B, i.e., for all values of $x > 1$ ft, which is term ②. But this upward uniform load has now annulled the actual downward uniform load in region DE. We will return to this shortly.

III. Immediately to the right of B the upward reaction R_B has a shear effect of 350 lb upward so that it tends to produce displacement such as shown in Fig. 6-7(*c*), which we term positive, hence the positive sign in term ③.

IV. The applied couple of 150 lb·ft has no force effect in any direction, hence does not appear in Eq. (*4*).

V. Immediately to the right of D the upward reaction R_D has a shear effect of 250 lb upward so that it tends to produce displacement such as shown in Fig. 6-7(*c*), which we term positive, hence the positive sign in term ④.

VI. As mentioned in (II), the true downward uniform load in DE has temporarily been annulled, hence we must introduce the term ⑤ to return it and make the external loading correct.

Equation (*4*) in terms of singularity functions now correctly specifies the vertical shear at all points on the beam from O to E. A plot of this is given below in Fig. 6-31(*a*).

In a nearly comparable manner, the bending moment from O to E may be written, except that now account must be taken of the applied moment of 150 lb·ft at C. The moment equation is given in (*5*) and a plot of it from O to E appears in Fig. 6-31(*b*).

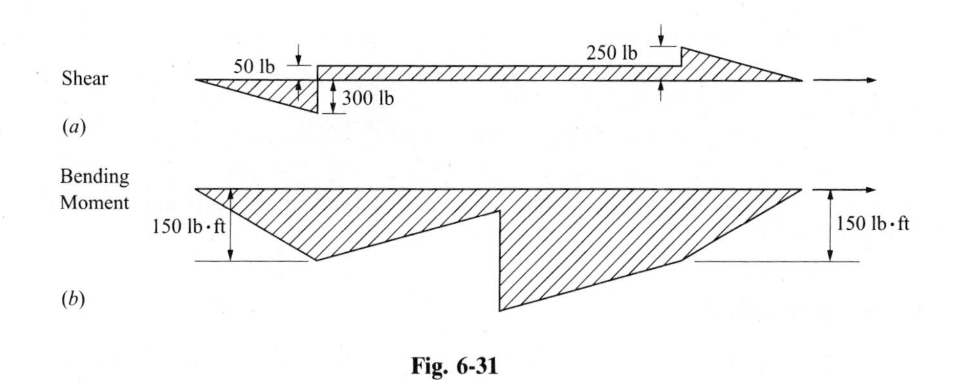

Fig. 6-31

6.13. The simply supported beam AD is subject to a uniform load over the segment BC together with a concentrated force applied at C as shown in Fig. 6-32. Using singularity functions, write equations for the shearing force and bending moment at any point in the beam and plot shear and moment diagrams.

The vertical reactions at A and D must first be determined from statics:

$$+\circlearrowleft \Sigma M_A = 4.5R_D - 12\text{ kN}(3.5\text{ m}) - (20\text{ kN})(3.5\text{ m}) = 0$$

$$R_D = 24.89\text{ kN}$$

$$\Sigma F_y = R_A + 24.89\text{ kN} - 12\text{ kN} - 20\text{ kN} = 0$$

$$R_A = 7.11\text{ kN}$$

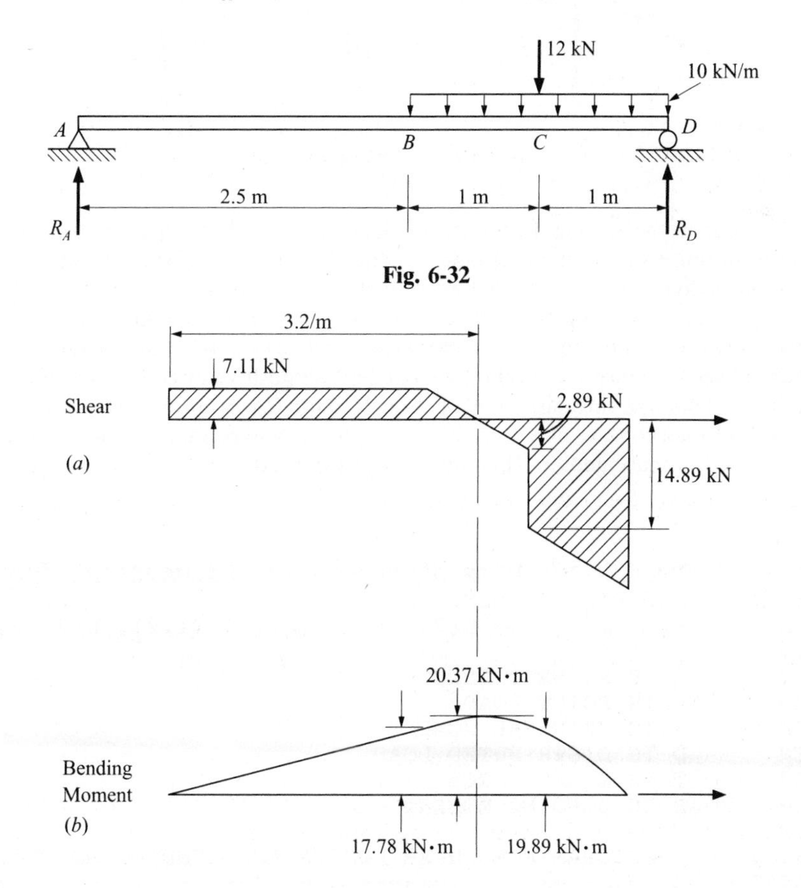

Fig. 6-32

Fig. 6-33

Introducing the coordinate system shown in Fig. 6-30 we can proceed as in Problem 6.12 and write

$$V = 7.11 - 10\langle x - 2.5\rangle^1 - 12\langle x - 3.5\rangle^0$$

$$M = 7.11\langle x\rangle^1 - 10\langle x - 2.5\rangle^1 \frac{\langle x - 2.5\rangle^1}{2} - 12\langle x - 3.5\rangle^1$$

From these equations the shear and moment diagrams may be plotted as shown in Figs. 6-33(a) and (b).

Computer Implementation

6.14. Consider a straight beam simply supported at any two points. Loading is by a system of concentrated forces, couples, and distributed loads that may (a) be uniform along a portion of the beam length, or (b) increase (or decrease) linearly. Write a computer program in BASIC to determine shearing force and bending moment at significant locations in the beam.

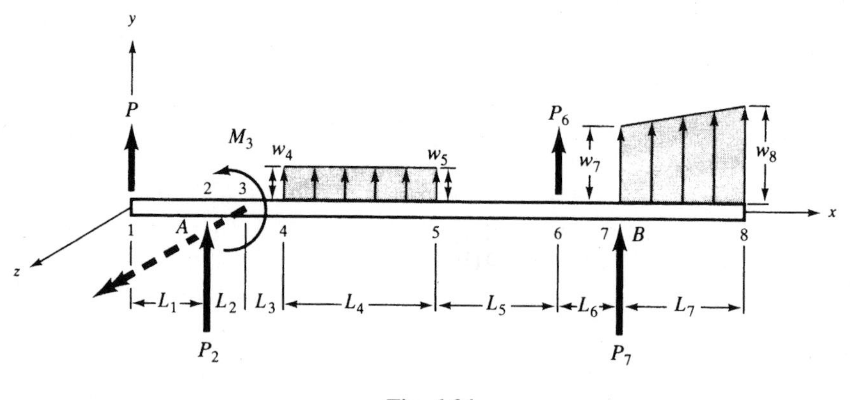

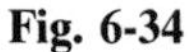

Fig. 6-34

Let us represent the loadings by the terminology of Fig. 6-34. It is first necessary to employ equations of statics to determine the reactions at points A and B. Next, we introduce numbers $1, 2, \ldots$ to designate points of application of concentrated forces (including reactions), moments, and left and right end coordinates of distributed loads. Positive directions of all such loads are indicated in Fig. 6-34. The applied moment M_3 is taken positive in the direction indicated because its vector representation (shown by the double-headed vector) is parallel to the z-axis and in the positive direction of that axis.

Use of the method of singularity functions leads to the BASIC program listed below. If more detailed information is needed concerning values of shearing forces and bending moment between number points, one may merely introduce additional points wherever desired.

```
00100 REM     THIS PROGRAM IS DEVELOPED TO EVALUATE THE SHEAR FORCES
00110 REM     AND BENDING MOMENTS.
00120 DIM     S(20), P(21), E(21), D(20,2), T(21,2), B(21,2)
00130 REM
00140 REM     S IS SEGMENT LENGTH
00150 REM     P IS POINT LOAD
00160 REM     E IS EXTERNAL MOMENT
00170 REM     D IS DISTRIBUTED LOAD
00180 REM     T IS SHEAR FORCE
00190 REM     B IS BENDING MOMENT
00200 REM
00210 PRINT   " PROGRAM FOR SHEAR FORCES AND BENDING MOMENTS "
00220 PRINT   " ---------------------------------------------- "
00230 PRINT
```

```
00240 PRINT     " PLEASE ENTER THE NUMBER OF SEGMENTS: "
00245 INPUT       N
00250 PRINT
00260 PRINT     " PLEASE ENTER THE LENGTH OF EACH SEGMENT FROM LEFT TO RIGHT.
00270 FOR       I=1 TO N
00280 INPUT     S(I)
00290 NEXT      I
00300 PRINT
00310 PRINT     " PLEASE ENTER THE NUMBER OF POINT LOADS: "
00315 INPUT       N1
00320 PRINT
00330 FOR       I=1 TO N1
00340 PRINT     " LOCATIONS AND LOADS: "
00345 INPUT       I1, P(I1)
00350 NEXT      I
00360 PRINT
00370 PRINT     " ENTER THE NUMBER OF EXTERNAL MOMENTS: "
00375 INPUT       N2
00380 PRINT
00390 FOR       I=1 TO N2
00400 PRINT     " ENTER THE LOCATIONS AND MOMENTS: "
00405 INPUT       L, E(L)
00410 NEXT      I
00420 PRINT
00430 PRINT     " ENTER THE NO. OF DISTRIBUTED LOADED SEGMENTS: "
00435 INPUT      N3
00440 PRINT
00450 FOR       I=1 TO N3
00460 PRINT     " ENTER THE SEGMENT NO., LOADLEFT, LOADRIGHT "
00465 INPUT       N4, D(N4,1), D(N4,2)
00470 NEXT      I
00480 PRINT
00490 LET       T(1,2)=P(1)
00500 LET       B(1,2)=-E(1)
00510 FOR       I=1 TO N
00520 LET       T(I+1,1)=T(I,2)+(D(I,1)+D(I,2))*S(I)/2
00530 LET       T(I+1,2)=T(I+1,1)+P(I+1)
00540 LET       T2=((2*D(I,1)+D(I,2))*S(I)^2)/6
00550 LET       B(I+1,1)=B(I,2)+T(I,2)*S(I)+T2
00560 LET       B(I+1,2)=B(I+1,1)-E(I+1)
00570 NEXT      I
00580 PRINT
00590 PRINT "LOCATION","SHEARLEFT","SHEARRIGHT","MOMENTLEFT","MOMENTRIGHT"
00600 FOR       I=1 TO N+1
00610 PRINT I,T(I,1),T(I,2),B(I,1),B(I,2)
00620 NEXT      I
00630 END
```

"Adapted from a program in *Basic Problems for Applied Mechanics: Statics*, William Weaver, Jr., McGraw-Hill, New York, 1972."

6.15. Use the BASIC program of Problem 6.14 to determine significant shearing forces and bending moments in the simply supported beam shown in Fig. 6-35.

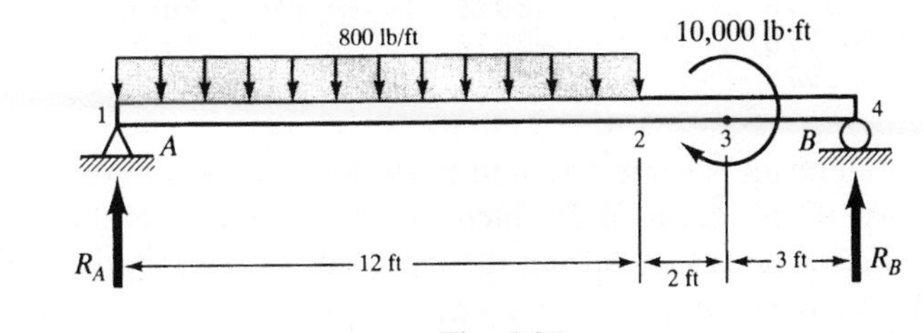

Fig. 6-35

It is first necessary to determine the reactions. From statics,

$$+\uparrow\Sigma M_A = -(9600\ \text{lb})\,(6\ \text{ft}) - 10{,}000\ \text{lb}\cdot\text{ft} + R_B(17\ \text{ft}) = 0$$

$$\Sigma F_y = R_A + R_B - 9600\ \text{lb} = 0$$

Solving,

$$R_A = 5624\ \text{lb} \qquad R_4 = 3976\ \text{lb}$$

Input to the program is

```
Number of segments:  3
Length of each segment:  12,2,3
Number of point loads (the reactions):  2
Location and magnitude of point loads:  1, 5624
                                        4, 3976
Number of external moments:  1
Location and magnitude of moments:  3, -10,000
Number of segments loaded by distributed load:  1
Segment number, load left, load right:  1, -800, -800
```

The computer output is shown below.

```
 PLEASE ENTER THE NUMBER OF SEGMENTS:
? 3

 PLEASE ENTER THE LENGTH OF EACH SEGMENT FROM LEFT TO RIGHT.
? 12
? 2
? 3

 PLEASE ENTER THE NUMBER OF POINT LOADS:
? 2

 LOCATIONS AND LOADS:
? 1,5624
 LOCATIONS AND LOADS:
? 4,3976

 ENTER THE NUMBER OF EXTERNAL MOMENTS:
? 1

 ENTER THE LOCATIONS AND MOMENTS:
? 3,-10000

 ENTER THE NO. OF DISTRIBUTED LOADED SEGMENTS:
? 1

 ENTER THE SEGMENT NO., LOADLEFT, LOADRIGHT
? 1,-800,-800
```

LOCATION	SHEARLEFT	SHEARRIGHT	MOMENTLEFT	MOMENTRIGHT
1	0	5624	0	0
2	-3976	-3976	9888	9888
3	-3976	-3976	1936	11936
4	-3976	0	8	8

6.16. A simply supported beam is subject to a uniform load of 2 kN/m over the region shown in Fig. 6-36. Use the BASIC program of Problem 6.14 to determine shearing forces and bending moments at significant points, including the midpoint of the length of the beam.

First, we must determine the end reactions from use of the statics equations. These are readily found to be 2 kN at each end.

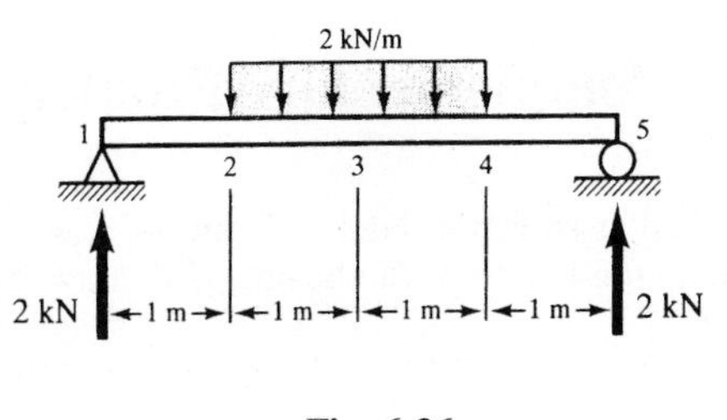

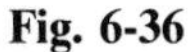

Fig. 6-36

For use of the program, it is necessary to number significant points along the length. These are usually points of application of applied loads. However, here we are asked for the shear and moment at the midpoint of the distributed load. Thus, we introduce an additional numbered point there with the result indicated in Fig. 6-36.

The input and output of the computer program are shown below.

```
 PLEASE ENTER THE NUMBER OF SEGMENTS:
? 4

 PLEASE ENTER THE LENGTH OF EACH SEGMENT FROM LEFT TO RIGHT.
? 1
? 1
? 1
? 1

 PLEASE ENTER THE NUMBER OF POINT LOADS:
? 2

 LOCATIONS AND LOADS:
? 1,2000
 LOCATIONS AND LOADS:
? 5,2000

 ENTER THE NUMBER OF EXTERNAL MOMENTS:
? 0

 ENTER THE NO. OF DISTRIBUTED LOADED SEGMENTS:
? 2

 ENTER THE SEGMENT NO., LOADLEFT, LOADRIGHT
? 2,-2000,-2000
 ENTER THE SEGMENT NO., LOADLEFT, LOADRIGHT
? 3,-2000,-2000
```

LOCATION	SHEARLEFT	SHEARRIGHT	MOMENTLEFT	MOMENTRIGHT
1	0	2000	0	0
2	2000	2000	2000	2000
3	0	0	3000	3000
4	-2000	-2000	2000	2000
5	-2000	0	0	0

```
SRU        0.129 UNTS.

 RUN COMPLETE.
```

Supplementary Problems

For the cantilever beams loaded as shown in Figs. 6-37 and 6-38, write equations for the shearing force and bending moment at any point along the length of the beam. Also, draw the shearing force and bending moment diagrams.

6.17.

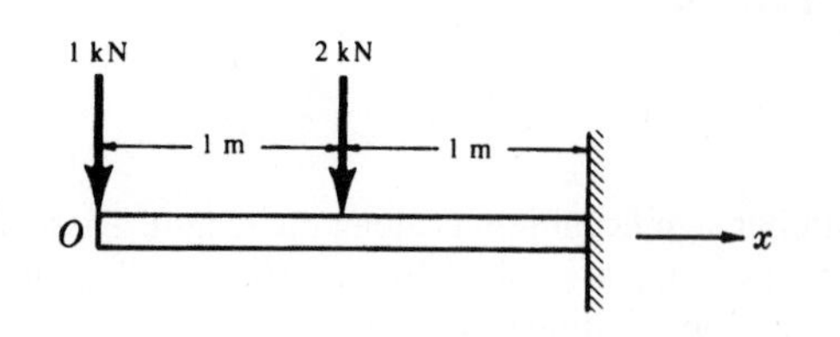

Ans.

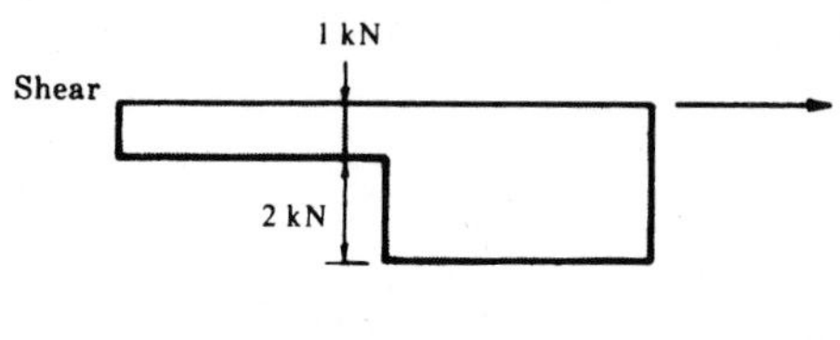

$V = -1\text{ kN}$ for $0 < x < 1\text{ m}$

$V = -3\text{ kN}$ for $1 < x < 2\text{ m}$

$M = -x\text{ kN}\cdot\text{m}$ for $0 < x < 1\text{ m}$

$M = -x - 2(x-1)\text{ kN}\cdot\text{m}$ for $1 < x < 2\text{ m}$

Bending Moment

1 kN·m

4 kN·m

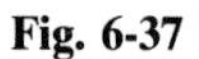

Fig. 6-37

6.18.

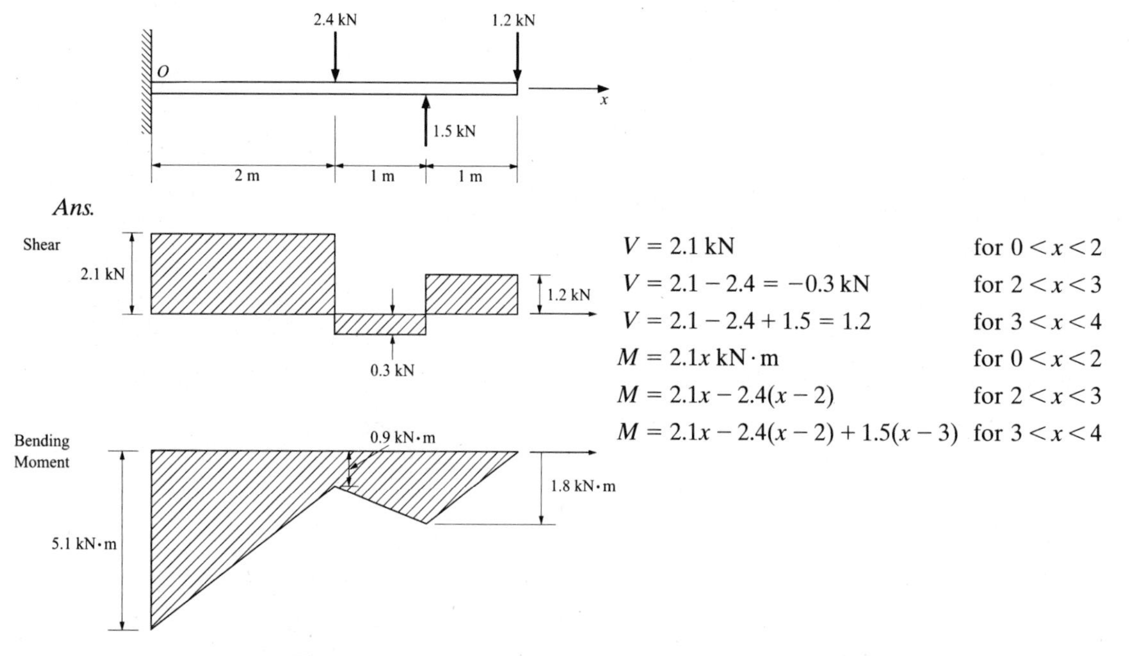

$V = 2.1\text{ kN}$ for $0 < x < 2$

$V = 2.1 - 2.4 = -0.3\text{ kN}$ for $2 < x < 3$

$V = 2.1 - 2.4 + 1.5 = 1.2$ for $3 < x < 4$

$M = 2.1x\text{ kN}\cdot\text{m}$ for $0 < x < 2$

$M = 2.1x - 2.4(x-2)$ for $2 < x < 3$

$M = 2.1x - 2.4(x-2) + 1.5(x-3)$ for $3 < x < 4$

Fig. 6-38

For the beams of Problems 6.19 through 6.25 simply supported at the ends and loaded as shown, write equations for the shearing force and bending moment at any point along the length of the beam. Also, draw the shearing force and bending moment diagrams.

6.19.

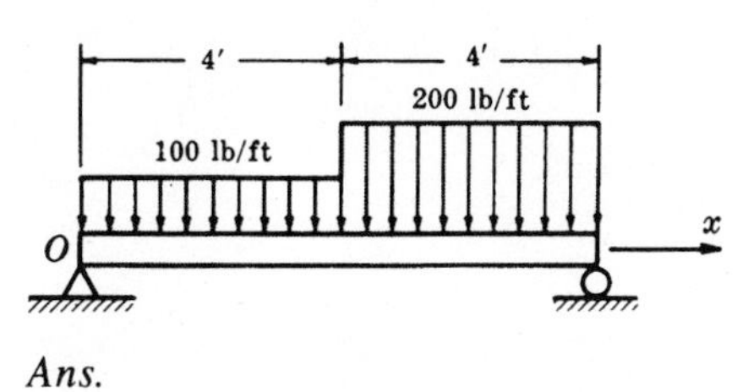

Ans.

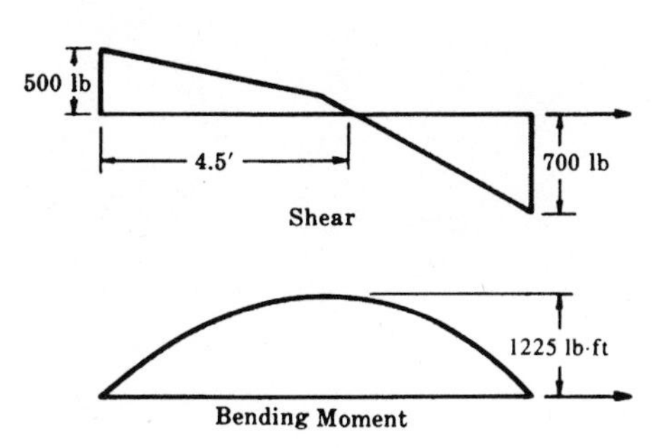

$V = 500 - 100x$ lb for $0 < x < 4$ ft

$V = 100 - 200(x - 4)$ lb for $4 < x < 8$ ft

$M = 500x - 50x^2$ lb · ft for $0 < x < 4$ ft

$M = 500x - 400(x - 2) - 100(x - 4)^2$ lb · ft for $4 < x < 8$ ft

Fig. 6-39

6.20.

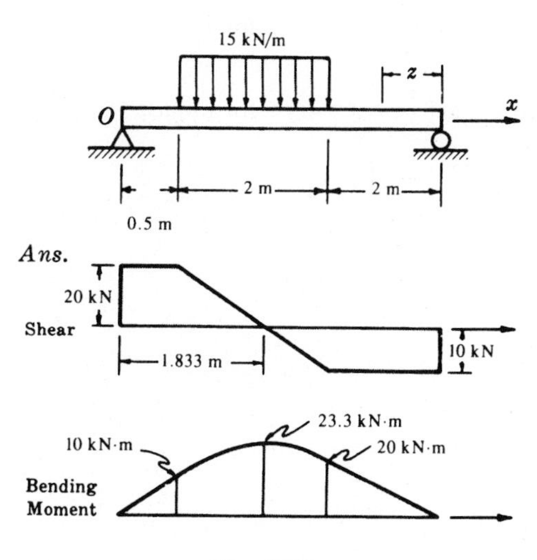

$V = 20$ kN for $0 < x < 0.5$ m

$V = 20 - 15(x - 0.5)$ for $0.5 < x < 2.5$ m

$V = -10$ kN for $2.5 < x < 4.5$ m

$M = 20x$ kN · m for $0 < x < 0.5$ m

$M = 20x - 7.5(x - 0.5)^2$ kN · m for $0.5 < x < 2.5$ m

$M = 10z$ kN · m for $0 < z < 2$ m

Fig. 6-40

6.21.

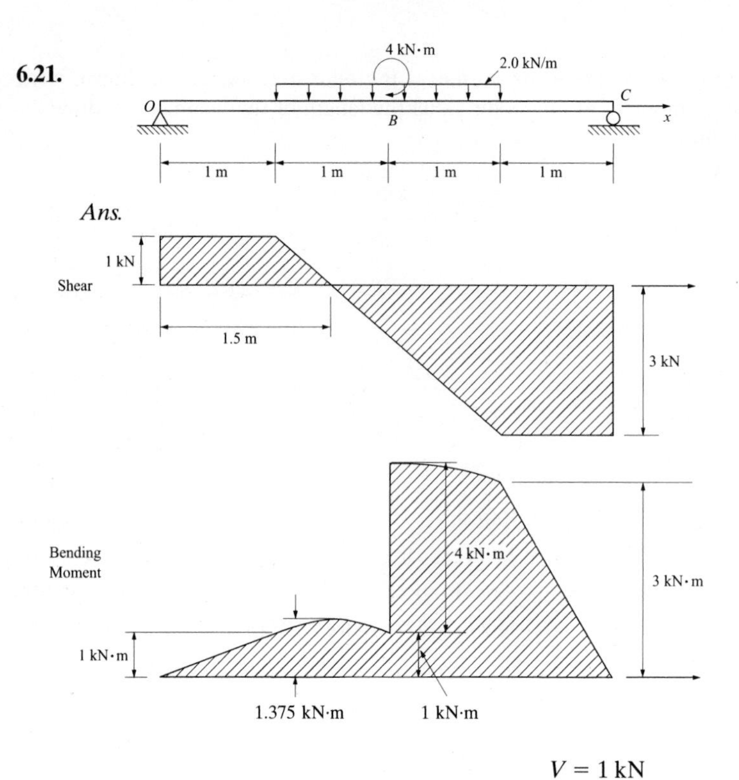

Fig. 6-41

$V = 1\text{ kN}$	for $0 < x < 1\text{ m}$
$V = 1 - 2(x - 1)\text{ kN}$	for $1 < x < 3$
$V = 3\text{ kN}$	for $3 < x < 4$
$M = 1x\text{ kN}\cdot\text{m}$	for $0 < x < 1\text{ m}$
$M = 1x - (x - 1)\left(\dfrac{x-1}{2}\right)$	for $1 < x < 2$
$M = 1x - 2(x - 1)\left(\dfrac{x-1}{2}\right) + 4$	for $2 < x < 3$
$M = 1x - 2(x - 1)\left(\dfrac{x-1}{2}\right) + 4$	for $3 < x < 4$

6.22.

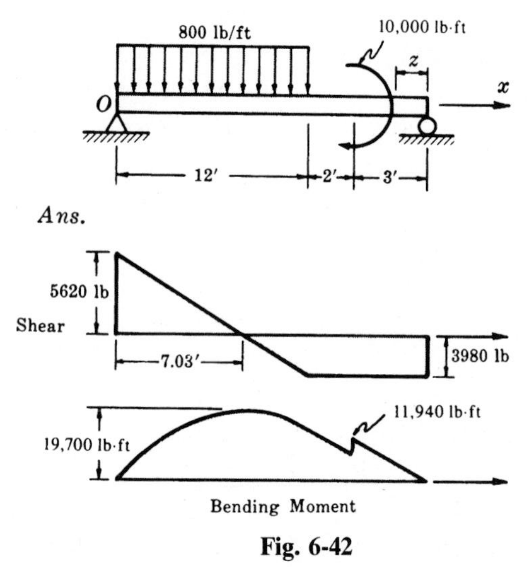

Fig. 6-42

$V = 5620 - 800x\text{ lb}$	for $0 < x < 12\text{ ft}$
$V = -3980\text{ lb}$	for $12 < x < 17\text{ ft}$
$M = 5620x - 400x^2\text{ lb}\cdot\text{ft}$	for $0 < x < 12\text{ ft}$
$M = 5620x - 9600(x - 6)\text{ lb}\cdot\text{ft}$	for $12 < x < 14\text{ ft}$
$M = 3980z$	for $0 < z < 3\text{ ft}$

6.23.

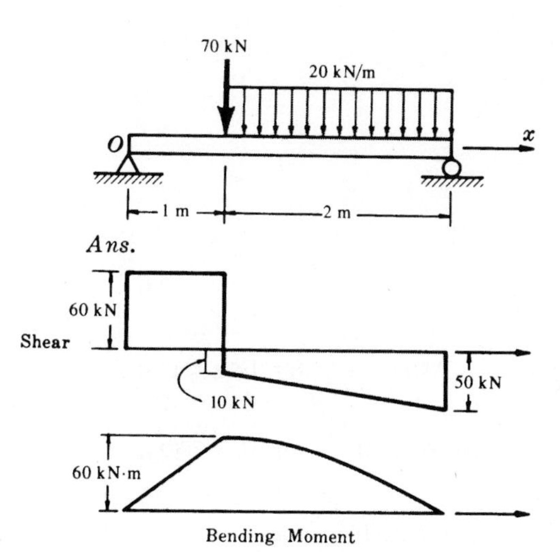

$V = 60 \text{ kN}$	for $0 < x < 1$ m
$V = 60 - 70 - 20(x - 1) \text{ kN}$	for $1 < x < 3$ m
$M = 60x \text{ kN} \cdot \text{m}$	for $0 < x < 1$ m
$M = 60x - 70(x - 3) - 10(x - 1)^2 \text{ kN} \cdot \text{m}$	for $1 < x < 3$ m

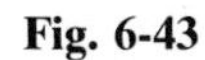
Fig. 6-43

6.24.

4000 lb·ft

3500 lb

O

x

2 ft 14 ft 2 ft

Ans.

Shear

187.5 lb

3312.5 lb

4000 lb·ft

6625 lb·ft

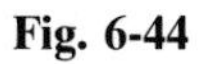
Fig. 6-44

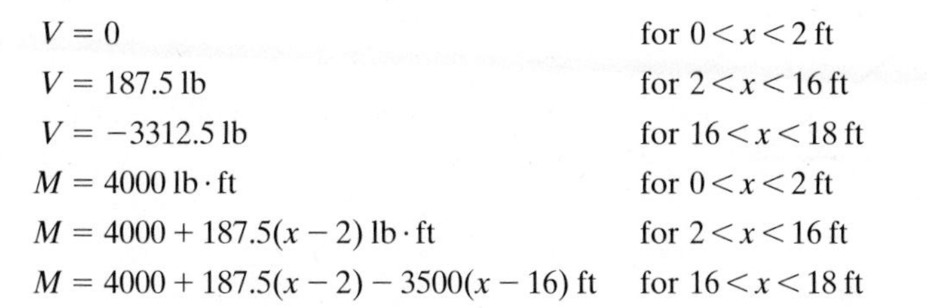

$V = 0$	for $0 < x < 2$ ft
$V = 187.5 \text{ lb}$	for $2 < x < 16$ ft
$V = -3312.5 \text{ lb}$	for $16 < x < 18$ ft
$M = 4000 \text{ lb} \cdot \text{ft}$	for $0 < x < 2$ ft
$M = 4000 + 187.5(x - 2) \text{ lb} \cdot \text{ft}$	for $2 < x < 16$ ft
$M = 4000 + 187.5(x - 2) - 3500(x - 16) \text{ ft}$	for $16 < x < 18$ ft

6.25.

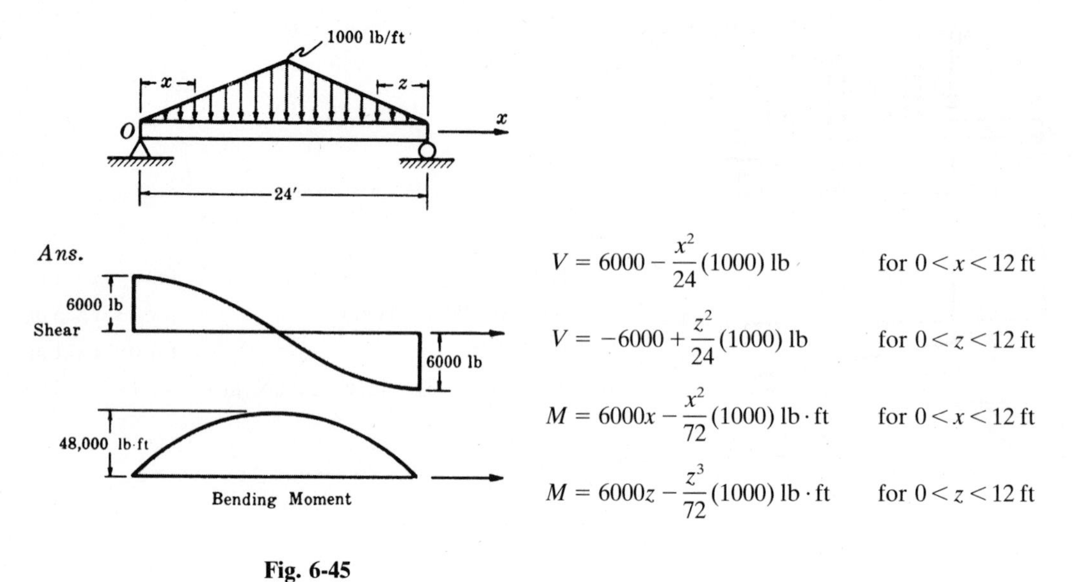

$$V = 6000 - \frac{x^2}{24}(1000)\ \text{lb} \qquad \text{for } 0 < x < 12\ \text{ft}$$

$$V = -6000 + \frac{z^2}{24}(1000)\ \text{lb} \qquad \text{for } 0 < z < 12\ \text{ft}$$

$$M = 6000x - \frac{x^2}{72}(1000)\ \text{lb}\cdot\text{ft} \qquad \text{for } 0 < x < 12\ \text{ft}$$

$$M = 6000z - \frac{z^3}{72}(1000)\ \text{lb}\cdot\text{ft} \qquad \text{for } 0 < z < 12\ \text{ft}$$

Fig. 6-45

For Problems 6.26 through 6.29 use singularity functions to write the equations for shearing force and bending moment at any point in the beam. Plot the corresponding diagrams.

6.26.

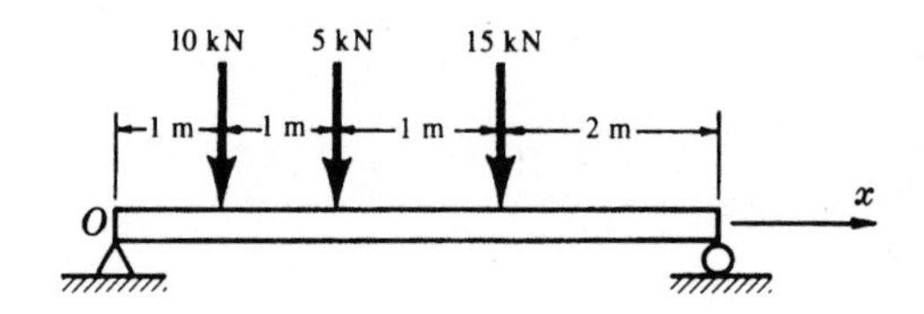

Ans.

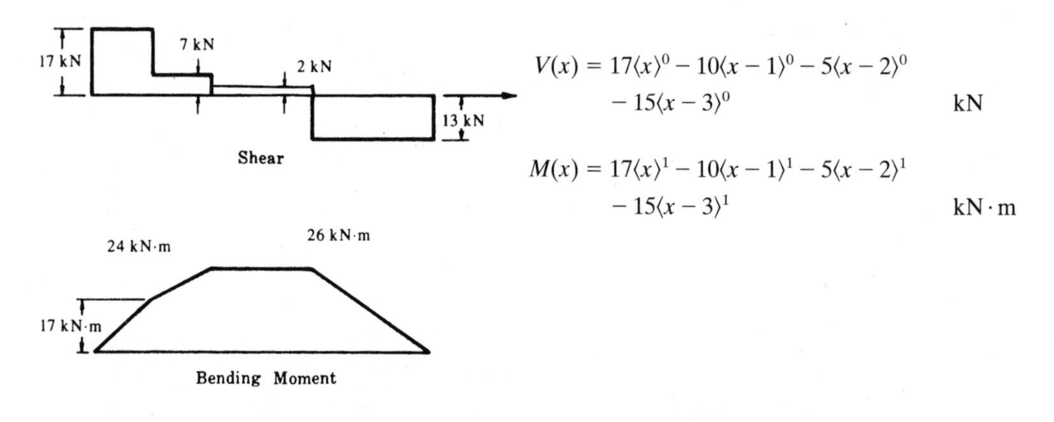

$$V(x) = 17\langle x\rangle^0 - 10\langle x-1\rangle^0 - 5\langle x-2\rangle^0 - 15\langle x-3\rangle^0 \qquad \text{kN}$$

$$M(x) = 17\langle x\rangle^1 - 10\langle x-1\rangle^1 - 5\langle x-2\rangle^1 - 15\langle x-3\rangle^1 \qquad \text{kN}\cdot\text{m}$$

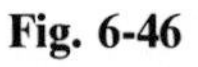

Fig. 6-46

6.27.

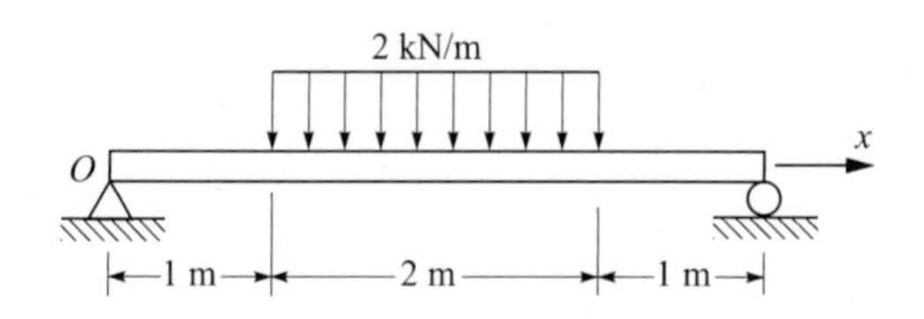

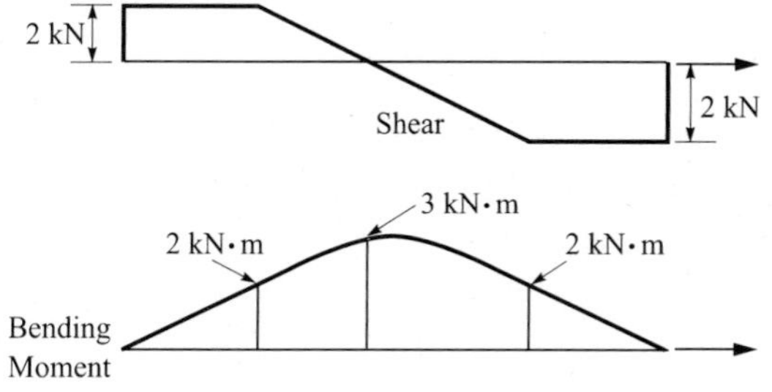

$$V(x) = 2\langle x\rangle^0 - 2\langle x-1\rangle^1 + 2\langle x-3\rangle^1 + 2\langle x-4\rangle^0 \text{ kN}$$
$$M(x) = 2\langle x\rangle^1 - 1\langle x-1\rangle^2 + 1\langle x-3\rangle^2 + 2\langle x-4\rangle^1 \text{ kN}\cdot\text{m}$$

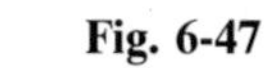

Fig. 6-47

6.28.

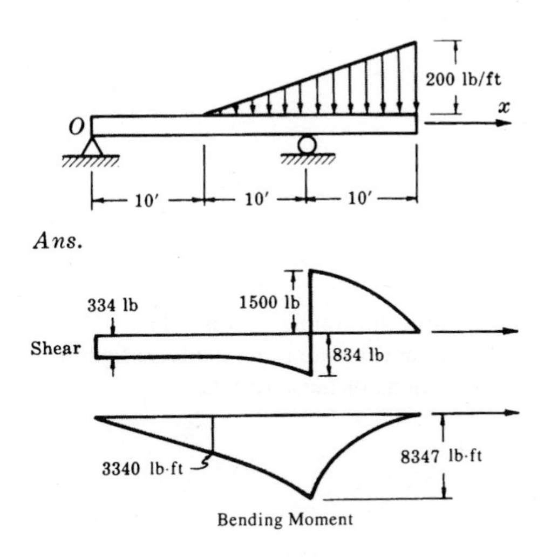

$$V\langle x\rangle = -334\langle x\rangle^0 - 5\langle x-10\rangle^2 + 2334\langle x-20\rangle^0$$
$$M\langle x\rangle = -334\langle x\rangle^1 - \tfrac{5}{3}\langle x-10\rangle^3 + 2334\langle x-20\rangle^1$$

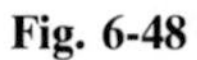

Fig. 6-48

6.29.

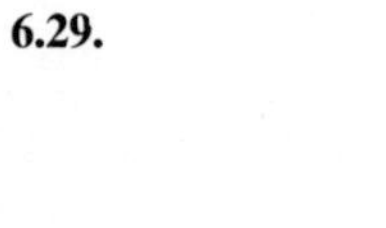

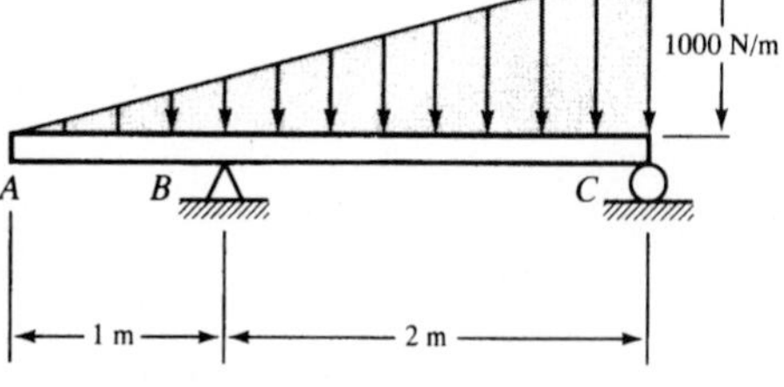

Ans.

(*a*)

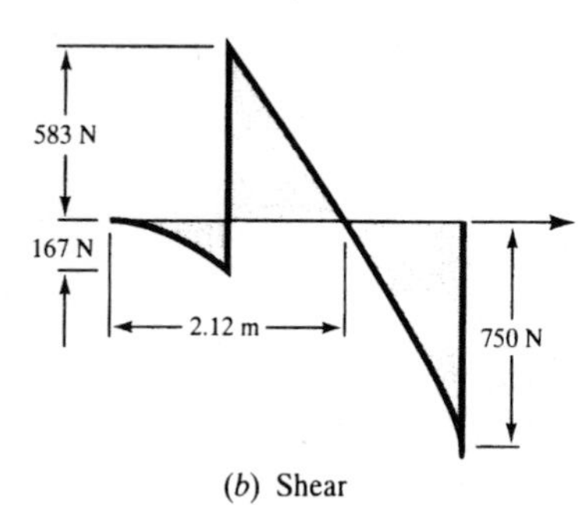

(*b*) Shear

$$V = -166.7\langle x\rangle^2 + 750\langle x-1\rangle^0$$
$$M = -55.6\langle x\rangle^3 + 750\langle x-1\rangle^1$$

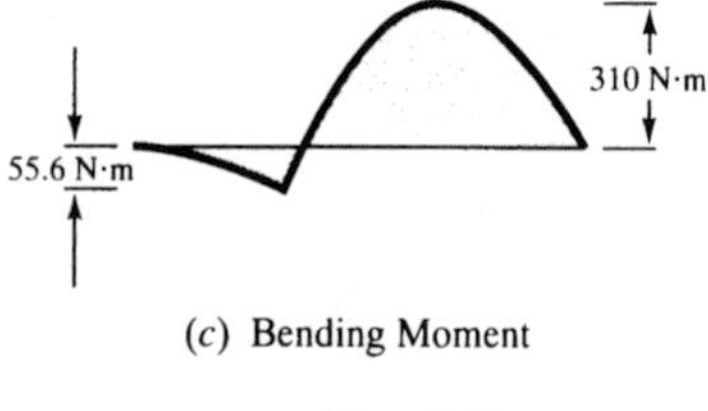

(*c*) Bending Moment

Fig. 6-49

6.30. A simply supported beam is subject to the uniform load together with the couple shown in Fig. 6-50. Use the BASIC program of Problem 6.14 to determine shearing forces and bending moments at significant points along the length of the beam. Draw approximate representations of these results.

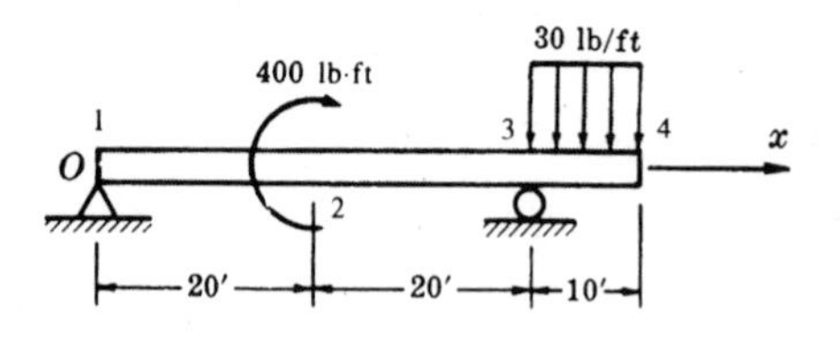

Fig. 6-50

Ans.

LOCATION	SHEARLEFT	SHEARRIGHT	MOMENTLEFT	MOMENTRIGHT
1	0	-47.5	0	0
2	-47.5	-47.5	-950	-550
3	-47.5	300	-1500	-1500
4	0	0	0	0

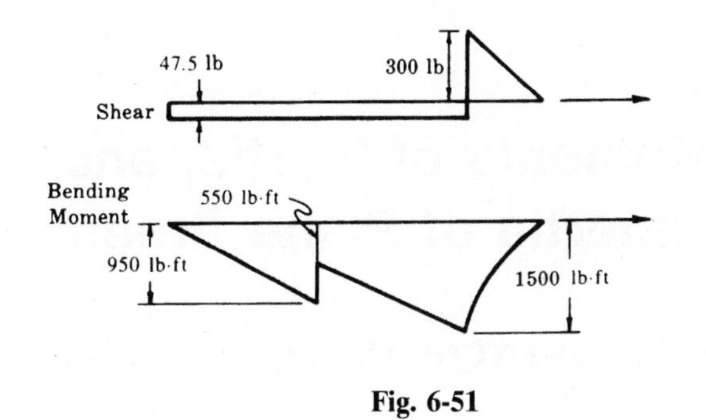

Fig. 6-51

6.31. A simply supported beam is subject to the uniform load together with the couple shown in Fig. 6-52. Use the BASIC program of Problem 6.14 to determine shearing forces and bending moments at significant points along the length of the beam.

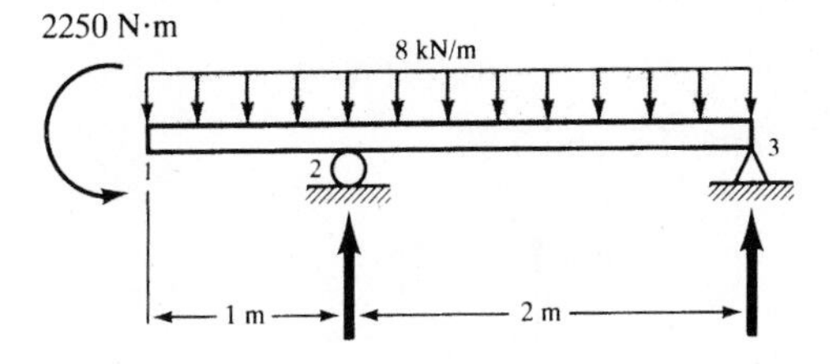

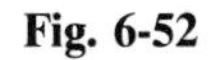

Fig. 6-52

Ans.

LOCATION	SHEARLEFT	SHEARRIGHT	MOMENTLEFT	MOMENTRIGHT
1	0	0	0	-2250
2	-8000	11125	-6250	-6250
3	-4875	0	0	0

Chapter 7

Centroids, Moments of Inertia, and Products of Inertia of Plane Areas

FIRST MOMENT OF AN ELEMENT OF AREA

The first moment of an element of area about any axis in the plane of the area is given by the product of the area of the element and the perpendicular distance between the element and the axis. For example, in Fig. 7-1 the first moment dQ_x of the element da about the x-axis is given by

$$dQ_x = y\,da$$

About the y-axis the first moment is

$$dQ_y = x\,da$$

For applications, see Problems 7.2 and 7.12.

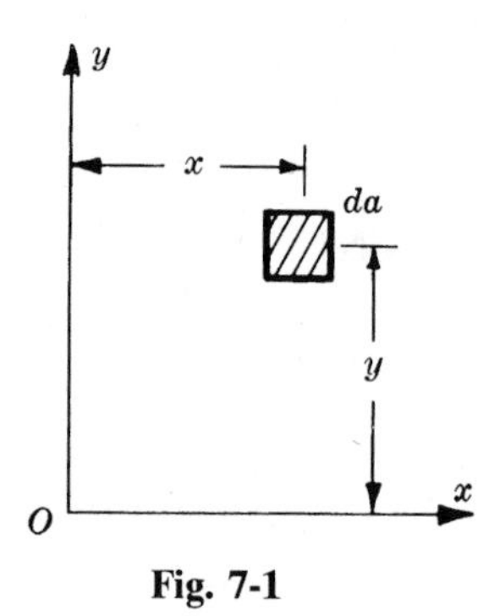

Fig. 7-1

FIRST MOMENT OF A FINITE AREA

The first moment of a finite area about any axis in the plane of the area is given by the summation of the first moments about that same axis of all the elements of area contained in the finite area. This is frequently evaluated by means of an integral. If the first moment of the finite area is denoted by Q_x, then

$$Q_x = \int dQ_x$$

For applications, see Problems 7.1 and 7.3.

CENTROID OF AN AREA

The centroid of an area is defined by the equations

$$\bar{x} = \frac{\int x\,da}{A} = \frac{Q_y}{A} \qquad \bar{y} = \frac{\int y\,da}{A} = \frac{Q_x}{A}$$

where A denotes the area. For a plane area composed of N subareas A_i each of whose centroidal coordinates $\bar{x}_i$ and $\bar{y}_i$ are known, the integral is replaced by a summation

$$\bar{x} = \frac{\sum_{i=1}^{N} \bar{x}_i A_i}{\sum_{i=1}^{N} A_i} \tag{7.1}$$

$$\bar{y} = \frac{\sum_{i=1}^{N} \bar{y}_i A_i}{\sum_{i=1}^{N} A_i} \tag{7.2}$$

For applications see Problems 7.2, 7.3, and 7.12.

The centroid of an area is the point at which the area might be considered to be concentrated and still leave unchanged the first moment of the area about any axis. For example, a thin metal plate will balance in a horizontal plane if it is supported at a point directly under its center of gravity.

The centroids of a few areas are obvious. In a symmetrical figure such as a circle or square, the centroid coincides with the geometric center of the figure.

It is common practice to denote a centroid distance by a bar over the coordinate distance. Thus $\bar{x}$ indicates the x-coordinate of the centroid.

SECOND MOMENT, OR MOMENT OF INERTIA, OF AN ELEMENT OF AREA

The second moment, or moment of inertia, of an element of area about any axis in the plane of the area is given by the product of the area of the element and the square of the perpendicular distance between the element and the axis. In Fig. 7-1, the moment of inertia dI_x of the element about the x-axis is

$$dI_x = y^2\, da$$

About the y-axis the moment of inertia is

$$dI_y = x^2\, da$$

SECOND MOMENT, OR MOMENT OF INERTIA, OF A FINITE AREA

The second moment, or moment of inertia, of a finite area about any axis in the plane of the area is given by the summation of the moments of inertia about that same axis of all of the elements of area contained in the finite area. This, too, is frequently found by means of an integral. If the moment of inertia of the finite area about the x-axis is denoted by I_x, then we have

$$I_x = \int dI_x = \int y^2\, da \tag{7.3}$$

$$I_y = \int dI_y = \int x^2\, da \tag{7.4}$$

For a plane area composed of N subareas A_i each of whose moment of inertia is known about the x- and y-axes, the integral is replaced by a summation

$$I_x = \sum_{i=1}^{N} (I_x)_i \qquad I_y = \sum_{i=1}^{N} (I_y)_i$$

For applications, see Problems 7.4, 7.6, 7.7, 7.8, 7.9, and 7.10.

UNITS

The units of moment of inertia are the fourth power of a length, in^4 or m^4.

PARALLEL-AXIS THEOREM FOR MOMENT OF INERTIA OF A FINITE AREA

The parallel-axis theorem for moment of inertia of a finite area states that the moment of inertia of an area about any axis is equal to the moment of inertia about a parallel axis through the centroid of the area plus the product of the area and the square of the perpendicular distance between the two axes. For the area shown in Fig. 7-2, the axes x_G and y_G pass through the centroid of the plane area. The x- and y-axes are parallel axes located at distances x_1 and y_1 from the centroidal axes. Let A denote the area of the figure, I_{x_G} and I_{y_G} the moments of inertia about the axes through the centroid, and I_x and I_y the moments of inertia about the x- and y-axes. Then we have

$$I_x = I_{x_G} + A(y_1)^2 \tag{7.5}$$

$$I_y = I_{y_G} + A(x_1)^2 \tag{7.6}$$

This relation is derived in Problem 7.5. For applications, see Problems 7.6, 7.8, 7.11, and 7.12.

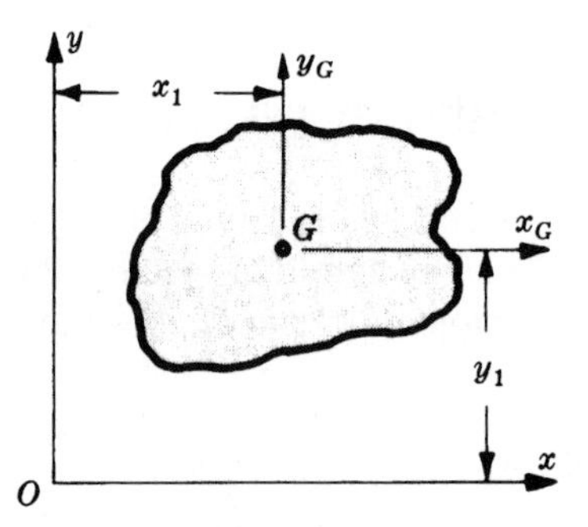

Fig. 7-2

RADIUS OF GYRATION

If the moment of inertia of an area A about the x-axis is denoted by I_x, then the radius of gyration r_x is defined by

$$r_x = \sqrt{\frac{I_x}{A}} \tag{7.7}$$

Similarly, the radius of gyration with respect to the y-axis is given by

$$r_y = \sqrt{\frac{I_y}{A}} \tag{7.8}$$

Since I is in units of length to the fourth power, and A is in units of length to the second power, then the radius of gyration has the units of length, say in or m. It is frequently useful for comparative purposes but has no physical significance. See Problems 7.10 and 7.11.

PRODUCT OF INERTIA OF AN ELEMENT OF AREA

The product of inertia of an element of area with respect to the x- and y-axes in the plane of the area is given by

$$dI_{xy} = xy\,da$$

where x and y are coordinates of the elemental area as shown in Fig. 7-1.

PRODUCT OF INERTIA OF A FINITE AREA

The product of inertia of a finite area with respect to the x- and y-axes in the plane of the area is given by the summation of the products of inertia about those same axes of all elements of area contained within the finite area. Thus

$$I_{xy} = \int xy\,da \tag{7.9}$$

From this, it is evident that I_{xy} may be positive, negative, or zero. For a plane area composed of N subareas A_i each of whose product of inertia is known with respect to specified x- and y-axes, the integral is replaced by the summation

$$I_{xy} = \sum_{i=1}^{N} (I_{xy})_i \tag{7.10}$$

For applications see Problems 7.13 and 7.15.

PARALLEL-AXIS THEOREM FOR PRODUCT OF INERTIA OF A FINITE AREA

The parallel-axis theorem for product of inertia of a finite area states that the product of inertia of an area with respect to the x- and y-axes is equal to the product of inertia about a set of parallel axes passing through the centroid of the area plus the product of the area and the two perpendicular distances from the centroid to the x- and y-axes. For the area shown in Fig. 7.2, the axes x_G and y_G pass through the centroid of the plane area. The x- and y-axes are parallel axes located at distances x_1 and y_1 from the centroidal axes. Let A represent the area of the figure and $I_{x_G y_G}$ be the product of inertia about the axes through the centroid. Then we have

$$I_{xy} = I_{x_G y_G} + Ax_1 y_1 \tag{7.11}$$

This relation is derived in Problem 7.14. For applications see Problems 7.15 and 7.16.

PRINCIPAL MOMENTS OF INERTIA

At any point in the plane of an area there exist two perpendicular axes about which the moments of inertia of the area are maximum and minimum for that point. These maximum and minimum values of moment of inertia are termed *principal moments of inertia* and are given by

$$(I_{x_1})_{\max} = \left(\frac{I_x + I_y}{2}\right) + \sqrt{\left(\frac{I_x - I_y}{2}\right)^2 + (I_{xy})^2} \tag{7.12}$$

$$(I_{x_1})_{\min} = \left(\frac{I_x + I_y}{2}\right) - \sqrt{\left(\frac{I_x - I_y}{2}\right)^2 + (I_{xy})^2} \tag{7.13}$$

These expressions are derived in Problem 7.17. For application, see Problem 7.18.

PRINCIPAL AXES

The pair of perpendicular axes through a selected point about which the moments of inertia of a plane area are maximum and minimum are termed *principal axes*. For application, see Problem 7.16.

The product of inertia vanishes if the axes are principal axes. Also, from the integral defining product of inertia of a finite area, it is evident that if either the x-axis, or the y-axis, or both, are axes of symmetry, the product of inertia vanishes. Thus, axes of symmetry are principal axes.

Type of section	Area	Location of centroid
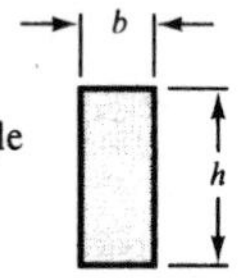 (a) Rectangle	bh	Geometric center
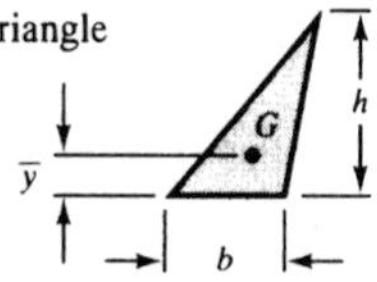 (b) Triangle	$\frac{1}{2}bh$	$\bar{y} = \frac{h}{3}$
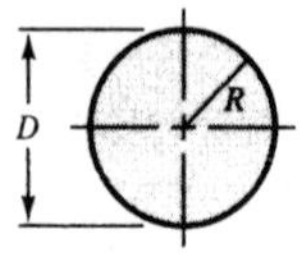 (c) Circle	πR^2 or $\frac{\pi}{4}D^2$	Geometric center
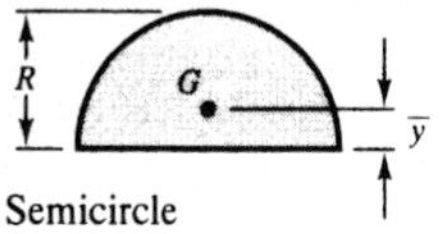 (d) Semicircle	$\frac{1}{2}\pi R^2$ or $\frac{\pi}{8}D^2$	$\bar{y} = \frac{4R}{3\pi}$
(e) Quadrant of circle 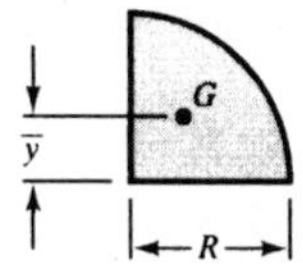	$\frac{\pi R^2}{4}$	$\bar{y} = \frac{4R}{3\pi}$
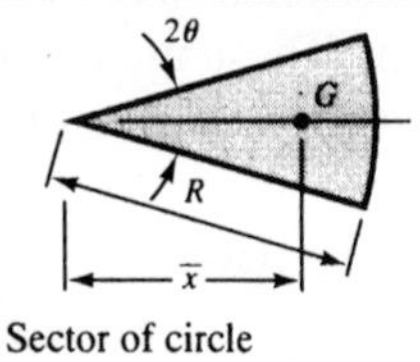 (f) Sector of circle	θR^2	$\bar{x} = \frac{2R \sin \theta}{3\theta}$

Fig. 7-3

INFORMATION FROM STATICS

Most texts on statics develop the properties of plane cross-sectional areas shown in Fig. 7-3 that will be needed in the present chapter. Those areas include (*a*) the rectangle, (*b*) the triangle, (*c*) the circle, (*d*) the semicircle, (*e*) the quadrant of a circle, and (*e*) the sector of a circle.

Solved Problems

7.1. The shaded area shown in Fig. 7-4 is bounded by the curves

$$y_1 = \sqrt[3]{x}$$

and

$$y_2 = x^3$$

Determine the y-coordinate of the centroid of this area which ends at (1,1).

We select an element that is horizontal (thus all points in this element have the same "y") and

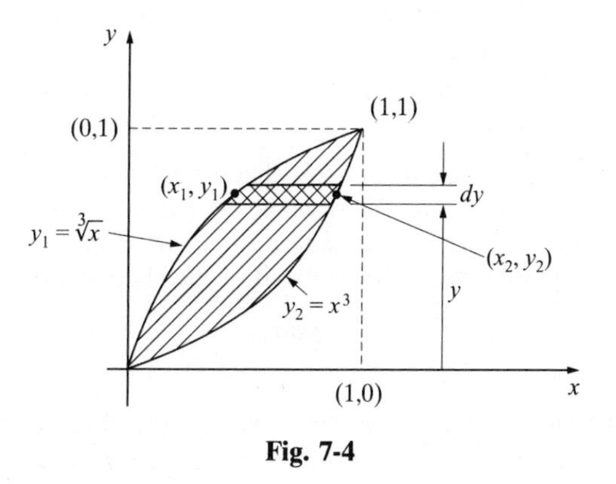

Fig. 7-4

extending from curve y_1 to y_2 as shown in Fig. 7-4. The height of the element is dy. From the definition of the location of the centroid,

$$\bar{y} = \frac{\int y\,da}{A}$$

we can write

$$da = (x_2 - x_1)\,dy$$

in which case we have

$$\bar{y} = \frac{\int_0^1 (x_2 - x_1)(y)(dy)}{\int_0^1 (x_2 - x_1)\,dy}$$

$$= \frac{\int_0^1 (y^{1/3} - y^3)(y)(dy)}{\int_0^1 (y^{1/3} - y^3)\,dy} = \frac{16}{70} = 0.229$$

Although the integrations involved in this problem are simple, for more complex problems one should resort to computers. A number of symbolic operations are available on proprietary software that permit easy and rapid treatments of such computations.

7.2. A circular cross section has a sector having a central angle 2θ removed as shown in Fig. 7-5. Locate the y-coordinate of the centroid of the shaded area.

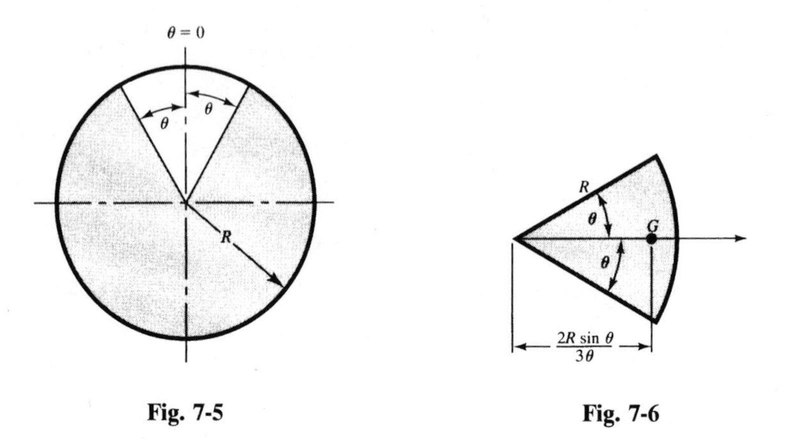

Fig. 7-5 **Fig. 7-6**

From the summary at the beginning of this chapter, we have for a sector of central angle 2θ the area and centroid given by θR^2 and $2R\sin\theta/3\theta$, respectively (see Fig. 7-6). The area of the entire circle having its centroid at its geometric center is also given in that summary.

By definition the y-coordinate of the centroid of the shaded area in Fig. 7-4 is given by

$$\bar{y} = \frac{\int y\,da}{A} \quad \text{or} \quad \frac{\Sigma\, y\,da}{A}$$

Here we consider the shaded area to be composed of the three components consisting of the lower semicircle ①, the upper semicircle ②, and the sector that has been removed ③. Thus the net shaded area is represented as shown in Fig. 7-7.

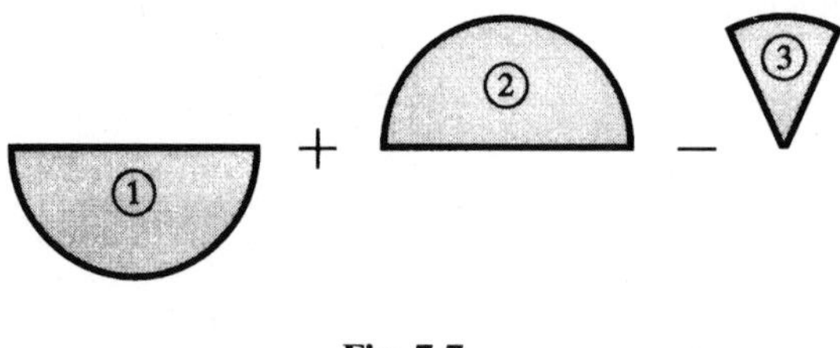

Fig. 7-7

Using these components in the finite summation (*7.1*), we have

$$\bar{y} = \frac{\overset{①}{\frac{\pi}{2}R^2\left(-\frac{4R}{3\pi}\right)} + \overset{②}{\frac{\pi}{2}R^2\left(\frac{4R}{3\pi}\right)} - \overset{③}{\theta R^2\left(\frac{2R}{3\theta}\sin\theta\right)}}{\pi R^2 - \theta R^2}$$

$$= -\frac{\frac{2}{3}(R\sin\theta)}{(\pi - \theta)}$$

7.3. A thin sheet of metal 600 mm by 1000 mm has its two upper corners folded over along the inclined lines AC and DF as shown in Fig. 7-8. In the regions bounded by the dotted lines, the metal thus becomes doubly thick. Determine the y-coordinate of the centroid of the folded sheet.

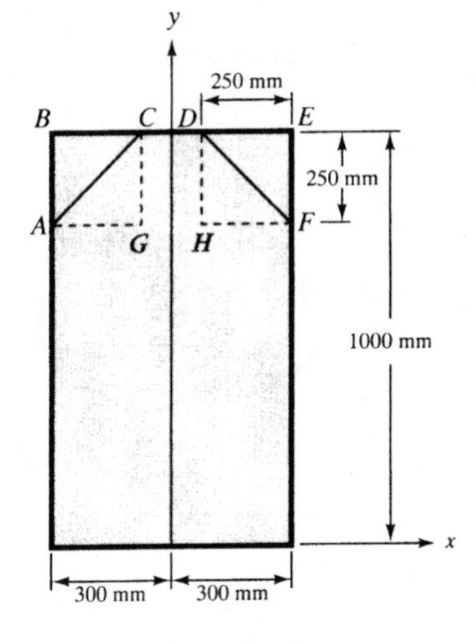

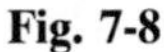
Fig. 7-8

By definition, the y-coordinate of the centroid is

$$\bar{y} = \frac{\int y\,da}{A} \quad \text{or} \quad \frac{\Sigma y_i A_i}{A}$$

where the numerator in each expression represents the first moment of the area about the x-axis. In the numerical evaluation, the triangles ABC and DEF have been removed but replaced by triangles ACG and DFH accounting for the double thickness. Thus we have

$$\bar{y} = \frac{(600)(1000)(500) - \overbrace{2\{\tfrac{1}{2}(250)(250)[1000 - \tfrac{250}{3}]\}}^{\triangle BCA} + \overbrace{2\{\tfrac{1}{2}(250)(250)[750 + \tfrac{250}{3}]\}}^{\triangle AGC}}{(600)(1000)}$$

$$= 491.3 \text{ mm}$$

7.4. Determine the moment of inertia of a rectangle about an axis through the centroid and parallel to the base.

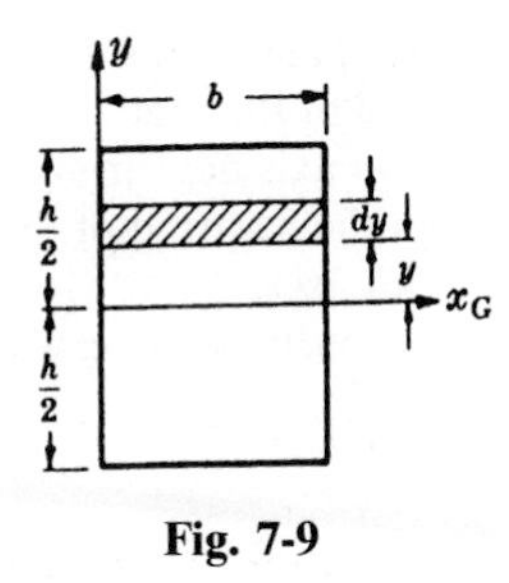

Fig. 7-9

Let us introduce the coordinate system shown in Fig. 7-9. The moment of inertia I_{x_G} about the x-axis passing through the centroid is given by $I_{x_G} = \int y^2\,da$. For convenience it is logical to select an element such that y is constant for all points in the element. The shaded area shown has this characteristic.

$$I_{x_G} = \int_{-h/2}^{h/2} y^2 b\,dy = b\left[\frac{y^3}{3}\right]_{-h/2}^{h/2} = \frac{1}{12}bh^3$$

This quantity has the dimension of a length to the fourth power, perhaps in^4 or m^4.

7.5. Derive the parallel-axis theorem for moments of inertia of a plane area.

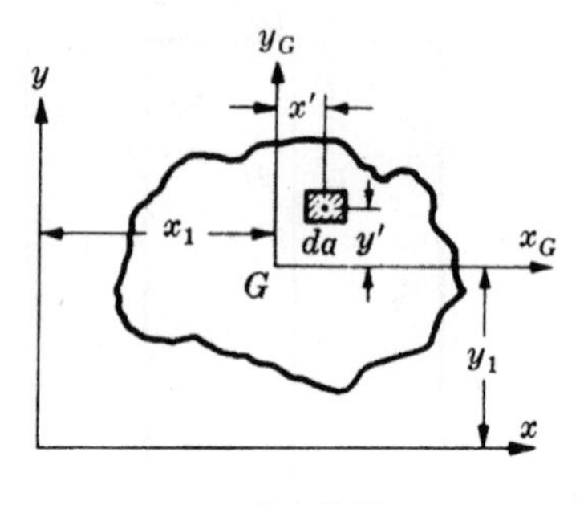

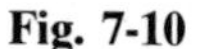
Fig. 7-10

Let us consider the plane area A shown in Fig. 7-10. The axes x_G and y_G pass through its centroid, whose location is presumed to be known. The axes x and y are located at known distances y_1 and x_1, respectively, from the axes through the centroid.

For the element of area da the moment of inertia about the x-axis is given by

$$dI_x = (y_1 + y')^2\,da$$

For the entire area A the moment of inertia about the x-axis is

$$I_x = \int dI_x = \int (y_1 + y')^2\,da = \int (y_1)^2\,da + 2\int y_1 y'\,da + \int (y')^2\,da$$

The first integral on the right is equal to $y_1^2 \int da = y_1^2 A$ because y_1 is a constant. The second integral on the right is equal to $2y_1 \int y'\,da = 2y_1(0) = 0$ because the axis from which y' is measured passes through the centroid of the area. The third integral on the right is equal to I_{x_G}, i.e., the moment of inertia of the area about the horizontal axis through the centroid. Thus

$$I_x = I_{x_G} + A(y_1)^2$$

A similar consideration in the other direction would show that

$$I_y = I_{y_G} + A(x_1)^2$$

This is the parallel-axis theorem for plane areas. It is to be noted that one of the axes involved in each equation must pass through the centroid of the area. In words, this may be stated as follows: The moment of inertia of an area with reference to an axis not through the centroid of the area is equal to the moment of inertia about a parallel axis through the centroid of the area plus the product of the same area and the square of the distance between the two axes.

The moment of inertia always has a positive value, with a minimum value for axes through the centroid of the area in question.

7.6. Find the moment of inertia of a rectangle about an axis coinciding with the base.

The coordinate system shown in Fig. 7-11 is convenient. By definition the moment of inertia about the x-axis is given by $I_x = \int y^2\,da$. For the element shown y is constant for all points in the element. Hence

$$I_x = \int_0^h y^2 b\,dy = b\left[\frac{y^3}{3}\right]_0^h = \frac{1}{3}bh^3$$

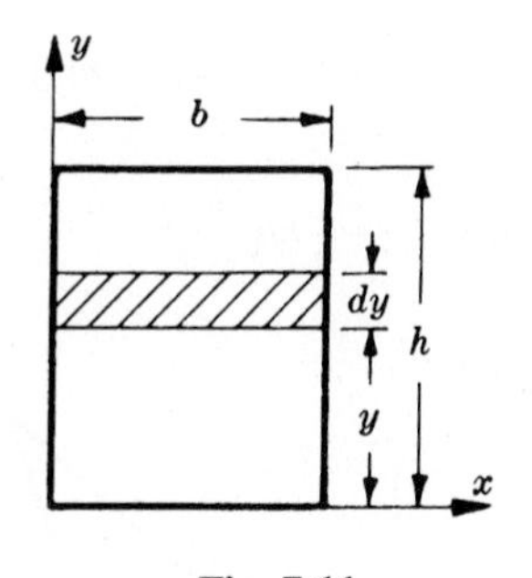

Fig. 7-11

This solution could also have been obtained by applying the parallel-axis theorem to the result obtained in Problem 7-4. This states that the moment of inertia about the base is equal to the moment of inertia about the horizontal axis through the centroid plus the product of the area and the square of the distance between these two axes. Thus

$$I_x = I_{x_G} + A(y_1)^2 = \frac{1}{12}bh^3 + bh\left(\frac{h}{2}\right)^2 = \frac{1}{3}bh^3$$

7.7. Determine the moment of inertia of a triangle about an axis coinciding with the base.

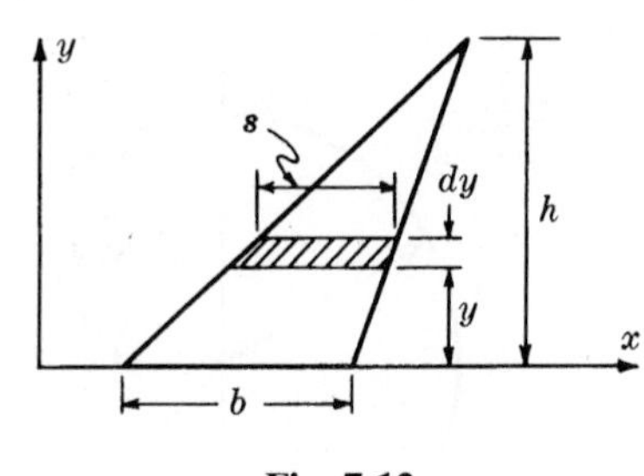

Fig. 7-12

Let us introduce the coordinate system shown in Fig. 7-12. The moment of inertia about the horizontal base is

$$I_x = \int y^2\, da$$

For the shaded element shown the quantity y is constant for all points in the element. Thus

$$I_x = \int_0^h y^2 s\, dy$$

By similar triangles, $s/b = (h - y)/h$, so that

$$I_x = \int_0^h y^2 \frac{b}{h}(h - y)\, dy = \frac{b}{h}\left[h\int_0^h y^2\, dy - \int_0^h y^3\, dy\right] = \frac{1}{12}bh^3$$

7.8. Determine the moment of inertia of a triangle about an axis through the centroid and parallel to the base.

Let the x_G-axis pass through the centroid and take the x-axis to coincide with the base as shown in Fig. 7-13.

From Fig. 7-3(b) the x_G-axis is located a distance of $h/3$ above the base. Also, the parallel-axis theorem tells us that

$$I_x = I_{x_G} + A(y_1)^2$$

But I_x was determined in Problem 7.7, and A and y_1 $(= h/3)$ are known. Hence we may solve for the desired unknown, I_{x_G}. Substituting,

$$\frac{1}{12}bh^3 = I_{x_G} + \frac{1}{2}bh\left(\frac{h}{3}\right)^2 \quad \text{or} \quad I_{x_G} = \frac{1}{36}bh^3$$

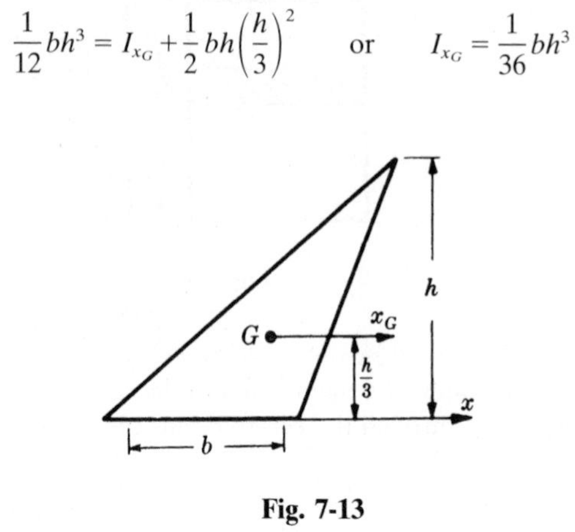

Fig. 7-13

7.9. Determine the moment of inertia of a circle about a diameter.

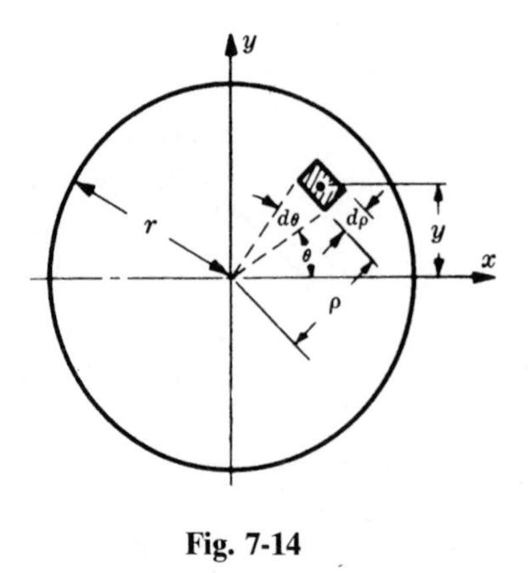

Fig. 7-14

Let us select the shaded element of area shown in Fig. 7-14, and work with the polar coordinate system. The radius of the circle is r.

To find I_x we have the definition $I_x = \int y^2\, da$.

But $y = \rho \sin\theta$ and $da = \rho\, d\theta\, d\rho$. Hence

$$I_x = \int_0^{2\pi}\int_0^r \rho^2 \sin^2\theta \rho\, d\theta\, d\rho = \int_0^{2\pi} \sin^2\theta\, d\theta \left[\frac{1}{4}\rho^4\right]_0^r$$

$$= \frac{r^4}{4}\int_0^{2\pi} \sin^2\theta\, d\theta = \frac{\pi r^4}{4}$$

If D denotes the diameter of the circle, then $D = 2r$ and $I_x = \pi D^4/64$. This is half the value of the polar moment of inertia of a solid circular area (see Problem 5.1).

The moment of inertia of a semicircular area about an axis coinciding with its base is

$$I_x = \frac{1}{2}\frac{\pi D^4}{64} = \frac{\pi D^4}{128}$$

7.10. Determine the moment of inertia about both the x- and y-axes as well as the corresponding radii of gyration of the plane area shown in Fig. 7-15.

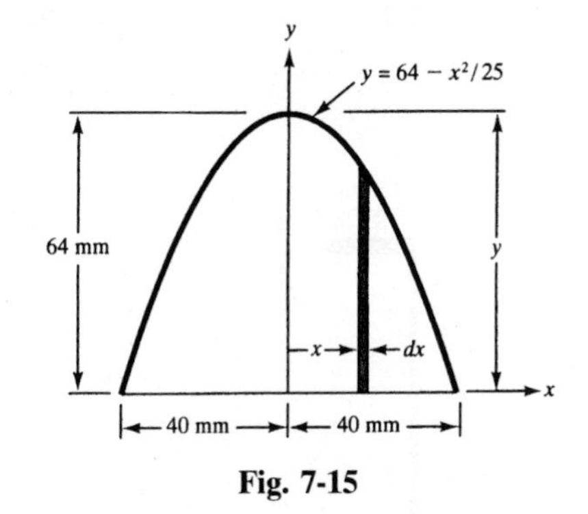

Fig. 7-15

Let us select the shaded element of width dx and altitude y shown in Fig. 7-15. From Problem 7.6 we have the moment of inertia of this element about the x-axis as

$$dI_x = \tfrac{1}{3}bh^3 = \tfrac{1}{3}(dx)y^3$$

Now, we must integrate over all values of x from -40 mm to $+40$ mm to account for all such elements. Thus,

$$\begin{aligned} I_x &= \int dI_x = \frac{1}{3}\int_{x=-40}^{x=40} y^3\,dx \\ &= \frac{2}{3}\int_{x=0}^{x=40}\left[64 - \frac{x^2}{25}\right]^3 dx \\ &= 3.197 \times 10^6\ \text{mm}^4 \end{aligned}$$

The same element may be employed to determine the moment of inertia of the entire area about the y-axis. By definition we have

$$dI_y = x^2\,da$$

which becomes

$$\begin{aligned} I_y &= \int dI_y = \int_{x=-40}^{x=40} x^2 y\,dx \\ &= 2\int_{x=0}^{x=40} x^2\left(64 - \frac{x^2}{25}\right) dx \\ &= 1.092 \times 10^6\ \text{mm}^4 \end{aligned}$$

To determine the radii of gyration, it is first necessary to find the area under the curve. It is given by

$$\begin{aligned} A &= \int y\,dx \\ &= 2\int_{x=0}^{x=40}\left(64 - \frac{x^2}{25}\right) dx = 3413\ \text{mm}^2 \end{aligned}$$

from which we have

$$r_x = \sqrt{\frac{I_x}{A}} = \sqrt{\frac{3.197 \times 10^6\ \text{mm}^4}{3413\ \text{mm}^2}} = 30.6\ \text{mm}$$

$$r_y = \sqrt{\frac{I_y}{A}} = \sqrt{\frac{1.092 \times 10^6\ \text{mm}^4}{3413\ \text{mm}^2}} = 17.9\ \text{mm}$$

7.11. Two channel sections are attached to a cover plate 16 in long by $\frac{1}{2}$ in thick, as indicated in Fig. 7-16. Locate the centroid of the cross section and determine the moment of inertia and radius of gyration about an axis parallel to the x-axis and passing through the centroid.

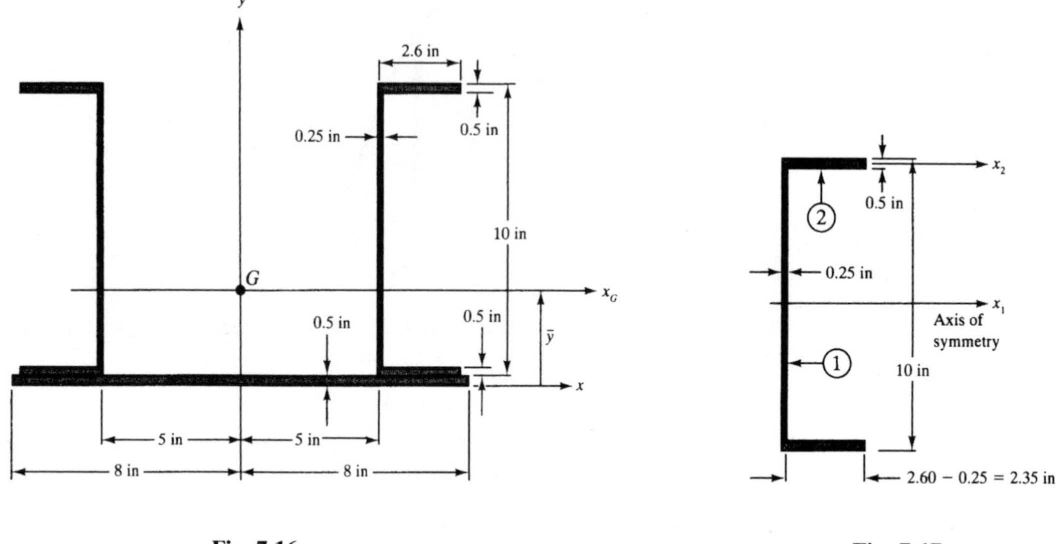

Fig. 7-16 **Fig. 7-17**

Let us first consider a single channel section, as shown in Fig. 7-17. The area of the cross section is

$$A = 2(\tfrac{1}{2})(2.60 - 0.25) + 10(\tfrac{1}{4}) = 4.85 \text{ in}^2$$

and from Problem 7.4 together with the parallel-axis theorem we have the moment of inertia of the channel about an axis parallel to the x-axis and passing through the centroid of the channel (the x_1-axis) as

$$\begin{aligned} I_{\text{ch}} &= \underset{①}{\tfrac{1}{12}(\tfrac{1}{4})(10)^3} + 2\{\underset{②}{\tfrac{1}{12}(2.35)(\tfrac{1}{2})^3} + \underset{③}{(2.35)(\tfrac{1}{2})(5 - \tfrac{1}{4})^2}\} \\ &= 73.90 \text{ in}^4 \end{aligned}$$

where term ① corresponds to the moment of inertia of the vertical rectangle about the x_1-axis, term ② corresponds to the moment of inertia of one horizontal rectangle about the x_2-axis through the centroid of the horizontal rectangle, and term ③ indicates the transfer term from the parallel axis theorem to pass from axis x_2 to axis x_1.

Now, we may write the moment of inertia of the entire assembly about the x-axis by applying the result of Problem 7.6 to the cover plate and applying the parallel axis theorem to I_{ch} to obtain

$$I_x = \tfrac{1}{3}(16)(\tfrac{1}{2})^3 + 2\{73.87 + 4.85(5.5)^2\} = 441.8 \text{ in}^4$$

The centroid of the cross section of the entire assembly is determined from the definition

$$\bar{y} = \frac{\Sigma\, y\, da}{A}$$

$$= \frac{\underset{③}{(16)(\tfrac{1}{2})(\tfrac{1}{4})} + \underset{④}{2[(4.85)(5.5)]}}{(16)(\tfrac{1}{2}) + 2[4.85]} = 3.13 \text{ in}$$

where the terms represented by ③ correspond to the horizontal cover plate and the terms numbered ④ correspond to the channels.

Now that we have located the centroidal axis x_G of the assembly, we may employ the parallel-axis theorem to transfer from the x- to the x_G-axis:

$$I_x = I_{x_G} + A(\bar{y})^2$$

$$441.8 \text{ in}^4 = I_{x_G} + (17.76 \text{ in}^2)(3.13 \text{ in})^2$$

$$I_{x_G} = 268.48 \text{ in}^4$$

The corresponding radius of gyration is

$$r_{x_G} = \sqrt{\frac{I_{x_G}}{A}} = \sqrt{\frac{268.48}{17.76}} = 3.89 \text{ in}$$

7.12. A plane section is in the form of an equilateral triangle, 200 mm on a side. From it is removed another equilateral triangle in such a manner that the width of the remaining section is 30 mm measured perpendicular to the sides of both equilateral triangles, as shown in Fig. 7-18. Determine the location of the centroid of the remaining (shaded) area as well as the moment of inertia about the axis through the centroid and parallel to the x-axis.

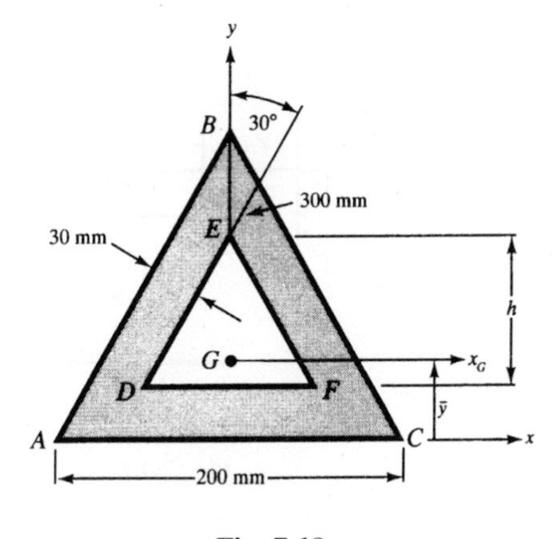

Fig. 7-18

It is necessary to determine the size of the inner triangle that has been removed. From the geometry of Fig. 7-18 it is evident that $BE = 60$ mm because of the 30° angle between BE and BC. Thus the altitude h of the "removed" triangle DEF is

$$h = 200 \cos 30 - 30 - 60 = 83.21 \text{ mm}$$

The length of a side of this triangle is

$$DF = \frac{83.21}{0.866} = 96.08 \text{ mm}$$

From symmetry the centroid lies on the y-axis and its location is found by the definition

$$\bar{y} = \frac{\int y\, da}{A} \quad \text{or} \quad \frac{\Sigma y\, dA}{A}$$

where the numerator represents the first moment of the area about the x-axis. Using the known location of the centroid of a triangle and its area, as given in the summary at the beginning of this chapter, we have

$$\bar{y} = \frac{\frac{1}{2}(200)(200 \cos 30)(\frac{200}{3} \cos 30) - \frac{1}{2}(96.08)(83.21)\{30 + 83.21/3\}}{\frac{1}{2}(200)(200 \cos 30) - \frac{1}{2}(96.08)(83.21)}$$

$$= 57.72 \text{ mm}$$

To determine the moment of inertia of the shaded area in Fig. 7-18, we begin by finding the moment of inertia of that area about the x-axis. This is accomplished by taking the moment of inertia of the outer triangle ABC about the x-axis using the result of Problem 7.7, then subtracting the moment of inertia of the inner triangle DEF about that same axis. This latter value is calculated by first determining the moment of inertia of DEF about an axis through the centroid of DEF using the result of Problem 7.8, then employing the parallel-axis theorem to transfer that value to the x-axis. Thus,

$$I_x = \tfrac{1}{12}(200)(200\cos 30)^3 - \{\tfrac{1}{36}(96.08)(83.21)^3 + \tfrac{1}{2}(96.08)(83.21)[30 + 83.21/3]^2\}$$
$$= 71.74 \times 10^6 \text{ mm}^4$$

Utilizing the parallel-axis theorem, we have

$$I_x = I_{x_G} + A(\bar{y})^2$$
$$71.74 \times 10^6 \text{ mm}^4 = I_{x_G} + \{\tfrac{1}{2}(200)(200\cos 30) - \tfrac{1}{2}(96.08)(83.21)\}(57.72 \text{ mm})^2$$
$$I_{x_G} = 27.35 \times 10^6 \text{ mm}^4$$

7.13. Determine the product of inertia of a rectangle with respect to the x- and y-axes indicated in Fig. 7-19.

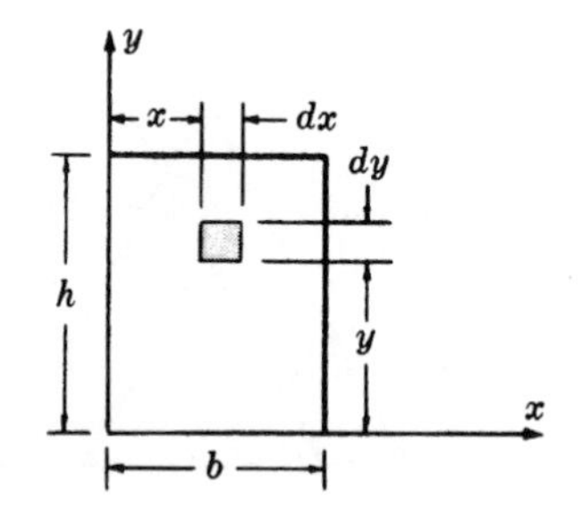

Fig. 7-19

We employ the definition $I_{xy} = \int xy\,da$ and consider the shaded element shown. Integrating,

$$I_{xy} = \int_{y=0}^{y=h} \int_{x=0}^{x=b} xy\,dx\,dy = \int_{y=0}^{y=h} \left[\frac{x^2}{2}\right]_0^b y\,dy$$
$$= \frac{b^2}{2}\left[\frac{y^2}{2}\right]_0^h = \frac{b^2h^2}{4} \qquad (1)$$

7.14. Derive the parallel-axis theorem for product of inertia of a plane area.

In Fig. 7-20, the axes x_G and y_G pass through the centroid of the area A. The axes x and y are located the known distances y_1 and x_1, respectively, from the axes through the centroid.

For the element of area da the product of inertia with respect to the x- and y-axes is given by

$$dI_{xy} = (x_1 + x')(y_1 + y')\,dx\,dy$$

For the entire area the product of inertia with respect to the x- and y-axes becomes

$$I_{xy} = \int dI_{xy} = \iint (x_1 + x')(y_1 + y')\,dx\,dy$$
$$= \iint x_1 y_1\,dx\,dy + \iint x' y_1\,dx\,dy + \iint x_1 y'\,dx\,dy + \iint x' y'\,dx\,dy$$

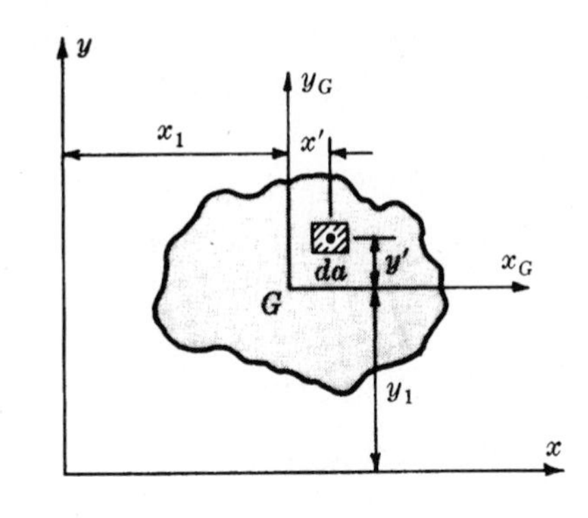

Fig. 7-20

The first integral on the right side equals $x_1 y_1 A$ since x_1 and y_1 are constants. The second and third integrals vanish because x' and y' are measured from the axes through the centroid of the area A. The fourth integral is equal to $I_{x_G y_G}$, that is, the product of inertia of the area with respect to axes through its centroid and parallel to the x- and y-axes. Thus, we have

$$I_{xy} = x_1 y_1 A + I_{x_G y_G} \tag{1}$$

This is the parallel-axis theorem for product of inertia of a plane area. It is to be noted that the x_G- and y_G-axes must pass through the centroid of the area. Also, x_1 and y_1 are positive only when the x- and y-coordinates have the location relative to the x_G-y_G system indicated in Fig. 7-20. Thus, care must be taken with regard to the algebraic signs of x_1 and y_1.

7.15. Determine I_{xy} for the angle section indicated in Fig. 7-21.

The area may be divided into the component rectangles as shown. For rectangle 1 we have, from (*1*) of Problem 7.13,

$$(I_{xy})_1 = \tfrac{1}{4}(10)^2(125)^2 = 39 \times 10^4 \text{ mm}^4$$

For rectangle 2 we employ (*1*) of Problem 7.14. The product of inertia of rectangle 2 about axes through its centroid and parallel to the x- and y-axes vanishes because these are axes of symmetry. Thus, for rectangle 2, $I_{x_G} = 0$. The parallel-axis theorem of Problem 7.14 thus becomes

$$(I_{xy})_2 = (42.5)(5)(65)(10) = 13.8 \times 10^4 \text{ mm}^4$$

For the entire angle section we thus have

$$I_{xy} = 39 \times 10^4 + 13.8 \times 10^4 = 52.8 \times 10^4 \text{ mm}^4$$

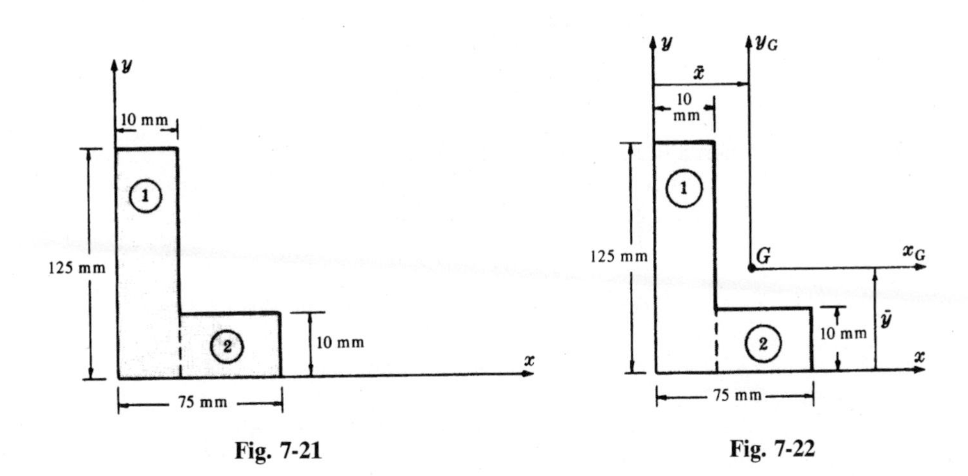

Fig. 7-21

Fig. 7-22

7.16. Determine the product of inertia of the angle section of Problem 7.15 with respect to axes parallel to the x- and y-axes and passing through the centroid of the angle section. See Fig. 7-22.

It is first necessary to locate the centroid of the area, that is, we must find $\bar{x}$ and $\bar{y}$. We have

$$\bar{x} = \frac{125(10)(5) + 65(10)(42.5)}{125(10) + 65(10)} = 17.8 \text{ mm}$$

$$\bar{y} = \frac{125(10)(62.5) + 65(10)(5)}{125(10) + 65(10)} = 42.8 \text{ mm}$$

Now we employ the parallel-axis theorem of Problem 7.13; that is,

$$I_{xy} = x_1 y_1 A + I_{x_G y_G}$$

In Problem 7.15 we found $I_{xy} = 52.8 \times 10^4 \text{ mm}^4$. Thus

$$52.8 \times 10^4 = 17.8(42.8)(1900) + I_{x_G y_G}$$

whence

$$I_{x_G y_G} = -92 \times 10^4 \text{ mm}^4$$

7.17. Consider a plane area A and assume that I_x, I_y, and I_{xy} are known. Determine the moments of inertia I_{x_1} and I_{y_1} as well as the product of inertia $I_{x_1 y_1}$ for the set of orthogonal axes x_1-y_1 oriented as shown in Fig. 7-23. Determine also the maximum and minimum values of I_{x1}.

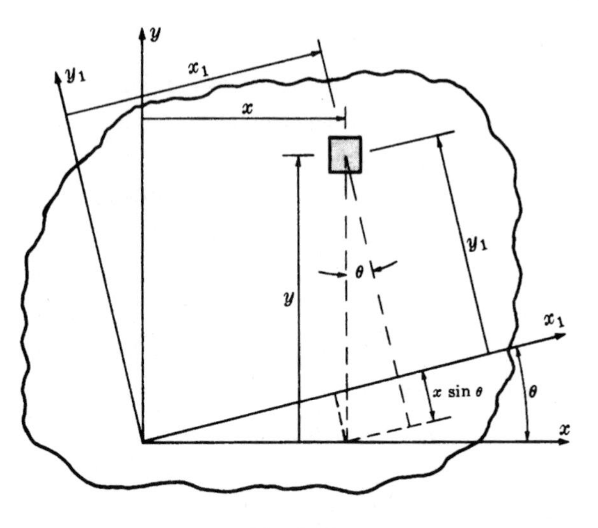

Fig. 7-23

The moment of inertia of the area with respect to the x_1-axis is

$$I_{x_1} = \int y_1^2 \, da = \int (y \cos\theta - x \sin\theta)^2 \, da$$

$$= \cos^2\theta \int y^2 \, da + \sin^2\theta \int x^2 \, da - 2 \sin\theta \cos\theta \int xy \, da$$

$$= I_x \cos^2\theta + I_y \sin^2\theta - 2I_{xy} \sin\theta \cos\theta$$

$$= I_x \left(\frac{1 + \cos 2\theta}{2} \right) + I_y \left(\frac{1 - \cos 2\theta}{2} \right) - I_{xy} \sin 2\theta$$

Or

$$I_{x_1} = \left(\frac{I_x + I_y}{2}\right) + \left(\frac{I_x - I_y}{2}\right)\cos 2\theta - I_{xy}\sin 2\theta \tag{1}$$

Analogously, I_{y_1} may be obtained from (*1*) by replacing θ by $\theta + \pi/2$ to yield

$$I_{y_1} = \left(\frac{I_x + I_y}{2}\right) - \left(\frac{I_x - I_y}{2}\right)\cos 2\theta + I_{xy}\sin 2\theta \tag{2}$$

The value of θ that renders I_{x_1} maximum or minimum is found by setting the derivative of Eq. (*1*) with respect to θ equal to zero. Thus, since I_x, I_y, and I_{xy} are constants we have from (*1*)

$$\frac{dI_{x_1}}{d\theta} = -(I_x - I_y)\sin 2\theta - 2I_{xy}\cos 2\theta = 0$$

Solving,

$$\tan 2\theta = -\frac{I_{xy}}{\left(\frac{I_x - I_y}{2}\right)} \tag{3}$$

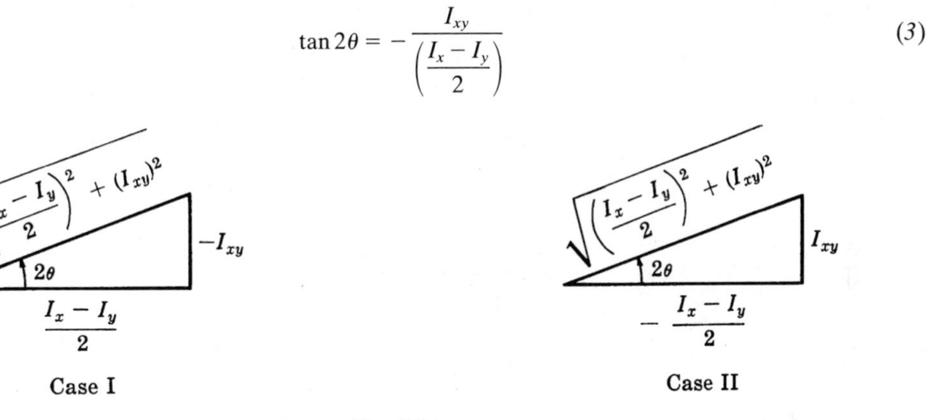

Fig. 7-24

Equation (*3*) has the convenient graphical interpretation shown in Cases I and II of Fig. 7-24.

If now the values of 2θ given by (*3*) are substituted into (*1*), we obtain

$$(I_{x_1})_{\substack{\max \\ \min}} = \left(\frac{I_x + I_y}{2}\right) \pm \sqrt{\left(\frac{I_x - I_y}{2}\right)^2 + (I_{xy})^2} \tag{4}$$

where the positive sign refers to Case I and the negative sign to Case II. These maximum and minimum values of moment of inertia correspond to axes defined by (*3*). The maximum and minimum values of moment of inertia are termed *principal moments of inertia* and the corresponding axes are termed *principal axes*.

We may now determine $I_{x_1y_1}$ from

$$\begin{aligned}
I_{x_1y_1} &= \int x_1 y_1\, da \\
&= \int (x\cos\theta + y\sin\theta)(y\cos\theta - x\sin\theta)\, da \\
&= \cos^2\theta \int xy\, da - \sin^2\theta \int xy\, da \\
&\qquad + \sin\theta\cos\theta \int y^2\, da - \sin\theta\cos\theta \int x^2\, da \\
&= I_{xy}(\cos^2\theta - \sin^2\theta) + (I_x - I_y)\sin\theta\cos\theta \\
&= \left(\frac{I_x - I_y}{2}\right)\sin 2\theta + I_{xy}\cos 2\theta
\end{aligned} \tag{5}$$

From (5), $I_{x_1y_1}$ vanishes if

$$\tan 2\theta = -\frac{I_{xy}}{\left(\frac{I_x - I_y}{2}\right)}$$

which is identical to condition (3). Since (3) defined principal axes, it follows that the product of inertia vanishes for principal axes.

7.18. A structural aluminum 6 Z 5.42 section has the nominal dimensions indicated in Fig. 7-25. Determine I_x, I_y, I_{xy} and also the maximum and minimum values of the moment of inertia with respect to axes through the point O.

The section may be divided into the component rectangles ①, ②, and ③ as indicated. The result obtained in Problem 7.4, together with the parallel-axis theorem given in Problem 7.5, may be used to determine I_x and I_y:

$$I_x = \tfrac{1}{12}(\tfrac{3}{8})(6)^3 + 2[\tfrac{1}{12}(3\tfrac{1}{8})(\tfrac{3}{8})^3 + (3\tfrac{1}{8})(\tfrac{3}{8})(2\tfrac{13}{16})^2] = 25.27 \text{ in}^4$$

$$I_y = \tfrac{1}{12}(6)(\tfrac{3}{8})^3 + 2[\tfrac{1}{12}(\tfrac{3}{8})(3\tfrac{1}{8})^3 + (\tfrac{3}{8})(3\tfrac{1}{8})(1\tfrac{3}{4})^2] = 9.08 \text{ in}^4$$

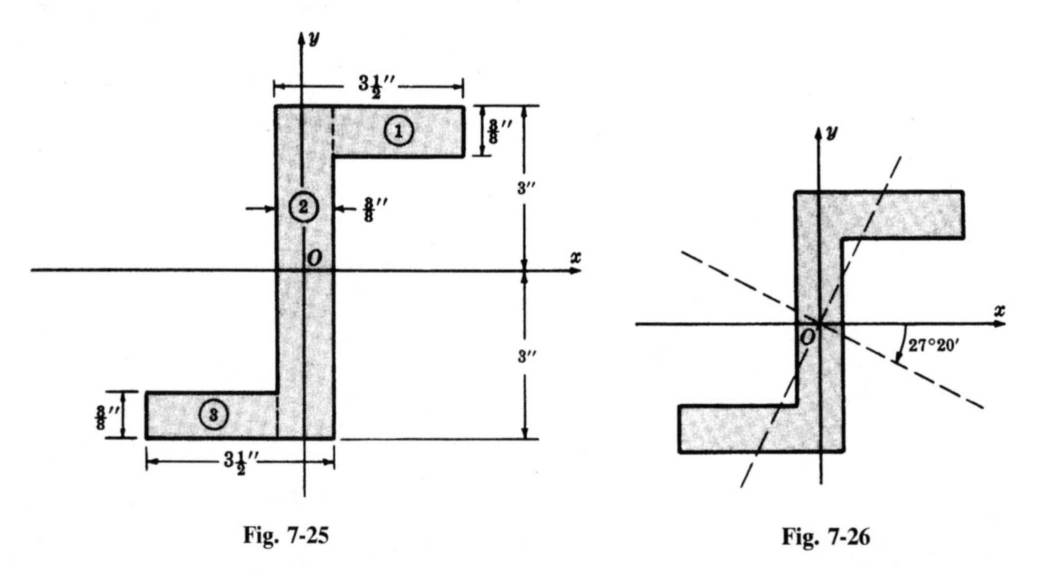

Fig. 7-25 Fig. 7-26

The product of inertia with respect to the x- and y-axes may be determined through use of the parallel-axis theorem for product of inertia as given in Problem 7.14. It is to be noted that the product of inertia of each of the component rectangles about axes through the centroid of each component and parallel to the x- and y-axes vanishes because these are axes of symmetry. Hence, from (1) of Problem 7.14 we have for the entire Z-section

$$I_{xy} = 2[(\tfrac{7}{4})(2\tfrac{13}{16})(3\tfrac{1}{8})(\tfrac{3}{8})] = 11.6 \text{ in}^4$$

The maximum and minimum values of moment of inertia with respect to axes through the point O may be found from (4) of Problem 7.17. From that equation

$$(I_{x_1})_{\substack{\max \\ \min}} = \left(\frac{I_x + I_y}{2}\right) \pm \sqrt{\left(\frac{I_x - I_y}{2}\right)^2 + (I_{xy})^2}$$

$$= \left(\frac{25.27 + 9.08}{2}\right) \pm \sqrt{\left(\frac{25.27 - 9.08}{2}\right)^2 + (11.6)^2}$$

$$(I_{x_1})_{\max} = 31.38 \text{ in}^4 \qquad (1)$$

$$(I_{x_1})_{\min} = 2.98 \text{ in}^4 \qquad (2)$$

The orientation of these principal moments of inertia is found from (3) of Problem 7.17 to be

$$\tan 2\theta = -\frac{I_{xy}}{\left(\frac{I_x - I_y}{2}\right)}$$

$$= -\frac{11.6}{\left(\frac{25.27 - 9.08}{2}\right)}$$

$$\theta = -27°20', \qquad -117°20' \tag{3}$$

The principal moments of inertia given in (1) and (2) correspond to the principal axes given by (3). These principal axes are represented by the dashed lines in Fig. 7-26.

Supplementary Problems

7.19. The structural channel section has welded to it a horizontal reinforcing plate as shown in cross section in Fig. 7-27. Determine the y-coordinate of the centroid of the composite section. *Ans.* $\bar{y} = 4.56$ in

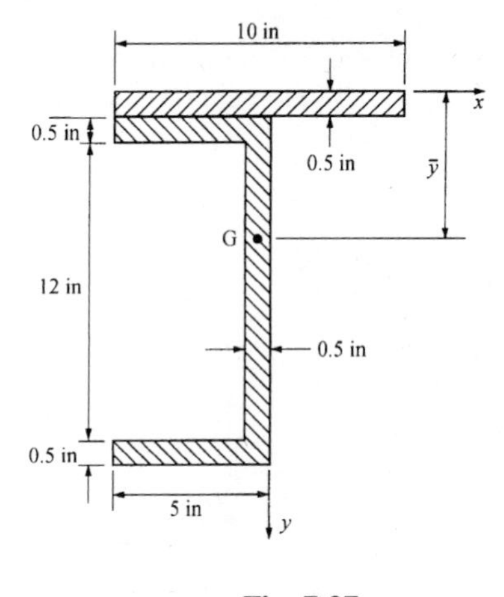

Fig. 7-27

7.20. The shaded area shown in Fig. 7-28 is bounded by a circular arc and a chord. Determine the location of the centroid of the area with respect to the center of the circular arc.

Ans. $\bar{y} = \frac{4R}{3} \cdot \frac{(\sin^3 \theta)}{(2\theta - \sin 2\theta)}$

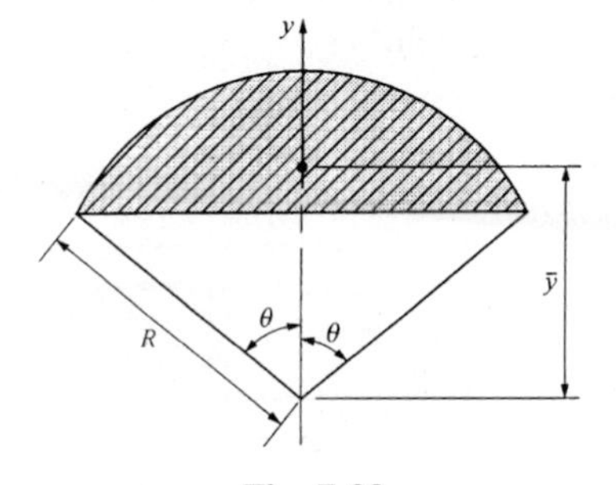

Fig. 7-28

7.21. An area consists of a circle of radius R from which a rectangle of dimensions $a \times 3a$ has been removed, as shown in Fig. 7-29. Determine the moment of inertia of the shaded area about the x- and also the y-axes.

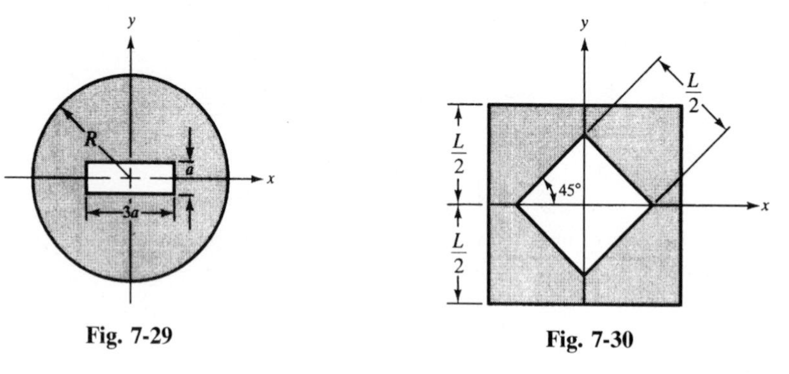

Fig. 7-29

Fig. 7-30

Ans. $I_x = \dfrac{\pi R^4}{4} - \dfrac{a^4}{4}, \qquad I_y = \dfrac{\pi R^4}{4} - \dfrac{9a^4}{4}$

7.22. The shaded area in Fig. 7-30 results from removing the central square from the outer square. Determine the moment of inertia of the net area about the x-axis. *Ans.* $I_x = 0.0781L^4$

7.23. A thin rectangular sheet has semicircular and also triangular areas removed, as shown in Fig. 7-31. Locate the centroid of the sheet and determine the moment of inertia about the horizontal axis passing through the centroid. *Ans.* $\bar{y} = 370.8$ mm, $I_{x_G} = 9937 \times 10^6$ mm^4

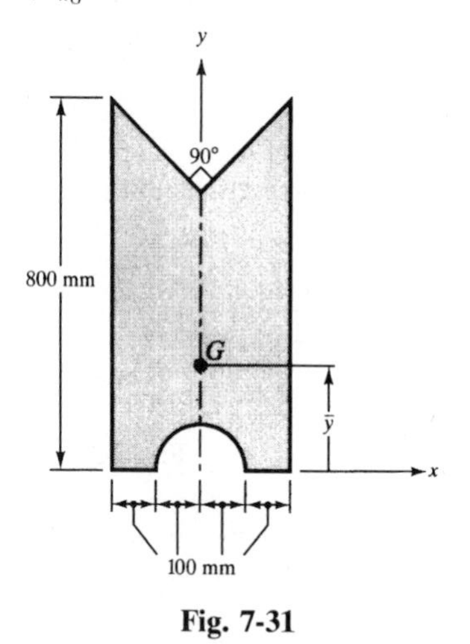

Fig. 7-31

7.24. A trapezoidal area has the dimensions indicated in Fig. 7-32. Determine the location of the centroid as well as the moment of inertia about an axis through the centroid and parallel to the x-axis.
Ans. $\bar{y} = 44.4$ mm, $I_{x_G} = 24.14 \times 10^6$ mm^4

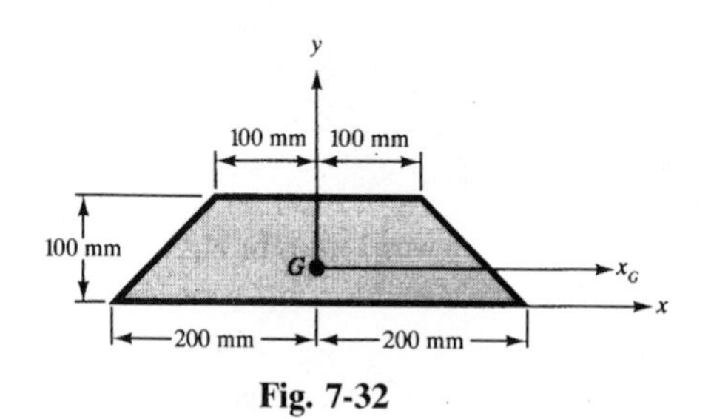

Fig. 7-32

7.25. A thin-walled section ($t \ll a$) has the configuration indicated in Fig. 7-33. Locate the centroid of the cross section and determine the moment of inertia of the area about an axis passing through the centroid and parallel to the x-axis. *Ans.* $\bar{y} = a$, $I_{x_G} = 5.33a^3 t + at^3/6$

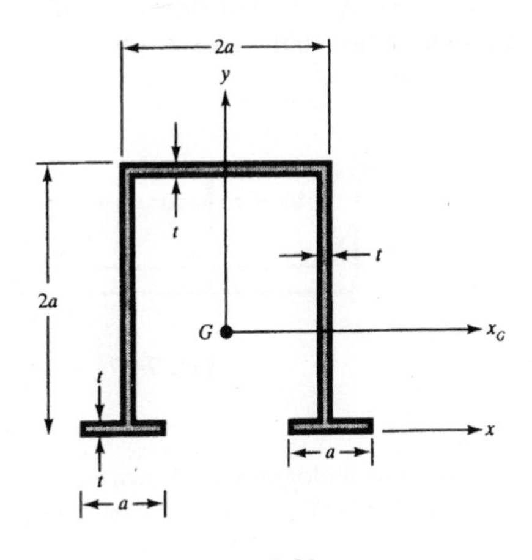

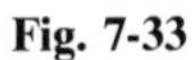

Fig. 7-33

7.26. An area of circular cross section from which three circular holes have been removed is shown in Fig. 7-34. Determine the location of the centroid of the section and the moment of inertia of an axis passing through the centroid and parallel to the x-axis. *Ans.* $\bar{y} = -R/10$, $I_{x_G} = 0.737R^4$

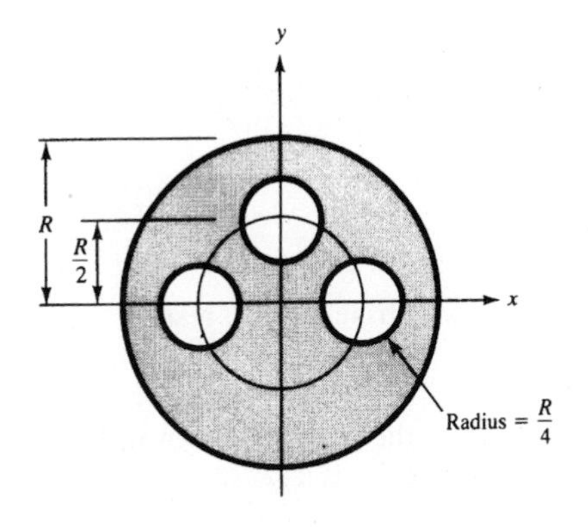

Fig. 7-34

7.27. Determine the moment of inertia of the diamond-shaped figure shown in Fig. 7-35 with respect to the horizontal axis of symmetry. *Ans.* $I_{x_G} = 85.4\ \text{in}^4$

Fig. 7-35

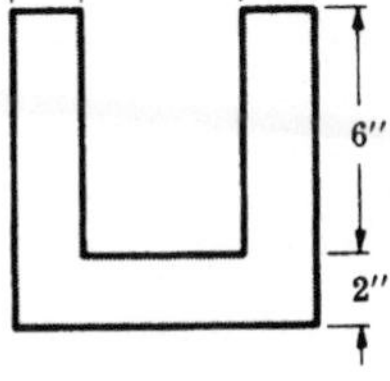

Fig. 7-36

7.28. Determine the moment of inertia of a channel-type section about a horizontal axis through the centroid. Refer to Fig. 7-36. What is the radius of gyration about this same axis?
Ans. $I_{x_G} = 231\ \text{in}^4$, $r_{x_G} = 2.40\ \text{in}$

7.29. Locate the centroid of the channel-type section shown in Fig. 7-37 and determine the moment of inertia of the cross-sectional area about a horizontal axis through the centroid.
Ans. $\bar{y} = 38.33\ \text{mm}$, $I_{x_G} = 33 \times 10^6\ \text{mm}^4$

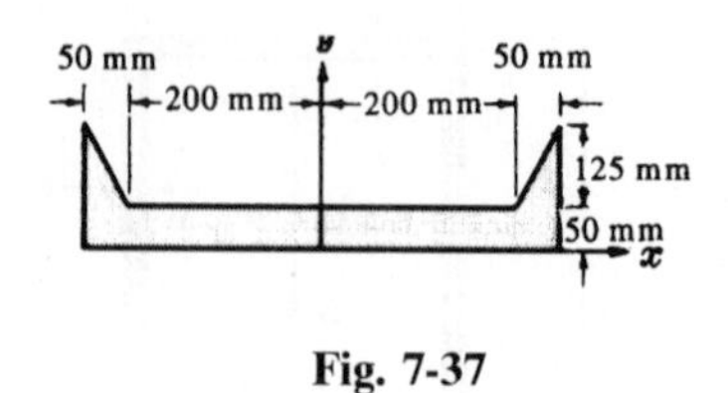

Fig. 7-37

7.30. A plane area has the shape of a parallelogram as shown in Fig. 7-38. The y- and z-axes pass through the centroid of the area. Determine I_y and I_z. *Ans.* $I_y = \frac{1}{12}bh^3$, $I_z = \frac{1}{12}hb(b^2 + c^2)$

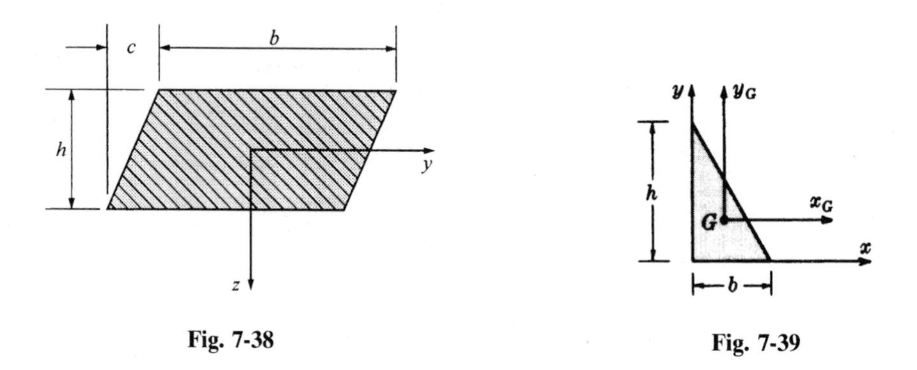

Fig. 7-38 **Fig. 7-39**

7.31. Determine the product of inertia of a triangle with respect to the x- and y-axes indicated in Fig. 7-39.
Ans. $b^2h^2/24$

7.32. Determine the product of inertia of the triangle shown in Fig. 7-39 with respect to the axes x_G and y_G passing through the centroid. *Ans.* $-b^2h^2/72$

7.33. For the plane area in Fig. 7-40 determine the moments of inertia and product of inertia with respect to the x_G- and y_G-axes passing through the centroid. Also, determine the principal second moments of area with respect to the centroid.
Ans. $I_{x_G} = 400 \times 10^6\ \text{mm}^4$; $I_{y_G} = 147 \times 10^6\ \text{mm}^4$; $I_{x_G y_G} = -58 \times 10^6\ \text{mm}^4$; $(I_{x_1})_{\max} = 805 \times 10^6\ \text{mm}^4$; $(I_{x_1})_{\min} = 142 \times 10^6\ \text{mm}^4$

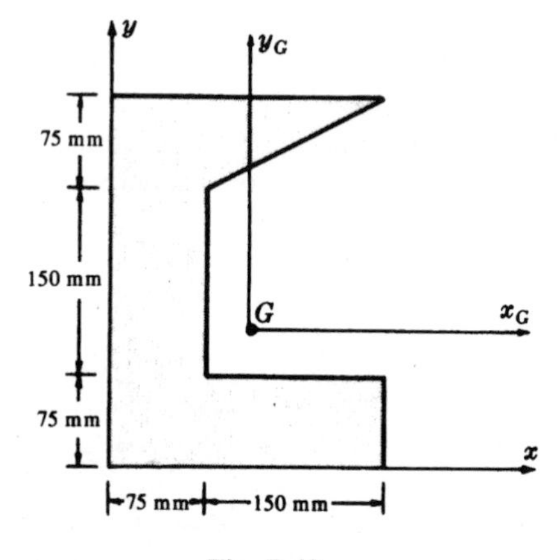

Fig. 7-40

Chapter 8

Stresses in Beams

TYPES OF LOADS ACTING ON BEAMS

Either forces or couples that lie in a plane containing the longitudinal axis of the beam may act upon the member. The forces are understood to act perpendicular to the longitudinal axis, and the plane containing the forces is assumed to be a plane of symmetry of the beam.

EFFECTS OF LOADS

The effects of these forces and couples acting on a beam are (*a*) to impart deflections perpendicular to the longitudinal axis of the bar and (*b*) to set up both normal and shearing stresses on any cross section of the beam perpendicular to its axis. Beam deflections will be considered in Chaps. 9, 10, and 11.

TYPES OF BENDING

If couples are applied to the ends of the beam and no forces act on the bar, then the bending is termed *pure bending*. For example, in Fig. 8-1 the portion of the beam between the two downward forces is subject to pure bending. Bending produced by forces that do not form couples is called *ordinary bending*. A beam subject to pure bending has only normal stresses with no shearing stresses set up in it; a beam subject to ordinary bending has both normal and shearing stresses acting within it.

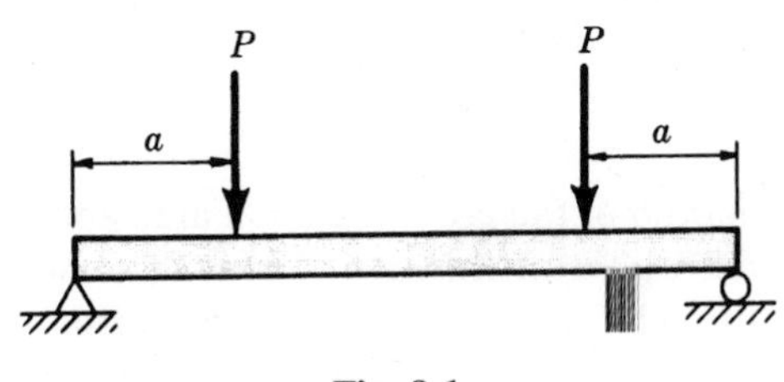

Fig. 8-1

NATURE OF BEAM ACTION

It is convenient to imagine a beam to be composed of an infinite number of thin longitudinal rods or fibers. Each longitudinal fiber is assumed to act independently of every other fiber, i.e., there are no lateral pressures or shearing stresses between the fibers. The beam of Fig. 8-1, for example, will deflect downward and the fibers in the lower part of the beam undergo extension, while those in the upper part are shortened. These changes in the lengths of the fibers set up stresses in the fibers. Those that are extended have tensile stresses acting on the fibers in the direction of the longitudinal axis of the beam, while those that are shortened are subject to compressive stresses.

NEUTRAL SURFACE

There always exists one surface in the beam containing fibers that do not undergo any extension or compression, and thus are not subject to any tensile or compressive stress. This surface is called the *neutral surface* of the beam.

NEUTRAL AXIS

The intersection of the neutral surface with any cross section of the beam perpendicular to its longitudinal axis is called the *neutral axis*. All fibers on one side of the neutral axis are in a state of tension, while those on the opposite side are in compression.

BENDING MOMENT

The algebraic sum of the moments of the external forces to one side of any cross section of the beam about an axis through that section is called the *bending moment* at that section. This concept was discussed in Chap. 6.

ELASTIC BENDING OF BEAMS

The following remarks apply *only if* all fibers in the beam are acting within the elastic range of action of the material.

Normal Stresses in Beams

For any beam having a longitudinal plane of symmetry and subject to a bending moment M at a certain cross section, the normal stress acting on a longitudinal fiber at a distance y from the neutral axis of the beam (see Fig. 8-2) is given by

$$\sigma = \frac{My}{I} \tag{8.1}$$

where I denotes the moment of inertia of the cross-sectional area about the neutral axis. This quantity was discussed in Chap. 7. The derivation of this equation is discussed in detail in Problem 8.1. For applications see Problems 8.2 through 8.18. These stresses vary from zero at the neutral axis of the beam to a maximum at the outer fibers as shown. The stresses are tensile on one side of the neutral axis, compressive on the other. These stresses are also called *bending*, *flexural*, or *fiber stresses*.

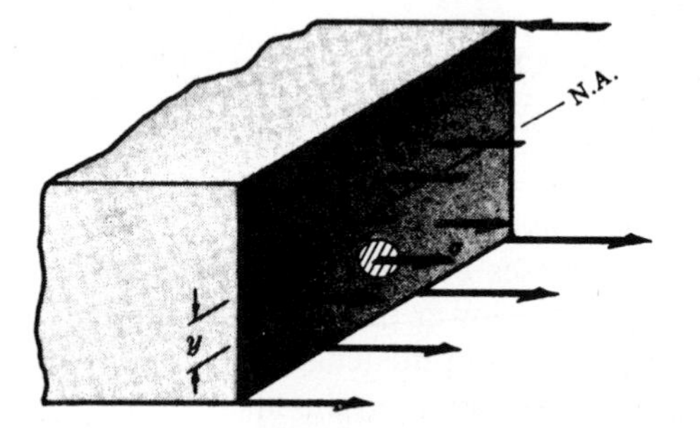

Fig. 8-2

Location of the Neutral Axis

When the beam action is entirely elastic the neutral axis passes through the centroid of the cross section. Hence, the moment of inertia I appearing in the above equation for normal stress is the moment of inertia of the cross-sectional area about an axis through the centroid of the cross section of the beam.

Section Modulus

At the outer fibers of the beam the value of the coordinate y is frequently denoted by the symbol c. In that case the maximum normal stresses are given by

$$\sigma = \frac{Mc}{I} \quad \text{or} \quad \sigma = \frac{M}{I/c} \tag{8.2}$$

The ratio I/c is called the *section modulus* and is usually denoted by the symbol Z. The units are in^3 or m^3. The maximum bending stresses may then be represented as

$$\sigma = \frac{M}{Z} \tag{8.3}$$

This form is convenient because values of Z are available in handbooks for a wide range of standard structural steel shapes. See Problems 8.5, 8.9, and 8.12.

Assumptions

In the derivation of the above expression for normal stresses it is assumed that a plane section of the beam normal to its longitudinal axis prior to loading remains plane after the forces and couples have been applied. Further, it is assumed that the beam is initially straight and of uniform cross section and that the moduli of elasticity in tension and compression are equal. Again, it is to be emphasized that no fibers of the beam are stressed beyond the proportional limit.

Shearing Force

The algebraic sum of all the vertical forces to one side of any cross section of the beam is called the shearing force at that section. This concept was discussed in Chap. 6.

Shearing Stresses in Beams

For any beam subject to a shearing force V (expressed in pounds) at a certain cross section, both vertical and horizontal shearing stresses τ are set up. The magnitudes of the vertical shearing stresses at any cross section are such that these stresses have the shearing force V as a resultant. In the cross section of the beam shown in Fig. 8-3, the vertical plane of symmetry contains the applied forces and the neutral axis passes through the centroid of the section. The coordinate y is measured from the neutral axis. The moment of inertia of the *entire* cross-sectional area about the neutral axis is denoted by I. The shearing stress on all fibers a distance y_0 from the neutral axis is given by the formula

$$\tau = \frac{V}{Ib} \int_{y_0}^{c} y \, da \tag{8.4}$$

where b denotes the width of the beam at the location where the shearing stress is being calculated. This expression is derived in Problem 8.19. For applications see Problems 8.20 through 8.23. The integral in (*8.4*) represents the first moment of the shaded area of the cross section about the neutral axis. This quantity was discussed in detail in Chap. 7. More generally, the integral always represents

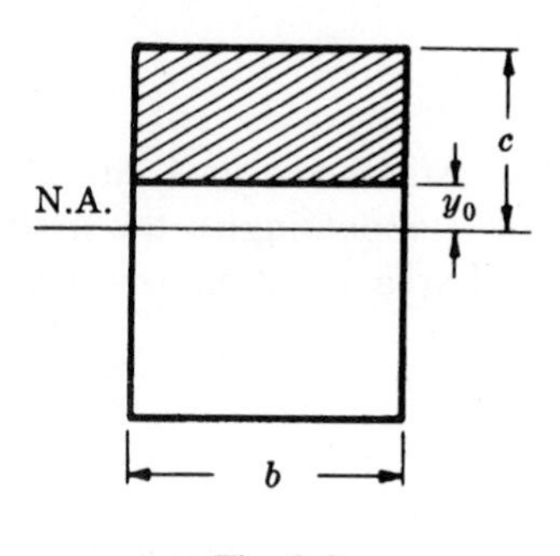

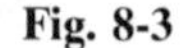
Fig. 8-3

the first moment about the neutral axis of that part of the cross-sectional area of the beam between the horizontal plane on which the shearing stress τ occurs and the outer fibers of the beam, i.e., the area between y_0 and c.

From (*8.4*) it is evident that the maximum shearing stress always occurs at the neutral axis of the beam, whereas the shearing stress at the outer fibers is always zero. In contrast, the normal stress varies from zero at the neutral axis to a maximum at the outer fibers.

In a beam of rectangular cross section the above equation for shearing stress becomes

$$\tau = \frac{V}{2I}\left(\frac{h^2}{4} - y_0^2\right) \tag{8.5}$$

where τ denotes the shearing stress on a fiber at a distance y_0 from the neutral axis and h denotes the depth of the beam. The distribution of vertical shearing stress over the rectangular cross section is thus parabolic, varying from zero at the outer fibers to a maximum at the neutral axis. For application see Problems 8.20 through 8.23.

Both the above equations for shearing stress give the vertical and also the horizontal shearing stresses at a point, as discussed in Problem 8.19, since the intensities of shearing stresses in these two directions are always equal.

PLASTIC BENDING OF BEAMS

The following remarks apply if some or all of the fibers of the beam are stressed to the yield point of the material.

We shall consider a simplified stress-strain curve such as that of Fig. 8-4, where it is assumed that the proportional limit and the yield point coincide. The yield region, i.e., the horizontal plateau of the curve, is assumed to extend indefinitely. This conventionalized representation of ductile material behavior is termed *elastic-perfectly plastic* behavior. Here, σ_{yp} denotes the yield point of the material and ϵ_{yp} represents the strain corresponding to that stress. We shall assume that material properties are identical in tension and compression.

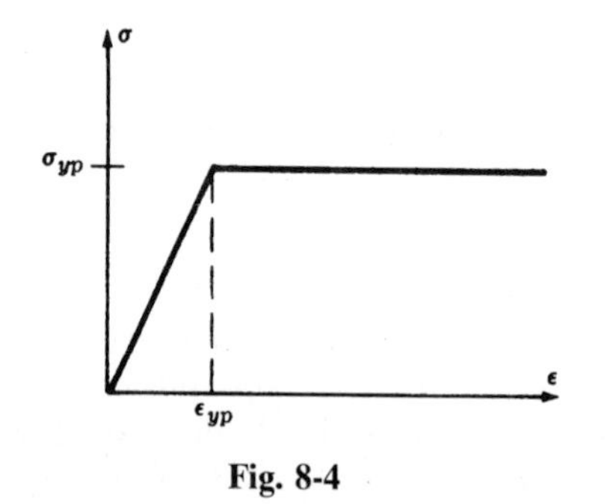

Fig. 8-4

Elastoplastic Action

For sufficiently large bending moments in a beam the interior fibers will be stressed in the elastic range of action, whereas the outer fibers will have reached the yield point of the material. Such a stress distribution may be as indicated in Fig. 8-5.

Fully Plastic Action

As bending moments continue to increase, a limiting case is approached in which all fibers are stressed to the yield point of the material. This stress distribution appears in Fig. 8-6.

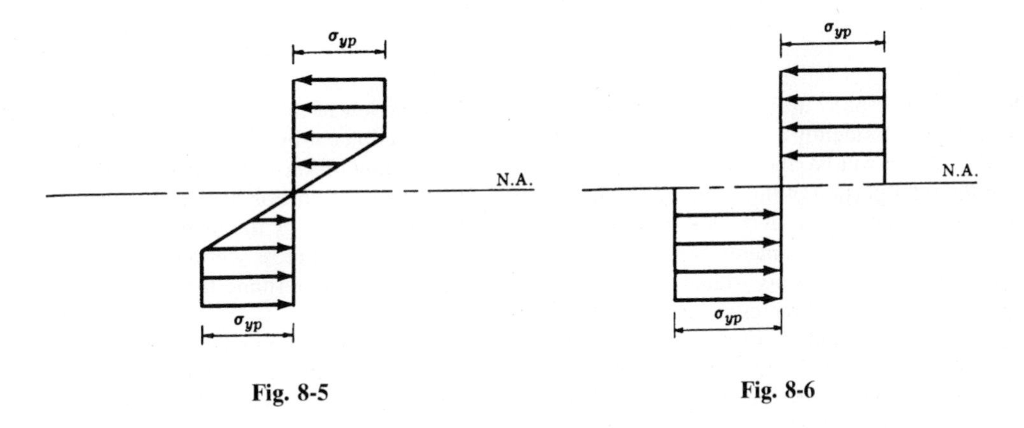

Fig. 8-5 Fig. 8-6

Location of Neutral Axis

When beam action is entirely elastic, the neutral axis passes through the centroid of the cross section. However, as plastic action spreads from the outer fibers inward, the neutral axis shifts from this location to another, which is determined by realizing that the resultant normal force over any cross section vanishes. In the limiting case of fully plastic action, the neutral axis assumes a position such that the total cross-sectional area is divided into two equal parts. This is discussed in Problem 8.29.

Fully Plastic Moment

The bending moment corresponding to fully plastic action is termed the *fully plastic moment* and will be denoted by M_p. For the stress-strain diagram assumed here no greater moment can be developed.

For a beam of rectangular cross section the fully plastic moment is shown in Problem 8.25 to be $M_p = bh^2\sigma_{yp}/4$ where b represents the width of the beam and h its depth.

Solved Problems

Elastic Bending of Beams

8.1. Derive an expression for the relationship between the bending moment acting at any section in a beam and the bending stress at any point in this same section. Assume Hooke's law holds.

The beam shown in Fig. 8-7(a) is loaded by the two couples M and consequently is in static equilibrium. Since the bending moment has the same value at all points along the bar, the beam is said to be in a condition of *pure bending*. To determine the distribution of bending stress in the beam, let us cut the beam by a plane passing through it in a direction perpendicular to the geometric axis of the bar. In this manner the forces under investigation become external to the new body formed, even though they were internal effects with regard to the original uncut body.

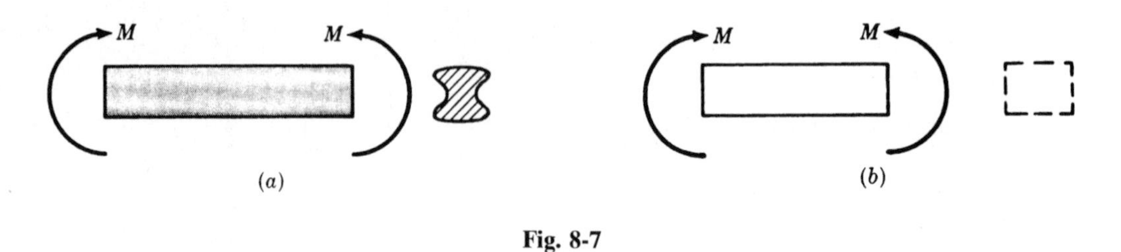

Fig. 8-7

The free-body diagram of the portion of the beam to the left of this cutting plane now appears as in Fig. 8-7(*b*). Evidently a moment M must act over the cross section cut by the plane so that the left portion of the beam will be in static equilibrium. The moment M acting on the cut section represents the effect of the right portion of the beam on the left portion. Since the right portion has been removed, it must be replaced by its effect on the left portion and this effect is represented by the moment M. This moment is the resultant of the moments of forces acting perpendicular to the cut cross section and in the plane of the page. It is now necessary to make certain assumptions in order to determine the nature of the variation of these forces over the cross section.

It is convenient to consider the beam to be composed of an infinite number of thin longitudinal rods or fibers. It is assumed that every longitudinal fiber acts independently of every other fiber; that is, there are no lateral pressures or shearing stresses between adjacent fibers. Thus each fiber is subject only to axial tension or compression. Further, it is assumed that a plane section of the beam normal to its axis before loads are applied remains plane and normal to the axis after loading. Finally, it is assumed that the material follows Hooke's law and that the moduli of elasticity in tension and compression are equal.

Let us next consider two adjacent cross sections *aa* and *bb* marked on the side of the beam, as shown in Fig. 8-8. Prior to loading, these sections are parallel to each other. After the applied moments have acted on the beam, these sections are still planes but they have rotated with respect to each other to the positions shown, where O represents the center of curvature of the beam. Evidently the fibers on the upper surface of the beam are in a state of compression, while those on the lower surface have been extended slightly and are thus in tension. The line *cd* is the trace of the surface in which the fibers do not undergo any strain during bending and this surface is called the *neutral surface*, and its intersection with any cross section is called the *neutral axis*. The elongation of the longitudinal fiber at a distance y (measured positive downward) may be found by drawing line *de* parallel to *aa*. If ρ denotes the radius of curvature of the bent beam, then from the similar triangles *cOd* and *edf* we find the strain of this fiber to be

$$\epsilon = \frac{\overline{ef}}{\overline{cd}} = \frac{\overline{de}}{\overline{cO}} = \frac{y}{\rho} \tag{1}$$

Thus, the strains of the longitudinal fibers are proportional to the distance y from the neutral axis.

Since Hooke's law holds, and therefore $E = \sigma/\epsilon$, or $\sigma = E\epsilon$, it immediately follows that the stresses existing in the longitudinal fibers are proportional to the distance y from the neutral axis, or

$$\sigma = \frac{Ey}{\rho} \tag{2}$$

Let us consider a beam of rectangular cross section, although the derivation actually holds for any cross section which has a longitudinal plane of symmetry. In this case, these longitudinal, or bending, stresses appear as in Fig. 8-9.

Let *da* represent an element of area of the cross section at a distance y from the neutral axis. The stress acting on *da* is given by the above expression and consequently the force on this element is the product of the stress and the area *da*, that is,

$$dF = \frac{Ey}{\rho}\,da \tag{3}$$

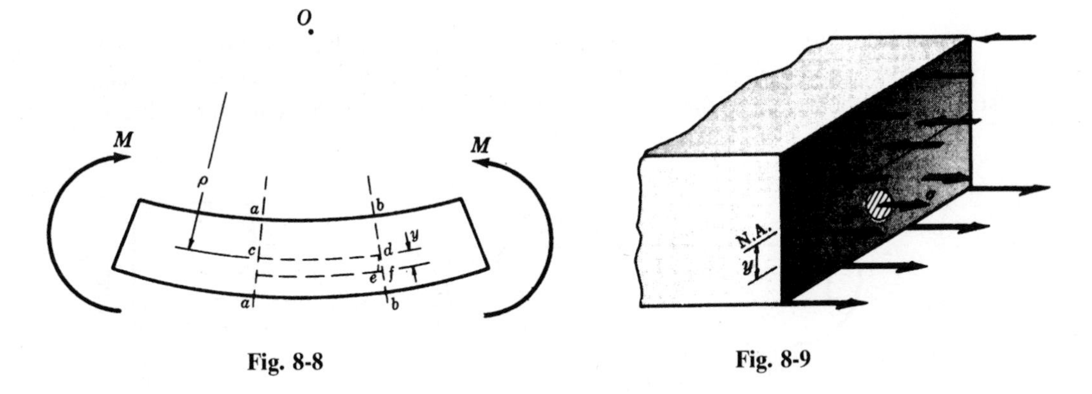

Fig. 8-8

Fig. 8-9

However, the resultant longitudinal force acting over the cross section is zero (for the case of pure bending) and this condition may be expressed by the summation of all forces dF over the cross section. This is done by integration:

$$\int \frac{Ey}{\rho} da = \frac{E}{\rho} \int y\, da = 0 \tag{4}$$

Evidently $\int y\, da = 0$. However, this integral represents the first moment of the area of the cross section with respect to the neutral axis, since y is measured from that axis. But, from Chap. 7 we may write $\int y\, da = \bar{y}A$, where $\bar{y}$ is the distance from the neutral axis to the centroid of the cross-sectional area. From this, $\bar{y}A = 0$; and since A is not zero, then $\bar{y} = 0$. Thus the neutral axis always passes through the centroid of the cross section, provided Hooke's law holds.

The moment of the elemental force dF about the neutral axis is given by

$$dM = y\, dF = y\left(\frac{Ey}{\rho} da\right) \tag{5}$$

The resultant of the moments of all such elemental forces summed over the entire cross section must be equal to the bending moment M acting at that section and thus we may write

$$M = \int \frac{Ey^2}{\rho} da \tag{6}$$

But $I = \int y^2\, da$ and thus we have

$$M = \frac{EI}{\rho} \tag{7}$$

It is to be carefully noted that this moment of inertia of the cross-sectional area is computed with respect to the axis through the centroid of the cross section. But previously we had

$$\sigma = \frac{Ey}{\rho} \tag{8}$$

Eliminating ρ from these last two equations, we obtain

$$\sigma = \frac{My}{I} \tag{9}$$

This formula gives the so-called bending or flexural stresses in the beam. In it, M is the bending moment at any section, I the moment of inertia of the cross-sectional area about an axis through the centroid of the cross section, and y the distance from the neutral axis (which passes through the centroid) to the fiber on which the stress σ acts.

The value of y at the outer fibers of the beam is frequently denoted by c. At these fibers the bending stresses are maximum and there we may write

$$\sigma = \frac{Mc}{I} \tag{10}$$

8.2. A beam is loaded by a couple of 12,000 lb · in at each of its ends, as shown in Fig. 8-10. The beam is steel and of rectangular cross section 1 in wide by 2 in deep. Determine the maximum bending stress in the beam and indicate the variation of bending stress over the depth of the beam.

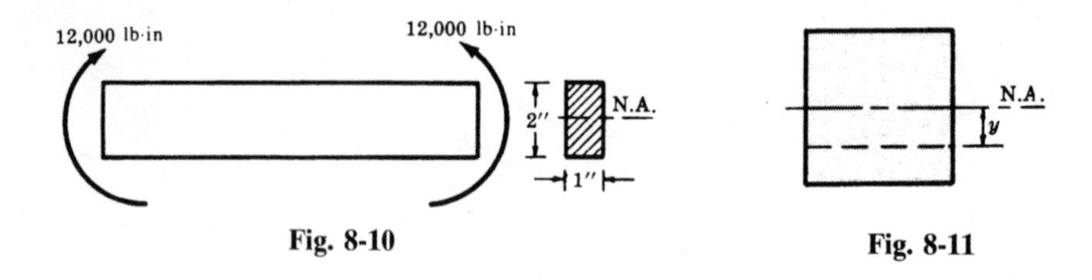

Fig. 8-10 **Fig. 8-11**

From Problem 8.1, bending takes place about the horizontal neutral axis denoted by N.A. This axis passes through the centroid of the cross section. The moment of inertia of the shaded rectangular cross section about this axis is found by the methods of Chap. 7 to be

$$I = \tfrac{1}{12}bh^3 = \tfrac{1}{12}(1)(2)^3 = 0.667 \text{ in}^4$$

Also from Problem 8.1, the bending stress at a distance y from the neutral axis is given by $\sigma = My/I$, where y is illustrated in Fig. 8-11. Thus, all longitudinal fibers of the beam at the distance y from the neutral axis are subject to the same bending stress given by the above formula.

Since M and I are constant along the length of the bar, evidently the maximum bending stress occurs on those fibers where y takes on its maximum value. These are the fibers along the upper and lower surfaces of the beam, and from inspection it is obvious that for the direction of loading shown the upper fibers are in compression and the lower fibers in tension. For the lower fibers, $y = 1$ in and the maximum bending stress is

$$\sigma = \frac{12{,}000(1)}{0.667} = 18{,}000 \text{ lb/in}^2$$

For the fibers along the upper surface y may be considered to be negative and we have

$$\sigma = \frac{12{,}000(-1)}{0.667} = -18{,}000 \text{ lb/in}^2$$

Thus the peak stresses are 18,000 lb/in^2 in tension for all fibers along the lower surface of the beam and 18,000 lb/in^2 in compression for all fibers along the upper surface. According to the formula $\sigma = My/I$, the bending stress varies linearly from zero at the neutral axis to a maximum at the outer fibers and hence the variation over the depth of the beam may be plotted as in Fig. 8-12.

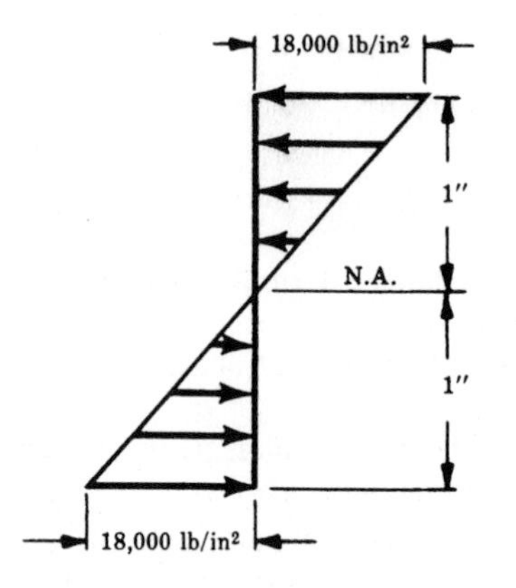

Fig. 8-12

8.3. A beam of circular cross section is 7 in in diameter. It is simply supported at each end and loaded by two concentrated loads of 20,000 lb each, applied 12 in from the ends of the beam. Determine the maximum bending stress in the beam.

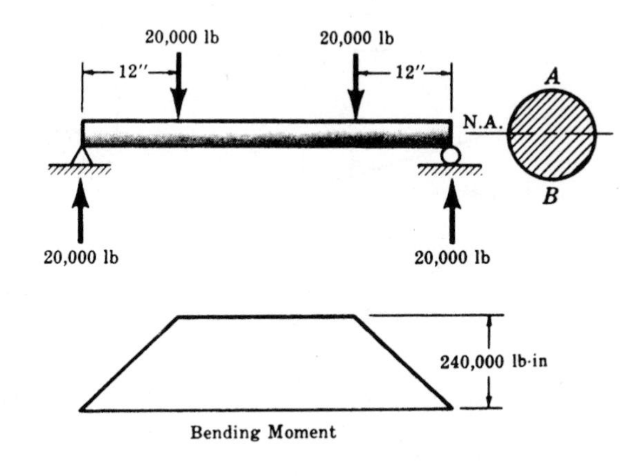

Fig. 8-13

Here the moment is not constant along the length of the beam, as it was in Problem 8.2. The loading is illustrated in Fig. 8-13 together with the bending moment diagram obtained by the methods of Chap. 6. It is to be noted that the portion of the beam between the two downward loads of 20,000 lb is in a condition termed *pure bending* and everywhere in that region the bending moment is equal to $20{,}000(12) = 240{,}000\ \text{lb} \cdot \text{in}$.

From Problem 7.9 the moment of inertia of the shaded circular cross section about the neutral axis, which passes through the centroid of the circle, is $I = \pi D^4/64 = \pi(7)^4/64 = 118\ \text{in}^4$.

The bending stress at a distance y from the horizontal neutral axis shown is $\sigma = My/I$. Evidently the maximum bending stresses occur along the fibers located at the ends of a vertical diameter and designated as A and B. This maximum stress is the same at all such points between the applied loads. At point B, $y = 3.5$ in and the stress becomes

$$\sigma = \frac{240{,}000(3.5)}{118} = 7120\ \text{lb/in}^2\ \text{tension}$$

At point A the stress is $7120\ \text{lb/in}^2$ compression.

8.4. A steel cantilever beam 16 ft 8 in in length is subjected to a concentrated load of 320 lb acting at the free end of the bar. The beam is of rectangular cross section, 2 in wide by 3 in deep. Determine the magnitude and location of the maximum tensile and compressive bending stresses in the beam.

The bending moment diagram for this type of loading, determined by the techniques of Chap. 6, is triangular with a maximum ordinate at the supporting wall, as shown below in Fig. 8-14(a). The maximum bending moment is merely the moment of the 320-lb force about an axis through point B and perpendicular to the plane of the page. It is $-320(200) = -64{,}000\ \text{lb} \cdot \text{in}$.

The bending stress at a distance y from the neutral axis, which passes through the centroid of the cross section, is $\sigma = My/I$ where y is illustrated in Fig. 8-14(b). In this expression I denotes the moment of inertia of the cross-sectional area about the neutral axis and is given by

$$I = \tfrac{1}{12}bh^3 = \tfrac{1}{12}(2)(3)^3 = 4.50\ \text{in}^4$$

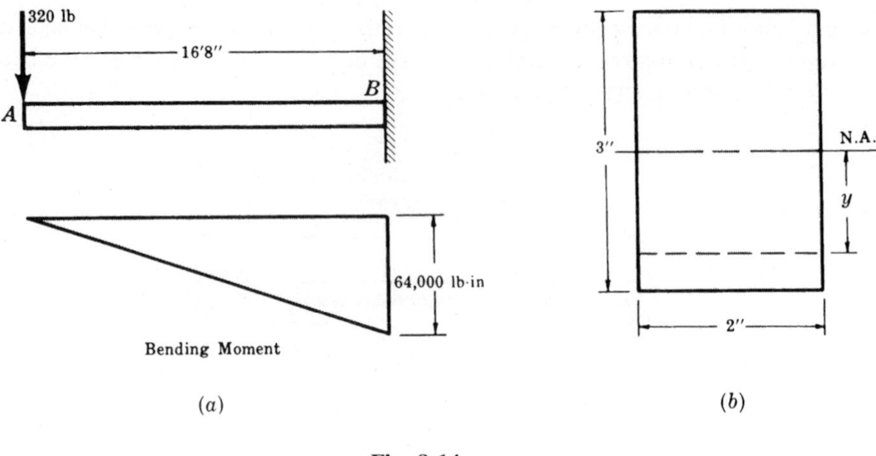

Fig. 8-14

Thus at the supporting wall, where the bending moment is maximum, the peak tensile stress occurs at the upper fibers of the beam and is

$$\sigma = \frac{My}{I} = \frac{(-64{,}000)(-1.5)}{4.50} = 21{,}400 \text{ lb/in}^2$$

It is evident that this stress must be tension because all points of the beam deflect downward. At the lower fibers adjacent to the wall the peak compressive stress occurs and is equal to 21,400 lb/in^2.

8.5. Let us reconsider Problem 8.4 for the case where the rectangular beam is replaced by a commercially available rolled steel section, designated as a W6 $\times$ $15\frac{1}{2}$. This standard manner of designation indicates that the depth of the section is 6 in, that it is a so-called wide-flange section, and that it weighs $15\frac{1}{2}$ lb per ft of length. Determine the maximum tensile and compressive bending stresses.

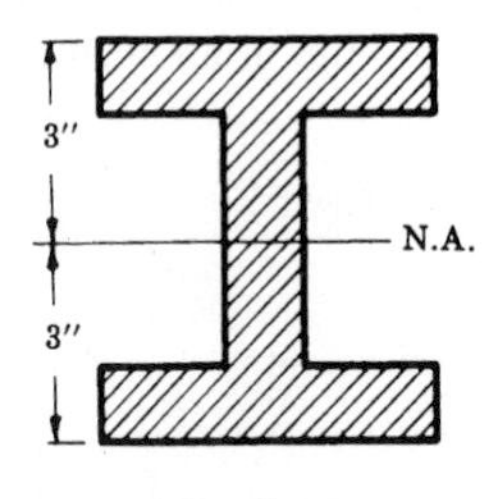

Fig. 8-15

Such a beam has the symmetric cross section shown in Fig. 8-15 and bending takes place about the horizontal neutral axis passing through the centroid. Extensive handbooks listing properties of all available rolled steel shapes are available to designers and abridged tables are presented at the end of this chapter. From that table the moment of inertia about the neutral axis is found to be 28.1 in^4.

The bending stress at a distance y from the neutral axis is given by $\sigma = My/I$. At the outer fibers, $y = c$ and

$$\sigma = \frac{Mc}{I} = \frac{M}{I/c}$$

The ratio I/c is designated as the *section modulus* and is usually denoted by the symbol Z. The units are obviously in^3. From the abridged table we find Z to be 9.7 in^3. Thus if one is concerned only with bending stresses occurring at the outer fibers, which is frequently the case since we are often interested only in maximum stresses, then the section modulus is a convenient quantity to work with, particularly for standard structural shapes.

The stresses in the extreme fibers at the section of the beam immediately adjacent to the wall are thus given by

$$\sigma = \frac{M}{I/c} = \frac{M}{Z} = \frac{64{,}000}{9.7} = 6600 \text{ lb/in}^2$$

Again, since the fibers along the top of the beam are stretching, the stress there will be tension. Along the lower face of the beam the fibers are shortening and there the stress is compressive.

8.6. A cantilever beam 3 m long is subjected to a uniformly distributed load of 30 kN per meter of length. The allowable working stress in either tension or compression is 150 MPa. If the cross section is to be rectangular, determine the dimensions if the height is to be twice as great as the width.

The bending moment diagram for a uniform load acting over a cantilever beam was determined in Problem 6.2. It was found to be parabolic, varying from zero at the free end of the beam to a maximum at the supporting wall. The loaded beam and the accompanying bending moment diagram are shown in Fig. 8-16. The maximum moment at the wall is given by

$$M_{x=3} = -30(3)(1.5) = -135 \text{ kN}\cdot\text{m}$$

It is to be noted that this problem involves the design of a beam, whereas all previous problems in this chapter called for the analysis of stresses acting in beams of known dimensions and subject to various loadings. The only cross section that need be considered for design purposes is the one where the bending moment is a maximum, i.e., at the supporting wall. Thus we wish to design a rectangular beam to resist a bending moment of 135 kN·m with a maximum bending stress of 150 MPa.

Since the cross section is to be rectangular it will have the appearance shown in Fig. 8-17, where the width is denoted by b and the height by $h = 2b$, in accordance with the specifications. The moment of inertia about the neutral axis, which passes through the centroid of the action, is given by

$$I = \tfrac{1}{12}bh^3 = \tfrac{1}{12}b(2b)^3 = \tfrac{2}{3}b^4$$

At the cross section of the beam adjacent to the supporting wall the bending stress in the beam is given by $\sigma = My/I$. The maximum bending stress in tension occurs along the upper surface of the beam, since these fibers elongate slightly, and at this surface $y = -b$ and $\sigma = 150$ MPa. Then

$$\sigma = \frac{My}{I} \qquad \text{or} \qquad 150 = \frac{-135 \times 10^3(10^3)(-b)}{\tfrac{2}{3}b^4}$$

from which $b = 110$ mm and $h = 2b = 220$ mm.

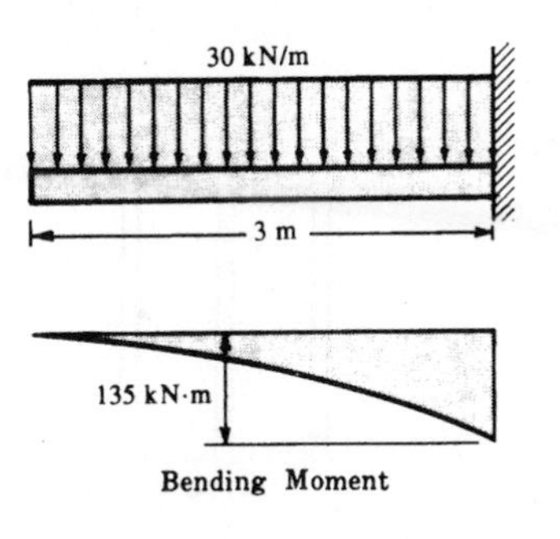

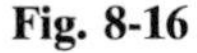

Fig. 8-16

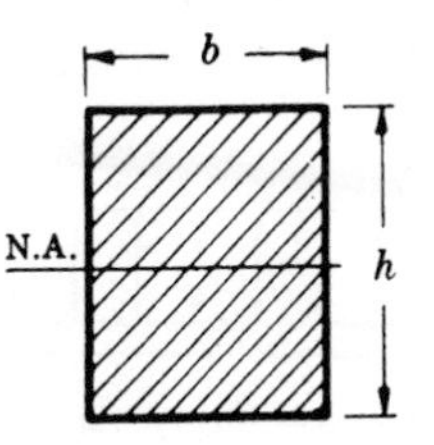

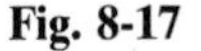

Fig. 8-17

8.7. A cantilever beam is of length 1.5 m, loaded by a concentrated force P at its tip as shown in Fig. 8-18(a), and is of circular cross section ($R = 100$ mm), having two symmetrically placed longitudinal holes as indicated. The material is titanium alloy, having an allowable working stress in bending of 600 MPa. Determine the maximum allowable value of the vertical force P.

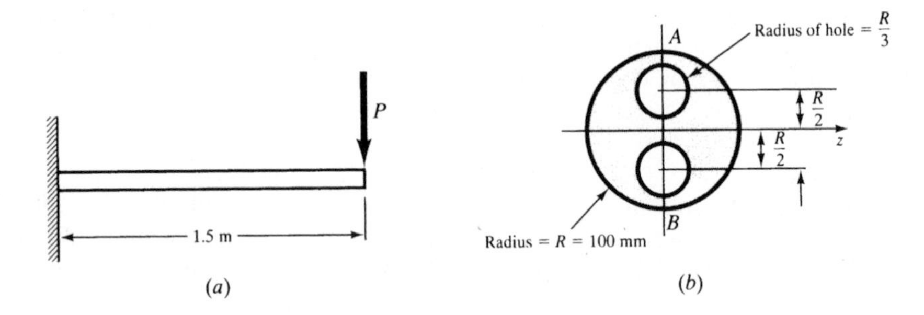

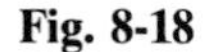

Fig. 8-18

It is first necessary to determine the section modulus of the beam. From Chap. 7, Problem 7.9, the moment of inertia of a solid circular cross section about a diametral axis z is $\pi R^4/4$. Using this value for the solid section and subtracting the moments of inertia of each of the holes about the same diametral axis z (from the parallel-axis theorem of Chap. 7), we have

$$I = \frac{\pi R^4}{4} - 2\left\{\frac{\pi}{4}\left(\frac{R}{3}\right)^4 + \pi\left(\frac{R}{3}\right)^2\left(\frac{R}{2}\right)^2\right\} = 0.592R^4$$

The section modulus from Eq. (8.3) is

$$Z = \frac{I}{c} = \frac{0.592R^4}{R} = 0.592R^3$$

The bending stresses in the uppermost and lowermost fibers, denoted by points A and B, respectively, in Fig. 8-18(b) are, from Eq. (8.3) and using $R = 0.1$ m,

$$\sigma_{\max} = \frac{M}{Z}$$

$$600 \times 10^6\ \text{N/m}^2 = \frac{P(1.5\ \text{m})}{0.592R^3}$$

Solving, $P = 237 \times 10^3$ N, or 237 kN.

8.8. The extruded beam shown in Fig. 8-19 is made of 6061-T6 aluminum alloy having an allowable working stress in either tension or compression of 90 MPa. The beam is a cantilever, subject to a uniform vertical load. Determine the allowable intensity of uniform loading.

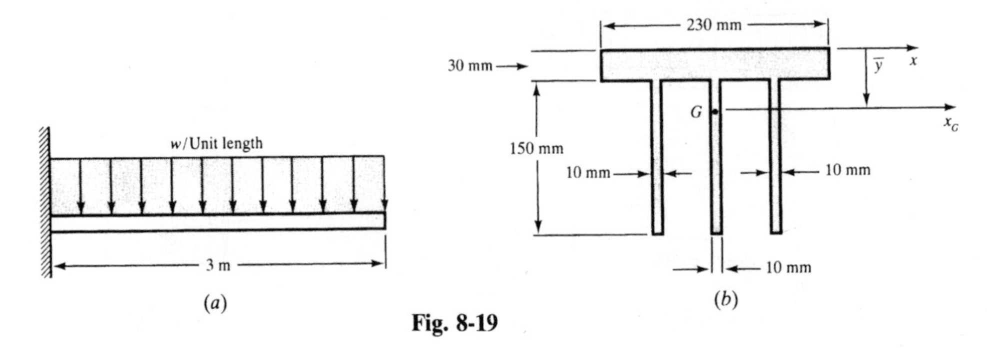

Fig. 8-19

It is first necessary to locate the centroid of the cross section. From the methods of Chap. 7, we have

$$\bar{y} = \frac{(200)(30)(15) + 3(180)(10)(90)}{(200)(30) + 3(180)(10)} = 50.5 \text{ mm}$$

It is next necessary to determine the moment of inertia of the cross section. Let us first work with the x-axis through the top of the flange. From Chap. 7 the moment of inertia of the entire section about that axis is

$$I_x = \tfrac{1}{3}(200 \text{ mm})(30 \text{ mm})^3 + 3\{\tfrac{1}{3}(10 \text{ mm})(180 \text{ mm})^3\}$$
$$= 60.12 \times 10^6 \text{ mm}^4$$

and from the parallel axis theorem of Chap. 7 we may now transfer to the x_G axis through the centroid of the cross section to find

$$I_{x_G} = 60.12 \times 10^6 \text{ mm}^4 - (11{,}400 \text{ mm}^2)(50.5 \text{ mm})^2$$
$$= 31.05 \times 10^6 \text{ mm}^4$$

The peak bending moment occurs at the supporting wall and was found in Problem 6.2 to be

$$M_{\text{max}} = \frac{wL^2}{2}$$

Next, applying Eq. (*8.1*) to the lowermost fibers (A) of the beam since those are the most distant from the neutral axis through G, we have

$$90 \times 10^6 \text{ N/m}^2 = \frac{[w(3 \text{ m})^2][(180 - 50.5) \text{ mm}](1 \text{ m}/1000 \text{ mm})}{(2)(31.05 \times 10^6 \text{ mm}^4)(1 \text{ m}/1000 \text{ mm})^4}$$

Solving,

$$w = 4.80 \text{ kN/m}$$

8.9. The simply supported beam AD is loaded by a concentrated force of 80 kN together with a couple of magnitude 30 kN · m, as shown in Fig. 8-20. From Table 8-2 at the end of this chapter select a commercially available steel wide-flange beam capable of carrying these loads if the peak allowable working stress in tension as well as compression is 160 MPa.

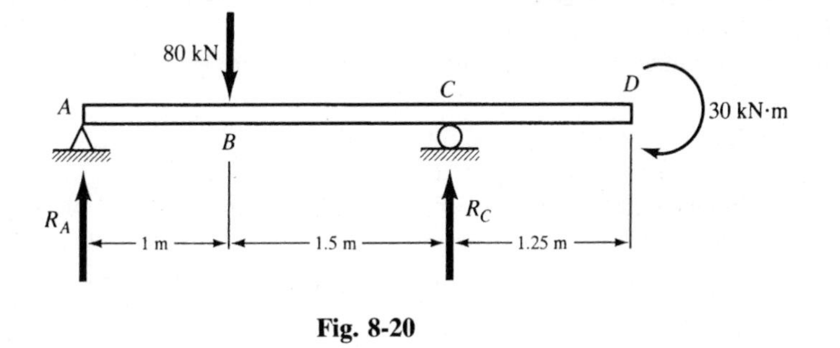

Fig. 8-20

It is first necessary to determine the reactions at A and C from statics. We have

$$+\circlearrowleft \Sigma M_A = -(80 \text{ kN})(1 \text{ m}) + R_C(2.5 \text{ m}) - 30 \text{ kN} \cdot \text{m} = 0$$
$$R_C = 44 \text{ kN}$$
$$\Sigma F_y = R_A + 44 - 80 = 0$$
$$R_A = 36 \text{ kN}$$

From the methods of Chap. 6, we can now construct the moment diagram which appears as in Fig. 8-21.

From Eq. (*8.3*) we have $\sigma_{max} = M/Z$. Substituting,

$$160 \times 10^6\ \text{N/m}^2 = \frac{36 \times 10^3\ \text{N} \cdot \text{m}}{Z}$$

Solving,

$$Z = 225 \times 10^{-6}\ \text{m}^3 \quad \text{or} \quad 225 \times 10^3\ \text{mm}^3$$

as the minimum acceptable value of section modulus. From Table 8-2 we see that the W203 × 28 section has a Z value of $262 \times 10^3\ \text{mm}^3$, which is adequate. Undoubtedly a more complete beam listing would indicate other sections with a Z value more nearly equal to the required minimum of $225 \times 10^3\ \text{mm}^3$. Only typical beams are listed in Table 8-2 for the sake of brevity.

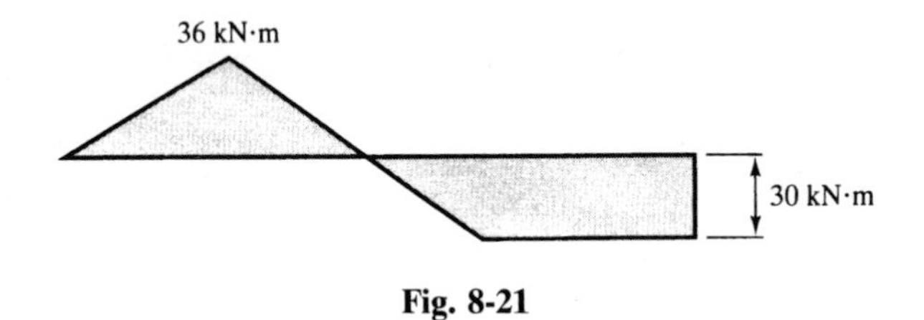

Fig. 8-21

8.10. If a steel wire 0.5 mm in diameter is coiled around a pulley 400 mm in diameter, determine the maximum bending stress set up in the wire. Take $E = 200$ GPa.

Since the radius of curvature of the wire is constant, 200 mm, it is evident from (*7*) of Problem 8.1, namely $M = EI/R$, that the bending moment M must be constant everywhere in the wire. Thus the wire acts as a beam subject to pure bending. An enlarged sketch of a portion of the wire is shown in Fig. 8-22. For any fiber in the wire at a distance y from the neutral axis, the normal strain was found in (*1*) of Problem 8.1 to be

$$\epsilon = \frac{y}{R}$$

where R denotes the radius of curvature of the beam at that point.

The maximum strain occurs at the fibers where y assumes its maximum value, that is, $\frac{1}{2}(0.5)$ mm from the neutral axis. The radius of curvature is approximately 200 mm. More accurately, this radius should be measured to the neutral surface of the wire, but the value in that case would only differ from 200 mm by 0.25 mm and this quantity may reasonably be neglected.

Thus the maximum strain at the outer fibers of the wire is

$$\epsilon = \frac{\frac{1}{2}(0.5)}{200} = 0.00125$$

The longitudinal fibers are subject to tensile stresses on one side of the wire and compressive on the other, with no other stresses acting. Hooke's law may then be used to find the stress:

$$\sigma = E\epsilon = (200 \times 10^9)(0.00125) = 250\ \text{MPa}$$

This is the maximum stress in the wire.

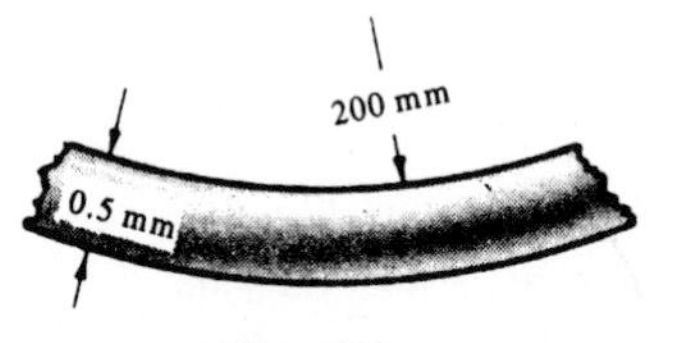

Fig. 8-22

8.11. The simply supported beam shown in Fig. 8-23(a) is subject to a uniformly varying load having a maximum intensity of w N per meter of length at the right end of the bar. If the beam is a wide-flange section having the dimensions shown in Fig. 8-23, determine the maximum load intensity w that may be applied if the working stress is 125 MPa in either tension or compression. Neglect the weight of the beam.

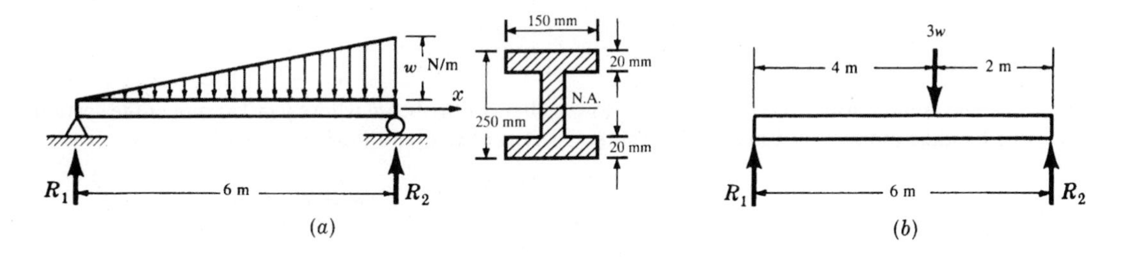

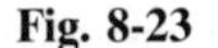
Fig. 8-23

The reactions R_1 and R_2 may readily be determined in terms of the unknown w by replacing the distributed load by its resultant. Since the average value of the distributed load is $w/2$ N/m acting over a length of 6 m, the resultant is a force of magnitude $6(w/2) = 3w$ N acting through the centroid of the triangular loading diagram, that is, 4 m to the right of R_1. This resultant thus appears as in Fig. 8-23(b). From statics we immediately have $R_1 = w$ N and $R_2 = 2w$ N.

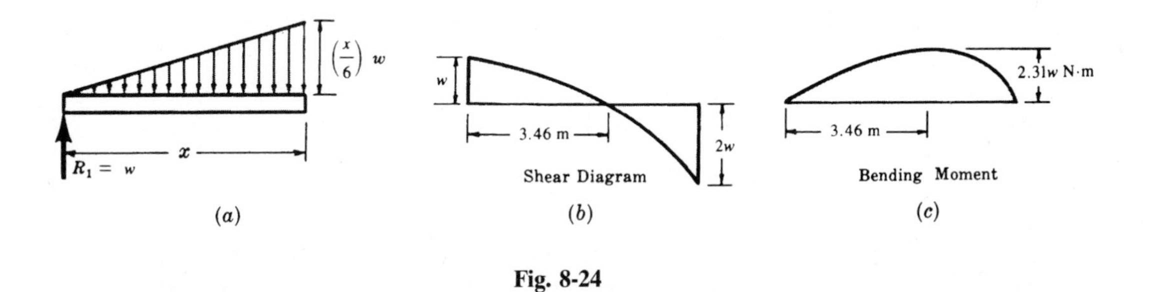

Fig. 8-24

The shearing force and bending moment diagrams for this type of loading were discussed in Problem 6.5. Let us introduce an x-axis coinciding with the beam and having its origin at the left support. Then at a distance x to the right of the left reaction, the intensity of load is found from similar-triangle relationships to be $(x/6)w$ N/m. This portion of the loaded beam between R_1 and the section x appears in Fig. 8-24(a). In accordance with the procedure explained in Problem 6.5, the shearing force V at the section a distance x from the left support is given by

$$V = w - \frac{1}{2}\left(\frac{x}{6}\right)wx = w - \frac{1}{12}wx^2$$

This equation holds for all values of x and from it the shear diagram is readily plotted, as shown in Fig. 8-24(b). The point of zero shear is found by setting

$$w - \tfrac{1}{12}wx^2 = 0 \qquad \text{from which} \qquad x = \sqrt{12} = 3.46\ \text{m}$$

This is also the point where the bending moment assumes its maximum value.

The bending moment M at the section a distance x from the left support is given by

$$M = wx - \frac{1}{2}\left(\frac{x}{6}\right)w\frac{x^2}{3} = wx - \frac{1}{36}wx^3$$

Again, this equation holds for all values of x and from it the bending moment diagram may be plotted as in Fig. 8-24(c). At the point of zero shear, $x = 3.46$ m, the bending moment is found by substitution in the above equation to be

$$M_{x=3.46} = 3.46w - \tfrac{1}{36}w(3.46)^3 = 2.31w \text{ N} \cdot \text{m}$$

This is the maximum bending moment in the beam.

The bending stress on any fiber a distance y from the neutral axis of the beam is given by $\sigma = My/I$. The moment of inertia I of the beam is found from

$$I_x = \frac{150(250)^3}{12} - 2\left[\frac{65(210)^3}{12}\right] = 95 \times 10^6 \text{ mm}^4$$

The maximum tensile stress occurs at the lower fibers of the beam where $y = 125$ mm at the section where the bending moment is a maximum. This stress is 125 MPa, and thus $\sigma = My/I$ becomes

$$125 \times 10^6 = \frac{(2.31w)(0.125)}{95 \times 10^6(10^{-12})} \qquad \text{or} \qquad w = 41 \text{ kN/m}$$

8.12. Determine the section modulus of a beam of rectangular cross section.

Let h denote the depth of the beam and b its width. Bending is assumed to take place about the neutral axis through the centroid of the cross section. The moment of inertia about the neutral axis is $I = bh^3/12$.

At the outer fibers the distance to the neutral axis is $h/2$, and this is commonly denoted by c. The maximum bending stresses at these outer fibers are given by

$$\sigma_{\max} = \frac{Mc}{I} = \frac{M}{I/c}$$

The ratio I/c is called the *section modulus* and is usually denoted by Z. Then $\sigma_{\max} = M/Z$. For the beam of rectangular cross section,

$$Z = \frac{L}{c} = \frac{bh^3/12}{h/2} = \frac{bh^2}{6}$$

The section modulus Z has units of m^3 or in^3.

8.13. A beam is loaded by one couple at each of its ends, the magnitude of each couple being 5 kN · m. The beam is steel and of T-type cross section with the dimensions indicated in Fig. 8-25(b). Determine the maximum tensile stress in the beam and its location, and the maximum compressive stress and its location.

It is first necessary to locate the centroid of the cross-sectional area since the neutral axis is known to pass through the centroid. To do this we introduce the x-y coordinate system shown and use the methods of Chap. 7. The y-coordinate of the centroid is defined by

$$\bar{y} = \frac{\int y\,da}{A}$$

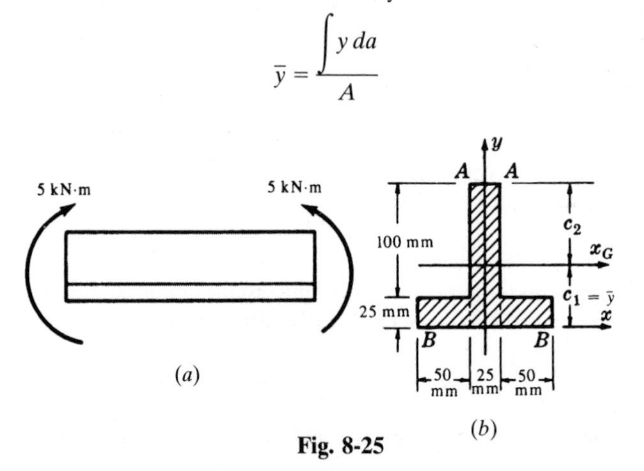

Fig. 8-25

where the numerator of the right side represents the first moment of the entire area about the x-axis. The T-section may be considered to consist of the three rectangles indicated by the dashed lines and this expression becomes

$$\bar{y} = \frac{125(25)(62.5) + 2[50(25)(12.5)]}{125(25) + 2[25(50)]} = 40.3 \text{ mm}$$

Thus, the centroid is located 40.3 mm above the x-axis. The horizontal axis passing through this point is denoted by x_G as shown.

The moment of inertia about the x-axis is given by the sum of the moments of inertia about this same axis of each of the three component rectangles comprising the cross section. Thus

$$I_x = \tfrac{1}{3}(25)(125)^3 + 2[\tfrac{1}{3}50(25)^3] = 16.8 \times 10^6 \text{ mm}^4$$

The moment of inertia about the x_G-axis may now be found by use of the parallel-axis theorem. Thus

$$I_x = I_{x_G} + A(\bar{y})^2 \qquad 16.8 \times 10^6 = I_{x_G} + 5625(40.3)^2 \qquad \text{and} \qquad I_{x_G} = 7.7 \times 10^6 \text{ mm}^4$$

Evidently for the loading shown, the fibers below the x_G-axis are in tension, while the fibers above this axis are in compression. Let c_1 and c_2 denote the distances of the extreme fibers from the neutral axis (x_G) as shown. Obviously $c_1 = 40.3$ mm and $c_2 = 84.7$ mm. The maximum tensile stress occurs in those fibers along B-B and is given by $\sigma = Mc_1/I$, where I denotes the moment of inertia of the entire cross section about the neutral axis passing through the centroid of the cross section. Thus the maximum tensile stress is given by

$$\sigma = \frac{Mc_1}{I} = 5 \times 10^3(10^3)(40.3)/7.7 \times 10^6 = 26.2 \text{ MPa}$$

The maximum compressive stress occurs in those fibers along A-A and is given by $\sigma = Mc_2/I$. To provide a consistent system of algebraic signs, it is necessary to assign a negative value to c_2 since it lies on the side of the x_G-axis opposite to that of c_1. Hence

$$\sigma = \frac{Mc_2}{I} = 5 \times 10^3(10^3)(-84.7)/7.7 \times 10^6 = -55 \text{ MPa}$$

The negative sign indicates that the stress is compressive.

8.14. A simply supported beam is loaded by the couple of 1000 lb · ft as shown in Fig. 8-26. The beam has a channel-type cross section as illustrated. Determine the maximum tensile and compressive stresses in the beam.

The bending moment diagram for this particular loading has been determined in Problem 6.11, where it was found to appear as in Fig. 8-27.

The techniques of Chap. 7 may be employed to locate the centroid as lying 1.5 in above the x-axis and the moment of inertia of the entire cross section about the x_G-axis as 41.6 in^4.

In this problem it is necessary to distinguish carefully between positive and negative bending moments. One method of attack is to consider a cross section of the beam slightly to the left of point B where the 1000 lb · ft couple is applied. According to the bending moment diagram the moment there is -600 lb · ft

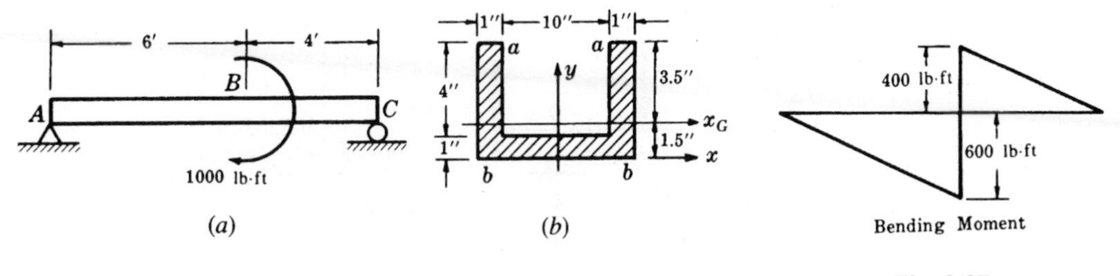

Fig. 8-26

Fig. 8-27

and, according to the sign convention adopted in Chap. 6, since the moment is negative the beam is concave downward at that section, as shown in Fig. 8-28. Thus the upper fibers are in tension and the lower fibers in compression. Along the upper fibers *a-a* the bending stress is given by $\sigma = My/I$. Then

$$\sigma_a = \frac{(-600)(12)(-3.5)}{41.6} = 605 \text{ lb/in}^2$$

Along the lower fibers *b-b* the value of *y* in the above formula for bending stress must be taken to be positive, and there we have

$$\sigma_b = \frac{(-600)(12)(+1.5)}{41.6} = -260 \text{ lb/in}^2$$

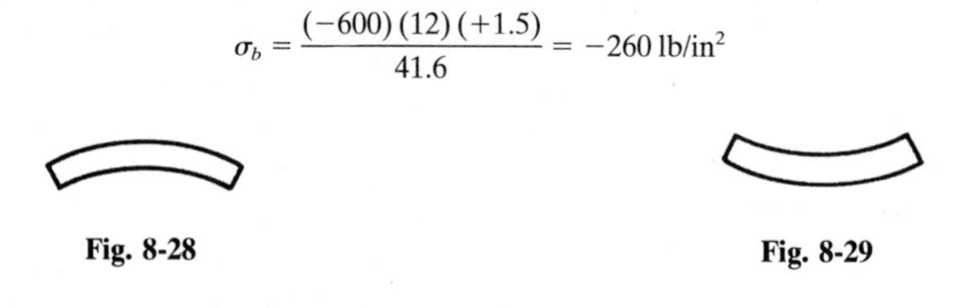

Fig. 8-28 **Fig. 8-29**

It is next necessary to investigate the bending stresses at a section slightly to the right of point *B*. There the bending moment is 400 lb · ft and according to the usual sign convention the beam is concave upward at that section, as shown in Fig. 8-29. Here the upper fibers are in compression and the lower fibers in tension. Along the upper fibers *a-a* the bending stress is

$$\sigma_a' = \frac{400(12)(-3.5)}{41.6} = -400 \text{ lb/in}^2$$

Along the lower fibers *b-b* we have

$$\sigma_b' = \frac{400(12)(1.5)}{41.6} = 170 \text{ lb/in}^2$$

The maximum tensile and compressive stresses must now be selected from the above four values. Evidently the maximum tension is 605 lb/in² occurring in the upper fibers just to the left of point *B*; the maximum compression is 400 lb/in² occurring in the upper fibers also but just to the right of point *B*.

8.15. Consider the beam with overhanging ends loaded by the three concentrated forces shown in Fig. 8-30. The beam is simply supported and of T-type cross section as shown. The material is gray cast iron having an allowable working stress in tension of 35 MPa and in compression of 150 MPa. Determine the maximum allowable value of *P*.

From symmetry each of the reactions denoted by *R* is equal to *P*/2. The bending moment diagram consists of a series of straight lines connecting the ordinates representing bending moments at the points *A*, *B*, *C*, *D*, and *E*. At *B* the bending moment is given by the moment of the force *P*/4 acting at *A* about an axis through *B*. Thus

$$M_B = -\left(\frac{P}{4}\right)(1) = \frac{-P}{4} \text{ N} \cdot \text{m}$$

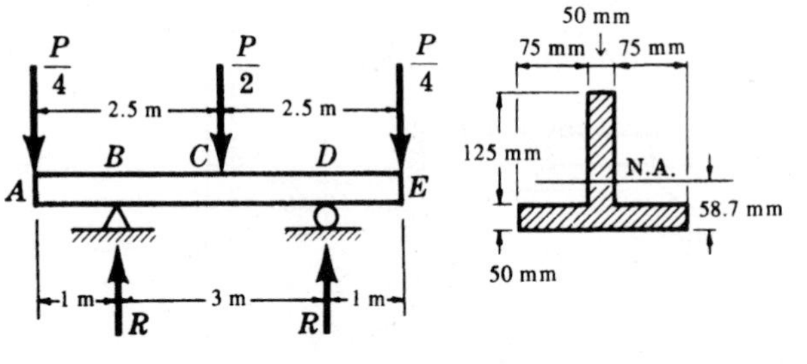

Fig. 8-30

At C the bending moment is given by the sum of the moments of the forces $P/4$ and $R = P/2$ about an axis through C. Thus

$$M_C = -\left(\frac{P}{4}\right)(2.5) + \left(\frac{P}{2}\right)(1.5) = \frac{P}{8}\,\text{N}\cdot\text{m}$$

The bending moment at D is equal to that at B by symmetry and the moment at each of the ends A and E is zero. Hence, the bending moment diagram plots as in Fig. 8-31.

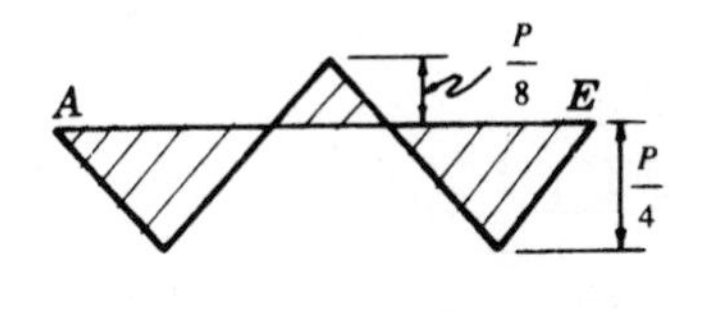

Fig. 8-31

Using the techniques described in Problem 8.13, we find the distance from the lower fibers of the flange to the centroid to be 58.7 mm and the moment of inertia of the area about the neutral axis passing through the centroid to be $40 \times 10^6\,\text{mm}^4$.

It is perhaps simplest to calculate four values of P based upon the various maximum tensile and compressive stresses that may exist at each of the points B and C and then select the minimum of these values. Let us first examine point B. Since the bending moment there is negative, the beam is concave downward at that point, as shown in Fig. 8-32. Evidently the upper fibers are in tension and the lower fibers are subject to compression. We shall first calculate a value of P, assuming that the allowable tensile stress of 35 MPa is realized in the upper fibers. Applying the flexure formula $\sigma = My/I$ to these upper fibers, we find

$$35 \times 10^6 = \frac{(-P/4)(0.116)}{40 \times 10^6(10^{-12})} \qquad \text{or} \qquad P = 48.3\,\text{kN}$$

Next we shall calculate a value of P, assuming that the allowable compressive stress of 150 MPa is set up in the lower fibers. Again applying the flexure formula, we find

$$-150 \times 10^6 = \frac{(-P/4)(0.0587)}{40 \times 10^6(10^{-12})} \qquad \text{or} \qquad P = 410\,\text{kN}$$

Fig. 8-32 **Fig. 8-33**

We shall now examine point C. Since the bending moment there is positive, the beam is concave upward at that point and appears as in Fig. 8-33. Here, the upper fibers are in compression and the lower fibers are subject to tension. First we will calculate a value of P, assuming that the allowable tension of 35 MPa is set up in the lower fibers. From the flexure formula we find

$$(35 \times 10^6) = \frac{(P/8)(0.0587)}{40 \times 10^6(10^{-12})} \qquad \text{or} \qquad P = 191\,\text{kN}$$

Last, we shall assume that the allowable compression of 150 MPa is set up in the upper fibers. Applying the flexure formula, we have

$$-150 \times 10^6 = \frac{(P/8)(-0.116)}{40 \times 10^6(10^{-12})} \qquad \text{or} \qquad P = 414\,\text{kN}$$

The minimum of these four values is $P = 48.3$ kN. Thus the tensile stress at the points B and D is the controlling factor in determining the maximum allowable load.

8.16. The cantilever beam ABC supports a uniform load over its right half and is of rectangular cross section with a square cutout as shown in Fig. 8-34. If the maximum permissible stress in either tension or compression is 140 MPa, determine the allowable uniform load w per unit length of the beam.

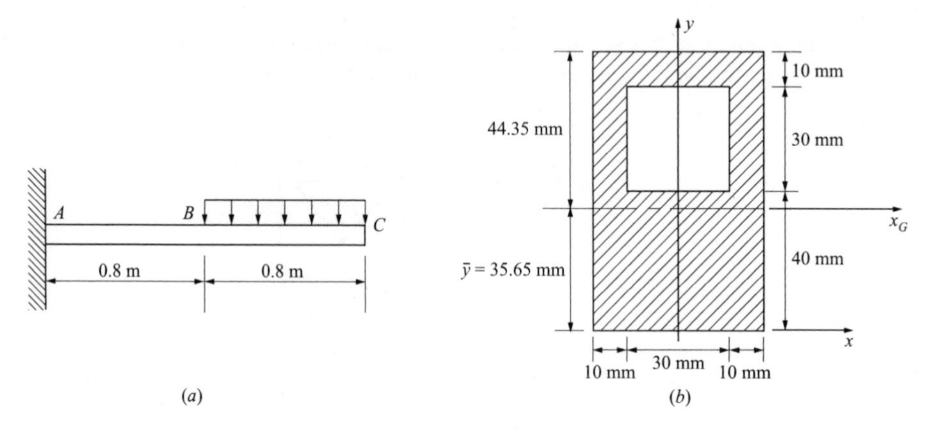

Fig. 8-34

It is first necessary to locate the neutral axis (N.A.) of the beam. For entirely elastic action this passes through the cross section of the beam and is given by (see Chap. 7)

$$\bar{y} = \frac{(80)(50)(40) - (30)(30)(55)}{(80)(50) - (30)(30)} = 35.65 \text{ mm}$$

Also, by the methods of Chap. 7, the moment of inertia about the x-axis is

$$I_x = \tfrac{1}{3}(50)(80)^3 - [\tfrac{1}{12}(30)(30)^3 + (900)(55)^2]$$
$$= (8193.25)(10)^3 \text{ mm}^4$$

Use of the parallel-axis theorem of Chap. 7 leads to the moment of inertia about an axis parallel to x but passing through the centroid, i.e., the x_G axis:

$$I_{x_G} = (8193.25)(10^3) \text{ mm}^4 - [(3100)(35.65)^2] = 4253.39 \times 10^3 \text{ mm}^2$$

The tensile fibers along the top surface of the beam are at a greater distance (44.35 mm) than the compressive fibers along the lower surface (35.65 mm). For these extreme fibers in tension we have

$$\sigma = \frac{Mc}{I}$$

$$140 \times 10^6 \text{ N/m}^2 = \frac{M(44.35 \text{ mm})(\text{m}/1000 \text{ mm})}{4253.39 \times 10^3 \text{ mm}^2 (\text{m}/1000 \text{ mm})^4}$$

Solving,

$$M_{max} = 13{,}372 \text{ N} \cdot \text{m}$$

From the loading conditions, $M_A = M_{max}$, so

$$M_A = M_{max} = (0.8 \text{ m} + 0.4 \text{ m}) w (0.8 \text{ m})$$

Solving,

$$w = \frac{(13{,}372 \text{ N} \cdot \text{m})}{(1.2 \text{ m})(0.8 \text{ m})}$$
$$= 13{,}929.6 \text{ N/m}$$

or

$$w = 13.93 \text{ kN/m}$$

8.17. The beam shown in Fig. 8-35 is of constant width b but the depth varies in the x-direction and further the depth is symmetric about the x-axis. Loading is due to a vertical force at the tip of the beam where $x = L$ and $y = 0$. Determine the equation of the beam contour $y = h(x)$ so that outer fiber bending stresses are equal to σ_0 at all points on the contour of the beam.

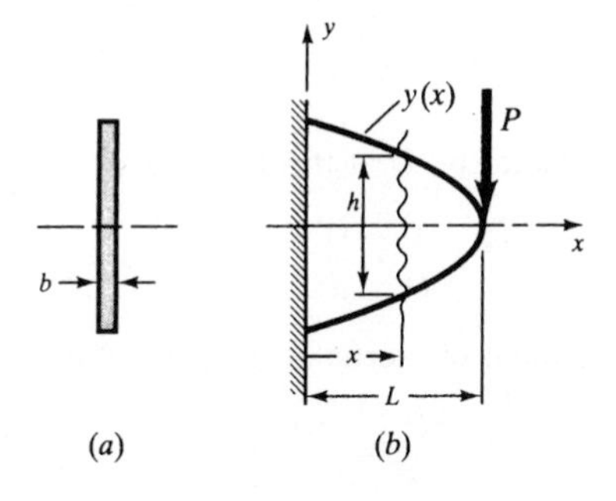

Fig. 8-35

The bending moment equation due to the concentrated load is $-P(L-x)$. From Problem 8.12, the section modulus of any cross section is given by $bh^2/6$. The outer fiber bending stresses along the top surface are, from Eq. (*8.3*),

$$\sigma = \frac{|M|}{Z} = \frac{P(L-x)}{(bh^2/6)} = \frac{6P(L-x)}{bh^2}$$

Since it is specified that this stress must be equal to σ_0 everywhere along the top surface, we have

$$\frac{6P(L-x)}{bh^2} = \sigma_0$$

Solving,

$$h = \sqrt{\frac{6P(L-x)}{b\sigma_0}}$$

This determines the beam contour for constant strength at all points along the length of the beam. This solution neglects the effect of the singular point $(L, 0)$ at the point of load application on stress distribution in the immediate vicinity of the force P.

8.18. A cantilever beam of circular cross section has the dimensions shown in Fig. 8-36. Determine the peak bending stress in the beam due to the concentrated force applied at the tip A.

To express the moment of inertia of the cross section at any point along the length of the beam in terms of the given geometry, we must first determine where the extensions of the top and bottom fibers would meet on the x-axis. From Fig. 8-36 we immediately have from similar triangles:

$$\frac{x_1}{d} = \frac{x_1 + L}{2.5d}$$

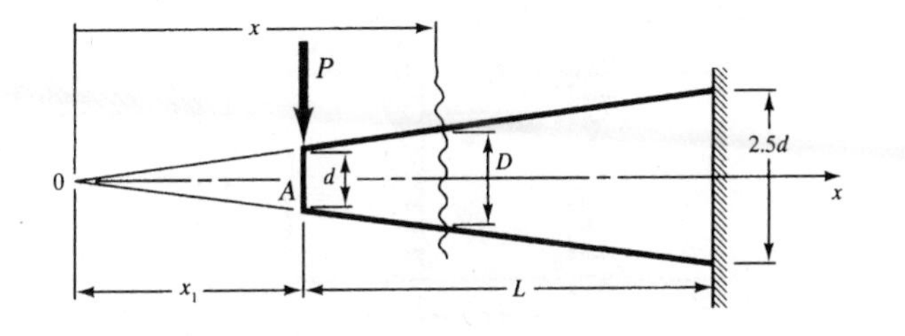

Fig. 8-36

Solving,

$$x_1 = \frac{2L}{3} \tag{1}$$

The bending moment at any station located a distance x from this fictitious point of intersection is

$$M = -P\left[x - \frac{2L}{3}\right] \tag{2}$$

If we designate the beam diameter by D at this location x, we have from geometry

$$\frac{x}{D} = \frac{x_1}{d} \quad \text{so} \quad D = \frac{3xd}{2L} \tag{3}$$

so that the cross-sectional moment of inertia at the general location x is

$$I = \frac{\pi D^4}{64} = \frac{\pi}{64}\left[\frac{3d}{2L}x\right]^4 = \left(\frac{81\pi d^4}{(64)(16)L^4}\right)x^4 \tag{4}$$

From Eq. (*8.2*) we find the outer fiber bending stresses to be

$$\sigma = \frac{M_c}{I} = \frac{P(x - 2L/3)(3xd/2L)}{[81\pi d^4/(64)(16)L^4]x^4} = \frac{256PL^3}{9\pi d^3}\left[\frac{x - 2L/3}{x^3}\right] \tag{5}$$

Note that Eq. (*5*) indicates that the peak bending stress does *not* occur at the clamped end $x = L$.

To find where the outer fiber stresses reach a maximum value, we take the derivative $d\sigma/dx$ and set it equal to zero to locate the critical value of x. Thus,

$$\frac{d\sigma}{dx} = \left(\frac{256PL^3}{9\pi d^3}\right)\left[\frac{x^3(1) - (x - 2L/3)3x^2}{x^6}\right] = 0 \tag{6}$$

Solving, $x = L$ measured from point 0. Substituting this value of x in Eq. (*5*), we find the peak outer fiber bending stress to be

$$\sigma = \frac{256PL^3}{9\pi d^3}\left[\frac{L - 2L/3}{L^3}\right] = \left(\frac{256PL^3}{9\pi d^3}\right)\left(\frac{1}{3L^2}\right) = 3.02\frac{PL}{d^3}$$

Note that from Eq. (*5*) the outer fiber bending stress at the clamped end $x = (L + 2L/3)$ is $1.96\ PL/d^3$, which is less than the peak value.

8.19. In a beam loaded by transverse forces acting perpendicular to the axis of the beam, not only are bending stresses parallel to the axis of the bar produced but shearing stresses also act over cross sections of the beam perpendicular to the axis of the bar. Express the intensity of these shearing stresses in terms of the shearing force at the section and the properties of the cross section.

The theory to be developed applies only to a cross section of rectangular shape. However, the results of this analysis are commonly used to give approximate values of the shearing stress in other cross sections having a plane of symmetry.

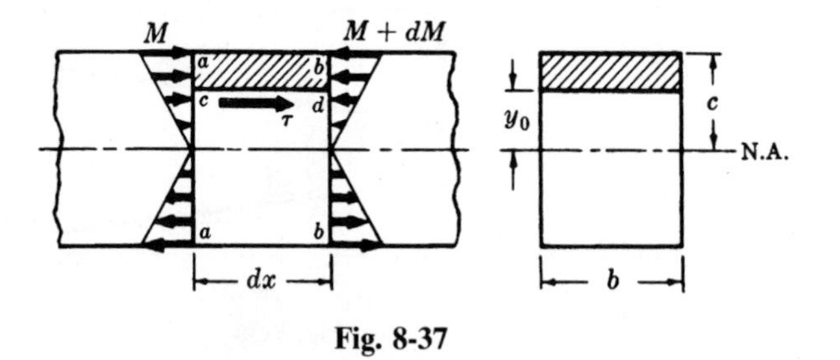

Fig. 8-37

Let us consider an element of length dx cut from a beam as shown in Fig. 8-37. We shall denote the bending moment at the left side of the element by M and that at the right side by $M + dM$, since in general the bending moment changes slightly as we move from one section to an adjacent section of the beam. If y is measured upward from the neutral axis, then the bending stress at the left section a-a is given by

$$\sigma = \frac{My}{I}$$

where I denotes the moment of inertia of the entire cross section about the neutral axis. This stress distribution is illustrated above. Similarly, the bending stress at the right section b-b is

$$\sigma' = \frac{(M + dM)y}{I}$$

Let us now consider the equilibrium of the shaded element $acdb$. The force acting on an area da of the face ac is merely the product of the intensity of the force and the area; thus

$$\sigma\, da = \frac{My}{I} da$$

The sum of all such forces over the left face ac is found by integration to be

$$\int_{y_0}^{c} \frac{My}{I} da$$

Likewise, the sum of all normal forces over the right face bd is given by

$$\int_{y_0}^{c} \frac{(M + dM)y}{I} da$$

Evidently, since these two integrals are unequal, some additional horizontal force must act on the shaded element to maintain equilibrium. Since the top face ab is assumed to be free of any externally applied horizontal forces, then the only remaining possibility is that there exists a horizontal shearing force along the lower face cd. This represents the action of the lower portion of the beam on the shaded element. Let us denote the shearing stress along this face by τ as shown. Also, let b denote the width of the beam at the position where τ acts. Then the horizontal shearing force along the face cd is $\tau b\, dx$. For equilibrium of the element $acdb$ we have

$$\Sigma F_h = \int_{y_0}^{c} \frac{My}{I} da - \int_{y_0}^{c} \frac{(M + dM)y}{I} da + \tau b\, dx = 0$$

Solving,

$$\tau = \frac{1}{Ib} \frac{dM}{dx} \int_{y_0}^{c} y\, da$$

But from Problem 6.1 we have $V = dM/dx$, where V represents the shearing force (in pounds or Newtons) at the section a-a. Substituting,

$$\tau = \frac{V}{Ib} \int_{y_0}^{c} y\, da \qquad (1)$$

The integral in this last equation represents the first moment of the shaded cross-sectional area about the neutral axis of the beam. This area is always the portion of the cross section that is above the level at which the desired shear acts. This first moment of area is sometimes denoted by Q in which case the above formula becomes

$$\tau = \frac{VQ}{Ib} \qquad (2)$$

The units of $\int y\, da$ or of Q are in^3 or m^3.

The shearing stress τ just determined acts horizontally as shown in Fig. 8-37. However, let us consider the equilibrium of a thin element $mnop$ of thickness t cut from any body and subject to a shearing stress

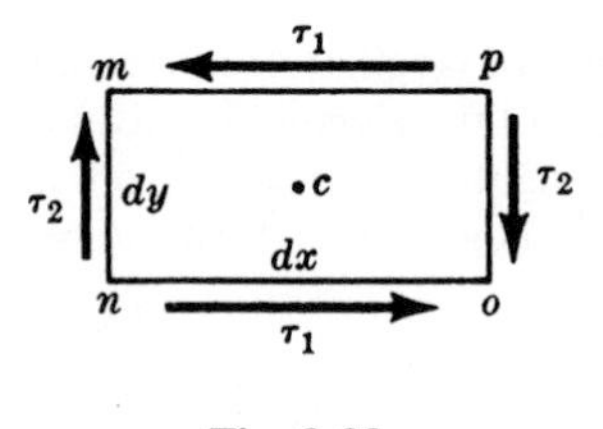

Fig. 8-38

τ_1 on its lower face, as shown in Fig. 8-38. The total horizontal force on the lower face is $\tau_1 t\,dx$. For equilibrium of forces in the horizontal direction, an equal force but acting in the opposite direction must act on the upper face, hence the shear stress intensity there too is τ_1. These two forces give rise to a couple of magnitude $\tau_1 t\,dx\,dy$. The only way in which equilibrium of the element can be maintained is for another couple to act over the vertical faces. Let the shear stress intensity on these faces be denoted by τ_2. The total force on either vertical face is $\tau_2 t\,dy$. For equilibrium of the moments about the center of the element we have

$$\Sigma M_c = \tau_1 t\,dx\,dy - \tau_2 t\,dy\,dx = 0 \qquad \text{or} \qquad \tau_1 = \tau_2$$

Thus we have the interesting conclusion that the shearing stresses on any two perpendicular planes through a point on a body are equal. Consequently, not only are there shearing stresses τ acting horizontally at any point in the beam, but shearing stresses of an equal intensity also act vertically at that same point.

In summary, when a beam is loaded by transverse forces, both horizontal and vertical shearing stresses arise in the beam. The vertical shearing stresses are of such magnitudes that their resultant at any cross section is exactly equal to the shearing force V at that same section.

8.20. A beam of rectangular cross section is simply supported at the ends and subject to the single concentrated force shown in Fig. 8-39(a). Determine the maximum shearing stress in the beam. Also, determine the shearing stress at a point 1 in below the top of the beam at a section 1 ft to the right of the left reaction.

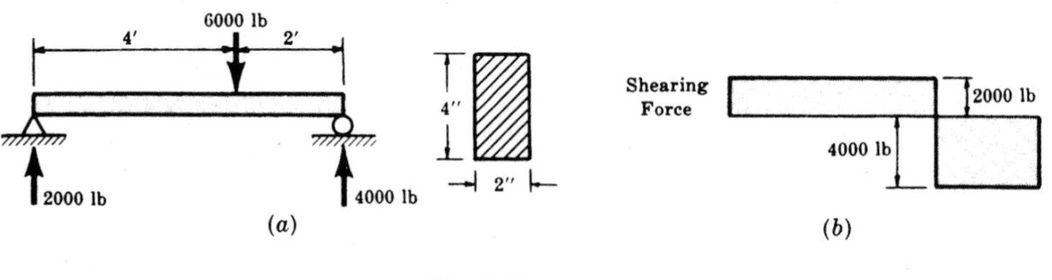

Fig. 8-39

The reactions are readily found from statics to be 2000 lb and 4000 lb as shown. The shearing force diagram for this type of loading appears in Fig. 8-39(b).

From the shear diagram, the shearing force acting at a section 1 ft to the right of the left reaction is 2000 lb. The shearing stress τ at any point in this section a distance y_0 from the neutral axis was shown in Problem 8.19 and also Eq. (*8.5*) to be

$$\tau = \frac{V}{2I}\left(\frac{h^2}{4} - y_0^2\right) \tag{1}$$

At a point 1 in below the top fibers of the beam, $y_0 = 1$ in. Also, we have $h = 4$ in and $I = bh^3/12 = 2(4)^3/12 = 10.67\ \text{in}^4$. Substituting,

$$\tau_{y_0=1} = \frac{2000}{2(10.67)}\left(\frac{4^2}{4} - 1\right) = 280\ \text{lb/in}^2$$

From Eq. (*1*) it is clear that the peak shearing stress occurs at the neutral axis where $y_0 = 0$. Thus,

$$\tau_{\max} = \frac{4000}{2(10.67)}\left(\frac{4^2}{4} - 1\right) = 750\ \text{lb/in}^2$$

Note that for a rectangular cross section this peak shearing stress is 50 percent greater than the average shearing stress, which is given by

$$\tau_{\text{mean}} = \frac{4000}{(4)(2)} = 500\ \text{lb/in}^2$$

8.21. Consider the cantilever beam subject to the concentrated load shown in Fig. 8-40. The cross section of the beam is of T-shape. Determine the maximum shearing stress in the beam and also determine the shearing stress 25 mm from the top surface of the beam at a section adjacent to the supporting wall.

The shear force has a constant value of 50 kN at all points along the length of the beam. Because of this simple, constant value the shear diagram need not be drawn.

The location of the centroid and the moment of inertia about the centroidal axis for this particular cross section were determined in Problem 8.15. The centroid was found to be 58.7 mm above the lower surface of the beam and the moment of inertia about a horizontal axis through the centroid was found to be $40 \times 10^6\ \text{mm}^4$.

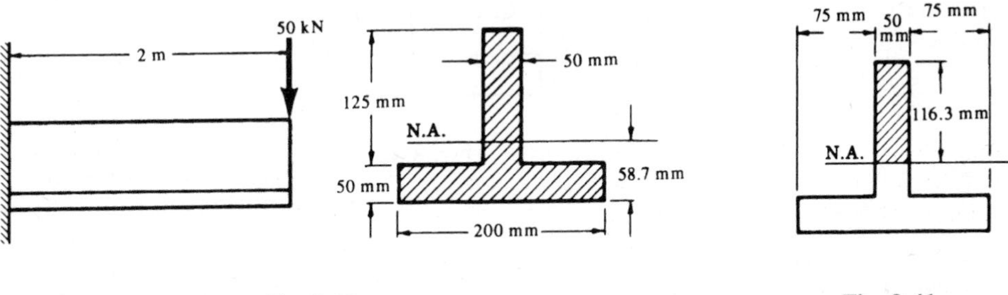

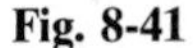

Fig. 8-40 **Fig. 8-41**

The shearing stress at a distance y_0 from the neutral axis through the centroid was found in Problem 8.19 to be

$$\tau = \frac{V}{Ib}\int_{y_0}^{c} y\,da$$

Inspection of this equation reveals that the shearing stress is a maximum at the neutral axis, since at that point $y_0 = 0$ and consequently the integral assumes the largest possible value. It is not necessary to integrate, however, since the integral is known in this case to represent the first moment of the area between the neutral axis and the outer fibers of the beam about the neutral axis. This area is represented by the shaded region in Fig. 8-41. The value of the integral could also, of course, be found by taking the first moment of the unshaded area below the neutral axis about the line, but that calculation would be somewhat more difficult.

Thus the first moment of the shaded area about the neutral axis is

$$50(116.3)(58.15) = 3.38 \times 10^5\ \text{mm}^3$$

and the shearing stress at the neutral axis, where $b = 50$ mm, is found by substitution in the above general formula to be

$$\tau = \frac{50 \times 10^3}{50(40 \times 10^6)}(3.38 \times 10^5) = 8.45\ \text{MPa}$$

In this formula b was taken to be 50 mm, since that is the width of the beam at the point where the shearing stress is being calculated. Thus the maximum shearing stress is 8.45 MPa and it occurs at all points on the neutral axis along the entire length of the beam, since the shearing force has a constant value along the entire length of the beam.

The shearing stress 25 mm from the top surface of the beam is again given by the formula

$$\tau = \frac{V}{Ib}\int_{y_0}^{c} y\,da$$

Now, the integral represents the first moment of the new shaded area shown in Fig. 8-42, about the neutral axis. Again it is not necessary to integrate to evaluate the integral, since the coordinate of the centroid of this shaded area is known. It is 103.8 mm above the neutral axis. Thus the first moment of this shaded area about the neutral axis is $50(25)(103.8) = 1.3 \times 10^5\ \text{mm}^3$, and the shearing stress 25 mm below the top fibers is

$$\tau = \frac{50 \times 10^3}{50(40 \times 10^6)}(1.3 \times 10^5) = 3.25\ \text{MPa}$$

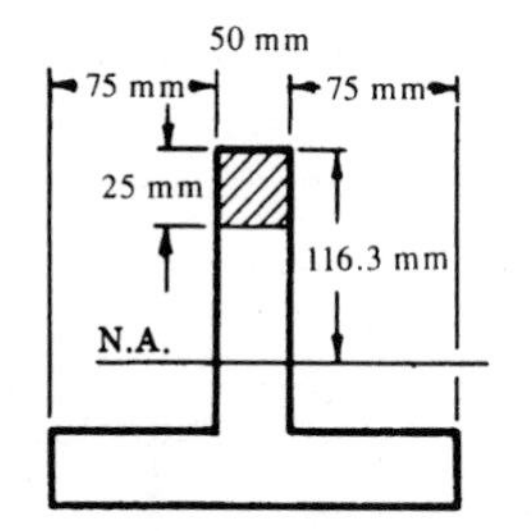

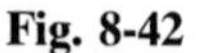
Fig. 8-42

Again, b was taken to be 50 mm since that is the width of the beam at the point where the shearing stress is being evaluated. Since the shearing force is equal to 50 kN everywhere along the length of the beam, the shearing stress 25 mm below the top fibers is 3.25 MPa everywhere along the beam.

8.22. The vertically oriented wide-flange section shown in Fig. 8-43 is loaded by a single horizontal concentrated force of 6.5 kN directed parallel to the z-axis. Determine the horizontal shear stress distribution on a flange at a section 3 m above the lower clamped end in the x-z plane.

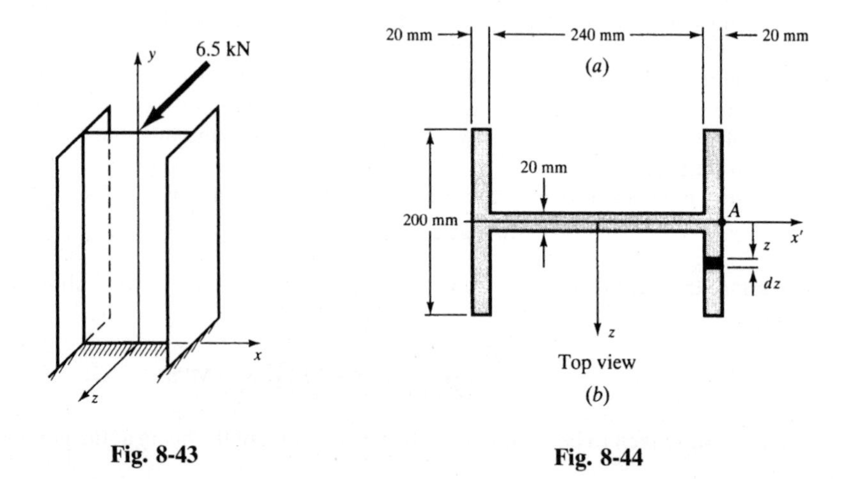

Fig. 8-43 **Fig. 8-44**

Figure 8-44 shows a typical horizontal cross section parallel to the x-z plane as well as dimensions of the web and flange. The shear stress τ in this plane acts in the z-direction and at a distance z from the x-axis. The specification of 3 m above the x-z plane is unimportant: all that matters is that the equation for shear stress derived in Problem 8.19 does not apply at horizontal sections near either the bottom or top of the vertically oriented bar. To apply Eq. (*1*) of Problem 8.19 to find τ we must first determine the moment of inertia of the cross section about the x-axis. From the methods of Chap. 7, we find

$$I = \tfrac{1}{12}(2)\,(20\text{ mm})\,(200\text{ mm})^3 + \tfrac{1}{12}(240\text{ mm})\,(20\text{ mm})^2 = 2683 \times 10^4\text{ mm}^4 \qquad (1)$$

We next introduce a coordinate z running from the x-y plane in the direction of the z-axis, and appearing as in Fig. 8-44. From Problem 8.19 we have here $V = 6500$ N, and the flange thickness b here is 0.02 m. The integral in Problem 8.19 represents the first moment of the area extending from z to the extreme fibers of the flange—that area is shaded in Fig. 8-44. Thus, we need not integrate and we may evaluate the first moment of the shaded area about the x'-axis by taking the product of the area and the distance of the centroid of the area from the x'-axis: that is,

$$[(0.1 - z)\,(0.04\text{ m})]\left(\frac{0.1+z}{2}\text{m}\right) \qquad \text{or} \qquad (0.02)\,[(0.1)^2 - z^2]\text{ m}^3$$

Equation (*2*) of Problem 8.19 now yields the desired shearing stress as

$$\begin{aligned}\tau &= \frac{6500\text{ N}}{(26.83\times 10^{-6}\text{ m}^4)\,(0.02\text{ m})\,(2)}\{(0.02)\,[(0.1)^2 - z^2]\text{ m}^3\}\\ &= 121.1[(0.1)^2 - z^2]\,(10^6)\end{aligned} \qquad (2)$$

At the point A where the value of z is zero, the peak shearing stress is found from Eq. (*2*) to be

$$\tau_A = (121.1)\,[(0.1)^2 - 0]\,(10^6) = 1.21 \times 10^6\text{ N/m}^2 \qquad \text{or} \qquad 1.21\text{ MPa}$$

8.23. Consider a beam having an I-type cross section as shown in Fig. 8-45. A shearing force V of 150 kN acts over the section. Determine the maximum and minimum values of the shearing stress in the vertical web of the section.

The shearing stress at any point in the cross section is given by

$$\tau = \frac{V}{Ib}\int_{y_0}^{c} y\,da$$

as derived in Problem 8.19. Here, y_0 represents the location of the section on which τ acts, and is measured from the neutral axis as shown. In this expression, I represents the moment of inertia of the entire cross section about the neutral axis, which passes through the centroid of the section. I is readily calculated by dividing the section into rectangles, as indicated by the dashed lines, and we have

$$I = \tfrac{1}{12}(10)\,(350)^3 + 2[\tfrac{1}{12}(200)\,(25)^3 + 200(25)\,(187.5)^2] = 389 \times 10^6\text{ mm}^4$$

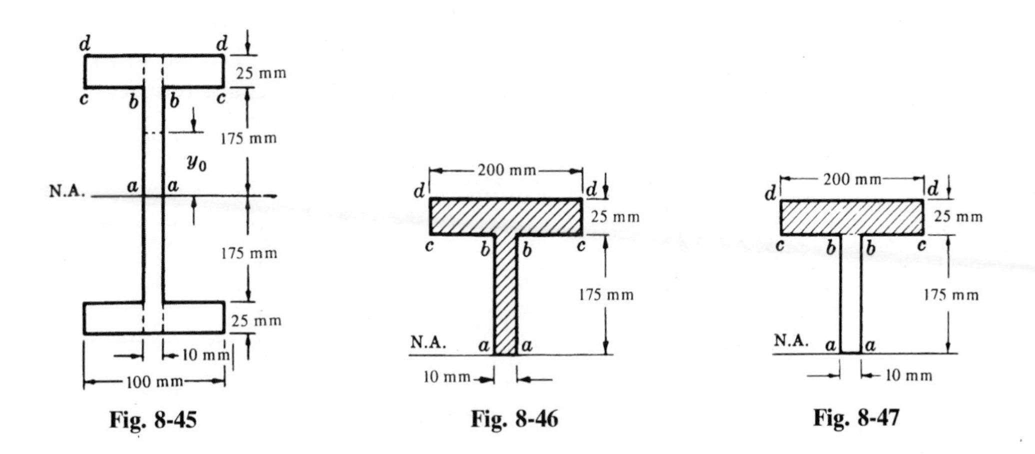

Fig. 8-45 **Fig. 8-46** **Fig. 8-47**

Inspection of the general formula for shearing stress reveals that this stress has a maximum value when $y_0 = 0$, that is, at the neutral axis, since at that point the integral takes on its largest possible value. It is not necessary to integrate to obtain the value of $\int_{y_0}^{c} y\,da$, since this integral is shown to represent the first moment of the area between $y_0 = 0$ (that is, the neutral axis) and the outer fibers of the beam. This area is shaded in Fig. 8-46. For this area we have, taking its first moment about the neutral axis,

$$\int_0^{200} y\,da = 175(10)(87.5) + 200(25)(187.5) = 1.1 \times 10^6\ \text{mm}^3$$

Consequently the maximum shearing stress in the web occurs at the section *a-a* along the neutral axis and by substituting in the general formula for shearing stress is found to be

$$\tau_{\max} = \frac{150 \times 10^3}{10(389 \times 10^6)}(1.1 \times 10^6) = 42.4\ \text{MPa}$$

The minimum shearing stress in the web occurs at that point in the web farthest from the neutral axis, i.e., across the section *b-b*. To calculate the shearing stress there, it is necessary to evaluate $\int_{y_0}^{c} y\,da$ for the area between *b-b* and the outer fibers of the beam. This is the shaded area shown in Fig. 8-47. Again, it is not necessary to integrate, since this integral merely represents the first moment of this shaded area about the neutral axis. It is

$$\int_{175}^{200} y\,da = 200(25)(187.5) = 9.375 \times 10^5\ \text{mm}^3$$

The value of b is still 10 mm, since that is the width of the beam at the position where the shearing stress is being calculated. Substituting in the general formula

$$\tau_{\min} = \frac{150 \times 10^3}{10(389 \times 10^6)} = (9.375 \times 10^5) = 36.2\ \text{MPa}$$

It is to be noted that there is not too great a difference between the maximum and minimum values of shearing stress in the web of the beam. In fact, it is customary to calculate only an approximate value of the shearing stress in the web of such an I-beam. This value is obtained by dividing the total shearing force V by the cross-sectional area of the web alone. This approximate value becomes

$$\tau_{\text{av}} = \frac{100 \times 10^3}{(400)(10)} = 37.5\ \text{MPa}$$

A more advanced analysis of shearing stresses in an I-beam reveals that the vertical web resists nearly all of the shearing force V and that the horizontal flanges resist only a small portion of this force. The shear stress in the web of an I-beam is specified by various codes at rather low values. Thus some codes specify 70 MPa, others 90 MPa.

Plastic Bending of Beams

8.24. Consider a beam of arbitrary doubly symmetric cross section, as in Fig. 8-48(a), subject to pure bending. The material is considered to be elastic-perfectly plastic, i.e., the stress-strain diagram has the appearance shown in Fig. 8-48(b) and stress-strain characteristics in tension and

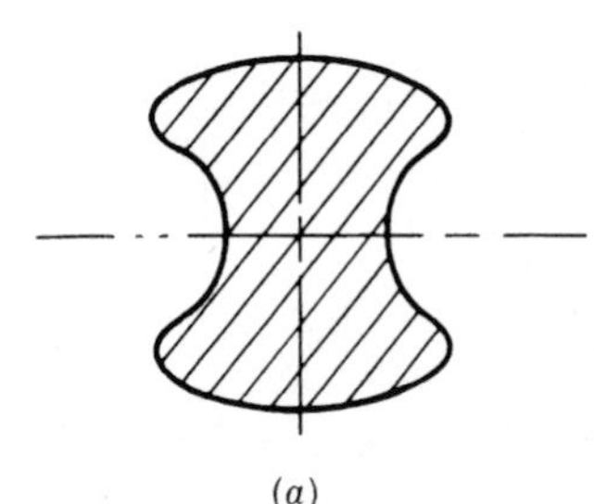

(a)

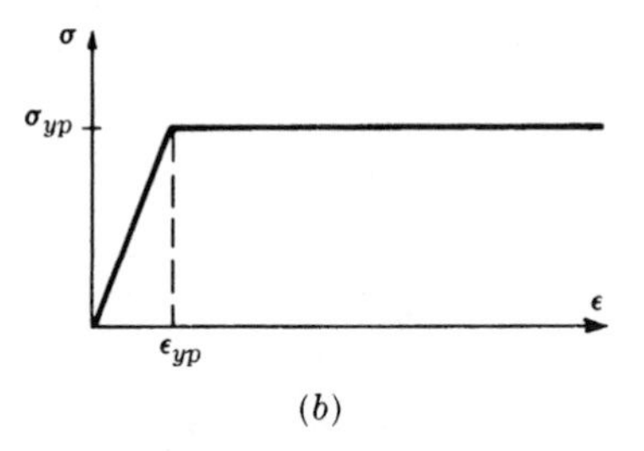

(b)

Fig. 8-48

compression are identical. Determine the moment acting on the beam when all fibers a distance y_1 from the neutral axis have reached the yield point of the material.

Even though bending of the beam has caused the outer fibers to have yielded it is still realistic to assume that plane sections of the beam normal to the axis before loads are applied remain plane and normal to the axis after loading. Consequently, normal strains of the longitudinal fibers of the beam still vary linearly with the distance of the fiber from the neutral axis.

As the value of the applied moment is increased, the extreme fibers of the beam are the first to reach the yield point of the material and the normal stresses on all interior fibers vary linearly as the distance of the fiber from the neutral axis, as indicated in Fig. 8-49(a). A further increase in the value of the moment puts interior fibers at the yield point, with yielding progressing from the outer fibers inward, as indicated in Fig. 8-49(b). In the limiting case when all fibers (except those along the neutral axis) are stressed to the yield point the normal stress distribution appears as in Fig. 8-49(c). The bending moment corresponding to Fig. 8-49(c) is termed a *fully plastic moment.* For the type of stress-strain curve shown in Fig. 8-48(b), no greater moment is possible.

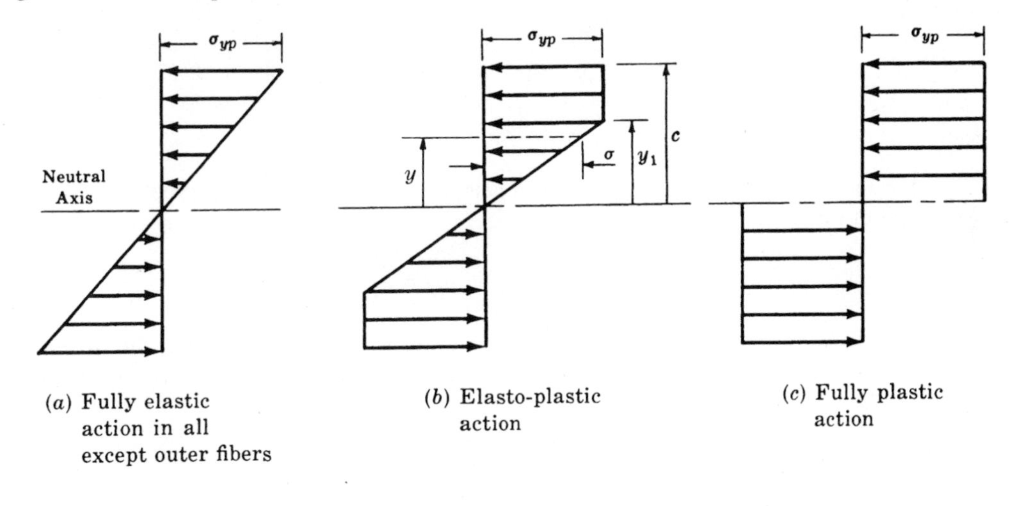

(a) Fully elastic action in all except outer fibers

(b) Elasto-plastic action

(c) Fully plastic action

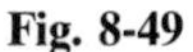

Fig. 8-49

For a beam in pure bending, the sum of the normal forces over the cross section must vanish. Hence, for the doubly symmetric section under consideration, it is evident from inspection of Fig. 8-49(b) that the neutral axis must pass through the centroid of such a section; i.e., the area above the neutral axis must be equal to the area below that axis. However, in Problem 8.29 it will be found that for a more general, nonsymmetric cross section the location of the neutral axis after certain of the fibers have yielded is not the same as that found for purely elastic action where the neutral axis passes through the centroid of the cross section.

From Fig. 8-48(b) we have for $y < y_1$:

$$\frac{\sigma}{y} = \frac{\sigma_{yp}}{y_1} \quad \text{or} \quad \sigma = \frac{y}{y_1}\sigma_{yp}$$

and for $y > y_1 : \sigma = \sigma_{yp}$ = constant. Thus the bending moment is

$$\begin{aligned} M &= \int \sigma y\, da = 2\int_0^{y_1} \frac{y}{y_1}\sigma_{yp}\, y\, da + 2\int_{y_1}^{c} \sigma_{yp}\, y\, da \\ &= \frac{2\sigma_{yp}}{y_1}\int_0^{y_1} y^2\, da + 2\sigma_{yp}\int_{y_1}^{c} y\, da \end{aligned}$$

8.25. For a beam of rectangular cross section determine the moment acting when all fibers a distance y_1 from the neutral axis have reached the yield point of the material.

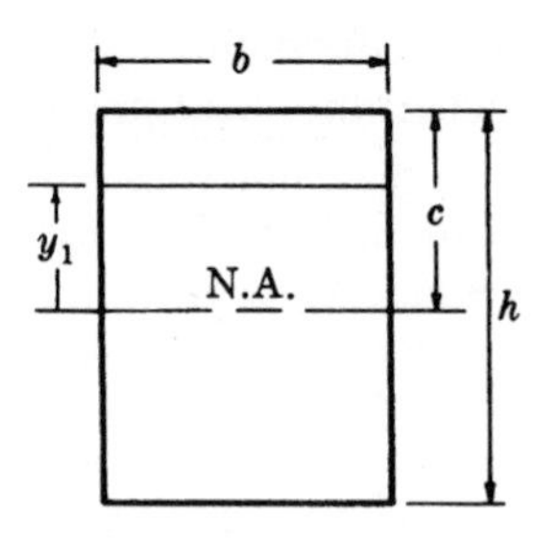

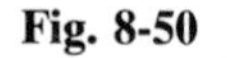
Fig. 8-50

From the result of Problem 8-24 for the geometry indicated in Fig. 8-50 we have

$$M = \frac{2\sigma_{yp}}{y_1}\left(\frac{1}{3}by_1^3\right) + 2\sigma_{yp}b(c - y_1)\left(\frac{c + y_1}{2}\right)$$

$$= \left(bc^2 - \frac{b}{3}y_1^2\right)\sigma_{yp}$$

For the limiting case when $y_1 = 0$ which is indicated by Fig. 8-49(c) of Problem 8.24 the fully plastic moment of this rectangular beam is

$$M_p = bc^2\sigma_{yp} = \frac{bh^2}{4}\sigma_{yp} \qquad (1)$$

It is to be noted that the maximum possible elastic moment, i.e., when the extreme fibers have just reached the yield point but all interior fibers are in the elastic range of action as indicated by Fig. 8-49(a), is

$$M_e = \frac{bh^2}{6}\sigma_{yp} \qquad (2)$$

Thus, for a rectangular cross section, the fully plastic moment is 50 percent greater than the maximum possible elastic moment.

8.26. Determine the fully plastic moment of a rectangular beam, 1×2 in in cross section, of steel with a yield point of 38,000 lb/in^2. Compare this with the maximum possible elastic moment that this same section may carry.

From (1) of Problem 8.25, the fully plastic moment is

$$M_p = \frac{1(2)^2}{4}(38{,}000) = 38{,}000 \text{ lb} \cdot \text{in}$$

From (2) of that same problem, the maximum possible elastic moment is

$$M_e = \frac{1(2)^2}{6}(38{,}000) = 25{,}400 \text{ lb} \cdot \text{in}$$

It is evident that M_p is 50 percent greater than M_e.

8.27. For a beam of rectangular cross section (Fig. 8-51) determine the relation between the bending moment and the radius of curvature when all fibers at a distance y_1 from the neutral axis have reached the yield point of the material.

As in Problem 8.25, we assume that plane sections before loading remain plane and normal to the beam axis after loading. Because of this, normal strains of the longitudinal fibers vary linearly as the

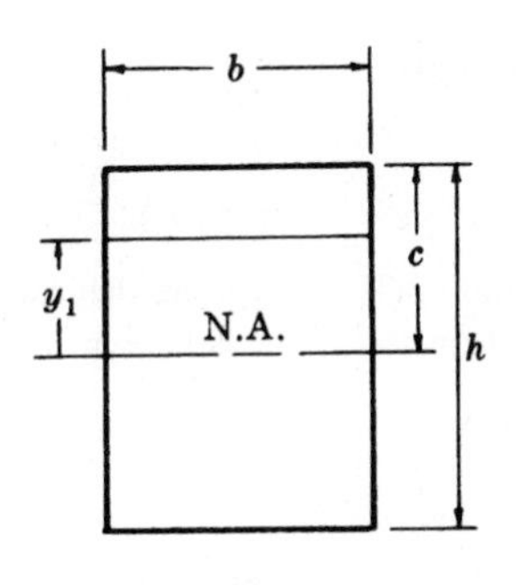

Fig. 8-51

distance of the fibers from the neutral axis. Thus, if ϵ_{yp} denotes the strain of the fibers at a distance y_1 from the neutral axis and ϵ_c represents the outer fiber strain, we have

$$\frac{\epsilon_c}{c} = \frac{\epsilon_{yp}}{y_1} \tag{1}$$

Consideration of the geometry of an originally rectangular element of length dx along the beam axis, as shown in Fig. 8-52(a), reveals that after bending it assumes the configuration indicated in Fig. 8-52(b). From that sketch we have

$$\frac{1}{R} = \frac{d\theta}{dx} = \frac{\epsilon_c}{c} \tag{2}$$

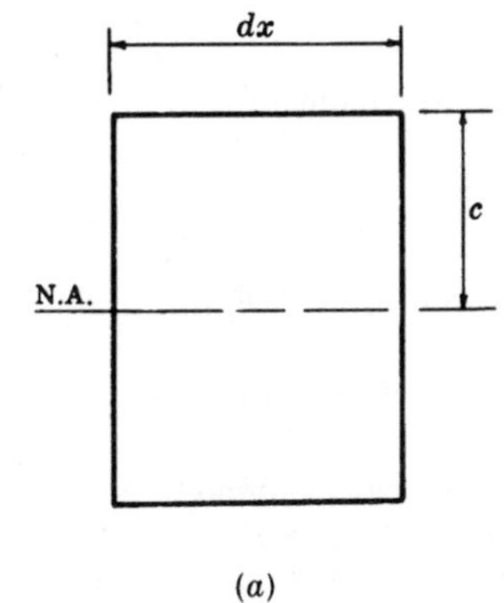

(a)

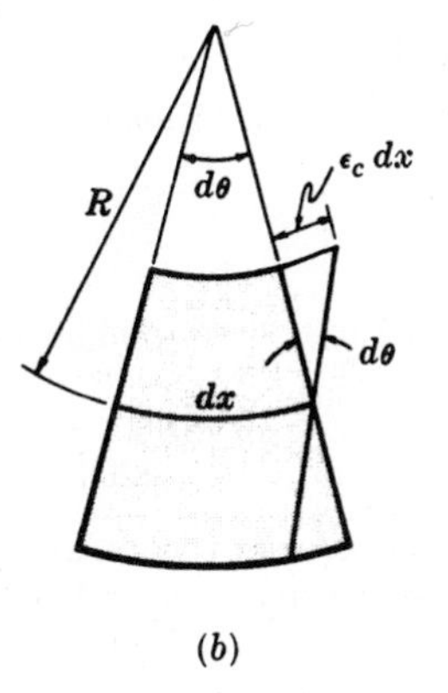

(b)

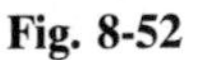

Fig. 8-52

Thus
$$\frac{d\theta}{dx} = \frac{\epsilon_{yp}}{y_1} = \frac{\sigma_{yp}}{Ey_1} \tag{3}$$

since the fibers a distance y_1 from the neutral axis obey Hooke's law: $\sigma_{yp} = E\epsilon_{yp}$. From Problem 8.25, the moment corresponding to these strains is

$$M = \left(bc^2 y_1 - \frac{b}{3} y_1^3\right)\frac{\sigma_{yp}}{y_1} \tag{4}$$

Thus, from (3) and (4),

$$\frac{d\theta}{dx} = \frac{M}{Eby_1(c^2 - \frac{1}{3}y_1^2)} \tag{5}$$

Finally, from (*2*) and (*5*) we have

$$\frac{1}{R} = \frac{M}{EI(M/M_e)\sqrt{3 - 2M/M_e}} \tag{6}$$

where $M_e = bh^2\sigma_{yp}/6$ as in Problem 8.25. This is the desired relation between the bending moment M and the radius of curvature R. Equation (*6*) plots as shown in Fig. 8-53.

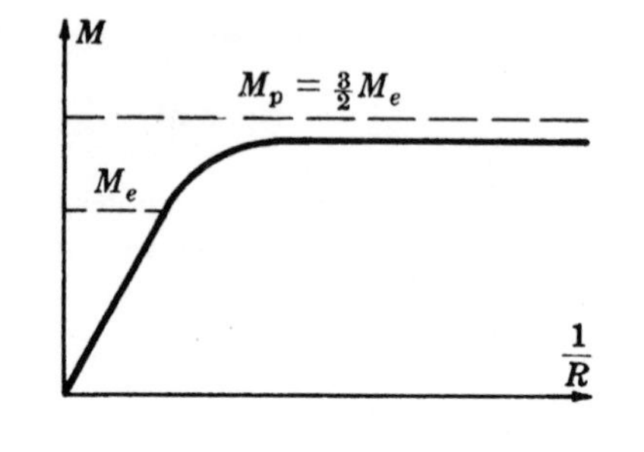

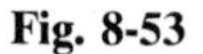
Fig. 8-53

8.28. Consider a beam of rectangular cross section where $b = 25$ mm, $h = 10$ mm. The material is steel for which $\sigma_{yp} = 200$ MPa and $E = 200$ GPa. Determine the radius of curvature corresponding to the maximum possible elastic moment and also the radius of curvature for a moment of 100 N · m.

From (*2*) of Problem 8.25, the maximum possible elastic moment is

$$M_e = \frac{0.025(0.01)^2}{6}(200 \times 10^6) = 83 \text{ N} \cdot \text{m}$$

The curvature corresponding to this moment is found from (*6*) of Problem 8.27 to be

$$\frac{1}{R} = \frac{83}{(200 \times 10^9)[(0.025)(0.01)^3/12]\sqrt{3 - 2}} = 0.2 \qquad \text{or} \qquad R = 5 \text{ m}$$

The value of y_1 corresponding to a moment of 100 N · m may be found from Problem 8.25 to be 4 mm. The curvature corresponding to this is found from (*6*) of Problem 8.27 to be

$$\frac{1}{R} = \frac{100}{(200 \times 10^9)[(0.025)(0.01)^3/12]\sqrt{3 - 200/83}} = 0.312 \qquad \text{or} \qquad R = 3.2 \text{ m}$$

8.29. Consider the more general case of a beam with a cross section symmetric only about the vertical axis, as shown in Fig. 8-54(*a*). For fully plastic bending [Fig. 8-54(*b*)], determine the location of the neutral axis.

Although the location of the neutral axis is unknown, let us denote the area of that portion of the cross section lying below that axis by A_1 and the area of the portion above that axis by A_2. As shown by Fig. 8-54(*b*), all fibers in A_1 are subject to a tensile stress equal to the yield point of the material and all fibers in A_2 are subject to the same magnitude compressive stress. For horizontal equilibrium of these forces, we have

$$\sigma_{yp}A_1 - \sigma_{yp}A_2 = 0 \tag{1}$$

from which

$$A_1 = A_2 = \frac{A}{2} \tag{2}$$

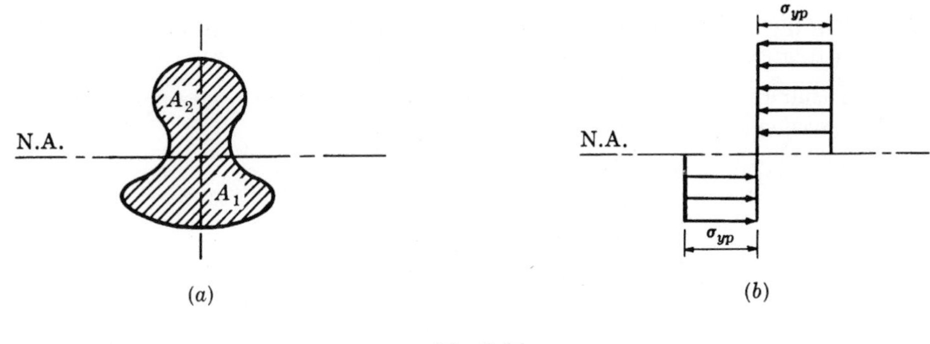

Fig. 8-54

where A is the area of the entire cross section. Thus, for fully plastic action, the neutral axis divides the cross section into two equal parts. This is in contrast to the situation for fully elastic action, where the neutral axis was found in Problem 8.1 to pass through the centroid of the cross section.

Also, the sum of the moments of the tensile and compressive stresses must equal the applied moment M_p, the fully plastic moment. If $\bar{y}_1$ and $\bar{y}_2$ denote the distances from the neutral axis to the centroids of the areas A_1 and A_2, respectively, then from statics

$$\sigma_{yp} A_1 \bar{y}_1 + \sigma_{yp} A_2 \bar{y}_2 = M_p \tag{3}$$

From (2) this becomes

$$\sigma_{yp} \frac{A}{2} (\bar{y}_1 + \bar{y}_2) = M_p \tag{4}$$

or

$$\sigma_{yp} = \frac{M_p}{(A/2)(\bar{y}_1 + \bar{y}_2)} \tag{5}$$

This is frequently written in the form

$$\sigma_{yp} = \frac{M_p}{Z_p} \tag{6}$$

where $Z_p = (A/2)(\bar{y}_1 + \bar{y}_2)$ is termed the *plastic section modulus.*

8.30. For a W8 × 40 wide-flange section of steel having a yield point of 38,000 lb/in^2, determine the fully plastic moment. Compare this with the maximum possible elastic moment that the same section can carry.

From Problem 8.29, the fully plastic moment M_p is given by

$$M_p = \sigma_{yp} Z_p$$

where Z_p is the plastic section modulus. For selected wide-flange sections Z_p is tabulated at the end of this chapter. In particular, for this section it is found to be 39.9 in^3. Thus

$$M_p = 38{,}000(39.9) = 1{,}520{,}000 \text{ lb} \cdot \text{in}$$

The maximum possible elastic moment is $M_e = \sigma_{yp} Z$ where Z is the usual (elastic) section modulus. Thus

$$M_e = 38{,}000(35.5) = 1{,}350{,}000 \text{ lb} \cdot \text{in}$$

The plastic moment is only 12.6 percent greater than the maximum elastic moment for this particular section. In fact, the fully plastic moment usually exceeds the maximum possible elastic moment by approximately 12 to 15 percent for most wide-flange sections.

8.31. Consider the T-section shown in Fig. 8-55(a) in which all fibers in the vertical web at a distance y_1 from the neutral axis have reached the yield point of the material, whereas all other fibers are still in the elastic range of action. Determine the location of the neutral axis and also the moment that corresponds to this stress distribution.

The neutral axis (described by the unknown c_1) may be located by investigating the normal forces over the cross section as shown in Fig. 8-55(b). From geometry

$$\frac{\sigma_0}{5-c_1} = \frac{\sigma_{yp}}{y_1} \quad \text{or} \quad \sigma_0 = \frac{5-c_1}{y_1}\sigma_{yp}$$

$$\frac{\sigma_0'}{4-c_1} = \frac{\sigma_{yp}}{y_1} \quad \text{or} \quad \sigma_0' = \frac{4-c_1}{y_1}\sigma_{yp}$$

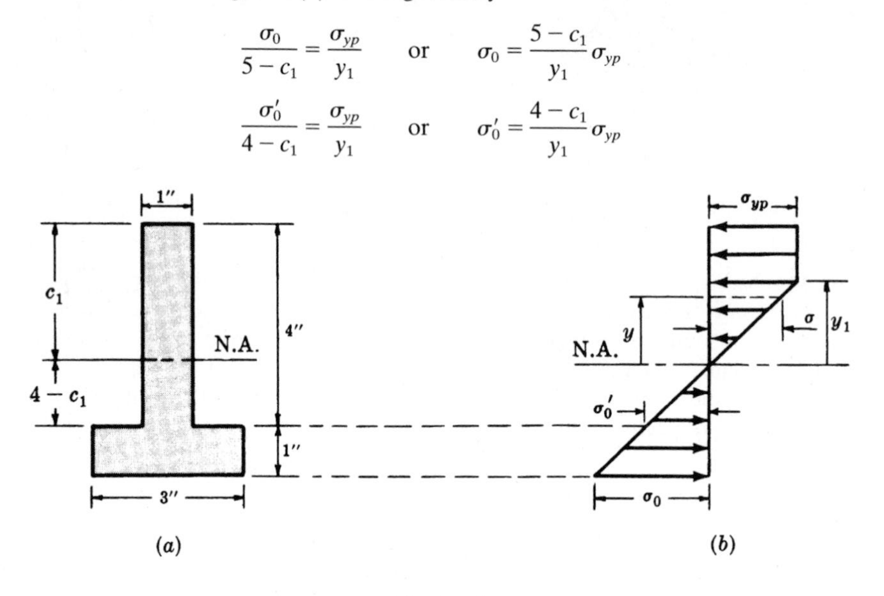

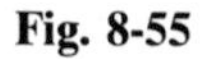
Fig. 8-55

For the resultant normal force to vanish

$$\Sigma F_N = (c_1 - y_1)(1)\sigma_{yp} + y_1(1)\left(\frac{\sigma_{yp}}{2}\right)$$
$$-\left\{\left[\frac{5-c_1}{2y_1}(5-c_1)(3)\sigma_{yp}\right] - \left[\frac{4-c_1}{2y_1}(4-c_1)(2)\sigma_{yp}\right]\right\} = 0$$

from which we obtain the quadratic equation

$$c_1^2 - (2y_1 + 14)c_1 + (y_1^2 + 43) = 0 \tag{1}$$

which determines c_1 for any specified value of y_1. This locates the neutral axis. Note that since y_1 occurred in the denominator in the above derivation, the equation should not be used to locate the neutral axis if $y_1 = 0$. Thus, when the action is entirely elastic the neutral axis passes through the centroid of the cross section. As plastification increases (i.e., as y_1 decreases), the neutral axis shifts to the location indicated by (1).

The moment corresponding to the stresses in Fig. 8-55(b) may be found from

$$M = \int \sigma y \, da$$
$$= \int_0^{y_1} \frac{y}{y_1}\sigma_{yp}(y)(1)\,dy + \int_{y_1}^{c_1} \sigma_{yp}(y)(1)\,dy$$
$$+ \int_0^{(5-c_1)} \left(\frac{y}{5-c_1}\right)\left(\frac{5-c_1}{y_1}\right)(\sigma_{yp})(y)(3)\,dy$$
$$- \int_0^{(4-c_1)} \left(\frac{y}{4-c_1}\right)\left(\frac{4-c_1}{y_1}\right)(\sigma_{yp})(y)(2)\,dy$$

or

$$M = \frac{\sigma_{yp}}{y_1}\left[\frac{y_1^3}{3} + \frac{y_1}{2}(c_1^2 - y_1^2) + (5-c_1)^3 - \frac{2}{3}(4-c_1)^3\right] \tag{2}$$

8.32. For the T-section of Problem 8.31 determine the location of the neutral axis when the action is fully plastic over the entire cross section. For fully plastic action determine the moment-carrying capacity and compare this with the maximum possible elastic moment.

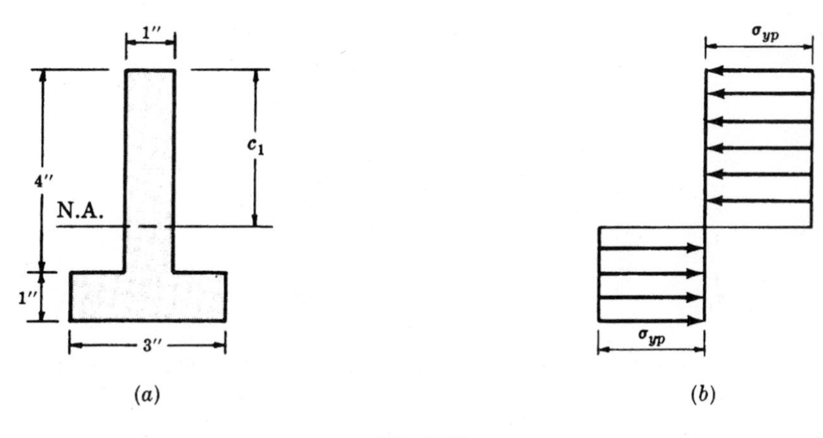

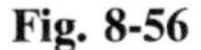
Fig. 8-56

In this case, the normal forces appear as indicated in Fig. 8-56(*b*). For equilibrium of normal forces over the cross section, we have

$$-\sigma_{yp}(1)(c_1) + [\sigma_{yp}(5 - c_1)(3) - \sigma_{yp}(4 - c_1)(2)] = 0$$

from which $c_1 = 3.5$ in. Thus, as mentioned in Problem 8.29, for fully plastic action the neutral axis divides the cross section into two equal parts.

The moment corresponding to this fully plastic action is

$$\begin{aligned} M_p &= \int \sigma y \, da \\ &= \int_0^{c_1} \sigma_{yp}(y)(1)\,dy + \int_0^{(5-c_1)} \sigma_{yp}(y)(3)\,dy - \int_0^{(4-c_1)} \sigma_{yp}(y)(2)\,dy \\ &= \sigma_{yp}[c_1^2 - 7c_1 + 21.5] \end{aligned}$$

For $c_1 = 3.5$ this becomes

$$M_p = 9.25\sigma_{yp}$$

By setting $y_1 = c_1$ in (*1*) of Problem 8.31, the neutral axis is located for the case of the maximum possible elastic moment. This location is found to be $c_1 = 3.07$ in (i.e., the neutral axis passes through the centroid of the cross section). The maximum possible elastic moment is found from (*2*) of Problem 8.31 to be

$$M_e = 5.32\sigma_{yp}$$

The fully plastic moment exceeds this value by 74 percent.

8.33. A beam is of square cross section, oriented as shown in Fig. 8-57, and carries a vertical load. If only the extreme top and bottom fibers reach the yield point, determine the maximum allowable elastic bending moment. Also, if the stress reaches yield at all fibers, determine the fully plastic moment.

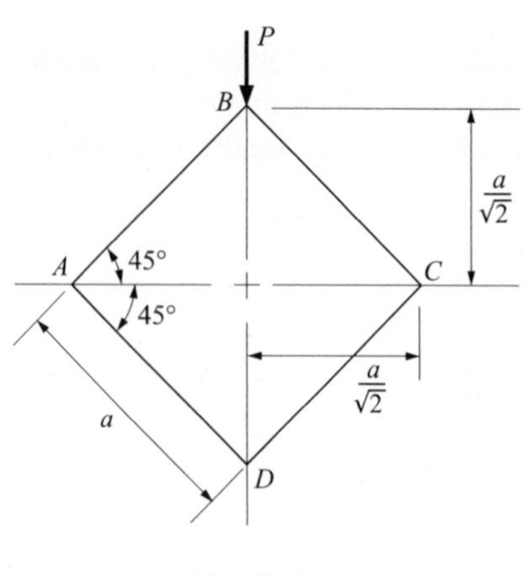

Fig. 8-57

If tensile yield is reached at fiber B and compressive yield at fiber D, the stress distribution over the cross section is given by (see Problem 8.1)

$$\sigma = \frac{Mc}{I} \tag{1}$$

To determine I, we consider the cross section to consist of triangles ABC and ADC. For each of these we have, from Problem 7.7,

$$I' = \tfrac{1}{12}bh^3 \tag{2}$$

So for the entire cross section the moment of inertia is

$$I = 2\left\{\frac{1}{12}\left(\frac{2a}{\sqrt{2}}\right)\left(\frac{a}{\sqrt{2}}\right)^3\right\}$$

$$= \frac{a^4}{24} \tag{3}$$

So from Eq. (*1*) at the extreme fibers we have

$$\sigma_{yp} = \frac{M_e\left(\dfrac{a}{\sqrt{2}}\right)}{\left(\dfrac{a^4}{24}\right)}$$

$$\therefore M_e = \frac{(\sigma_{yp})a^3\sqrt{2}}{12}$$

For fully plastic action over the entire cross section we have the stress distribution shown in Fig. 8-58.

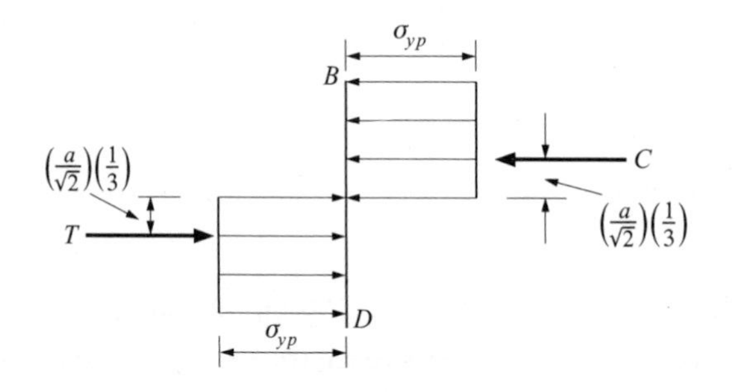

Fig. 8-58

The resultant of the compressive stresses above AC is

$$C = \frac{1}{2}\left(\frac{2a}{\sqrt{2}}\right)\left(\frac{a}{\sqrt{2}}\right)\sigma_{yp} = \frac{a^2}{2}\sigma_{yp}$$

which acts at the centroid of triangle ABC, and the resultant of the tensile stresses below AC is

$$T = \frac{1}{2}\left(\frac{2a}{\sqrt{2}}\right)\left(\frac{a}{\sqrt{2}}\right)\sigma_{yp} = \frac{a^2}{2}\sigma_{yp}$$

acting at the centroid of triangle ADC. These forces form a couple of magnitude

$$M_p = \left(\frac{a^2}{2}\sigma_{yp}\right)\left[2\left(\frac{a}{3\sqrt{2}}\right)\right] = \sigma_{yp}\frac{a^3}{3\sqrt{2}}$$

It is also of interest to form the ratio M_p/M_e:

$$\frac{M_p}{M_e} = \frac{\sigma_{yp}\left(\dfrac{a^3}{3\sqrt{2}}\right)}{\sigma_{yp}\left(\dfrac{a^3\sqrt{2}}{12}\right)} = 2$$

Supplementary Problems

8.34. A beam made of titanium, type Ti-6Al-4V, has a yield point of 120,000 lb/in^2. The beam has 1-in $\times$ 2-in rectangular cross section and bends about an axis parallel to the 1-in face. If the maximum bending stress is 90,000 lb/in^2, find the corresponding bending moment. *Ans.* 60,000 lb $\cdot$ in

8.35. A cantilever beam 3 m long carries a concentrated force of 35 kN at its free end. The material is structural steel and the maximum bending stress is not to exceed 125 MPa. Determine the required diameter if the bar is to be circular. *Ans.* 204 mm

8.36. Two $\frac{1}{2}$-in $\times$ 8-in cover plates are welded to two channels 10 in high to form the cross section of the beam shown in Fig. 8-59. Loads are in a vertical plane and bending takes place about a horizontal axis. The moment of inertia of each channel about a horizontal axis through the centroid is 78.5 in^4. If the maximum allowable elastic bending stress is 18,000 lb/in^2, determine the maximum bending moment that may be developed in the beam. *Ans.* 1,232,000 lb $\cdot$ in

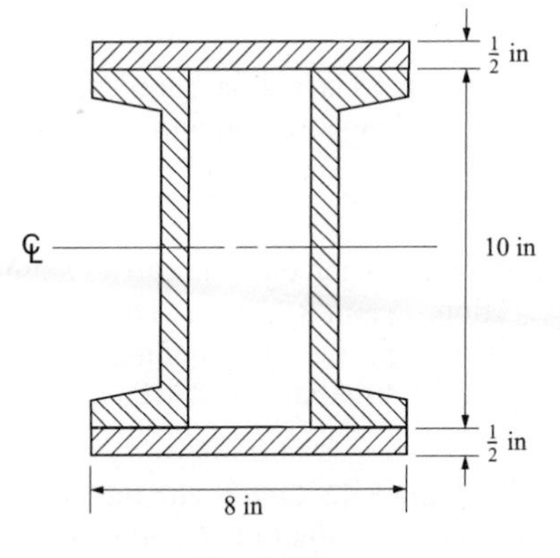

Fig. 8-59

8.37. A 250 mm deep wide-flange section with $I = 61 \times 10^6\ \text{mm}^4$ is used as a cantilever beam. The beam is 2 m long and the allowable bending stress is 125 MPa. Determine the maximum allowable intensity of uniform load that may be carried along the entire length of the beam. *Ans.* 30.5 kN/m

8.38. The beam shown in Fig. 8-60 is simply supported at the ends and carries the two symmetrically placed loads of 60 kN each. If the working stress in either tension or compression is 125 MPa, what is the required moment of inertia of area required for a 250-mm-deep beam? *Ans.* $60 \times 10^6\ \text{mm}^4$

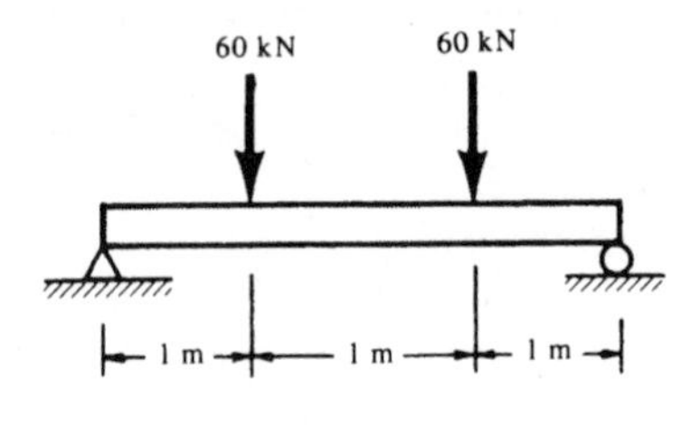

Fig. 8-60

8.39. Consider the simply supported beam subject to the two concentrated forces (60 kN each) shown in Fig. 8-60. Now, the beam is of hollow circular cross section as shown in Fig. 8-61, with an allowable working stress in either tension or compression of 125 MPa. Determine the necessary outer diameter of the beam. *Ans.* 17.4 mm

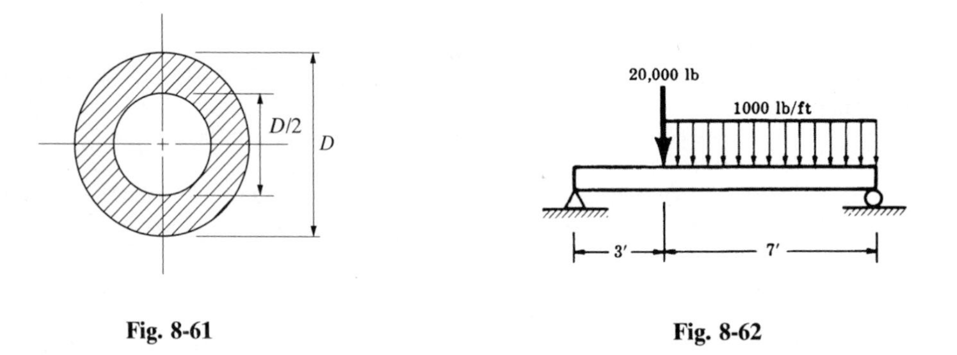

Fig. 8-61

Fig. 8-62

8.40. Consider a simply supported beam carrying the concentrated and uniform loads shown in Fig. 8-62. Select a suitable wide-flange section to resist these loads based upon a working stress in either tension or compression of $20{,}000\ \text{lb/in}^2$. *Ans.* $W12 \times 25$

8.41. Select a suitable wide-flange section to act as a cantilever beam 3 m long that carries a uniformly distributed load of 30 kN/m. The working stress in either tension or compression is 150 MPa. *Ans.* $W305 \times 66$

8.42. A beam 3 m long is simply supported at each end and carries a uniformly distributed load of 10 kN/m. The beam is of rectangular cross section, 75 mm $\times$ 150 mm. Determine the magnitude and location of the peak bending stress. Also, find the bending stress at a point 25 mm below the upper surface at the section midway between supports. *Ans.* 40 MPa, -26.8 MPa

8.43. Reconsider the steel beam of Problem 8-42. Determine the maximum bending stress if now the weight of the beam is considered in addition to the load of 10 kN/m. The weight of steel is $77.0\ \text{kN/m}^3$. *Ans.* 43.6 MPa

8.44. The two distributed loads are carried by the simply supported beam as shown in Fig. 8-63. The beam is a W8 × 28 section. Determine the magnitude and location of the maximum bending stress in the beam. *Ans.* 9000 lb/in^2, 5.5 ft from the right support

8.45. A T-beam having the cross section shown in Fig. 8-64 projects 2 m from a wall as a cantilever beam and carries a uniformly distributed load of 8 kN/m, including its own weight. Determine the maximum tensile and compressive bending stresses. *Ans.* +38.5 MPa, −81 MPa

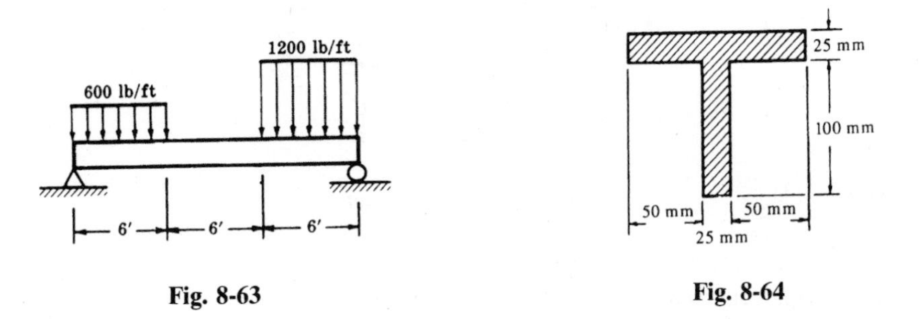

Fig. 8-63 **Fig. 8-64**

8.46. The simply supported beam *AC* shown in Fig. 8-65(*a*) supports a concentrated load *P*. The beam section is rectangular, 60 mm by 100 mm, with two square cutouts as shown in Fig. 8-65(*b*). If the allowable working stress is 120 MPa, determine the maximum value of *P*. *Ans.* 1.80 kN

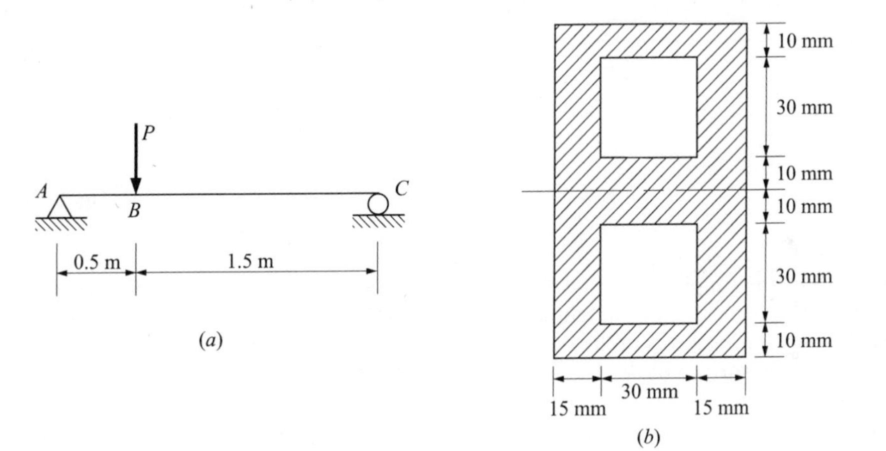

Fig. 8-65

8.47. A simply supported steel beam of channel-type cross section is loaded by both the uniformly distributed load and the couple shown in Fig. 8-66. Determine the maximum tensile and compressive stresses. *Ans.* 31.2 MPa, −56.8 MPa

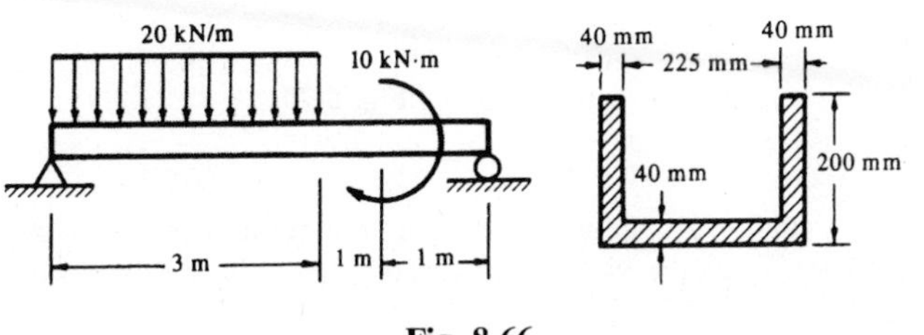

Fig. 8-66

8.48. A beam of circular cross section has the geometry shown in Fig. 8-67 and is subjected to a single concentrated vertical force at its midpoint. Determine the location of the point of maximum bending stress and the value of that stress. *Ans.* $x = L/4$, $\sigma_{max} = 0.377\ PL/d^3$

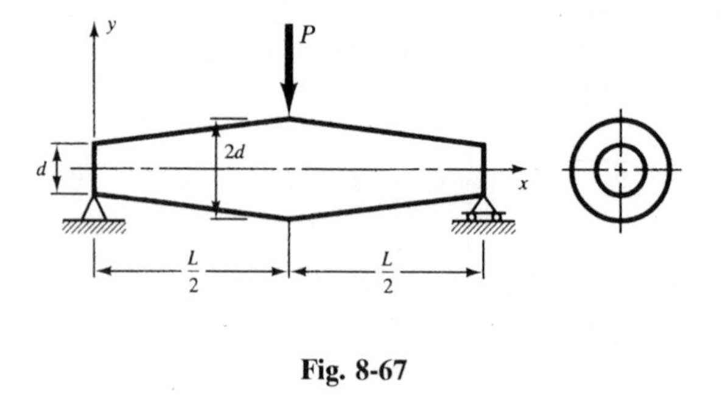

Fig. 8-67

8.49. A channel-shape beam with an overhanging end is loaded as shown in Fig. 8-68. The material is gray cast iron having an allowable working stress of 5000 lb/in² in tension and 20,000 lb/in² in compression. Determine the maximum allowable value of P. *Ans.* 2400 lb

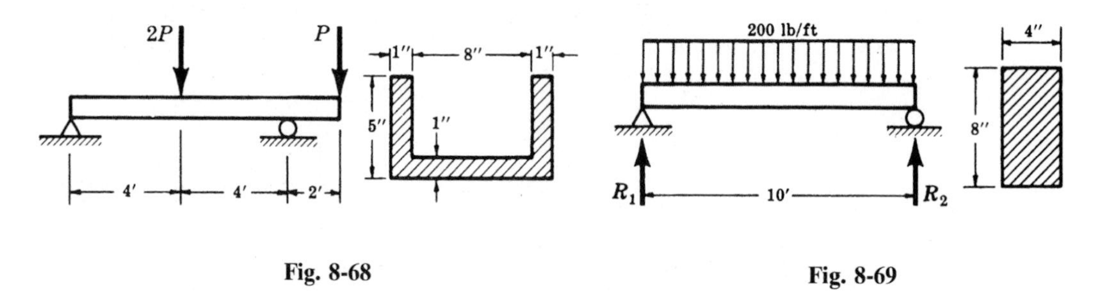

Fig. 8-68 **Fig. 8-69**

8.50. In Fig. 8-69 the simply supported beam of length 10 ft and cross section 4 in × 8 in carries a uniform load of 200 lb/ft. Neglecting the weight of the beam, find (*a*) the maximum normal stress in the beam, (*b*) the maximum shearing stress in the beam, and (*c*) the shearing stress at a point 2 ft to the right of R_1 and 1 in below the top surface of the beam. *Ans.* (*a*) 705 lb/in², (*b*) 47 lb/in², (*c*) 12.3 lb/in²

8.51. Determine (*a*) the maximum bending stress and (*b*) the maximum shearing stress in the simply supported beam shown in Fig. 8-70. *Ans.* (*a*) 22,000 lb/in², (*b*) 1660 lb/in²

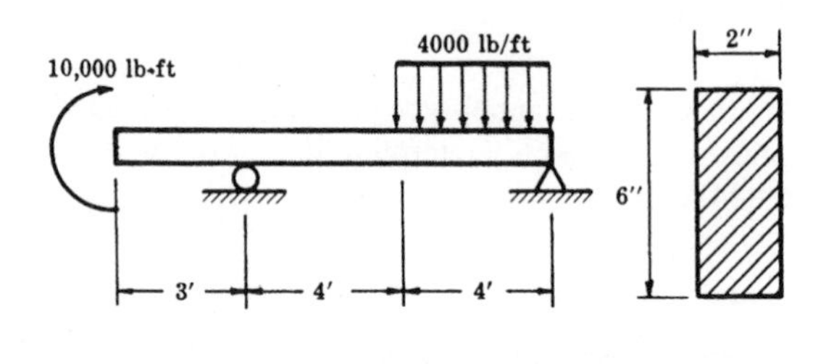

Fig. 8-70

8.52. For a bar of solid circular cross section, determine the amount by which the fully plastic moment exceeds the moment that just causes the yield point to be reached in the extreme fibers. *Ans.* 69.6 percent

8.53. Consider bending of a bar of isosceles triangular cross section (Fig. 8-71). The loads lie in the vertical plane of symmetry. Determine the ratio of the fully plastic moment to the moment that just causes yielding of the extreme fibers. *Ans.* 2.48

8.54. For the T-section shown in Fig. 8-72, determine the location of the neutral axis for fully plastic action. *Ans.* 137.5 mm above the lowest fibers of the section

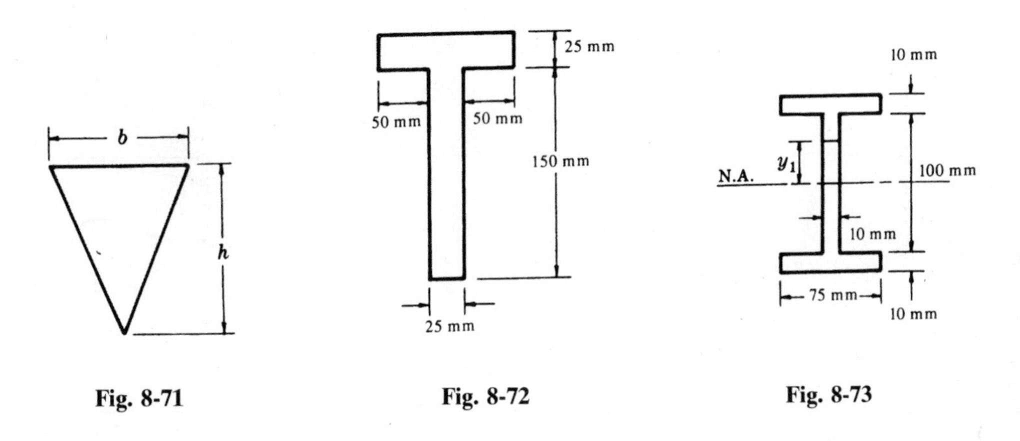

Fig. 8-71 **Fig. 8-72** **Fig. 8-73**

8.55. A bar of solid circular cross section of radius r is subject to bending. By what percent does the bending moment required to cause plastic action at the distance $r/2$ from the neutral axis exceed that required to just cause the yield point to be reached in the extreme fibers? *Ans.* 49.2 percent

8.56. For the section shown in Fig. 8-73 determine the value of y_1 which represents the point where elastic action terminates and plastic flow begins, when the beam is subject to a bending moment of 20 kN·m. Also determine the radius of curvature. Take the yield point of the material to be 200 MPa, and $E = 200$ GPa. *Ans.* $y_1 = 47.4$ mm, $R = 52.6$ m

8.57. A wide-flange section 600 mm high has welded to each of its flanges a 25 mm thick cover plate (see Fig. 8-74). The moment of inertia of the section is 1000×10^6 mm^4. At a particular location along the length of the beam, the transverse shear force is 300 kN. Determine the shear force per unit length existing in each of the four welds. *Ans.* 146 N/mm

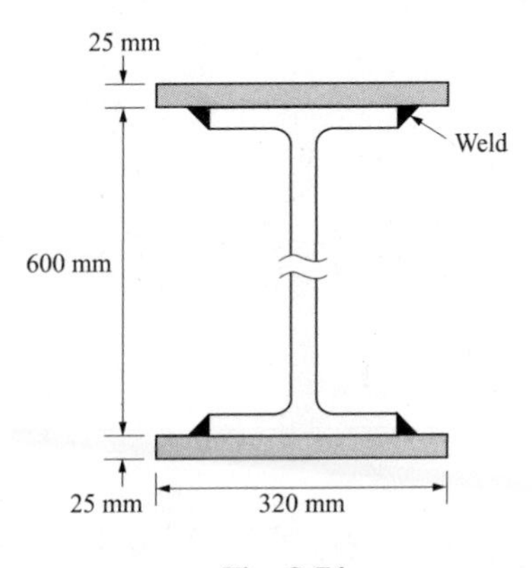

Fig. 8-74

Table 8-1. Properties of Selected Wide-Flange Sections, USCS Units

Designation*	Weight per foot, lb/ft	Area, in^2	I (about x-x axis), in^4	Z, in^3	I (about y-y axis), in^4	Z_p (plastic section modulus), in^3
W 18 × 70	70.0	20.56	1153.9	128.2	78.5	144.7
W 18 × 55	55.0	16.19	889.9	98.2	42.0	111.6
W 12 × 72	72.0	21.16	597.4	97.5	195.3	108.1
W 12 × 58	58.0	17.06	476.1	78.1	107.4	86.5
W 12 × 50	50.0	14.71	394.5	64.7	56.4	72.6
W 12 × 45	45.0	13.24	350.8	58.2	50.0	64.9
W 12 × 40	40.0	11.77	310.1	51.9	44.1	57.6
W 12 × 36	36.0	10.59	280.8	45.9	23.7	51.4
W 12 × 32	32.0	9.41	246.8	40.7	20.6	45.0
W 12 × 25	25.0	7.39	183.4	30.9	14.5	35.0
W 10 × 89	89.0	26.19	542.4	99.7	180.6	114.4
W 10 × 54	54.0	15.88	305.7	60.4	103.9	67.0
W 10 × 49	49.0	14.40	272.9	54.6	93.0	60.3
W 10 × 45	45.0	13.24	248.6	49.1	53.2	55.0
W 10 × 37	37.0	10.88	196.9	39.9	42.2	45.0
W 10 × 29	29.0	8.53	157.3	30.8	15.2	34.7
W 10 × 23	23.0	6.77	120.6	24.1	11.3	33.7
W 10 × 21	21.0	6.19	106.3	21.5	9.7	24.1
W 8 × 40	40.0	11.76	146.3	35.5	49.0	39.9
W 8 × 35	35.0	10.30	126.5	31.1	42.5	34.7
W 8 × 31	31.0	9.12	109.7	27.4	37.0	30.4
W 8 × 28	28.0	8.23	97.8	24.3	21.6	27.1
W 8 × 27	27.0	7.93	94.1	23.4	20.8	23.9
W 8 × 24	24.0	7.06	82.5	20.8	18.2	23.1
W 8 × 19	19.0	5.59	64.7	16.0	7.9	17.7
W 6 × $15\frac{1}{2}$	15.5	4.62	28.1	9.7	9.7	11.3

*The first number after the W is the nominal depth of the section in inches. The second number is the weight in pounds per foot of length.

Table 8-2. Properties of Selected Wide-Flange Sections, SI Units

Designation*	Mass per meter, kg/m	Area, mm^2	I (about x-x axis), 10^6 mm^4	Z, 10^3 mm^3	I (about y-y axis), 10^6 mm^4	Z_p (plastic section modulus), 10^3 mm^3
W 460 × 103	102.9	13,200	479	2100	32.6	2370
W 460 × 81	80.9	10,400	369	1610	17.4	1820
W 305 × 106	105.8	13,600	248	1590	81.0	1770
W 305 × 85	85.3	11,000	198	1280	44.6	1410
W 305 × 74	73.5	9,480	164	1060	23.4	1190
W 305 × 66	66.2	8,530	146	952	20.7	1060
W 305 × 59	58.8	7,580	129	849	18.3	942
W 305 × 53	52.9	6,820	117	750	9.83	840
W 305 × 47	47.0	6,060	102	665	8.55	736
W 305 × 37	36.8	4,760	76.1	505	6.02	572
W 254 × 131	130.8	16,900	225	1630	74.9	1870
W 254 × 79	79.4	10,200	127	988	43.1	1100
W 254 × 72	72.0	9,280	113	893	38.6	986
W 254 × 66	66.2	8,530	103	803	22.1	899
W 254 × 54	54.4	7,010	81.7	652	17.5	736
W 254 × 43	42.6	5,490	65.3	504	6.31	567
W 254 × 34	33.8	4,360	50.0	394	4.69	551
W 254 × 31	30.9	3,990	44.1	352	4.02	394
W 203 × 59	58.8	7,580	60.7	580	20.3	652
W 203 × 51	51.4	6,630	52.5	508	17.6	567
W 203 × 46	45.6	5,870	45.5	448	15.4	497
W 203 × 41	41.2	5,300	40.6	397	8.96	443
W 203 × 40	39.7	5,110	39.0	383	8.63	391
W 203 × 35	35.3	4,550	34.2	340	7.55	378
W 203 × 28	27.9	3,600	26.8	262	3.28	290
W 152 × 23	22.8	2,980	11.7	159	4.02	185

*The first number after the W is the nominal depth of the section in millimeters. The second number is the mass in kilograms per meter of length.

Chapter 9

Elastic Deflection of Beams: Double-Integration Method

INTRODUCTION

In Chap. 8 it was stated that lateral loads applied to a beam not only give rise to internal bending and shearing stresses in the bar, but also cause the bar to deflect in a direction perpendicular to its longitudinal axis. The stresses were examined in Chap. 8 and it is the purpose of this chapter and also Chap. 10 to examine methods for calculating the deflections.

DEFINITION OF DEFLECTION OF A BEAM

The deformation of a beam is most easily expressed in terms of the deflection of the beam from its original unloaded position. The deflection is measured from the original neutral surface to the neutral surface of the deformed beam. The configuration assumed by the deformed neutral surface is known as the elastic curve of the beam. Figure 9-1 represents the beam in its original undeformed state and Fig. 9-2 represents the beam in the deformed configuration it has assumed under the action of the load.

Fig. 9-1 Fig. 9-2

The displacement y is defined as the deflection of the beam. Often it will be necessary to determine the deflection y for every value of x along the beam. This relation may be written in the form of an equation which is frequently called the equation of the deflection curve (or elastic curve) of the beam.

IMPORTANCE OF BEAM DEFLECTIONS

Specifications for the design of beams frequently impose limitations upon the deflections as well as the stresses. Consequently, in addition to the calculation of stresses as outlined in Chap. 8, it is essential that the designer be able to determine deflections. For example, in many building codes the maximum allowable deflection of a beam is not to exceed $\frac{1}{300}$ of the length of the beam. Components of aircraft usually are designed so that deflections do not exceed some preassigned value, else the aerodynamic characteristics may be altered. Thus, a well-designed beam must not only be able to carry the loads to which it will be subjected but it must not undergo undesirably large deflections. Also, the evaluation of reactions of statically indeterminate beams involves the use of various deformation relationships. These will be examined in detail in Chap. 11.

METHODS OF DETERMINING BEAM DEFLECTIONS

Numerous methods are available for the determination of beam deflections. The most commonly used are the following:

1. Double-integration method
2. Method of singularity functions
3. Elastic energy methods

The first method is described in this chapter, the use of singularity functions is discussed in Chap. 10, and elastic energy methods are treated in Chap. 15. It is to be carefully noted that all of these methods apply *only* if all portions of the beam are acting in the *elastic range of action*.

DOUBLE-INTEGRATION METHOD

The differential equation of the deflection curve of the bent beam is

$$EI\frac{d^2y}{dx^2} = M \tag{9.1}$$

where x and y are the coordinates shown in Fig. 9-2. That is, y is the deflection of the beam. This equation is derived in Problem 9.1. In the equation E denotes the modulus of elasticity of the beam and I represents the moment of inertia of the beam cross section about the neutral axis, which passes through the centroid of the cross section. Also, M represents the bending moment at the distance x from one end of the beam. This quantity was defined in Chap. 6 to be the algebraic sum of the moments of the external forces to one side of the section at a distance x from the end about an axis through this section. Usually, M will be a function of x and it will be necessary to integrate (*9.1*) twice to obtain an algebraic equation expressing the deflection of y as a function of x.

Equation (*9.1*) is the basic differential equation that governs the elastic deflection of all beams irrespective of the type of applied loading. For applications, see Problems 9.2 through 9.14 and 9.16 through 9.22.

THE INTEGRATION PROCEDURE

The double-integration method for calculating deflections of beams merely consists of integrating (*9.1*). The first integration yields the slope dy/dx at any point in the beam and the second integration gives the deflection y for any value of x. The bending moment M must, of course, be expressed as a function of the coordinate x before the equation can be integrated. For the cases to be studied here the integrations are extremely simple.

Since the differential equation (*9.1*) is of the second order, its solution must contain two constants of integration. These two constants must be evaluated from known conditions concerning the slope or deflection at certain points in the beam. For example, in the case of a cantilever beam the constants would be determined from the conditions of zero change of slope as well as zero deflection at the built-in end of the beam.

Frequently two or more equations are necessary to describe the bending moment in the various regions along the length of a beam. This was emphasized in Chap. 6. In such a case, (*9.1*) must be written for each region of the beam and integration of these equations yields two constants of integration for each region. These constants must then be determined so as to impose conditions of continuous deformations and slopes at the points common to adjacent regions. See Problems 9.17 through 9.19.

SIGN CONVENTIONS

The sign conventions for bending moment adopted in Chap. 6 will be retained here. The quantities E and I appearing in (*9.1*) are, of course, positive. Thus, from this equation, if M is positive for a certain value of x, then d^2y/dx^2 is also positive. With the above sign convention for bending moments, it is necessary to consider the coordinate x along the length of the beam to be positive to the right and the deflection y to be positive upward. This will be explained in detail in Problem 9.1. With these algebraic signs the integration of (*9.1*) may be carried out to yield the deflection y as a function of x, with the understanding that upward beam deflections are positive and downward deflections negative.

ASSUMPTIONS AND LIMITATIONS

In the derivation of (*9.1*) it is assumed that deflections caused by shearing action are negligible compared to those caused by bending action. Also, it is assumed that the deflections are small compared to the cross-sectional dimensions of the beam and that all portions of the beam are acting in the elastic range. Equation (*9.1*) is derived on the basis of the beam being straight prior to the application of loads. Beams with slight deviations from straightness prior to loading may be treated by modifying this equation as indicated in Problem 9.25.

Solved Problems

9.1. Obtain the differential equation of the deflection curve of a beam loaded by lateral forces.

In Problem 8.1 the relationship

$$M = \frac{EI}{\rho} \qquad (1)$$

was derived. In this expression M denotes the bending moment acting at a particular cross section of the beam, ρ the radius of curvature to the neutral surface of the beam at this same section, E the modulus of elasticity, and I the moment of the cross-sectional area about the neutral axis passing through the centroid of the cross section. In this book we will usually be concerned with those beams for which E and I are constant along the entire length of the beam, but in general both M and ρ will be functions of x.

Equation (*1*) may be written in the form

$$\frac{1}{\rho} = \frac{M}{EI} \qquad (2)$$

where the left side of Eq. (*2*) represents the curvature of the neutral surface of the beam. Since M will vary along the length of the beam, the deflection curve will be of variable curvature.

Let the heavy line in Fig. 9-3 represent the deformed neutral surface of the bent beam. Originally the beam coincided with the x-axis prior to loading and the coordinate system that is usually found to be most convenient is shown in the sketch. The deflection y is taken to be positive in the upward direction; hence for the particular beam shown, all deflections are negative.

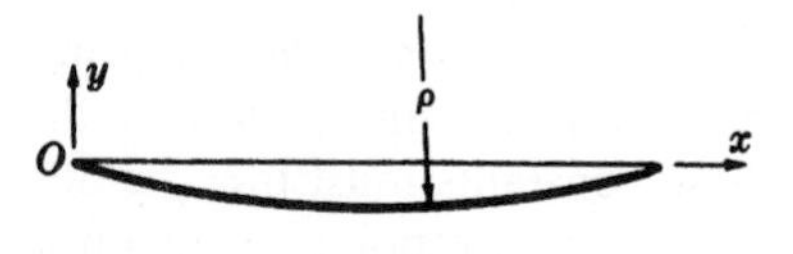

Fig. 9-3

An expression for the curvature at any point along the curve representing the deformed beam is readily available from differential calculus. The exact formula for curvature is

$$\frac{1}{\rho} = \frac{d^2y/dx^2}{[1 + (dy/dx)^2]^{3/2}} \tag{3}$$

In this expression, dy/dx represents the slope of the curve at any point; and for small beam deflections this quantity and in particular its square are small in comparison to unity and may reasonably be neglected. This assumption of small deflections simplifies the expression for curvature into

$$\frac{1}{\rho} \approx \frac{d^2y}{dx^2} \tag{4}$$

Hence for small deflections, (2) becomes $d^2y/dx^2 = M/EI$ or

$$EI\frac{d^2y}{dx^2} = M \tag{5}$$

This is the differential equation of the deflection curve of a beam loaded by lateral forces. In honor of its codiscoverers, it is called the Euler-Bernoulli equation of bending of a beam. In any problem it is necessary to integrate this equation to obtain an algebraic relationship between the deflection y and the coordinate x along the length of the beam. This will be carried out in the following problems.

9.2. Determine the deflection at every point of the cantilever beam subject to the single concentrated force P, as shown in Fig. 9-4.

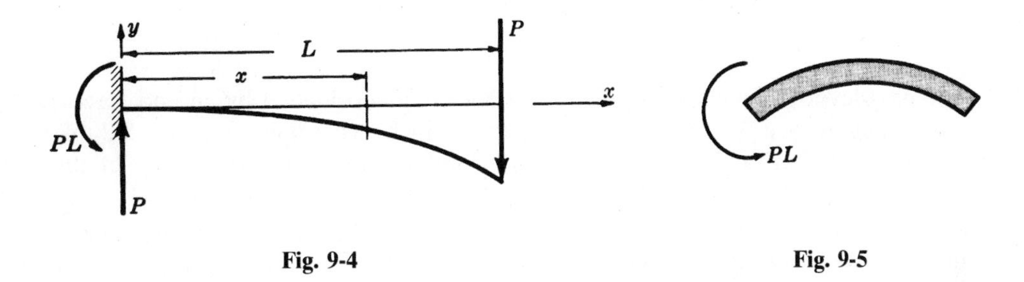

Fig. 9-4 **Fig. 9-5**

The x-y coordinate system shown is introduced, where the x-axis coincides with the original unbent position of the beam. The deformed beam has the appearance indicated by the heavy line. It is first necessary to find the reactions exerted by the supporting wall upon the bar, and these are easily found from statics to be a vertical force reaction P and a moment PL as shown.

The bending moment at any cross section a distance x from the wall is given by the sum of the moments of these two reactions about an axis through this section. Evidently the upward force P produces a positive bending moment Px, and the couple PL if acting alone would produce curvature of the bar as shown in Fig. 9-5. According to the sign convention of Chap. 6, this constitutes negative bending. Hence the bending moment M at the section x is

$$M = -PL + Px$$

The differential equation of the bent beam is

$$EI\frac{d^2y}{dx^2} = M$$

where E denotes the modulus of elasticity of the material and I represents the moment of inertia of the cross section about the neutral axis. Substituting,

$$EI\frac{d^2y}{dx^2} = -PL + Px \tag{1}$$

This equation is readily integrated once to yield

$$EI\frac{dy}{dx} = -PLx + \frac{Px^2}{2} + C_1 \tag{2}$$

which represents the equation of the slope, where C_1 denotes a constant of integration. This constant may be evaluated by use of the condition that the slope dy/dx of the beam at the wall is zero since the beam is rigidly clamped there. Thus $(dy/dx)_{x=0} = 0$. Equation (2) is true for all values of x and y, and if the condition $x = 0$ is substituted we obtain $0 = 0 + 0 + C_1$ or $C_1 = 0$.

Next, integration of (2) yields

$$EIy = -PL\frac{x^2}{2} + \frac{Px^3}{6} + C_2 \tag{3}$$

where C_2 is a second constant of integration. Again, the condition at the supporting wall will determine this constant. There, at $x = 0$, the deflection y is zero since the bar is rigidly clamped. Substituting $(y)_{x=0} = 0$ in Eq. (3), we find $0 = 0 + 0 + C_2$ or $C_2 = 0$.

Thus Eqs. (2) and (3) with $C_1 = C_2 = 0$ give the slope dy/dx and deflection y at any point x in the beam. The deflection is a maximum at the right end of the beam ($x = L$), under the load P, and from Eq. (3),

$$EIy_{max} = \frac{-PL^3}{3} \tag{4}$$

where the negative value denotes that this point on the deflection curve lies below the x-axis. If only the magnitude of the maximum deflection at $x = L$ is desired, it is usually denoted by Δ_{max} and we have

$$\Delta_{max} = \frac{PL^3}{3EI} \tag{5}$$

9.3. The cantilever beam shown in Fig. 9-4 is 3 m long and loaded by an end force of 20 kN. The cross section is a W203 × 59 steel section, which according to Table 8-2 of Chap. 8 has $I = 60.7 \times 10^{-6}\ \text{m}^4$ and $Z = 580 \times 10^{-6}\ \text{m}^3$. Find the maximum deflection of the beam. Take $E = 200$ GPa.

The maximum deflection occurs at the free end of the beam under the concentrated force and was found in Problem 9.2 to be, by Eq. (4),

$$y_{max} = -\frac{PL^3}{3EI} = -\frac{(20{,}000\ \text{N})(3\ \text{m})^3}{3(200 \times 10^9\ \text{N/m}^2)(60.7 \times 10^{-6}\ \text{m}^4)} = -0.0148\ \text{m} \quad \text{or} \quad 14.8\ \text{mm}$$

The negative sign of course indicates downward deflection. In the derivation of this deflection formula it was assumed that the material of the beam follows Hooke's law. Actually, from the above calculation alone there is no assurance that the material is not stressed beyond the proportional limit. If it were then the basic beam-bending equation $EI(d^2y/dx^2) = M$ would no longer be valid and the above numerical value would be meaningless. Consequently, in every problem involving beam deflections it is to be emphasized that it is necessary to determine that the maximum bending stress in the beam is below the proportional limit of the material. This is easily done by use of the flexure formula derived in Problem 8.1. According to this formula

$$\sigma = \frac{Mc}{I}$$

where σ denotes the bending stress, M the bending moment, c the distance from the neutral axis to the outer fibers of the beam, and I the second moment of area of the beam cross section about the neutral axis. The maximum bending moment in this problem occurs at the supporting wall and is given by $M_{max} = (20{,}000\ \text{N})(3\ \text{m}) = 60{,}000\ \text{N}\cdot\text{m}$. Using this in the formula for bending stress, we have

$$\sigma_{max} = \frac{M}{Z} = \frac{60{,}000\ \text{N}\cdot\text{m}}{580 \times 10^{-6}\ \text{m}^3} = 103\ \text{MPa}$$

Since this value is below the proportional limit of steel, which is approximately 200 MPa, the use of the beam deflection equation was justifiable.

9.4. Determine the slope of the right end of the cantilever beam loaded as shown in Fig. 9-4. For the beam described in Problem 9.3, determine the value of this slope.

In Problem 9.2 the equation of the slope was found to be

$$EI\frac{dy}{dx} = -PLx + \frac{Px^2}{2}$$

At the free end, $x = L$, and

$$EI\left(\frac{dy}{dx}\right)_{x=L} = -PL^2 + \frac{PL^2}{2}$$

The slope at the end is thus

$$\left(\frac{dy}{dx}\right)_{x=L} = \frac{-PL^2}{2EI}$$

For the beam described in Problem 9.3, this becomes

$$\left(\frac{dy}{dx}\right)_{x=L} = \frac{-(20{,}000\text{ N})(3\text{ m})^3}{2(200\times 10^9\text{ N/m}^2)(60.7\times 10^{-6}\text{ m}^4)}$$

$$= 0.0222\text{ rad} \qquad \text{or} \qquad 1.27°$$

9.5. Determine the deflection at every point of a cantilever beam subject to the uniformly distributed load w per unit length shown in Fig. 9-6.

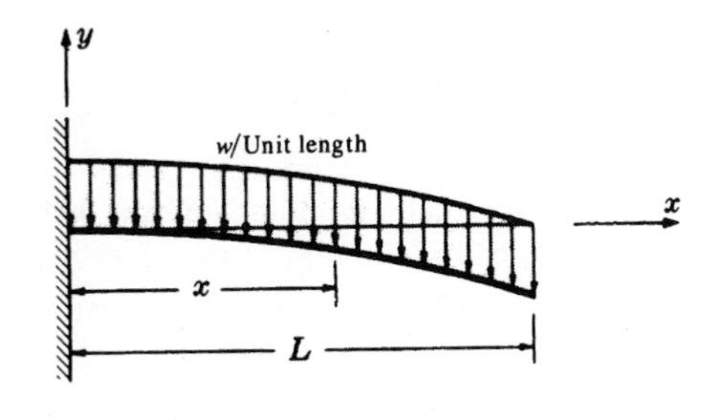

Fig. 9-6

The x-y coordinate system shown is introduced, where the x-axis coincides with the original unbent position of the beam. The deformed beam has the appearance indicated by the heavy line. The equation for the bending moment could be determined in a manner analogous to that used in Problem 9.2, but instead let us seek a slight simplification of that technique. Let us determine the bending moment at the section a distance x from the wall by considering the forces to the right of this section rather than those to the left.

The force of w/unit length acts over the length $L - x$ to the right of this section and hence the resultant force is $w(L - x)$ lb. This force acts at the midpoint of this length of beam to the right of x and thus its moment arm from x is $\frac{1}{2}(L - x)$. The bending moment at the section x is thus given by

$$M = -\frac{w}{2}(L - x)^2$$

the negative sign being necessary since downward loads produce negative bending.

The differential equation describing the bent beam is thus

$$EI\frac{d^2y}{dx^2} = -\frac{w}{2}(L-x)^2 \tag{1}$$

The first integration yields

$$EI\frac{dy}{dx} = \frac{w}{2}\frac{(L-x)^3}{3} + C_1 \tag{2}$$

where C_1 denotes a constant of integration.

This constant may be evaluated by realizing that the left end of the beam is rigidly clamped. At that point, $x = 0$, we have no change of slope and hence $(dy/dx)_{x=0} = 0$. Substituting these values in (2), we find $0 = wL^3/6 + C_1$ or $C_1 = -wL^3/6$. We thus have

$$EI\frac{dy}{dx} = \frac{w}{6}(L-x)^3 - \frac{wL^3}{6} \tag{2'}$$

The next integration yields

$$EIy = -\frac{w}{6}\frac{(L-x)^4}{4} - \frac{wL^3}{6}x + C_2 \tag{3}$$

where C_2 represents a second constant of integration.

At the clamped end, $x = 0$, of the beam the deflection is zero and since (3) holds for all values of x and y, it is permissible to substitute this pair of values in it. Doing this, we obtain

$$0 = \frac{-wL^4}{24} + C_2 \qquad \text{or} \qquad C_2 = \frac{wL^4}{24}$$

The final form of the deflection curve of the beam is thus

$$EIy = -\frac{w}{24}(L-x)^4 - \frac{wL^3}{6}x + \frac{wL^4}{24} \tag{3'}$$

The deflection is a maximum at the right end of the bar ($x = L$) and there we have from (3′)

$$EIy_{\text{max}} = -\frac{wL^4}{6} + \frac{wL^4}{24} = -\frac{wL^4}{8}$$

where the negative value denotes that this point on the deflection curve lies below the x-axis. The magnitude of the maximum deflection is

$$\Delta_{\text{max}} = \frac{wL^4}{8EI} \tag{4}$$

9.6. A cantilever beam carrying a parabolically distributed load is shown in Fig. 9-7. Determine the equation of the deflected beam as well as the deflection of the tip.

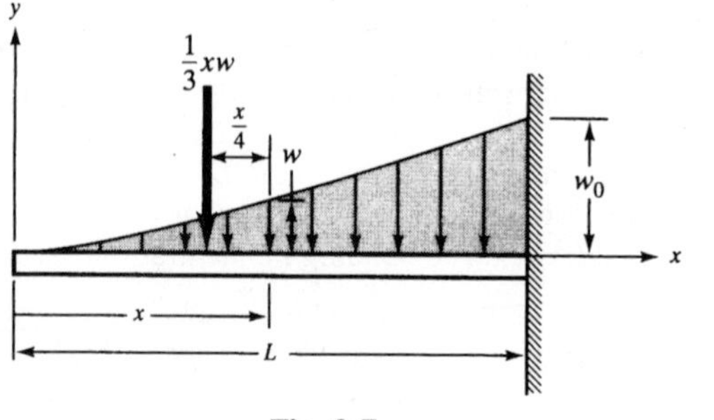

Fig. 9-7

Let us introduce a coordinate system having its origin at the tip of the beam. The intensity of loading at any point x to the right of the tip is, from the properties of a parabola,

$$w = w_0\left(\frac{x}{L}\right)^2 \tag{1}$$

From statics it is known that for any parabolic area such as shown in Fig. 9-8 the area is given by $A = \frac{1}{3}ah$ and the centroid C is located at $x = 3a/4$. Accordingly, it is now possible to determine the bending moment at the point x as the sum of the moments of all loads to the left of x about that point. The resultant of the loading to the left of x is $\frac{1}{3}xw$ and this resultant, shown by the solid arrow in Fig. 9-7, is located a

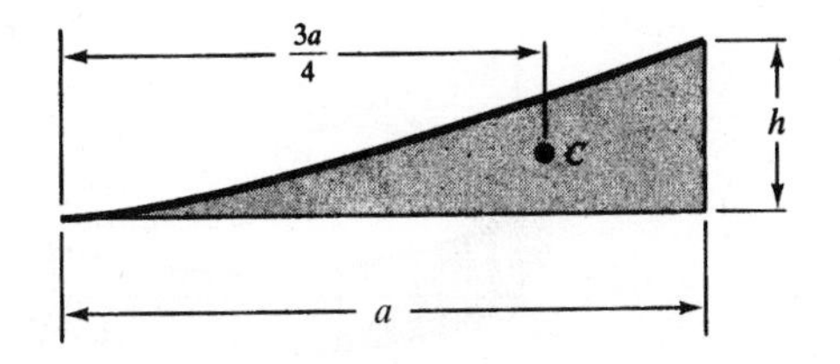

Fig. 9-8

distance $3x/4$ from the tip, or, alternatively, $(x/4)$ from position x. Thus, the bending moment at x is found, with the aid of Eq. (*1*), to be

$$-\frac{1}{3}xw\left(\frac{x}{4}\right) \quad \text{or} \quad -\frac{w_0x^4}{12L^2}$$

and the differential equation of the deflection curve is

$$EI\frac{d^2y}{dx^2} = -\frac{w_0x^4}{12L^2} \tag{2}$$

Integrating the first time, we find

$$EI\frac{dy}{dx} = -\frac{w_0}{12L^2}\cdot\frac{x^5}{5} + C_1 \tag{3}$$

When $x = L$, the slope $dy/dx = 0$, so from Eq. (*3*), we have

$$0 = -\frac{w_0L^3}{60} + C_1 \quad \text{and therefore } C_1 = \frac{w_0L^3}{60}$$

Integrating again, we have

$$EIy = -\frac{w_0}{60L^2}\cdot\frac{x^6}{6} + \frac{w_0L^3}{60}x + C_2 \tag{4}$$

When $x = L$, $y = 0$, so from Eq. (*4*), we have

$$0 = -\frac{w_0L^4}{360} + \frac{w_0L^4}{60} + C_2 \quad \text{and therefore } C_2 = -\frac{1}{72}w_0L^4$$

The desired equation of the deflected beam is

$$EIy = -\frac{w_0}{360L^2}x^6 + \frac{x_0L^3}{60}x - \frac{1}{72}w_0L^4$$

and the deflection at the tip is

$$EIy]_{x=0} = -\tfrac{1}{72}w_0L^4$$

9.7. Obtain an expression for the deflection curve of the simply supported beam of Fig. 9-9 subject to the uniformly distributed load w per unit length as shown.

The x-y coordinate system shown is introduced, where the x-axis coincides with the original unbent position of the beam. The deformed beam has the appearance indicated by the heavy line. The total load acting on the beam is wL and, because of symmetry, each of the end reactions is $wL/2$. Because of the symmetry of loading, it is evident that the deflected beam is symmetric about the midpoint $x = L/2$.

The equation for the bending moment at any section of a beam loaded and supported as this one is was discussed in Problem 6.3. According to the method indicated there, the portion of the uniform load to the left of the section a distance x from the left support is replaced by its resultant acting at the midpoint of the section of length x. The resultant is wx lb acting downward and hence giving rise to a negative bending moment.

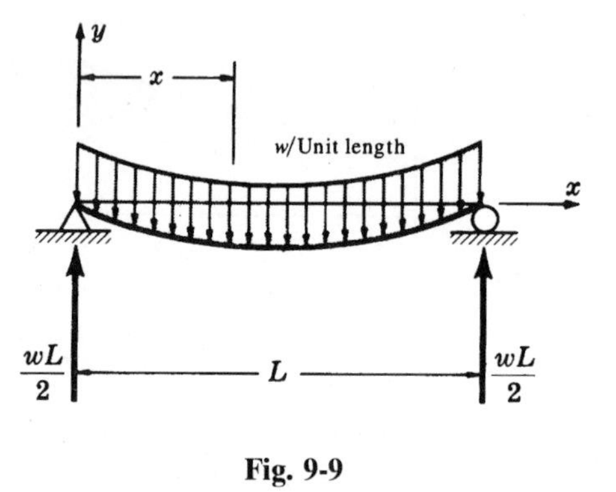

Fig. 9-9

The reaction $wL/2$ gives rise to a positive bending moment. Consequently, for any value of x, the bending moment is

$$M = \frac{wL}{2}x - wx\frac{x}{2}$$

The differential equation of the bent beam is $EI(d^2y/dx^2) = M$. Substituting,

$$EI\frac{d^2y}{dx^2} = \frac{wL}{2}x - \frac{wx^2}{2} \tag{1}$$

Integrating,

$$EI\frac{dy}{dx} = \frac{wL}{2}\frac{x^2}{2} - \frac{w}{2}\frac{x^2}{3} + C_1 \tag{2}$$

It is to be noted that dy/dx represents the slope of the beam. Since the deflected beam is symmetric about the center of the span, i.e., about $x = L/2$, it is evident that the slope must be zero there. That is, the tangent to the deflected beam is horizontal at the midpoint of the beam. This condition enables us to determine C_1. Substituting this condition in (2), we obtain $(dy/dx)_{x=L/2} = 0$,

$$0 = \frac{wL}{4}\frac{L^2}{4} - \frac{w}{6}\frac{L^3}{8} + C_1 \qquad \text{or} \qquad C_1 = -\frac{wL^3}{24}$$

The slope dy/dx at any point is thus given by

$$EI\frac{dy}{dx} = \frac{wL}{4}x^2 - \frac{w}{6}x^3 - \frac{wL^3}{24} \tag{2'}$$

Integrating again, we find

$$EIy = \frac{wL}{4}\frac{x^3}{3} - \frac{w}{6}\frac{x^4}{4} - \frac{wL^3}{24}x + C_2 \tag{3}$$

This second constant of integration C_2 is readily determined by the fact that the deflection y is zero at the left support. Substituting $y_{x=0} = 0$ in (3), we find $0 = 0 - 0 - 0 + C_2$ or $C_2 = 0$.

The final form of the deflection curve of the beam is thus

$$EIy = \frac{wL}{12}x^3 - \frac{w}{24}x^4 - \frac{wL^3}{24}x \qquad (3')$$

The maximum deflection of the beam occurs at the center because of symmetry. Substituting $x = L/2$ in (3′), we obtain

$$EIy_{max} = -\frac{5wL^4}{384}$$

Or, without regard to algebraic sign, we have for the maximum deflection of a uniformly loaded, simply supported beam

$$\Delta_{max} = \frac{5}{384}\frac{wL^4}{EI} \qquad (4)$$

9.8. A simply supported beam of length 10 ft and rectangular cross section 1 in × 3 in carries a uniform load of 200 lb/ft. The beam is titanium, type Ti-5Al-2.5Sn, having a yield strength of 115,000 lb/in^2 and $E = 16 \times 10^6$ lb/in^2. Determine the maximum deflection of the beam.

From Problem 9.7 the maximum deflection is

$$\Delta_{max} = \frac{5}{384}\frac{wL^4}{EI}$$

Substituting,

$$\Delta_{max} = \frac{5}{384}\frac{(200/12)(120)^4}{(16 \times 10^6)\frac{1}{12}(1)(3)^3} = 1.25 \text{ in}$$

Using the methods of Chap. 8, the maximum bending stress is found to be only 20,000 lb/in^2, well below the nonlinear range of action of the material. Thus the use of the deflection formula is justified.

9.9. Consider the simply supported beam subject to the two end couples M_1 and M_2 as shown in Fig. 9-10. Determine the equation of the deflection curve and locate the point of peak deflection if $M_1 = 0$.

For equilibrium the resultant of the applied couples, that is, $(M_1 - M_2)$, must be another couple corresponding to the vertical reactions at the ends R_L and R_R. From statics,

$$+\circlearrowleft \Sigma M_0 = -M_1 + M_2 + R_R L = 0$$

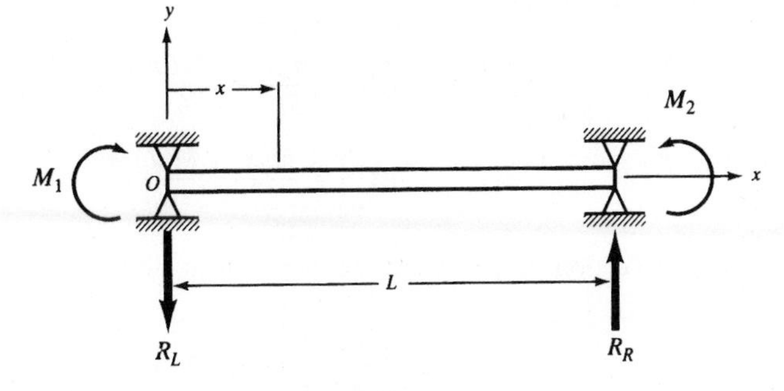

Fig. 9-10

Therefore,

$$R_R = \frac{M_1 - M_2}{L}(\uparrow)$$

$$\Sigma F_y = -R_L + R_R = 0$$

Therefore,

$$R_L = \frac{M_1 - M_2}{L}(\downarrow)$$

The differential equation describing the bent beam is thus

$$EI\frac{d^2y}{dx^2} = M_1 - R_L x \tag{1}$$

Integrating,

$$EI\frac{dy}{dx} = M_1 x - R_L \frac{x^2}{2} + C_1 \tag{2}$$

We have no information concerning the slope anywhere in the beam. Hence it is not possible to determine the constant of integration C_1 at this stage. Let us integrate again:

$$EIy = M_1\frac{x^2}{2} - \frac{R_L}{2}\cdot\frac{x^3}{3} + C_1 x + C_2 \tag{3}$$

We may now determine the two constants of integration through use of the fact that the beam deflection is zero at each end. Accordingly,

When $x = 0$, $y = 0$, so from Eq. (3) we have

$$0 = 0 - 0 + 0 + C_2 \qquad \text{and therefore } C_2 = 0$$

Next, when $x = L$, $y = 0$, so we have from Eq. (3)

$$0 = M_1\frac{L^2}{2} - \frac{R_L}{6}L^3 + C_1 L$$

from which

$$C_1 = -\frac{M_1 L}{3} - \frac{M_2 L}{6}$$

so that the desired equation of the deflection curve is

$$EIy = \frac{M_1}{2}x^2 - \left(\frac{M_1 - M_2}{6L}\right)x^3 - \left(\frac{M_1 L}{3} + \frac{M_2 L}{6}\right)x \tag{4}$$

If $M_1 = 0$, Eq. (4) becomes

$$EIy = \frac{M_2 x^3}{6L} - \frac{M_2 L x}{6} \tag{5}$$

and

$$EI\frac{dy}{dx} = \frac{M_2 x^2}{2L} - \frac{M_2 L}{6} \tag{6}$$

The point of peak deflection occurs when the slope given by Eq. (6) is zero. Solving Eq. (6) for this value of x,

$$x = \frac{L}{\sqrt{3}} \tag{7}$$

At this point (for $M_1 = 0$) the deflection is given by Eq. (5) to be

$$EIy_{max} = \frac{M_2}{6L}\left(\frac{L}{\sqrt{3}}\right)^3 - \frac{M_2 L}{6}\left(\frac{L}{\sqrt{3}}\right) = -\frac{M_2 L^2\sqrt{3}}{27} \tag{8}$$

Inspection of Eq. (4) for the case $M_1 = M_2 = M$ indicates that

$$EIy = \frac{M}{2}x^2 - \frac{ML}{2}x \tag{9}$$

which indicates a parabolic deflection curve. Yet, Eq. (2) of Problem 9.1 indicates that if M = constant along the length of the beam, the curvature $(1/\rho)$ is constant; i.e., the bar bends into a circular arc. The reason for the very slight discrepancy is that Eq. (5) of Problem 9.1, that is,

$$EI\frac{d^2y}{dx^2} = M$$

incorporates the approximation

$$\frac{1}{\rho} \approx \frac{d^2y}{dx^2}$$

as explained in Problem 9.1. In reality the numerical difference between the parabola and the circular arc is very small and in almost all cases may be neglected.

9.10. A simply supported beam is loaded by a couple M_2 as shown in Fig. 9-11. The beam is 2 m long and of square cross section 50 mm on a side. If the maximum permissible deflection in the beam is 5 mm, and the allowable bending stress is 150 MPa, find the maximum allowable load M_2. Take $E = 200$ GPa.

It is perhaps simplest to determine two values of M_2: one based upon the assumption that the deflection of 5 mm is realized, the other based on the assumption that the maximum bending stress in the bar is 150 MPa. The true value of M_2 is then the minimum of these two values.

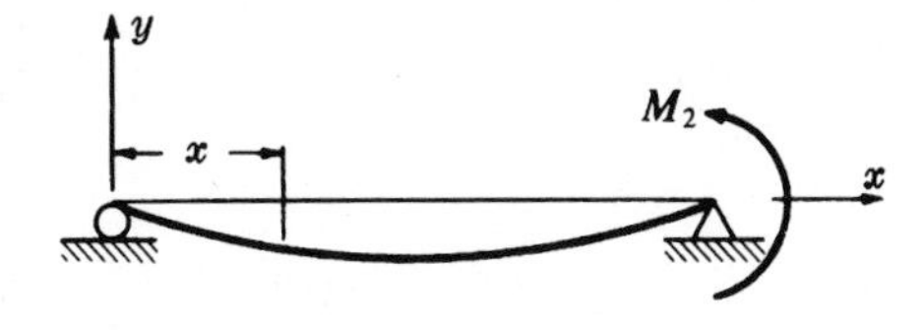

Fig. 9-11

Let us first consider that the maximum deflection in the beam is 5 mm. According to Eq. (8), Problem 9.9, we have

$$0.005 = \frac{M_2(2)^2\sqrt{3}}{27(200\times10^9)(\frac{1}{12})(0.05)(0.05)^3} \qquad \text{or} \qquad M_2 = 2.03\ \text{kN}\cdot\text{m}$$

We shall now assume that the allowable bending stress of 150 MPa is set up in the outer fibers of the beam at the section of maximum bending moment. Referring to Problem 9.9, since $M_1 = 0$, we find the reactions at the ends of the beam are

$$|R| = \frac{M_2}{L}$$

so that they have the appearance shown in Fig. 9-12, and the bending moment diagram for the beam is as shown in Fig. 9-13.

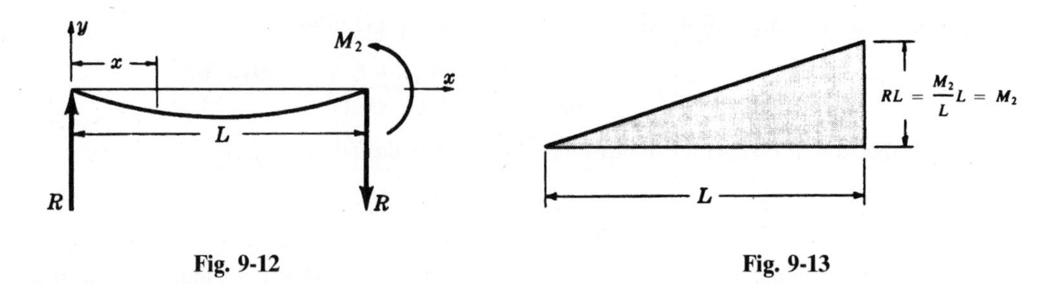

Fig. 9-12 Fig. 9-13

The maximum bending moment in the beam is M_2. Using the usual flexure formula, $\sigma = Mc/I$, we have at the outer fibers of the bar at the right end, i.e., at the section of maximum bending moment,

$$150 \times 10^6 = \frac{M_2(0.025)}{(\frac{1}{12})(0.05)(0.05)^3} \quad \text{or} \quad M_2 = 3.125 \text{ kN} \cdot \text{m}$$

Thus the maximum allowable moment is $M_2 = 2.03$ kN · m.

9.11. A simply supported beam is subjected to the sinusoidal loading shown in Fig. 9-14. Determine the deflection curve of the beam as well as the peak deflection.

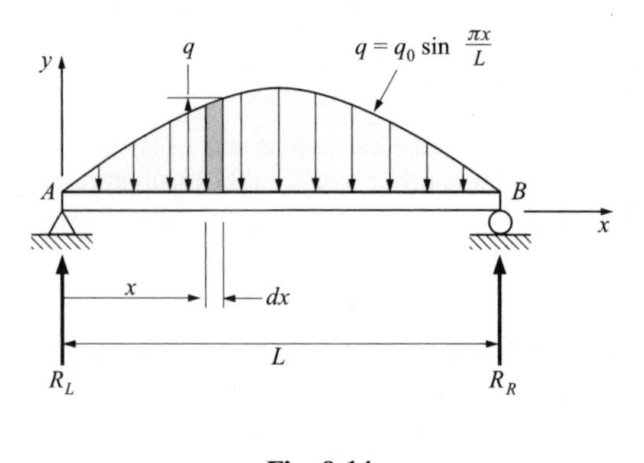

Fig. 9-14

It is first necessary to determine the total load on the beam. Let us consider the shaded element a distance x from the end A and of width dx. If q denotes load per unit length, then the load corresponding to the shaded element is $q\,dx$ and the load on the entire beam is found by integrating:

$$\text{Load} = \int_{x=0}^{x=L} q\,dx = \int_0^L q_0 \sin \frac{\pi x}{L} dx = \frac{2q_0 L}{\pi}$$

From statics, half of this load is carried at each end reaction. Thus,

$$R_L = R_R = \frac{q_0 L}{\pi}$$

The bending moment at the point denoted by x is found as the sum of the moments of all forces to the left of that point. To determine the moment about x of the portion of the sinusoidal load to the left of x, it is necessary to introduce another variable of integration, u, corresponding to a second vertical shaded element of width du, as shown in Fig. 9-15. The variable u must run from $u = 0$ to $u = x$ so as to yield the bending moment due to the sinusoidal load to the left of x.

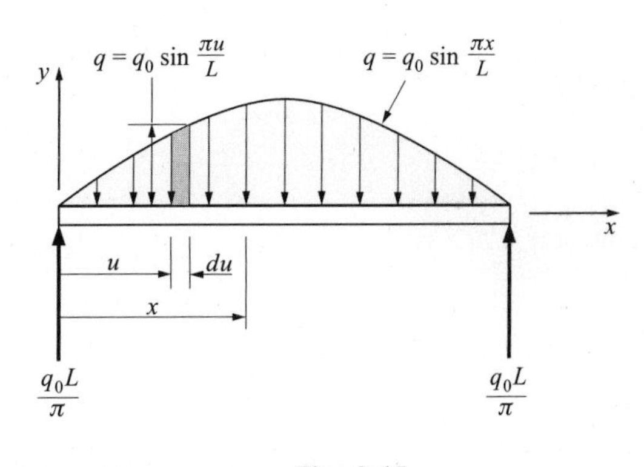

Fig. 9-15

Remembering the contribution that the left support makes to the bending moment, we have

$$M = \frac{q_0 L}{\pi} x - \int_{u=0}^{u=x} q_0 \left[\sin \frac{\pi u}{L} \right] (du)(x - u)$$

$$= \frac{q_0 L}{\pi} x - q_0 \int_{u=0}^{u=x} x \sin \frac{\pi u}{L} du + q_0 \overset{\circledast}{\int_{u=0}^{u=x}} u \sin \frac{\pi u}{L} du \tag{1}$$

In this integration u is a variable and x is to be (temporarily) regarded as a constant. The last integral $\circledast$ in Eq. (*1*) must be integrated by parts, remembering that

$$\int \theta(\sin \theta)\, d\theta = \sin \theta - \theta \cos \theta \tag{2}$$

Here,

$$\theta = \frac{\pi u}{L} \qquad d\theta = \frac{\pi}{L} du$$

so that the last integral $\circledast$ becomes

$$\int_{u=0}^{u=x} u \sin \frac{\pi u}{L} du = \frac{L^2}{\pi^2} \left[\sin \frac{\pi u}{L} - \frac{\pi u}{L} \cos \frac{\pi u}{L} \right]_{u=0}^{u=x}$$

$$= \frac{L^2}{\pi^2} \left[\sin \frac{\pi x}{L} \right] - \frac{Lx}{\pi} \cos \frac{\pi x}{L} \tag{3}$$

The bending moment, Eq. (*1*), is thus

$$M = \frac{q_0 L x}{\pi} - q_0 x \left(\frac{L}{\pi} \right) \left[-\cos \frac{\pi u}{L} \right]_{u=0}^{u=x} + \frac{q_0 L^2}{\pi^2} \left[\sin \frac{\pi x}{L} - \frac{\pi x}{L} \cos \frac{\pi x}{L} \right]$$

$$= \frac{q_0 L^2}{\pi^2} \sin \frac{\pi x}{L} \tag{4}$$

The differential equation of the deflected beam is thus

$$EI \frac{d^2 y}{dx^2} = \frac{q_0 L^2}{\pi^2} \sin \frac{\pi x}{L} \tag{5}$$

Integrating the first time, we have

$$EI \frac{dy}{dx} = -\frac{q_0 L^2}{\pi^2} \left(\frac{L}{\pi} \right) \cos \frac{\pi x}{L} + C_1 \tag{6}$$

As the first boundary condition, from symmetry, when $x = L/2$, $dy/dx = 0$. Substituting in Eq. (*6*), we find $C_1 = 0$. Integrating again,

$$EIy = -\frac{q_0 L^3}{\pi^3} \left(\frac{L}{\pi} \right) \sin \frac{\pi x}{L} + C_2 \tag{7}$$

The second boundary condition is that when $x = 0$, $y = 0$. Substituting in Eq. (7), we have $C_2 = 0$. The equation of the deflected beam is

$$EIy = -\frac{q_0 L^4}{\pi^4}\sin\frac{\pi x}{L} \tag{8}$$

and the peak deflection, at $x = L/2$, is

$$EIy]_{\max} = -\frac{q_0 L^4}{\pi^4}$$

9.12. Determine the deflection curve of a simply supported beam subject to the concentrated force P applied as shown in Fig. 9-16.

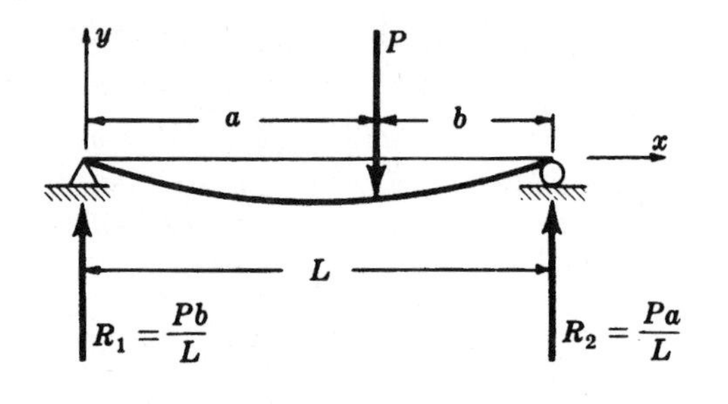

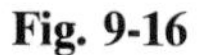
Fig. 9-16

The x-y coordinate system is introduced as shown. The heavy line indicates the configuration of the deformed beam. From statics the reactions are found to be $R_1 = Pb/L$ and $R_2 = Pa/L$.

This problem presents one feature that distinguishes it from the other problems solved thus far in this chapter. Namely, it is essential to consider two different equations describing the bending moment in the beam. One equation is valid to the left of the load P, the other holds to the right of this force. The integration of each equation gives rise to two constants of integration and thus there are four constants of integration to be determined. All problems met thus far have offered only two constants.

In the region to the left of the force P we have the bending moment $M = (Pb/L)x$ for $0 < x < a$. The differential equation of the bent beam thus becomes

$$EI\frac{d^2y}{dx^2} = \frac{Pb}{L}x \qquad \text{for} \qquad 0 < x < a \tag{1}$$

The first integration yields

$$EI\frac{dy}{dx} = \frac{Pb}{L}\frac{x^2}{2} + C_1 \tag{2}$$

No numerical information is available about the slope dy/dx at any point in this region. Since the load is not applied at the center of the beam, there is no reason to believe that the slope is zero at $x = L/2$. However, for the slope of the beam under the point of application of the force P we can write

$$EI\left(\frac{dy}{dx}\right)_{x=a} = \frac{Pba^2}{2L} + C_1 \tag{3}$$

The next integration of (2) yields

$$EIy = \frac{Pb}{2L}\frac{x^3}{3} + C_1 x + C_2 \tag{4}$$

At the left support, $y = 0$ when $x = 0$. Substituting these values in *(4)* we immediately find $C_2 = 0$. It is to be noted that it is not permissible to use the condition $y = 0$ at $x = L$ in *(4)* since *(1)* is not valid in that region. We have for the deflection under the point of application of the force P

$$EIy_{x=a} = \frac{Pba^3}{6L} + C_1 a \tag{5}$$

In the region to the right of the force P the bending moment equation is $M = (Pb/L)x - P(x-a)$ for $a<x<L$. Thus

$$EI\frac{d^2y}{dx^2} = \frac{Pb}{L}x - P(x-a) \qquad \text{for} \qquad a<x<L \tag{6}$$

The first integration of this equation yields

$$EI\frac{dy}{dx} = \frac{Pb}{L}\frac{x^2}{2} - \frac{P(x-a)^2}{2} + C_3 \tag{7}$$

Although nothing definite may be said about the slope in this portion of the beam, we have for the slope under the point of application of the force P

$$EI\left(\frac{dy}{dx}\right)_{x=a} = \frac{Pba^2}{2L} + C_3 \tag{8}$$

Under the concentrated load P the slope as given by *(3)* must be equal to that given by *(8)*. Consequently the right sides of these two equations must be equal and we have

$$\frac{Pba^2}{2L} + C_1 = \frac{Pba^2}{2L} + C_3 \qquad \text{or} \qquad C_1 = C_3$$

Equation *(7)* may now be integrated to give

$$EIy = \frac{Pb}{2L}\frac{x^3}{3} - \frac{P(x-a)^3}{6} + C_3 x + C_4 \tag{9}$$

We may write for the deflection under the concentrated load

$$EIy_{x=a} = \frac{Pba^3}{6L} + C_3 a + C_4 \tag{10}$$

The deflection at $x = a$ given by (5) must equal that given by *(10)*. Thus the right sides of these two equations are equal and we have

$$\frac{Pba^3}{6L} + C_1 a = \frac{Pba^3}{6L} + C_3 a + C_4$$

Since $C_1 = C_3$, we have $C_4 = 0$.

The condition that $y = 0$ when $x = L$ may now be substituted in *(9)*, yielding

$$0 = \frac{PbL^2}{6} - \frac{Pb^3}{6} + C_3 L \qquad \text{or} \qquad C_3 = \frac{Pb}{6L}(b^2 - L^2)$$

In this manner all four constants of integration are determined. These values may now be substituted in Eqs. *(4)* and *(9)* to give

$$EIy = \frac{Pb}{6L}[x^3 - (L^2 - b^2)x] \qquad \text{for} \qquad 0<x<a \tag{4'}$$

$$EIy = \frac{Pb}{6L}\left[x^3 - \frac{L}{b}(x-a)^3 - (L^2 - b^2)x\right] \qquad \text{for} \qquad a<x<L \tag{9'}$$

These two equations are necessary to describe the deflection curve of the bent beam. Each equation is valid only in the region indicated.

If the load P acts at the center of the beam, the peak deflection which occurs at $x = L/2$ by symmetry is given by Eq. (4′) as

$$EIy]_{x=L/2} = \frac{P(L/2)}{6L}\left[\left(\frac{L}{2}\right)^3 - \left(L^2 - \left\{\frac{L}{2}\right\}^2\right)\frac{L}{2}\right]$$

$$= -\frac{PL^3}{48} \qquad (11)$$

9.13. The simply supported beam described in Problem 9.12 is 14 ft long and of circular cross section 4 in in diameter. If the maximum permissible deflection is 0.20 in, determine the maximum value of the load P if $a = b = 7$ ft. The material is steel for which $E = 30 \times 10^6$ lb/in².

The maximum deflection, given by (11) of Problem 9.12, is $\Delta_{max} = PL^3/48EI$. For a circular cross section (see Problem 7.9), $I = \pi D^4/64 = \pi 4^4/64 = 12.6$ in⁴. Also, $L = 14$ ft $= 168$ in. Thus,

$$0.20 = \frac{P(168)^3}{48(30 \times 10^6)(12.6)} \qquad \text{or} \qquad P = 765 \text{ lb}$$

With this load applied at the center of the beam the reaction at each end is 383 lb and the bending moment at the center of the beam is $383(7) = 2681$ lb·ft. This is the maximum bending moment in the beam and the maximum bending stress occurs at the outer fibers at this central section. The maximum bending stress is $\sigma = Mc/I$. Then $\sigma_{max} = 2681(12)(2)/12.6 = 5100$ lb/in². This is below the proportional limit of the material; hence the use of the deflection equation was permissible.

9.14. Consider the simply supported beam described in Problem 9.12. If the cross section is rectangular, 50×100 mm and $P = 20$ kN with $a = 1$ m, $b = 0.5$ m, determine the maximum deflection of the beam. The beam is steel, for which $E = 200$ GPa.

Since $a > b$, it is evident that the maximum deflection must occur to the left of the load P. It occurs at that point where the slope of the beam is zero.

Differentiating Eq. (4′) of Problem 9.12, we find that the slope in this region is given by

$$EI\frac{dy}{dx} = \frac{Pb}{6L}[3x^2 - (L^2 - b^2)]$$

Setting the slope equal to zero, we find $x = \sqrt{L^2 - b^2/3}$ for the point where the deflection is maximum. The deflection at this point is found by substituting this value of x in (4′):

$$EIy_{max} = \frac{Pb\sqrt{3}}{27L}(L^2 - b^2)^{3/2}$$

For the rectangular section $I = 50(100)^3/12 = 4.167 \times 10^6$ mm⁴. Substituting,

$$y_{max} = \frac{20 \times 10^3(0.5 \times 10^3)[(1.5 \times 10^3)^2 - (0.5 \times 10^3)^2]^{3/2}(\sqrt{3})(10^6)}{27(1.5 \times 10^3)(4.167 \times 10^6)(200 \times 10^9)} = -1.45 \text{ mm}$$

The negative sign indicates that this point on the bent beam lies below the x-axis.

From $\sigma = Mc/I$ the maximum bending stress, which occurs under the load P, is 80 MPa. This is below the proportional limit of steel, so the above deflection equations are valid.

9.15. The beam AC is simply supported at A and at C is pinned to a cantilever beam CD as shown in Fig. 9-17(a). Both beams have identical flexural rigidities EI. The vertical load of 8 kN acts at point B. Determine the deflection of point B.

Free-body diagrams of the flexible beams AC and CD appear as in Figs. 9-17(b) and 9-17(c), respectively. For AC, because of symmetry the reaction at C is 4 kN and by Newton's law the equal and opposite force must be exerted at the end C of beam CD as shown in Fig. 9.17(c).

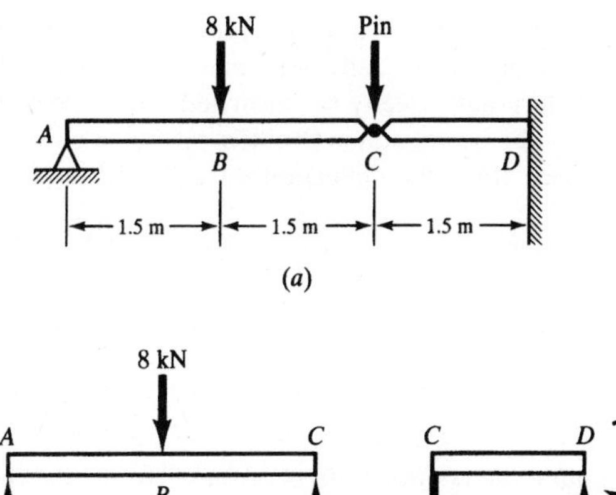

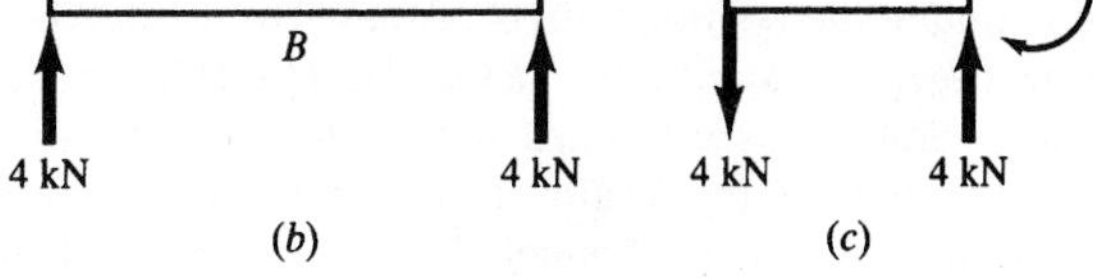

Fig. 9-17

From Problem 9.2 the downward deflection of point C regarded as the tip of beam CD is

$$\Delta_C = \frac{PL^3}{3EI} = \frac{(4\text{ kN})(1.5\text{ m})^3}{3EI} = \frac{4.5}{EI}$$

This same deflection must describe the downward displacement of C regarded as the right end of beam AC. Prior to the deformation of AC due to the 8-kN load, the displacement of point C (on AC) imparts a downward displacement of half that, namely $2.25/EI$ to point B, since the bar during this stage will rotate as a rigid body about A. Then, the deflection of point B due to the 8-kN load must be considered. From Problem 9.12 this is

$$\frac{PL^3}{48EI} = \frac{(8\text{ kN})(3\text{ m})^3}{48EI} = \frac{4.5}{EI}$$

The resultant deflection at point B is thus

$$\Delta_B = \frac{4.5}{EI} + \frac{2.25}{EI} = \frac{6.75}{EI}\ (\downarrow)$$

9.16. Determine the equation of the deflection curve for a cantilever beam loaded by a uniformly distributed load w per unit length, as well as by a concentrated force P at the free end. See Fig. 9-18.

The deformed beam has the configuration indicated by the heavy line. The x-y coordinate system is introduced as shown. One logical approach to this problem is to determine the reactions at the wall, then

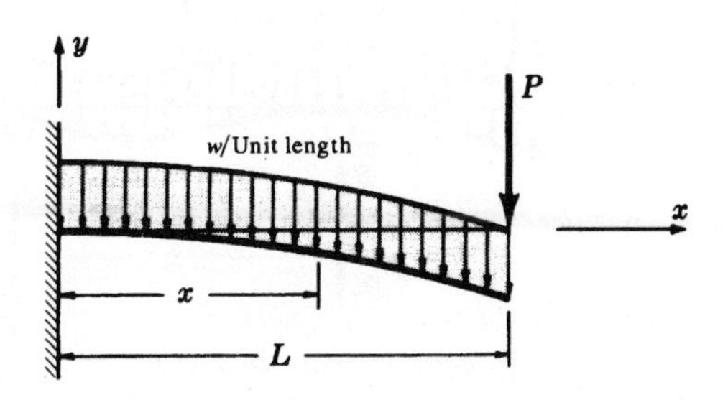

Fig. 9-18

write the differential equation of the bent beam, integrate this equation twice, and determine the constant of integration from the conditions of zero slope and zero deflection at the wall.

Actually this procedure has already been carried out in Problem 9.2 for the case in which only the concentrated load acts on the beam, and in Problem 9.5 when only the uniformly distributed load is acting. For the concentrated force alone the deflection y was found in (*3*) of Problem 9.2 to be

$$EIy = -PL\frac{x^2}{2} + \frac{Px^3}{6} \tag{1}$$

For the uniformly distributed load alone the deflection y was found in (*3'*) of Problem 9.5 to be

$$EIy = -\frac{w}{24}(L-x)^4 - \frac{wL^3}{6}x + \frac{wL^4}{24} \tag{2}$$

It is possible to obtain the resultant effect of these two loads when they act simultaneously merely by adding together the effects of each as they act separately. This is called the *method of superposition*. It is useful in determining deflections of beams subject to a combination of loads, such as we have here. Essentially it consists in utilizing the results of simpler beam-deflection problems to build up the solutions of more complicated problems. Thus it is not an independent method of determining beam deflections.

According to this method the deflection at any point of a beam subject to a combination of loads can be obtained as the sum of the deflections produced at this point by each of the loads acting separately. The final deflection equation resulting from the combination of loads is then obtained by adding the deflection equations for each load.

For the present beam the final deflection equation is given by adding Eqs. (*1*) and (*2*):

$$EIy = -PL\frac{x^2}{2} + \frac{Px^3}{6} - \frac{w}{24}(L-x)^4 - \frac{wL^3}{6}x + \frac{wL^4}{24} \tag{3}$$

The slope dy/dx at any point in the beam is merely found by differentiating both sides of (*3*) with respect to x.

The method of superposition is valid in all cases where there is a linear relationship between each separate load and the separate deflection which it produces.

9.17. Determine the deflection curve of an overhanging beam subject to a uniform load w per unit length and supported as shown in Fig. 9-19.

We replace the distributed load by its resultant of wL acting at the midpoint of the length L. Taking moments about the right reaction, we have

$$\Sigma M_C = R_1 b - \frac{wL^2}{2} = 0 \qquad \text{or} \qquad R_1 = \frac{wL^2}{2b}$$

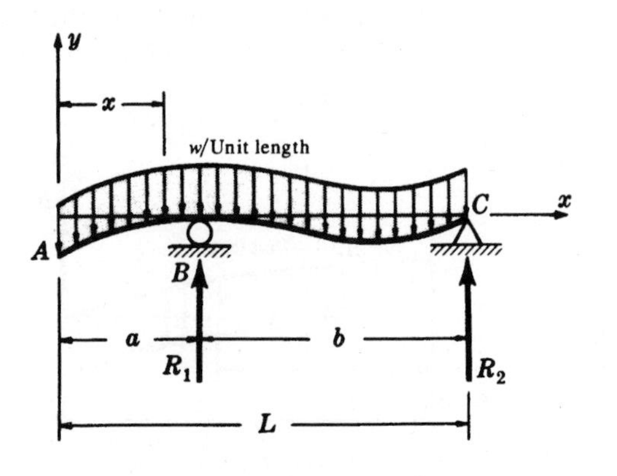

Fig. 9-19

Summing forces vertically, we find

$$\Sigma F_v = \frac{wL^2}{2b} + R_2 - wL = 0$$

or

$$R_2 = wL - \frac{wL^2}{2b}$$

The bending moment equation in the left overhanging region is $M = -wx^2/2$ for $0<x<a$. Consequently the differential equation of the bent beam in that region is

$$EI\left(\frac{d^2y}{dx^2}\right) = \frac{-wx^2}{2} \quad \text{for} \quad 0<x<a \tag{1}$$

Two successive integrations yield

$$EI\frac{dy}{dx} = -\frac{w}{2}\frac{x^3}{3} + C_1 \tag{2}$$

$$EIy = -\frac{w}{6}\frac{x^4}{4} + C_1x + C_2 \tag{3}$$

The bending moment equation in the region between supports is $M = -wx^2/2 + R_1(x-a)$. The differential equation of the bent beam in that region is thus

$$EI\frac{d^2y}{dx^2} = -\frac{wx^2}{2} + \frac{wL^2}{2b}(x-a) \quad \text{for} \quad a<x<L \tag{4}$$

Two integrations of this equation yield

$$EI\frac{dy}{dx} = -\frac{w}{2}\frac{x^3}{3} + \frac{wL^2}{2b}\frac{(x-a)^2}{2} + C_3 \tag{5}$$

$$EIy = -\frac{w}{6}\frac{x^4}{4} + \frac{wL^2}{4b}\frac{(x-a)^3}{3} + C_3x + C_4 \tag{6}$$

Since we started with two second-order differential equations, (1) and (4), and two constants of integration arose from each, we have four constants C_1, C_2, C_3, and C_4 to evaluate from known conditions concerning slopes and deflections. These conditions are the following:

1. When $x = a$, $y = 0$ in the overhanging region.
2. When $x = a$, $y = 0$ in the region between supports.
3. When $x = L$, $y = 0$ in the region between supports.
4. When $x = a$, the slope given by (2) must be equal to that given by (5); consequently the right sides of these equations must be equal when $x = a$.

Substituting condition (1) in (3), we obtain

$$0 = \frac{-wa^4}{24} + C_1a + C_2 \tag{7}$$

Substituting condition (2) in (6), we find

$$0 = \frac{-wa^4}{24} + C_3a + C_4 \tag{8}$$

Substituting condition (3) in (6), we get

$$0 = \frac{-wL^4}{24} + \frac{wL^2b^2}{12} + C_3L + C_4 \tag{9}$$

Finally, equating slopes at the left reaction by substituting $x = a$ in the right sides of equations (2) and (5), we obtain

$$\frac{-wa^3}{6} + C_1 = \frac{-wa^3}{6} + C_3 \tag{10}$$

Note that there is no reason for assuming the slope to be zero at the left support, $x = a$.

These last four Eqs. (7), (8), (9), (10) may now be solved for the four unknown constants C_1, C_2, C_3, C_4. The solution is found to be

$$C_1 = C_3 = \frac{w(L^4 - a^4)}{24b} - \frac{wL^2 b}{12} \tag{11}$$

$$C_2 = C_4 = \frac{wa^4}{24} - \frac{w(L^4 - a^4)a}{24b} + \frac{wL^2 ab}{12} \tag{12}$$

The two equations describing the deflection curve of the bent bar are found by substituting these values of the constants in (3) and (6). These equations may be written in the final forms

$$EIy = -\frac{wx^4}{24} + \frac{w(L^4 - a^4)x}{24b} - \frac{wL^2 bx}{12} + \frac{wa^4}{24} - \frac{w(L^4 - a^4)a}{24b} + \frac{wL^2 ab}{12} \quad \text{for } 0 < x < a \tag{3'}$$

$$EIy = -\frac{wx^4}{24} + \frac{wL^2(x-a)^3}{12b} + \frac{w(L^4 - a^4)x}{24b} - \frac{wL^2 bx}{12} + \frac{wa^4}{24} - \frac{w(L^4 - a^4)a}{24b} + \frac{wL^2 ab}{12} \quad \text{for } a < x < L \tag{6'}$$

Problem 9.17, although involving relatively simple geometry and loading, is obviously very tedious when solved by the method of double integration. Usually the method is well suited only to situations where a single equation describes the entire deflected beam. Chapter 10 will be based upon use of singularity functions (see Chap. 6) as a much-simplified approach to beam deflections far better adapted to more complex conditions of loading and support than is the straightforward double-integration approach. Also, the singularity function approach is very well adapted to computer implementation, as will be shown in Chap. 10.

9.18. Determine the equation of the deflection curve for the overhanging beam loaded by the two equal forces P shown in Fig. 9-20.

The x-y coordinate system is introduced as shown with the x-axis coinciding with the original unbent position of the bar. The fact that the left end of the bar deflects from the coordinate curve presents no difficulties. For the condition of symmetry it is evident that each support exerts a vertical force P upon the bar.

The bending moment in the left overhanging region is

$$M = -Px \quad \text{for} \quad 0 < x < a$$

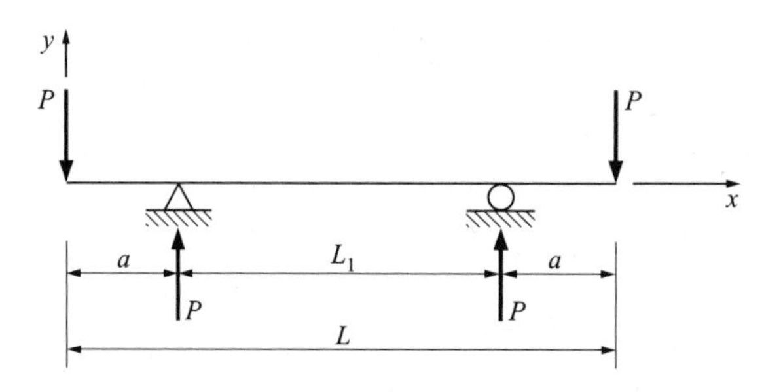

Fig. 9-20

and the differential equation of the bent beam in that region is

$$EI\frac{d^2y}{dx^2} = -Px \qquad \text{for} \qquad 0 < x < a \tag{1}$$

The first integration of this equation yields

$$EI\frac{dy}{dx} = -P_1\frac{x^2}{2} + C_1 \tag{2}$$

Nothing definite is known about the slope dy/dx in this region. In particular, it is to be emphasized that there is no justification for assuming the slope to be zero at the point of support $x = a$. We may denote the slope there by the notation

$$EI\left(\frac{dy}{dx}\right)_{x=a} = -P\left(\frac{a^2}{2}\right) + C_1 \tag{3}$$

The next integration yields

$$EIy = -\frac{P}{2}\left(\frac{x^3}{3}\right) + C_1x + C_2 \tag{4}$$

Since the beam is hinged at the support, it is known that the deflection y is 0 there. Thus, $(y)_{x=a} = 0$. Substituting $y = 0$ when $x = a$ in (*4*), we find

$$0 = -\frac{Pa^3}{6} + C_1a + C_2 \tag{5}$$

The bending moment in the central region of the beam between supports is $M = -Pa$ and the differential equation of the bent beam in the central region is

$$EI\frac{d^2y}{dx^2} = -Pa \qquad \text{for} \qquad a < x < (L - a) \tag{6}$$

Integrating, we obtain

$$EI\frac{dy}{dx} = -Pax + C_3 \tag{7}$$

Because of the symmetry of loading it is evident that the slope dy/dx must be zero at the midpoint of the bar. Thus $(dy/dx)_{x=L/2} = 0$. Substituting these values in Eq. (*7*), we find

$$0 = -Pa\left(\frac{L}{2}\right) + C_3 \qquad \text{or} \qquad C_3 = \frac{PaL}{2} \tag{8}$$

Also, from Eq. (*7*) we may say that the slope of the beam over the left support, $x = a$, is given by substituting $x = a$ in this equation. This yields

$$EI\left(\frac{dy}{dx}\right)_{x=a} = -Pa^2 + \frac{PaL}{2} \tag{9}$$

But the slope dy/dx as given by this expression must be equal to that given by Eq. (*3*), since the bent bar at that point must have the same slope, no matter which equation is considered. Equating the right sides of Eqs. (*3*) and (*9*), we obtain

$$-\frac{Pa^2}{2} + C_1 = -Pa^2 + \frac{PaL}{2} \tag{10}$$

or

$$C_1 = -\frac{Pa^2}{2} + \frac{PaL}{2} \tag{11}$$

Substituting this value of C_1 in Eq. (5), we find

$$0 = -\frac{Pa^3}{6} - \frac{Pa^3}{2} + \frac{Pa^2 L}{2} + C_2 \qquad (12)$$

or

$$C_2 = \frac{2Pa^3}{3} - \frac{Pa^2 L}{2}$$

The next integration of Eq. (7) yields

$$EIy = -Pa\frac{x^2}{2} + \frac{PaL}{2}(x) + C_4 \qquad (13)$$

Again, it may be said that the deflection y is zero at the left support, where $x = a$. Although this same condition was used previously in obtaining Eq. (5), there is no reason why it should not be used again. In fact, it is essential to use it in order to solve for the constant C_4 in Eq. (13). Thus, substituting the values $(y)_{x=a} = 0$ in Eq. (13), we obtain

$$0 = -\frac{Pa^3}{2} + \frac{Pa^2 L}{2} + C_4 \qquad \text{or} \qquad C_4 = \frac{Pa^3}{2} - \frac{Pa^2 L}{2} \qquad (14)$$

Thus two equations were required to define the bending moment in the left and central regions of the beam. Each equation was used in conjunction with the second-order differential equation describing the bent beam, and thus two constants of integration arose from the solution of each of these two equations. It was necessary to utilize four conditions concerning slope and deflection in order to determine these four constants. These conditions were:

(*a*) When $x = a$, $y = 0$ for the overhanging portion of the beam.
(*b*) When $x = a$, $y = 0$ for the central portion of the beam.
(*c*) When $x = L/2$, $dy/dx = 0$ for the central portion of the beam.
(*d*) When $x = a$, the slope dy/dx is the same for the deflection curve on either side of the support.

Finally, the equations of the bent beam may be written in the forms

$$EIy = -\frac{Px^3}{6} - \frac{Pa^2 x}{2} + \frac{PaLx}{2} + \frac{2Pa^3}{3} - \frac{Pa^2 L}{2} \qquad \text{for} \qquad 0 < x < a \qquad (15)$$

$$EIy = -\frac{Pax^2}{2} + \frac{PaLx}{2} + \frac{Pa^3}{2} - \frac{Pa^3 L}{2} \qquad \text{for} \qquad a < x < (L - a) \qquad (16)$$

Because of the symmetry there is no need to write the equation for the deformed beam in the right overhanging region.

9.19. For the overhanging beam of Problem 9.18, each force P is 4000 lb. The distance a is 3 ft and the length L is 16 ft. The bar is steel and of circular cross section 4 in in diameter. Determine the deflection under each load and also the deflection at the center of the beam. Take $E = 30 \times 10^6$ lb/in^2.

The moment of inertia is given by $I = \pi(4)^4/64 = 12.6$ in^4, according to Problem 7.9 in Chap. 7. Also, we have $a = 3$ ft $= 36$ in, $L = 16$ ft $= 192$ in. The deflection anywhere in the left overhanging region is given by Eq. (15) of Problem 9.18. Under the concentrated force P we have $x = 0$, and substituting these values in Eq. (15) we obtain

$$30 \times 10^6 (12.6)\,(y)_{x=0} = \frac{2(4000)\,(36)^3}{3} - \frac{4000(36)^2(192)}{2}$$

or

$$(y)_{x=0} = -0.96 \text{ in}$$

The deflection anywhere in the central portion between supports is given by Eq. (*16*) of Problem 9.18. At the center of the beam we have $x = 8\text{ ft} = 96\text{ in}$ and, as before $a = 36\text{ in}$, $L = 192\text{ in}$, and $P = 4000\text{ lb}$. Substituting in Eq. (*16*), we find

$$(30 \times 10^6)(12.6)(y)_{x=8\,\text{ft}} = \frac{-4000(36)(96)^2}{2} + \frac{(4000)(36)(192)(96)}{2} + \frac{4000(36)^3}{2} - \frac{4000(36)^2(192)}{2}$$

Solving

$$y_{x=8\,\text{ft}} = 0.69\text{ in}$$

The maximum bending stress occurs at the outer fibers of the bar everywhere between the supports, since the bending moment has the constant value of $4000(3) = 12{,}000\text{ lb}\cdot\text{ft}$ in this region. This maximum stress is given by

$$\sigma = \frac{M_c}{I} = \frac{(12{,}000)(12)(2)}{12.6} = 22{,}800\text{ lb/in}^2$$

This is less than the proportional limit of the material.

9.20. A cantilever beam Fig. 9-21(*a*) lying in a horizontal plane when viewed from the top has the triangular plan form shown in Fig. 9-21(*b*). The side view, Fig. 9-21(*c*), shows the constant thickness h of the beam. Determine the deflection curve of the beam and also the deflection of the tip due to the weight of the beam, which is γ per unit volume.

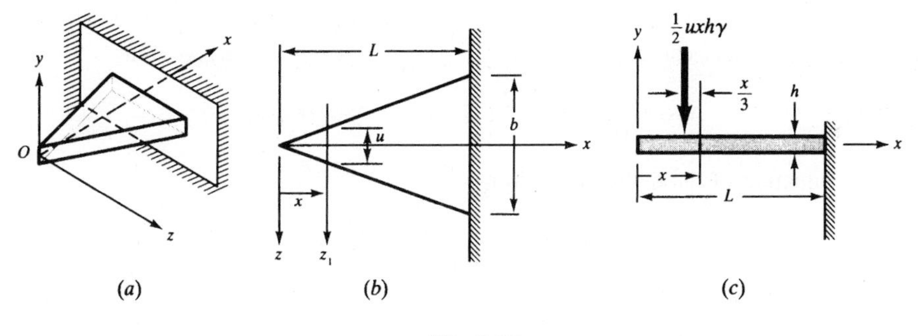

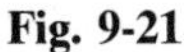
Fig. 9-21

We introduce an x-y-z coordinate system having its origin at point O, the tip of the beam. The location of an arbitrary cross section is denoted by x and the width there is u, as shown in Fig. 9-21(*b*). The overall beam length and base width are denoted by L and b, respectively. From geometry we have

$$u = b\left(\frac{x}{L}\right)$$

and the bending moment at section x is due to the weight of the portion of the triangular beam to the left of x. That weight is

$$\tfrac{1}{2}uxh\gamma$$

and the resultant force corresponding to this weight acts at a distance $x/3$ from the cross-section x, as shown in Fig. 9-21(*c*). Thus, the bending moment at x due to the weight of material to the left of x is

$$M = -\frac{uxh\gamma}{2}\cdot\frac{x}{3} = -\frac{x^2h\gamma}{6}\left(\frac{bx}{L}\right) = -\frac{bh\gamma x^3}{6L} \tag{1}$$

so that the differential equation of the deflected beam is

$$EI\frac{d^2y}{dx^2} = -\frac{bh\gamma x^3}{6L} \tag{2}$$

However, I is a function of x. Consideration of the cross-section x indicates that I (about an axis z_1 parallel to the z-axis) is

$$I = \frac{1}{12}uh^3 = \frac{1}{12}b\left(\frac{x}{L}\right)h^3$$

so that the differential equation of the beam becomes

$$E\left[\frac{1}{12}b\left(\frac{x}{L}\right)h^3\right]\frac{d^2y}{dx^2} = -\frac{bh\gamma x^3}{6L} \tag{3}$$

or

$$\frac{d^2y}{dx^2} = -\left(\frac{2\gamma}{Eh^2}\right)x^2 \tag{4}$$

Integrating the first time, we obtain

$$\frac{dy}{dx} = -\left(\frac{2\gamma}{Eh^2}\right)\frac{x^3}{3} + C_1 \tag{5}$$

and when $x = L$, $dy/dx = 0$; hence substituting in Eq. (5), we have

$$0 = -\frac{2\gamma L^3}{3Eh^2} + C_1 \qquad \text{and therefore } C_1 = \frac{2\gamma L^3}{3Eh^2}$$

Integrating again, we find

$$y = -\left(\frac{2\gamma}{3Eh^2}\right)\frac{x^4}{4} + \frac{2\gamma L^3}{3Eh^2}x + C_2 \tag{6}$$

As a second boundary condition, when $x = L$, $y = 0$, so from Eq. (6) we find

$$0 = -\frac{2\gamma}{3Eh^2}\cdot\frac{L^4}{4} + \frac{2\gamma L^4}{3Eh^2} + C_2 \qquad \text{and therefore } C_2 = -\frac{\gamma L^4}{2Eh^2}$$

Thus, the equation of the deflected beam is

$$y = -\frac{\gamma}{6Eh^2}x^4 + \frac{2\gamma L^3}{3Eh^2}x - \frac{\gamma L^4}{2Eh^2}$$

which at the tip becomes

$$y]_{x=0} = -\frac{\gamma L^4}{2Eh^2}$$

9.21. A cantilever beam is in the form of a circular truncated cone, of length L, diameter d at the small end, and $2d$ at the large end, as shown in Fig. 9-22. The beam is loaded only by its own weight, which is γ per unit volume. Determine the deflection at the free end.

From the geometry, we may extend the sloping sides until they intersect at distance x_0 from the left end. By similar triangles we have

$$\frac{d}{x_0} = \frac{2d}{x_0 + L}$$

from which $x_0 = L$. Also,

$$\frac{y}{x} = \frac{d}{2L}$$

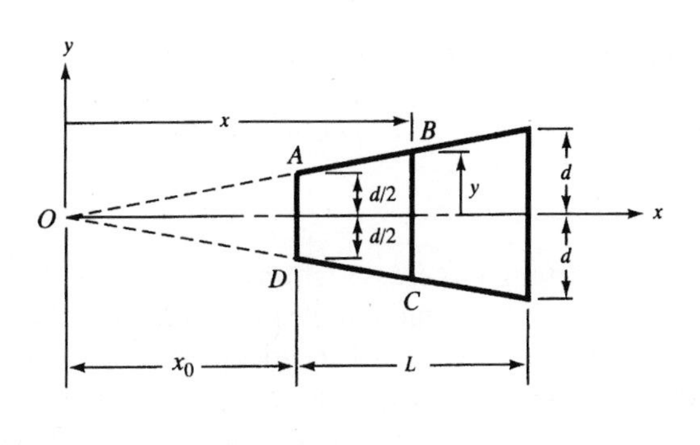

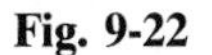

Fig. 9-22

so

$$y = \left(\frac{d}{2L}\right)x$$

The moment of inertia of any circular cross section a distance x from the point O is

$$I = \frac{\pi y^4}{4} = \frac{\pi}{4}\left(\frac{d^4}{16L^4}\right)x^4$$

The differential equation of the deflected beam is given by employing Eq. (*5*) of Problem 9.1 and using as the bending moment at x the moment of the weight of the solid region $ABCD$ which is found as the moment of the weight of the complete solid cone $OBCO$ about x minus the moment of the cone OAD about that same section. Remembering that the volume of a complete cone is $\frac{1}{3}$ (base) (altitude) and that the center of mass of a solid cone lies $\frac{1}{4}$ the altitude above the base, we have for the equation of the bent beam

$$E\left\{\frac{\pi d^4}{64L^4}\cdot x^4\right\}\frac{d^2y}{dx^2} = -\left\{\frac{1}{3}\pi y^2 x\gamma\left(\frac{x}{4}\right) - \frac{1}{3}\gamma\pi\left(\frac{d}{2}\right)^2 L\left(x - \frac{3}{4}L\right)\right\} \tag{1}$$

This simplifies to the form

$$\frac{d^2y}{dx^2} = \frac{16L^4\gamma\pi}{3\pi d^4 E}\left\{-\frac{d^2}{4L^2} + \frac{Ld^2}{x^3} - \frac{3L^2d^2}{4x^4}\right\} \tag{2}$$

The first integration leads to

$$\frac{dy}{dx} = \frac{16L^4\gamma}{3d^4E}\left\{-\frac{d^2}{4L^2}x + Ld^2\left(-\frac{1}{2x^2}\right) - \frac{3L^2d^2}{4}\left(-\frac{1}{3x^3}\right)\right\} + C_1 \tag{3}$$

As the first boundary condition, when $x = 2L$, $dy/dx = 0$. Substituting in (*3*), we find

$$C_1 = \frac{19L^3\gamma}{6d^2E}$$

The next integration gives us

$$y = \frac{16L^4\gamma}{3d^4E}\left\{-\frac{d^2}{4L^2}\cdot\frac{x^2}{2} - \frac{Ld^2}{2}\left(-\frac{1}{x}\right) + \frac{L^2d^2}{4}\left(-\frac{1}{2x^2}\right)\right\} + \frac{19L^3\gamma}{6d^2E}x + C_2 \tag{4}$$

and the second boundary condition is that when $x = 2L$, $y = 0$. From Eq. (*4*) we have

$$C_2 = -\frac{29}{6}\frac{L^4\gamma}{d^2E}$$

The equation of the deflected beam is thus

$$y = \frac{16L^4\gamma}{3d^4E}\left\{-\frac{d^2}{8L^2}x^2 + \frac{Ld^2}{2}\left(\frac{1}{x}\right) - \frac{L^2d^2}{8}\left(\frac{1}{x^2}\right)\right\} + \frac{19L^3\gamma}{6d^2E}x - \frac{29L^4\gamma}{6d^2E} \tag{5}$$

The deflection of the tip is found by setting $x = L$ in Eq. (5) and is

$$y]_{x=L} = -\frac{\gamma L^4}{3d^2 E}$$

9.22. The beam of variable rectangular cross section shown in Fig. 9-23 is simply supported at the ends and loaded by equal magnitude end couples each equal to PL as well as symmetrically placed transverse forces each equal to $1.5P$. The thickness h of the beam is constant. Determine the manner in which the width must vary so that all outer fibers are stressed to the same value σ_0 in both tension and compression. Also determine the central deflection of the beam.

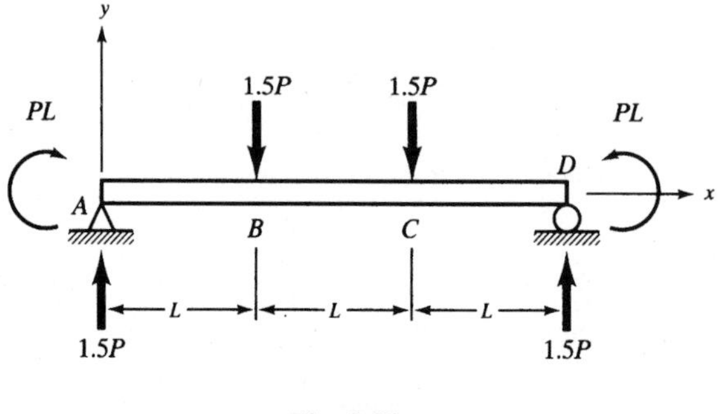

Fig. 9-23

The end reactions are easily found from statics to each be $1.5P$, as shown. The bending moment diagrams corresponding to the force loadings and to the end couples are found by the methods of Chap. 6 and are illustrated in Figs. 9-24(a) and 9-24(b), respectively. The resultant bending moment diagram is found by superposition of these two to be that shown in Fig. 9-24(c).

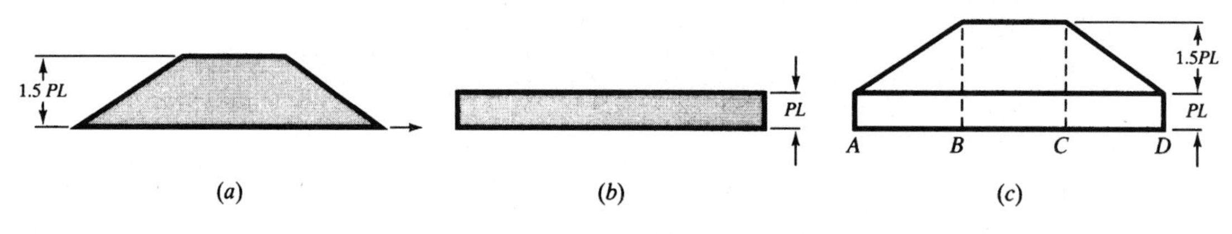

Fig. 9-24

The outer fiber bending stresses in each of the regions AB and BC are found for the rectangular cross section through use of the results of Problems 8.1 and 8.12 to be

$$\sigma_Z = \frac{M_c}{I} = \frac{M}{I/c} = \frac{M}{Z} = \frac{6M}{bh^2} \tag{1}$$

where for the rectangular bar

$$Z = \frac{bh^2}{6} \tag{2}$$

Figure 9-24(c) together with Eq. (1) indicates that in the region BC (since the bending moment is constant) the beam width must also be constant. In that region the cross section must withstand a maximum bending moment of $2.5PL$ and the value of the outer fiber bending stresses is

$$\sigma_0 = \frac{6(2.5PL)}{b_{max} h^2} \tag{3}$$

Solving, we find the maximum width everywhere in BC to be

$$b_{\max} = \frac{15PL}{\sigma_0 h^2} \tag{4}$$

In the end region AB, the bending moment from Fig. 9-24(c) is

$$M = PL + 1.5PL\left(\frac{x}{L}\right) \qquad \text{for } 0 < x < L \tag{5}$$

where x is measured positive to the right from the support at A. Since $x = 0$ at A, the width of the beam there must be sufficient to withstand the bending moment PL. Thus, for the outer fiber bending stresses at $x = 0$ to have the magnitude σ_0, we have

$$\sigma_0 = \frac{6M}{b_{\min} h^2} = \frac{6PL}{b_{\min} h^2}$$

Solving,

$$b_{\min} = \frac{6PL}{\sigma_0 h^2} \tag{6}$$

The same width $b_{\min}$ must also exist at the right end $x = 3L$ by symmetry. Equation (5) indicates a linear variation of bending moment between A and B so that the width increases linearly from A to B. The resulting constant outer fiber bending stress beam thus appears as shown in Fig. 9-25.

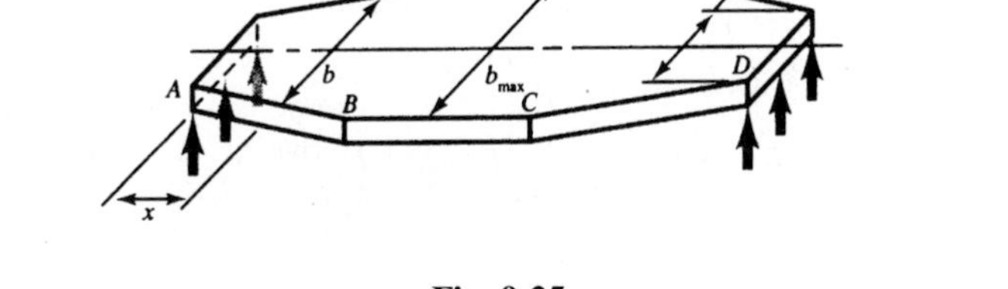

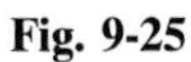
Fig. 9-25

To find the peak deflection, which, because of symmetry, obviously occurs at the midpoint of BC where $x = 3L/2$, we must write the differential equations for bending in regions AB and BC. Because of symmetry of loading and support, there is no need to consider CD since its behavior is symmetric to that of AB. First,

In AB:

$$M = 1.5Px + PL$$

and

$$\sigma_0 = \frac{Mc}{I} = \frac{(PL + 1.5Px)(h/2)}{\frac{1}{12}bh^3} \tag{7}$$

Thus,

$$b = \frac{(PL + 1.5Px)(6)}{\sigma_0 h^2} \tag{8}$$

where b denotes the width of the bar at a distance x from A as indicated in Fig. 9-25. The moment of inertia of the cross section a distance x from A is thus

$$\frac{1}{12}\left[\frac{(PL + 1.5Px)(6)}{\sigma_0 h^2}\right]h^3 \tag{9}$$

The differential equation of the bent beam in AB is

$$E\left[\frac{(PL + 1.5Px)h}{2\sigma_0}\right]\frac{d^2y}{dx^2} = 1.5Px + PL \tag{10}$$

or

$$\frac{d^2y}{dx^2} = \frac{2\sigma_0}{Eh} = \text{constant} \tag{11}$$

Integrating

$$\frac{dy}{dx} = \left(\frac{2\sigma_0}{Eh}\right)x + C_1 \tag{12}$$

Integrating a second time

$$y = \frac{2\sigma_0}{Eh}\cdot\frac{x^2}{2} + C_1 x + C_2 \tag{13}$$

As a boundary condition, when $x = 0$, $y = 0$; hence $C_2 = 0$ from Eq. (*13*). Also, when $x = L$, the deflection from Eq. (*13*) is

$$y]_{x=L} = \frac{2\sigma_0}{Eh}\cdot\frac{L^2}{2} + C_1 L \tag{14}$$

and the slope at $x = L$ is, from (*12*)

$$\left.\frac{dy}{dx}\right]_{x=L} = \frac{2\sigma_0 L}{Eh} + C_1 \tag{15}$$

In BC, $M = 2.5PL$, and since the width b_{max} in BC is constant, the moment of inertia anywhere in BC is

$$\tfrac{1}{12}b_{max}h^3 \tag{16}$$

so the bent beam in BC is described by the equation

$$E\left[\frac{b_{max}h^3}{12}\right]\frac{d^2y}{dx^2} = 2.5PL \tag{17}$$

or

$$\frac{d^2y}{dx^2} = \frac{30PL}{Eb_{max}h^3} = \text{constant} \tag{18}$$

Integrating,

$$\frac{dy}{dx} = \frac{30PLx}{Eb_{max}h^2} + C_3 \tag{19}$$

As a boundary condition, from symmetry we know that at $x = 3L/2$, $dy/dx = 0$. Hence from (*19*) we have

$$C_3 = -\frac{45PL^2}{Eb_{max}h^3}$$

Integrating again,

$$y = \left(\frac{30PL}{Eb_{max}h^3}\right)\frac{x^2}{2} - \left(\frac{45PL^2}{Eb_{max}h^3}\right)x + C_4 \tag{20}$$

When $x = L$, the deflections are represented by Eqs. (*14*) and (*20*), leading to

$$\frac{2\sigma_0 L^2}{2Eh} + C_1 L = -\frac{30PL^3}{Eb_{max}h^3} + C_4 \tag{21}$$

Finally, equating slopes at $x = L$ as given by Eqs. (*15*) and (*19*), we have

$$\frac{2\sigma_0 L}{Eh} + C_1 = \frac{30PL^2}{Eb_{max}h^3} - \frac{45PL^2}{Eb_{max}h^3} \tag{22}$$

Solving Eqs. (*21*) and (*22*), we find

$$C_1 = -\frac{45PL^2}{Eb_{max}h^2} \quad \text{and therefore } C_4 = 0$$

Hence in the region BC from Eq. (*20*), we have

$$y_{max}\big]_{x=3L/2} = -\frac{33.75PL^3}{Eb_{max}h^3}$$

9.23. Consider the bending of a cantilever beam which remains in contact with a rigid cylindrical surface as it deflects. The tangent to the cantilever is horizontal at point A in Fig. 9-26. Determine the deflection of the tip B due to the load P.

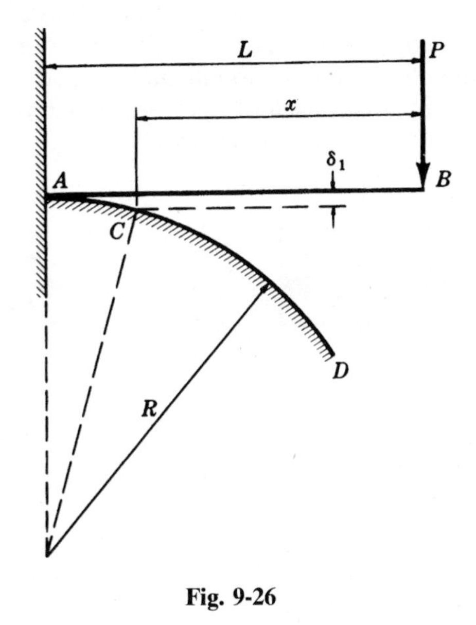

Fig. 9-26

If the curvature of the cantilever at A is less than the curvature of the rigid cylindrical surface, then the cantilever touches the surface only at point A and the deflection is exactly as found in Problem 9.2. From Problem 9.1, the curvature of the beam at A is given by

$$\frac{1}{\rho} = \frac{M}{EI} = \frac{PL}{EI}$$

and thus this curvature must be less than the curvature of the rigid surface, which is $1/R$.

If, however, $1/R = PL/EI$, then the beam comes into contact with the surface to the right of point A. We shall denote by P^* the limiting value of the load given by $P^* = EI/RL$. For $P > P^*$ some region AC of the beam will be in contact with the surface and at point C the curvature of the rigid surface $1/R$ is equal to the curvature of the beam, that is, $Px/EI = 1/R$ from which $x = EI/PR$.

The deflection at the tip B may now be found as the sum of

1. The deflection of C from the tangent at A, which is given by δ_1 in the diagram and is found from the relation

$$(R + \delta_1)^2 = R^2 + (L - x)^2$$

to be approximately

$$\delta_1 = \frac{(L - x)^2}{2R}$$

2. The deflection of the portion of the beam of length x acting as a simple cantilever, given by

$$\delta_2 = \frac{Px^3}{3EI} = \frac{(EI)^2}{3P^2R^3}$$

3. The deflection owing to the rotation at point C, given by

$$\delta_3 = \frac{x(L-x)}{R} = \frac{EI}{PR^2}\left(L - \frac{EI}{PR}\right)$$

The desired deflection at the tip is thus

$$\delta = \delta_1 + \delta_2 + \delta_3 = \frac{L^3}{2R} - \frac{(EI)^2}{6P^2R^3}$$

9.24. A thermostat consists of two strips of different materials of equal thickness bonded together at their interface. Frequently this configuration takes the form of a cantilever beam, as in Fig. 9-27. If E_1 and E_2 denote the Young's moduli and α_1 and α_2 denote the coefficients of linear expansion of the two materials, each of thickness h, determine the deflection of the end of the cantilever assembly due to a temperature rise T.

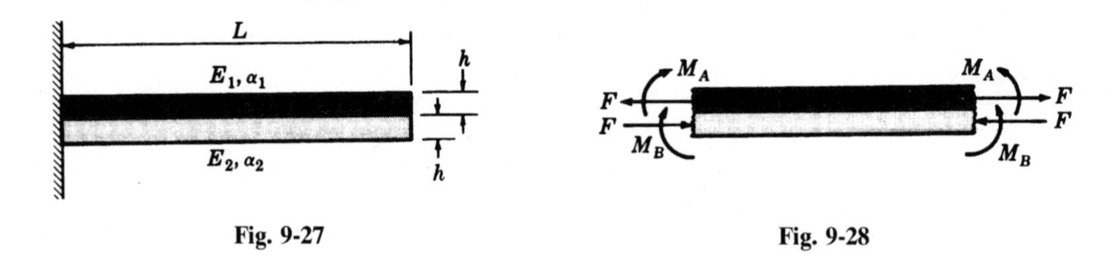

Fig. 9-27 **Fig. 9-28**

Let b represent the width of the assembly. As in Problem 8.1, we shall assume that a plane section prior to deformation remains plane after deformation. The resultant normal forces F acting over each strip must be numerically equal since no external forces are applied along the length of the beam. Thus a cross section at any station along the length has Fig. 9-28 as its free-body representation.

The normal strain in the lower fibers of the top strip is found as the sum of (a) the strain due to the normal load, F/E_1bh; (b) the strain due to bending, which is $M_A(h/2)/E_1I$ from Problem 8.1; and (c) the strain due to the temperature rise, which is $\alpha_1 T$ as mentioned in Chap. 1. The sum of these strains must be the same as the strain in the upper fibers of the lower strip. Thus

$$\frac{F}{E_1bh} + \frac{M_A(h/2)}{E_1I} + \alpha_1 T = \frac{-F}{E_2bh} - \frac{M_B(h/2)}{E_2I} + \alpha_2 T \qquad (1)$$

The curvatures at this interface must also be equal. Thus, from Problem 9.1,

$$\frac{1}{R_1} = \frac{M_A}{E_1I} \quad \text{and} \quad \frac{1}{R_2} = \frac{M_B}{E_1I} \qquad (2)$$

and since $R_1 = R_2$, we have

$$M_A = \left(\frac{E_1}{E_2}\right) M_B \qquad (3)$$

From statics it is evident that

$$M_A + M_B = Fh \qquad (4)$$

from which

$$M_B = \frac{Fh}{1 + (E_1/E_2)} \qquad M_A = \frac{Fh}{1 + (E_2/E_1)} \qquad (5)$$

Substituting (5) in (1), we find

$$F = \frac{(\alpha_2 - \alpha_1)\, TbhE_1 E_2(E_1 + E_2)}{E_1^2 + E_2^2 + 14E_1 E_2} \tag{6}$$

and from (5) we get

$$M_A = \frac{(\alpha_2 - \alpha_1)\, Tbh^2 E_1^2 E_2}{E_1^2 + E_2^2 + 14E_1 E_2} \tag{7}$$

We may now use the result obtained in Problem 9.23 for the deflection δ of a point on a cylindrical surface (which represents the interface, since in pure bending the assembly deforms into a circular configuration according to Problem 9.1) and express the deflection δ of the end of the assembly as

$$\delta = \frac{L^2}{2R} \tag{8}$$

Substituting from Eq. (2),

$$\delta = \frac{M_A L^2}{2E_1 I}$$

From (7) we then get

$$\delta = \frac{6(\alpha_2 - \alpha_1)\, TE_1 E_2 L^2}{h(E_1^2 + E_2^2 + 14E_1 E_2)}$$

9.25. A beam has a slight initial curvature such that the initial configuration (which is stress free) is described by the relation $y_0 = Kx^3$. The beam is rigidly clamped at the origin and is subjected to a concentrated force at its extreme end, as shown in Fig. 9-29. As the force is increased, the beam deflects downward and the region near the clamped end comes in contact with the rigid horizontal plane. If the value of the applied force is P, determine the length of the beam in contact with the horizontal plane and the vertical distance of the extreme end from the plane.

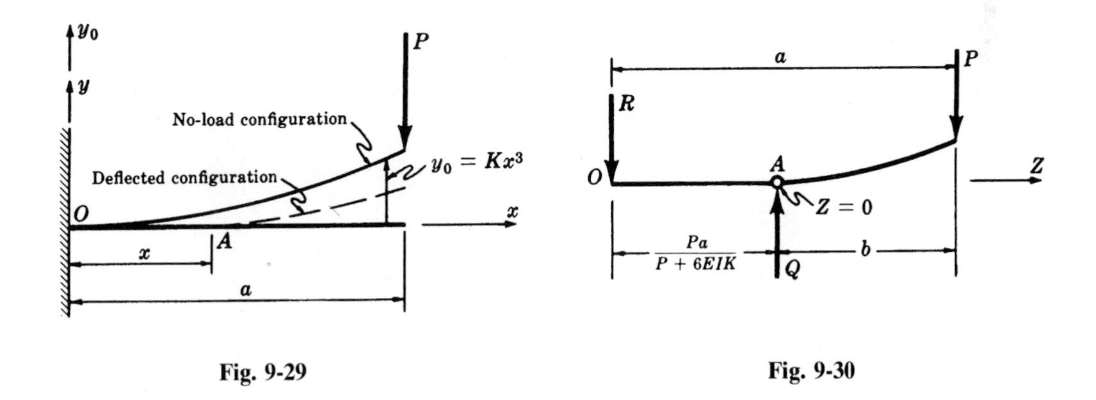

Fig. 9-29

Fig. 9-30

The initial curvature may be determined from the expression $y_0 = Kx^3$ so that the bending moment arising from straightening the portion of the beam near the support is readily found to be $EI(d^2y_0/dx^2) = 6EIKx$, where x is the length of beam in contact with the horizontal plane. If this expression for moment is equated to the moment of the applied load about the point of contact, that is, $P(a - x)$, we have

$$6EIKx = P(a - x) \qquad \text{whence } x = \frac{Pa}{P + 6EIK}$$

Since the beam is considered to be weightless, there is no normal force between the beam and the rigid horizontal plane between the clamp at O and the extreme point of contact at A. The beam is flat between O and A. A free-body diagram of the deformed beam thus appears as in Fig. 9.30. A simple statics equation for equilibrium of moments about point A indicates that the clamp exerts a downward force equal to $6EIK$. For vertical equilibrium there is a concentrated force reaction $Q = P + 6EIK$ acting on the beam at the extreme point of contact A.

We now seek the equation of the deflection curve in the region to the right of point A. In Problem 9.1, Eq. (5) indicated that for an initially straight beam bending moment M is proportional to the curvature, d^2y/dx^2. However, in the present problem it is necessary to modify (5) to say that the bending moment M is proportional to the *change of curvature* since the beam is not initially straight. Thus, the Euler-Bernoulli equation for the portion of the beam to the right of point A is

$$EI\left(\frac{d^2y_0}{dx^2} - \frac{d^2y}{dZ^2}\right) = P(b - Z)$$

where a new coordinate Z has been introduced. This coordinate runs along the x-axis but has its origin at point A. It is important to note that, as the beam deflects, the curvature decreases from its original value; hence the quantity in parentheses on the left side of the equation is positive. Accordingly, the right side must be written as positive. This does not contradict our previous sign convention of downward forces giving negative moments since it was applied to *initially straight* beams. If we substitute $EI(d^2y_0/dx^2) = 6EIKx$, the last equation becomes

$$EI\frac{d^2y}{dZ^2} = 6EIK\left[\frac{Pa}{P + 6EIK} + Z\right] - Pb + PZ$$

Integrating twice and imposing the boundary conditions that $y = dy/dZ = 0$ at $Z = 0$, we obtain the desired deflection

$$EIy_{Z=b} = \frac{36(EIKa)^3}{(P + 6EIK)^2}$$

9.26. The bar ABC in Fig. 9-31 has flexural rigidity $E(3I)$ in region AB and flexural rigidity EI in region BC. The bar is pinned at A, supported by a roller at B, and subject to an applied bending moment M_0 at the free end. Determine the vertical deflection at B.

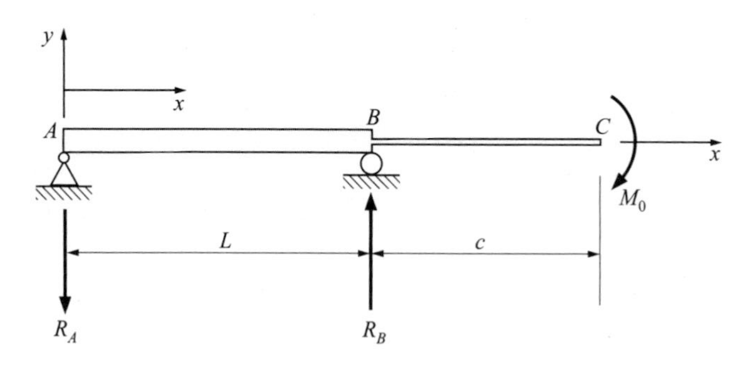

Fig. 9-31

Let us introduce the x-y coordinate system shown, where x may designate a cross section in either AB or BC. It is first necessary to determine the reactions from statics, viz.,

$$+\circlearrowleft \Sigma M_B = -M_0 + R_A L = 0 \qquad \therefore R_A = \frac{M_0}{L}(\downarrow)$$

$$\Sigma F_y = -R_A + R_B = 0 \qquad \therefore R_B = \frac{M_0}{L}(\uparrow)$$

We first write the differential equation of the deflected bar in region AB:

$$E(3I)\frac{d^2y}{dx^2} = -R_A x \qquad \text{for} \qquad 0 < x < L$$

Integrating,

$$E(3I)\frac{dy}{dx} = -R_A\frac{x^2}{2} + C_1 \tag{1}$$

Integrating again,

$$E(3I)y = -\frac{R_A}{2}\cdot\frac{x^3}{3} + C_1x + C_2 \tag{2}$$

As the first boundary condition we have: When $x = 0$, $y = 0$. Substituting in Eq. (*2*), we have

$$C_2 = 0$$

As a second boundary condition we have: When $x = L$, $y = 0$, and using $R_A = M_0/L$ we have

$$0 = -\frac{M_0}{L}\cdot\frac{L^3}{6} + C_1L + C_2$$

Thus,

$$C_1 = \frac{M_0L}{6}$$

Next, we write the differential equation of the deflected beam in region BC:

$$EI\frac{d^2y}{dx^2} = -R_Ax + R_B(x - L) \qquad \text{for} \qquad L < x < (L + c)$$

$$= -\frac{M_0x}{L} + \frac{M_0x}{L} - R_BL$$

$$= -M_0$$

This result could also have been obtained by taking moments of applied loads to the right of any section designated by "x" in BC.

Integrating,

$$EI\frac{dy}{dx} = -M_0x + C_3 \tag{3}$$

Integrating again,

$$EIy = -M_0\cdot\frac{x^2}{2} + C_3x + C_4 \tag{4}$$

As a third boundary condition at $x = L$, $y = 0$ in Eq. (*4*), so from (*4*)

$$0 = -\frac{M_0L^2}{2} + C_3L + C_4 \tag{5}$$

As the fourth boundary condition at $x = L$ the slopes dy/dx as given by Eqs. (*1*) and (*3*) must be equal. This leads to

$$\frac{1}{3EI}\left[\frac{R_AL^2}{2} + \frac{M_0L}{6}\right] = \frac{1}{EI}[-M_0L + C_3] \tag{6}$$

Solving Eq. (*6*) for C_3, then (*5*) for C_4, we find

$$C_3 = \tfrac{8}{9}M_0L; \qquad C_4 = -\tfrac{7}{18}M_0L^2$$

The equations of the deflected beam are thus

$$E(3I)y = -\frac{M_0}{6L}x^3 + \frac{M_0 L}{6}x \qquad \text{for} \qquad 0<x<L \tag{7}$$

$$EIy = -\frac{M_0}{2}x^2 + \frac{8}{9}M_0 Lx - \frac{7}{18}M_0 L^2 \qquad \text{for} \qquad L<x<(L+C) \tag{8}$$

When $x = (L + C)$, we have from Eq. (8) the desired tip deflection:

$$[y]_{x=L+C} = \frac{M_0}{EI}\left[\frac{(L+C)^2}{2} + \frac{8}{9}L(L+C) - \frac{7}{18}L^2\right]$$

$$= -\frac{M_0 C}{EI}\left(\frac{L}{9} + \frac{C}{2}\right) \tag{9}$$

Supplementary Problems

9.27. The cantilever beam loaded as shown in Problem 9.2 is made of a titanium alloy, having $E = 105$ GPa. The load P is 20 kN, $L = 4$ m, and the moment of inertia of the beam cross section is $104 \times 10^6\ \text{mm}^4$. Find the maximum deflection of the beam. *Ans.* -39 mm

9.28. Consider the simply supported beam loaded as shown in Problem 9.12. The length of the beam is 20 ft, $a = 15$ ft, the load $P = 1000$ lb, and $I = 150\ \text{in}^4$. Determine the deflection at the center of the beam. Take $E = 30 \times 10^6\ \text{lb/in}^2$. *Ans.* -0.044 in

9.29. Refer to Fig. 9-32. Determine the deflection at every point of the cantilever beam subject to the single moment M_1 shown. *Ans.* $EIy = -M_1x^2/2$

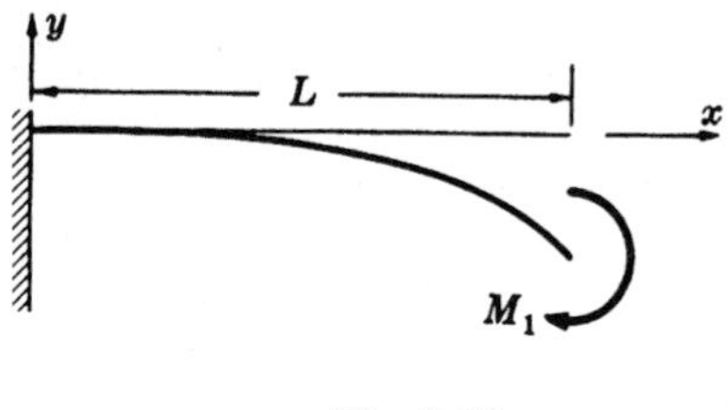

Fig. 9-32

9.30. The cantilever beam described in Problem 9.29 is of circular cross section, 5 in in diameter. The length of the beam is 10 ft and the applied moment is 5000 lb · ft. Determine the maximum deflection of the beam. Take $E = 30 \times 10^6\ \text{lb/in}^2$. *Ans.* -0.469 in

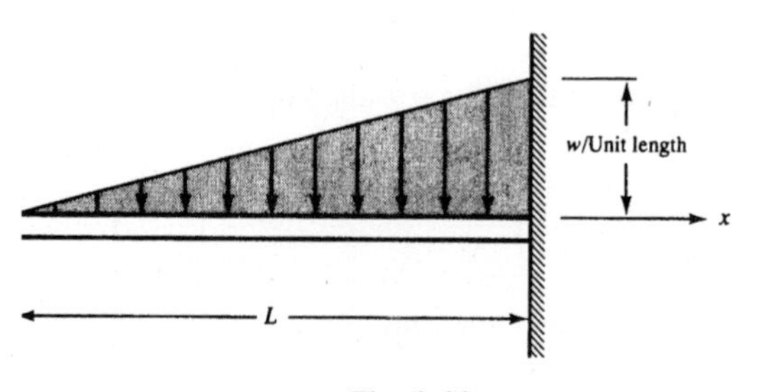

Fig. 9-33

9.31. Refer to Fig. 9-33. Find the equation of the deflection curve for the cantilever beam subject to the uniformly varying load shown.

Ans. $EIy = -\dfrac{wx^5}{120L} + \dfrac{wL^3x}{24} - \dfrac{wL^4}{30}$

9.32. A cantilever beam is loaded by the sinusoidal load indicated in Fig. 9-34. Determine the deflection of the tip of the beam. *Ans.* $EIy]_{x=0} = -0.07385q_0L^4$

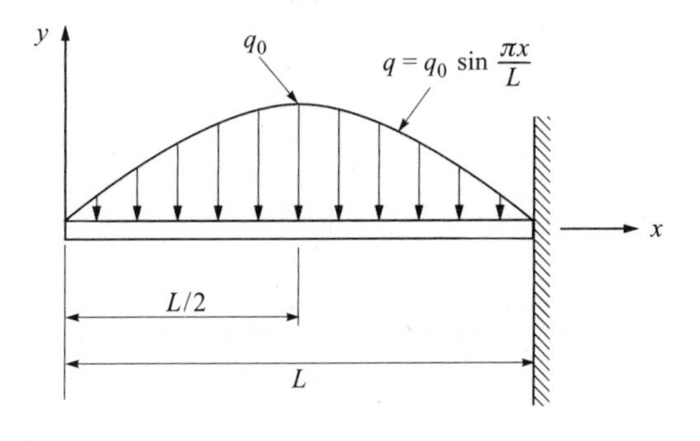

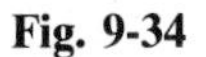

Fig. 9-34

9.33. A cantilever beam carrying a parabolically distributed load is shown in Fig. 9-35. Determine the equation of the deflected beam as well as the deflection at the tip.

Ans. $y]_{x=0} = -\dfrac{56}{945}w_0L^4, EIy = -\dfrac{16}{945}\dfrac{w_0}{L^{1/2}}x^{9/2} + \dfrac{8}{105}x_0L^3x - \dfrac{56}{945}w_0L^4$

9.34. The cross section of the cantilever beam loaded as shown in Fig. 9-33 is rectangular, 50 × 75 mm. The bar, 1 m long, is aluminum for which $E = 65$ GPa. Determine the permissible maximum intensity of loading if the maximum deflection is not to exceed 5 mm and the maximum stress is not to exceed 50 MPa.
Ans. $w = 14.1$ kN/m

9.35. Refer to Fig. 9-36. Determine the equation of the deflection curve for the simply supported beam supporting the load of uniformly varying intensity.

Ans. $EIy = \dfrac{wL}{2}\left(-\dfrac{x^5}{60L^2} + \dfrac{x^3}{18} - \dfrac{7L^2x}{180}\right)$

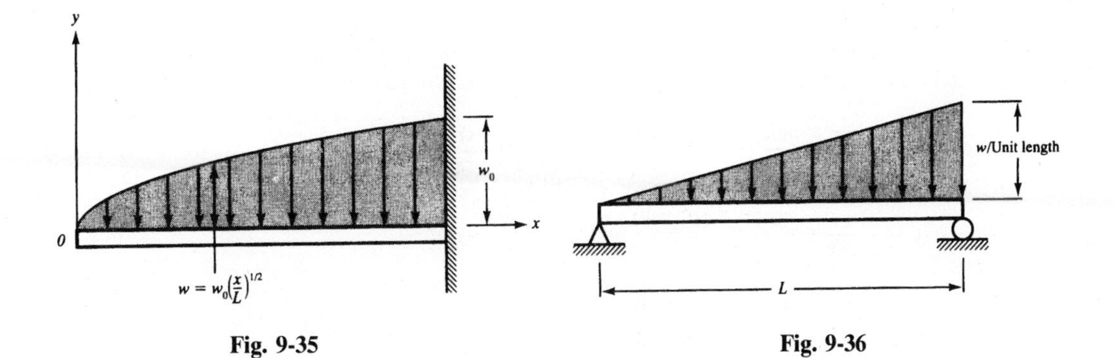

Fig. 9-35 **Fig. 9-36**

9.36. Determine the equation of the deflection curve for the cantilever beam loaded by the concentrated force P as shown in Fig. 9-37.

Ans. $EIy = -\frac{P}{6}(a-x)^3 - \frac{Pa^2}{2}x + \frac{Pa^3}{6}$ for $0 < x < a$; $EIy = -\frac{Pa^2}{2}x + \frac{Pa^3}{6}$ for $a < x < L$

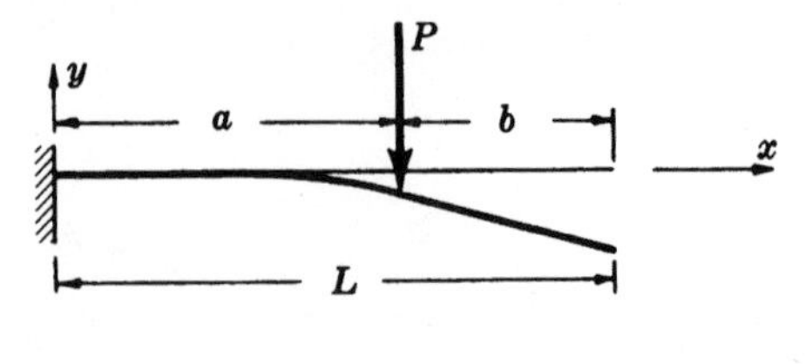

Fig. 9-37

9.37. For the cantilever beam of Fig. 9-37, take $P = 5$ kN, $a = 2$ m, and $b = 1$ m. The beam is of equilateral triangular cross section, 150 mm on a side, with a vertical axis of symmetry. Determine the maximum deflection of the beam. Take $E = 200$ GPa. *Ans.* -12.8 mm

9.38. The cantilever beam shown in Fig. 9-38 is subjected to a uniform load w per unit length over its right half BC. Determine the equations of the deflection curve as well as the maximum deflection.

Ans. $$EIy = \frac{wLx^3}{12} - \frac{3wL^2x^2}{16} \quad \text{for} \quad 0 < x < \frac{L}{2}$$

$$EIy = -\frac{w(L-x)^4}{24} - \frac{7wL^3x}{48} + \frac{15wL^4}{384} \quad \text{for} \quad \frac{L}{2} < x < L$$

$$\Delta_{\max} = \frac{41}{384}\left(\frac{wL^4}{EI}\right)$$

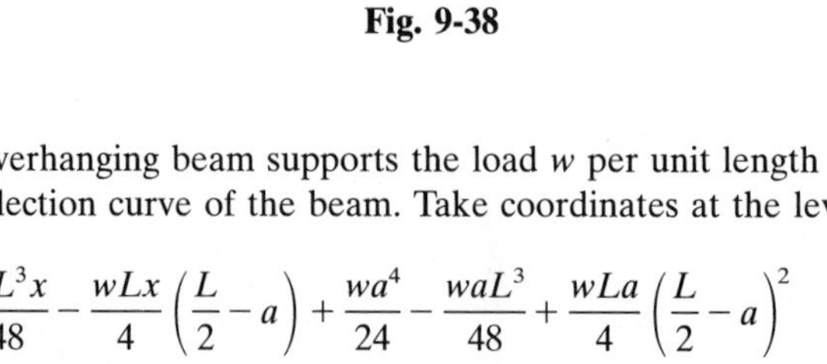

Fig. 9-38

9.39. The simply supported overhanging beam supports the load w per unit length as shown in Fig. 9-39. Find the equations of the deflection curve of the beam. Take coordinates at the level of the supports.

Ans. $$EIy = -\frac{wx^4}{24} + \frac{wL^3x}{48} - \frac{wLx}{4}\left(\frac{L}{2} - a\right) + \frac{wa^4}{24} - \frac{waL^3}{48} + \frac{wLa}{4}\left(\frac{L}{2} - a\right)^2 \quad \text{for} \quad 0 < x < a$$

$$EIy = -\frac{wx^4}{24} + \frac{wL(x-a)^3}{12} + \frac{wL^3x}{48} - \frac{wLx}{4}\left(\frac{L}{2} - a\right)^2$$

$$+\frac{wa^4}{24} - \frac{waL^3}{48} + \frac{wLa}{4}\left(\frac{L}{2} - a\right)^2 \quad \text{for} \quad a < x < (a+b)$$

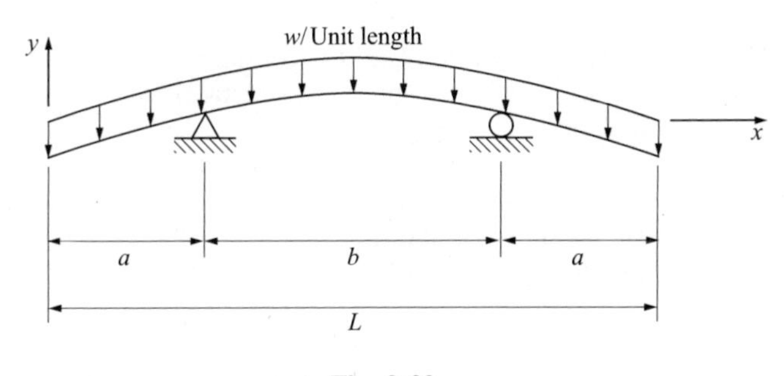

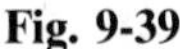

Fig. 9-39

9.40. A simply supported beam with overhanging ends is loaded by the uniformly distributed loads shown in Fig. 9-40. Determine the deflection of the midpoint of the beam with respect to an origin at the level of the supports.

Ans. $\dfrac{wa^2(L-2a)^2}{16EI}$ (above level of supports)

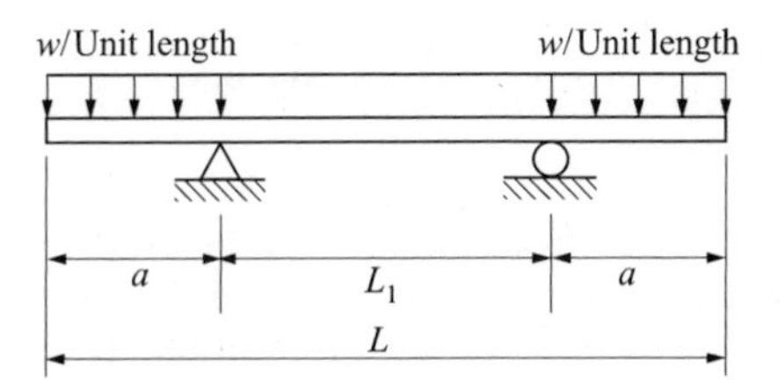

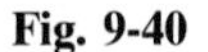

Fig. 9-40

9.41. For the beam described in Problem 9.40, determine the deflection of one end of the beam with respect to an origin at the level of the supports.

Ans. $\dfrac{wa^3L}{4EI} - \dfrac{3wa^4}{8EI}$ (below level of supports)

9.42. The overhanging beam is loaded by the uniformly distributed load as well as the concentrated force shown in Fig. 9-41. Determine the deflection of point A of the beam.

Ans. $\dfrac{-wa^3b}{3EI} + \dfrac{Pab^2}{4EI} - \dfrac{wa^4}{8EI}$ (below level of supports)

9.43. Figure 9-42 shows a cantilever beam in the form of a circular cone whose length L is large compared to the base diameter D. If the only force acting is its own weight, which is γ per unit volume, determine the equation of the deflection curve.

Ans. $y = -\dfrac{2\gamma L^2}{45ED^2}(x^3 + 2L^3 - 3L^2x)$

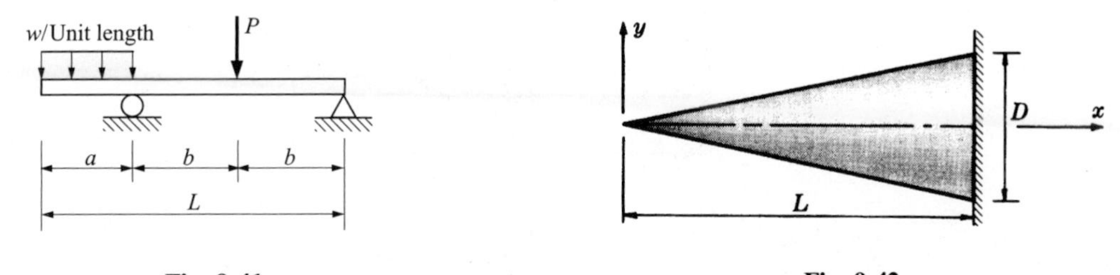

Fig. 9-41 **Fig. 9-42**

9.44. For the overhanging beam treated in Problem 9.17 consider the uniform load to be 120 lb/ft, $a = 3$ ft, and $b = 12$ ft. The bar has a 3-in $\times$ 4-in rectangular cross section. Determine the maximum deflection of the beam. Take $E = 30 \times 10^6$ lb/in^2. *Ans.* -0.10 in at $x = 110.4$ in

9.45. A cantilever beam when viewed from the top [see Fig. 9-43(*a*)] has a triangular configuration. The thickness h of the beam is constant, as shown in the side view Fig. 9-43(*b*). Determine the deflection of the beam due to a concentrated load P at the tip. Neglect the weight of the beam. *Ans.* $y]_{x=0} = -6PL^3/Ebh^3$

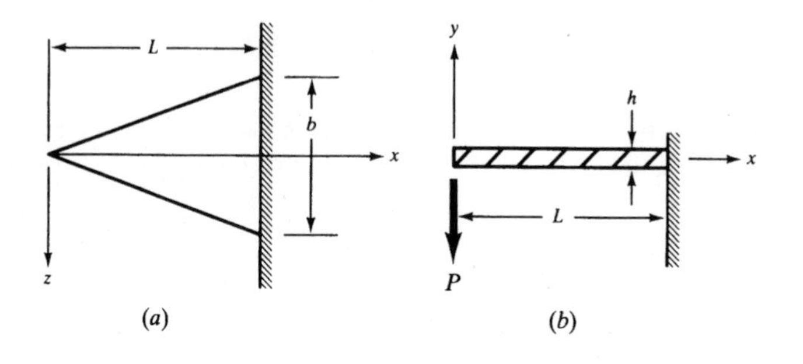

Fig. 9-43

9.46. A cantilever beam when viewed from the top has the configuration indicated in Fig. 9-44(*a*) and is of constant thickness h, as indicated in Fig. 9-44(*b*). Find the equation of the deflection curve as the beam bends under the action of the concentrated force P at the tip. Neglect the weight of the beam.

$$\textit{Ans.}\quad y = \left[-\frac{16P(L-x)^{11/4}}{77} - \frac{4}{9}PL^{7/4}x + \frac{16PL^{11/4}}{77}\right]\left(\frac{6L^{1/4}}{Eh^3a^{1/2}}\right)$$

$$y]_{x=L} = -\frac{24PL^3}{11Eh^3a^{1/2}}$$

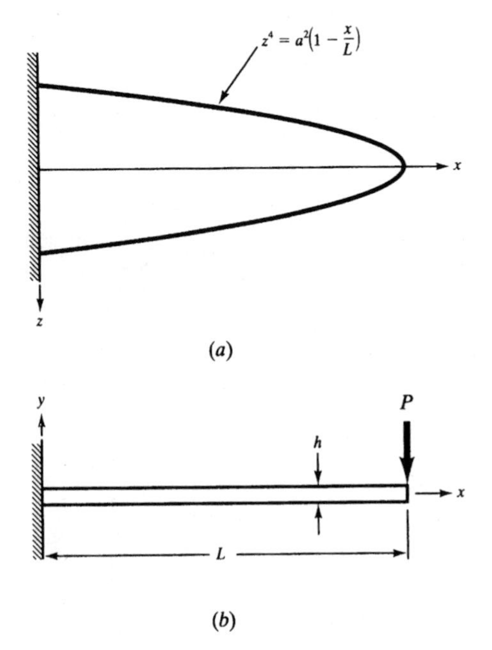

Fig. 9-44

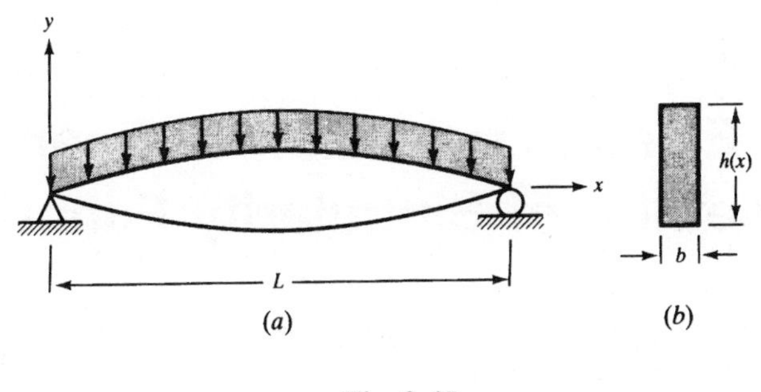

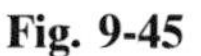

Fig. 9-45

9.47. A simply supported beam of length L is subjected to a uniformly distributed loading w per unit length. The width b of the beam is constant and the height varies in such a manner that all outer fibers along both the top and lower surfaces are subject to the same magnitude normal stress σ_0. Determine the variation of height of the beam as a function of x, as shown in Fig. 9-45(b). Also determine the maximum deflection of the beam.

Ans. $h = \dfrac{2h_{\text{max}}\sqrt{Lx - x^2}}{L}$, $y_{\text{max}} = -0.0178\dfrac{wL^4}{Eb(h_{\text{max}})^3}$

9.48. The cantilever beam of variable cross section shown in Fig. 9-46 is in the form of a wedge of constant width b. The midplane of the wedge lies in the horizontal plane x-z. Find the deflection of the tip of the beam due to its own weight γ per unit volume. *Ans.* $y]_{x=0} = -\gamma L^4/Eh^2$

9.49. Two solid rigid cylinders I and II have their geometric axes in a horizontal plane spaced a distance L apart, as shown in Fig. 9-47. A beam of flexural rigidity EI is then placed across the tops of the cylinders and loaded by a centrally applied vertical force P. The beam deflects (dotted line) and is tangent to each of the cylinders at the points designated as A. Determine the angle θ describing this point of contact.

Ans. $\theta = \dfrac{PL^2}{16EI}\left(1 - \dfrac{PLR}{4EI}\right)$

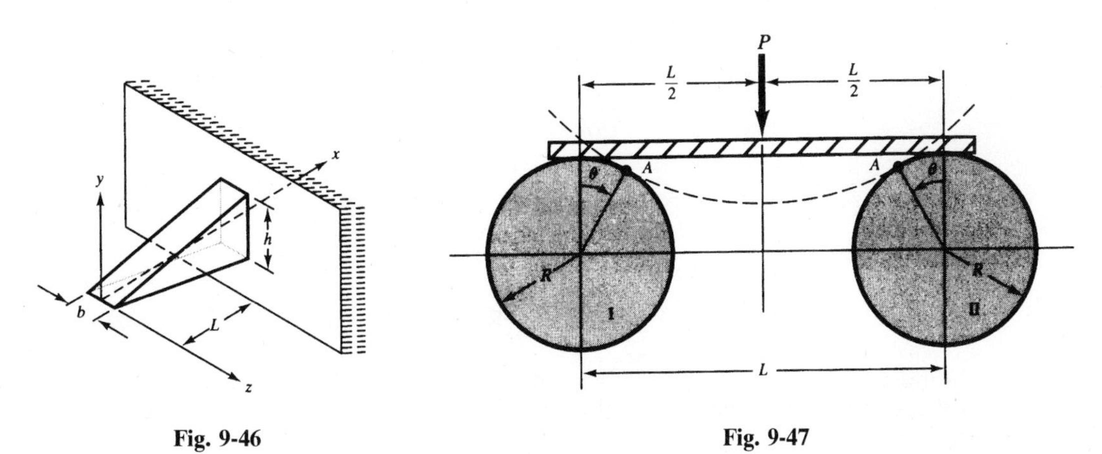

Fig. 9-46

Fig. 9-47

Chapter 10

Elastic Deflection of Beams: Method of Singularity Functions

In Chap. 9 we found the elastic deflections of transversely loaded beams through direct integration of the second-order Euler-Bernoulli equation. As we saw, the approach is direct but may become very lengthy even for relatively simple engineering situations.

A more expedient approach is based upon the use of the singularity functions introduced in Chap. 6. The method is direct and may be applied to a beam subject to any combination of concentrated forces, moments, and distributed loads. One must only remember the definition of the singularity function given in Chap. 6; i.e., the quantity $\langle x - a \rangle$ vanishes if $x < a$ but is equal to $(x - a)$ if $x > a$.

There are several possible approaches for using singularity functions for the determination of beam deflections. Perhaps the simplest is to employ the approach of Chap. 6 in which the bending moment is written in terms of singularity functions in the form of one equation valid along the entire length of the beam. Two integrations of this equation lead to the equation for the deflected beam in terms of two constants of integration which must be determined from boundary conditions. As noted in Chap. 6, integration of the singularity functions proceeds directly and in the same manner as simple power functions. Thus, the approach is direct and avoids the problem of the determination of a pair of constants corresponding to each region of the beam (between loads) as in the case of double integration exemplified in Chap. 9.

Most important, the singularity function approach leads directly into a computerized approach for the determination of beam deflections. See Problems 10.16, 10.17, and 10.18.

Solved Problems

10.1. Using singularity functions, determine the deflection curve of the cantilever beam subject to the loads shown in Fig. 10-1.

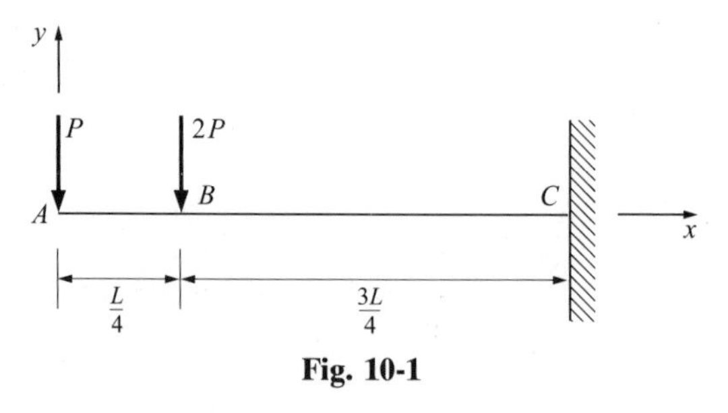

Fig. 10-1

In this case it is not necessary to determine the reactions of the wall supporting the beam at C. From the techniques of Chap. 6 we find the bending moment along the entire length of the beam to be given by

$$M = -P\langle x \rangle^1 - 2P\left\langle x - \frac{L}{4} \right\rangle^1 \qquad (1)$$

where the angular brackets have the meanings given in the section "Singularity Functions" of Chap. 6, pages 135–136. Thus, the differential equation for the bent beam is

$$EI\frac{d^2y}{dx^2} = -P\langle x \rangle^1 - 2P\left\langle x - \frac{L}{4} \right\rangle^1 \qquad (2)$$

The first integration yields

$$EI\frac{dy}{dx} = -P\frac{\langle x\rangle^2}{2} - 2P\frac{\left\langle x - \frac{L}{4}\right\rangle^2}{2} + C_1 \tag{3}$$

where C_1 is a constant of integration. The next integration leads to

$$EIy = -\frac{P}{2}\frac{\langle x\rangle^3}{3} - 2P\frac{\left\langle x - \frac{L}{4}\right\rangle^3}{2(3)} + C_1\langle x\rangle + C_2 \tag{4}$$

where C_2 is a second constant of integration. These two constants may be determined from the boundary conditions:

(*a*) When $x = L$, $dy/dx = 0$, so from (*3*):

$$0 = -\frac{PL^2}{2} - P\left(\frac{3L}{4}\right)^2 + C_1 \tag{5}$$

(*b*) When $x = L$, $y = 0$, so from (*4*):

$$0 = -\frac{PL^3}{6} - \frac{P}{3}\left(\frac{3L}{4}\right)^3 + C_1 L + C_2 \tag{6}$$

Solving (*5*) and (*6*),

$$C_1 = \frac{17}{16}PL^2; \qquad C_2 = -\frac{145}{192}PL^3 \tag{7}$$

The desired deflection curve is thus

$$EIy = -\frac{P}{6}\langle x\rangle^3 - \frac{P}{3}\left\langle x - \frac{L}{4}\right\rangle^3 + \frac{17}{16}PL^2\langle x\rangle - \frac{145}{192}PL^3 \tag{8}$$

For example, the deflection at point B where $x = L/4$ is found from (*8*) to be

$$EIy]_{x=L/4} = -\frac{P}{6}\left(\frac{L}{4}\right)^3 - 0 + \frac{17}{16}PL^2\left(\frac{L}{4}\right) - \frac{145}{192}PL^3$$

or

$$y]_{x=L/4} = -\frac{94.5PL^3}{192EI} \quad \text{or} \quad -\frac{0.492PL^3}{EI}$$

10.2. The cantilever beam ABC shown in Fig. 10-2 is subject to a uniform load w per unit length distributed over its right half, together with a concentrated couple $wL^2/2$ applied at C. Using singularity functions, determine the deflection curve of the beam.

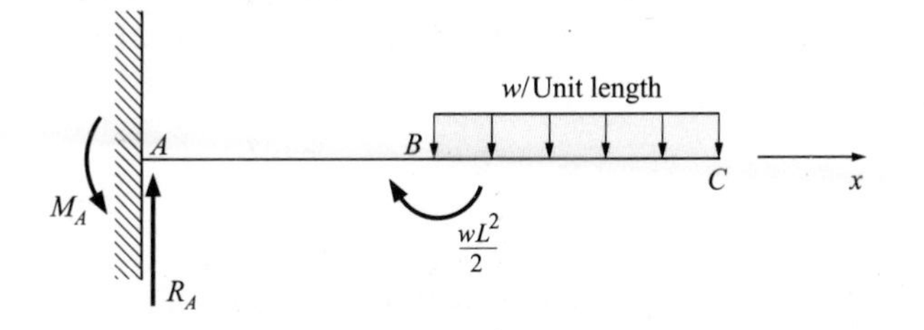

Fig. 10-2

It is first necessary to find from statics the shear and moment reactions exerted by the wall on the beam at A. From statics we have

$$+\circlearrowright \Sigma M_A = M_A - \frac{wL^2}{2} - w\left(\frac{L}{2}\right)\left(\frac{3L}{4}\right) = 0$$

$$M_A = \frac{7wL^2}{8}$$

$$\Sigma F_y = R_A - w\left(\frac{L}{2}\right) = 0; \qquad R_A = \frac{wL}{2}$$

By the singularity function approach we may write the bending moment along the entire length of the beam as

$$M = \frac{wL}{2}\langle x\rangle^1 - \frac{7wL^2}{8}\langle x\rangle^0 + \frac{wL^2}{2}\left\langle x - \frac{L}{2}\right\rangle^0 - w\left\langle x - \frac{L}{2}\right\rangle^1 \frac{\left\langle x - \frac{L}{2}\right\rangle^1}{2} \tag{1}$$

where, again, the singularity functions are as defined in Chap. 6. Thus the differential equation of the bent beam is

$$EI\frac{d^2y}{dx^2} = \frac{wL}{2}\langle x\rangle^1 - \frac{7wL^2}{8}\langle x\rangle^0 + \frac{wL^2}{2}\left\langle x - \frac{L}{2}\right\rangle^0 - w\left\langle x - \frac{L}{2}\right\rangle^1 \frac{\left\langle x - \frac{L}{2}\right\rangle^1}{2} \tag{2}$$

Integrating,

$$EI\frac{dy}{dx} = \frac{wL}{2}\frac{\langle x\rangle^2}{2} - \frac{7wL^2}{8}\langle x\rangle + \frac{wL^2}{2}\frac{\left\langle x - \frac{L}{2}\right\rangle^1}{1} - \frac{w}{2}\frac{\left\langle x - \frac{L}{2}\right\rangle^3}{3} + C_1 \tag{3}$$

The first boundary condition is: When $x = 0$, $dy/dx = 0$. Substituting in (*3*), we find $C_1 = 0$.
Integrating again,

$$EIy = \frac{wL}{4}\frac{\langle x\rangle^3}{3} - \frac{7wL^2}{8}\frac{\langle x\rangle^2}{2} + \frac{wL^2}{2}\frac{\left\langle x - \frac{L}{2}\right\rangle^2}{2} - \frac{w}{6}\frac{\left\langle x - \frac{L}{2}\right\rangle^4}{4} + C_2 \tag{4}$$

The second boundary condition is: When $x = 0$, $y = 0$. Substituting in (*4*), we find $C_2 = 0$.
Thus, the desired deflection equation is

$$EIy = \frac{wL}{12}\langle x\rangle^3 - \frac{7wL^2}{16}\langle x\rangle^2 + \frac{wL^2}{4}\left\langle x - \frac{L}{2}\right\rangle^2 - \frac{w}{24}\left\langle x - \frac{L}{2}\right\rangle^4 \tag{5}$$

This yields the deflection at the tip to be

$$EIy]_{x=L} = \frac{wL^4}{12} - \frac{7wL^4}{16} + \frac{wL^2}{4}\left(\frac{L}{2}\right)^2 - \frac{w}{24}\left(\frac{L}{2}\right)^4$$

or

$$y]_{x=L} = -\frac{113}{384EI}$$

10.3. Consider a simply supported beam subject to a uniform load distributed over a portion of its length, as indicated in Fig. 10-3. Use singularity functions to determine the deflection curve of the beam.

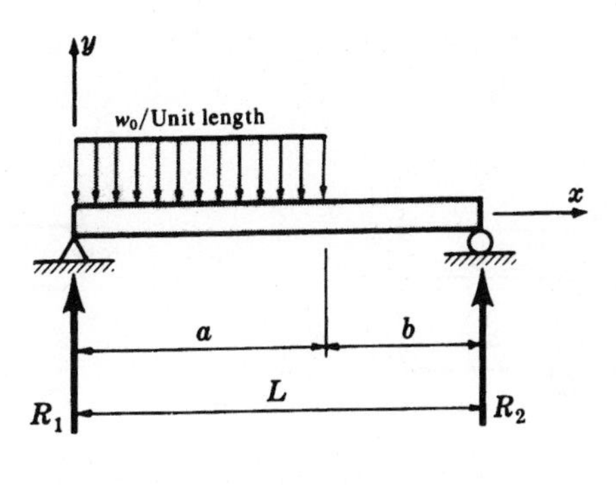

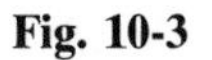
Fig. 10-3

From statics the reactions are found to be

$$R_1 = \frac{w_0}{2L}(L^2 - b^2)$$

$$R_2 = w_0 a - \frac{w_0}{2L}(L^2 - b^2)$$

The bending moment at any point x along the length of the beam is

$$M = R_1 x - \frac{w_0}{2}\langle x\rangle^2 + \frac{w_0}{2}\langle x - a\rangle^2 \tag{1}$$

Note that the last term on the right is required to cancel the distributed load represented by the term

$$-\frac{w_0}{2}\langle x\rangle^2$$

for all values of x greater than $x = a$. Thus

$$EI\frac{d^2y}{dx^2} = M = R_1\langle x\rangle^1 - \frac{w_0}{2}\langle x\rangle^2 + \frac{w_0}{2}\langle x - a\rangle^2 \tag{2}$$

Integrating,

$$EI\frac{dy}{dx} = \frac{R_1}{2}\langle x\rangle^2 - \frac{w_0}{6}\langle x\rangle^3 + \frac{w_0}{6}\langle x - a\rangle^3 + C_1 \tag{3}$$

Finally,

$$EIy = \frac{R_1}{6}\langle x\rangle^3 - \frac{w_0}{24}\langle x\rangle^4 + \frac{w_0}{24}\langle x - a\rangle^4 + C_1 x + C_2 \tag{4}$$

To determine C_1 and C_2, we impose the boundary conditions that $y = 0$ at $x = 0$ and $x = L$. From (4) we thus find

$$C_1 = \frac{w_0 L^3}{24} - \frac{w_0 b^4}{24L} - \frac{w_0 L}{12}(L^2 - b^2)$$

$$C_2 = 0$$

The deflection curve is accordingly

$$EIy = \frac{w_0}{12L}(L^2 - b^2)\langle x\rangle^3 - \frac{w_0}{24}\langle x\rangle^4 + \frac{w_0}{24}\langle x - a\rangle^4 + \left[-\frac{w_0 L^3}{24} - \frac{w_0 b^4}{24L} + \frac{w_0 L b^2}{12}\right]x \tag{5}$$

10.4. Consider the overhanging beam shown in Fig. 10-4. Determine the equation of the deflection curve using singularity functions.

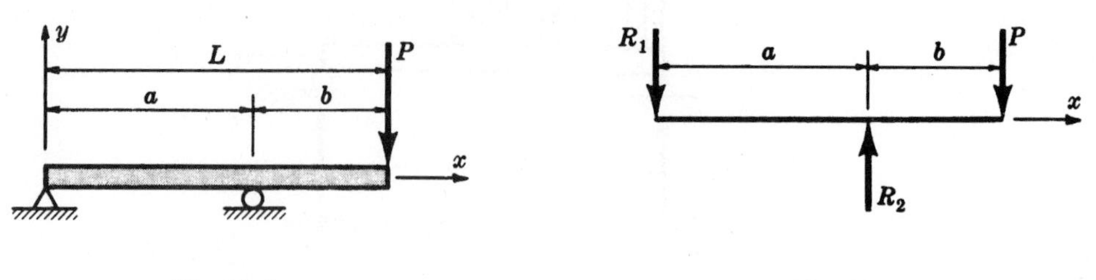

Fig. 10-4

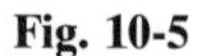

Fig. 10-5

From statics the reactions are first found to be $R_1 = Pb/a$ and $R_2 = P[1 + (b/a)]$, acting as indicated in Fig. 10-5. The bending moment at any point x along the entire length of the beam is

$$M(x) = -R_1\langle x\rangle^1 + R_2\langle x-a\rangle^1 \tag{1}$$

Thus

$$EI\frac{d^2y}{dx^2} = M = -R_1\langle x\rangle^1 + R_2\langle x-a\rangle^1 \tag{2}$$

from which

$$EI\frac{dy}{dx} = -\frac{R_1}{2}\langle x\rangle^2 + \frac{R_2}{2}\langle x-a\rangle^2 + C_1 \tag{3}$$

$$EIy = -\frac{R_1}{6}\langle x\rangle^3 + \frac{R_2}{6}\langle x-a\rangle^3 + C_1x + C_2 \tag{4}$$

The boundary conditions are $y = 0$ at $x = 0$ and $x = a$. From these conditions, C_1 and C_2 are found from (*4*) to be

$$C_1 = \frac{Pab}{6} \qquad C_2 = 0$$

The deflection curve is thus

$$EIy = -\frac{Pb}{6a}\langle x\rangle^3 + \frac{P}{6}\left(1+\frac{b}{a}\right)\langle x-a\rangle^3 + \frac{Pabx}{6} \tag{5}$$

10.5. Through the use of singularity functions determine the equation of the deflected cantilever beam subject to the triangular loading together with the couple indicated in Fig. 10-6.

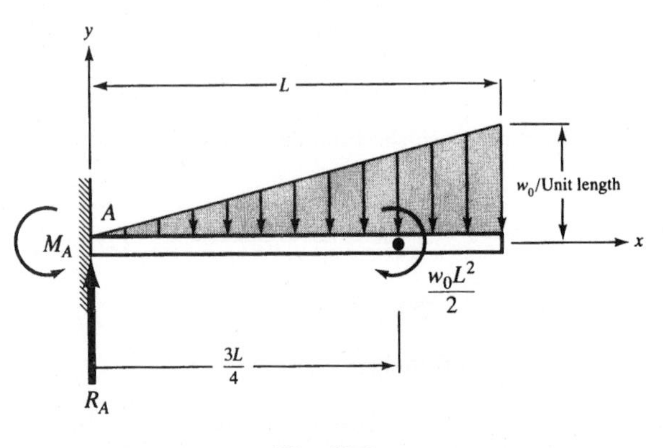

Fig. 10-6

We must first determine the reactions at point A through the use of statics. There will be a vertical shear reaction R_A as well as a moment M_A to prevent angular rotation at point A. From statics

$$\circlearrowright \Sigma M_A = M_A - \frac{w_0 L^2}{2} - \frac{w_0}{2}(L)\left(\tfrac{2}{3}L\right) = 0$$

Therefore

$$M_A = \tfrac{5}{6} w_0 L^2$$

$$\Sigma F_y = R_A - \frac{w_0 L}{2} = 0$$

Therefore

$$R_A = \frac{w_0 L}{2}$$

To write the expression for bending moment, let us first examine the contribution from the distributed loading. At any position x to the right of point A, the load intensity from geometry is $w = w_0(x/L)$ and the resultant (shown by the dotted vector in Fig. 10-7) is of magnitude

$$\frac{wx}{2} = w_0 \frac{x^2}{2L}$$

and acts at a point distance $\frac{2}{3}x$ from A. Thus, the moment at x due *only* to the triangular loading is

$$-w_0 \frac{x^2}{2L}\left(\frac{L}{3}\right) \quad \text{or} \quad -\frac{w_0 x^3}{6L}$$

where the negative sign is inserted because this downward loading gives negative bending moment.

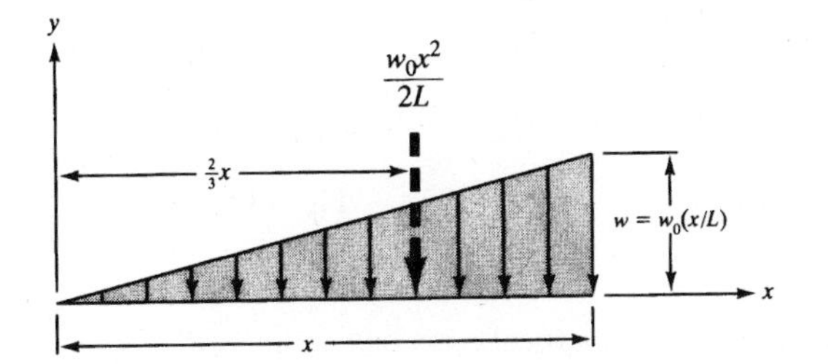

Fig. 10-7

Due to *all* loadings, that is, M_A, R_A, and the triangular load, the bending moment at any location x is

$$M = -\frac{5}{6}w_0 L^2 + \frac{w_0 L}{2}x - \frac{w_0 x^2}{2L}\cdot\frac{x}{3} + \frac{w_0 L^2}{2}\left\langle x - \frac{3L}{4}\right\rangle^0 \qquad (1)$$

so that the differential equation of the deflected beam is

$$EI\frac{d^2y}{dx^2} = -\frac{5}{6}w_0 L^2 + \frac{w_0 L}{2}x - \frac{w_0 x^3}{6L} + \frac{w_0 L^2}{2}\left\langle x - \frac{3L}{4}\right\rangle^0 \qquad (2)$$

Integrating the first time, we obtain

$$EI\frac{dy}{dx} = -\frac{5}{6}w_0 L^2 x + \frac{w_0 L}{2}\cdot\frac{x^2}{2} - \frac{w_0}{6L}\cdot\frac{x^4}{4} + \frac{w_0 L^2}{2}\left\langle x - \frac{3L}{4}\right\rangle^1 + C_1 \qquad (3)$$

As the first boundary condition, we have $dy^2/dx = 0$ at $x = 0$ which when substituted in Eq. (*3*) yields $C_1 = 0$. Integrating a second time, we obtain

$$EIy = -\frac{5}{12}w_0L^2x^2 + \frac{w_0L}{4}\frac{x^3}{3} - \frac{w_0}{24L}\cdot\frac{x^5}{5} + \frac{w_0L^2}{4}\left\langle x - \frac{3L}{4}\right\rangle^2 + C_2 \tag{4}$$

The second boundary condition, $y = 0$ at $x = 0$, leads, upon substitution in Eq. (*4*), to $C_2 = 0$. Thus the beam deflection equation is

$$EIy = -\frac{5}{12}w_0L^2x^2 + \frac{w_0L}{12}x^3 - \frac{w_0}{120L}x^5 + \frac{w_0L^2}{4}\left\langle x - \frac{3L}{4}\right\rangle^2 \tag{5}$$

The deflection at the tip, $x = L$, is found from Eq. (*5*) to be

$$EIy]_{x=L} = -0.326w_0L^4$$

10.6. Using singularity functions, determine the equation of the deflection curve of the beam simply supported at points B and C and subject to the triangular loading shown in Fig. 10-8.

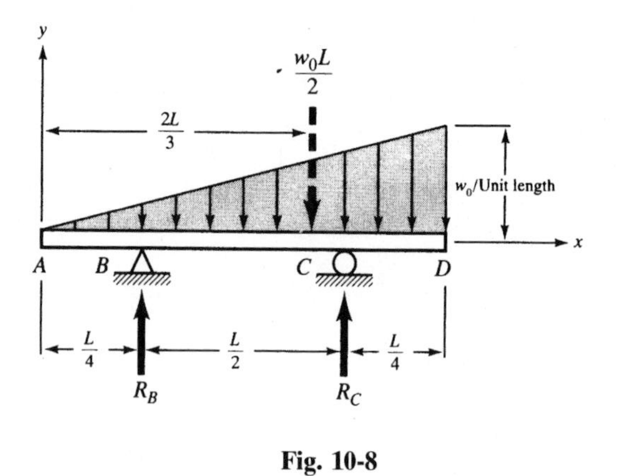

Fig. 10-8

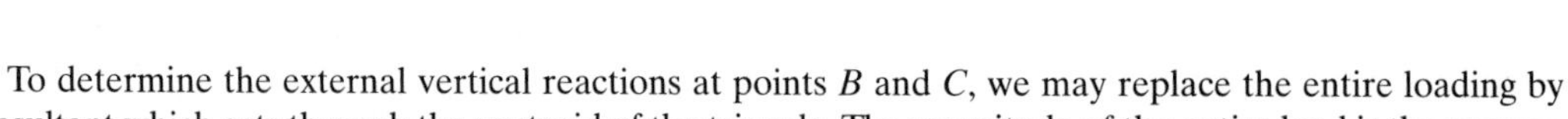

To determine the external vertical reactions at points B and C, we may replace the entire loading by its resultant which acts through the centroid of the triangle. The magnitude of the entire load is the average load per unit length, $w_0/2$, multiplied by the beam length L, or $w_0L/2$. This acts at a distance $2L/3$ from the left end A and is shown by the dotted vector in Fig. 10-8. From statics

$$+\circlearrowleft \Sigma M_B = R_C\cdot\frac{L}{2} - \frac{w_0L}{2}\left(\frac{2}{3}L - \frac{L}{4}\right) = 0$$

Therefore

$$R_C = \frac{5w_0L}{12}$$

$$\Sigma F_y = R_B + \frac{5w_0L}{12} - \frac{w_0L}{2} = 0$$

Therefore

$$R_B = \frac{w_0L}{12}$$

At any station x measured from the origin at A, the bending moment in terms of singularity functions is given as the sum of the moments of all forces to the left of that station. Let us examine a portion of the triangular load of horizontal length x. The resultant of that much of the loading is shown by the dotted vector in Fig. 10-9 and the resultant is of magnitude

$$\frac{w}{2} \cdot x = w_0 \frac{x}{L} \cdot \frac{x}{2}$$

and acts at a point distance $\frac{2}{3}x$ from A. Thus, the moment at x due *only* to the triangular loading is

$$-w_0 \frac{x^2}{2L} \cdot \frac{x}{3} \qquad \text{or} \qquad -\frac{w_0 x^3}{6L}$$

where the minus sign is inserted because according to our bending moment sign conventions in Chap. 6 downward loads give rise to negative bending moment.

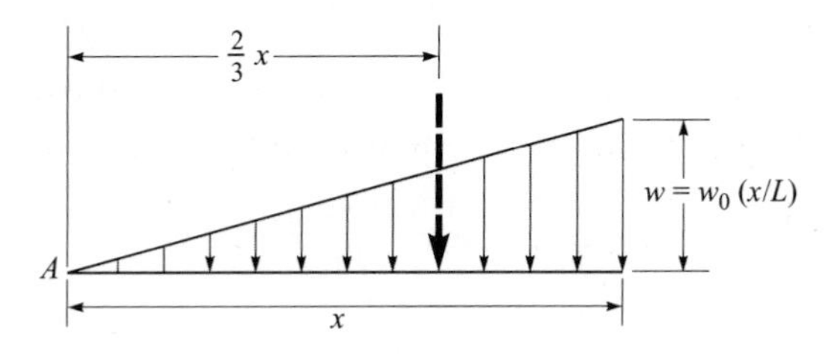

Fig. 10-9

In terms of singularity functions, the bending moment at any station x due to all loadings (including reactions) is

$$M = -\frac{w_0\langle x\rangle^3}{6L} + \frac{w_0 L}{12}\left\langle x - \frac{L}{4}\right\rangle + \frac{5w_0 L}{12}\left\langle x - \frac{3L}{4}\right\rangle \tag{1}$$

so that the differential equation of the bent beam is

$$EI\frac{d^2y}{dx^2} = -\frac{w_0\langle x\rangle^3}{6L} + \frac{w_0 L}{12}\left\langle x - \frac{L}{4}\right\rangle + \frac{5w_0 L}{12}\left\langle x - \frac{3L}{4}\right\rangle \tag{2}$$

Integrating the first time, we obtain

$$EI\frac{dy}{dx} = -\frac{w_0}{24L}\langle x\rangle^4 + \frac{w_0 L}{24}\left\langle x - \frac{L}{4}\right\rangle^2 + \frac{5w_0 L}{24}\left\langle x - \frac{3L}{4}\right\rangle^2 + C_1 \tag{3}$$

and integrating again, we find

$$EIy = -\frac{w_0}{120L}\langle x\rangle^5 + \frac{w_0 L}{72}\left\langle x - \frac{L}{4}\right\rangle^3 + \frac{5}{72}w_0 L\left\langle x - \frac{3L}{4}\right\rangle^3 + C_1 x + C_2 \tag{4}$$

As boundary conditions, when $x = L/4$, $y = 0$, so substituting in Eq. (*4*) we obtain

$$0 = -\frac{w_0}{120L}\left(\frac{L}{4}\right)^5 + C_1\frac{L}{4} + C_2 \tag{5}$$

Also, when $x = 3L/4$, $y = 0$, and substitution in Eq. (*4*) yields

$$0 = -\frac{w_0}{120L}\left(\frac{3L}{4}\right)^5 + \frac{w_0 L}{72}\left(\frac{L}{2}\right)^3 + C_1 \cdot \frac{3L}{4} + C_2 \tag{6}$$

Solving Eqs. (*5*) and (*6*), we obtain

$$C_1 = 0.0004666 w_0 L^3$$

$$C_2 = -0.0001085 w_0 L^4$$

so that the equation of the deflected beam is

$$EIy = -\frac{w_0}{120L}\langle x\rangle^5 + \frac{w_0 L}{72}\left\langle x - \frac{L}{4}\right\rangle^3 + \frac{5w_0 L}{72}\left\langle x - \frac{3L}{4}\right\rangle^3 + 0.0004666w_0 L^3 x - 0.0001085w_0 L^4 \quad (7)$$

10.7. If the beam subject to triangular loading in Problem 10.6 is a W203 × 40 steel section, of length $L = 4$ m, $I = 39 \times 10^6$ mm^4, and $w_0 = 80$ kN/m, determine the deflection at the point D.

Using Eq. (7) of Problem 10.6, we have

$$\begin{aligned} EI[y]_{x=L} &= -\frac{w_0 L^4}{120} + \frac{w_0 L}{72}\left(\frac{3L}{4}\right)^3 + \frac{5w_0 L}{72}\left(\frac{L}{4}\right)^3 \\ &\quad + 0.0004666w_0 L^4 - 0.0001085w_0 L^4 \\ &= -0.001031w_0 L^4 \\ y]_{x=L} &= -\frac{0.001031w_0 L^4}{EI} \\ &= -\frac{(0.001031)(80{,}000\ \text{N/m})(4\ \text{m})^4}{(200 \times 10^9\ \text{N/m}^2)(39 \times 10^{-6}\ \text{m}^4)} \\ &= -0.0027\ \text{m} \quad \text{or} \quad -2.7\ \text{mm} \end{aligned}$$

10.8. The beam AD in Fig. 10-10 is simply supported at A and C, loaded by a uniform load from B to D, and also by a couple applied as shown at D. Determine the equation of the deflection curve through the use of singularity functions.

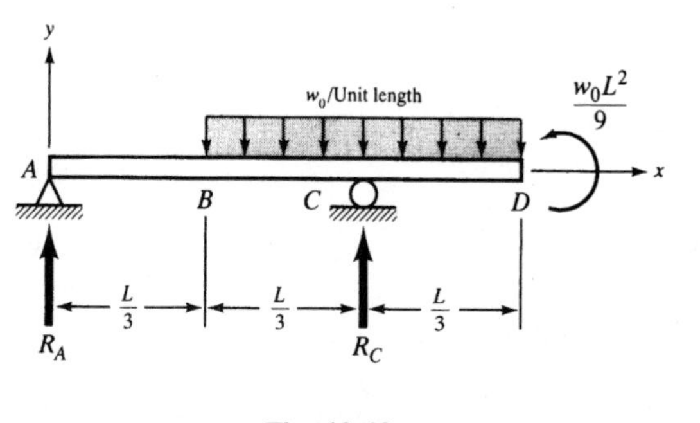

Fig. 10-10

The reactions at A and C are assumed to be positive in the directions shown and are found from the two statics equations to be

$$\circlearrowright \Sigma M_A = R_C\left(\frac{2L}{3}\right) - w_0\left(\frac{2L}{3}\right)^2 + \frac{w_0 L^2}{9} = 0 \quad (1)$$

$$\Sigma F_y = R_A + R_C - w_0\left(\frac{2L}{3}\right) = 0 \quad (2)$$

Solving,

$$R_A = \frac{w_0 L}{6} \qquad R_C = \frac{w_0 L}{2}$$

The singularity approach lets us write the equation of the entire deflected beam in the form

$$EI\frac{d^2y}{dx^2} = \frac{1}{6}w_0L\langle x\rangle^1 - \frac{w_0\left\langle x-\frac{L}{3}\right\rangle^2}{2} + \frac{w_0L}{2}\left\langle x-\frac{2L}{3}\right\rangle^1 \tag{3}$$

The applied couple does not appear directly in this equation but its effect is incorporated in the statics equations (*1*) and (*2*). Integrating the first time

$$EI\frac{dy}{dx} = \frac{w_0L}{6}\cdot\frac{\langle x\rangle^2}{2} - \frac{w_0}{2}\frac{\left\langle x-\frac{L}{3}\right\rangle^3}{3} + \frac{w_0L}{2}\frac{\left\langle x-\frac{2L}{3}\right\rangle^2}{2} + C_1\langle x\rangle \tag{4}$$

Integrating the second time

$$EIy = \frac{w_0L}{12}\frac{\langle x\rangle^3}{3} - \frac{w_0}{6}\frac{\left\langle x-\frac{L}{3}\right\rangle^4}{4} + \frac{w_0L}{4}\frac{\left\langle x-\frac{2L}{3}\right\rangle^3}{3} + C_1\frac{\langle x\rangle^2}{2} + C_2 \tag{5}$$

As boundary conditions we have: when $x = 0$, $y = 0$, from which Eq. (*5*) leads to $C_2 = 0$. Also, when $x = 2L/3$, $y = 0$, from which Eq. (*5*) gives us

$$C_1 = -0.03472w_0L^2$$

The required equation of the deflected beam is thus

$$EIy = \frac{w_0L\langle x\rangle^3}{36} - \frac{w_0}{24}\left\langle x-\frac{L}{3}\right\rangle^4 + \frac{w_0L}{12}\left\langle x-\frac{2L}{3}\right\rangle^3 - 0.01736w_0L^2x^2 \tag{6}$$

10.9. In Problem 10-8 if the beam is a steel wide-flange section W203 × 51 (having $I = 52.5 \times 10^6\ \text{mm}^4$ from Table 8-2 of Chap. 8), of length 6 m, and subject to a uniform load over *BD* of intensity 22 kN/m, determine the deflection at point *B*.

From the general equation of the deflection curve, Eq. (*6*) of Problem 10.8, we may write the expression for the deflection at $y = L/3$ as

$$EIy]_{x=L/3} = \frac{w_0L}{36}\cdot\frac{L^3}{27} - 0 + 0 - 0.01736w_0L^2\left(\frac{L^2}{9}\right)$$
$$= -0.0009w_0L^4$$

Substituting,

$$y]_{x=L/3} = \frac{\dfrac{(22{,}000\ \text{N/m})(6\ \text{m})}{36}\cdot\dfrac{(6\ \text{m})^3}{27} - 0.01736\left(\dfrac{22{,}000\ \text{N}}{\text{m}}\right)(6\ \text{m})^2\left(\dfrac{6\ \text{m}}{9}\right)^2}{(200\times10^9\ \text{N/m}^2)(52.5\times10^{-6}\ \text{m}^4)}$$
$$= -2.44\times10^{-2}\ \text{m} \quad \text{or} \quad -24.4\ \text{mm}$$

10.10. The cantilever beam *AD* is loaded by the applied couples M_1 and $M_1/3$, as shown in Fig. 10-11. Use the method of singularities to determine the equation of the deflected beam.

For static equilibrium, there must be a reactive couple M_A acting at point *A*, as well as possibly a shear-type reactive force R_A. From statics we find

$$+\circlearrowright \Sigma M_A = M_A - M_1 + \frac{M_1}{3} = 0 \qquad \text{and therefore } M_A = \frac{2}{3}M_1$$

$$\Sigma F_y = R_A = 0$$

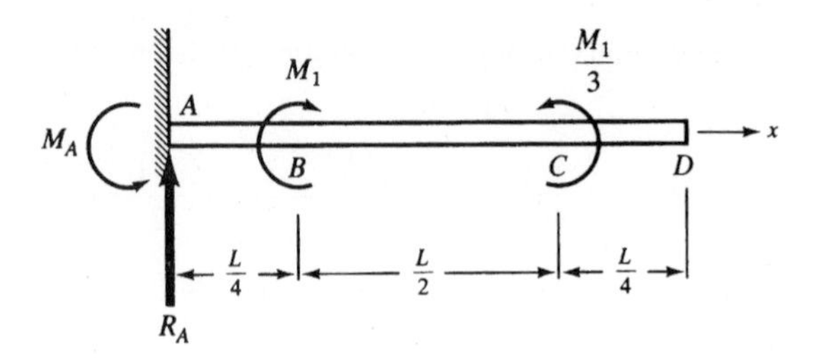

Fig. 10-11

The bending moment for any value of x is

$$M = -\frac{2}{3}M_1\langle x\rangle^0 + M_1\left\langle x - \frac{L}{4}\right\rangle^0 - \frac{M_1}{3}\left\langle x - \frac{3L}{4}\right\rangle^0 \tag{1}$$

so that the differential equation of the deflected beam is

$$EI\frac{d^2y}{dx^2} = -\frac{2}{3}M_1\langle x\rangle^0 + M_1\left\langle x - \frac{L}{4}\right\rangle^0 - \frac{M_1}{3}\left\langle x - \frac{3L}{4}\right\rangle^0 \tag{2}$$

Integrating the first time, we obtain

$$EI\frac{dy}{dx} = -\frac{2}{3}M_1\langle x\rangle^1 + M_1\left\langle x - \frac{L}{4}\right\rangle^1 - \frac{M_1}{3}\left\langle x - \frac{3L}{4}\right\rangle^1 + C_1 \tag{3}$$

and the first boundary condition is that $dy/dx = 0$ when $x = 0$. Hence, $C_1 = 0$.
Integrating a second time

$$EIy = -\frac{2}{3}M_1\frac{\langle x\rangle^2}{2} + \frac{M_1}{2}\left\langle x - \frac{L}{4}\right\rangle^2 - \frac{M_1}{6}\left\langle x - \frac{3L}{4}\right\rangle^2 + C_2 \tag{4}$$

and the second boundary condition is that $y = 0$ when $x = 0$. Hence $C_2 = 0$.
The equation describing the deflected beam is finally

$$EIy = -\frac{M_1\langle x\rangle^2}{3} + \frac{M_1}{2}\left\langle x - \frac{L}{4}\right\rangle^2 - \frac{M_1}{6}\left\langle x - \frac{3L}{4}\right\rangle^2 \tag{5}$$

10.11. The cantilever beam in Problem 10-10 is a steel wide-flange section W254 × 31, having $I = 44.1 \times 10^{-6}\ \text{m}^4$ and a length of 2 m. Determine M_1 if the deflection at point D is to be 3 mm.

We employ Eq. (5) of Problem 10.10 and simplify it for the deflection at $x = L$ to find

$$EIy]_{x=L} = -\frac{M_1L^2}{16}$$

Substituting the given numerical values, we find the tip deflection to be

$$y]_{x=L} = -\frac{M_1(2\ \text{m})^2}{(16)(200 \times 10^9\ \text{N/m}^2)(44.1 \times 10^{-6}\ \text{m}^4)} = 0.003\ \text{m}$$

Solving,

$$M_1 = 106\ \text{kN}\cdot\text{m}$$

10.12. Through the use of singularity functions determine the equation of the deflection curve of the simply supported beam of Fig. 10-12 subject to the couple applied at B plus the linearly varying load in CD.

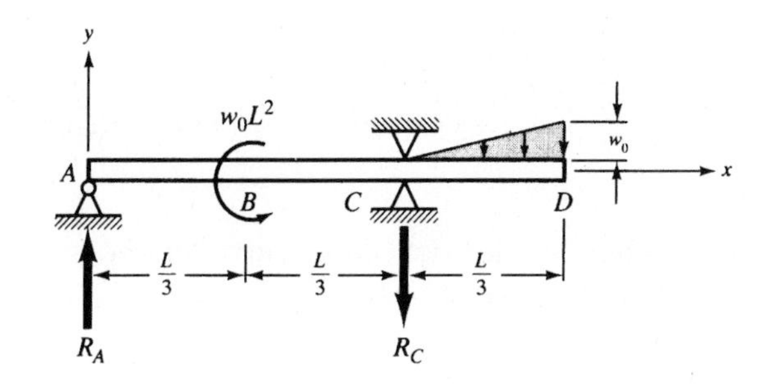

Fig. 10-12

Denoting the reactions at A and C by R_A and R_C assumed positive in the directions indicated and writing the two statics equations for this parallel force system, we obtain

$$+\circlearrowleft \Sigma M_A = w_0 L^2 - R_C\left(\frac{2L}{3}\right) - \frac{w_0}{2}\left(\frac{L}{3}\right)\left(\frac{2L}{3} + \frac{2}{3}\cdot\frac{L}{3}\right) = 0 \tag{1}$$

$$\Sigma F_y = R_A - R_C - \frac{w_0}{2}\cdot\frac{L}{3} = 0 \tag{2}$$

Solving, $R_A = \frac{13}{9} w_0 L$ and $R_C = \frac{23}{18} w_0 L$. Since each of these is positive, the assumed directions are correct.

In terms of singularity functions, the differential equation of the deflected beam is

$$EI\frac{d^2y}{dx^2} = \frac{13}{9} w_0 L\langle x\rangle^1 - w_0 L^2\left\langle x - \frac{L}{3}\right\rangle^0 - \frac{23}{18} w_0 L\left\langle x - \frac{2L}{3}\right\rangle^1$$

$$- w_0 \frac{\left\langle x - \frac{2L}{3}\right\rangle\left\langle x - \frac{2L}{3}\right\rangle\left(\frac{1}{2}\right)}{\left(\frac{L}{3}\right)} \cdot \frac{\left\langle x - \frac{2L}{3}\right\rangle}{3} \tag{3}$$

where the effect of the triangular loading in CD is represented as the last term in Eq. (3) using the technique for triangular load discussed in Problem 10.6 and illustrated in Fig. 10-9.

Integrating the first time, we obtain

$$EI\frac{dy}{dx} = \frac{13}{9} w_0 L\frac{\langle x\rangle^2}{2} - w_0 L^2\left\langle x - \frac{L}{3}\right\rangle^1 - \frac{23}{18} w_0 L\frac{\left\langle x - \frac{2L}{3}\right\rangle^2}{2} - \frac{w_0}{2L}\frac{\left\langle x - \frac{2L}{3}\right\rangle^4}{4} + C_1 \tag{4}$$

We have no boundary conditions on slope; hence we are unable to determine C_1 at this time. Integrating the second time

$$EIy = \frac{13}{18} w_0 L\frac{\langle x\rangle^3}{3} - w_0 L^2\frac{\left\langle x - \frac{L}{3}\right\rangle^2}{2} - \frac{23}{36} w_0 L\frac{\left\langle x - \frac{2L}{3}\right\rangle^3}{3} - \frac{w_0}{8L}\frac{\left\langle x - \frac{2L}{3}\right\rangle^5}{5} + C_1 x + C_2 \tag{5}$$

As boundary conditions, we have $x = 0$ at $y = 0$, so from Eq. (5) we find $C_2 = 0$. Also, when $x = 2L/3$, $y = 0$, from which we have from Eq. (5)

$$0 = \frac{13}{54} w_0 L\left(\frac{8L^3}{27}\right) - \frac{w_0 L^2}{2}\cdot\frac{L^2}{9} - 0 - 0 + C_1\left(\frac{2L}{3}\right)$$

Solving,

$$C_1 = -0.02366 w_0 L^3$$

The deflection curve of the bent beam is thus

$$EIy = \frac{13}{54}w_0L\langle x\rangle^3 - \frac{w_0L^2}{2}\left\langle x - \frac{L}{3}\right\rangle^2 - \frac{23}{108}w_0L\left\langle x - \frac{2L}{3}\right\rangle^3 - \frac{w_0}{40L}\left\langle x - \frac{2L}{3}\right\rangle^5 - 0.02366w_0L^3\langle x\rangle$$

10.13. Determine the equation of the deflection curve of the simply supported beam shown in Fig. 10-13(a). Use singularity functions.

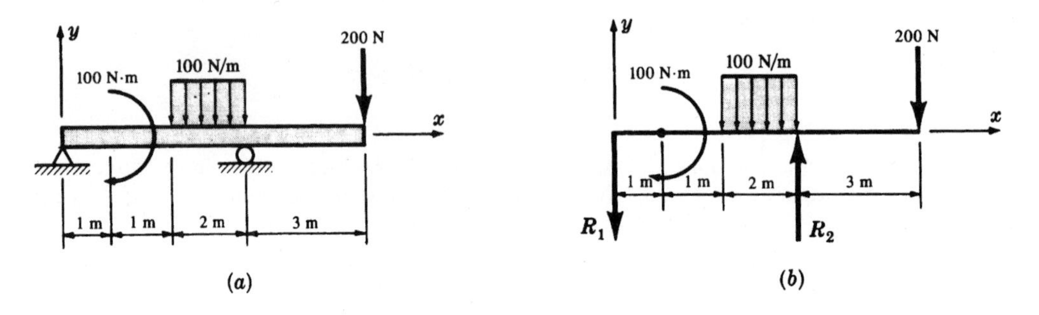

Fig. 10-13

The free-body diagram is shown in Fig. 10-13(b). From statics the reactions are readily found to be $R_1 = 225$ N, $R_2 = 525$ N.

Writing the bending moment corresponding to Fig. 10-13(b) in terms of singularity functions, we have

$$EI\frac{d^2y}{dx^2} = M = -225\langle x\rangle^1 + 100\langle x-1\rangle^0 - \frac{100\langle x-2\rangle^2}{2} + \overset{\circledast}{\frac{100\langle x-4\rangle^2}{2}} + 525\langle x-4\rangle^1 \qquad (1)$$

where the term denoted by ⊛ is necessary to annul the effect of the 100 N/M load to the right of $x = 4$ m.

Integrating,

$$EI\frac{dy}{dx} = -\frac{225}{2}\langle x\rangle^2 + 100\langle x-1\rangle^1 - \frac{50}{3}\langle x-2\rangle^3 + \frac{50}{3}\langle x-4\rangle^3 + \frac{525}{2}\langle x-4\rangle^2 + C_1 \qquad (2)$$

$$EIy = -\frac{225}{6}\langle x\rangle^3 + \frac{100}{2}\langle x-1\rangle^2 - \frac{50}{12}\langle x-2\rangle^4 + \frac{50}{12}\langle x-4\rangle^4 + \frac{525}{6}\langle x-4\rangle^3 + C_1x + C_2 \qquad (3)$$

The boundary conditions are $y = 0$ at $x = 0$, $x = 4$ m. Using these conditions in (3) to determine C_1 and C_2, we find $C_1 = 504$, $C_2 = 0$.

The desired deflection curve is thus

$$EIy = -\frac{225}{6}\langle x\rangle^3 + \frac{100}{2}\langle x-1\rangle^2 - \frac{50}{12}\langle x-2\rangle^4 + \frac{50}{12}\langle x-4\rangle^4 + \frac{525}{6}\langle x-4\rangle^3 + 504x \qquad (4)$$

10.14. The elastic beam AD shown in Fig. 10-14 is simply supported at B and C and subject to an applied couple M_1 at point A together with a uniformly distributed load in the overhanging region CD. Find the equation of the deformed beam as well as the deflection at point A.

From statics the reactions R_B and R_C are found to be

$$R_B = \tfrac{73}{48}wL\,(\downarrow) \qquad R_C = \tfrac{91}{48}wL\,(\uparrow)$$

Using the method of singularity functions, we find that the differential equation of the bent beam is

$$EI\frac{d^2y}{dx^2} = \frac{1}{2}wL^2 - \frac{73}{48}wL\left\langle x - \frac{L}{4}\right\rangle + \frac{91}{48}wL\left\langle x - \frac{5}{8}L\right\rangle - \frac{w}{2}\left\langle x - \frac{5}{8}L\right\rangle^2 \qquad (1)$$

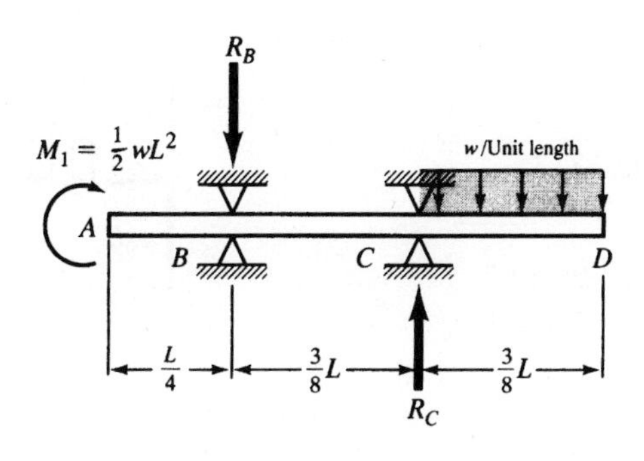

Fig. 10-14

Integrating the first time, we have

$$EI\frac{dy}{dx} = \frac{1}{2}wL^2\langle x\rangle - \frac{73}{96}wL\left\langle x - \frac{L}{4}\right\rangle^2 + \frac{91}{96}wL\left\langle x - \frac{5}{8}L\right\rangle^2 - \frac{w}{6}\left\langle x - \frac{5}{8}L\right\rangle^3 + C_1 \tag{2}$$

Integrating a second time

$$EIy = \frac{1}{2}wL\frac{\langle x\rangle^2}{2} - \frac{73}{288}wL\left\langle x - \frac{L}{4}\right\rangle^3 + \frac{91}{288}wL\left\langle x - \frac{5}{8}L\right\rangle^3 - \frac{w}{24}\left\langle x - \frac{5}{8}L\right\rangle^4 + C_1x + C_2 \tag{3}$$

As boundary conditions to determine C_1 and C_2, we have

First: When $x = L/4$, $y = 0$. Substituting in Eq. (*3*), we have

$$0 = \frac{wL}{4}\cdot\frac{L^2}{16} - 0 + 0 - 0 + C_1\frac{L}{4} + C_2 \tag{4}$$

Second: When $x = 5L/8$, $y = 0$. Substituting in Eq. (*3*), we have

$$0 = \frac{wL^2}{4}\cdot\frac{25}{64}L^2 - \frac{73}{288}wL\cdot\frac{27L^3}{512} + 0 + \frac{5}{8}LC_1 + C_2 \tag{5}$$

Solving Eqs. (*4*) and (*5*), we obtain

$$C_1 = -0.1831wL^3 \qquad C_2 = 0.03015wL^4$$

The equation of the deflected beam, for all values of x, is

$$EIy = -\frac{wL^2}{4}\langle x\rangle^2 - \frac{73}{288}wL\left\langle x - \frac{L}{4}\right\rangle^3 + \frac{91}{288}wL\left\langle x - \frac{5}{8}L\right\rangle^3 - \frac{w}{24}\left\langle x - \frac{5}{8}L\right\rangle^4 - 0.1831wL^3x + 0.03015wL^4 \tag{6}$$

At the left end, $x = 0$, and the deflection there is

$$EIy]_{x=0} = 0.03015w_0L^4$$

10.15. Use singularity functions to determine the equation of the deflection curve of the simply supported beam subject to a uniformly varying load as in Fig. 10-15(*a*). What is the central deflection of the beam?

The free-body diagram with the reactions found from statics is shown in Fig. 10-15(*b*).

If we refer to Problem 10.6, we can write the bending moment at any location x in the form

$$M(x) = +\frac{w_0L}{4}\langle x\rangle^1 - \frac{w_0}{3L}\langle x\rangle^3 + \frac{2w_0}{3L}\left\langle x - \frac{L}{2}\right\rangle^3 \tag{1}$$

where the second term on the right side of (*1*) represents a uniformly varying load extending completely across the beam as indicated by the triangle *OAB* in Fig. 10-16. To remove the portion of this loading represented by triangle *ABD*, we add the third term on the right side, which leaves the true load represented by triangle *ODB*.

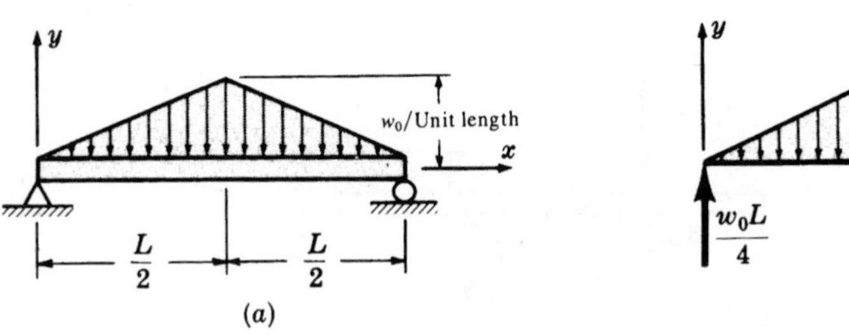

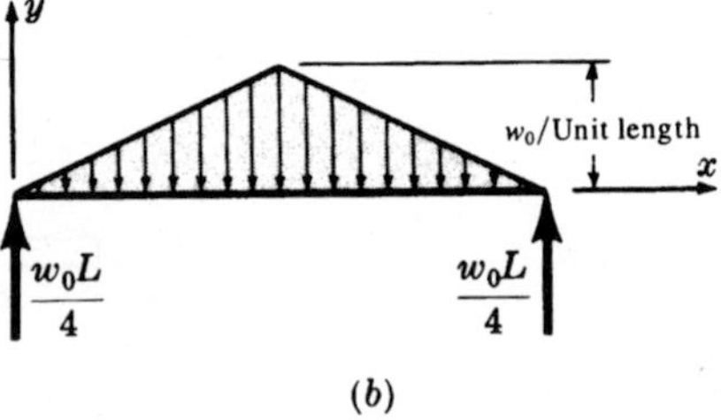

Fig. 10-15

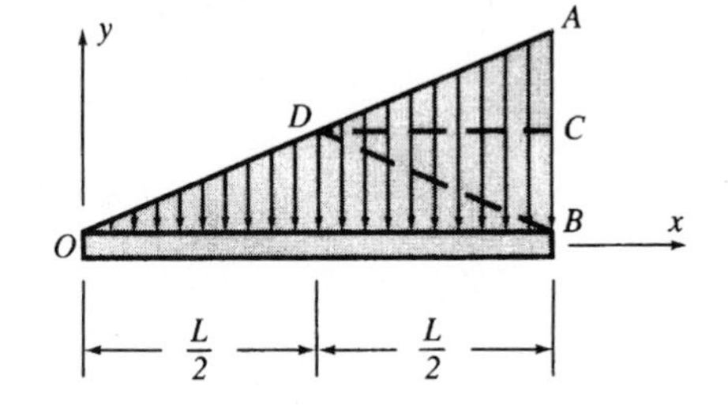

Fig. 10-16

Thus

$$EI\frac{d^2y}{dx^2} = M = +\frac{w_0L}{4}\langle x\rangle^1 - \frac{w_0}{3L}\langle x\rangle^3 + \frac{2w_0}{3L}\left\langle x - \frac{L}{2}\right\rangle^3 \tag{2}$$

from which

$$EI\frac{dy}{dx} = +\frac{w_0L}{8}\langle x\rangle^2 - \frac{w_0}{12L}\langle x\rangle^4 + \frac{w_0}{6L}\left\langle x - \frac{L}{2}\right\rangle^4 + C_1 \tag{3}$$

From symmetry we have as a boundary condition $dy/dx = 0$ at $x = L/2$. From (3) we find that $C_1 = -5w_0L^3/192$. Integrating again we get the desired deflection curve.

$$EIy = \frac{w_0L}{24}\langle x\rangle^3 - \frac{w_0}{60L}\langle x\rangle^5 + \frac{w_0}{30L}\left\langle x - \frac{L}{2}\right\rangle^5 - \frac{5}{192}w_0L^3x + C_2 \tag{4}$$

Since $y = 0$ at $x = 0$, it follows that $C_2 = 0$. The central deflection is found from (4) to be

$$y = -\frac{w_0L^4}{120EI}$$

Statically Determinate Beams—Computerized Solutions

Problems 10.1 through 10.15 have demonstrated the efficiency of the method of singularity functions for the determination of beam deflections. The technique is very well suited to computer implementation because there is a direct correspondence between the singularity function $\langle x - a\rangle$ defined as

$$\langle x - a\rangle = \begin{cases} 0 & \text{if } x < a \\ (x - a) & \text{if } x > a \end{cases}$$

and the "if" statement in FORTRAN. This feature is utilized extensively in the computerized approach in Problem 10.16.

10.16. Write a FORTRAN program for determination of slope and deflection at selected points along the length of a beam of constant cross section, simply supported at two arbitrary points, and loaded by arbitrary concentrated forces, moments, and uniformly distributed loads.

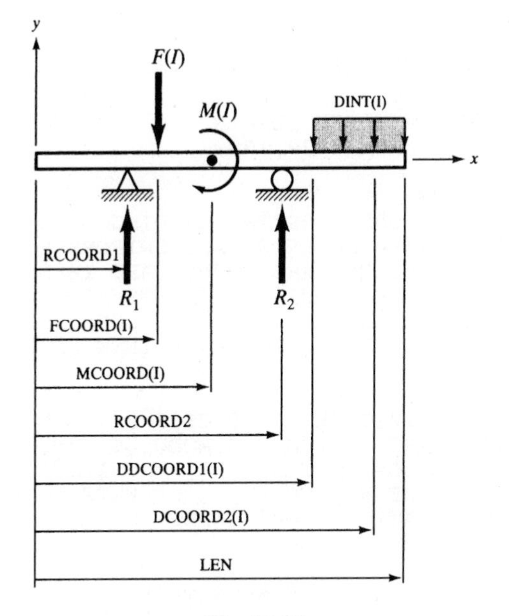

Fig. 10-17

Let us employ the terminology shown in Fig. 10-17. See Table 10-1.

A complete listing of the program based upon numerical solution of the beam bending equation

$$EI\frac{d^2y}{dx^2} = M$$

utilizing singularity functions follows. One must introduce all parameters of beam loading, geometry, and elastic properties. The program will then print out the slope and deflection (with appropriate algebraic sign) at each of the (NUM + 1) points along the length of the beam as well as values of the reactions R_1 and R_2.

Table 10-1

Units	USCS or SI
E	Young's modulus
I	Moment of inertia of beam cross section about the neutral axis
LEN	Length of beam
NF	Number of applied concentrated forces (not including reactions)
NM	Number of applied moments
ND	Number of uniformly distributed loads
NUM	Number of segments into which length of beam is divided for purpose of analysis
RCOORD1	Coordinate locating reaction R
RCOORD2	Coordinate locating reaction R
FCOORD(I)	Coordinate locating applied concentrated force I
FMAG(I)	Magnitude of concentrated force I
MCOORD(I)	Coordinate of locating moment I
MMAG(I)	Magnitude of moment I
DDCOORD1(I)	Left coordinate of distributed load I
DDCOORD2(I)	Right coordinate of distributed load I
MCOORD(I)	Magnitude (load/unit length) of uniformly distributed load I

```
00010*************************************************************************
00020         PROGRAM BEND (INPUT,OUTPUT)
00030*************************************************************************
00040*
00050*          AUTHOR: KATHLEEN DERWIN
00060*          DATE  : JANUARY 29,1989
00070*
00080*  BRIEF DESCRIPTION:
00090*      THIS PROGRAM CONSIDERS THE BENDING OF BEAMS DUE TO CONCENTRATED
00100*  FORCES, CONCENTRATED MOMENTS, AND UNIFORMLY DISTRIBUTED LOADS. FIRST,
00110*  THE PIN REACTION FORCES ARE FOUND, AND THEN THE SLOPE AND DEFLECTION OF
00120*  THE LOADED BEAM AT VARIOUS INCREMENTS ALONG ITS LENGTH ARE DETERMINED.
00130*  NOTE, THIS PROGRAM WAS DEVELOPED TO CONSIDER GENERAL LOADING, AND THE
00140*  PINS DO NOT HAVE TO BE AT THE ENDPOINTS OF THE BEAM.
00150*
00160*  INPUT:
00170*    THE USER MUST FIRST ENTER IF USCS OF SI UNITS ARE DESIRED. THEN,
00180*  THE MOMENT OF INERTIA, YOUNG'S MODULUS, AND THE LENGTH OF THE BEAM
00190*  ARE ENTERED. FINALLY, THE NUMBER, MAGNITUDE, AND LOCATION OF ALL
00200*  LOAD TYPES, AND THE NUMBER OF INCREMENTS TO PERFORM THE SLOPE AND
00210*  DEFLECTION CALCULATIONS  ARE INPUTTED.
00220*
00230*  OUTPUT:
00240*      THE PROGRAM PRINTS THE MAGNITUDE AND SENSE OF THE TWO REACTION
00250*  FORCES, AS WELL AS THE SLOPE AND DEFLECTION AT SUCCESSIVE INTERVALS
00260*  ALONG THE BEAM.
00270*
00280*  VARIABLES:
00290*     E,INER,LEN       ---    YOUNG'S MODULUS, MOMENT OF INERTIA, LENGTH
00300*                             OF BEAM
00310*     NUM              ---    NUMBER OF INCREMENTS TO DO CALCULATIONS ON
00320*  RCOORD1,RCOORD2      ---    LOCATION OF THE PINS
00330*     R1,R2            ---    MAGNITUDE OF THE PIN REACTION FORCES
00340*  FCOORD(I),FMAG(I)    ---    LOCATION AND MAGNITUDE OF CONCENTRATED FORCE
00350*  DDCOORD1(I),DCOORD2(I)-    LOCATION OF DISTRIBUTED LOADS
00360*    DINT(I)            ---    INTENSITY OF DISTRIBUTED LOADS
00370*  MCOORD(I),MMAG(I)    ---    LOCATION AND MAGNITUDE OF MOMENTS
00380*      DX               ---    INCREMENTAL STEP ALONG BEAM (LENGTH/NUM)
00390*  VV1,VV2,...VV6       ---    THE 'BRACKET TERMS' OF THE SINGULARITY FNCTS
00400*  SLF(I),DF(I),SLM(I),
00410*  DM(I),SLD(I),DD(I)---      THE SUMMING ARRAYS FOR SLOPE AND DEFLECTION
00420*                             DUE TO EACH APPLIED FORCE AT A PARTICULAR PT
00430*  SLR1,SLR2,DR1,DR2 ---      THE EFFECTS OF THE REACTION FORCES AT A POIN
00440*  SLFX,SLMX,SLDX       ---    THE TOTAL SLOPE AND DEFLECTION DUE TO BOTH
00450*  DFX,DMX,DDX                APPLIED AND REACTIVE FORCES AT A POINT
00460*     C1,C2            ---    THE CONSTANTS OF INTEGRATION
00470*  SL(I),D(I)           ---    THE FINAL SLOPE AND DEFLECTION AT ANY POINT
00480*    NF,NM,ND           ---    THE NUMBER OF CONCENTRATED FORCES (NOT
00490*                             INCLUDING REACTIONS), APPLIED MOMENTS, AND
00500*                             UNIFORMLY DISTRIBUTED LOADS
00510*   FSUM,MSUM           ---    THE SUM OF THE FORCES AND MOMENTS, USED TO
00520*                             COMPUTE THE REACTIVE FORCES
00530*  DDIST(I),LOAD(I)     ---    THE DISTANCE EACH DISTRIBUTED LOAD SPANS, AN
00540*                             THE MAGNITUDE OF THE RESULTING FORCE
00550*      BIG              ---    GIVES THE LARGEST NUMBER OF ALL FORCE TYPES
00560*      ANS              ---    DENOTES IF USCS OR SI UNITS ARE DESIRED
00570*
00580*
00590**************************************************************************
00600*********                     MAIN PROGRAM                    ************
00610**************************************************************************
00620*
00630*          VARIABLE DECLARATIONS
00640*
00650       REAL E,INER,LEN,NUM,RCOORD1,RCOORD2,FCOORD(10),FMAG(10),MCOORD(10)
00660       REAL MMAG(10),DCOORD1(10),DCOORD2(10),DINT(10),DX,X,XX,VV1,VV2
```

```
00670       REAL VV3,VV4,VV5,VV6,SLF(10),SLD(10),SLM(10),DF(10),DD(10),DM(10)
00680       REAL SLR1,SLR2,DR1,DR2,R1,R2,SLFX,SLDX,SLMX,DFX,DDX,DMX
00690       REAL C1,C2,FSUM,MSUM,DDIST(10),LOAD(10),SL(100),D(100)
00700       INTEGER NF,ND,NM,BIG,ANS
00710*
00720*            INITIALIZING VARIABLES TO ZERO
00730*
00740       FCOORD(10)=0.0
00750       FMAG(10)=0.0
00760       MCOORD(10)=0.0
00770       MMAG(10)=0.0
00780       DCOORD1(10)=0.0
00790       DCOORD2(10)=0.0
00800       DINT(10)=0.0
00810       SLF(10)=0.0
00820       SLD(10)=0.0
00830       SLM(10)=0.0
00840       DF(10)=0.0
00850       DD(10)=0.0
00860       DM(10)=0.0
00870       SL(100)=0.0
00880       D(100)=0.0
00890       SLFX=0.0
00900       SLDX=0.0
00910       SLMX=0.0
00920       DFX =0.0
00930       DDX =0.0
00940       DMX =0.0
00950*
00960******      USER INPUT      *****
00970*
00980       PRINT*,'PLEASE INDICATE YOUR CHOICE OF UNITS:'
00990       PRINT*,'1 - USCS'
01000       PRINT*,'2 - SI'
01010       PRINT*,' '
01020       PRINT*,'ENTER 1,2:'
01030       READ*,ANS
01040       IF (ANS.EQ.1) THEN
01050          PRINT*,'PLEASE INPUT ALL DATA IN UNITS OF POUND AND/OR INCH...'
01060       ELSE
01070          PRINT*,'PLEASE INPUT ALL DATA IN UNITS OF NEWTON AND/OR METER...
01080       ENDIF
01090*
01100
01110       PRINT*,' '
01120       PRINT*,'ENTER THE VALUES FOR E,I,LEN,NF,ND,NM,NUM:'
01130       READ(*,*)E,INER,LEN,NF,ND,NM,NUM
01140       PRINT*,' '
01150       PRINT*,'ENTER THE COORDINATES OF THE ALL FORCE TYPES AS DISTANCES'
01160       PRINT*,'FROM THE LEFT END OF THE BEAM...ALSO, CONSIDER FORCES'
01170       PRINT*,'DIRECTED DOWNWARD, AND MOMENTS ACTING CLOCKWISE AS POSITIVE
01180       PRINT*,' '
01190       PRINT*,'ENTER THE COORDINATES OF THE REACTION POINTS:'
01200       READ(*,*)RCOORD1,RCOORD2
01210       IF (NF.GT.0) THEN
01220       PRINT*,'ENTER THE COORDINATE AND MAGNITUDE OF ALL CONCENTRATED '
01230       PRINT*,'FORCES:'
01240       READ(*,*)(FCOORD(I),FMAG(I),I=1,NF)
01250       ENDIF
01260       IF (NM.GT.0) THEN
01270       PRINT*,'ENTER THE COORDINATE AND MAGNITUDE OF ALL CONCENTRATED '
01280       PRINT*,'MOMENTS:'
01290       READ(*,*)(MCOORD(I),MMAG(I),I=1,NM)
01300       ENDIF
01310       IF (ND.GT.0) THEN
01320       PRINT*,'ENTER THE FIRST AND SECOND COORDINATE AND THEN INTENSITY '
```

```
01330      PRINT*,'OF ALL DISTRIBUTED LOADS:'
01340      READ(*,*)(DCOORD1(I),DCOORD2(I),DINT(I),I=1,ND)
01350      ENDIF
01360*
01370******      END USER INPUT      *****
01380*
01390      PRINT*,'  '
01400      PRINT*,' THE MAGNITUDES OF THE TWO REACTIVE FORCES (LB OR  NEWTONS)
01410*
01420******       CALCULATIONS       ******
01430*
01440*
01450*           CALCULATING THE MAGNITUDE AND DIRECTION OF THE PIN REACTION
01460*           FORCES
01470*
01480      FSUM=0.0
01490      MSUM=0.0
01500      DO 15 I=1,ND
01510         DDIST(I)= DCOORD2(I) - DCOORD1(I)
01520         LOAD(I) = DINT(I)*DDIST(I)
01530         FSUM = LOAD(I) + FSUM
01540         MSUM =(((0.5*DDIST(I) + DCOORD1(I)) - RCOORD1) * LOAD(I)) + MSUM
01550 15   CONTINUE
01560      DO 20 I = 1,NF
01570         FSUM = FSUM + FMAG(I)
01580         MSUM =((FCOORD(I) - RCOORD1)*FMAG(I)) + MSUM
01590 20   CONTINUE
01600      DO 30 I = 1,NM
01610         MSUM = MSUM + MMAG(I)
01620 30   CONTINUE
01630      R2 = -(MSUM/(RCOORD2-RCOORD1))
01640      R1 = -(FSUM+R2)
01650*
01660*           PRINTING THE REACTION FORCES
01670*
01680      PRINT*,' '
01690      PRINT*,'R1 = ',R1,'        R2 = ',R2
01700      PRINT*,' '
01710*
01720*           CALCULATING THE LARGEST NUMBER OF EITHER FORCES, DISTRIBUTED
01730*           LOADS, OR MOMENTS
01740*
01750      IF (NF.GE.ND) THEN
01760         IF (NF.GE.NM) THEN
01770            BIG=NF
01780         ELSE
01790            BIG=NM
01800         ENDIF
01810      ELSE
01820         IF (ND.GE.NM) THEN
01830            BIG=ND
01840         ELSE
01850            BIG=NM
01860         ENDIF
01870      ENDIF
01880*
01890*
01900*      THE FOLLOWING SECTION OF THIS PROGRAM PERFORMS THE CALCULATIONS
01910*      THAT DETERMINE THE SLOPE AND DEFLECTION AT SEVERAL INTERVALS ALONG
01920*      THE BEAM. THE METHOD OF SINGULARITY FUNCTIONS AND INTEGRATION IS
01930*      EMPLOYED, AND THE PRINCIPAL OF SUPERPOSITION ALLOWS EACH TYPE OF
01940*      FORCE TO BE CONSIDERED SEPARATELY AND THEN SUMMED TO PRODUCE
01950*      THE NET EFFECT ON THE BEAM.
01960*
01970      DX=LEN/NUM
01980      J=1
01990 10   DO 50 XX=0,LEN,DX
```

```
02000          X=XX
02010*
02020*        THE FUNCTIONS ARE FIRST SOLVED FOR THE INITIAL CONDITIONS OF ZERO
02030*        DISPLACEMENT AT THE TWO PIN REACTION POINTS, RCOORD1 AND RCOORD2,
02040*        THAT THE CONSTANTS OF INTEGRATION MAY BE DETERMINED.
02050*
02060          IF (J.EQ.1) X=RCOORD1
02070          IF (J.EQ.2) X=RCOORD2
02080*
02090*      EVALUATING THE 'BRACKET TERMS' USED WITH THE SINGULARITY FUNCTIONS
02100*
02110          DO 60 I=1,BIG
02120             VV1=X-FCOORD(I)
02130             VV2=X-DCOORD1(I)
02140             VV3=X-DCOORD2(I)
02150             VV4=X-RCOORD1
02160             VV5=X-RCOORD2
02170             VV6=X-MCOORD(I)
02180*
02190*          RECALL, WITH SINGULARITY FUNCTIONS IF THE QUANTITY IN THE
02200*          BRACKETS IS LESS THAN OR EQUAL TO ZERO, THAT TERM MAKES NO
02210*          CONTRIBUTION TO THE SLOPE AND/OR DEFLECTION AT THAT POINT.
02220*
02230             IF (VV1.LE.0) VV1=0
02240             IF (VV2.LE.0) VV2=0
02250             IF (VV3.LE.0) VV3=0
02260             IF (VV4.LE.0) VV4=0
02270             IF (VV5.LE.0) VV5=0
02280             IF (VV6.LE.0) VV6=0
02290*
02300*         DETERMINING THE SLOPE AND DISPLACEMENT DUE TO EACH FORCE AT A
02310*         PARTICULAR POINT ON THE BEAM
02320*
02330             SLF(I) = FMAG(I)/2*(VV1**2)
02340             DF(I)  = FMAG(I)/6*(VV1**3)
02350
02360             SLD(I) = (DINT(I)/6*(VV2**3)) - (DINT(I)/6*(VV3**3))
02370             DD(I) = (DINT(I)/24*(VV2**4)) - (DINT(I)/24*(VV3**4))
02380
02390             SLM(I) = MMAG(I)*VV6
02400             DM(I)  = MMAG(I)/2*(VV6**2)
02410 60       CONTINUE
02420*
02430*         DETERMINING THE SLOPE AND DISPLACEMENT DUE TO THE REACTION FORCE
02440*         AT A PARTICULAR POINT ON THE BEAM
02450*
02460          SLR1 = R1/2 * (VV4**2)
02470          SLR2 = R2/2 * (VV5**2)
02480          DR1  = R1/6 * (VV4**3)
02490          DR2  = R2/6 * (VV5**3)
02500*
02510*         SUMMING THE EFFECTS OF ALL FORCE CONTRIBUTIONS OF THE SLOPE AND
02520*         DISPLACEMENT AT A PARTICULAR POINT ON THE BEAM
02530*
02540          DO 40 I=1,BIG
02550             SLFX= SLFX+ SLF(I)
02560             SLDX= SLDX+ SLD(I)
02570             SLMX= SLMX+ SLM(I)
02580             DFX = DFX + DF(I)
02590             DDX = DDX + DD(I)
02600             DMX = DMX + DM(I)
02610 40       CONTINUE
02620
02630          SL(J) = SLFX + SLDX + SLMX + SLR1 + SLR2
02640          D(J)  = DFX + DDX + DMX + DR1 + DR2
02650          J =J+1
```

```
02660*
02670*        SETTING THE SLOPE AND DISPLACEMENT SUMS BACK TO ZERO BEFORE
02680*        MOVING TO NEXT POINT ON BEAM
02690*
02700         SLFX=0.0
02710         SLDX=0.0
02720         SLMX=0.0
02730         DFX =0.0
02740         DDX =0.0
02750         DMX =0.0
02760         IF (J.EQ.3) GO TO 10
02770*
02780*        REPEAT THIS PROCEDURE FOR NEXT POINT ON BEAM
02790 50   CONTINUE
02800*
02810*           CALCULATING THE CONSTANTS OF INTEGRATION FROM THE INITIAL
02820*           CONDITIONS OF ZERO DISPLACEMENT AT THE PINS.
02830*
02840      C1 = (D(2) - D(1))/(RCOORD1 - RCOORD2)
02850      C2 =(-D(1) - (C1*RCOORD1))
02860
02870      X=0.0
02880*
02890*           FINALLY, DETERMINING THE SLOPE AND DISPLACEMENT AT EVERY POIN
02900*           BY CONSIDERING ALL THE FORCE CONTRIBUTIONS AT EACH RESPECTIVE
02910*           POINT, AND THE CONSTANTS OF INTEGRATION.
02920*
02930      DO 80 I=3,J-1
02940           SL(I) =(SL(I) + C1)/(E*INER)
02950           D(I)  =(D(I) + (C1*X) + C2)/(E*INER)
02960*          PRINT*,SL(I),D(I)
02970           X=X+DX
02980 80   CONTINUE
02990*
03000*          PRINTING THE SLOPE AND DELECTION AT INCREMENTS ALONG THE BEAM
03010*
03020      PRINT 82,'NODE','LOCATION','SLOPE','DEFLECTION'
03030      IF (ANS.EQ.1) THEN
03040         PRINT 83
03050      ELSE
03060         PRINT 84
03070      ENDIF
03080      X=0.0
03090*
03100      DO 85 I=3,J-1
03110         PRINT 90,I-2,X,SL(I),D(I)
03120         X=X+DX
03130 85   CONTINUE
03140*
03150*        FORMAT STATEMENTS
03160*
03170 82   FORMAT(//,2X,A4,5X,A8,5X,A5,6X,A10)
03180 83   FORMAT(3X,'NO',9X,'IN',8X,'IN/IN',10X,'IN')
03190 84   FORMAT(3X,'NO',9X,'M',9X,' M/M ',10X,'M')
03200 90   FORMAT(3X,I2,6X,F8.3,3X,E10.3,4X,E10.3)
03210      STOP
03220      END
```

10.17. A beam 12 m long is supported at knife edge reactions and loaded by a concentrated moment of $8000\,\mathrm{N\cdot m}$ together with a concentrated force of 8500 N as shown in Fig. 10-18. Use the FORTRAN program of Problem 10.16 to determine the deflection by considering 25 segments along the length of the beam. The beam is of rectangular cross section 60 mm wide and 280 mm high and $E = 200$ GPa.

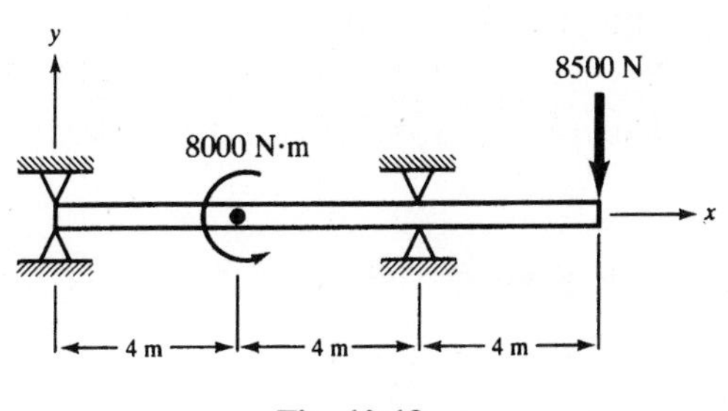

Fig. 10-18

The input into the program is shown in Table 10-2.

Input of these parameters into the program leads to the following output:

```
 PLEASE INDICATE YOUR CHOICE OF UNITS:
 1 — USCS
 2 — SI

 ENTER 1.2:
? 2
 PLEASE INPUT ALL DATA IN UNITS OF NEWTON AND/OR METER...

 ENTER THE VALUES FOR E,I,LEN,NF,ND,NM,NUM:
? 200E+9,109E-6,12,1,0,1,25

 ENTER THE COORDINATES OF ALL THE FORCE TYPES AS DISTANCES
 FROM THE LEFT END OF THE BEAM...ALSO, CONSIDER FORCES
 DIRECTED DOWNWARD, AND MOMENTS ACTING CLOCKWISE AS POSITIVE.

 ENTER THE COORDINATES OF THE REACTION POINTS:
? 0,8
```

Table 10-2

Units	SI
E	200×10^9
I	$\frac{1}{12}$ (0.06 m) $(0.28\ \text{m})^3 = 109 \times 10^{-6}\ \text{m}^4$
LEN	12
NF	1
ND	0
NM	1
NUM	25
RCOORD1	0
RCOORD2	8
FCOORD(I)	12
FMAG(I)	8500
MCOORD(I)	4
MMAG(I)	8000
DCOORD1(I)	0
DCOORD2(I)	0
DMAG(I)	0

```
 ENTER THE COORDINATE AND MAGNITUDE OF ALL CONCENTRATED
 FORCES:
? 12,8500
 ENTER THE COORDINATE AND MAGNITUDE OF ALL CONCENTRATED
 MOMENTS:
? 4,8000

  THE MAGNITUDES OF THE TWO REACTIVE FORCES (LB OR NEWTONS):

 R1 = 5250.          R2 = -13750.

  NODE      LOCATION       SLOPE       DEFLECTION
   NO          M            M/M            M
    1         .000      -.294E-02      .000E+00
    2         .480      -.291E-02     -.140E-02
    3         .960      -.282E-02     -.278E-02
    4        1.440      -.269E-02     -.411E-02
    5        1.920      -.249E-02     -.535E-02
    6        2.400      -.224E-02     -.649E-02
    7        2.880      -.194E-02     -.750E-02
    8        3.360      -.158E-02     -.834E-02
    9        3.840      -.116E-02     -.900E-02
   10        4.320      -.571E-03     -.943E-02
   11        4.800       .132E-03     -.954E-02
   12        5.280       .891E-03     -.929E-02
   13        5.760       .171E-02     -.867E-02
   14        6.240       .257E-02     -.765E-02
   15        6.720       .350E-02     -.619E-02
   16        7.200       .448E-02     -.428E-02
   17        7.680       .552E-02     -.188E-02
   18        8.160       .660E-02      .103E-02
   19        8.640       .763E-02      .445E-02
   20        9.120       .856E-02      .833E-02
   21        9.600       .941E-02      .127E-01
   22       10.080       .102E-01      .174E-01
   23       10.560       .108E-01      .224E-01
   24       11.040       .114E-01      .277E-01
   25       11.520       .119E-01      .333E-01
   26       12.000       .123E-01      .391E-01

 SRU       1.284 UNTS.

 RUN COMPLETE.
```

From the printout we note that the deflection under the 8500-N force is 0.0391 m or 39.1 mm and under the 8000-N · m moment located between nodes 9 and 10 it is approximately −0.0092 m or −9.2 mm.

10.18. A beam 100 in long and of rectangular cross section with $I = 3.375\ \text{in}^4$ is loaded and supported as shown in Fig. 10-19. Use the FORTRAN program of Problem 10.16 to determine the deflections if the beam is represented by 50 segments along its length. Take $E = 30 \times 10^6\ \text{lb/in}^2$.

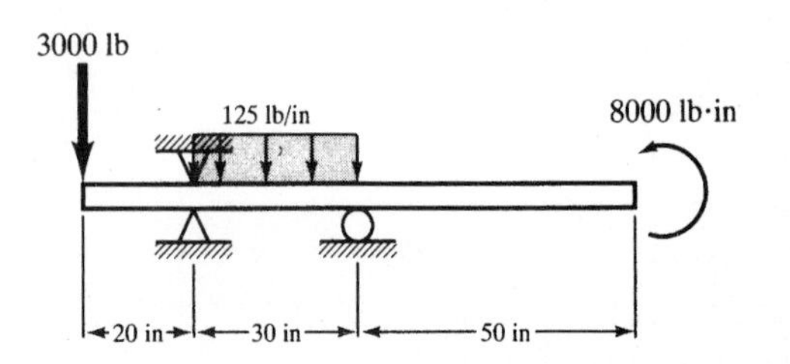

Fig. 10-19

The input to the program is shown in Table 10-3.

Table 10-3

Units	USCS
E	30×10^6
I	3.375
LEN	100
NF	1
ND	1
NM	1
NUM	50
RCOORD1	20
RCOORD2	50
FCOORD(I)	0
FMAG(I)	3000
MCOORD(I)	100
MMAG(I)	-8000
DCOORD1(I)	20
DCOORD2(I)	50
DMAG(I)	125

Input of these parameters into the program leads to the following output:

```
run
 PLEASE INDICATE YOUR CHOICE OF UNITS:
 1 — USCS
 2 — SI

 ENTER 1.2:
? 1
 PLEASE INPUT ALL DATA IN UNITS OF POUND AND/OR INCH...

 ENTER THE VALUES FOR E,I,LEN,NF,ND,NM,NUM:
? 30E6,3.375,100,1,1,1,50

 ENTER THE COORDINATES OF ALL THE FORCE TYPES AS DISTANCES
 FROM THE LEFT END OF THE BEAM...ALSO, CONSIDER FORCES
 DIRECTED DOWNWARD, AND MOMENTS ACTING CLOCKWISE AS POSITIVE.

 ENTER THE COORDINATES OF THE REACTION POINTS:
? 20,50
 ENTER THE COORDINATE AND MAGNITUDE OF ALL CONCENTRATED
 FORCES:
? 0,3000
 ENTER THE COORDINATE AND MAGNITUDE OF ALL CONCENTRATED
 MOMENTS:
? 100,-8000
 ENTER THE FIRST AND SECOND COORDINATE AND THEN MAGNITUDE
 OF ALL DISTRIBUTED LOADS:
? 20,50,125
```

THE MAGNITUDES OF THE TWO REACTIVE FORCES (LB OR NEWTONS):

R1 = -7141.666666667 R2 = 391.6666666667

NODE NO	LOCATION IN	SLOPE IN/IN	DEFLECTION IN
1	.000	-.101E-01	.162E+00
2	2.000	-.100E-01	.142E+00
3	4.000	-.983E-02	.122E+00
4	6.000	-.953E-02	.103E+00
5	8.000	-.912E-02	.838E-01
6	10.000	-.859E-02	.661E-01
7	12.000	-.793E-02	.496E-01
8	14.000	-.716E-02	.345E-01
9	16.000	-.628E-02	.210E-01
10	18.000	-.527E-02	.943E-02
11	20.000	-.414E-02	.000E+00
12	22.000	-.304E-02	-.715E-02
13	24.000	-.209E-02	-.123E-01
14	26.000	-.128E-02	-.156E-01
15	28.000	-.605E-03	-.175E-01
16	30.000	-.556E-04	-.181E-01
17	32.000	.380E-03	-.178E-01
18	34.000	.710E-03	-.166E-01
19	36.000	.946E-03	-.150E-01
20	38.000	.110E-02	-.129E-01
21	40.000	.117E-02	-.106E-01
22	42.000	.119E-02	-.826E-02
23	44.000	.114E-02	-.592E-02
24	46.000	.106E-02	-.371E-02
25	48.000	.933E-03	-.172E-02
26	50.000	.784E-03	.000E+00
27	52.000	.626E-03	.141E-02
28	54.000	.468E-03	.250E-02
29	56.000	.310E-03	.328E-02
30	58.000	.152E-03	.374E-02
31	60.000	-.617E-05	.389E-02
32	62.000	-.164E-03	.372E-02
33	64.000	-.322E-03	.323E-02
34	66.000	-.480E-03	.243E-02
35	68.000	-.638E-03	.131E-02
36	70.000	-.796E-03	-.123E-03
37	72.000	-.954E-03	-.187E-02
38	74.000	-.111E-02	-.394E-02
39	76.000	-.127E-02	-.632E-02
40	78.000	-.143E-02	-.902E-02
41	80.000	-.159E-02	-.120E-01
42	82.000	-.174E-02	-.154E-01
43	84.000	-.190E-02	-.190E-01
44	86.000	-.206E-02	-.230E-01
45	88.000	-.222E-02	-.273E-01
46	90.000	-.238E-02	-.319E-01
47	92.000	-.253E-02	-.368E-01
48	94.000	-.269E-02	-.420E-01
49	96.000	-.285E-02	-.475E-01
50	98.000	-.301E-02	-.534E-01
51	100.000	-.317E-02	-.596E-01

SRU 1.305 UNTS.

RUN COMPLETE.

Supplementary Problems

10.19. The cantilever beam ABC is loaded by a uniformly distributed load w per unit length over the right half BC as shown in Fig. 10-20. Use singularity functions to determine the deflection curve of the bent beam. Also, determine the deflection at the tip C.

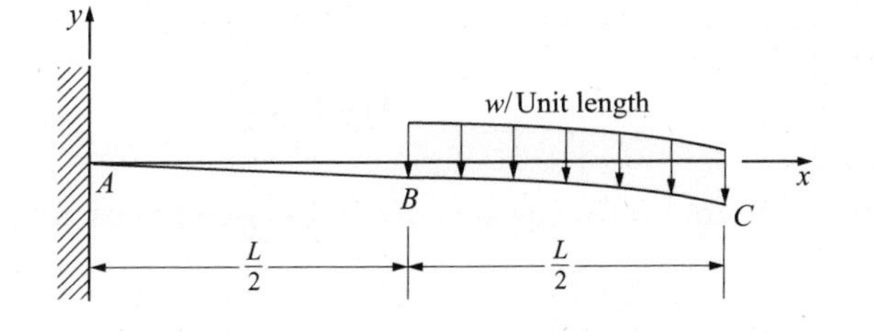

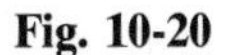

Fig. 10-20

Ans. $$EIy = \frac{wL}{12}\langle x\rangle^3 - \frac{3}{8}wL^2\frac{\langle x\rangle^2}{2} - \frac{w}{24}\left\langle x - \frac{L}{2}\right\rangle^4$$

$$EIy]_{x=L} = -\frac{41}{384}$$

10.20. Consider a simply supported beam subject to a uniform load acting over a portion of the beam as indicated in Fig. 10-21. Use singularity functions to determine the equation of the deflection curve.

Ans. $$EIy = \frac{wb}{6L}\left(\frac{b}{2} + c\right)\langle x\rangle^3 - \frac{w}{24}\langle x - a\rangle^4 + \frac{w}{24}\langle x - a - b\rangle^4$$

$$+ \left\{\frac{w}{24L}[(L-a)^4 - (L-c)^4] - \frac{wbL}{6}\left(\frac{b}{2} + c\right)\right\}\langle x\rangle$$

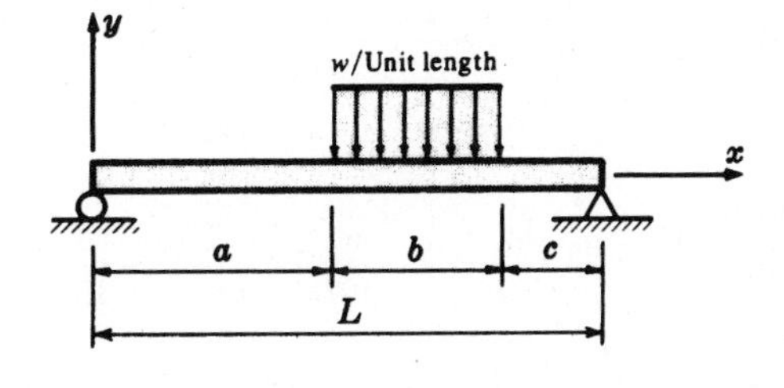

Fig. 10-21

10.21. The beam $ABCD$ is pinned at B, rests on a roller at C, and is subjected to the tip loads each of magnitude P as shown in Fig. 10-22. Use the method of singularity functions to determine the deflection curve of the beam, which is symmetric about the midlength of the beam. Also, determine the deflection at point A.

Ans. $$EIy = -\frac{P}{6}\langle x\rangle^3 + \frac{P}{6}\langle x - a\rangle^3 + \frac{P}{6}\langle x - (a + L_1)\rangle^3 + \left(\frac{PLa}{2} - \frac{Pa^2}{2}\right)\langle x\rangle$$

$$EIy]_{x=0} = \frac{2}{3}Pa^3 - \frac{PLa^2}{2}$$

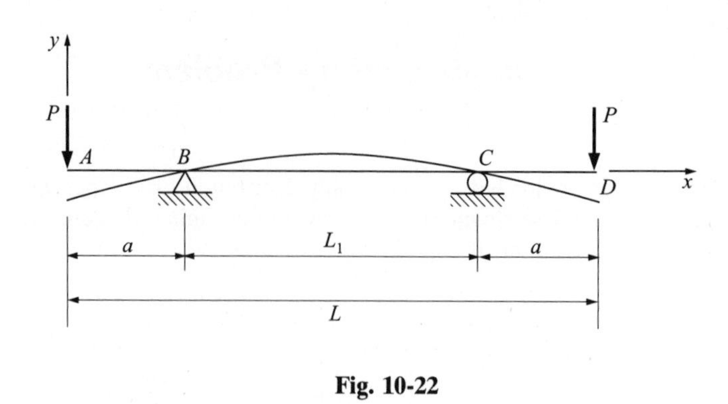

Fig. 10-22

Use singularity functions to determine the equation of the deflected beams in Problems 10.22 through 10.25.

10.22. See Fig. 10-23.

$$\textit{Ans.}\quad EIy = -\frac{wa}{24}\langle x\rangle^3 - \frac{w}{24}\langle x\rangle^4 + \frac{w}{24}\langle x-a\rangle^4 + \frac{wa^2}{2}\langle x-a\rangle^2 + \frac{9}{24}wa\langle x-2a\rangle^3 - \frac{wa}{6}\langle x-3a\rangle^3 + \frac{11}{48}wa^3\langle x\rangle^1$$

10.23. See Fig. 10-24.

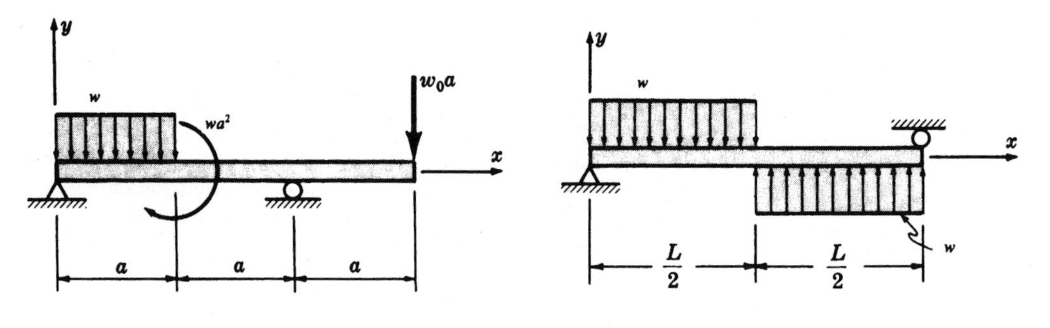

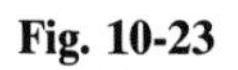

Fig. 10-23

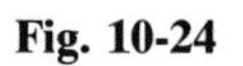

Fig. 10-24

$$\textit{Ans.}\quad EIy = \frac{wL}{24}\langle x\rangle^3 - \frac{w}{24}\langle x\rangle^4 + \frac{w}{12}\left\langle x - \frac{L}{2}\right\rangle^4 - \frac{wL^3}{192}\langle x\rangle^1$$

10.24. See Fig. 10-25.

$$\textit{Ans.}\quad EIy = \frac{w_0 L}{24}\langle x\rangle^3 - \frac{w_0}{24}\langle x\rangle^4 + \frac{w_0}{60L}\langle x\rangle^5 - \frac{w_0}{10L}\left\langle x - \frac{L}{2}\right\rangle^5 - \frac{3}{192}w_0 L^3 x$$

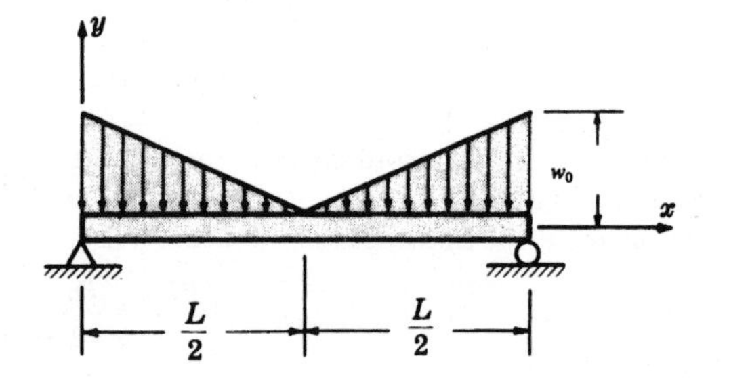

Fig. 10-25

10.25. See Fig. 10-26.

Ans. $EIy = -\frac{850}{3}\langle x\rangle^3 + 3300\langle x-3\rangle^2 - \frac{500}{12}\langle x-6\rangle^4 + \frac{500}{12}\langle x-9\rangle^4 + \frac{2350}{3}\langle x-9\rangle^3 + 10{,}175x$

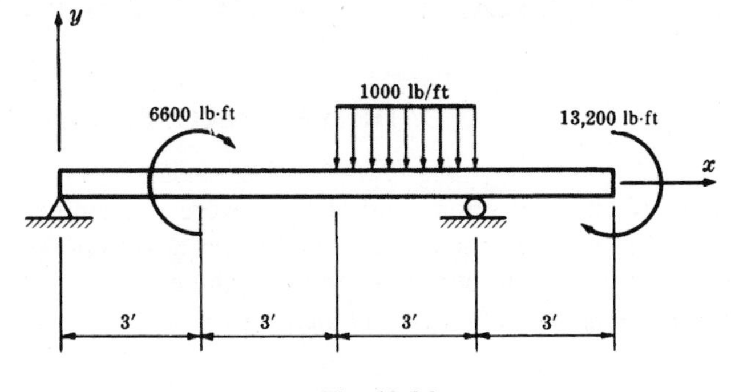

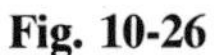
Fig. 10-26

10.26. The beam AC in Fig. 10-27 is 15 ft long, 3 in × 4 in in rectangular cross section, is subject to a uniform load of 120 lb/ft, and has $E = 30 \times 10\ \text{lb/in}^2$. Use the FORTRAN program of Problem 10.16 to determine (*a*) the deflection at the left end of the beam and (*b*) the maximum deflection of the beam.
Ans. (*a*) 0.065 in, (*b*) −0.10 in at $x = 110$ in

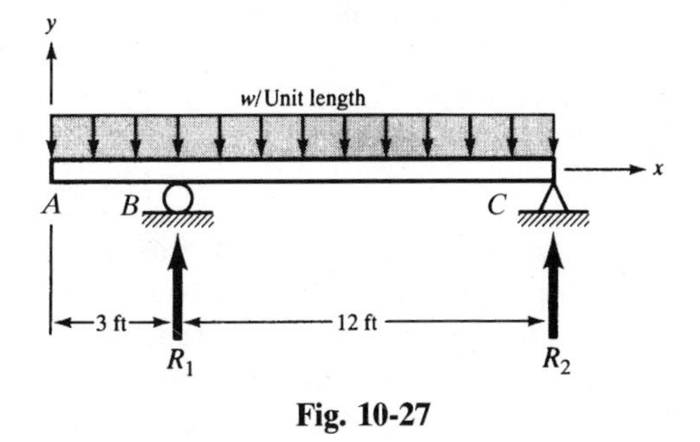

Fig. 10-27

10.27. Through the use of singularity functions, determine the equation of the deflection curve of the beam simply supported at B and C and subject to the triangular loading shown in Fig. 10-28.

$$\textit{Ans.}\quad EIy = -\frac{w_0\langle x\rangle^5}{180L} + \frac{w_0 L}{16}\left\langle x - \frac{L}{2}\right\rangle^3 - 0.02050 w_0 L^3 x + 0.01042 w_0 L^4$$

10.28. The beam shown in Fig. 10-29 is simply supported and subject to a concentrated force, the moment, and the uniformly distributed load indicated. The material has $E = 200$ GPa and the beam cross section has $I = 20 \times 10^{-6}\ \text{m}^4$. Use the FORTRAN program of Problem 10.16 to determine the deflection under the point of application of the 4200-N force. *Ans.* 19.8 mm

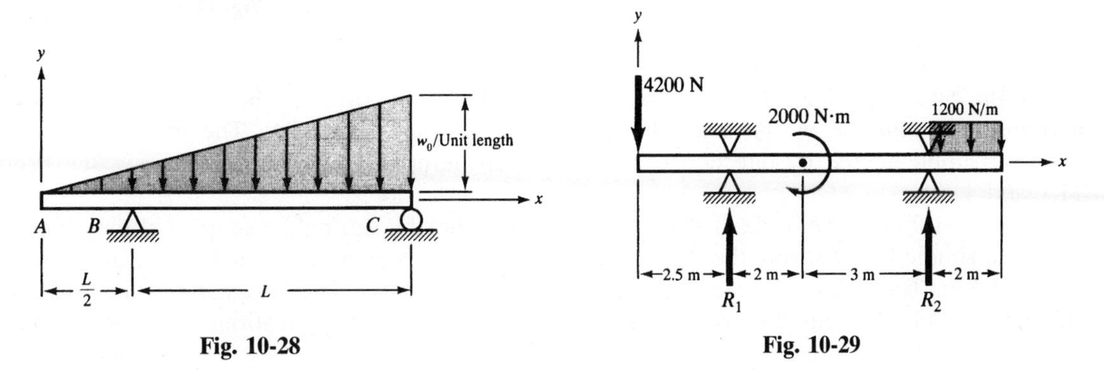

Fig. 10-28

Fig. 10-29

Chapter 11

Statically Indeterminate Elastic Beams

STATICALLY DETERMINATE BEAMS

In Chaps. 8, 9, and 10 the deflections and stresses were determined for beams having various conditions of loading and support. In the cases treated it was always possible to completely determine the reactions exerted upon the beam merely by applying the equations of static equilibrium. In these cases the beams are said to be *statically determinate*.

STATICALLY INDETERMINATE BEAMS

In this chapter we shall consider those beams where the number of unknown reactions exceeds the number of equilibrium equations available for the system. In such a case it is necessary to supplement the equilibrium equations with additional equations stemming from the deformations of the beam. In these cases the beams are said to be *statically indeterminate*.

TYPES OF STATICALLY INDETERMINATE BEAMS

Several common types of statically indeterminate beams are illustrated below. Although a wide variety of such structures exists in practice, the following four diagrams will illustrate the nature of an indeterminate system. For the beams shown below the reactions of each constitute a parallel force system and hence there are two equations of static equilibrium available. Thus the determination of the reactions in each of these cases necessitates the use of additional equations arising from the deformation of the beam.

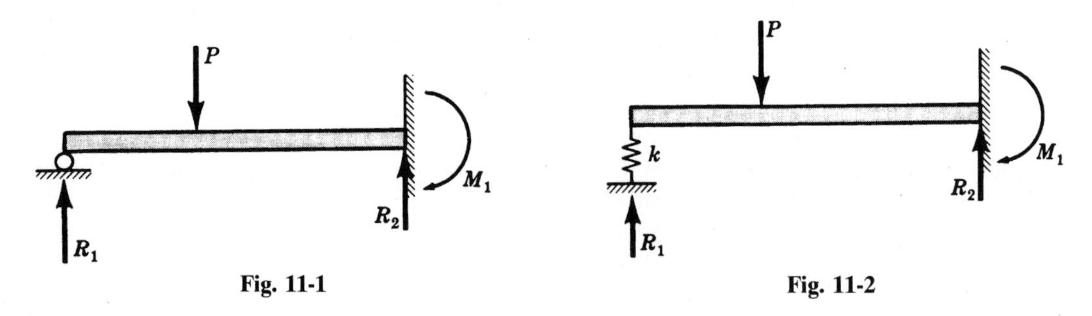

Fig. 11-1 **Fig. 11-2**

In the case (Fig. 11-1) of a beam fixed at one end and supported at the other, sometimes termed a *supported cantilever*, we have as unknown reactions R_1, R_2, and M_1. The two statics equations must be supplemented by one equation based upon deformations. For applications, see Problems 11.1 and 11.3.

In Fig. 11-2 the beam is fixed at one end and has a flexible springlike support at the other. In the case of a simple linear spring the flexible support exerts a force proportional to the beam deflection at that point. The unknown reactions are again R_1, R_2, and M_1. The two statics equations must be supplemented by one equation stemming from deformations. For applications see Problems 11.2 and 11.16.

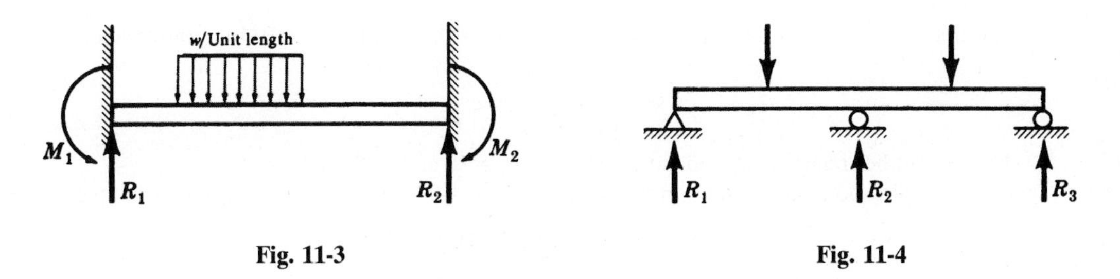

Fig. 11-3 **Fig. 11-4**

As shown in Fig. 11-3, a beam fixed or clamped at both ends has the unknown reactions R_1, R_2, M_1, and M_2. The two statics equations must be supplemented by two equations arising from the deformations. For applications, see Problems 11.4, 11.6, and 11.12.

In Fig. 11-4 the beam is supported on three supports at the same level. The unknown reactions are R_1, R_2, and R_3. The two statics equations must be supplemented by one equation based upon deformations. A beam of this type that rests on more than two supports is called a *continuous beam*.

Solved Problems

11.1. A beam is clamped at A, simply supported at B, and subject to the concentrated force shown in Fig. 11-5. Determine all reactions.

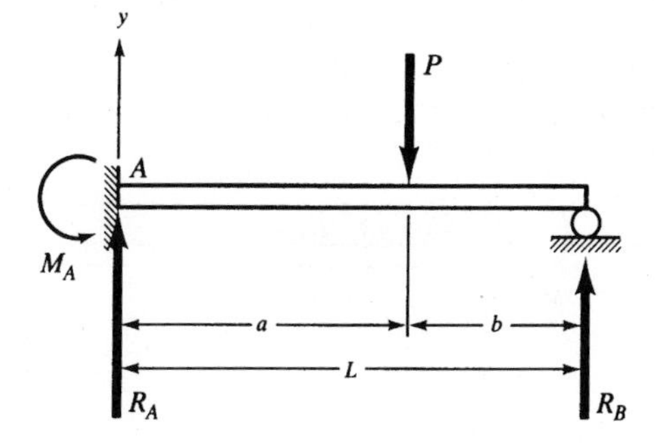

Fig. 11-5

The reactions are R_A, R_B, and M_A. From statics we have

$$+\circlearrowright \Sigma M_A = M_A - Pa + R_b L = 0 \tag{1}$$

$$\Sigma F_y = R_A + R_B - P = 0 \tag{2}$$

Thus there are two equations in the three unknowns R_A, R_B, and M_A. We can supplement the statics equations with an equation stemming from deformations using the method of singularity functions to describe the bent beam. This is

$$EI\frac{d^2y}{dx^2} = R_A\langle x\rangle - M_A\langle x\rangle^0 - P\langle x-a\rangle \tag{3}$$

Integrating the first time, we have

$$EI\frac{dy}{dx} = R_A\frac{\langle x\rangle^2}{2} - M_A\langle x\rangle - \frac{P}{2}\langle x-a\rangle^2 + C_1 \tag{4}$$

The first boundary condition is that at $x = 0$, $dy/dx = 0$, and thus $C_1 = 0$. Integrating again,

$$EIy = \frac{R_A}{2}\frac{\langle x\rangle^3}{3} - M_A\frac{\langle x\rangle^2}{2} - \frac{p}{2}\frac{\langle x-a\rangle^3}{3} + C_2 \tag{5}$$

The second boundary condition is that at $x = 0$, $y = 0$, and we find $C_2 = 0$.

The third boundary condition is that at $x = L$, $y = 0$. Substituting in Eq. (*5*), we have

$$0 = \frac{R_A L^3}{6} - \frac{M_A L^2}{2} - \frac{Pb^3}{6} \tag{6}$$

Simultaneous solution of the three equations (*1*), (*2*), and (*6*) leads to

$$R_A = \frac{Pb}{2L^3}(3L^2 - b^2)$$

$$R_B = \frac{Pa^2}{2L^3}(2L + b)$$

$$M_A = \frac{Pb}{2L^2}(L^2 - b^2)$$

11.2. The beam AB in Fig. 11-6 is clamped at A, spring supported at B, and loaded by the uniformly distributed load w per unit length. Prior to application of the load, the spring is stress free. The spring constant is 345 kN/m. To determine the flexural rigidity EI of the beam, an experiment is conducted without the uniform load w and also without the spring being present. In this experiment it is found that a vertical force of 10,000 N applied at end B deflects that point 50 mm. The spring is then attached to the beam at B and a uniform load of magnitude 5 kN/m is applied between A and B. Determine the deflection of point B under these conditions.

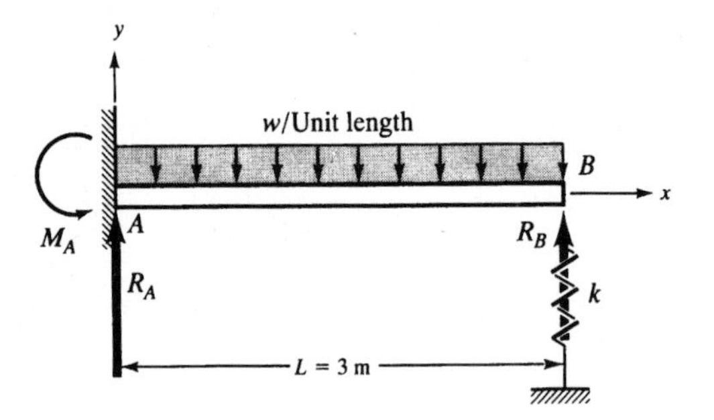

Fig. 11-6

The forces acting on the beam when it is uniformly loaded as well as spring supported at its tip are shown in Fig. 11-6. The force R_B represents the force exerted by the spring on the beam. The differential equation of the bent beam in terms of singularity functions is

$$EI\frac{d^2y}{dx^2} = -M_A\langle x\rangle^0 + R_A\langle x\rangle^1 - \frac{w}{2}\langle x\rangle^2 \tag{1}$$

Integrating the first time, we find

$$EI\frac{dy}{dx} = -M_A\langle x\rangle^1 + \frac{R_A}{2}\langle x\rangle^2 - \frac{w}{6}\langle x\rangle^3 + C_1 \tag{2}$$

Now, invoking the boundary condition that when $x = 0$, $dy/dx = 0$, we find from Eq. (*2*) that $C_1 = 0$. The second integration yields

$$EIy = -\frac{M_A}{2}\langle x\rangle^2 + \frac{R_A}{6}\langle x\rangle^3 - \frac{w}{24}\langle x\rangle^4 + C_2 \tag{3}$$

and the second boundary condition is that $x = 0$ when $y = 0$, so from Eq. (*3*) we have $C_2 = 0$. From Eq. (*3*) we have the deflection at B due to the uniform load plus the presence of the spring to be given by

$$EI[y]_{x=L} = -\frac{M_A L^2}{2} + \frac{R_A L^3}{6} - \frac{wL^4}{24} \tag{4}$$

But for linear action of the spring we have the usual relation

$$R_B = -k[y]_{x=L} = +k\,\Delta_B \tag{5}$$

Also, from statics for this parallel force system we have the two equilibrium equations

$$+\circlearrowleft \Sigma M_A = M_A + R_B L - \frac{wL^2}{2} = 0 \tag{6}$$

$$\Sigma F_y = R_A + R_B - (5000\ \text{N/m})(3\ \text{m}) = 0 \tag{7}$$

Simultaneous solution of Eqs. (*4*), (*6*), and (*7*) indicates that

$$R_A\left(\frac{EI}{k} + \frac{L^3}{3}\right) = \frac{EIwL}{k} + \frac{5wL^4}{24} \tag{8}$$

The flexural rigidity EI is easily found by consideration of the experimental evidence. The tip deflection of a tip-loaded cantilever beam is

$$\frac{PL^3}{3EI}$$

which becomes, for this experiment,

$$0.050\ \text{m} = \frac{(10{,}000\ \text{N})(3\ \text{m})^3}{3EI}$$

from which

$$EI = 1.8 \times 10^6\ \text{N}\cdot\text{m}^2 \tag{9}$$

If this value together with the spring constant of 345,000 N/m is substituted in Eq. (*8*), we find that $R_A = 11{,}440$ N. From Eq. (*7*) we find that $R_B = 3560$ N, so that the spring equation (*5*) indicates the displacement of point B to be

$$\Delta_B = \frac{3560\ \text{N}}{345{,}000\ \text{N/m}} = 0.01032\ \text{m} \quad \text{or} \quad 10.3\ \text{mm} \tag{10}$$

11.3. Consider the overhanging beam shown in Fig. 11-7. Determine the magnitude of the supporting force at B.

There are two statics equations

$$\circlearrowleft \Sigma M_A = M_1 + R_2 a - \frac{w(a+b)^2}{2} = 0 \tag{1}$$

$$\Sigma F_v = R_1 + R_2 - w(a+b) = 0 \tag{2}$$

Let us employ the method of singularity functions to write the differential equation of the bent beam

$$EI\frac{d^2y}{dx^2} = -M_1\langle x\rangle^0 + R_1\langle x\rangle^1 - \frac{w}{2}\langle x\rangle^2 + R_2\langle x-a\rangle^1 \tag{3}$$

Note that in (*1*) a negative sign is assigned to M_1 since, as we work from left to right starting at the origin A, the reactive moment M_1 tends to bend the portion of the beam to the right of A into a configuration having curvature concave downward, which is negative according to the bending moment sign convention given in Chap. 6.

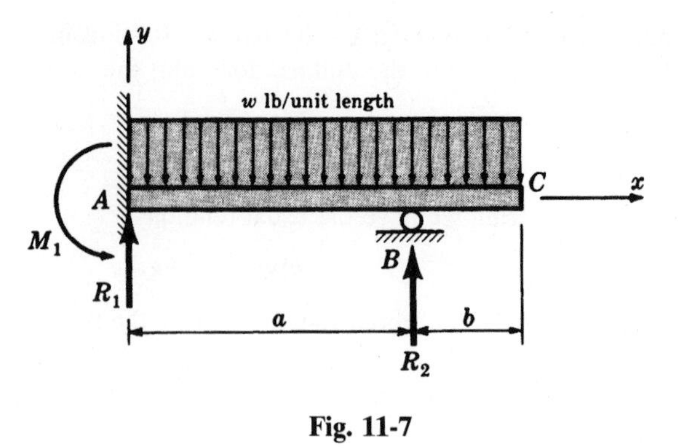

Fig. 11-7

Integrating

$$EI\frac{dy}{dx} = -M_1\langle x\rangle^1 + \frac{R_1}{2}\langle x\rangle^2 - \frac{w}{6}\langle x\rangle^3 + \frac{R_2}{2}\langle x-a\rangle^2 + C_1 \tag{4}$$

But when $x = 0$, $dy/dx = 0$; hence $C_1 = 0$. Integrating again,

$$EIy = -\frac{M_1}{2}\langle x\rangle^2 + \frac{R_1}{6}\langle x\rangle^3 - \frac{w}{24}\langle x\rangle^4 + \frac{R_2}{6}\langle x-a\rangle^3 + C_2 \tag{5}$$

But when $x = 0$, $y = 0$, so that $C_2 = 0$.

Since the support at point B is unyielding, y must vanish in (5) when $x = a$. Substituting, we find

$$0 = -\frac{M_1 a^2}{2} + \frac{R_1 a^3}{6} - \frac{wa^4}{24} \qquad \text{from which } M_1 = R_1\frac{a}{3} - \frac{wa^2}{12}$$

Solving this in conjunction with the statics equations, we find

$$R_1 = \frac{5}{8}wa - \frac{3wb^2}{4a} \qquad R_2 = \frac{3}{8}wa + wb + \frac{3wb^2}{4a}$$

11.4. The clamped end beam is loaded as shown in Fig. 11-8 by a couple M_0. Determine all reactions.

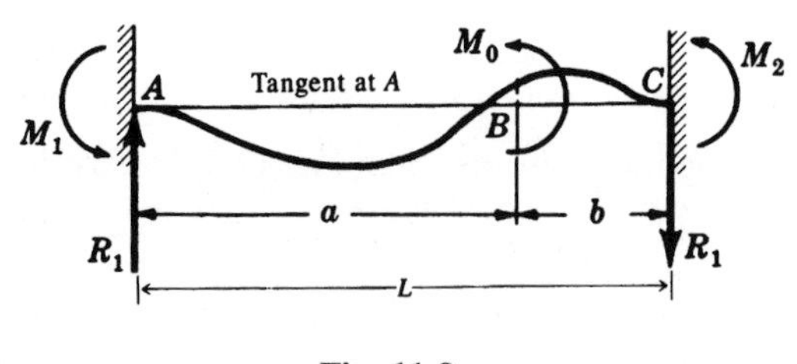

Fig. 11-8

Under the action of the couple, the initially straight beam bends into the configuration shown by the curved line. Tangents to the deformed configuration remain horizontal at ends A and B and of course there is zero vertical displacement at each of these ends. This gives rise to the reactions shown in which the vertical (shear) reactions are of equal magnitude for vertical equilibrium. This leaves only one equation from statics, namely,

$$+\circlearrowright \Sigma M_A = -M_1 - M_2 - M_0 + R_1(a+b) = 0 \tag{1}$$

This equation contains R_1, M_1, and M_2 as unknowns. Since there are no more statics equations available, we must supplement Eq. (1) with two additional equations stemming from deformations of the system. We employ the method of singularity functions and write the bending moment at any point along the length of the beam as

$$M = -M_1\langle x\rangle^0 + R_1\langle x\rangle - M_0\langle x-a\rangle^0 \tag{2}$$

The differential equation of the bent beam is thus

$$EI\frac{d^2y}{dx^2} = -M_1\langle x\rangle^0 + R_1\langle x\rangle - M_0\langle x-a\rangle^0 \tag{3}$$

Integrating the first time, we obtain

$$EI\frac{dy}{dx} = -M_1\langle x\rangle + R_1\frac{\langle x\rangle^2}{2} - M_0\frac{\langle x-a\rangle^1}{1} + C_1 \tag{4}$$

As the first boundary condition, when $x = 0$, $dy/dx = 0$; hence from (*4*) we have $C_1 = 0$. Integrating again

$$EIy = -M_A\frac{\langle x\rangle^2}{2} + \frac{R_1}{2}\cdot\frac{\langle x\rangle^3}{3} - M_0\frac{\langle x-a\rangle^2}{2} + C_2 \tag{5}$$

The second boundary condition states that when $x = 0$, $y = 0$. Substituting these values in Eq. (*5*), we find $C_2 = 0$.

The third boundary condition is that when $x = L$, $dy/dx = 0$. Thus from Eq. (*4*) we have

$$0 = -M_1L + \frac{R_1L^2}{2} - M_0b \tag{6}$$

The fourth and last boundary condition is that when $x = L$, $y = 0$. From Eq. (*5*) we obtain

$$0 = -\frac{M_1}{2}L^2 + \frac{R_1}{2}\cdot\frac{L^3}{6} - M_0\frac{b^2}{2} \tag{7}$$

It is now possible to solve Eqs. (*1*), (*6*), and (*7*) simultaneously to obtain the desired reactions

$$R_1 = \frac{6M_0ab}{L^3}$$

$$M_1 = \frac{M_0(2ab-b^2)}{L^2} \tag{8}$$

$$M_2 = \frac{M_0(2ab-a^2)}{L^2}$$

There may have been a temptation to say that the deflection under the point of application of the couple, at B, is zero. There is no reason for making such an assumption and, in fact, we may now return to the deflection Eq. (*5*) and calculate the deflection at $x = a$ and find that it is

$$EI[y]_{x=a} = \frac{M_0a^2(2ab-b^2)}{2L^2} + \frac{M_0a^4b}{L^3} \tag{9}$$

which is clearly nonzero.

11.5. The horizontal beam shown in Fig. 11-9(*a*) is simply supported at the ends and is connected to a composite elastic vertical rod at its midpoint. The supports of the beam and the top of the copper rod are originally at the same elevation, at which time the beam is horizontal. The temperature of both vertical rods is then decreased 40°C. Find the stress in each of the vertical rods. Neglect the weight of the beam and of the rods. The cross-sectional area of the copper rod is $500\ \text{mm}^2$, $E_{cu} = 100$ GPa, and $\alpha_{cu} = 20 \times 10^{-6}$/°C. The cross-sectional area of the aluminum rod is $1000\ \text{mm}^2$, $E_{al} = 70$ GPa, and $\alpha_{al} = 25 \times 10^{-6}$/°C. For the beam, $E = 10$ GPa and $I = 400 \times 10^6\ \text{mm}^4$.

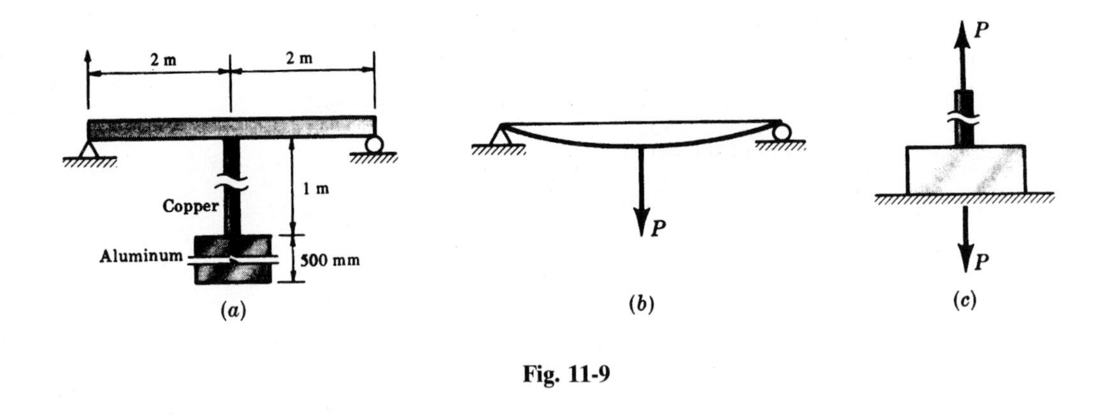

Fig. 11-9

A free-body diagram of the horizontal beam appears as in Fig. 11-9(*b*). Here, P denotes the force exerted upon the beam by the copper rod. Since this force is initially unknown, there are three forces acting upon the beam, but only two equations of equilibrium for a parallel force system; hence the problem is statically indeterminate. It will thus be necessary to consider the deformations of the system.

A free-body diagram of the two vertical rods appears as in Fig. 11-9(*c*). The simplest procedure is temporarily to cut the connection between the beam and the copper rod, and then allow the vertical rods to contract freely because of the decrease in temperature. If the horizontal beam offers no restraint, the copper rod will contract an amount

$$\Delta_{cu} = (20 \times 10^{-6})(10^3)(40) = 0.8 \text{ mm}$$

and the aluminum rod will contract by an amount

$$\Delta_{al} = (25 \times 10^{-6})(500)(40) = 0.5 \text{ mm}$$

However, the beam exerts a tensile force P upon the copper rod and the same force acts in the aluminum rod as shown in Fig. 11-9(*c*). These axial forces elongate the vertical rods and this elongation (see Problem 1.1) is

$$\frac{P(10^3)(10^6)}{500(100 \times 10^9)} + \frac{P(500)(10^6)}{10^3(70 \times 10^6)}$$

The downward force P exerted by the copper rod upon the horizontal beam causes a vertical deflection of the beam. In Problem 9.12 this central deflection was found to be $\Delta = PL^3/48EI$.

Actually, of course, the connection between the copper rod and the horizontal beam is not cut in the true problem and we realize that the resultant shortening of the vertical rods is exactly equal to the downward vertical deflection of the midpoint of the beam. This change of length of the vertical rods is caused partially by the decrease in temperature and partially by the axial force acting in the rods. For the shortening of the rods to be equal to the deflection of the beam we must have

$$(0.8 + 0.5) - \left[\frac{P(10^3)(10^6)}{500(100 \times 10^9)} + \frac{P(500)(10^6)}{10^3(70 \times 10^9)}\right] = \frac{P(4 \times 10^3)^3(10^6)}{48(10 \times 10^9)(400 \times 10^6)}$$

Solving, $P = 3.61$ kN; then,

$$\sigma_{cu} = 3.61 \times 10^3/500 = 7.22 \text{ MPa} \quad \text{and} \quad \sigma_{al} = 3.61 \times 10^3/1000 = 3.61 \text{ MPa}$$

11.6. The beam of flexural rigidity EI shown in Fig. 11-10 is clamped at both ends and subjected to a uniformly distributed load extending along the region BC of length $0.6L$. Determine all reactions.

At end A as well as C the supporting walls exert bending moments M_A and M_C plus shearing forces R_A and R_C as shown. For such a plane, parallel force system there are two equations of static equilibrium

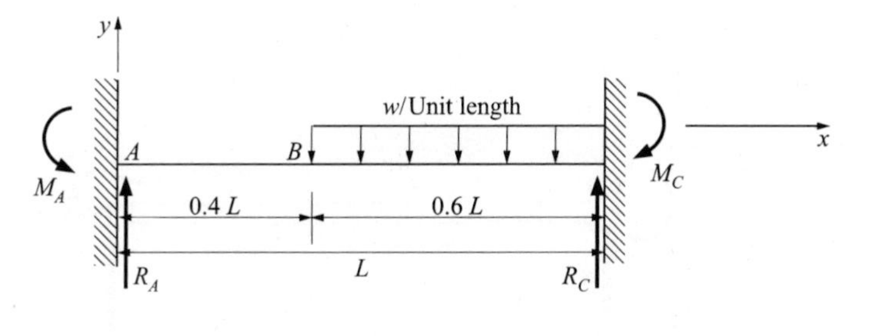

Fig. 11-10

and we must supplement these equations with additional relations stemming from beam deformations. The bending moment along the length ABC is conveniently written in terms of singularity functions:

$$EI\frac{d^2y}{dx^2} = -M_A\langle x\rangle^0 + R_A\langle x\rangle - \frac{w\langle x-0.4L\rangle^2}{2} \tag{1}$$

Integrating,

$$EI\frac{dy}{dx} = -M_A\langle x\rangle^1 + R_A\frac{\langle x\rangle^2}{2} - \frac{w}{2}\frac{\langle x-0.4L\rangle^3}{3} + C_1 \tag{2}$$

where C_1 is a constant of integration. As the first boundary condition, we have: when $x = 0$, the slope $dy/dx = 0$. Substituting in Eq. (2), we have

$$0 = -0 + 0 - 0 + C_1 \qquad \text{for} \qquad C_1 = 0$$

As the second boundary condition, when $x = L$, $dy/dx = 0$. Substituting in Eq. (2), we find

$$0 = -M_A L + \frac{R_A L^3}{2} - \frac{w}{6}(0.6L)^3 \tag{3}$$

Next, integrating Eq. (2), we find

$$EIy = -M_A\frac{\langle x\rangle^2}{2} + \frac{R_A}{2}\frac{\langle x\rangle^3}{3} - \frac{w}{6}\frac{\langle x-0.4L\rangle^4}{4} + C_2 \tag{4}$$

The third boundary condition is: when $x = 0$, $y = 0$, so from Eq. (4) we have $C_2 = 0$. The fourth boundary condition is: when $x = L$, $y = 0$, so from Eq. (4) we have

$$0 = -\frac{M_A L^2}{2} + \frac{R_A L^3}{6} - \frac{w}{24}\langle 0.6L\rangle^4 \tag{5}$$

The expressions for M_A given in Eqs. (3) and (5) may now be equated to obtain a single equation containing R_A as an unknown. Solving this equation, we find

$$R_A = wL\left\{(0.6)^3 - \frac{(0.6)^4}{2}\right\}$$

$$= 0.1512wL$$

Substituting this value in Eq. (3), we find $M_A = 0.0396wL^2$.

From statics we have

$$\Sigma F_y = -(0.6L)w + 0.1512wL + R_C = 0 \qquad \therefore R_C = 0.4488wL$$

and

$$+\circlearrowright \Sigma M_A = -0.0396wL^2 - M_C + (0.4488wL)(L) - [w(0.6L)](0.7L) = 0$$

$$\therefore M_C = 0.0684wL^2$$

11.7. The beam in Fig. 11-11 of flexural rigidity EI is clamped at A, supported between knife edges at B, and loaded by a vertical force P at the unsupported tip C. Determine the deflection at C.

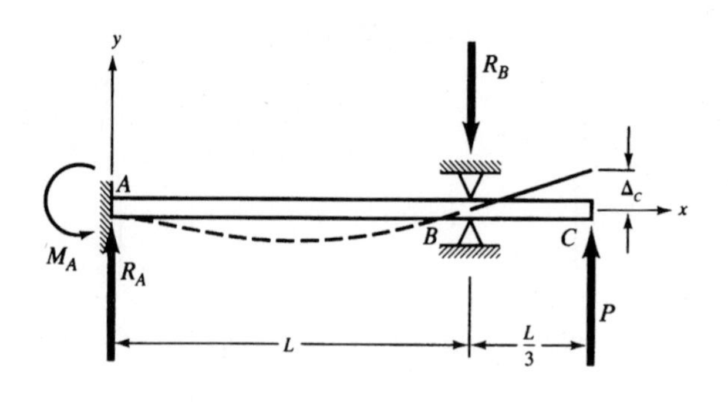

Fig. 11-11

The reactions at A are the moment M_A and shear force R_A as shown in Fig. 11-11. From statics we have

$$+\circlearrowleft\ \Sigma M_A = M_A + P\left(\frac{4L}{3}\right) - R_B(L) = 0 \tag{1}$$

$$\Sigma F_y = R_A + P - R_B = 0 \tag{2}$$

These two equations contain the three unknowns M_A, R_A, and R_B. Thus, we must supplement these two statics equations with another equation arising from deformation of the beam. Using the x-y coordinate system shown, the differential equation of the deformed beam in terms of singularity functions is

$$EI\frac{d^2y}{dx^2} = -M_A\langle x\rangle^0 + R_A\langle x\rangle^1 - R_B\langle x - L\rangle^1 \tag{3}$$

The first integration yields

$$EI\frac{dy}{dx} = -M_A\langle x\rangle^1 + R_A\frac{\langle x\rangle^2}{2} - R_B\frac{\langle x - L\rangle^2}{2} + C_1 \tag{4}$$

where C_1 is a constant of integration. The first boundary condition is that when $x = 0$, $dy/dx = 0$; hence from (*4*), $C_1 = 0$. The next integration yields

$$EIy = -M_A\frac{\langle x\rangle^2}{2} + \frac{R_A}{2}\frac{\langle x\rangle^3}{2} - \frac{R_B}{2}\frac{\langle x - L\rangle^3}{3} + C_2 \tag{5}$$

where the constant C_2 is determined from the second boundary condition $x = 0$, $y = 0$, leading to $C_2 = 0$. The third boundary condition arises from the fact that there is no deflection at B; that is, when $x = L$, $y = 0$. Substituting in Eq. (*5*), we find

$$0 = -\frac{M_A L^2}{2} + \frac{R_A L^3}{6} - 0 \tag{6}$$

Solving Eqs. (*1*), (*2*), and (*6*) simultaneously, we have

$$R_A = \frac{3M_A}{L} = \frac{P}{2} \qquad M_A = \frac{PL}{6} \qquad R_B = \frac{3P}{2} \tag{7}$$

If we now introduce these values into Eq. (*5*) and also set $x = 4L/3$ (point C), we have

$$EI\Delta_C = 0.0401PL^3 \tag{8}$$

11.8. In Problem 11.7 if the beam is a $W6 \times 15\frac{1}{2}$ steel wide-flange section of length 10 ft, determine the force P required to deflect the tip C 0.2 in.

From Eq. (*8*) of Problem 11.7, we have the tip deflection Δ_C as

$$EI\Delta_C = 0.0401PL^3$$

For this structural shape, we have from Table 8-1 that $I = 28.1\ \text{in}^4$. Substituting

$$\left(30 \times 10^6 \frac{\text{lb}}{\text{in}^2}\right)(28.1\ \text{in}^4)(0.2\ \text{in}) = 0.0401P(120\ \text{in})^3$$

Solving, $P = 2430$ lb.

11.9. The beam of flexural rigidity EI in Fig. 11-12 is clamped at end A, supported at C, and loaded by the couple at B together with the load uniformly distributed over the region BC. Determine all reactions.

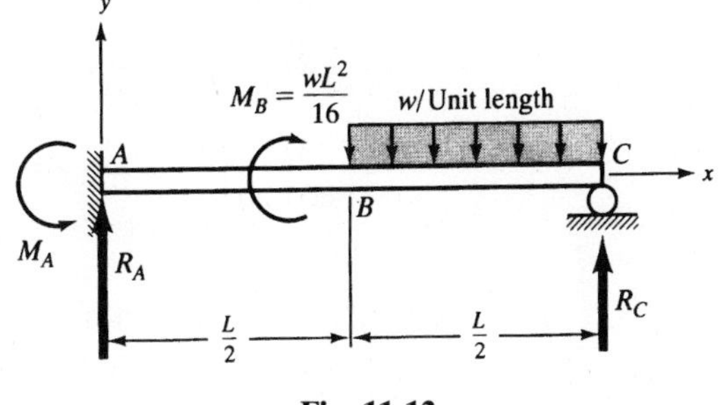

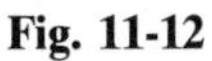
Fig. 11-12

The reactions at the left support A consist of the moment M_A plus the shear force R_A. From statics, for this parallel force system, we have two equations of equilibrium

$$+\circlearrowright\ \Sigma M_A = M_A - \frac{wL^2}{16} - \left(w\frac{L}{2}\right)\left(\frac{3L}{4}\right) + R_C(L) = 0 \tag{1}$$

$$\Sigma F_y = R_A + R_C - \frac{wL}{2} = 0 \tag{2}$$

These two equations contain the three unknowns M_A, R_A, and R_C. Accordingly we must supplement the two statics equations with another equation stemming from deformations of the system.

For the x-y coordinate system shown, the differential equation of the bent beam written in terms of singularity functions is

$$EI\frac{d^2y}{dx^2} = -M_A\langle x\rangle^0 + R_A\langle x\rangle^1 + M_B\left\langle x - \frac{L}{2}\right\rangle^0 - \frac{w}{2}\left\langle x - \frac{L}{2}\right\rangle^2 \tag{3}$$

Integrating the first time, this becomes

$$EI\frac{dy}{dx} = -M_A\langle x\rangle^1 + R_A\frac{\langle x\rangle^2}{2} + M_B\left\langle x - \frac{L}{2}\right\rangle - \frac{w}{2}\frac{\langle x - L/2\rangle^3}{3} + C_1 \tag{4}$$

where C_1 is a constant of integration. As the first boundary condition, when $x = 0$, $dy/dx = 0$. Substituting these values in Eq. (4), we find $C_1 = 0$. Integrating the second time, we find

$$EIy = -M_A\frac{\langle x\rangle^2}{2} + \frac{R_A}{2}\frac{\langle x\rangle^3}{3} + M_B\frac{\langle x - L/2\rangle^2}{2} - \frac{w}{6}\frac{\langle x - L/2\rangle^4}{4} + C_2 \tag{5}$$

where C_2 is the second constant of integration. As a second boundary condition, we have at point A, $x = 0$, $y = 0$, and so from Eq. (5) we see that $C_2 = 0$. The third boundary condition is that at point C when $x = L$, $y = 0$. Substituting these values in Eq. (5), we have

$$-\frac{M_AL^2}{2} + \frac{R_AL^3}{6} + \frac{M_B}{2}\cdot\frac{L^2}{4} - \frac{w}{24}\cdot\left(\frac{L}{2}\right)^4 = 0 \tag{6}$$

Solving Eqs. (*1*), (*2*), and (*6*) simultaneously, we find

$$M_A = \tfrac{3}{64}wL^2 \qquad R_A = \tfrac{7}{64}wL \qquad R_C = \tfrac{25}{64}wL \tag{7}$$

11.10. In Problem 11.9 if the beam is titanium having a Young's modulus of 110 GPa, with a rectangular cross section 20 mm × 30 mm, is 2 m long, and carries the uniform load in BC of 960 N/m, determine the deflection at the midpoint B.

From Eq. (*5*) of Problem 11.9 we have the deflection at the midpoint B as

$$EIy]_{x=L/2} = -\frac{M_A}{2}\left(\frac{L}{2}\right)^2 + \frac{R_A}{6}\left(\frac{L}{2}\right)^3$$

$$= -\frac{3}{64}wL^2\left(\frac{L^2}{8}\right) + \frac{7}{64}wL\left(\frac{L^3}{48}\right)$$

$$= -\frac{11}{(48)(64)}wL^4 = -0.00358wL^4 \tag{1}$$

For this beam

$$I = \tfrac{1}{12}(0.020\ \text{m})(0.030\ \text{m})^3 = 0.045 \times 10^{-6}\ \text{m}^4$$

so that Eq. (*1*) becomes

$$(110 \times 10^9\ \text{N/m}^2)(0.045 \times 10^{-6}\ \text{m}^4)[y]_{x=L/2} = -0.00358(960\ \text{N/m})(2\ \text{m})^4$$

Solving,

$$y]_{x=L/2} = -11.1\ \text{mm}$$

11.11. The beam AB of flexural rigidity EI is simply supported at A, rigidly clamped at end B, and subject to the load of uniformly varying intensity shown in Fig. 11-13. Determine the reactions developed at A and B by the use of the method of singularity functions.

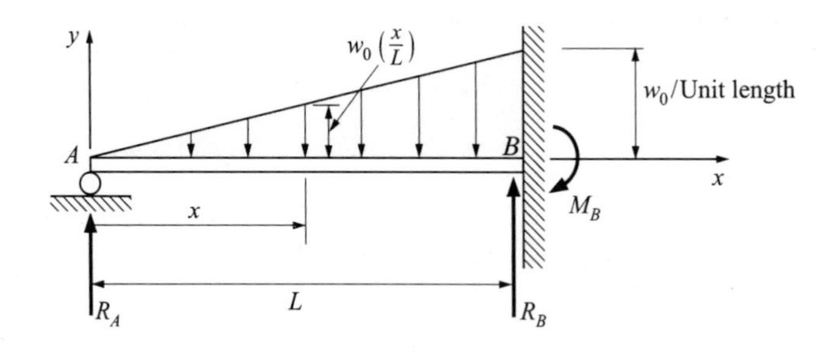

Fig. 11-13

Let us denote the vertical force reaction at A by R_A, that at B by R_B, and the moment exerted by the wall on the beam at B by M_B, as shown in Fig. 11-13. A related problem is 10.5 in this book. Following the procedure discussed there, we write the contribution to bending moment of the distributed loading at any point a distance x to the right of A:

$$M = R_A\langle x\rangle - w_0\left(\frac{x}{L}\right)(x)\left(\frac{1}{2}\right)\left(\frac{x}{3}\right)$$

Thus,

$$EI\frac{d^2y}{dx^2} = R_A\langle x\rangle - \frac{w_0\langle x\rangle^3}{6L} \tag{1}$$

Integrating the first time,

$$EI\frac{dy}{dx} = R_A\frac{\langle x\rangle^2}{2} - \frac{w_0}{6L}\cdot\frac{\langle x\rangle^4}{4} + C_1 \tag{2}$$

When $x = L$, $dy/dx = 0$, so from Eq. (2)

$$0 = R_A\frac{L^2}{2} - \frac{w_0L^3}{24} + C_1 \tag{3}$$

Integrating a second time,

$$EIy = \frac{R_A}{2}\frac{\langle x\rangle^3}{3} - \frac{w_0}{24L}\frac{\langle x\rangle^5}{5} + C_1x + C_2 \tag{4}$$

When $x = L$, $y = 0$, so we have from Eq. (4)

$$0 = \frac{R_AL^3}{6} - \frac{w_0L^4}{120} + C_1L + C_2 \tag{5}$$

Also, when $x = 0$, $y = 0$, so from Eq. (4), $C_2 = 0$.

From Eqs. (3) and (5) we have

$$C_1 = \frac{w_0L^3}{24} - \frac{R_AL^2}{2} = -\frac{R_AL^2}{6} + \frac{w_0L^3}{120} \tag{6}$$

Solving,

$$R_A = \tfrac{1}{10}w_0L \tag{7}$$

The two statics equations for such a force system are

$$\Sigma F_y = R_A + R_B - \frac{w_0L}{2} = 0$$

$$+\circlearrowleft \Sigma M_B = -R_AL - M_0 + \left(\frac{w_0}{2}\right)(L)\left(\frac{L}{3}\right) = 0$$

Solving,

$$R_B = \tfrac{2}{5}w_0L$$
$$M_B = \tfrac{1}{15}w_0L^2$$

11.12. The beam AC in Fig. 11-14 is rigidly clamped at both ends and loaded by a concentrated force P at point B. Determine all reactions, the deflection at B, and the maximum deflection occurring to the left of point B. Take $a > b$.

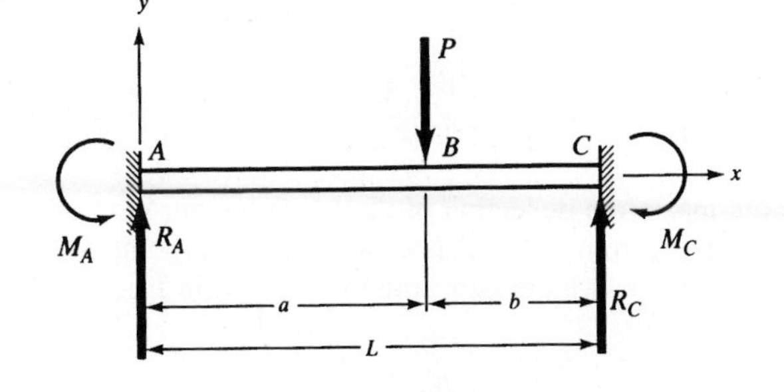

Fig. 11-14

The end moment and shear reactions are shown in Fig. 11-14. From statics we have the two equations

$$+\circlearrowleft \Sigma M_A = M_A - Pa + R_C L - M_C = 0 \tag{1}$$

$$\Sigma F_y = R_A + R_C - P = 0 \tag{2}$$

Next, writing the differential equation of the deflected beam in terms of singularity functions,

$$EI\frac{d^2y}{dx^2} = -M_A\langle x\rangle^0 + R_A\langle x\rangle^1 - P\langle x-a\rangle \tag{3}$$

Integrating the first time, we obtain

$$EI\frac{dy}{dx} = -M_A\langle x\rangle^1 + R_A\frac{\langle x\rangle^2}{2} - \frac{P\langle x-a\rangle^2}{2} + C_1 \tag{4}$$

As the first boundary condition, when $x = 0$, the slope $dy/dx = 0$. Substituting these values in Eq. (*4*), we obtain $C_1 = 0$.

Integrating again, we find

$$EIy = -M_A\frac{\langle x\rangle^2}{2} + \frac{R_A}{2}\frac{\langle x\rangle^3}{3} - \frac{P}{2}\frac{\langle x-a\rangle^3}{3} + C_2 \tag{5}$$

The second boundary condition is that when $x = 0$, $y = 0$. Substituting these values in Eq. (*5*), we find $C_2 = 0$. The equation of the deflected beam is consequently

$$EIy = -\frac{M_A}{2}\langle x\rangle^2 + \frac{R_A}{6}\langle x\rangle^3 - \frac{P}{6}\langle x-a\rangle^3 \tag{6}$$

Now, apply the boundary conditions at point C. The slope there is zero; hence from Eq. (*4*) we obtain the equation

$$-M_A L + \frac{R_A}{2}L^2 - \frac{Pb^2}{2} = 0 \tag{7}$$

The deflection $y = 0$ at $x = L$; hence we have from Eq. (*6*) the relation

$$-\frac{M_A}{2}L^2 + \frac{R_A}{6}L^3 - \frac{Pb^3}{6} = 0 \tag{8}$$

We may now solve Eqs. (*1*), (*2*), (*7*), and (*8*) simultaneously to find the end reactions

$$R_A = \frac{Pb^2}{L^3}(3a+b) \qquad M_A = \frac{Pab^2}{L^2}$$
$$R_C = \frac{Pa^2}{L^3}(a+3b) \qquad M_C = \frac{Pa^2b}{L^2} \tag{9}$$

The deflection at B under the point of application of the load P is found by setting $x = a$ in Eq. (*6*):

$$EI[y]_{x=a} = -\frac{M_A}{2}a^2 + \frac{R_A}{6}a^3 = -\frac{Pa^3b^3}{3L^3} \tag{10}$$

To determine the maximum deflection of the beam for our case of $a > b$, we consider the deflected bar as shown in Fig. 11-15, from which it is evident that the point of horizontal tangency to the beam occurs to the left of B; that is, we are concerned with $x < a$ in Eq. (*4*) so that the slope in region AB is given by

$$EI\frac{dy}{dx} = -M_A\langle x\rangle^1 + \frac{R_A}{2}\langle x\rangle^2 \tag{11}$$

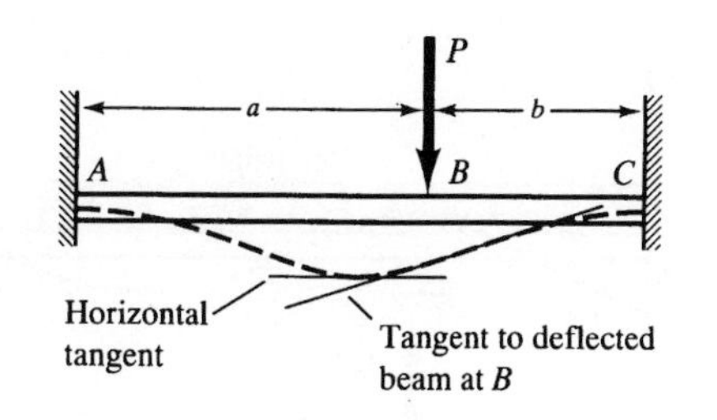

Fig. 11-15

which we set equal to zero to find the value x_1. This leads to a horizontal tangent at the value of x_1 given by

$$x_1 = \frac{2aL}{(3a+b)} \tag{12}$$

Substituting this x_1 in Eq. (*6*) and remembering that $x < a$, we obtain

$$EI[y]_{\max} = -\frac{2Pa^3b^2}{3(3a+b)^2} \tag{13}$$

11.13. In Problem 11.12 the beam has $a = 6$ ft, $b = 3$ ft, and is of circular cross section 2.5 in in diameter. The applied load is $P = 6000$ lb. Determine the deflection under the point of application of the load as well as the maximum deflection of the beam. Take $E = 30 \times 10^6$ lb/in^2.

The moment of inertia of the cross section is

$$I = \frac{\pi}{64}D^4 = \frac{\pi}{64}(2.5\text{ in})^4 = 1.917\text{ in}^4$$

The deflection under the point of application of the load is given by Eq. (*10*) of Problem 11.12 to be

$$y]_{x=a} = -\frac{Pa^3b^3}{3EIL^3}$$

Substituting,

$$y]_{x=a} = \frac{-(6000\text{ lb})(72\text{ in})^3(36\text{ in})^3}{3(30\times10^6\text{ lb/in}^2)(1.917\text{ in}^4)(108\text{ in})^3} = -0.480\text{ in}$$

The location of the point of maximum deflection is given by Eq. (*12*) to be

$$x_1 = \frac{2aL}{3a+b} = \frac{2(6\text{ ft})(9\text{ ft})}{18\text{ ft}+3\text{ ft}} = 5.14\text{ ft}$$

and the desired maximum deflection is found from Eq. (*13*) to be

$$\begin{aligned} y]_{\max} &= -\frac{2Pa^3b^2}{3(3a+b)^2EI} \\ &= \frac{2(6000\text{ lb})(72\text{ in})^3(36\text{ in})^3}{2[(3)(72\text{ in})+(36\text{ in})]^2(30\times10^6\text{ lb/in}^2)(1.917\text{ in}^4)} \\ &= -0.522\text{ in} \end{aligned}$$

11.14. The initially horizontal beam ABC in Fig. 11-16 is clamped at C and supported on a smooth roller at B. A uniform load w per unit length acts over the entire length of the beam. After

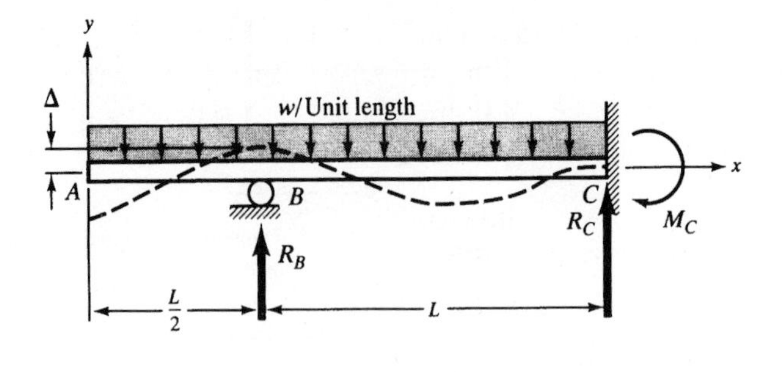

Fig. 11-16

application of the load, the reaction at B is mechanically displaced upward an amount Δ so that the beam then has the configuration shown by the dotted line. Determine the reaction R_B after this displacement has been imposed.

The beam reactions are R_B, R_C, and a moment M_C. Using the method of singularity functions, we have the equation of the bent beam,

$$EI\frac{d^2y}{dx^2} = -w\frac{\langle x\rangle^2}{2} + R_B\left\langle x - \frac{L}{2}\right\rangle \tag{1}$$

Integrating the first time, we obtain

$$EI\frac{dy}{dx} = -\frac{w}{2}\frac{\langle x\rangle^3}{3} + \frac{R_B}{2}\left\langle x - \frac{L}{2}\right\rangle^2 + C_1 \tag{2}$$

For the first boundary condition, we know that, when $x = 3L/2$, $dy/dx = 0$. Substituting in Eq. (2)

$$0 = -\frac{w}{6}\left(\frac{27L^3}{8}\right) + R_B\frac{L^2}{8} + C_1$$

from which

$$C_1 = \frac{9}{16}wL^3 - \frac{R_BL^2}{8} \tag{3}$$

Integrating a second time,

$$EIy = -\frac{w}{6}\frac{\langle x\rangle^4}{4} + \frac{R_B}{2}\frac{\langle x - L/2\rangle^3}{3} + \left(\frac{9}{16}wL^3 - R_B\frac{L^2}{8}\right)\langle x\rangle + C_2 \tag{4}$$

For the second boundary condition, when $x = 3L/2$, $y = 0$. Substituting in Eq. (4), we have

$$wL^4\left[-\frac{27}{(8)(16)} + \frac{27}{32}\right] + R_B\left[\frac{L^3}{6} - \frac{3L^3}{16}\right] + C_2 = 0$$

from which

$$C_2 = -\frac{81}{128}wL^4 + \frac{1}{48}R_BL^3$$

The third and last boundary condition stems from the imposed displacement at point b; that is, when $x = L/2$, $y = \Delta$. Substituting these values in Eq. (4), we have

$$EI\Delta = -\frac{w}{24}\left(\frac{L^4}{16}\right) + 0 + \left(\frac{9}{16}wL^3 - R_B\frac{L^2}{8}\right)\left(\frac{L}{2}\right) - \frac{81}{128}wL^4 + \frac{1}{48}R_BL^3$$

Solving for R_B, we obtain

$$R_B = \frac{3EI\Delta}{L^3} - 272w_0L$$

11.15. The horizontal beam AB shown in Fig. 11-17 is clamped at A, subjected to a uniformly distributed load w per unit length, and supported at B in such a manner that it is free to deflect vertically but is completely restrained against rotation at that point. Determine the vertical deflection at B after the beam has deflected as shown by the dotted line.

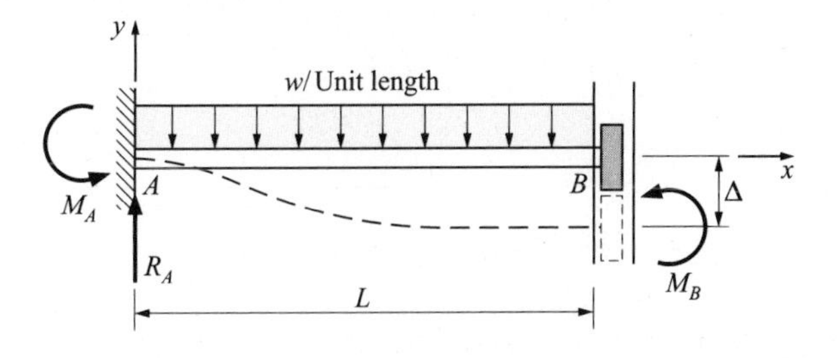

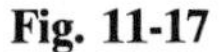
Fig. 11-17

The equation of the deflected beam is

$$EI\frac{d^2y}{dx^2} = -M_A\langle x\rangle^0 + R_A\langle x\rangle^1 - \frac{w\langle x\rangle^2}{2} \tag{1}$$

Integrating the first time, we find

$$EI\frac{dy}{dx} = -M_A\langle x\rangle^1 + R_A\frac{\langle x\rangle^2}{2} - \frac{w}{2}\frac{\langle x\rangle^3}{3} + C_1 \tag{2}$$

The first boundary condition is that when $x = 0$, $dy/dx = 0$. Substituting these values in Eq. (2), we find that $C_1 = 0$. Integrating again,

$$EIy = -M_A\frac{\langle x\rangle^2}{2} + \frac{R_A}{2}\frac{\langle x\rangle^3}{3} - \frac{w}{6}\frac{\langle x\rangle^4}{4} + C_2 \tag{3}$$

Imposing the boundary condition that $x = 0$ at $y = 0$, we have $C_2 = 0$.

The third boundary condition is that, when $x = L$, $dy/dx = 0$. Substituting these values in Eq. (2), we obtain the equation

$$0 = -M_A L + \frac{R_A L^2}{2} - \frac{wL^3}{6} \tag{4}$$

From statics, we have the two equilibrium equations

$$+\circlearrowleft\ \Sigma M_A = M_A + M_B - \frac{wL^2}{2} = 0 \tag{5}$$

$$\Sigma F_y = R_A - wL = 0 \tag{6}$$

Solving Eqs. (4), (5), and (6) simultaneously, we have

$$R_A = wL$$

$$M_A = \tfrac{2}{3}wL^2$$

$$M_B = \tfrac{1}{6}wL^2$$

Substitution of these values in Eq. (3) leads to

$$EI[y]_{x=L} = -\frac{wL^2}{3}\cdot\frac{L^2}{2} + \frac{wL}{2}\cdot\frac{L^3}{3} - \frac{wL^4}{24}$$

or

$$y]_{x=L} = -\frac{wL^4}{24EI}$$

11.16. The cantilever beam AB in Fig. 11-18 is clamped at B and supported through a hinge by a partially submerged (in water) pontoon at A. The beam is of flexural rigidity EI and length L. It is loaded by a vertical concentrated force F at A. Determine the reactive moment at B.

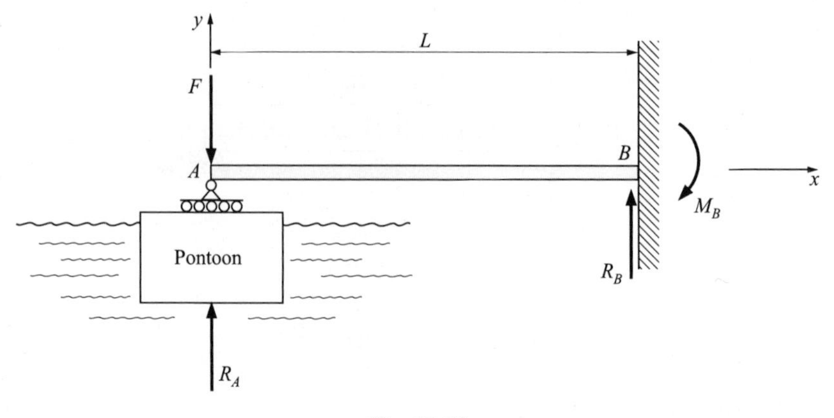

Fig. 11-18

When the force F is applied, the pontoon submerges a distance Δ. According to the law of Archimedes, the pontoon is buoyed up by a force R_A of magnitude equal to the weight of the additional water displaced during the movement through Δ. If the cross-sectional area of the pontoon is A_0 and the weight of the water per unit volume is γ, then

$$A_0 \Delta \gamma = -R_A \tag{1}$$

For the coordinate system shown in Fig. 11-18, we have

$$EI\frac{d^2y}{dx^2} = R_A x - Fx \tag{2}$$

Integrating the first time, we have

$$EI\frac{dy}{dx} = R_A\frac{x^2}{2} - F\frac{x^2}{2} + C_1 \tag{3}$$

As a boundary condition, we have $dy/dx = 0$ when $x = L$, so from Eq. (3)

$$C_1 = \frac{FL^2}{2} - \frac{R_A L^2}{2}$$

Integrating a second time,

$$EIy = \frac{R_A}{2}\cdot\frac{x^3}{3} = \frac{F}{2}\cdot\frac{x^3}{3} + \left(\frac{FL^2}{2} - \frac{R_A L^2}{2}\right)x + C_2 \tag{4}$$

As a second boundary condition we have $y = 0$ when $x = L$, so from Eq. (4)

$$C_2 = \frac{R_A L^3}{2} - \frac{FL^3}{3}$$

The equation of the deflected beam AB is thus

$$EIy = \frac{R_A}{6}x^3 - \frac{F}{6}x^3 + \left(\frac{FL^2}{2} - \frac{R_A L^2}{2}\right)x + \frac{R_A L^3}{3} - \frac{FL^3}{3} \tag{5}$$

We seek the deflection y at $x = 0$. From Eq. (1), it is

$$y = \Delta = -\frac{R_A}{A_0 \gamma}$$

Accordingly at $x = 0$ from Eq. (5) we have

$$-\frac{R_A EI}{A_0 \gamma} = \frac{R_A L^3}{3} - \frac{FL^3}{3}$$

Solving,

$$R_A = \frac{\left(\frac{FL^3}{3}\right)}{\frac{L^3}{3} + \frac{EI}{A_0 \gamma}} \tag{6}$$

From statics,

$$+\circlearrowright \Sigma M_B = R_A L - FL + M_B = 0 \tag{7}$$

Solving Eqs. (6) and (7) simultaneously we have

$$M_B = \frac{3FLEI}{L^3 A_0 \gamma + 3EI}$$

Supplementary Problems

11.17. A clamped-end beam is supported at the right end, clamped at the left, and carries the two concentrated forces shown in Fig. 11-19. Determine the reaction at the wall and the reaction at the right end of the beam.
Ans. $4P/3$ acting upward at left end, $PL/3$ acting counterclockwise at left end, $2P/3$ acting upward at right end

11.18. Determine the deflection under the point of application of the force P located a distance $L/3$ from the right end of the beam described in Problem 11.17. *Ans.* $7PL^3/486EI$

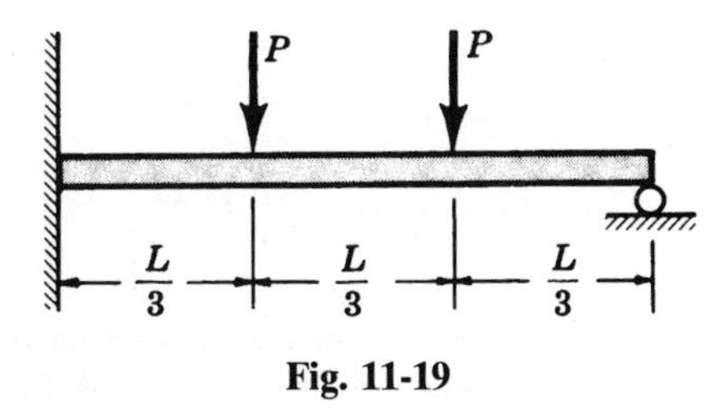

Fig. 11-19

11.19. The beam of Problem 11.17 is of titanium Ti-4Al-3Mo-IV (STA) with a tensile ultimate strength of 175,000 lb/in^2 at room temperature. If the cross section is 2 in × 5 in and a safety factor of 1.4 is employed, determine the maximum allowable value of each load P. *Ans.* 17,400 lb

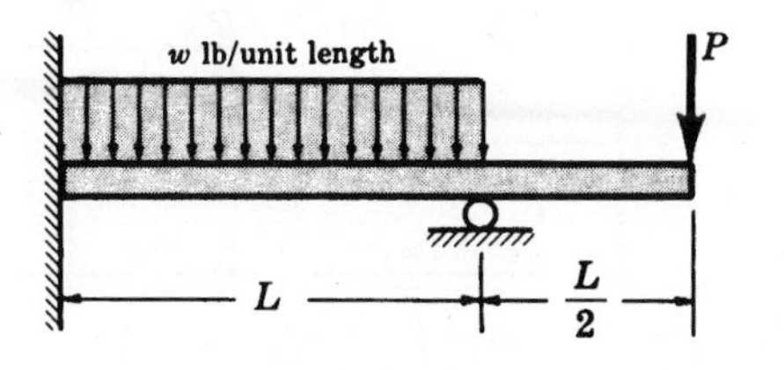

Fig. 11-20

11.20. A clamped-end beam is supported at an intermediate point and loaded as shown in Fig. 11-20. Determine the various reactions.
Ans. $\frac{5}{8}wL - \frac{3}{4}P$ upward at left end, $\frac{1}{8}wL^2 - \frac{1}{4}PL$ counterclockwise at left end, $\frac{3}{8}wL + \frac{7}{4}P$ upward at support

11.21. A clamped-end beam is supported at the right end, clamped at the left, and carries the load of uniformly varying intensity, as indicated in Fig. 11-21. Determine the moment exerted by the support on the beam. *Ans.* $7wL^2/120$

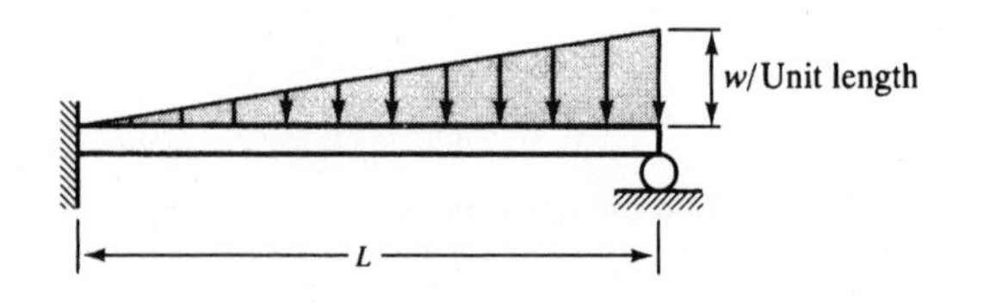

Fig. 11-21

11.22. The beam shown in Fig. 11-22 is clamped at the left end, supported at the right, and loaded by a couple M_0. Determine the reaction at the right support. *Ans.* $3M_0a(a + 2b)/2(a + b)^3$

11.23. For the beam shown in Fig. 11-22, determine the deflection under the point of application of the applied moment M_0. *Ans.* $M_0a^2b(a^2 - 2b^2)/4(a + b)^3 EI$

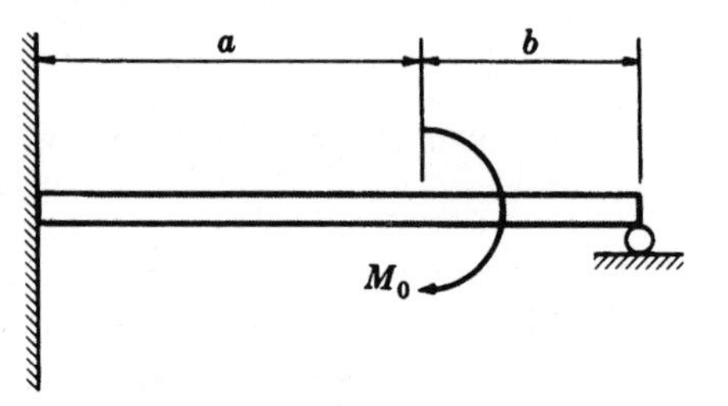

Fig. 11-22

11.24. In Fig. 11-23 AB and CD are cantilever beams with a roller E between their end points. A load of 5 kN is applied as shown. Both beams are made of steel for which $E = 200$ GPa. For beam AB, $I = 20 \times 10^6\ \text{mm}^4$; for CD, $I = 30 \times 10^6\ \text{mm}^4$. Find the reaction at E. *Ans.* 398 N

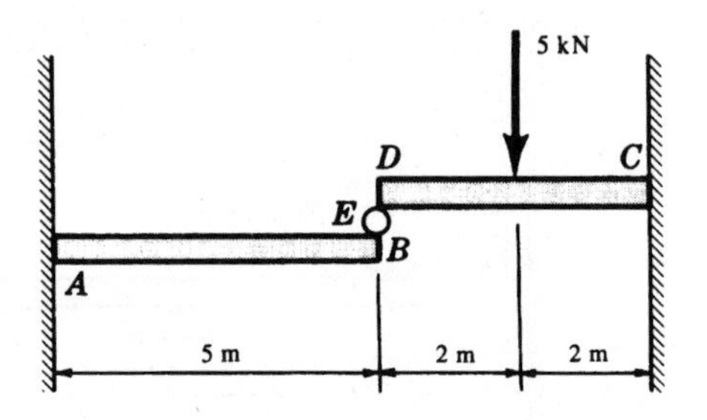

Fig. 11-23

11.25. The straight elastic beam AB in Fig. 11-24 is a W152 × 23 wide-flange section having $I = 11.7 \times 10^6$ mm^4. Member CD is a vertical steel wire of 3-mm-diameter circular cross section and length 4 m. Both the beam and the wire are steel for which $E = 200$ GPa. Prior to the application of any load to the beam, due to a fabrication error, the end D of the wire is 5 mm above the tip B of the beam. The end D of the wire and the tip B of the beam are then mechanically pulled together and joined. Determine the axial stress in the bar prior to the application of any load to the beam. *Ans.* 106 MPa

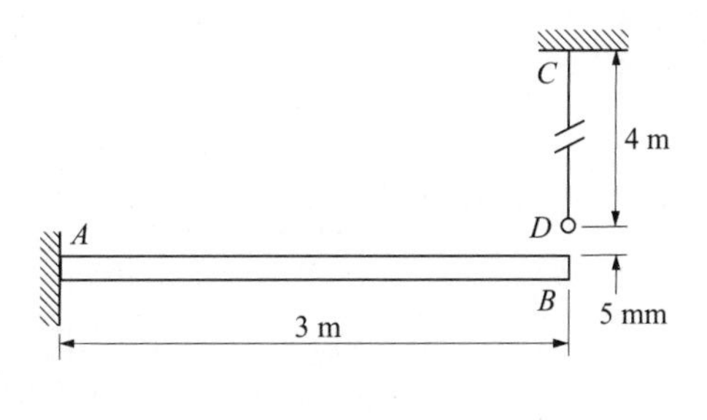

Fig. 11-24

11.26. A beam is clamped at both ends and supports a uniform load over its right half, as shown in Fig. 11-25. Determine all reactions.
Ans. $3wL/32$ acting upward at left end, $5wL^2/192$ acting counterclockwise at left end, $13wL/32$ acting upward at right end, $11wL^2/192$ acting clockwise at right end

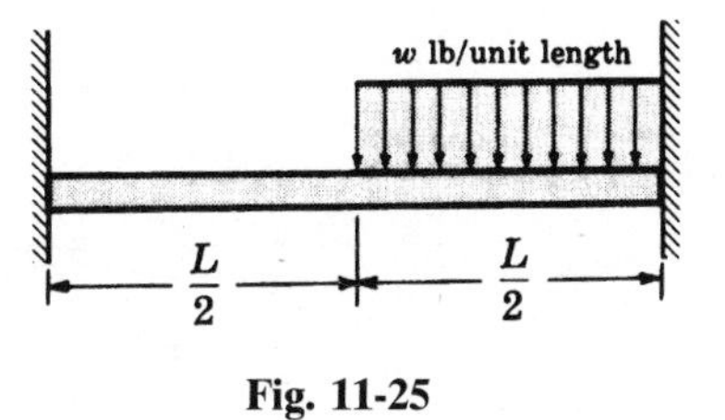

Fig. 11-25

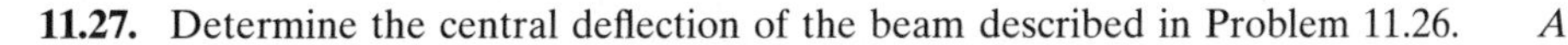
11.27. Determine the central deflection of the beam described in Problem 11.26. *Ans.* $wL^4/768EI$

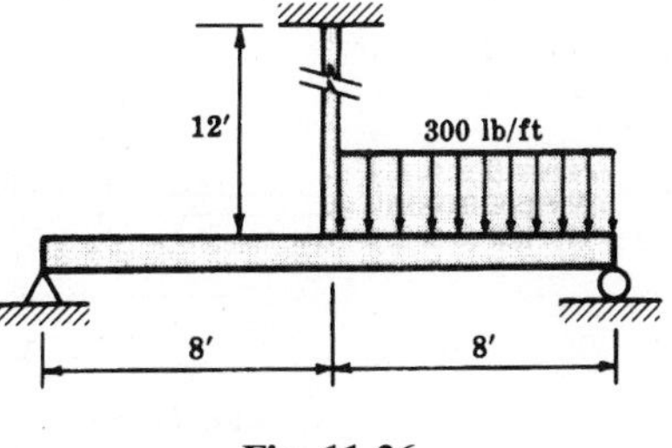

Fig. 11-26

11.28. A 16-ft beam carries a uniform load over the right half of its span and is supported at the center of the span by a vertical rod, as shown in Fig. 11-26. The rod is steel, 12 ft in length, 0.5 in^2 in cross-sectional area, and $E_s = 30 \times 10^6$ lb/in^2. The beam is wood 4 in × 8 in in cross section and $E_w = 1.5 \times 10^6$ lb/in^2. Determine the stress in the vertical steel rod. *Ans.* 2960 lb/in^2

11.29. The beam of flexural rigidity EI in Fig. 11-27 is clamped at A, supported between knife edges at B, and subjected to the couple M_0 at its unsupported tip C. Determine the deflection of point C.
Ans. $M_0L^2/4EI$

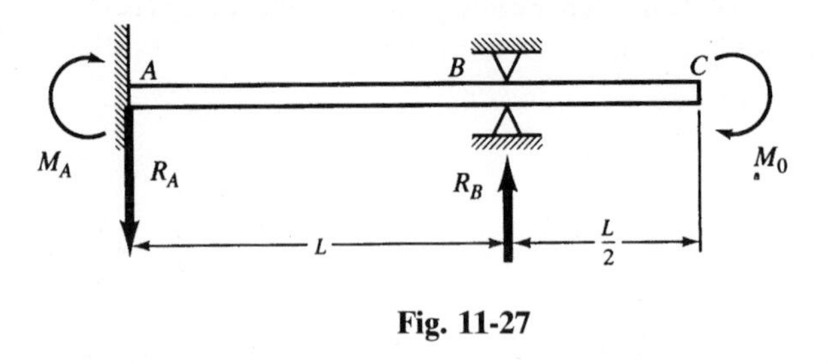

Fig. 11-27

11.30. The cantilever beam in Fig. 11-28 of length 3 m and rectangular cross section 100 mm × 200 mm has its free end (at no load) 3 mm above the top of a spring whose constant is 150 kN/m. The material is titanium alloy, which has $E = 110$ GPa and a yield point of 900 MPa. A downward force P of 7000 N is applied to the tip of the beam. Find the deformation of the top of the spring under this load. *Ans.* 4.72 mm

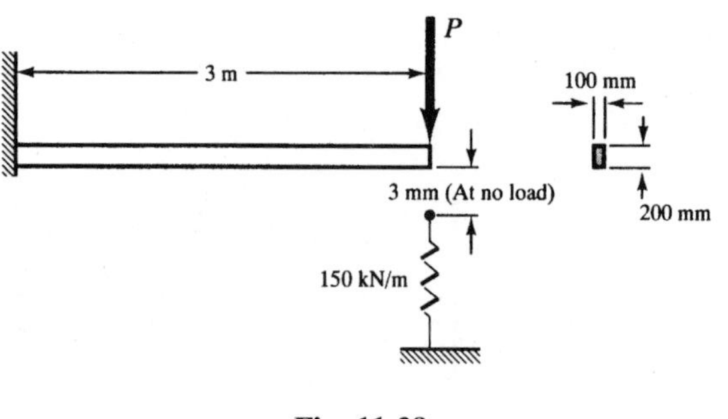

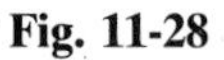

Fig. 11-28

11.31. A beam AB is clamped at each end and subject to a load of uniformly varying intensity as shown in Fig. 11-29. Determine the moment reactions developed at each end of the beam.
Ans. $wL^2/30$ counterclockwise at A, $wL^2/20$ clockwise at B

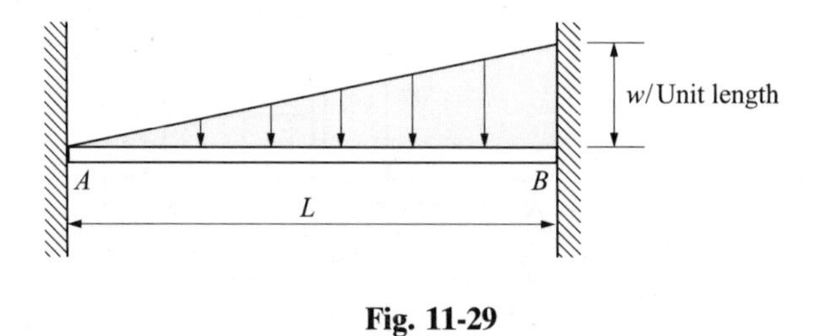

Fig. 11-29

11.32. The beam AB is pinned at its left end, clamped at the right end, and subjected to the uniformly varying vertical load shown in Fig. 11-30. Determine the vertical reaction at the support at A.

Ans. $R_1 = \dfrac{w_0L_2}{240L^3}\,[\{10L_2L_3 + 5L_3^2 + 15(L_2 + L_3)^2\}L - L_3^3 - 2L_3^2(L_2 + L_3) - 3L_3^2(L_2 + L_3) - 4(L_2 + L_3)^3]$

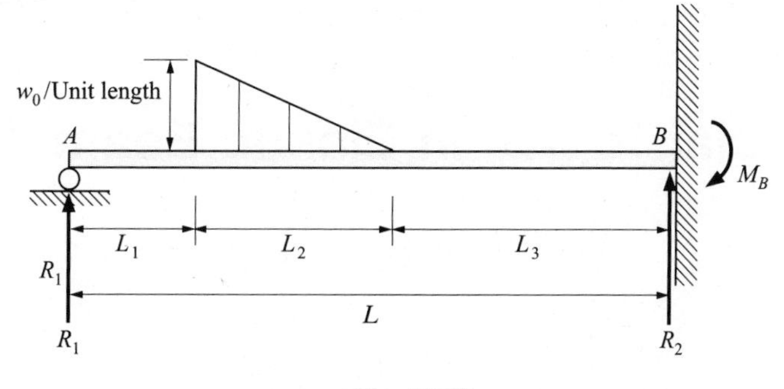

Fig. 11-30

11.33. The two-span continuous beam shown in Fig. 11-31 supports the two concentrated loads shown. Determine the various reactions. *Ans.* $R_A = 668$ lb, $R_B = 12{,}061$ lb, $R_C = 7271$ lb

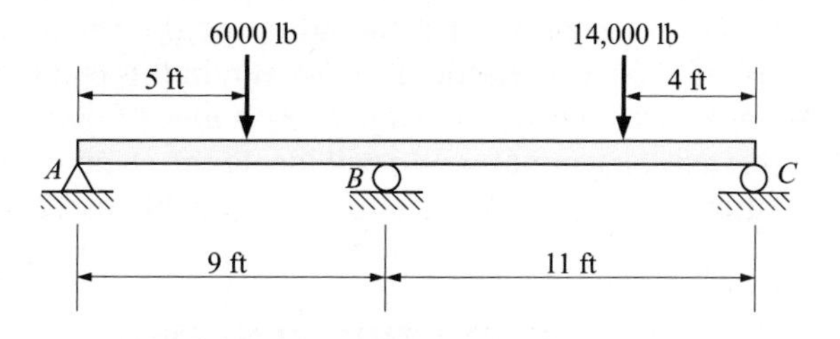

Fig. 11-31

11.34. The three-span continuous beam shown in Fig. 11-32 supports a uniformly distributed load in the left and central span, but is unloaded in the right span. Determine the reactions at A, B, C, and D.
Ans. $R_A = 0.383wL(\uparrow)$, $R_B = 1.20wL\ (\uparrow)$, $R_C = 0.450wL\ (\uparrow)$, $R_D = -0.033wL\ (\downarrow)$

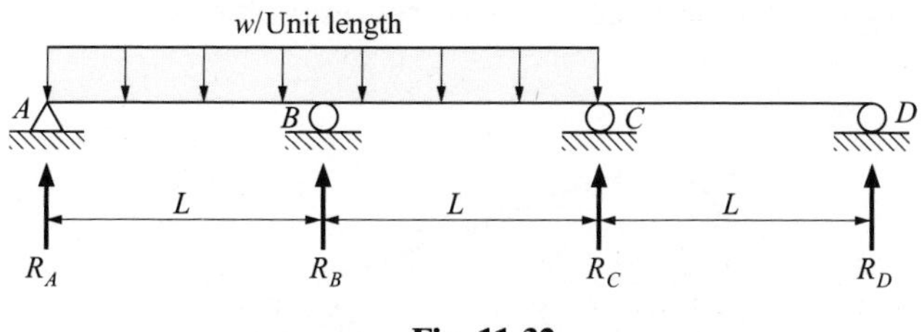

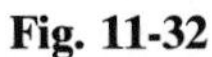
Fig. 11-32

11.35. The beam shown in Fig. 11-33 is simply supported at the left and right ends and spring supported at the center. Determine the spring constant so that the bending moment will be zero at the point where the spring supports the beam. *Ans.* $k = 16EI/L^3$

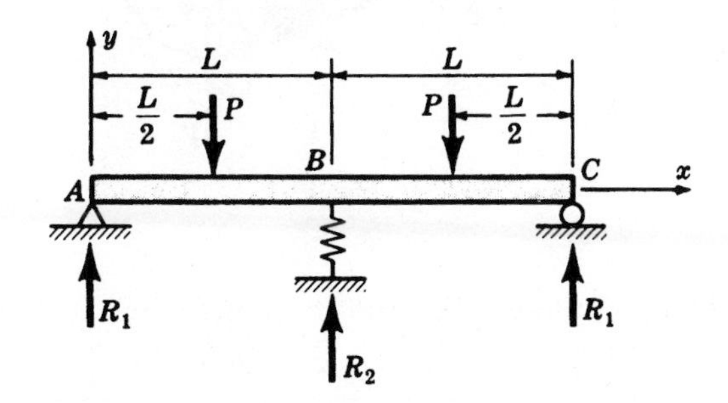

Fig. 11-33

Chapter 12

Special Topics in Elastic Beam Theory

SHEAR CENTER

The simple flexure formula $\sigma = My/I$ determined in Problem 8.1 is valid only if the transverse loads which give rise to bending act in a plane of symmetry of the beam cross section. In this type of loading there is obviously no torsion of the beam. However, in more general cases the beam cross section will have no axes of symmetry and the problem of where to apply transverse loads so that the action is entirely bending with no torsion arises. Every elastic beam cross section possesses a point through which transverse forces may be applied so as to produce *bending only* with *no torsion* of the beam. This point is called the *shear center*. In general, determination of the shear center location is extremely difficult and requires use of the theory of elasticity. However, in this chapter we will be concerned only with beams of *thin-walled open* cross section having a *single axis of symmetry*, with the loads acting in a plane perpendicular to this axis of symmetry. We will locate the shear center of the open cross section on the axis of symmetry of the beam. For applications, see Problems 12.1 through 12.4.

UNSYMMETRIC BENDING

Frequently beams are of unsymmetric cross section, or even if the cross section is symmetric the plane of the applied loads may not be one of the planes of symmetry. In either of these cases the expression $\sigma = My/I$ derived in Problem 8.1 is not valid for determination of the bending stress. It is convenient to resolve the bending moment into components along the y- and z-axes of the cross section, as indicated by the double-headed vector representations of these moments in Fig. 12-1.

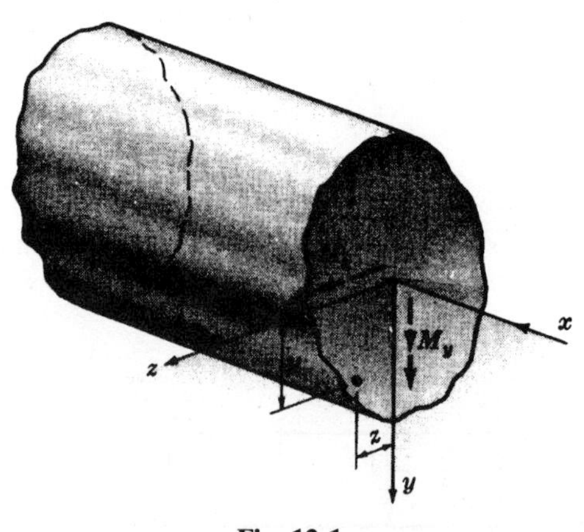

Fig. 12-1

The bending stress at a point located by the coordinates y, z is shown in Problem 12.5 to be

$$\sigma = \frac{(M_z I_y + M_y I_{yz})y + (-M_y I_z - M_z I_{yz})z}{I_y I_z - I_{yz}^2} \tag{12.1}$$

where I_y and I_z denote the moments of inertia about the y- and z-axes, respectively, and I_{yz} is the product of inertia. These quantities are determined by the methods of Chap. 7. There exists a *neutral axis* and those longitudinal fibers lying on the neutral axis are not subject to any normal stress. However, the neutral axis is usually not perpendicular to the plane of the applied loads nor does it coincide with either of the principal axes. For applications, see Problems 12.6 and 12.7. A computerized approach for determination of bending stresses is offered in Problem 12.8 and examples are offered in Problems 12.9 and 12.10.

CURVED BEAMS

Occasionally initially curved beams are encountered in machine design and other areas. Here we consider only those elastic beams for which the plane of curvature is also a plane of symmetry of every cross section and the bending loads act in this plane of symmetry. Unlike the case of the initially straight beam, the neutral axis no longer passes through the centroid of the cross section but instead shifts toward the center of curvature of the beam by a distance denoted by $\bar{y}$. The bending stress distribution over the cross section is hyperbolic in nature and in Problem 12.11 it is shown that these stresses are given by

$$\sigma = \frac{My}{A\bar{y}(r+y)} \tag{12.2}$$

where M is the bending moment, A is the cross-sectional area, r is the radius of curvature of the neutral axis, and y denotes the distance of any fiber from the neutral axis. For applications see Problem 12.12. Because of the tedious nature of calculations associated with bending of curved beams, the problem is well suited to computer implementation and a FORTRAN program is developed in Problem 12.13 together with examples in Problems 12.14 and 12.15.

Solved Problems

Shear Center

12.1. Determine the shear center of half of a thin-walled cylindrical section oriented as shown in Fig. 12-2 and subject to a vertical load.

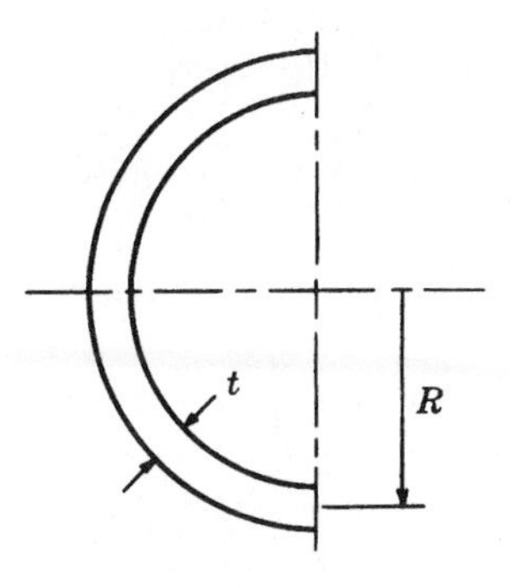

Fig. 12-2

Since the beam action is one of bending only with no torsion, it follows that normal stresses are distributed over the cross section in accordance with the flexure formula $\sigma = My/I$. Consequently, according to Problem 8.19, page 198, horizontal shearing stresses acting perpendicular to the plane of the cross section are generated and are determined by the relation

$$\tau = \frac{V}{Ib}\int_{y_0}^{c} y\,da$$

As indicated in Problem 8.19, the presence of these horizontal shearing stresses necessitates the presence of equal intensity shear stresses acting over the vertical cross section. In Fig. 12-3(a) these shear stresses have been shown as acting tangential to the center line of the cross section and further, for a thin-walled section, it is customary to assume a uniform distribution of the shear stresses across the thickness t. Finally, it is assumed that shearing stresses perpendicular to the circular centerline of the section are negligible. In Fig. 12-3(a), V denotes the resultant of the distributed shearing stresses and it, of course, acts vertically, since the horizontal components of the various stress vectors above and below the axis of symmetry annul one another.

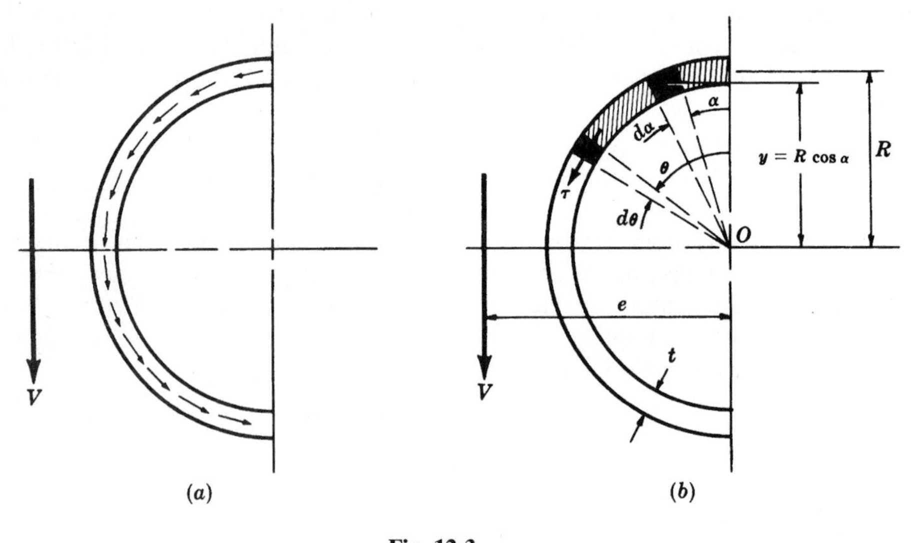

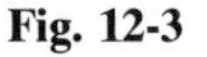
Fig. 12-3

Let us examine the shearing stress τ at an arbitrary point denoted by the angle θ, as indicated in Fig. 12-3(b). Determination of this stress from the relation

$$\tau = \frac{V}{Ib}\int_{y_0}^{c} y\,da \tag{a}$$

necessitates evaluation of I as well as the integral, which, as explained in Chap. 7, represents the first moment of the shaded area about the axis of symmetry. This is accomplished by introducing an auxiliary variable α $(0 < \alpha < \theta)$ as shown in Fig. 12-3(b) so that

$$\int_{y_0}^{c} y\,da = \int_{0}^{\theta} (R\cos\alpha)t(R\,d\alpha) = R^2 t \sin\theta \tag{b}$$

Next, the moment of inertia of the entire cross section about the axis of symmetry is given by

$$I = \int y^2\,da = \int_{0}^{\pi} (R\cos\theta)^2 tR\,d\theta = \frac{\pi R^3 t}{2} \tag{c}$$

The shearing stress at any point represented by θ is now found from (a), (b), and (c) to be

$$\tau = \frac{V}{(\pi R^3 t/2)t}[R^2 t\sin\theta] = \frac{2V}{\pi R t}\sin\theta \tag{d}$$

The moment of these distributed shearing stresses about any point, say O, must be equal to the moment of the resultant V about that same point. Thus since τ acts over an area $t(R\,d\theta)$ we have

$$\int_{\theta=0}^{\theta=\pi}\left(\frac{2V}{\pi Rt}\sin\theta\right)(Rt\,d\theta)R = Ve$$

Thus $$e = \frac{4R}{\pi}$$

gives the location of the shear center.

12.2. Determine the shear center of the "hat"-type thin-walled section indicated in Fig. 12-4. The thickness t is constant throughout the beam.

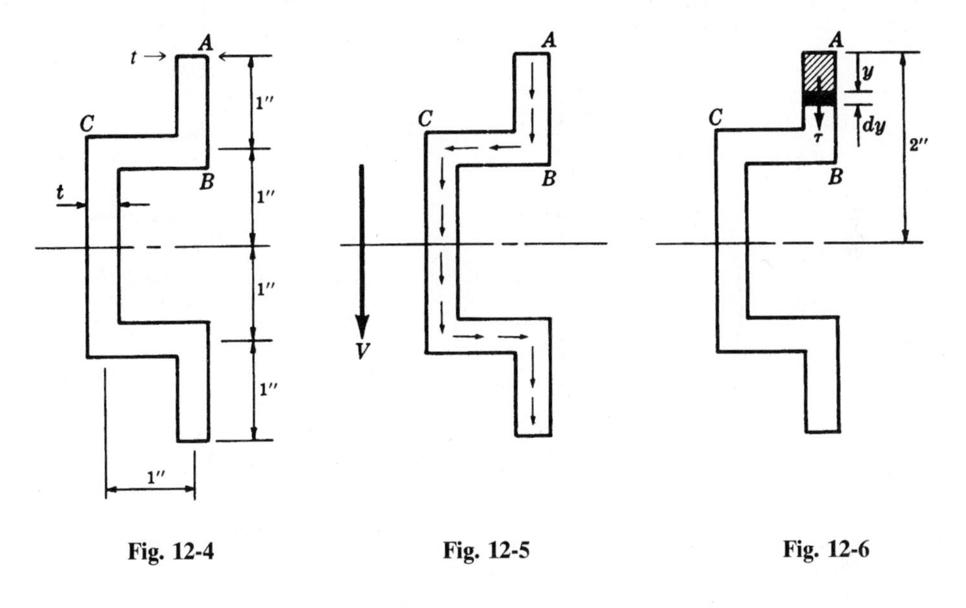

Fig. 12-4 Fig. 12-5 Fig. 12-6

In accordance with the reasoning given in Problem 12.1, the distribution of shear stresses over the cross section appears as in Fig. 12-5. The resultant of the distributed shearing stresses, denoted V, acts vertically because the net horizontal effect of the shearing stresses in the two horizontal portions of the "hat" is zero. Let us first examine the shearing stress in the upper vertical member AB. At a distance y below the extreme point A, as shown in Fig. 12-6, the shearing stress is given by

$$\tau = \frac{V}{It}\int_{y_0}^{c} y\,da \qquad (a)$$

The integral represents the first moment of the shaded area about the axis of symmetry and may be readily evaluated as the product of the area, that is, yt, and the distance from the centroid of the area to the axis of symmetry, that is, $2 - y/2$. The shear stress at y is thus

$$\tau = \frac{V}{It}\left(2 - \frac{y}{2}\right)yt \qquad (b)$$

where it is to be remembered that V and I pertain to the shear force acting over the *entire* cross section and the moment of inertia of the *entire* cross section, respectively. The resultant shear force V_1 acting over the vertical region AB, as indicated in Fig. 12-7, is found by integration to be

$$V_1 = \int_{y=0}^{y=1} \tau t\,dy = \frac{Vt}{I}\int_0^1\left(2y - \frac{y^2}{2}\right)dy = \frac{5}{6}\frac{Vt}{I} \qquad (c)$$

Let us next examine the shearing stress in the upper horizontal member BC. At a distance x from point B, as indicated in Fig. 12-8, the shearing stress is given by Eq. (a), where now the integral represents the first moment of the shaded area in Fig. 12-8 about the axis of symmetry. By inspection the integral has the value $(1)(t)(1.5) + (x)(t)(1)$ and the shear stress at x is thus

$$\tau = \frac{V}{It}[1.5t + xt] \qquad (d)$$

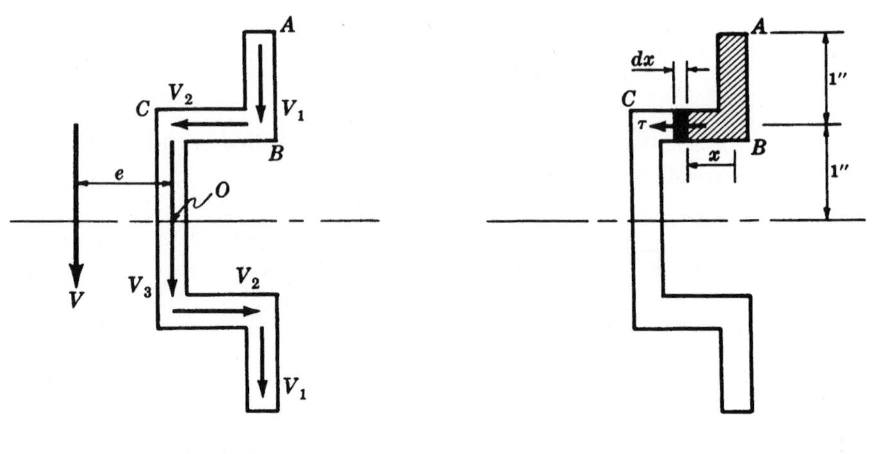

Fig. 12-7 **Fig. 12-8**

where V and I again pertain to the resultant shear over the *entire* cross section and the moment of inertia of the *entire* cross section, respectively. The resultant shear force V_2, as indicated in Fig. 12-7, is found to be

$$V_2 = \int_{x=0}^{x=1} \tau t\, dx = \frac{Vt}{I}\int_0^1 (1.5 + x)\, dx = \frac{3Vt}{2I} \qquad (e)$$

Since the entire section is thin walled it is customary to use only nominal dimensions and thus neglect any slight duplication of areas at the intersections of the various members.

Because of symmetry the forces on the lower members are identical to those just found. The sum of the moments of these forces about any point, such as O in Fig. 12-7, must equal the moment of the resultant V about that same point. Thus, we have $-2V_1(1) + 2V_2(1) = Ve$ or

$$e = \frac{4t}{3I} \qquad (f)$$

Finally, I may be calculated by the methods of Chap. 7 to be

$$I = \frac{1}{12}(t)(4)^3 + 2[(1)(t)(1)^2] = \frac{22t}{3} \qquad (g)$$

The shear center from (f) thus becomes

$$e = \frac{4t}{3(22t/3)} = \frac{4}{22} = 0.182 \text{ in} \qquad (h)$$

Note that by choosing the moment center at O it is not necessary to determine V_3.

12.3. Determine the shear center of a thin-walled rectangular section in which there is a narrow longitudinal slit (see Fig. 12-9). The thickness t is constant.

Observe that this section corresponds to the "hat" section of Problem 12-2 except that the outstanding flanges of the "hat" are turned toward the axis of symmetry here. The distribution of shear stresses appears

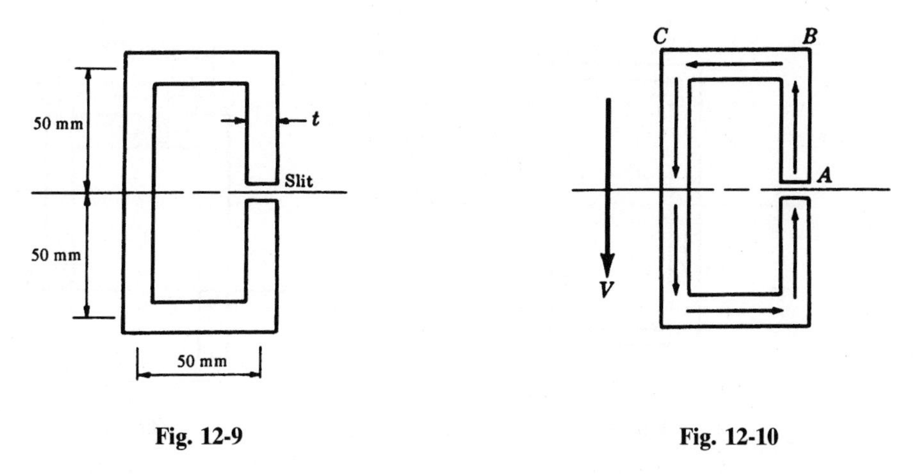

Fig. 12-9 **Fig. 12-10**

as indicated in Fig. 12-10 and the vertical force V denotes the resultant of these distributed shearing stresses. Let us first examine the shearing stress in the vertical member AB. See Fig. 12-11. At a distance z above the axis of symmetry (assuming the slit to be of negligible thickness) the shearing stress is again given by

$$\tau = \frac{V}{Ib} \int_{y_0}^{c} y\, da \tag{a}$$

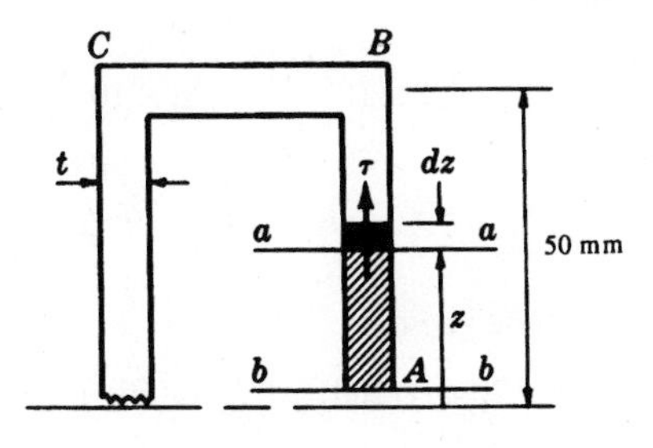

Fig. 12-11

where it is of utmost importance to observe that the integral represents the first moment of the area lying *between* the section a-a where the shear stress is desired and the extreme fibers b-b of the section. This is true even though fibers b-b lie *closer* to the axis of symmetry than a-a. This statement follows from the derivation of the above equation as given in Chap. 7. The integral is evaluated as the product of the area, that is, zt, and the distance from the centroid of the area to the axis of symmetry, that is, $z/2$. The shear stress at z is thus

$$\tau = \frac{V}{It}\left[zt\frac{z}{2}\right] = \frac{Vz^2}{2I} \tag{b}$$

The resultant shear force V_1 acting over the vertical region AB, indicated in Fig. 12-12, is found by integration to be

$$V_1 = \int_{z=0}^{z=50} \tau t\, dz = \int_0^{50} \frac{Vz^2}{2I} t\, dz = 2.08 \times 10^4 \frac{Vt}{I} \tag{c}$$

Let us next examine the shearing stress in the upper horizontal member BC. At a distance x from point B, as indicated in Fig. 12-13, the shearing stress is given by Eq. (a) where the integral represents the first moment of the shaded area in Fig. 12-13 about the axis of symmetry. From (a),

$$\tau = \frac{V}{It}[(x)(t)50 + (50)(t)(25)] = \frac{50V}{I}(x + 25) \tag{d}$$

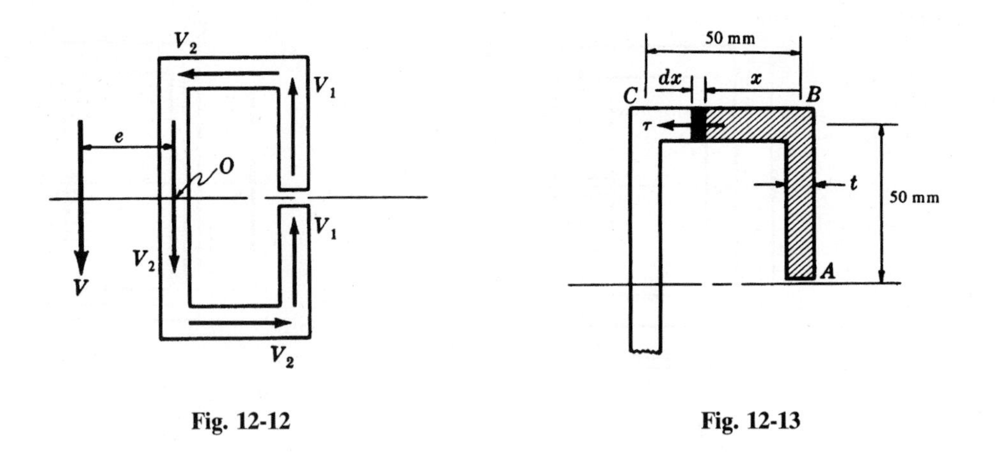

Fig. 12-12

Fig. 12-13

The resultant shear force V_2 acting over the horizontal member BC, as indicated in Fig. 12-12, is found by integration to be

$$V_2 = \int_{x=0}^{x=50} \tau t\, dx = \int_0^{50} \frac{50V}{I}(x+25)\, dx = 1.25 \times 10^5 \frac{Vt}{I} \tag{e}$$

From Fig. 12-12 the sum of the moments of the forces V_1, V_2, and V_3 about any point, such as O, must equal the moment of the resultant about that point. Thus $2(50V_1) + 2(50V_2) = Ve$.

Substituting from (c) and (e),

$$2.1 \times 10^5 \frac{Vt}{I} + 1.25 \times 10^7 \frac{Vt}{I} = Ve \tag{f}$$

$$e = 1.46 \times 10^7 \frac{t}{I} \tag{g}$$

The second moment of area is given by

$$I = 2[\tfrac{1}{12}(t)(100)^3] + 2[50t(50)^2] = 4.167 \times 10^5 t$$

Thus

$$e = \frac{1.46 \times 10^7 t}{4.167 \times 10^5 t} = 35 \text{ mm}$$

which locates the shear center.

12.4. Determine the shear center of the thin-walled section indicated in Fig. 12-14. The thickness t is constant.

The distribution of shear stresses appears as in Fig. 12-15 where the vertical force V denotes the resultant of these distributed shearing stresses. Let us first determine the shearing stress in the horizontal member AB. At a distance x from point A, as indicated in Fig. 12-16, the shearing stress is found to be

$$\tau = \frac{V}{Ib}\int_{y_0}^{c} y\, da \tag{a}$$

or

$$\tau = \frac{V}{It}[(x)(t)(3)] = \frac{3Vx}{I} \tag{b}$$

The resultant shear force V_1 acting over AB, as indicated in Fig. 12-17, is found by integration to be

$$V_1 = \int_{x=0}^{x=2} \tau t\, dx = \int_0^2 \frac{3V\,xt}{I}\, dx = \frac{6Vt}{I} \tag{c}$$

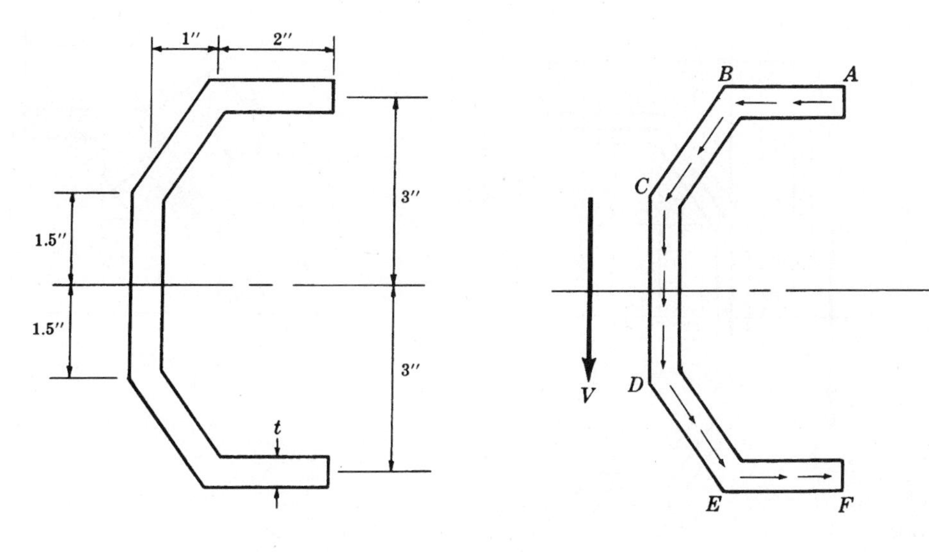

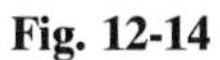
Fig. 12-14

Fig. 12-15

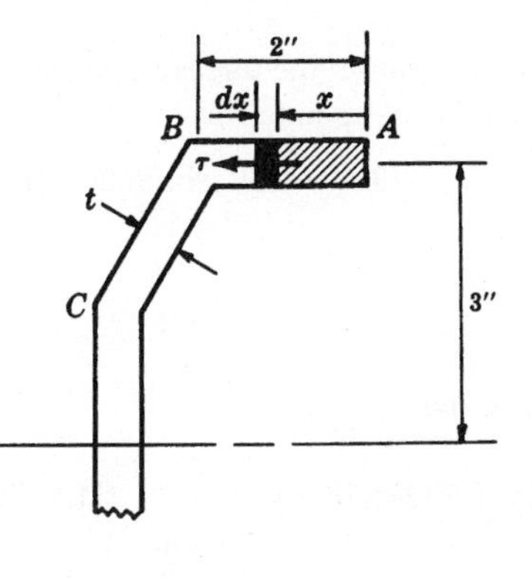

Fig. 12-16

The shearing stress in the inclined member BC at a distance y from point B, as indicated in Fig. 12-18, is again given by Eq. (a), where the integral represents the first moment of the shaded area in Fig. 12-18 about the axis of symmetry. For the inclined portion of that area, it is simplest to integrate through introduction of an auxiliary variable u as indicated. Thus

$$\tau = \frac{V}{It}\left[(2)(t)(3) + \int_{u=0}^{u=y} [1.5 + (1.80 - u)\sin 56^\circ 20']t\,du\right]$$

$$= \frac{V}{I}(6 + 3y - 0.416y^2) \qquad (d)$$

The resultant shear force V_2 acting over the inclined member BC in Fig. 12-17 is found by integration to be

$$V_2 = \int_{y=0}^{y=1.80} \tau t\,dy$$

$$= \int_0^{1.80} \frac{Vt}{I}(6 + 3y - 0.416y^3)\,dy = \frac{14.85Vt}{I} \qquad (e)$$

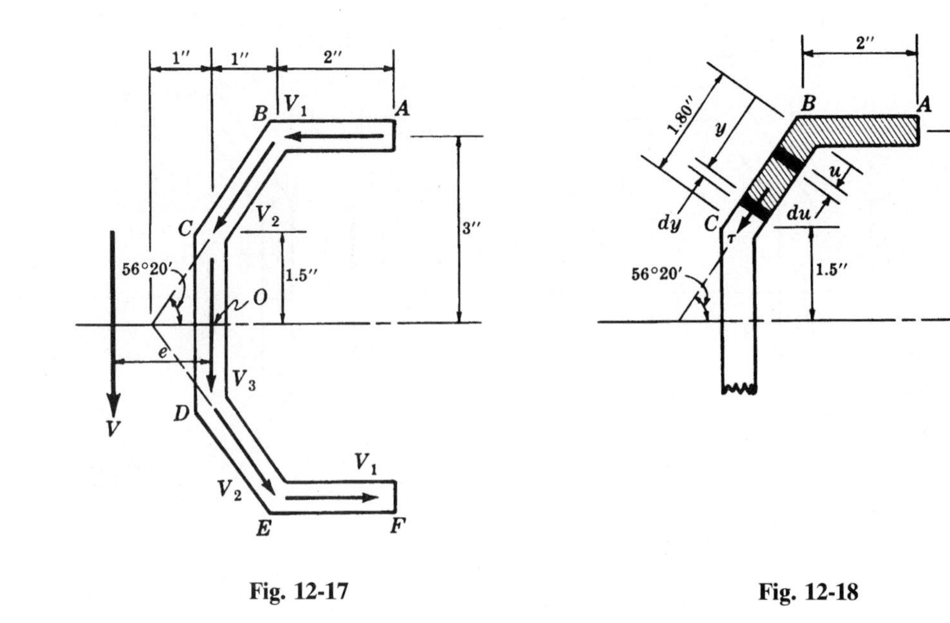

Fig. 12-17 Fig. 12-18

From Fig. 12-17 the sum of the moments of the forces V_1, V_2, and V_3 about any point, such as O, must equal the moment of the resultant about that point. Thus

$$2(3V_1) + 2(V_2 \sin 56°20')(1) = Ve$$

Substituting from (c) and (e),

$$e = \frac{60.8t}{I} \tag{f}$$

The moment of inertia is given by

$$I = \frac{1}{12}(t)(3)^3 + 2[(2)(t)(3)^2] + 2\int_{u=0}^{u=1.80} [1.5 + (1.80 - u)\sin 56°20']^2 t\,du$$

We then have

$$e = \frac{60.8t}{57.2t} = 1.06 \text{ in}$$

which locates the shear center.

Unsymmetric Bending

12.5. Consider a beam of arbitrary unsymmetric cross section subject to pure bending, as indicated in Fig. 12-19(a). Derive an expression for the relationship between the bending moment and the bending stress at any point in this section. Assume Hooke's law holds.

It is convenient to resolve the moment M, which acts in a plane oblique to the y- and z-axes (through the centroid), into moment components about those axes. These components are designated as M_y and M_z and their positive directions are indicated by the double-headed vectors in Fig. 12-19(b).

As in Problem 8.1 it is reasonable to assume that cross sections that were plane prior to bending remain plane after application of the loads. However, in the general case being considered here there is one radius of curvature ρ_z in the x-y plane and another ρ_y in the x-z plane. Thus, for a longitudinal fiber of area da as indicated in Fig. 12-19(b) the normal strain, analogous to (1) of Problem 8.1, is given by

$$\varepsilon = \frac{y}{\rho_z} + \frac{z}{\rho_y} \tag{1}$$

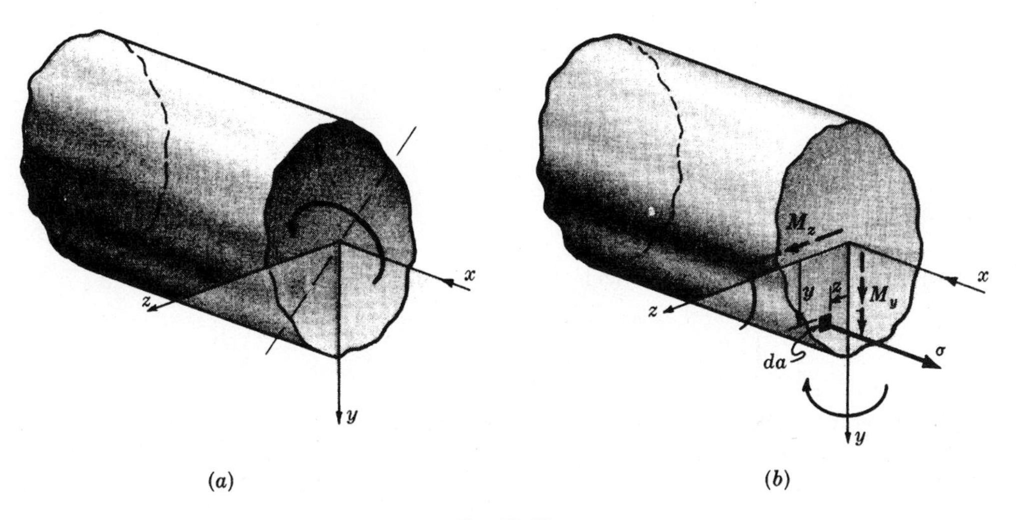

Fig. 12-19

Since Hooke's law holds, we immediately have

$$\sigma = \frac{Ey}{\rho_z} + \frac{Ez}{\rho_y} \tag{2}$$

and this longitudinal, or bending, stress is indicated in the figure.

The resultant longitudinal force acting over the cross section is zero (for the case of pure bending) and this condition may be expressed as

$$\int_A \sigma\, da = 0 \qquad \text{or} \qquad \int_A \left(\frac{Ey}{\rho_z} + \frac{Ez}{\rho_y}\right) da = 0$$

where the integration is extended over the cross-sectional area A. Since ρ_y and ρ_z are constant over the cross section, we have

$$\frac{E}{\rho_z}\int_A y\, da + \frac{E}{\rho_y}\int_A z\, da = 0 \tag{3}$$

This equation is satisfied if the integrals vanish. This implies taking the origin of the y-z coordinate system to coincide with the centroid of the cross section.

From Fig. 12-19(b) it is evident that

$$M_z = \int_A \sigma y\, da = \int_A \left(\frac{Ey^2}{\rho_z} + \frac{Eyz}{\rho_y}\right) da$$

$$= \frac{E}{\rho_z}\int_A y^2\, da + \frac{E}{\rho_y}\int_A yz\, da$$

where the first integral represents the moment of inertia of the cross-sectional area about the z-axis and the second integral (as mentioned in Chap. 7) represents the product of inertia of the same area about the y- and z-axes. Using the notation of Chap. 7, this last equation becomes

$$M_z = \frac{EI_z}{\rho_z} + \frac{EI_{yz}}{\rho_y} \tag{4}$$

Also from Fig. 12-19(b) we have

$$M_y = -\int_A \sigma z\, da = -\int_A \left(\frac{Eyz}{\rho_z} + \frac{Ez^2}{\rho_y}\right) da$$

$$= -\frac{EI_{yz}}{\rho_z} + \frac{EI_y}{\rho_y} \tag{5}$$

Equations (*4*) and (*5*) may be solved for ρ_y and ρ_z to yield

$$\frac{1}{\rho_y} = \frac{-M_y I_z - M_z I_{yz}}{E(I_y I_z - I_{yz}^2)} \tag{6}$$

$$\frac{1}{\rho_z} = \frac{M_z I_y + M_y I_{yz}}{E(I_y I_z - I_{yz}^2)} \tag{7}$$

Substituting (*6*) and (*7*) in (*2*) yields the bending stress

$$\sigma = \frac{(M_z I_y + M_y I_{yz})y + (-M_y I_z - M_z I_{yz})z}{I_y I_z - I_{yz}^2} \tag{8}$$

Equation (*8*) is termed the *generalized flexure formula* and holds for an elastic beam of arbitrary cross section with bending loads in an arbitrary plane. For the special case $M_y = I_{yz} = 0$ (implying that the y- and z-axes are principal axes and that bending takes place only about the z-axis) (*8*) reduces to $\sigma = M_z y / I_z$, which is equivalent to (*9*) of Problem 8.1.

The equation of the neutral axis is readily found by setting the stress from (*8*) equal to zero, since by definition the fibers along the neutral axis are free of longitudinal stress. Thus

$$\frac{y}{z} = \frac{M_y I_z + M_z I_{yz}}{M_z I_y + M_y I_{yz}} = \tan\alpha \tag{9}$$

where α denotes the angle of inclination of the neutral axis as indicated in Fig. 12-20. In general the neutral axis is *not* perpendicular to the plane of the applied moments nor does it coincide with either of the principal axes.

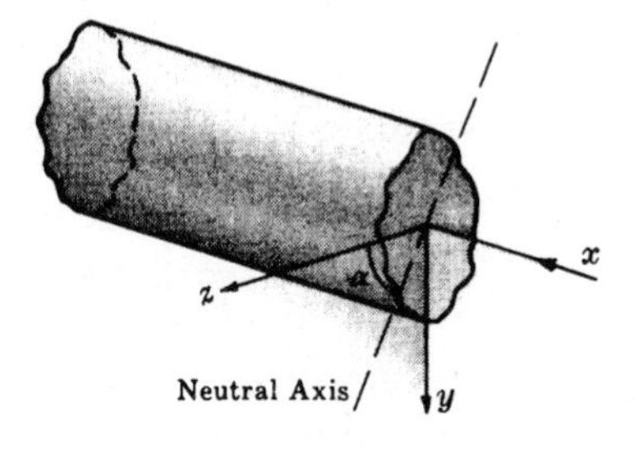

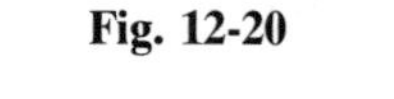
Fig. 12-20

12.6. The rectangular beam of Fig. 12-21 is subject to loads that create a bending moment of 2000 lb · ft acting in a plane oriented at 30° to the y-axis. Determine the peak tensile and compressive stresses in the beam.

The vector representation of the 2000 lb · ft moment is indicated by the solid double-headed vector in Fig. 12-22, together with its moment components (dashed vectors) in the y- and z-directions. This convenient vector representation enables us to find the components as

$$M_y = 2000 \sin 30^\circ = 1000 \text{ lb} \cdot \text{ft} \qquad M_z = 2000 \cos 30^\circ = 1732 \text{ lb} \cdot \text{ft}$$

From Problem 7.3, we have

$$I_y = \tfrac{1}{12}(6)(3)^3 = 13.5 \text{ in}^4 \qquad I_z = \tfrac{1}{12}(3)(6)^3 = 54 \text{ in}^4$$

Also, since the y- and z-axes are axes of symmetry, they are principal axes of the cross section and, from Chap. 7, the product of inertia with respect to these axes vanishes: $I_{yz} = 0$.

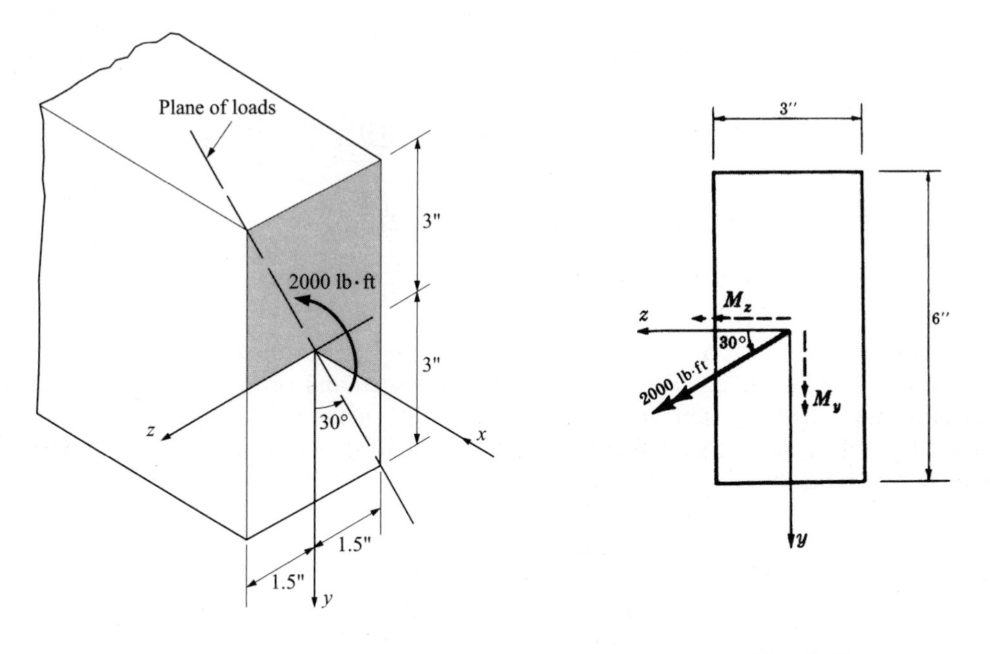

Fig. 12-21 Fig. 12-22

The angle of inclination of the neutral axis (which passes through the centroid) is given by (9) of Problem 12.5 to be

$$\begin{aligned}\tan\alpha &= \frac{M_y I_z + M_z I_{yz}}{M_z I_y + M_y I_{yz}} \\ &= \frac{(1000)(54) + (1732)(0)}{(1732)(13.5) + 1000(0)} = 2.31 \\ \alpha &= 66^\circ 40'\end{aligned}$$

As mentioned in Problem 12.5, there is no reason to expect the neutral axis, as indicated in Fig. 12-23, to be normal to the plane of the loads.

In Problem 12.5, it was assumed that plane sections remain plane during bending. The originally plane section rotates about the neutral axis indicated in Fig. 12-23 and since both strains as well as stresses vary as the distance from the neutral axis it is evident that the peak tensile stress occurs at point B and the peak compressive stress occurs at A, i.e., *at those points most remote from the neutral axis*. Substituting the

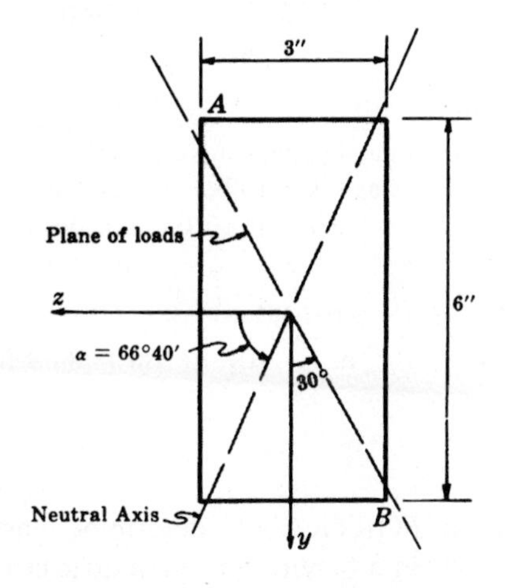

Fig. 12-23

coordinates of these points and the values of the moment components in (8) of Problem 12.5, we obtain

$$\sigma_B = \frac{[(1732)(12)(13.5) + 0](3) + [-(1000)(12)(54) - 0](-15)}{(13.5)(54) - 0} = 2480 \text{ lb/in}^2$$

$$\sigma_A = \frac{[(1732)(12)(13.5) + 0](-3) + [-(1000)(12)(54) - 0](1.5)}{(13.5)(54) - 0} = -2480 \text{ lb/in}^2$$

12.7. The structural angle section designated as L127 × 127 × 22.2 has the dimensions and centroidal axis indicated in Fig. 12.24. The values of the cross-sectional properties with respect to the centroidal axis of the section are $I_y = I_z = 7.41 \times 10^{-6}\,\text{m}^4$ and $I_{yz} = -4.201 \times 10^{-6}\,\text{m}^4$. For a loading $M_y = 0$, $M_z = 10\,\text{kN}\cdot\text{m}$, find the angle of inclination of the neutral axis and the bending stress at point A.

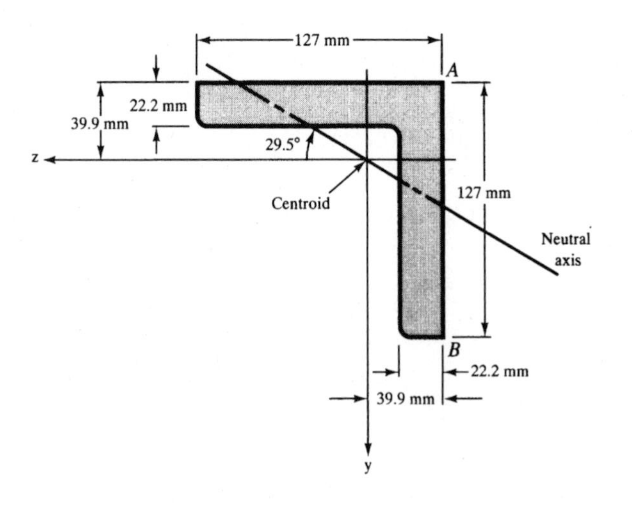

Fig. 12-24

The angle of inclination of the neutral axis is given by Eq. (9) of Problem 12.5 as

$$\tan\alpha = \frac{0 + M_z(-4.201 \times 10^{-6}\,\text{m}^4)}{M_z(7.41 \times 10^{-6}\,\text{m}^4) + 0}$$
$$= -0.567$$
$$\alpha = -29.5^\circ$$

which is shown in Fig. 12-24. The minus sign indicates clockwise rotation from the positive end of the z-axis because the positive direction of α was taken to be counterclockwise as indicated in Fig. 12-20.

Point A has coordinates $y = z = -39.9$ mm so that the desired stress at that point from Eq. (8) of Problem 12.5 is

$$\sigma = \frac{[(10{,}000\,\text{N}\cdot\text{m})(7.41 \times 10^{-6}\,\text{m}^4) - 0](0.0399\,\text{m}) + [0 - (10{,}000\,\text{N}\cdot\text{m})(-4.201 \times 10^{-6}\,\text{m}^4)(-0.0399\,\text{m})}{(7.41 \times 10^{-6}\,\text{m}^4)(7.41 \times 10^{-6}\,\text{m}^4) - (-4.201 \times 10^{-6}\,\text{m}^4)^2}$$
$$= -124\,\text{MPa}$$

12.8. Write a computer program in FORTRAN to determine elastic bending stresses as well as the orientation of the neutral axis in a beam of unsymmetric cross section subject to pure bending as shown in Fig. 12-19.

The desired stress is given by Eq. (*8*) in Problem 12.5 and the angular orientation of the neutral axis is indicated by Eq. (*9*) of that problem. The components of moment M_y and M_z have the positive directions shown in Fig. 12-19 and all other symbols are defined in Problem 12.5. The program is

```
00010*********************************************************************
00020                          PROGRAM BEND (INPUT,OUTPUT)
00030*********************************************************************
00040*
00050*           AUTHOR: KATHLEEN DERWIN
00060*           DATE  : JANUARY 27,1989
00070*
00080*  BRIEF DESCRIPTION:
00090*      THIS PROGRAM CONSIDERS A BEAM OF ARBITRARY UNSYMMETRIC CROSS
00100*  SECTION SUBJECTED TO PURE BENDING. THE GENERALIZED FLEXURE FORMULA
00110*  HOLDS FOR THIS CASE, AND PROVIDES A RELATIONSHIP BETWEEN THE BENDING
00120*  MOMENT AND THE BENDING STRESS AT ANY POINT IN THE SECTION. ALSO,
00130*  THE ANGLE OF INCLINATION OF THE NEUTRAL AXIS CAN BE CALCULATED AS A
00140*  FUNCTION OF THE BENDING MOMENTS.
00150*
00160*  INPUT:
00170*      THE USER IS FIRST ASKED IF USCS OR SI UNITS WILL BE USED. THEN,
00180*  THE SECTIONAL PROPERTIES (MOMENTS OF INERTIA IY,IZ,IYZ) ARE INPUTTED,
00190*  AS WELL AS THE BENDING MOMENTS. FINALLY, THE COORDINATES OF THE POINT
00200*  WHERE THE BENDING STRESS IS DESIRED ARE ENTERED.
00210*
00220*  OUTPUT:
00230*      THE BENDING STRESS AT ANY POINT ON THE CROSS SECTION MAY BE
00240*  OBTAINED, AS WELL AS THE ANGLE OF INCLINATION OF THE NEUTRAL AXIS.
00250*
00260*  VARIABLES:
00270*    IY,IZ,IYZ     ---     SECTIONAL PROPERTIES (MOMENTS OF INERTIA)
00280*     MY,MZ        ---     BENDING MOMENTS
00290*     SIGMA        ---     BENDING STRESS AT THE DESIRED POINT ON THE SECTI
00300*    TALPHA      ---     THE TANGENT OF THE ANGLE OF INCLINATION OF THE
00310*                        NEUTRAL AXIS
00320*     ALPHA      ---     THE ANGLE OF INCLINATION OF THE NEUTRAL AXIS
00330*     Y,Z          ---     COORDINATE OF THE POINT WHERE STRESS DETERMINATI
00340*                          IS DESIRED
00350*     ANS          ---     DENOTES IF USCS OR SI UNITS ARE TO BE USED
00360*     UNIT         ---     GIVES THE USCS OR SI UNIT FOR STRESS
00370*
00380*********************************************************************
00390******                       MAIN PROGRAM                       ******
00400*********************************************************************
00410*
00420*          VARIABLE DECLARATIONS
00430*
00440       REAL IY,IZ,IYZ,SIGMA,MY,MZ,TALPHA,ALPHA
00450       INTEGER ANS
00460       CHARACTER UNIT*4
00470*
00480*          USER INPUT
00490*
00500       PRINT*,'PLEASE INDICATE YOUR CHOICE OF UNITS:'
00510       PRINT*,'1 - USCS'
00520       PRINT*,'2 - SI'
00530       PRINT*,' '
00540       PRINT*,'ENTER 1,2'
00550       READ*,ANS
00560       PRINT*,' '
00570       PRINT*,' '
00580       PRINT*,'NOTE, THE COORDINATE SYSTEM USED HAS THE X-AXIS ORIENTED'
00590       PRINT*,'SO THAT IT IS POSITIVE INTO THE PAGE AND ACTING AS THE'
00600       PRINT*,'NEUTRAL AXIS OF THE SECTION. THE POSITIVE Y-AXIS IS DIRECTED'
00610       PRINT*,'DOWNWARD, WHILE THE POSITIVE Z-AXIS IS TO THE LEFT AS ONE'
```

```
00620       PRINT*,'FACES THE SECTION. (IT IS A RIGHT HANDED SYSTEM.)'
00630       PRINT*,' '
00640       PRINT*,' '
00650       IF (ANS.EQ.1) THEN
00660          PRINT*,'PLEASE ENTER THE SECTION PROPERTIES IY,IZ,IYZ ,(IN^4):'
00670          READ*,IY,IZ,IYZ
00680          PRINT*,' '
00690          PRINT*,'PLEASE ENTER THE MAGNITUDE OF THE BENDING MOMENTS MY,MZ'
00700          PRINT*,'FOLLOWING THE SIGN CONVENTION STATED (LB-FT):'
00710          READ*,MY,MZ
00720          MY = MY*12
00730          MZ = MZ*12
00740       ELSE
00750          PRINT*,'PLEASE ENTER THE SECTION PROPERTIES IY,IZ,IYZ ,(MM^4):'
00760          READ*,IY,IZ,IYZ
00770          PRINT*,' '
00780          PRINT*,'PLEASE ENTER THE MAGNITUDE OF THE BENDING MOMENTS MY,MZ'
00790          PRINT*,'FOLLOWING THE SIGN CONVENTION STATED (KN-M):'
00800          READ*,MY,MZ
00810          MY = MY*1E6
00820          MZ = MZ*1E6
00830       ENDIF
00840*
00850       PRINT*,' '
00860       PRINT*,'ENTER THE Y AND Z COORDINATES OF THE POINT WHERE STRESS '
00870       PRINT*,'DETERMINATION IS DESIRED.(FOLLOW THE SIGN CONVENTION STATED
00880       IF(ANS.EQ.1) THEN
00890          PRINT*,'Y AND Z ARE DISTANCES IN INCHES FROM THE NEUTRAL AXIS:'
00900       ELSE
00910          PRINT*,'Y AND Z ARE DISTANCES IN MILLIMETERS FROM NEUTRAL AXIS:'
00920       ENDIF
00930       READ*,Y,Z
00940*
00950*              END USER INPUT
00960***********************************************************
00970*              CALCULATIONS FOR BENDING STRESS AND THE ANGLE OF INCLINATION
00980*              AS FUNCTIONS OF THE APPLIED BENDING MOMENTS AND THE SECTION
00990*              PROPERTIES
01000*
01010       SIGMA=(((MZ*IY + MY*IYZ)*Y) +((-MY*IZ - MZ*IYZ)*Z))/(IY*IZ - IYZ**2
01020       TALPHA =((MY*IZ + MZ*IYZ)/(MZ*IY + MY*IYZ))
01030       ALPHA  = ATAN(TALPHA)
01040       ALPHA = ALPHA*180/3.14159
01050*
01060*             PRINTING OUTPUT
01070*
01080       IF (ANS.EQ.1) THEN
01090          UNIT = ' PSI'
01100       ELSE
01110          UNIT = ' MPA'
01120       ENDIF
01130      PRINT 10,'THE BENDING STRESS AT (',Y,',',Z,') IS',SIGMA,UNIT,'.'
01140      PRINT 20,'THE ANGLE OF INCLINATION OF THE NEUTRAL AXIS IS',ALPHA,'DEG.'
01150*
01160*             FORMAT STATEMENTS
01170*
01180 10    FORMAT(//,2X,A23,F6.1,A1,F6.1,A4,F10.1,A4,A1)
01190 20    FORMAT(/,2X,A,F8.2,1X,A)
01200*
01210       STOP
01220       END
```

12.9. Rework Problem 12.6 using the FORTRAN program in Problem 12.8.

The self-prompting program is utilized by entering the moment components and sectional properties from Problem 12.6. Consideration of the directions of moment components indicates that the peak tensile stress will occur at point B in Fig. 12-23 and the coordinates of that point are $y = 3$, $z = -1.5$. The printout is

```
run
 PLEASE INDICATE YOUR CHOICE OF UNITS:
 1 - USCS
 2 - SI

 ENTER 1,2
? 1


 NOTE, THE COORDINATE SYSTEM USED HAS THE X-AXIS ORIENTED
 SO THAT IT IS POSITIVE INTO THE PAGE AND ACTING AS THE
 NEUTRAL AXIS OF THE SECTION. THE POSITIVE Y-AXIS IS DIRECTED
 DOWNWARD, WHILE THE POSITIVE Z-AXIS IS TO THE LEFT AS ONE
 FACES THE SECTION. (IT IS A RIGHT HANDED SYSTEM.)


 PLEASE ENTER THE SECTION PROPERTIES IY,IZ,IYZ ,(IN^4):
? 13.5,54,0

 PLEASE ENTER THE MAGNITUDE OF THE BENDING MOMENTS MY,MZ
 FOLLOWING THE SIGN CONVENTION STATED (LB-FT):
? 1000,1732

 ENTER THE Y AND Z COORDINATES OF THE POINT WHERE STRESS
 DETERMINATION IS DESIRED.(FOLLOW THE SIGN CONVENTION STATED)
 Y AND Z ARE DISTANCES IN INCHES FROM THE NEUTRAL AXIS:
? 3,-1.5


  THE BENDING STRESS AT (    3.0,   -1.5) IS    2488.0 PSI.

  THE ANGLE OF INCLINATION OF THE NEUTRAL AXIS IS   66.59 DEG.

SRU        0.895 UNTS.

 RUN COMPLETE.
```

12.10. Rework Problem 12.7 using the FORTRAN program of Problem 12.8.

Enter the given cross-sectional properties, moment components, and coordinates of point A indicated in Problem 12.8 into the self-prompting program to obtain the following printout, which agrees with the results of Problem 12.7

```
run
 PLEASE INDICATE YOUR CHOICE OF UNITS:
 1 - USCS
 2 - SI

 ENTER 1,2
? 2


 NOTE, THE COORDINATE SYSTEM USED HAS THE X-AXIS ORIENTED
 SO THAT IT IS POSITIVE INTO THE PAGE AND ACTING AS THE
```

```
 NEUTRAL AXIS OF THE SECTION. THE POSITIVE Y-AXIS IS DIRECTED
 DOWNWARD, WHILE THE POSITIVE Z-AXIS IS TO THE LEFT AS ONE
 FACES THE SECTION. (IT IS A RIGHT HANDED SYSTEM.)


 PLEASE ENTER THE SECTION PROPERTIES IY,IZ,IYZ ,MM^4):
? 7.41E+6,7.41E+6,-4.201E+6

 PLEASE ENTER THE MAGNITUDE OF THE BENDING MOMENTS MY, MZ
 FOLLOWING THE SIGN CONVENTION STATED (KN-M):
? 0.10

 ENTER THE Y AND Z COORDINATES OF THE POINT WHERE STRESS
 DETERMINATION IS DESIRED. (FOLLOW THE SIGN CONVENTION STATED)
 Y AND Z ARE DISTANCES IN MILLIMETERS FROM NEUTRAL AXIS:
? -39.9,-39.9

  THE BENDING STRESS AT ( -39.9, -39.9) IS    -124.3 MPA.

  THE ANGLE OF INCLINATION OF THE NEUTRAL AXIS IS  -29.55 DEG.
```

Curved Beams

12.11. Consider the bending of an initially curved elastic beam for which the plane of curvature is also a plane of symmetry of every cross section. The bending loads act in this plane of symmetry. Derive an expression for the relationship between the bending moment and the bending stress at any point in the cross section. Assume Hooke's law holds.

The beam is illustrated in Fig. 12-25, where R denotes the distance from the center of curvature C to the axis through the centroid of the cross section. The bending moment M is taken to be positive in the direction indicated, i.e., when it tends to *increase* the curvature (*decrease* the radius of curvature).

Let us examine the behavior of a part of the beam corresponding to a central angle $d\theta$ before deformation. After deformation, this angle changes to $d\theta + \Delta\, d\theta$, as shown in Fig. 12-26. Just as in the case of the initially straight beam studied in Problem 8.1, we will assume that plane cross sections originally perpendicular to the geometric axis of the beam remain plane after bending. Thus, the normal section CD prior to loading moves to $C'D'$ after loading. For convenience we shall assume that AB remains fixed in space but this in no way influences the results we will obtain. It will still be assumed that there exists one axis, the neutral axis, for which the longitudinal fibers do not change length, and thus the section CD may be considered to rotate about this neutral axis as indicated in Fig. 12-26. However, there is no reason to believe that the neutral axis coincides with the centroid of the cross section as it did for the initially straight beam in Problem 8.1. In the present problem involving the curved beam, Fig. 12.26 indicates that the *total elongation* of a longitudinal fiber varies as the distance y of the fiber from the neutral axis. The coordinate y is measured positive away from the center of curvature. However, the lengths of these fibers prior to loading are obviously different; hence the *unit* elongations, i.e., normal strains, are *not* proportional to the distances from the neutral axis. This point constitutes the fundamental difference between behavior of a curved beam and behavior of the initially straight beam discussed in Problem 8.1. Since Hooke's law is assumed to hold for this curved beam, it follows that stresses on these fibers are *not* proportional to the distances from the neutral axis.

Let us consider the elongation of the fiber at a distance y from the neutral axis. From Fig. 12-26 this is $y(\Delta\, d\theta)$. Dividing this elongation by the original length of the fiber, $(r + y)\, d\theta$, yields the normal strain as

$$\epsilon = \frac{y(\Delta\, d\theta)}{(r + y)\, d\theta} \qquad (a)$$

where r denotes the radius of curvature of the neutral axis. Since Hooke's law holds, the normal stress is

$$\sigma = \frac{Ey(\Delta\, d\theta)}{(r + y)\, d\theta} \qquad (b)$$

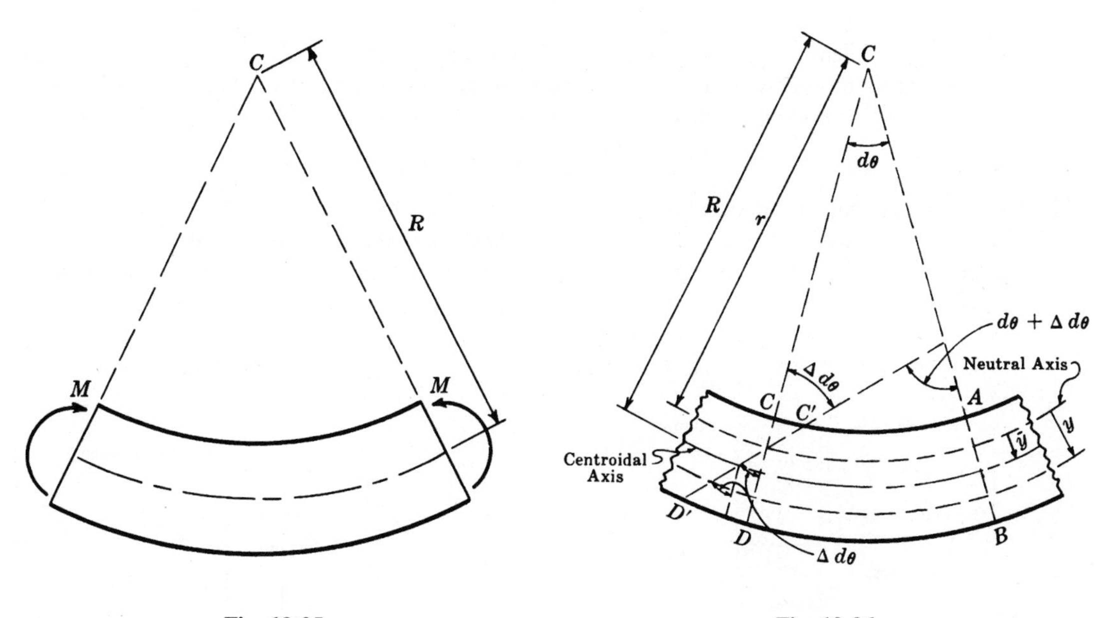

Fig. 12-25

Fig. 12-26

The neutral axis may now be located by requiring the resultant normal force over the cross section to vanish. Thus

$$\int_A \sigma\, da = \int_A \frac{Ey(\Delta\, d\theta)\, da}{(r+y)\, d\theta} = \frac{E(\Delta\, d\theta)}{d\theta}\int_A \frac{y\, da}{(r+y)} = 0 \tag{c}$$

where the integration is over the entire cross-section area A. If $u = r + y$ (i.e., the distance of any fiber from the center of curvature C) then (c) becomes

$$\int_A \frac{(u-r)\, da}{u} = 0 \qquad \text{or} \qquad r = \frac{A}{\displaystyle\int_A da/u} \tag{d}$$

where the integral in the denominator represents a mathematical property of the cross-sectional area and is analogous to the moment of inertia that arises in the case of bending of an initially straight beam.

The sum of the moments of the normal forces on the fibers must equal the bending moment:

$$M = \int_A \sigma y\, da = \int_A \frac{Ey^2(\Delta\, d\theta)\, da}{(r+y)\, d\theta} = \frac{E(\Delta\, d\theta)}{d\theta}\int_A \frac{y^2\, da}{r+y}$$

Simplifying,

$$\int_A \frac{y^2\, da}{r+y} = \int_A y\, da - r\int_A \frac{y\, da}{r+y}$$

The first integral represents the first or static moment of the cross-sectional area about the neutral axis, and the second according to (c) vanishes. Thus

$$M = \frac{E(\Delta\, d\theta)}{d\theta}[A\bar{y}] \tag{e}$$

where $\bar{y}$ denotes the distance from the neutral axis to the centroidal axis. Combining (b) and (e), we find the normal stress on any fiber to be

$$\sigma = \frac{My}{A\bar{y}(r+y)} \tag{f}$$

From (f) it is evident that the stress distribution across the depth of the curved beam is *hyperbolic*. The maximum stress always occurs at the outer fibers on the concave side of the beam. Further, the neutral axis always lies between the centroidal axis and the center of curvature.

12.12. The U-shaped bar of rectangular cross section is loaded by collinear, oppositely directed forces of 9680 N, as shown in Fig. 12-27. The cross-sectional dimensions are 40 mm × 60 mm. The action line of the forces lies 120 mm from the centroid of the cross section. Determine the normal stresses at points A and B.

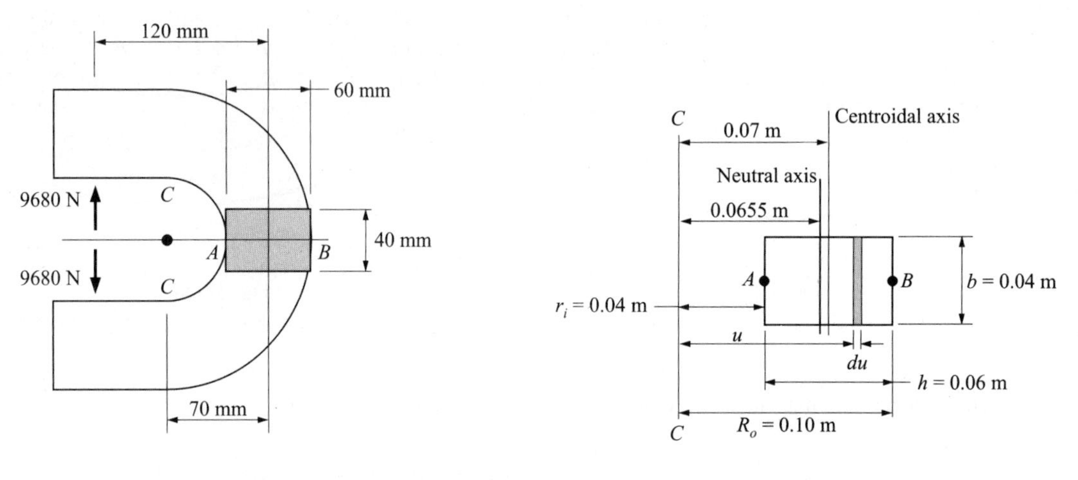

Fig. 12-27 **Fig. 12-28**

It is first necessary to use Eq. (d) of Problem 12.11 to locate the neutral axis. A horizontal cross section of the system coinciding with points A and B is shown in Fig. 12-28, where the variable u is introduced to carry out the integration in Eq. (d). We have

$$r = \frac{bh}{\int b(du)/u} = \frac{h}{(\ln u)_{r_i}^{R_o}} = \frac{0.6\text{ m}}{\ln(0.1\text{ m}/0.04\text{ m})} = 0.0655\text{ m}$$

as the distance from center of curvature to the neutral axis. The variable $\bar{y}$ is thus 0.07 m − 0.0655 m = 0.0045 m

The bending stresses are given by Eq. (f) of Problem 12.11, where $M = -(9680\text{ N})(0.12\text{ m}) = -1162\text{ N}\cdot\text{m}$ since the loading tends to decrease the curvature, and thus we must call it negative moment. At point A in Fig. 12-28, we have $y = -0.0255$ m and the bending stress at A is

$$\sigma_A = \frac{(-1162\text{ N}\cdot\text{m})(-0.0255\text{ m})}{(0.06\text{ m})(0.04\text{ m})(0.0045\text{ m})[0.0655\text{ m} - 0.0255\text{ m}]} = 68.6\text{ MPa}$$

At point B, we have $y = 0.0345$ m and the bending stress at B is

$$\sigma_B = \frac{(-1162\text{ N}\cdot\text{m})(0.0345\text{ m})}{(0.06\text{ m})(0.04\text{ m})(0.0045\text{ m})[0.0655\text{ m} + 0.0345\text{ m}]} = -37.1\text{ MPa}$$

In addition to these bending stresses, the tensile action of the applied loads on the cross section A–B sets up uniform tensile stresses given by

$$\sigma = \frac{P}{A} = \frac{9680\text{ N}}{(0.04\text{ m})(0.06\text{ m})} = 4.03\text{ MPa}$$

The resultant normal stress at point A is thus

$$\sigma'_A = 68.6\text{ MPa} + 4.03\text{ MPa} = 72.63\text{ MPa}$$

and at B it is

$$\sigma'_B = -37.1 \text{ MPa} + 4.03 \text{ MPa} = -33.07 \text{ MPa}$$

12.13. Develop a computer program in FORTRAN to determine extreme fiber bending stresses in the curved beam loaded in pure bending as shown in Fig. 12-25.

The general theory given in Problem 12.11 indicates that it is first necessary to determine the location of the neutral axis, which is a distance r from the center of curvature in Fig. 12-26. From Eq. (d) of Problem 12.11, r is seen to be a function of the shape of the cross section. From this general expression (d) we choose to write a computer program for the three common types of cross section: (a) rectangular, (b) circular, and (c) trapezoidal. The following program carries out the integration of Eq. (d) over each of these cross sections, then develops outer fiber stress according to Eq. (f) for a pure bending moment loading M as shown in Fig. 12-25, where it must be carefully noted that the moment is negative if it acts so as to reduce the curvature of the beam. The program is

```
00010***********************************************************************
00020                          PROGRAM CRVBEAM
00030***********************************************************************
00040*
00050*            AUTHOR: KATHLEEN DERWIN
00060*            DATE  : FEBRUARY 5, 1989
00070*
00080*    BRIEF DESCRIPTION:
00090*       THE FOLLOWING PROGRAM CONSIDERS THE BENDING OF AN INITIALLY
00100*    CURVED ELASTIC BEAM FOR WHICH THE PLANE OF CURVATURE IS ALSO A
00110*    PLANE OF SYMMETRY AT EVERY CROSS SECTION. THE BENDING LOAD ACTS IN
00120*    THIS PLANE OF SYMMETRY. THE MAXIMUM BENDING STRESS OCCURS AT THE
00130*    EXTREME FIBERS OF THE SECTION, AND CAN BE DETERMINED FOR A RECTANGULA
00140*    CIRCULAR, OR TRAPEZOIDAL CROSS SECTION. NOTE, THE RELATIONSHIP BETWEE
00150*    THE BENDING MOMENT AND BENDING STRESS INVOLVES TAKING THE NATURAL
00160*    LOGARITHM OF THE RATIO BETWEEN THE DISTANCE FROM THE CENTER OF CURV-
00170*    ATURE TO THE OUTER AND INNER EXTREME FIBERS. FOR EXTREMELY THIN
00180*    CROSS SECTIONS, THIS RATIO MAY BE QUITE CLOSE TO UNITY, IN WHICH
00190*    CASE THE CALCULATION REQUIRES PRECISION BEYOND THE CAPABILITIES OF
00200*    MOST COMPUTERS. TO AVOID THIS PROBLEM, A SERIES EXPANSION HAS BEEN
00210*    EMPLOYED TO APPROXIMATE THE LOGARITHMIC FUNCTION. FOR THE CASE OF
00220*    THE TRAPEZOIDAL CROSS SECTION, THE LOGARITHMIC FUNCTION IS USED,
00230*    ASSUMING THAT IF THE BEAM WERE SUFFICIENTLY THIN TO CAUSE PROBLEMS
00240*    IN THE CALCULATIONS, THE USER COULD APPROXIMATE THE CROSS SECTION
00250*    AS RECTANGULAR WITH CONSIDERABLE ACCURACY.
00260*
00270*    INPUT:
00280*       THE USER IS FIRST ASKED IF USCS OR SI UNITS ARE DESIRED, AND THEN
00290*    FOR THE SHAPE OF THE BEAM CROSS SECTION. THEN, DEPENDING ON THE SHAPE
00300*    OF THE SECTION, THE PHYSICAL DIMENSIONS AND THE DISTANCE FROM THE
00310*    CENTER OF CURVATURE TO THE INNER FIBERS OF THE SECTION ARE INPUTTED.
00320*    FINALLY, AFTER THE PROGRAM FINDS THE CENTRAL AXIS LOCATION, THE USER
00330*    MUST DETERMINE AND ENTER THE BENDING MOMENT BASED ON THE LOADING.
00340*
00350*    OUTPUT:
00360*       THE PROGRAM INITIALLY WILL DETERMINE THE LOCATION OF THE CENTRAL
00370*    AXIS FOR THE PARTICULAR CROSS SECTION. (FROM THIS INFORMATION, THE
00380*    USER THEN MUST DETERMINE THE BENDING MOMENT BASED ON THE LOADING.)
00390*    ULTIMATELY, THE BENDING STRESS AT THE EXTREME FIBERS OF THE CROSS
00400*    SECTION IS GIVEN.
00410*
00420*    VARIABLES:
00430*      ANS          ---   USER INPUT FOR CHOICE OF UNITS
00440*      SHAPE        ---   USER INPUT FOR CHOICE OF X-SECTIONAL SHAPE
```

```
00450*     B,H          ---  DIMENSIONS OF BASE, HEIGHT FOR RECTANGULAR SECTION
00460*     D            ---  DIAMETER OF CIRCULAR SECTION
00470*     B1,B2,H      ---  DIMENSIONS OF INNER BASE, OUTER BASE, AND HEIGHT
00480*                       FOR TRAPEZOIDAL SECTION
00490*     RI,RO        ---  DISTANCE FROM THE CENTER OF CURVATURE TO THE INNER
00500*                       AND OUTER FIBERS OF THE SECTION RESPECTIVELY
00510*     A            ---  AREA OF THE SECTION
00520*     RR           ---  DISTANCE FROM THE CENTER OF CURVATURE TO THE CENTRA
00530*                       AXIS OF THE SECTION
00540*     YBAR         ---  DISTANCE FROM THE CENTRAL AXIS TO THE NEUTRAL AXIS
00550*     R            ---  THE DISTANCE FROM THE CENTER OF CURVATURE TO THE
00560*                       NEUTRAL AXIS (THE DIFFERENCE BETWEEN RR AND YBAR)
00570*     K            ---  A CONSTANT USED FOR THE CASE OF THE CIRCULAR SECTIO
00580*     YI,YO        ---  THE DISTANCES FROM THE NEUTRAL AXIS TO THE INNER AN
00590*                       OUTER FIBERS RESPECTIVELY
00600*     M            ---  THE BENDING MOMENT ACTING ON THE SECTION
00610*   SIGMAI,SIGMAO---    THE BENDING STRESSES AT THE INNER AND OUTER FIBERS
00620*   A1,A2,YJ,YK, ---    VARIABLES USED TO FIND THE CENTROID OF TRAPEZOIDAL
00630*   SUMAY,SUMA,HOLD     SECTION
00640*    UNIT          ---  CHARACTER VARIABLE DENOTING THE APPROPRIATE UNITS
00650*
00660**********************************************************************
00670**********                     MAIN PROGRAM                ***********
00680**********************************************************************
00690*
00700*          VARIABLE DECLARATION
00710*
00720      REAL B,H,D,B1,B2,RI,RO,A,RR,YBAR,R,K,YI,YO,M,SIGMAI,SIGMAO
00730      REAL A1,A2,YJ,YK,SUMAY,SUMA,HOLD
00740      INTEGER ANS,SHAPE
00750      CHARACTER UNIT*7
00760*
00770*          USER INPUT
00780*
00790      PRINT*,'PLEASE INPUT YOUR CHOICE OF UNITS:'
00800      PRINT*,'1 - USCS'
00810      PRINT*,'2 - SI'
00820      PRINT*,' '
00830      PRINT*,'ENTER 1,2 :'
00840      READ*,ANS
00850      PRINT*,' '
00860*
00870      PRINT*,' '
00880 10   PRINT*,'PLEASE INPUT THE SHAPE OF THE BEAM CROSS SECTION:'
00890      PRINT*,'1 - RECTANGULAR'                 '
00900      PRINT*,'2 - CIRCULAR'
00910      PRINT*,'3 - TRAPEZOIDAL'
00920      PRINT*,' '
00930      PRINT*,'ENTER 1,2,3:'
00940      READ*,SHAPE
00950*
00960      IF(ANS.EQ.1) THEN
00970         PRINT*,'PLEASE INPUT THE FOLLOWING DIMENSIONS IN INCHES...'
00980         UNIT='INCHES.'
00990      ELSE
01000         PRINT*,'PLEASE INPUT THE FOLLOWING DIMENSIONS IN METERS...'
01010         UNIT ='METERS.'
01020      ENDIF
01030*
01040*               PROMPTS FOR THE DIMENSIONS OF THE APPROPRIATE SECTION
01050*
01060     PRINT*,' '
01070     IF (SHAPE.EQ.1) THEN
01080         PRINT*,'PLEASE INPUT THE DIMENSIONS OF THE BASE AND HEIGHT, '
01090         PRINT*,'AND THE DISTANCE FROM THE CENTER OF CURVATURE TO THE  '
```

```
01100          PRINT*,'INNER FIBERS OF THE X-SECTION: (B,H,RI)'
01110          READ*,B,H,RI
01120          PRINT*,' '
01130      ELSEIF (SHAPE.EQ.2) THEN
01140          PRINT*,'PLEASE INPUT THE DIAMETER AND DISTANCE FROM THE CENTER O
01150          PRINT*,'CURVATURE TO THE INNER FIBERS OF THE X-SECTION: (D,RI)'
01160          READ*,D,RI
01170          PRINT*,' '
01180      ELSEIF (SHAPE.EQ.3) THEN
01190         PRINT*,'PLEASE INPUT THE DIMENSIONS OF THE INSIDE, THEN OUTSIDE'
01200         PRINT*,'BASES, THE HEIGHT, AND THE DISTANCE FROM THE CENTER OF'
01210         PRINT*,'CURVATURE TO THE INNER FIBERS OF THE X-SECTION:(B1,B2,H,RI)'
01220         READ*,B1,B2,H,RI
01230         PRINT*,' '
01240      ELSE
01250         PRINT*,'YOU MUST ENTER A 1,2 OR 3!'
01260         GO TO 10
01270      ENDIF
01280*
01290*            END USER INPUT
01300*
01310*       CALCULATIONS --- IN EACH CASE, THE DISTANCE FROM THE CENTER OF
01320*                        CURVATURE TO THE CENTRAL AND NEUTRAL AXIS IS
01330*                        FOUND (RR AND  R) ,AND THEN THE DISTANCE FROM
01340*                        THE NEUTRAL AXIS TO THE EXTREME FIBERS (YI,YO) IS
01350*                        DETERMINED.
01360*
01370       IF (SHAPE.EQ.1) THEN
01380*
01390*                IF SHAPE EQUALS ONE, THEN THE SECTION IS RECTANGULAR
01400*
01410         A = B*H
01420         RO = RI + H
01430         RR = (H)/2 + RI
01440         YBAR = H**2/(12*RR)
01450         R = RR-YBAR
01460         YI = YBAR -(H/2)
01470         YO = YBAR +(H/2)
01480       ELSEIF (SHAPE.EQ.2) THEN
01490*
01500*                IF SHAPE EQUALS TWO, THEN THE SECTION IS CIRCULAR
01510*
01520         A = (3.14159/4)*D**2
01530         RR = RI + (D/2.)
01540         K = ((D/(2*RR))**2)/4  + ((D/(2*RR))**4)/8
01550         YBAR = (K*RR)/(1-K)
01560         R = RR - YBAR
01570         YI = YBAR -(D/2)
01580         YO = YBAR +(D/2)
01590       ELSEIF (SHAPE.EQ.3) THEN
01600*
01610*                IF SHAPE EQUALS THREE, THEN THE SECTION IS TRAPEZOIDAL
01620*
01630         A = ((B1 + B2)/2)*H
01640         RO = RI + H
01650         HOLD = 0.0
01660*
01670*                FIRST, THE CENTROID OF THE TRAPEZOIDAL SECTION IS FOUND
01680*
01690 20      IF (B1.GT.B2) THEN
01700            A1 = (H/4)*(B1-B2)
01710            A2 = B2*H
01720            YJ = H/3.
01730            YK = H/2.
01740            SUMA = (2*A1) + A2
```

```
01750              SUMAY = (A1*YJ*2) + (A2*YK)
01760           ELSE
01770              HOLD = B2
01780              B2 = B1
01790              B1 = HOLD
01800              GO TO 20
01810           ENDIF
01820           IF (HOLD.EQ.0.) THEN
01830              RR = RI + (SUMAY/SUMA)
01840           ELSE
01850              RR = RI + (H - (SUMAY/SUMA))
01860           ENDIF
01870*
01880           R = ((H**2)*(B1 + B2))/2.
01890           R = R/(((B1*RO) - (B2*RI))*(LOG(RO/RI)) - H*(B1 - B2))
01900           YBAR = RR-R
01910           YI = YBAR - (SUMAY/SUMA)
01920           YO = YBAR + (H-(SUMAY/SUMA))
01930        ENDIF
01940*
01950*                 ONCE THE CENTRAL AXIS HAS BEEN DETERMINED, THE USER
01960*                 IS PROMPTED FOR THE BENDING MOMENT WHICH THEY MUST
01970*                 CALCULATE BASED ON THIS DIMENSION AND THE GIVEN LOAD
01980*
01990        PRINT*,'THE DISTANCE FROM THE CENTER OF CURVATURE TO THE CENTRAL'
02000        PRINT 15,'AXIS OF THE CURVED SECTION IS:',RR,UNIT
02010        PRINT*,' '
02020        PRINT*,'GIVEN THIS DIMENSION, THE USER MUST NOW CALCULATE THE'
02030        PRINT*,'MOMENT ACTING ON THE CROSS SECTION...THE MOMENT IS THE'
02040        PRINT*,'PRODUCT OF THE APPLIED LOAD AND THE DISTANCE TO THE CENTRAL'
02050        PRINT*,'AXIS FROM THE POINT OF APPLICATION. NOTE, THE MOMENT IS '
02060        PRINT*,'NEGATIVE IF IT ACTS TO REDUCE THE CURVATURE!'
02070        PRINT*,' '
02080        PRINT*,'PLEASE ENTER THE MOMENT (IN N-M OR LB-IN):'
02090        READ*,M
02100        PRINT*,' '
02110*
02120*           CALCULATING THE BENDING STRESS AT THE INNER AND OUTER FIBERS
02130*
02140        SIGMAI = (M*YI)/(A*YBAR*(R+YI))
02150        SIGMAO = (M*YO)/(A*YBAR*(R+YO))
02160*
02170        IF (ANS.EQ.1) THEN
02180           UNIT = ' PSI.'
02190        ELSE
02200           SIGMAI = SIGMAI/1E6
02210           SIGMAO = SIGMAO/1E6
02220           UNIT = ' MPA.'
02230        ENDIF
02240*
02250*                 PRINTING OUTPUT
02260*
02270        PRINT*,' '
02280        PRINT 15,'THE BENDING STRESS AT THE INNER FIBERS IS :',SIGMAI,UNIT
02290        PRINT 15,'THE BENDING STRESS AT THE OUTER FIBERS IS :',SIGMAO,UNIT
02300*
02310 15     FORMAT(1X,A,F11.3,1X,A)
02320*
02330        STOP
02340        END
```

12.14. Return to Problem 12.12 and use the FORTRAN program of Problem 12.13 to determine the bending stress at point A.

Using the moment loading and geometry of Problem 12.12 we have 68.58, which is in good agreement with the value found in Problem 12.12. Note that the uniform normal stress of 4.03 MPa must be added to this value to obtain the resultant normal stress at A. The following computer program yields only the bending effect.

```
run
 PLEASE INPUT YOUR CHOICE OF UNITS:
 1 - USCS
 2 - SI

 ENTER 1,2 :
? 2


 PLEASE INPUT THE SHAPE OF THE BEAM CROSS SECTION:
 1 - RECTANGULAR
 2 - CIRCULAR
 3 - TRAPEZOIDAL

 ENTER 1,2,3:
? 1
 PLEASE INPUT THE FOLLOWING DIMENSIONS IN METERS...

 PLEASE INPUT THE DIMENSIONS OF THE BASE AND HEIGHT,
 AND THE DISTANCE FROM THE CENTER OF CURVATURE TO THE
 INNER FIBERS OF THE X-SECTION: (B,H,RI)
? 0.04,0.06,0.04

 THE DISTANCE FROM THE CENTER OF CURVATURE TO THE CENTRAL
 AXIS OF THE CURVED SECTION IS:          .070 METERS.

 GIVEN THIS DIMENSION, THE USER MUST NOW CALCULATE THE
 MOMENT ACTING ON THE CROSS SECTION...THE MOMENT IS THE
 PRODUCT OF THE APPLIED LOAD AND THE DISTANCE TO THE CENTRAL
 AXIS FROM THE POINT OF APPLICATION. NOTE, THE MOMENT IS
 NEGATIVE IF IT ACTS TO REDUCE THE CURVATURE!

 PLEASE ENTER THE MOMENT (IN N-M OR LB-IN):
? -1162


 THE BENDING STRESS AT THE INNER FIBERS IS :
 68.58
```

12.15. Consider a crane hook subject to a vertical load of 5000 lb. The cross section is trapezoidal, as shown in Fig. 12-29. Determine the tensile stress at point A using the computer program of Problem 12.13.

The theory of Problem 12.11 is applicable here but the evaluation of the integral in Eq. (d) of that problem is tedious; hence we employ the FORTRAN program of Problem 12.13 using as input the geometry indicated in Fig. 12-29. The printout first indicates that the distance from the center of curvature to the centroidal axis is 2.287 in and from that we can calculate the acting moment as

$$M = -(1.18 \text{ in} + 2.287 \text{ in})(5000 \text{ lb}) = -17{,}335 \text{ lb} \cdot \text{in}$$

Now, using this moment as input in the program, we have the stresses at inner and outer fibers as indicated in the final two lines of the printout.

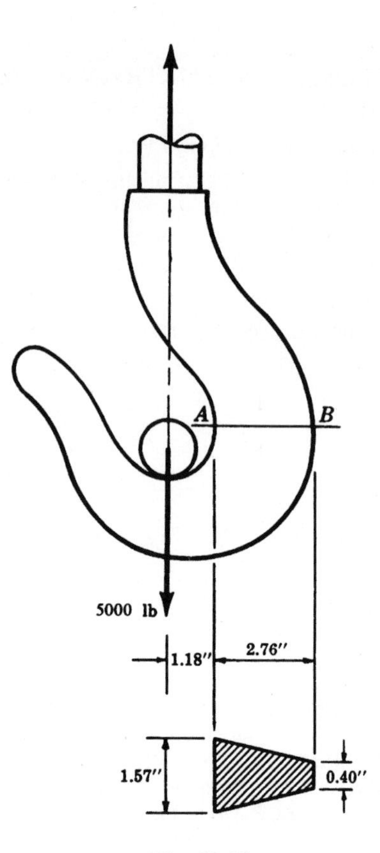

Fig. 12-29

```
run
 PLEASE INPUT YOUR CHOICE OF UNITS:
 1 - USCS
 2 - SI

 ENTER 1,2 :
? 1


 PLEASE INPUT THE SHAPE OF THE BEAM CROSS SECTION:
 1 - RECTANGULAR
 2 - CIRCULAR
 3 - TRAPEZOIDAL

 ENTER 1,2,3:
? 3
 PLEASE INPUT THE FOLLOWING DIMENSIONS IN INCHES...

 PLEASE INPUT THE DIMENSIONS OF THE INSIDE, THEN OUTSIDE
 BASES, THE HEIGHT, AND THE DISTANCE FROM THE CENTER OF
 CURVATURE TO THE INNER FIBERS OF THE X-SECTION:(B1,B2,H,RI)
? 1.57,0.40,2.76,1.18

 THE DISTANCE FROM THE CENTER OF CURVATURE TO THE CENTRAL
 AXIS OF THE CURVED SECTION IS:       2.287 INCHES.

 GIVEN THIS DIMENSION, THE USER MUST NOW CALCULATE THE
 MOMENT ACTING ON THE CROSS SECTION...THE MOMENT IS THE
 PRODUCT OF THE APPLIED LOAD AND THE DISTANCE TO THE CENTRAL
 AXIS FROM THE POINT OF APPLICATION. NOTE, THE MOMENT IS
 NEGATIVE IF IT ACTS TO REDUCE THE CURVATURE!
```

```
PLEASE ENTER THE MOMENT (IN N-M OR LB-IN):
? -17335


THE BENDING STRESS AT THE INNER FIBERS IS :  19878.782 PSI.
THE BENDING STRESS AT THE OUTER FIBERS IS : -12928.359 PSI.

SRU       1.134 UNTS.
```

In addition to these bending stresses, there is a uniformly distributed set of tensile stresses over the cross section AB due to the direct, tensile effect of the 5000-lb load. These stresses are given by

$$\sigma = \frac{P}{A} = \frac{5000 \text{ lb}}{[(1.57 + 0.40)/2 \text{ in}]\,(2.76 \text{ in})} = 1839 \text{ lb/m}^2$$

and must be added to the bending stresses found by the computer program. Thus, the true stress at point A is

$$\sigma'_A = 19.879 \text{ lb/in}^2 + 1839 \text{ lb/in}^2 = 21{,}718 \text{ lb/in}^2 \qquad \text{or} \qquad 21{,}700 \text{ lb/in}^2$$

Supplementary Problems

12.16. Locate the shear center of a thin-walled circular section with a longitudinal slit (Fig. 12-30).
Ans. $e = 2R$

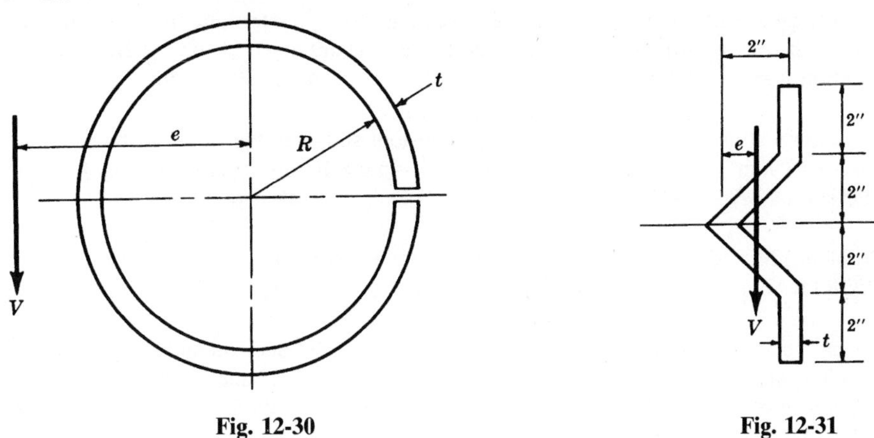

Fig. 12-30 **Fig. 12-31**

12.17. Determine the shear center of the thin-walled "hat" section shown in Fig. 12-31. *Ans.* $e = 0.51$ in

12.18. Determine the shear center of the thin-walled section indicated in Fig. 12-32. *Ans.* $e = 6.85$ mm

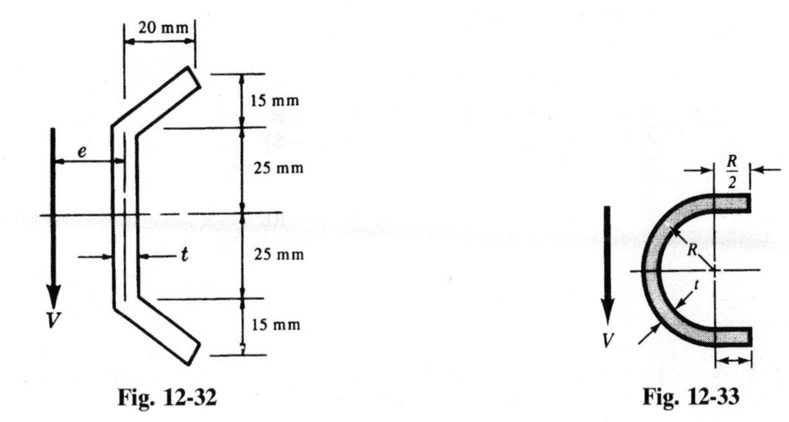

Fig. 12-32 **Fig. 12-33**

12.19. Determine the shear center of the thin-walled section shown in Fig. 12-33.
Ans. $0.747R$ measured from the centroid

12.20. Find the shear center of the thin-walled section shown in Fig. 12-34.
Ans. $0.703a$ measured from the centroid

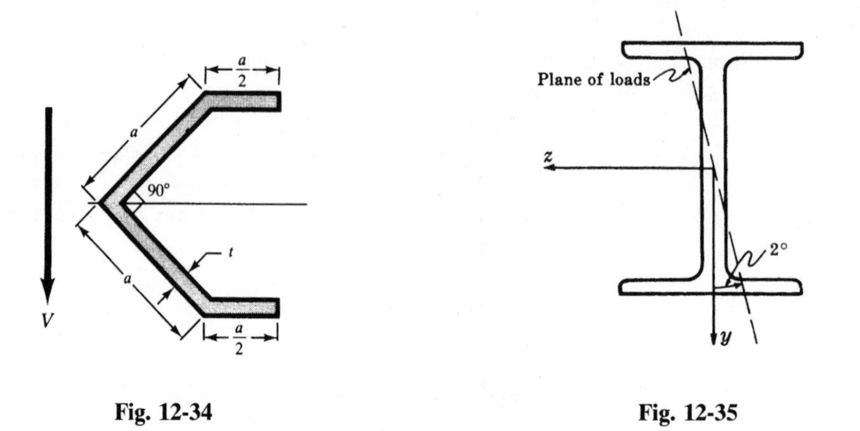

Fig. 12-34

Fig. 12-35

12.21. A structural steel I-beam 250 mm deep is subjected to a bending moment lying in a plane oriented at 2° to the vertical axis of symmetry of the beam (see Fig. 12-35). Determine the percentage increase in elastic tensile stress over the stress that would exist if the moment acted in the vertical plane of symmetry. For this section $I_z = 57 \times 10^6\ \text{mm}^4$ and $I_y = 3.3 \times 10^6\ \text{mm}^4$. *Ans.* 30 percent

12.22. The structural aluminum z-section has the dimensions shown in Fig. 12-36 with cross-sectional properties $I_y = 4.1 \times 10^6\ \text{mm}^4$, $I_z = 10.7 \times 10^6\ \text{mm}^4$, and $I_{yz} = 5.0 \times 10^6\ \text{mm}^4$. The loading has components $M_y = -2.235\ \text{kN} \cdot \text{m}$, $M_z = 4.47\ \text{kN} \cdot \text{m}$. Determine the bending stress at point A. *Ans.* -35.5 MPa

12.23. In Problem 12.7 find the bending stress at point B, neglecting the effect of the rounded corner there. Use the FORTRAN program of Problem 12.8 *Ans.* 153.3 MPa

12.24. A semicircular bar is of square cross section and is clamped at one end and subject to a load P at the other end, as indicated in Fig. 12-37. The cross section is 4 in on a side and the radius of the bar is 20 in. If the maximum tensile stress at the support is not to exceed 28,000 lb/in², determine the maximum allowable value of the load P. *Ans.* 6460 lb

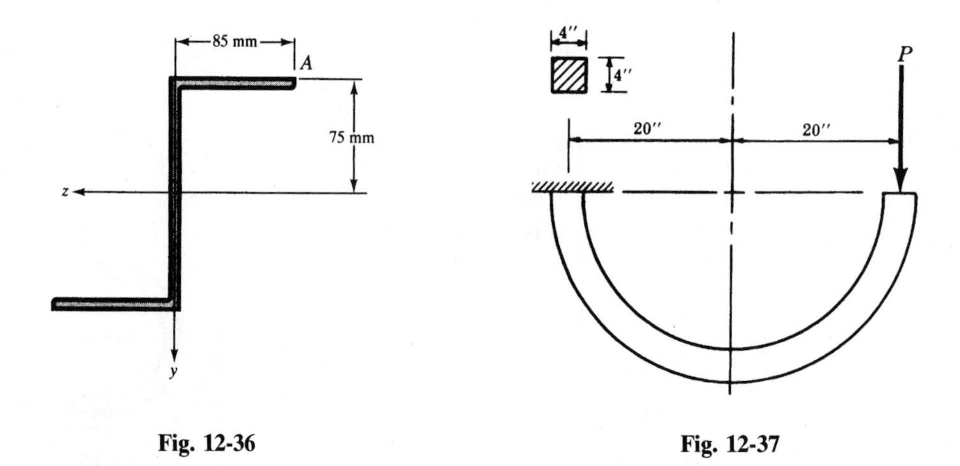

Fig. 12-36

Fig. 12-37

Chapter 13

Plastic Deformations of Beams

INTRODUCTION

In certain situations in structural design it is acceptable to permit a modest amount of permanent deformation of the structural element. If this is the case, then it is possible to permit loads greater than indicated by elastic theory, which permits no stress greater than the yield point of the material to develop at any point. This results in more efficient use of the material and is called *plastic design*. Fundamentally, this more efficient design is possible because of the ability of certain materials, such as structural steel, to undergo relatively large plastic deformations after the yield point has been reached. This is illustrated by the horizontal region of the stress-strain diagram shown in Fig. 1-5, page 3.

PLASTIC HINGE

As the transverse loads on a beam increase, yielding begins at the outer fibers at some critical station along the length of the beam and progresses rather rapidly toward the central fibers at this station. When finally all the fibers on one side of the neutral axis are in a state of tension corresponding to the yield point of the material and all those on the other side are in a state of compression, again at the yield point, then a flowing or *hinging* action occurs at that station and the bending moment transmitted across the *plastic hinge* remains constant. In this book a plastic hinge is denoted by a small, open circle.

FULLY PLASTIC MOMENT

The bending moment developed at a plastic hinge is termed a *fully plastic moment*. This concept was discussed in Chap. 8.

LOCATION OF PLASTIC HINGES

In general, plastic hinges form at points of maximum moment. For beams subject to concentrated forces and moments, the peak bending moment must always occur under one of these loadings or at some reaction and thus the plastic hinges must develop first at these points. In the case of distributed loads, the location of the plastic hinges is considerably more difficult to determine and often several possible points must be investigated. This is discussed in Problems 13.8 and 13.9.

COLLAPSE MECHANISM

When enough plastic hinges have formed in a structure to develop its full plastic load-carrying capacity, then portions of the structure (such as a beam or frame) between hinges may displace without any further increase of load; i.e., the portions between hinges behave as a *mechanism*. Essentially, the hinges allow a kinematic freedom of motion. Under these conditions the shape of the deformed body may be characterized as a straight line between any pair of hinges. Typical representations of collapse mechanisms are shown in Problems 13.2, 13.4, and 13.8 through 13.10.

LIMIT LOAD

The external load sufficient to cause the structure to behave as a mechanism is termed the *limit load* or *collapse load*. Any design based upon the concept of development of a mechanism is termed *limit design*. All problems in this chapter illustrate computation of the limit load.

Solved Problems

13.1. The simply supported beam *ABC* in Fig. 13-1(*a*) is loaded by a central vertical force of 1200 lb and made of steel having a yield point of 38,000 lb/in^2. The beam is of rectangular cross section, as shown in Fig. 13-1(*b*), with width b, depth $1.6b$, and length $L = 40$ in. Determine b for fully plastic action. Also determine the width b' when only the extreme fibers have reached yield.

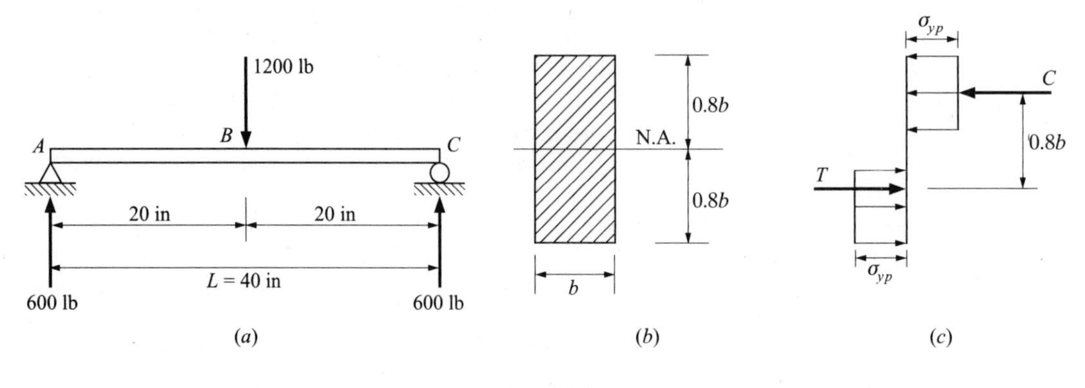

Fig. 13-1

The reactions at A and C are each 600 lb by symmetry. The peak bending moment at the midpoint B is given by

$$(600 \text{ lb})(20 \text{ in}) = 12{,}000 \text{ lb} \cdot \text{in}$$

At that time all fibers above the centrally located neutral axis (N.A.) are acting in compression C and those below that axis are in tension T, as shown in Fig. 13-1(*c*). The location of the action line of each of these forces is shown in Fig. 13-1(*c*). The moment resulting from the effect of T and C is

$$M_P = (\sigma_{yp})(0.8b)(b)[0.8b]$$
$$= 0.64b^3\sigma_{yp}$$
$$= 0.64b^3(38{,}000)$$

Thus,

$$0.64b^3(38{,}000) = 12{,}000$$
$$b = 0.79 \text{ in}$$
$$1.6b = 1.26 \text{ in}$$

so that the beam cross-sectional area is 0.995 in^2.

From Problem 8.25 for a rectangular cross section, the maximum possible fully elastic moment (i.e., when only the extreme outer fibers have reached the yield point) is given by

$$M_e = \frac{b'(h')^2}{6}\sigma_{yp}$$

Hence,

$$12{,}000 = \frac{b'(1.6b')^2}{6}(38{,}000)$$

Solving,

$$b' = 0.905 \text{ in}$$
$$1.6b' = 1.45 \text{ in}$$

Here, the cross-sectional area is 1.312 in^2. The fully elastic moment corresponds to an area of 1.312 in^2. Thus, allowing fully plastic action leads to a 24.2 percent reduction of beam weight for any given length. Suitable safety factors, usually specified by building codes, must be introduced into each of the above computations.

13.2. Determine the limit load of the simply supported beam shown in Fig. 13-2.

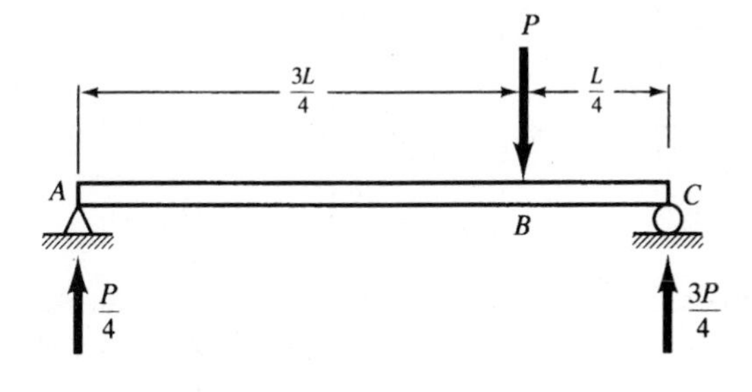

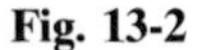
Fig. 13-2

The end reactions at A and C are readily found from statics to be $P/4$ and $3P/4$, respectively, irrespective of whether the beam is in the elastic or plastic state. The peak bending moment occurs under the point of application of P and is thus $(P/4)(3L/4) = 3PL/16$. When this bending moment reaches a value corresponding to fully plastic action of the section of the beam at B, which we term M_p, a plastic hinge forms at B and the beam continues to deflect without further increase of P. This collapse mechanism has the form shown in Fig. 13-3.

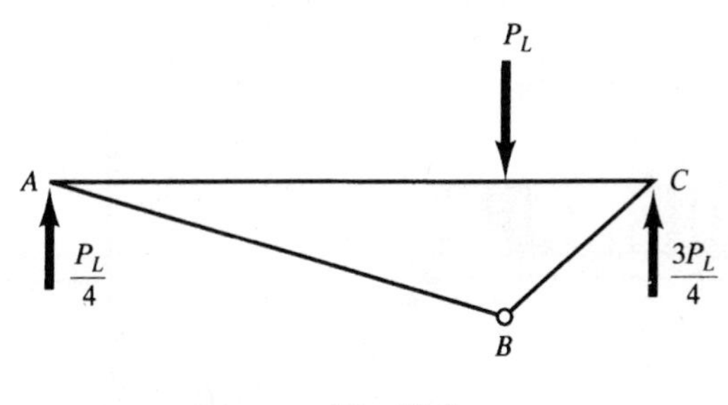

Fig. 13-3

The value of the load P corresponding to this condition is termed the *limit load* P_L. The reaction at A is then $(P_L/4)$ and thus the moment at B is

$$\left(\frac{P_L}{4}\right)\left(\frac{3L}{4}\right) = M_p$$

Solving, $P_L = 16M_p/3L$. Dividing P_L by some suitable safety factor gives an allowable working load. This procedure is called *limit design*.

13.3. The beam of Problem 13.2 is of rectangular cross section 1.75 in × 3 in. It is titanium, type Ti-8Mn, with a yield point stress of 115,000 lb/in². If the length of the beam is 5 ft, determine the central force P necessary to develop the plastic hinge at B.

From Problem 8.25 the fully plastic moment for a rectangular cross section is given by

$$M_p = \sigma_{yp}\frac{bh^2}{4}$$

Substituting,

$$M_p = (115{,}000\ \text{lb/in}^2)\frac{(1.75\ \text{in})(3\ \text{in})^2}{4} = 453{,}000\ \text{lb}\cdot\text{in}$$

Using the result of Problem 13.2,

$$P_L = \frac{16M_p}{3L} = \frac{16(453{,}000\ \text{lb}\cdot\text{in})}{3(60\ \text{in})} = 40{,}300\ \text{lb}$$

This is the limit load of the beam.

From Problem 8.25, the peak elastic moment that this beam could withstand is given by

$$M_e = \sigma_{yp}\frac{bh^2}{6} = 302{,}000\ \text{lb}\cdot\text{in}$$

from which the maximum allowable load P_e based on elastic design is

$$P_e = \frac{16M_e}{3L} = 26{,}850\ \text{lb}$$

Thus use of limit design permits a 50 percent greater load than elastic analysis. However, the designer would want to incorporate some safety factor into the above limit load.

13.4. Determine the limit load of a simply supported beam subject to a unformly distributed load. See Fig. 13-4.

According to the methods developed in Chap. 6, the peak bending moment occurs at the midpoint of the length of the beam and is given by $wL^2/8$. For fully plastic action at the midpoint, this moment is denoted by M_p. Thus, when the plastic hinge forms at the center, the uniform load has the value w_L (limit load) so that

$$\frac{w_L L^2}{8} = M_p \qquad \text{or} \qquad w_L = \frac{8M_p}{L^2}$$

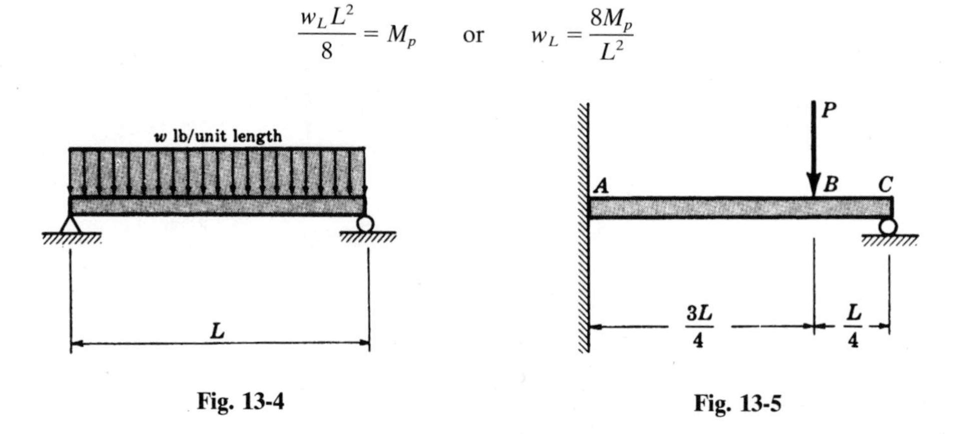

Fig. 13-4 Fig. 13-5

13.5. The beam shown in Fig. 13-5 is clamped at the left end, simply supported at the right, and subject to the concentrated load indicated. Determine the magnitude of the limit load P_L corresponding to plastic collapse.

This statically indeterminate beam cannot collapse plastically through formation of a single plastic hinge at B because the region AB is constrained to very small lateral deflections until another hinge forms somewhere along its length. It has been demonstrated in Chap. 6 that significant bending moments in a beam subject to concentrated forces always occur either at the points of application of these forces or where the reactions are applied. In the present case, this would imply the formation of another plastic hinge at A. With hinges at A and B, we have a so-called *kinematically admissible mechanism* of collapse. The order in which the plastic hinges are formed is of no consequence. The collapse mechanism appears in Fig. 13-6.

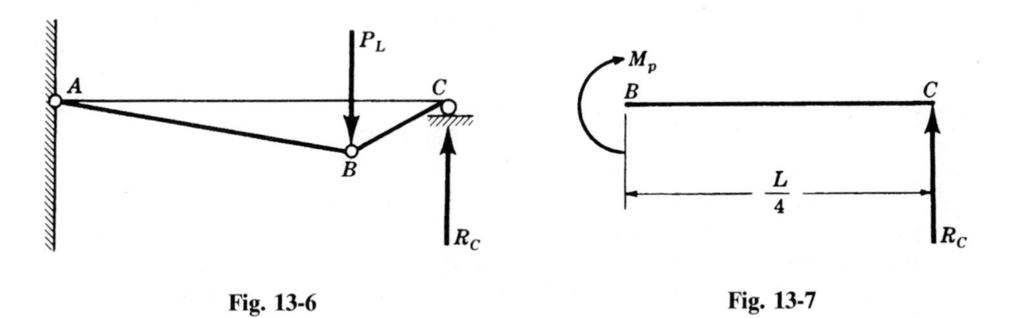

Fig. 13-6 **Fig. 13-7**

The free-body diagram of the right portion of the beam, extending from C to a point just to the right of the applied load P when that force is the limit load P_L, is shown in Fig. 13-7, in which M_p denotes the fully plastic moment at B. From statics,

$$M_p - \frac{R_C L}{4} = 0 \quad \text{or} \quad R_C = \frac{4M_p}{L} \tag{1}$$

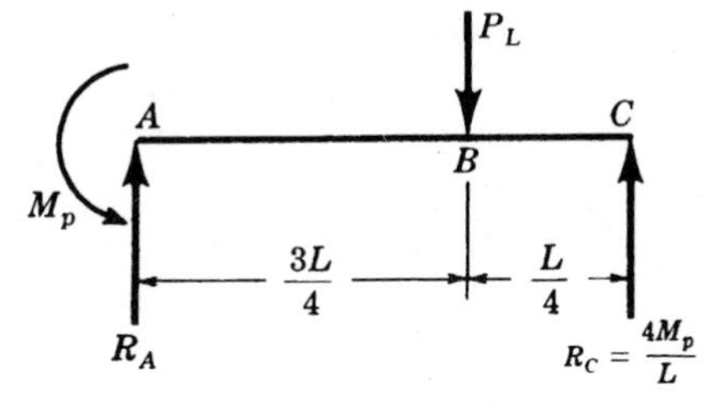

Fig. 13-8

Next, from the free-body diagram of the entire beam (Fig. 13-8), with plastic hinges at A and B, we have

$$\Sigma F_v = R_A + \frac{4M_p}{L} - P_L = 0$$

Hence

$$R_A = P_L - \frac{4M_p}{L} \tag{2}$$

$$\Sigma M_C = R_A L - M_p - P_L\left(\frac{L}{4}\right) = 0 \tag{3}$$

Substituting R_A from (2) in (3) yields

$$P_L = \frac{20}{3}\frac{M_p}{L}$$

as the limit load.

13.6. The beam described in Problem 13.5 is of hollow circular cross section, as shown in Fig. 13-9, and is of steel having a yield point of 200 MPa. Find the limit load that may be carried if $r = 20$ mm and $L = 2$ m.

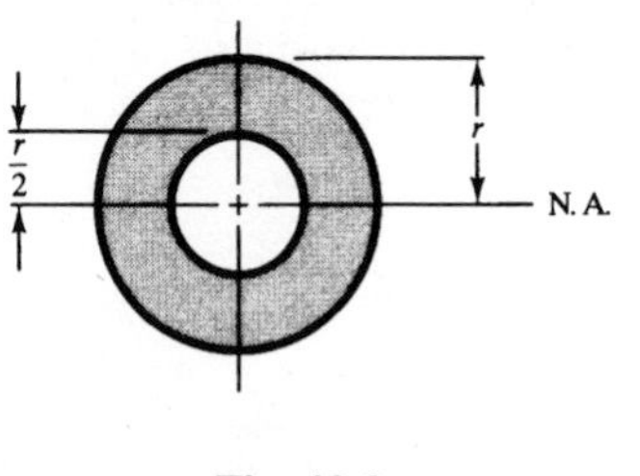

Fig. 13-9

For simplicity, let us first find the fully plastic moment M_p for a solid circular cross section of radius r. Above the neutral axis (N.A.) there is a uniform normal stress distribution equal to the yield point stress, and the resultant of these stresses acts at the centroid, which is at a distance $(4r/3\pi)$ above the N.A. A like situation exists below the N.A., where the normal stresses are oppositely directed from those above that axis. Thus,

$$M_p = 2\left[\sigma_{yp}\left(\frac{\pi r^2}{2}\right)\left(\frac{4r}{3\pi}\right)\right] = \frac{4r^3}{3}\sigma_{yp}$$

The fully plastic moment for the hollow circular cross section is now given by

$$M_p = \frac{4r^3}{3}\sigma_{yp} - (2)\left[\sigma_{yp}\frac{\pi(r/2)^2}{2}\left\{\frac{4(r/2)}{3\pi}\right\}\right] = \frac{7r^3}{6}\sigma_{yp} \qquad (1)$$

For our parameters,

$$M_p = \tfrac{7}{6}(0.02\text{ m})^3(200 \times 10^6\text{ N/m}^2) = 1867\text{ N}\cdot\text{m}$$

and from Problem 13.5 we have

$$P_L = \frac{20M_p}{3L} = \frac{20(1867\text{ N}\cdot\text{m})}{3(2\text{ m})} = 6225\text{ N}$$

as the limit load.

13.7. The beam described in Problem 13.5 is a wide-flange section having the dimensions indicated in Fig. 13-10. For this section, determine the limit load P_L. The material is structural steel with a yield point of 250 MPa and the length of the beam is 2 m.

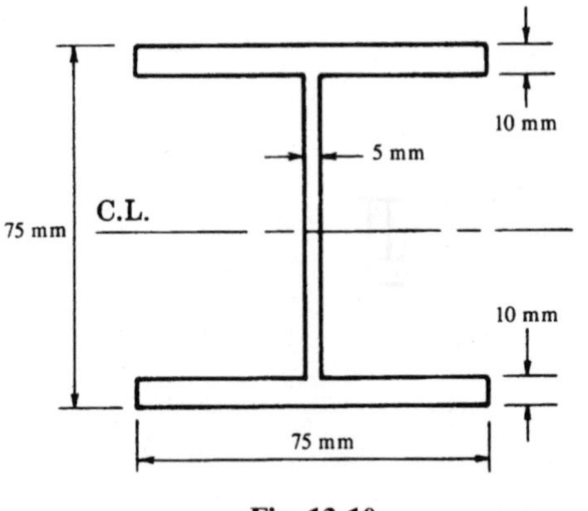

Fig. 13-10

As mentioned in Problem 8.29, for fully plastic action, the neutral axis divides the cross-sectional area into two parts of equal area. Here, because of the symmetry, the neutral axis coincides with the centerline (C.L.) and the centroidal distances from that line are

$$\bar{y}_1 = \bar{y}_2 = \frac{(75)(10)(37.5-5) + (27.5)(5)(27.5/2)}{(75)(10) + (27.5)(5)} = 29.6 \text{ mm}$$

The fully plastic moment is thus

$$M_p = \sigma_{yp}\frac{A}{2}(\bar{y}_1 + \bar{y}_2) = 250[(75)(10) + (27.5)(5)](29.6 + 29.6) = 13.13 \text{ kN}\cdot\text{m}$$

The limit load from Problem 13.5 is

$$P_L = \frac{20}{3}\frac{(13.13 \times 10^3)}{2} = 43.8 \text{ kN}$$

It is of interest to carry out an elastic analysis of this same beam. In this case the outer fibers are taken to be stressed to the yield point and, of course, the stresses vary linearly over the depth, being zero at the neutral axis. The second moment of area of the cross section is found by the methods of Chap. 7 to be

$$I = \tfrac{1}{12}(75)(75)^3 - \tfrac{1}{12}(70)(55)^3 = 1.67 \times 10^6 \text{ mm}^4$$

and the outer fiber stresses are found from

$$\sigma_{yp} = \frac{M_e c}{I} \qquad \text{or} \qquad 250 = \frac{M_e(37.5)}{1.67 \times 10^6}$$

and thus the maximum elastic moment M_e that the section can support is $M_e = 11.13$ kN · m. From Problem 11.1 the bending moment at point A is found to be $0.116PL$ while that at point B is $0.159PL$. Using the latter value we can find the maximum load that the beam can support for entirely elastic action to be

$$0.159P_eL = 11.13 \text{ kN}\cdot\text{m} \qquad \text{or} \qquad P_e = 35 \text{ kN}$$

The load P_L, corresponding to plastic collapse, exceeds this value by 25 percent.

13.8. Determine the limit load of a clamped-end beam carrying a uniformly distributed load (Fig. 13-11).

The collapse mechanism appears in Fig. 13-12, where plastic hinges have formed at points A, B, and C. By virtue of symmetry the shear is zero at the midpoint C; hence we may draw the free-body diagram of the left half of the beam as in Fig. 13-13. From statics,

$$\Sigma M_A = 2M_p - w_L\left(\frac{L}{2}\right)\left(\frac{L}{4}\right) = 0$$

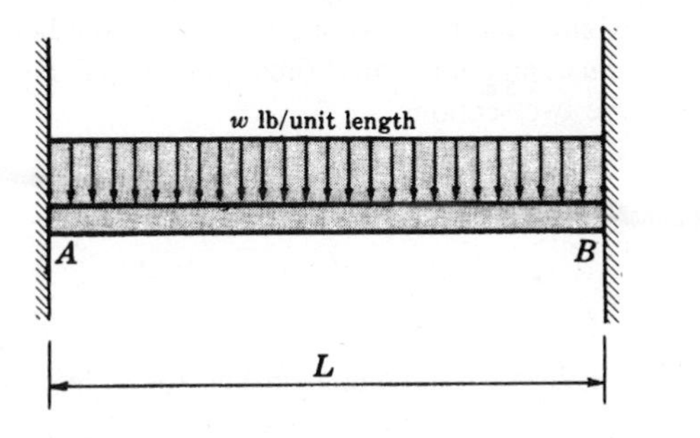

Fig. 13-11

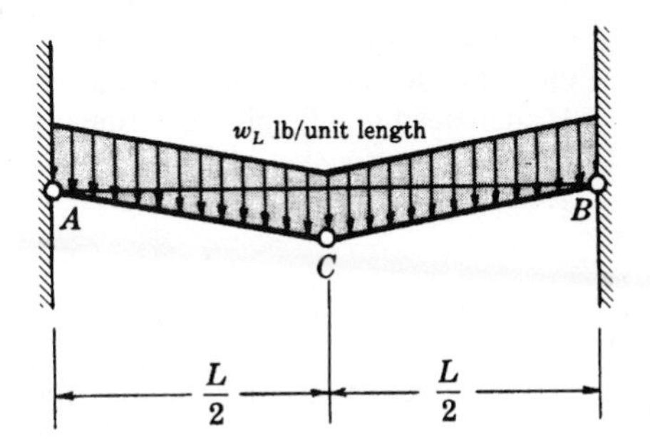

Fig. 13-12

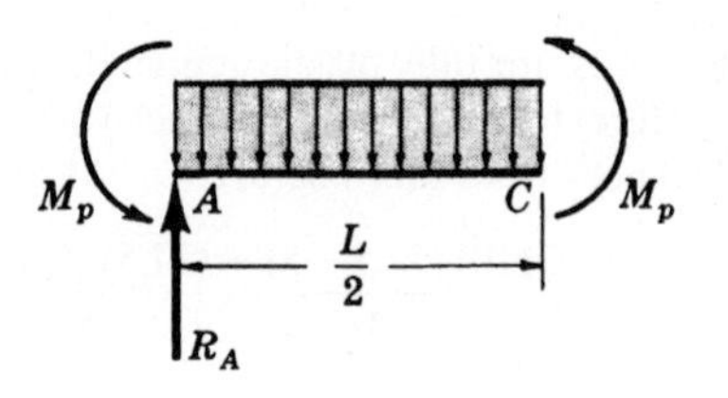

Fig. 13-13

The limit load is thus $w_L = 16M_p/L^2$. From considerations similar to Problem 11.6, the permissible load based upon the outer fibers being at the yield point and all interior fibers acting in the elastic range of action is $w_e = 12M_e/L^2$ so that in this case the ratio of limit load w_L to maximum elastic load w_e is $\frac{4}{3}M_p/M_e$. However, the ratio M_p/M_e itself may be significant. For a rectangular cross section it has the value $\frac{3}{2}$, as indicated in Problem 8.25. For such a rectangular bar we then have

$$\frac{w_L}{w_e} = \frac{4}{3}\frac{M_p}{M_e} = \frac{4}{3}\left(\frac{3}{2}\right) = 2$$

indicating that in this particular case, limit design permits application of twice the load permitted by elastic analysis. This rather large variation between the permissible loads is due partially to the indeterminate nature of this beam. It should be noted that there are exceptional cases where the limit load and maximum elastic load coincide even for an indeterminate system.

13.9. The beam shown in Fig. 13-14 is clamped at the left end, simply supported at the right, and subject to a uniformly distributed load. Determine the magnitude of this load corresponding to plastic collapse of the beam.

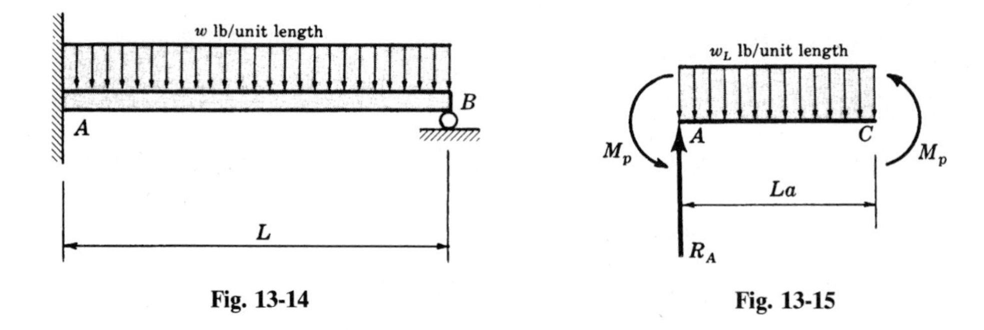

Fig. 13-14 Fig. 13-15

This problem is somewhat analogous to Problem 13.5 because the beam cannot collapse plastically through formation of a single plastic hinge but instead, two hinges must form. One of these is obviously at the clamped end A but the location of the other is not immediately apparent. It of course occurs at the position of relative maximum moment (excluding point A) but that point is not known. However, since the shear is known to be zero at the point of maximum moment, we may draw the free-body diagram of the left region of the beam of length La and regard a as an unknown. It thus appears as in Fig. 13-15, where M_p denotes the fully plastic moment at each of the two sections.

From statics,

$$\Sigma F_v = R_A - w_L La = 0 \qquad (1)$$

$$\Sigma M_A = 2M_p - \frac{w_L L^2 a^2}{2} = 0 \qquad (2)$$

Next, let us consider the free-body diagram of the entire beam, as in Fig. 13-16. From statics,

$$\Sigma M_B = -R_A L + \frac{w_L L^2}{2} + M_p = 0 \qquad (3)$$

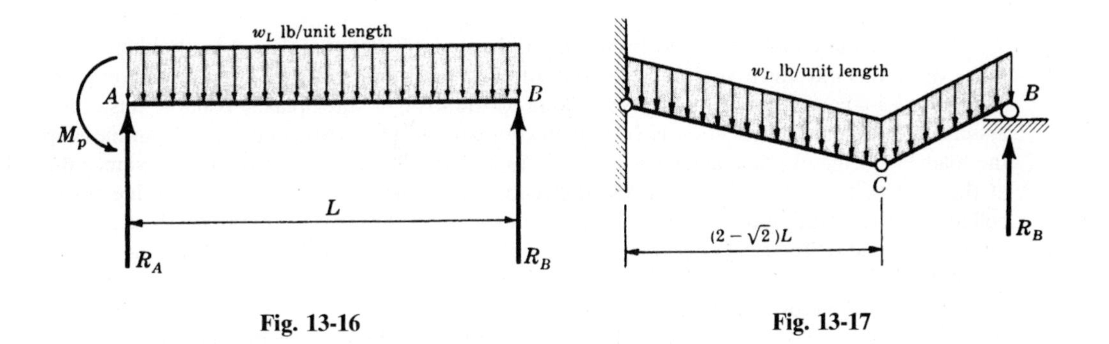

Fig. 13-16

Fig. 13-17

Solving (*1*), (*2*), and (*3*) simultaneously we arrive at the single equation

$$a^2 - 4a + 2 = 0$$

for determination of the point of relative maximum moment. Solving, we obtain $a = 2 - \sqrt{2}$, the other root of the quadratic being of no physical significance.

Substituting this value in (*2*), we find

$$w_L = (6 + 4\sqrt{2})\frac{M_p}{L^2}$$

as the limit load. The collapse mechanism appears in Fig. 13-17.

13.10. The clamped-end beam is subject to a concentrated force as shown in Fig. 13-18. Determine the magnitude of this load corresponding to plastic collapse of the beam.

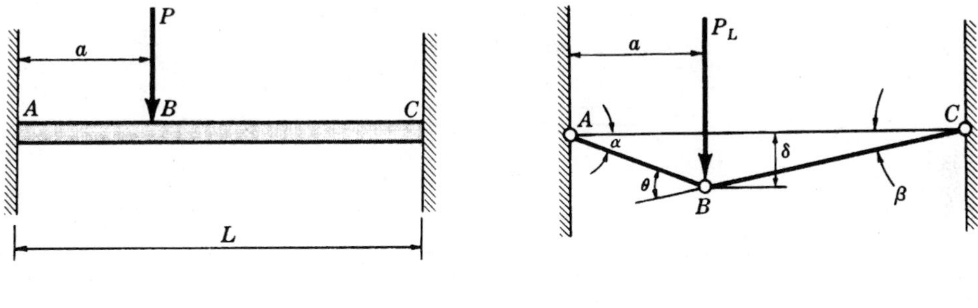

Fig. 13-18

Fig. 13-19

The only logical collapse mechanism is that of Fig. 13-19, where plastic hinges form at A, B, and C. From the geometry of triangle ABC we have

$$\alpha + \beta = \theta \tag{1}$$

or

$$\frac{\delta}{a} + \frac{\delta}{L-a} = \theta \tag{2}$$

since the deflection δ is still small compared to L even though plastic collapse has occurred. Solving (*2*) we obtain

$$\delta = \theta a\left(1 - \frac{a}{L}\right) \tag{3}$$

and from geometry we have

$$\alpha = \theta\left(1 - \frac{a}{L}\right) \qquad \beta = \frac{\theta a}{L} \tag{4}$$

This problem could be solved by use of statics equations as employed in Problems 13.5 and 13.8. However, let us introduce another technique which will be well suited to even more complex problems. This involves a consideration of the work done by the load P_L after plastic collapse has occurred. If we assume that the elastic deflection is very small compared to the plastic deflection, then the work done by the load P_L during plastic collapse is $P_L \delta$. It is to be carefully noted that the load assumes the value P_L at the start of the collapse through the deflection δ and maintains this constant value throughout the collapse process. During the collapse, the beam develops the fully plastic moment M_p at each of the hinge points A, B, and C. The total energy dissipated at these hinges is provided by and is equal to the work done by the load P_L.

The work done by the plastic hinge at A is given by $M_p \alpha$, at B it is given by $M_p \theta$ and at C by $M_p \beta$. Thus, equating work done by P_L to the net work done by these three plastic moments, and using (4) we have

$$P_L \delta = M_p \theta \left(1 - \frac{a}{L}\right) + M_p \theta + M_p \left(\frac{\theta a}{L}\right) \tag{5}$$

Substituting δ from (3) we have as the collapse load

$$P_L = \frac{2M_p L}{a(L - a)}$$

13.11. A horizontal beam of rectangular cross section 50 mm × 120 mm is 1.5 m long and hinged at its left end A as shown in Fig. 13-20. The right end C is supported by a vertical bar of the same material, of cross-sectional area 3 cm^2. The yield point of each material is 200 MPa. The beam is subject to a vertical force P applied at B. Determine the limit load P_L.

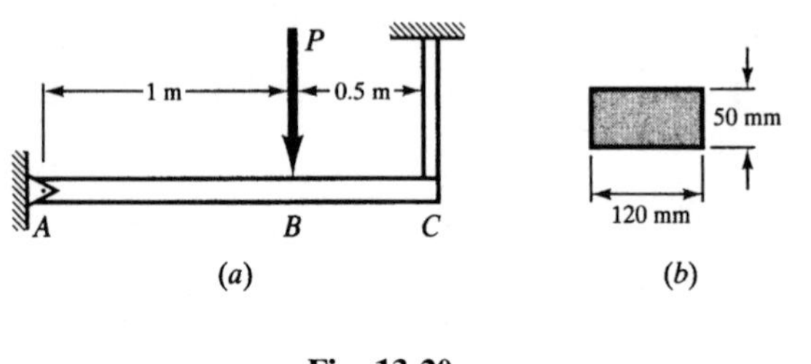

Fig. 13-20

It is not clear which yields first, the vertical bar or the horizontal beam AC. Let us assume that the vertical bar is the first to yield. The force in it is

$$F_1 = (200 \times 10^6 \text{ N/m}^2)(3 \text{ cm}^2)(1 \text{ m}/100 \text{ cm})^2 = 6 \times 10^4 \text{ N}$$

The free-body diagram of the beam is shown in Fig. 13-21.

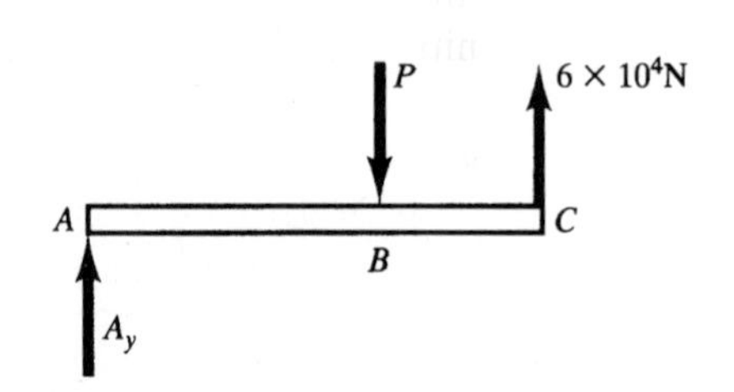

Fig. 13-21

For equilibrium,

$$+\uparrow \Sigma M_A = -P(1\text{ m}) + (6 \times 10^4\text{ N})(1.5\text{ m}) = 0$$

from which

$$P'_L = 9 \times 10^4\text{ N} \qquad \text{or} \qquad 90\text{ kN}$$

Next, assume that the beam develops a plastic hinge at B with the vertical bar still being entirely elastic. The free-body diagram of the left portion of the beam between A and a point just slightly to the left of B is shown in Fig. 13-22.

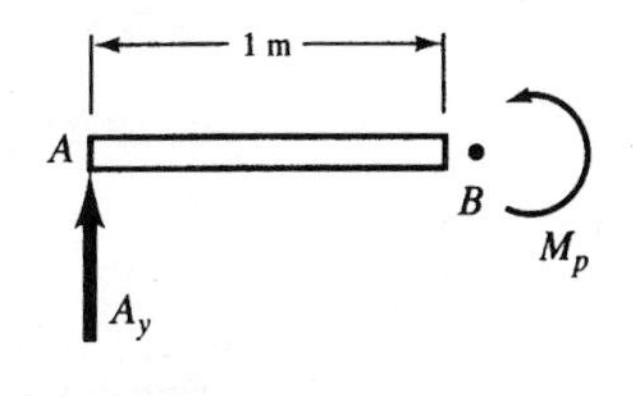

Fig. 13-22

For equilibrium of this portion of the beam,

$$+ \circlearrowright \Sigma M_B = M_p - A_y(1\text{ m}) = 0$$

For equilibrium of the entire beam AC about point C,

$$+ \circlearrowright \Sigma M_C = P''_L(0.5\text{ m}) - A_y(1\text{ m}) = 0$$

Solving,

$$P''_L = 3M_p$$

But for a bar of rectangular cross section, the fully plastic moment (see Problem 8.25) is given by

$$M_p = \sigma_{yp}\frac{bh^2}{4}$$

$$= (200 \times 10^6\text{ N/m}^2)\frac{(0.12\text{ m})(0.05\text{ m})^2}{4} = 15{,}000\text{ N}\cdot\text{m}$$

Thus,

$$P''_L = 3(15{,}000) = 45{,}000\text{ N} \qquad \text{or} \qquad 45\text{ kN}$$

Since this load of 45 kN is reached before the load of 90 kN (causing yield of the vertical bar), it is evident that the limit load is 45 kN, which will cause formation of a plastic hinge at B while the vertical bar is still elastic.

13.12. Consider the rectangular frame with both bases clamped subject to the two equal loads shown in Fig. 13-23. Determine the magnitude of the loads corresponding to plastic collapse of the frame.

In this situation there are three possible plastic collapse mechanisms. These are shown in Fig. 13-24, where Cases I and II correspond to individual actions of the applied loads and Case III is a composite mechanism formed as a combination of I and II so as to eliminate a plastic hinge at point B. We shall determine the collapse loads of each of these three cases and then select the minimum of the three loads as the correct one.

Case I can be treated by the methods of Problem 13.1, so that we immediately have $P_{L1} = 4M_p/L$. Case II can be treated by the same methods, so for it we have $P_{L2} = 4M_p/L$.

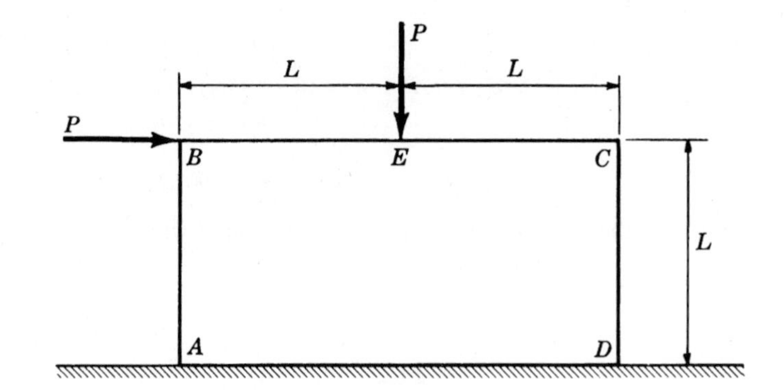

Fig. 13-23

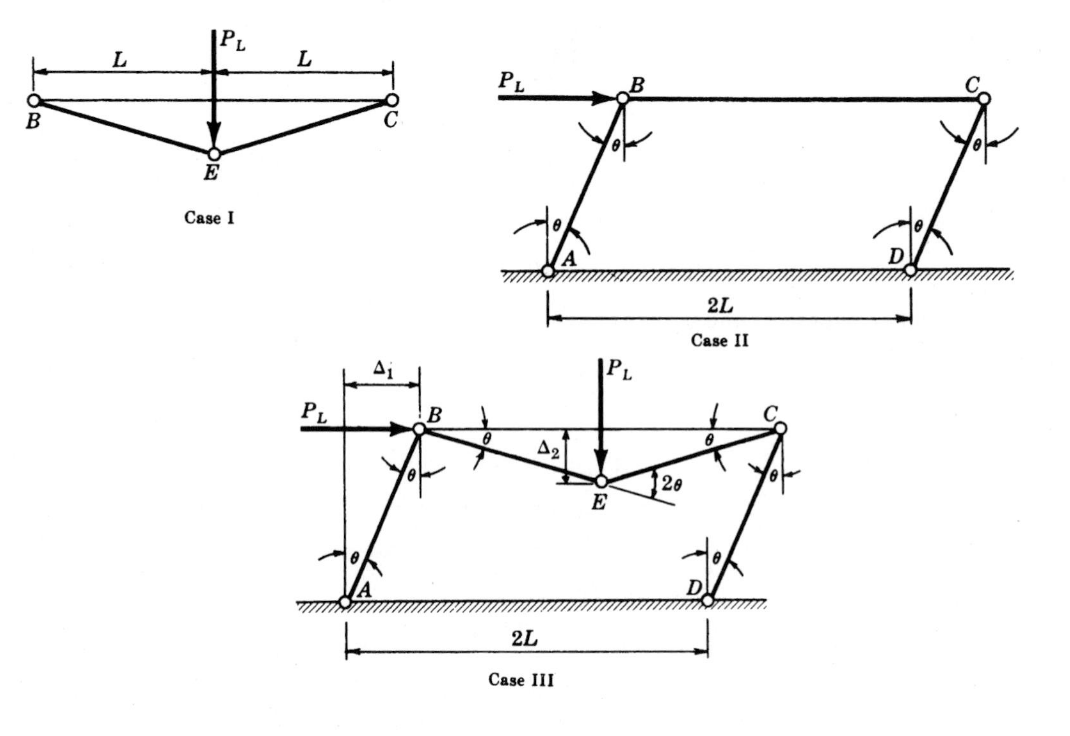

Fig. 13-24

For Case III there are plastic hinges at A, E, C, and D, with B constituting a rigid joint. Work-energy balance requires that

$$P_{L3}\Delta_1 + P_{L3}\Delta_2 = [M_p\,\theta]_A + [M_p(2\theta)]_E + [M_p(2\theta)]_C + [M_p\,\theta]_D$$

or

$$P_{L3}(L\theta) + P_{L3}(L\theta) = 6M_p\,\theta$$

from which $P_{L3} = 3M_p/L$.

Thus, the collapse load is $P_L = P_{L3} = 3M_p/L$ and collapse occurs as indicated by the sketch for Case III.

13.13. The continuous beam shown in Fig. 13-25(a) rests on three simple supports and is subject to the single concentrated load indicated. Determine the magnitude of this load for plastic collapse of the beam.

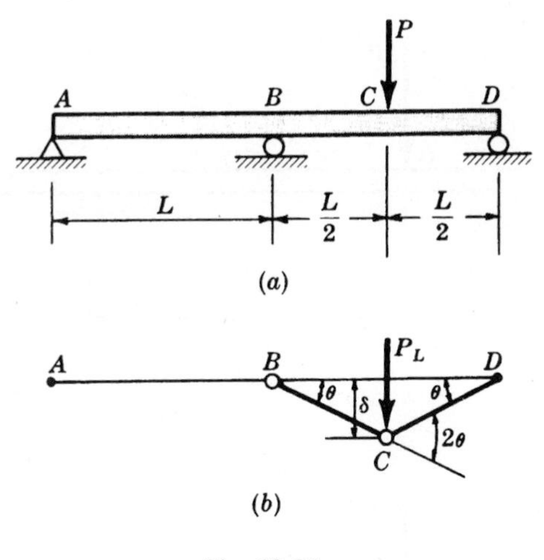

Fig. 13-25

The plastic collapse of such a beam usually occurs in only one of the spans and, in this case, collapse could occur by formation of a mechanism as indicated in Fig. 13-25(b), where plastic hinges form at points B and C.

The work done by the load P_L during plastic collapse is $P_L\,\delta$. The fully plastic moment M_p develops at each of the hinge points B and C. Work-energy balance requires that

$$P_L\,\delta = [M_p\,\theta]_B + [M_p(2\theta)]_C$$

or

$$P_L\left(\frac{L}{2}\theta\right) = 3M_p\,\theta$$

from which the collapse load is $P_L = 6M_p/L$.

13.14. A two-span continuous steel beam supports the concentrated forces indicated in Fig. 13-26(a). The beam is of rectangular cross section, 2 in wide by 4 in high, with the yield point of the steel being 38,000 lb/in^2. Determine the value of P to cause plastic collapse.

Let us first assume that collapse occurs in the span AC with the formation of the mechanism indicated in Fig. 13-26(b). Fully plastic moments develop at B and C and the work-energy balance requires that

$$2P_L(10\theta) = [M_p(2\theta)]_B + [M_p\,\theta]_C \qquad \text{or} \qquad P_L = \frac{3M_p}{20}$$

Next, consider the possibility of collapse in the span CE with the formation of the mechanism shown in Fig. 13-26(c). From the geometry of triangle CDE we have

$$\phi = \alpha + \beta$$

But since α is small compared to the span CE, this becomes

$$\frac{\delta_1}{8} + \frac{\delta_1}{2} = \phi$$

where δ_1 must of course be in consistent units (i.e., feet). Thus

$$\delta_1 = \tfrac{8}{5}\phi$$

and from geometry

$$\alpha = \tfrac{1}{5}\phi \qquad \beta = \tfrac{4}{5}\phi$$

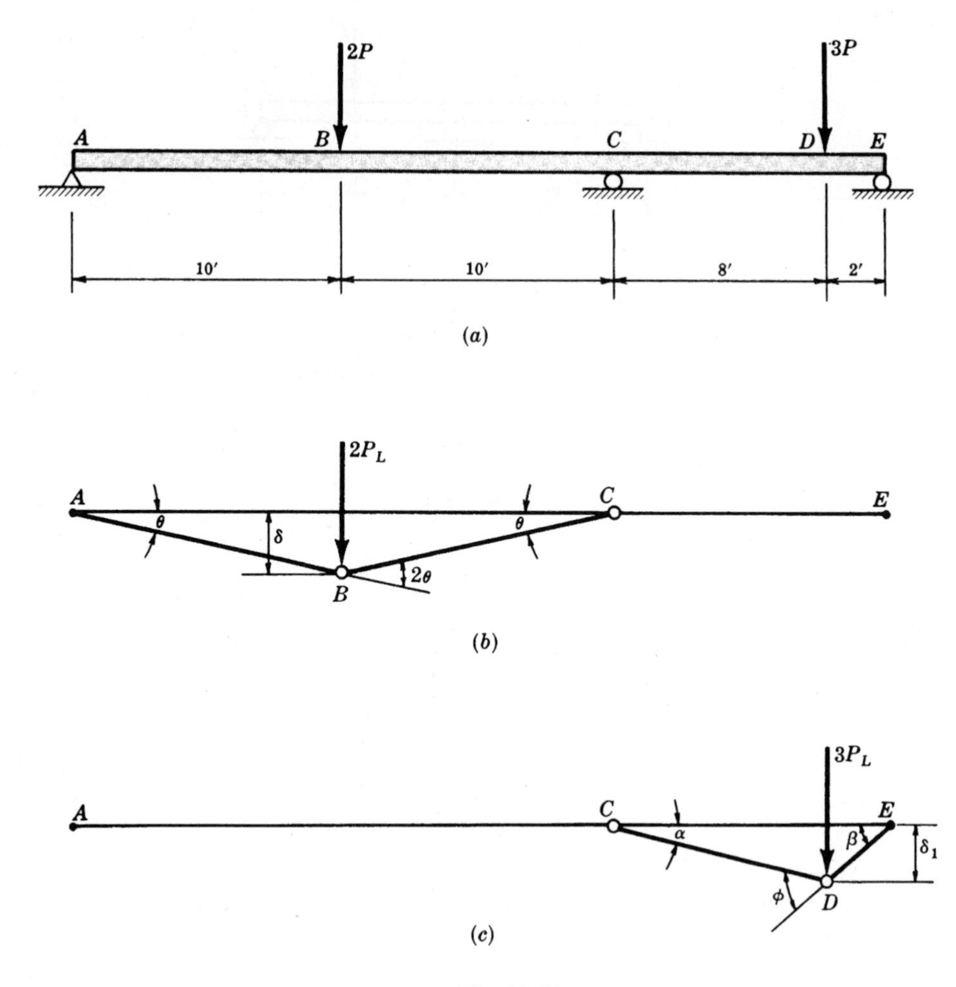

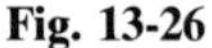
Fig. 13-26

In this case fully plastic moments develop at C and D and work-energy balance requires that

$$3P_L(8\alpha) = [M_p\phi]_D + [M_p\alpha]_C \qquad \text{or} \qquad P_L = \frac{M_p}{4}$$

Since this is larger than the P_L found for collapse of the left span, evidently collapse occurs with the formation of the mechanism shown for span AC.

Since the fully plastic moment for a rectangular cross section is given by

$$M_p = \sigma_{yp}\left(\frac{bh^2}{4}\right)$$

we find the collapse load to be

$$P_L = \frac{3}{20(12)}(38{,}000)\frac{(2)(4)^2}{4} = 3800 \text{ lb}$$

where the factor of 12 appears in the denominator to render the units consistent.

13.15. A simply supported beam of 50-mm × 75-mm rectangular cross section has a yield point stress of 250 MPa and carries the loads indicated in Fig. 13-27(a). Use the limit design criterion to determine the maximum load P.

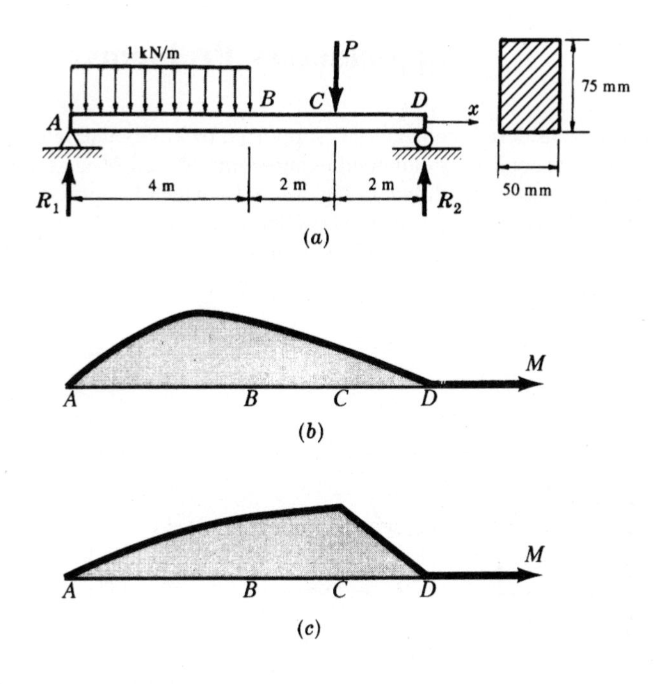

Fig. 13-27

From statics the reactions are $R_1 = 3 + (P/4)$ kN and $R_2 = 1 + (3P/4)$ kN.
Fully plastic action of this beam corresponds to a moment of

$$M_p = \sigma_{yp}\frac{bh^2}{4} = 250\frac{(50)(75)^2}{4} = 17.6 \text{ kN}\cdot\text{m}$$

In any problem involving several loads, the location of the first plastic hinge to form is usually not apparent. Here, two possibilities exist. In the first [Fig. 13-27(*b*)], the maximum moment would occur between points A and B. If this is the correct form of the moment diagram then the shear must vanish at some point for which $x < 4$. Thus, since

$$V = 3 + \frac{P}{4} - 1x$$

we must find P from the equation

$$0 = 3 + \frac{P}{4} - x \qquad \text{or} \qquad x = 3 + \frac{P}{4}$$

Since $x < 4$ in this consideration, this implies $P < 4$. A simple calculation indicates that $P = 4$ kN cannot develop the fully plastic moment of 17.6 kN · m.

For the second possibility [Fig. 13-27(*c*)], the maximum moment occurs at point C. The presence of a plastic hinge at C corresponds to a load P, given by

$$\left(1 + \frac{3P}{4}\right)(2) = 17.6 \text{ kN}\cdot\text{m} \qquad \text{or} \qquad P = 10.4 \text{ kN}$$

In this case the moment at B must be less than that at C, since the moment diagram must have a common tangent to the two branches meeting at B. Hence there is no need to investigate the moment at B. Thus $P = 10.4$ kN is the peak load that may be applied according to the limit design criterion.

Supplemetary Problems

13.16. In Problem 6.4 we considered the beam AD supported by knife-edge reactions at B and C as shown in Fig. 13-28(a). Loading was applied by end bending moments M_1 and $M_1/2$ as indicated in that figure. The beam has a T-shaped cross-section as shown in Fig. 13-28(b), which has previously been considered in Problem 8.32. If the material has a yield point of 39,000 lb/in², determine the maximum value of applied load for fully plastic action. *Ans.* $M_p = 360{,}750\ \text{lb} \cdot \text{in}$

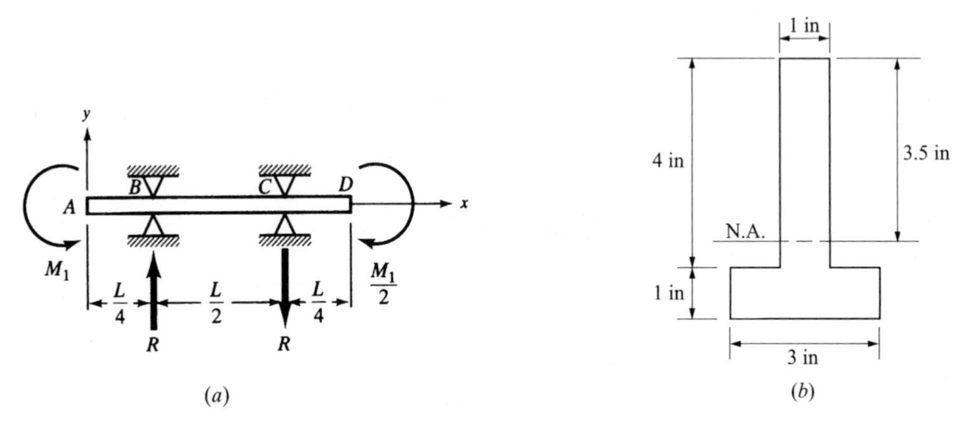

Fig. 13-28

13.17. Consider again the beam AD and loading shown in Fig. 13-28. The cross section is now a hollow rectangular shape as shown in Fig. 13-29. For a yield point of 39,000 lb/in², determine the maximum value of applied load for fully plastic action. *Ans.* 546,000 lb · in

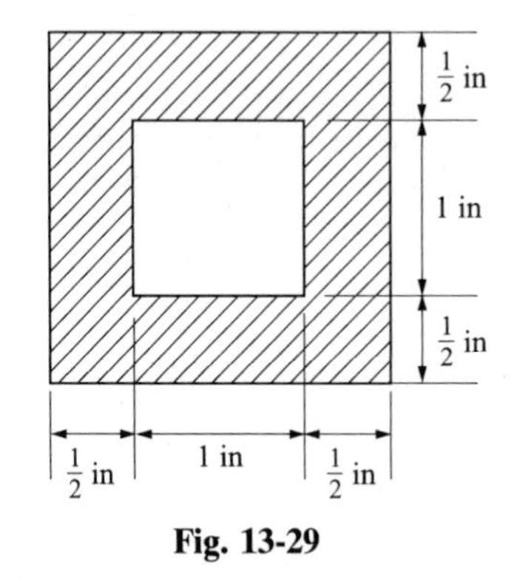

Fig. 13-29

13.18. Determine the limit load P of the simply supported beam of Fig. 13-30. *Ans.* $P_L = 4.5M_p/L$

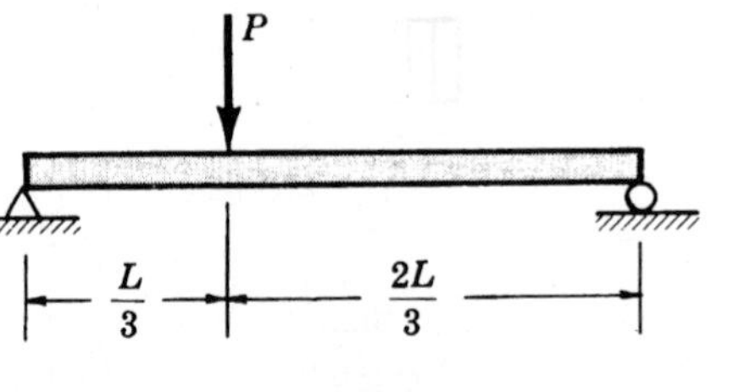

Fig. 13-30

13.19. The beam of Fig. 13-30 is of rectangular cross section, 25 mm × 50 mm. It is Hy-80 steel with a yield strength of 500 MPa. The length of the beam is 1 m. Determine the limit load when the loading is applied at the third point as indicated. *Ans.* $P_L = 35.2$ kN

13.20. The beam of Problem 13.4 is 2 m long and of square cross section 50 mm × 50 mm. It is structural steel with a yield stress of 250 MPa. Determine the limit load. *Ans.* $w_L = 15.6$ kN/m

13.21. Determine the magnitude of the limit load P_L for the beam clamped at one end and simply supported at the other (Fig. 13-31).

Ans. $P_L = M_p \dfrac{L+x}{(Lx - x^2)}$

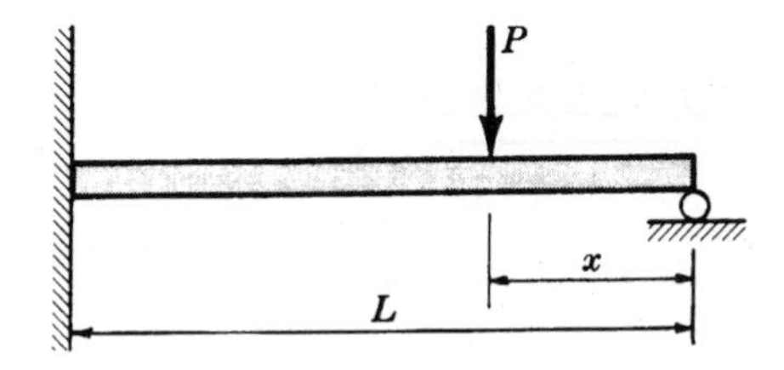

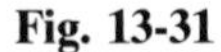
Fig. 13-31

13.22. In Problem 13.21 determine x so that P_L is a minimum. *Ans.* $x = 0.41L$, $(P_L)_{\min} = 5.64M_p/L$

13.23. The simply supported beam AC shown in Fig. 13-32 has a plastic moment M_p and carries the two concentrated loads shown. Determine the limit load P_L. *Ans.* $P_L = M_p/2L$

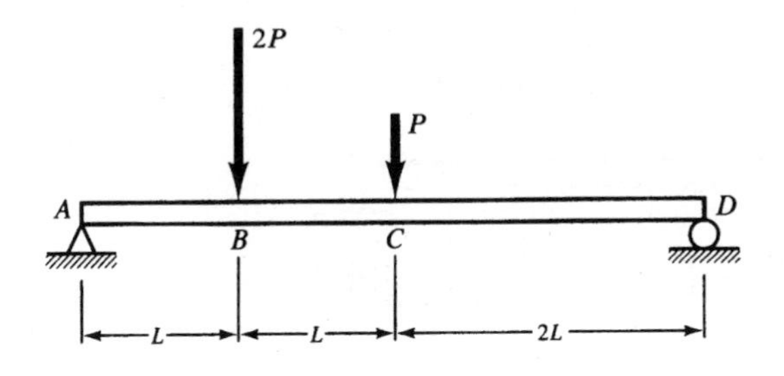

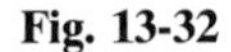
Fig. 13-32

Determine the magnitude of the load for plastic collapse of the systems shown in Figs. 13-33 and 13-34.

13.24. See Fig. 13-33.

Ans. $w_L = (6 + 4\sqrt{2})\dfrac{M_p}{L^2}$

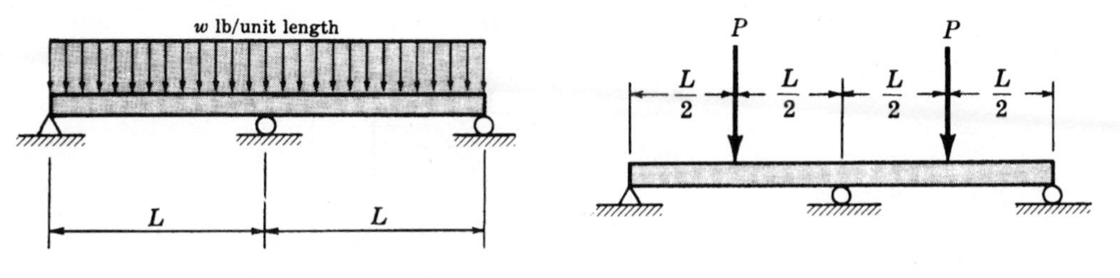

Fig. 13-33

Fig. 13-34

13.25. See Fig. 13-34.

Ans. $P_L = \dfrac{6M_p}{L}$

13.26. The continuous beam $ABCD$ is loaded as indicated in Fig. 13-35. Find the ratio $(w_L)L/P_L$ so that the limit load occurs in both AC and CD simultaneously. *Ans.* 2/3

13.27. The continuous beam shown in Fig. 13-36 rests on the three simple supports indicated. The span AC has a fully plastic moment $3M_p$ and the lighter span CD has a fully plastic moment M_p. A concentrated vertical force acts at the midpoint of AC. Find the limit load P_L. *Ans.* $P_L = 7M_p/L$

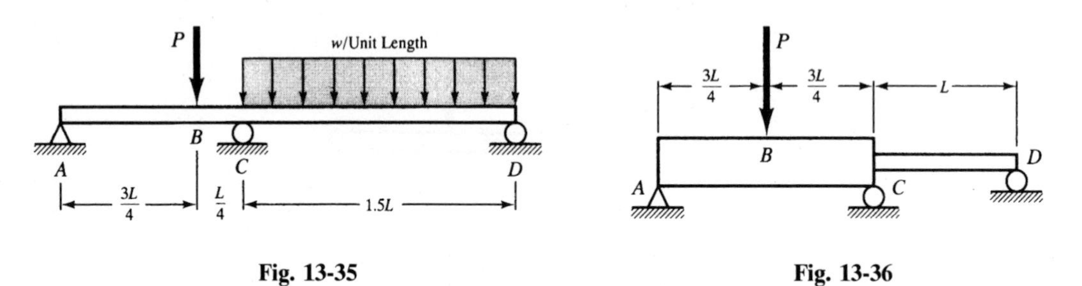

Fig. 13-35

Fig. 13-36

13.28. Determine the magnitude of the load P for plastic collapse of the beam shown in Fig. 13-37.

Ans. $P_L = \dfrac{6M_p}{L}$

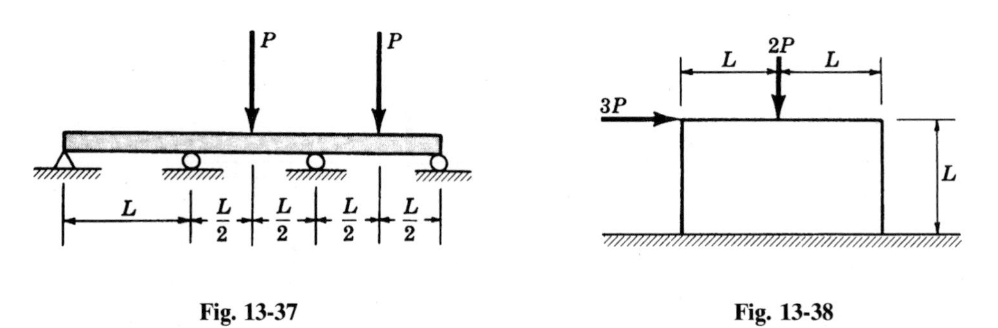

Fig. 13-37

Fig. 13-38

13.29. Determine the magnitude of P in Fig. 13-38 for plastic collapse of the rectangular frame having both bases clamped. *Ans.* $P_L = 1.2M_p/L$

13.30. Determine the magnitude of P for plastic collapse of the rectangular frame having both bases pinned (Fig. 13-39). *Ans.* $P_L = 4M_p/3L$

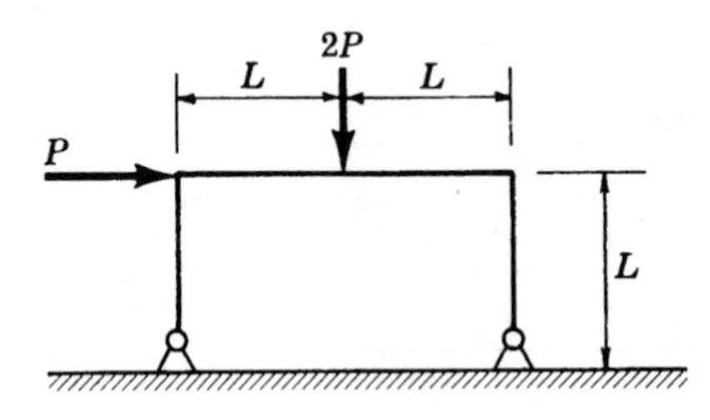

Fig. 13-39

13.31. Determine the magnitude of the force P for plastic collapse of the unsymmetric frame having both bases pinned (Fig. 13-40). *Ans.* $P_L = M_p(h_1 + h_2)/h_1 h_2$

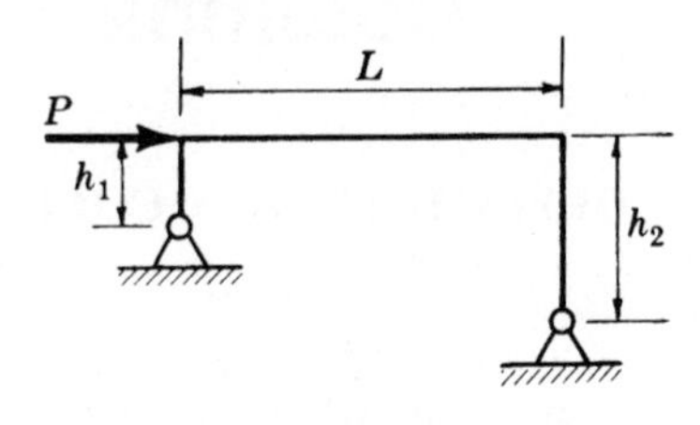

Fig. 13-40

13.32. See Fig. 13-41. Determine the value of P for plastic collapse of the system.

Ans. $P_L = \dfrac{2M_p}{L}$

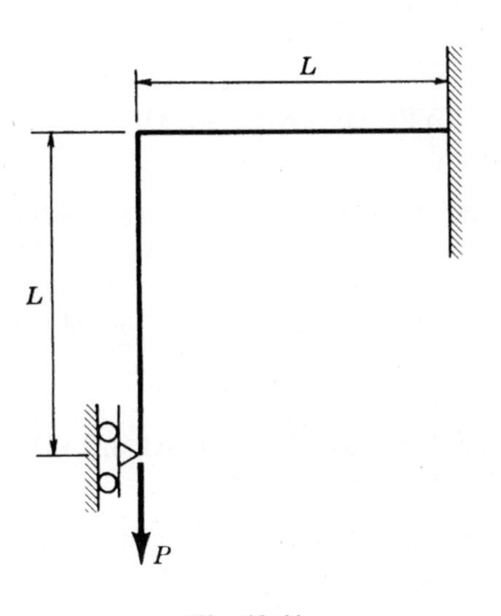

Fig. 13-41

Chapter 14

Columns

DEFINITION OF A COLUMN

A long slender bar subject to axial compression is called a *column*. The term "column" is frequently used to describe a vertical member, whereas the word "strut" is occasionally used in regard to inclined bars.

Examples

Many aircraft structural components, structural connections between stages of boosters for space vehicles, certain members in bridge trusses, and structural frameworks of buildings are common examples of columns.

TYPE OF FAILURE OF A COLUMN

Failure of a column occurs by buckling, i.e., by lateral deflection of the bar. In comparison it is to be noted that failure of a short compression member occurs by yielding of the material. Buckling, and hence failure, of a column may occur even though the maximum stress in the bar is less than the yield point of the material. Linkages in oscillating or reciprocating machines may also fail by buckling.

DEFINITION OF THE CRITICAL LOAD OF A COLUMN

The critical load of a slender bar subject to axial compression is that value of the axial force that is just sufficient to keep the bar in a slightly deflected configuration. Figure 14-1 shows a pin-ended bar in a buckled configuration due to the critical load P_{cr}.

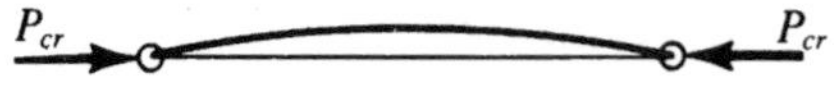

Fig. 14-1

SLENDERNESS RATIO OF A COLUMN

The ratio of the length of the column to the minimum radius of gyration of the cross-sectional area is termed the *slenderness ratio of the bar*. This ratio is of course dimensionless. The method of determining the radius of gyration of an area was discussed in Chap. 7.

If the column is free to rotate at each end, then buckling takes place about that axis for which the radius of gyration is a minimum.

CRITICAL LOAD OF A LONG SLENDER COLUMN

If a long slender bar of constant cross section is pinned at each end and subject to axial compression, the load P_{cr} that will cause buckling is given by

$$P_{cr} = \frac{\pi^2 EI}{L^2} \tag{14.1}$$

where E denotes the modulus of elasticity, I the minimum second moment of area of the cross-sectional area about an axis through the centroid, and L the length of the bar. The derivation of this formula is presented in Problem 14.1.

This formula was first obtained by the Swiss mathematician Leonhard Euler (1707–1783) and the load P_{cr} is called the *Euler buckling load*. As discussed in Problem 14.2, this expression is not immediately applicable if the corresponding axial stress, found from the expression $\sigma_{cr} = P_{cr}/A$, where A represents the cross-sectional area of the bar, exceeds the proportional limit of the material. For example, for a steel bar having a proportional limit of 210 MPa, the above formula is valid only for columns whose slenderness ratio exceeds 100. The value of P_{cr} represented by this formula is a failure load; consequently, a safety factor must be introduced to obtain a design load. Applications of this expression may be found in Problems 14.5 through 14.7.

INFLUENCE OF END CONDITIONS—EFFECTIVE LENGTH

Equation (*14.1*) may be modified to the form

$$P_{cr} = \frac{\pi^2 EI}{(KL)^2} \tag{14.2}$$

where KL is an effective length of the column. For a column pinned at both ends, $K = 1$. If both ends are clamped, $K = 0.5$; for one end clamped and the other pinned, $K = 0.7$. For a column clamped at one end and unsupported at the loaded end, $K = 2$. See Problems 14.1, 14.3, and 14.4.

DESIGN OF ECCENTRICALLY LOADED COLUMNS

The derivation of the expression leading to the Euler buckling load assumes that the column is loaded perfectly concentrically. If the axial force P is applied with an eccentricity e, the peak compressive stress in the bar occurs at the outer fibers at the midpoint of the length of the bar and is given by

$$\sigma_{max} = \frac{P}{A}\left[1 + \frac{ec}{r^2}\sec\left(\frac{L}{2}\sqrt{\frac{P}{AE}}\right)\right] \tag{14.3}$$

where c is the distance from the neutral axis to the outer fibers, r the radius of gyration, L the length of the column, and A the cross-sectional area. This is the *secant formula* for columns. It is discussed in detail in Problem 14.22.

INELASTIC COLUMN BUCKLING

The expression for the Euler buckling load may be extended into the inelastic range of action by replacing Young's modulus by the tangent modulus E_t. The resulting *tangent-modulus formula* is then

$$P_{cr} = \frac{\pi^2 E_t I}{L^2} \tag{14.4}$$

See Problem 14.9.

DESIGN FORMULAS FOR COLUMNS HAVING INTERMEDIATE SLENDERNESS RATIOS

The design of compression members having large values of the slenderness ratio proceeds according to the Euler formula presented above together with an appropriate safety factor. For the design of shorter compression members, it is customary to employ any one of the many semiempirical formulas giving a relationship between the yield stress and the slenderness ratio of the bar.

For steel columns, one commonly employed design expression is that due to the American Institute of Steel Construction (AISC), which states that the allowable (working) axial stress on a steel column having slenderness ratio L/r is

$$\sigma_a = \frac{[1-(KL/r)^2]\sigma_{yp}}{\left[\frac{5}{3} + \frac{3(KL/r)}{8C_c} - \frac{(KL/r)^3}{8C_c^3}\right]} \qquad \text{for } \frac{KL}{r} < C_c$$

$$\sigma_a = \frac{\pi^2 E}{(\frac{23}{12})(KL/r)^2} \qquad \text{for } \frac{KL}{r} > C_c \tag{14.5}$$

$$C_c = \sqrt{\frac{2\pi^2 E}{\sigma_{yp}}} \tag{14.6}$$

where σ_{yp} is the yield point of the material and E is Young's modulus. See Problems 14.11, 14.12, 14.13, and 14.14.

Another approach is in the use of the Structural Stability Research Council's (SSRC) equations which give mean axial compressive stress σ_u immediately prior to collapse:

$$\begin{aligned}
\sigma_u &= \sigma_{yp} && \text{for } 0 < \lambda < 0.15 \\
\sigma_u &= \sigma_{yp}(1.035 - 0.202\lambda - 0.222\lambda^2) && \text{for } 0.15 \leq \lambda \leq 1.0 \\
\sigma_u &= \sigma_{yp}(-0.111 + 0.636\lambda^{-1} + 0.087\lambda^{-2}) && \text{for } 1.0 \leq \lambda \leq 2.0 \\
\sigma_u &= \sigma_{yp}(0.009 + 0.877\lambda^{-2}) && \text{for } 2.0 \leq \lambda < 3.6 \\
\sigma_u &= \sigma_{yp}\lambda^{-2} \text{ (Euler's curve)} && \text{for } \lambda \geq 3.6
\end{aligned} \tag{14.7}$$

where

$$\lambda = \frac{L}{\pi r}\sqrt{\frac{\sigma_{yp}}{E}} \tag{14.8}$$

No safety factor is present in these equations but of course one must be introduced by the designer. See Problem 14.15.

COMPUTER IMPLEMENTATION

The design expression advanced by the AISC for allowable (working) stress on a steel column as well as the SSRC's equations giving mean axial compressive stress just prior to collapse are well suited to computer implementation. Problems 14.17 and 14.20, respectively, give FORTRAN programs for each of these recommendations. It is only necessary to input into the self-prompting programs the geometric and materials parameters of the column to obtain its resistance as indicated by each of these sets of relations. For application see Problems 14.18, 14.19, and 14.21.

BEAM-COLUMNS

Bars subjected to simultaneous axial compression and lateral loading are termed *beam-columns*. An example is given in Problem 14.25.

BUCKLING OF RIGID SPRING-SUPPORTED BARS

The columns discussed above are *flexible* members, i.e., capable of undergoing lateral bending immediately after buckling. A related type of buckling involves one or more *rigid* bars pinned to fixed supports or to each other and supported by one or more transverse springs. In certain cases the applied loads may cause the bar system to move suddenly to an alternate equilibrium position. This too is a form of instability of the system. See Problem 14.26.

Solved Problems

14.1. Determine the critical load for a long slender pin-ended bar loaded by an axial compressive force at each end. The line of action of the forces passes through the centroid of the cross section of the bar.

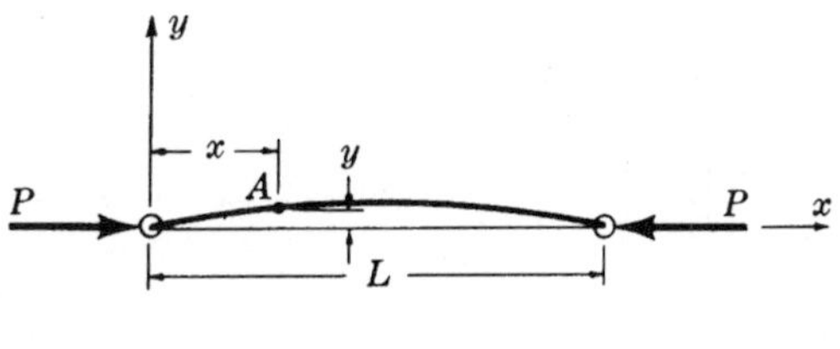

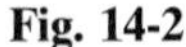

Fig. 14-2

The critical load is defined to be that axial force that is just sufficient to hold the bar in a slightly deformed configuration. Under the action of the load P the bar has the deflected shape shown in Fig. 14-2.

It is of course necessary that one end of the bar be able to move axially with respect to the other end in order that the lateral deflection may take place. The differential equation of the deflection curve is the same as that presented in Chap. 9, namely,

$$EI\frac{d^2y}{dx^2} = M \tag{1}$$

Here the bending moment at the point A having coordinates (x,y) is merely the moment of the force P applied at the left end of the bar about an axis through the point A and perpendicular to the plane of the page. It is to be carefully noted that this force produces curvature of the bar that is concave downward, which, according to the sign convention of Chap. 6, constitutes negative bending. Hence the bending moment is $M = -Py$. Thus we have

$$EI\frac{d^2y}{dx^2} = -Py \tag{2}$$

If we set

$$\frac{P}{EI} = k^2 \tag{3}$$

(*2*) becomes

$$\frac{d^2y}{dx^2} + k^2y = 0 \tag{4}$$

This equation is readily solved by any one of several standard techniques discussed in works on differential equations. However, the solution is almost immediately apparent. We need merely find a function which when differentiated twice and added to itself (times a constant) is equal to zero. Evidently either $\sin kx$ or $\cos kx$ possesses this property. In fact, a combination of these terms in the form

$$y = C \sin kx + D \cos kx \tag{5}$$

may also be taken to be a solution of (*4*). This may be readily checked by substitution of y as given by (*5*) into (*4*).

Having obtained y in the form given in (*5*), it is next necessary to determine C and D. At the left end of the bar, $y = 0$ when $x = 0$. Substituting these values in (*5*), we obtain

$$0 = 0 + D \qquad \text{or} \qquad D = 0$$

At the right end of the bar, $y = 0$ when $x = L$. Substituting these values in (*5*) with $D = 0$, we obtain

$$0 = C \sin kL$$

Evidently either $C = 0$ or $\sin kL = 0$. But if $C = 0$ then y is everywhere zero and we have only the trivial case of a straight bar which is the configuration prior to the occurrence of buckling. Since we are not interested in the solution, then we must take

$$\sin kL = 0 \tag{6}$$

For this to be true, we must have

$$kL = n\pi \text{ radians } (n = 1, 2, 3, \ldots) \tag{7}$$

Substituting $k^2 = P/EI$ in (*7*), we find

$$\sqrt{\frac{P}{EI}}L = n\pi \qquad \text{or} \qquad P = \frac{n^2\pi^2 EI}{L^2} \tag{8}$$

The smallest value of this load P evidently occurs when $n = 1$. Then we have the so-called first mode of buckling where the critical load is given by

$$P_{cr} = \frac{\pi^2 EI}{L^2} \tag{9}$$

This is called *Euler's buckling load for a pin-ended column*. The deflection shape corresponding to this load is

$$y = C \sin\left(\sqrt{\frac{P}{EI}}x\right) \tag{10}$$

Substituting in this equation from (*9*), we obtain

$$y = C \sin\frac{\pi x}{L} \tag{11}$$

Thus the deflected shape is in a sine curve. Because of the approximations introduced in the derivation of (*1*), it is not possible to obtain the amplitude of the buckled shape, denoted by C in (*11*).

As may be seen from (*9*), buckling of the bar will take place about that axis in the cross section for which I assumes a minimum value.

Equation (*9*) may be modified to the form

$$P_{cr} = \frac{\pi^2 EI}{(KL)^2} \tag{12}$$

where KL is an effective length of the column, defined to be a portion of the deflected bar between points corresponding to zero curvature. For example, for a column pinned at both ends, $K = 1$. If both ends are rigidly clamped, $K = 0.5$. For one end clamped and the other pinned, $K = 0.7$. In the case of a cantilever-type column loaded at its free end, $K = 2$.

14.2. Determine the axial stress in the column considered in Problem 14.1.

In the derivation of the equation $EI(d^2y/dx^2) = M$ used to determine the critical load in Problem 14.1, it was assumed that there is a linear relationship between stress and strain (see Chap. 9). Thus the critical load indicated by (*9*) of Problem 14.1 is correct only if the proportional limit of the material has not been exceeded.

The axial stress in the bar immediately prior to the instant when the bar assumes its buckled configuration is given by

$$\sigma_{cr} = \frac{P_{cr}}{A} \tag{1}$$

where A represents the cross-sectional area of the bar. Substituting for P_{cr} its value as given by (*9*) of Problem 14.1, we find

$$\sigma_{cr} = \frac{\pi^2 EI}{AL^2} \tag{2}$$

But from Chap. 7 we know that we may write

$$I = Ar^2 \tag{3}$$

where r represents the radius of gyration of the cross-sectional area. Substituting this value in (*2*), we find

$$\sigma_{cr} = \frac{\pi^2 EAr^2}{AL^2} = \pi^2 E\left(\frac{r}{L}\right)^2 \tag{4}$$

or

$$\sigma_{cr} = \frac{\pi^2 E}{(L/r)^2} \tag{5}$$

The ratio L/r is called the *slenderness ratio* of the column.

Let us consider a steel column having a proportional limit of 210 MPa and $E = 200$ GPa. The stress of 210 MPa marks the upper limit of stress for which (*5*) may be used. To find the value of L/r corresponding to these constants, we substitute in (*5*) and obtain

$$210 \times 10^6 = \frac{\pi^2(200 \times 10^9)}{(L/r)^2} \qquad \text{or} \qquad \frac{L}{r} \approx 100$$

Thus for this material the buckling load as given by (*9*) of Problem 14.1 and the axial stress as given by (*5*) are valid only for those columns having $L/r \geqslant 100$. For those columns having $L/r < 100$, the compressive stress exceeds the proportional limit before elastic buckling takes place and the above equations are not valid.

Equation (*5*) may be plotted as shown in Fig. 14-3. For the particular values of proportional limit and modulus of elasticity assumed above, the portion of the curve to the left of $L/r = 100$ is not valid. Thus for this material, point A marks the upper limit of applicability of the curve.

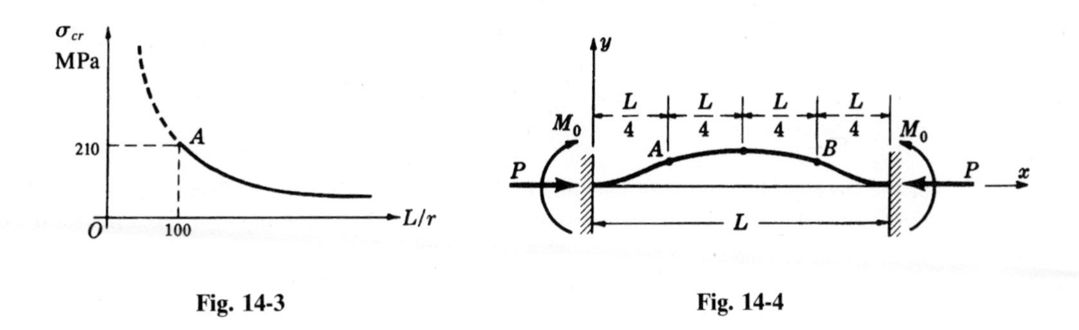

Fig. 14-3 **Fig. 14-4**

14.3. Determine the critical load of a long, slender bar clamped at each end and subject to axial thrust as shown in Fig. 14-4.

Let us introduce the x–y coordinate system shown in Fig. 14-4 and let (x, y) represent the coordinates of an arbitrary point on the bar. The bending moment at this point is found as the sum of the moments of the forces to the left of this section about an axis through this point and perpendicular to the plane of the page. Hence at this point we have $M = -Py + M_0$. The differential equation for the bending of the bar is then $EId^2y/dx^2 = -Py + M_0$, or

$$\frac{d^2y}{dx^2} + \frac{P}{EI}y = \frac{M_0}{EI} \tag{1}$$

As discussed in texts on differential equations, the solution to (*1*) consists of two parts. The first part is merely the solution of the so-called homogeneous equation obtained by setting the right-hand side of (*1*) equal to zero. We must then solve the equation

$$\frac{d^2y}{dx^2} + \frac{P}{EI}y = 0 \tag{2}$$

But the solution to this equation has already been found in Problem 14.1 to be

$$y = A_1 \cos\left(\sqrt{\frac{P}{EI}}x\right) + B_1 \sin\left(\sqrt{\frac{P}{EI}}x\right) \tag{3}$$

The second part of the solution of (*1*) is given by a so-called particular solution, i.e., any function satisfying (*1*). Evidently one such function is given by

$$y = \frac{M_0}{P} (= \text{constant}) \tag{4}$$

The general solution of (*1*) is given by the sum of the solutions represented by (*3*) and (*4*), or

$$y = A_1 \cos\left(\sqrt{\frac{P}{EI}}x + B_1 \sin\sqrt{\frac{P}{EI}}x\right) + \frac{M_0}{P} \tag{5}$$

Consequently

$$\frac{dy}{dx} = -A_1\sqrt{\frac{P}{EI}}\sin\left(\sqrt{\frac{P}{EI}}x\right) + B_1\sqrt{\frac{P}{EI}}\cos\left(\sqrt{\frac{P}{EI}}x\right) \tag{6}$$

At the left end of the bar we have $y = 0$ when $x = 0$. Substituting these values in (*5*), we find $0 = A_1 + M_0/P$. Also, at the left end of the bar we have $dy/dx = 0$ when $x = 0$; substituting in (*6*), we obtain $0 = 0 + B_1\sqrt{P/EI}$ or $B_1 = 0$.

At the right end of the bar we have $dy/dx = 0$ when $x = L$; substituting in (*6*), with $B_1 = 0$, we find

$$0 = -A_1\sqrt{\frac{P}{EI}}\sin\left(\sqrt{\frac{P}{EI}}L\right)$$

But $A_1 = -M_0/P$ and since this ratio is not zero, then $\sin(\sqrt{P/EI}\,L) = 0$. This occurs only when $\sqrt{P/EI}\,L = n\pi$ where $n = 1, 2, 3, \ldots$. Consequently

$$P_{cr} = \frac{n^2\pi^2 EI}{L^2} \tag{7}$$

For the so-called first mode of buckling illustrated in Fig. 14-4, the deflection curve of the bent bar has a horizontal tangent at $x = L/2$; that is, $dy/dx = 0$ there. Equation (*6*) now takes the form

$$\frac{dy}{dx} = \frac{M_0}{P}\left(\frac{n\pi}{L}\right)\sin\frac{n\pi x}{L} \tag{6'}$$

and since $dy/dx = 0$ at $x = L/2$, we find

$$0 = \frac{M_0}{P}\left(\frac{n\pi}{L}\right)\sin\frac{n\pi}{2}$$

The only manner in which this equation may be satisfied is for n to assume even values; that is, $n = 2, 4, 6, \ldots$.

Thus for the smallest possible value of $n = 2$, Eq. (7) becomes

$$P_{cr} = \frac{4\pi^2 EI}{L^2}$$

14.4. Determine the critical load for a long slender bar clamped at one end, free at the other, and loaded by an axial compressive force applied at the free end.

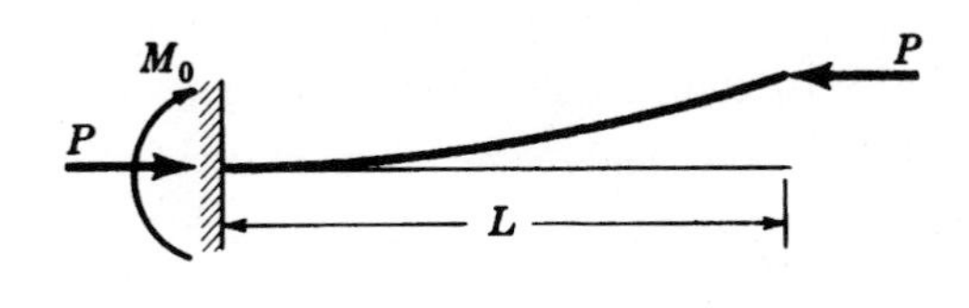

Fig. 14-5

The critical load is that axial compressive force P that is just sufficient to keep the bar in a slightly deformed configuration, as shown in Fig. 14-5. The moment M_0 represents the effect of the support in preventing any angular rotation of the left end of the bar.

Inspection of the above deflection curve for the buckled column indicates that the entire bar corresponds to one-half of the deflected pin-ended bar discussed in Problem 14.1. Thus for the column under consideration, the length L corresponds to $L/2$ for the pin-ended column. Hence the critical load for the present column may be found from Eq. (9), Problem 14.1, by replacing L by $2L$. This yields

$$P_{cr} = \frac{\pi^2 EI}{(2L)^2} = \frac{\pi^2 EI}{4L^2}$$

14.5. A steel bar of rectangular cross section 40 mm × 50 mm and pinned at each end is subject to axial compression. If the proportional limit of the material is 230 MPa and $E = 200$ GPa, determine the minimum length for which Euler's equation may be used to determine the buckling load.

The minimum second moment of area is $I = \frac{1}{12}bh^3 = \frac{1}{12}(50)(40)^3 = 2.67 \times 10^5\ \text{mm}^4$. Hence the least radius of gyration is

$$r = \sqrt{\frac{I}{A}} = \sqrt{\frac{2.67 \times 10^5}{(40)(50)}} = 11.5\ \text{mm}$$

The axial stress for such an axially loaded bar was found in Problem 14.2 to be

$$\sigma_{cr} = \frac{\pi^2 E}{(L/r)^2}$$

The minimum length for which Euler's equation may be applied is found by placing the critical stress in the above formula equal to 230 MPa. Doing this, we obtain

$$230 \times 10^6 = \frac{\pi^2(200 \times 10^9)}{(L/11.5)^2} \qquad \text{or} \qquad L = 1.065\ \text{m}$$

14.6. Consider again a rectangular steel bar 40 mm × 50 mm in cross section, pinned at each end and subject to axial compression. The bar is 2 m long and $E = 200$ GPa. Determine the buckling load using Euler's formula.

The *minimum* second moment of area of this cross section was found in Problem 14.5 to be $2.67 \times 10^5\ \text{mm}^4$. Applying the expression for buckling load given in (*9*) of Problem 14.1, we find

$$P_{cr} = \frac{\pi^2 EI}{L^2} = \frac{\pi^2(200 \times 10^9)(10^{-6})(2.67 \times 10^5)}{(2 \times 10^3)^2} = 132\ \text{kN}$$

The axial stress corresponding to this load is

$$\sigma_{cr} = \frac{P_{cr}}{A} = \frac{132 \times 10^3}{(40)(50)} = 66\ \text{MPa}$$

14.7. Determine the critical load for a W10 × 21 section acting as a pinned end column. The bar is 12 ft long and $E = 30 \times 10^6\ \text{lb/in}^2$. Use Euler's theory.

From Table 8-1 of Chap. 8 we find the minimum moment of inertia to be $9.7\ \text{in}^4$. Thus,

$$P_{cr} = \frac{\pi^2 EI}{L^2}$$

$$= \frac{\pi^2(30 \times 10^6\ \text{lb/in}^2)(9.7\ \text{in}^4)}{(144\ \text{in})^2} = 138{,}000\ \text{lb}$$

14.8. A long thin bar of length L and rigidity EI is pinned at end A, and at the end B rotation is resisted by a restoring moment of magnitude λ per radian of rotation at that end. Derive the equation for the axial buckling load P. Neither A nor B can displace laterally, but A is free to approach B.

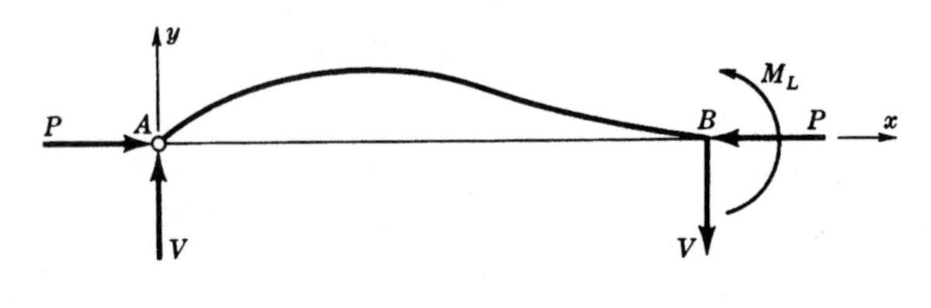

Fig. 14-6

The buckled bar is shown in Fig. 14-6, where M_L represents the restoring moment. The differential equation of the buckled bar is

$$EI\frac{d^2y}{dx^2} = Vx - Py$$

or

$$\frac{d^2y}{dx^2} + \frac{P}{EI}y = \frac{V}{EI}x$$

Let $\alpha^2 = P/EI$. Then

$$\frac{d^2y}{dx^2} + \alpha^2 y = \frac{V}{EI}x$$

The general solution of this equation is easily found to be

$$y = A\sin\alpha x + B\cos\alpha x + \frac{V}{P}x \qquad (1)$$

As the first boundary condition, when $x = 0$, $y = 0$; hence $B = 0$. As the second boundary condition, when $x = L$, $y = 0$; hence from (*1*) we obtain

$$0 = A \sin \alpha L + \frac{VL}{P} \qquad \text{or} \qquad \frac{V}{P} = -\frac{A}{L} \sin \alpha L$$

Thus

$$y = A\left[\sin \alpha x - \frac{x}{L} \sin \alpha L\right] \tag{2}$$

From (*2*) the slope at $x = L$ is found to be

$$\left[\frac{dy}{dx}\right]_{x=L} = A\left[\alpha \cos \alpha L - \frac{1}{L} \sin \alpha L\right] \tag{3}$$

The restoring moment at end B is thus

$$M_L = A\lambda\left[\alpha \cos \alpha L - \frac{1}{L} \sin \alpha L\right] \tag{4}$$

Also, since in general $M = EI(d^2y/dx^2)$, from (*2*) we have

$$M_L = -A\alpha^2 EI \sin \alpha L \tag{5}$$

Equating expressions (*4*) and (*5*) after carefully noting that as M_L increases dy/dx at that point decreases (necessitating the insertion of a negative sign), we have

$$-A\alpha^2 EI \sin \alpha L = -\left[A\lambda\alpha \cos \alpha L - \frac{A\lambda}{L} \sin \alpha L\right] \tag{6}$$

Simplifying, the equation for determination of the buckling load P becomes

$$\frac{PL}{\lambda} - \alpha L \cot \alpha L + 1 = 0 \tag{7}$$

This equation would have to be solved numerically for specific values of EI, L, and λ.

14.9. Discuss column behavior when the average applied axial stress in the bar exceeds the proportional limit of the material.

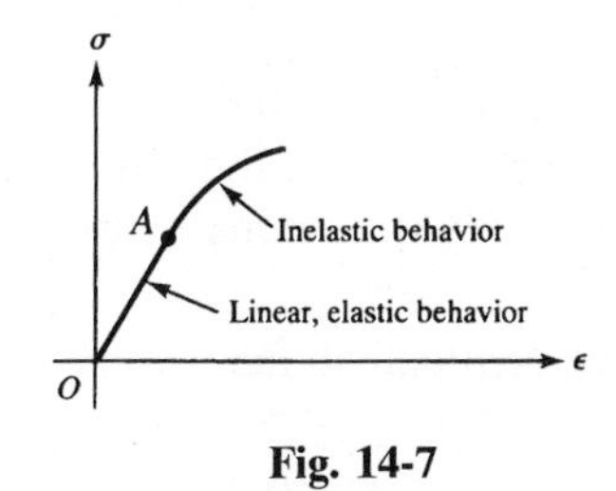

Fig. 14-7

The Euler buckling load determined in Problem 14.1 is based upon the assumption that the column everywhere is acting within the linear elastic range of action of the material, shown as OA in Fig. 14-7. In this range the modulus E is the slope of the straight line OA. When the stress-strain curve ceases to be linear, i.e., to the right of point A, the slope of the curve is called the *tangent modulus* E_t and it varies with strain. This parameter must be determined by materials tests. Under these conditions it is necessary to consider inelastic buckling. One of the earliest approaches to this, still used occasionally, is due to the German engineer Engesser who, in 1889, suggested replacing E in Euler's expression, Eq. (*9*) of Problem 14.1, by the *tangent modulus* E_t. In this case the axial stress immediately prior to buckling is given by

$$\sigma_{cr} = \frac{\pi^2 E_t}{(L/r)^2}$$

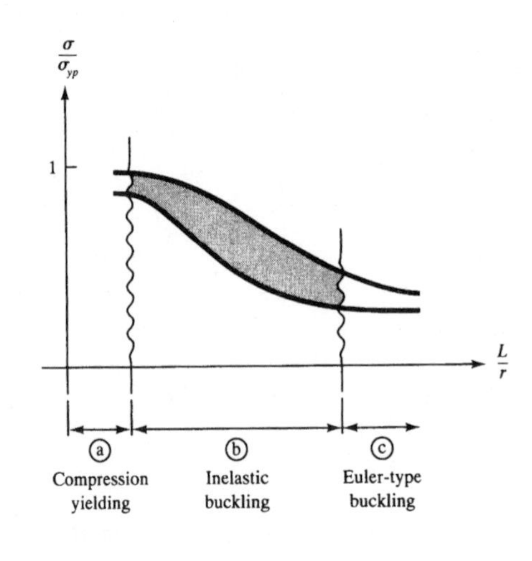

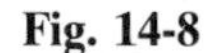

Fig. 14-8

This is the *tangent modulus formula* and the load $P_{cr} = A\sigma_{cr}$ is called the *Engesser load*. This approach is simple and easy to use—see Problem 14.10—and indicates a load only slightly less than the inelastic buckling load found experimentally. The theory has certain inconsistencies that will not be discussed here so it is not the best approach to rational column design.

Test results on axially compressed bars usually can be exhibited by the plot shown in Fig. 14-8, where the mean axial stress σ just before buckling (divided by the yield point of the material) is shown as a function of the slenderness ratio L/r. Experimental results indicate wide scatter, as shown by data points between the two solid curves. The scatter is due to initial geometric deviations from straightness of the bar as well as residual stresses incurred during fabrication. The plot indicates three modes of failure, depending on the value of L/r. The first is ⓐ, compressive yielding for very short columns; the second is ⓑ, inelastic buckling for intermediate length bars (which comprise many engineering applications); and the third is ⓒ, Euler-type buckling of very long slender bars. Failures of type ⓐ have been discussed in Chap. 1 and Euler column behavior was treated in Problems 14.1 through 14.7. The rational design of columns corresponding to condition ⓑ is based upon any one of a number of semiempirical approaches discussed in the following problems.

14.10. A pinned end column is 275 mm long and has a solid circular cross section. If it must support an axial load of 250 kN, determine the required radius of the rod if the tangent modulus theory is employed and the experimentally determined curve relating tangent modulus to axial stress is that shown in Fig. 14-9.

From Problem 14.9 the load, according to the tangent modulus theory, is given by

$$P_{cr} = (A)\frac{\pi^2 E_t}{(L/r)^2} = \frac{\pi^2 E_t I}{L^2} \tag{1}$$

For the solid circular cross section of radius R, we have $I = \pi R^4/4$ so that (*1*) becomes

$$E_t = \frac{(250{,}000\ \text{N})(0.275\ \text{m})^2}{\pi^2(\pi R^4/4)} = \frac{2439}{R^4}\ \text{N/m}^2 \tag{2}$$

For any assumed radius R it is easily possible to find the axial stress:

$$\sigma = \frac{P}{A} = \frac{250{,}000}{\pi R^2} \tag{3}$$

and for any value of σ from Fig. 14-9 we can ascertain the corresponding experimentally determined value of E_t. Thus, we can solve Eqs. (*2*) and (*3*) by trial and error.

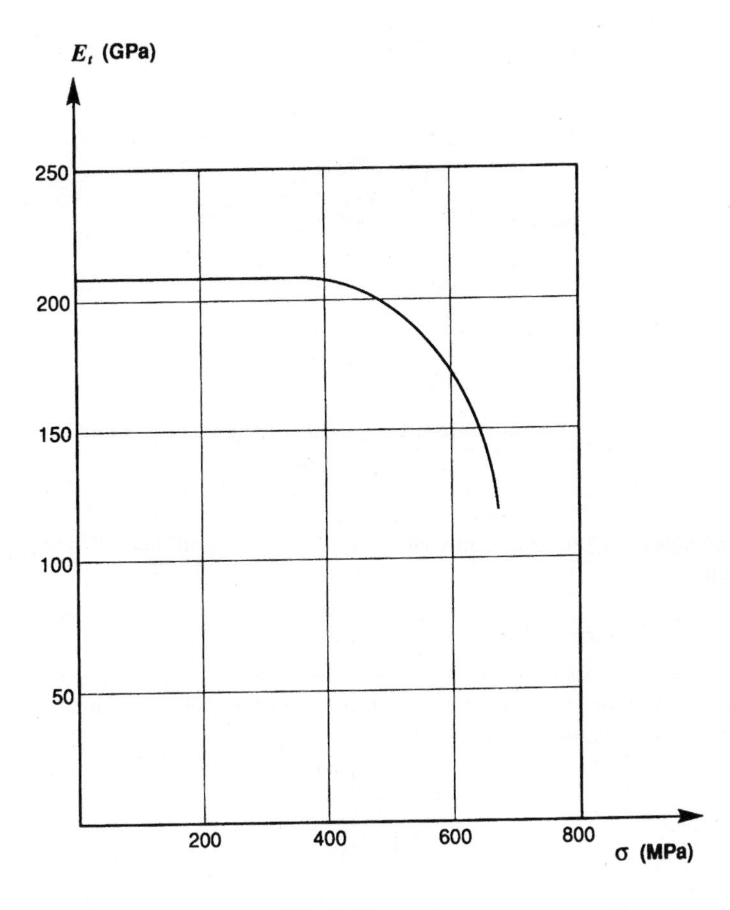

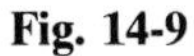
Fig. 14-9

Let us try $R = 0.012$ m. From Eq. (*3*)

$$\sigma = \frac{250{,}000}{\pi(0.012\text{ m})^2} = 553\text{ MPa}$$

For this value of σ from Fig. 14-9, we have $E_t = 175$ GPa. However, from Eq. (*2*) it is

$$E_t = \frac{2439}{(0.012\text{ m})^4} = 117\text{ GPa}$$

Clearly these values of E_t do not agree and the assumed radius is too large.

Next, let us try $R = 0.011$ m. From Eq. (*3*)

$$\sigma = \frac{250{,}000}{\pi(0.011\text{ m})^2} = 658\text{ MPa}$$

For this value of σ from Fig. 14-9, we have $E_t = 125$ GPa. However, from Eq. (*2*) it is

$$E_t = \frac{2439}{(0.011\text{ m})^4} = 167\text{ GPa}$$

It is instructive to plot these values as shown in Fig. 14-10. Clearly an acceptable value of radius lies between 0.011 and 0.012 m. Let us try $R = 0.0112$ m. From Eq. (*3*) we have

$$\sigma = \frac{250{,}000}{\pi(0.0112)^2} = 634\text{ MPa}$$

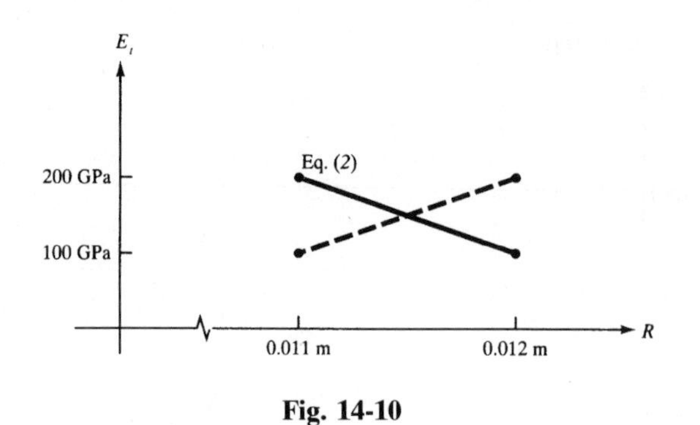

Fig. 14-10

and the corresponding value of E_t from Fig. 14-10 is 152 GPa. The value found from Eq. (2) is

$$E_t = \frac{2439}{(0.0112\ \text{m})^4} = 155\ \text{GPa}$$

These two values of E_t are sufficiently close that we may regard the radius of 0.0112 m as acceptable, that is, 11.2 mm.

14.11. Discuss design criteria for structural steel columns.

In one approach, advocated by the AISC, the allowable axial compressive stress σ_a on a steel column of length L, minimum radius of gyration of cross section r, material yield point σ_{yp}, and Young's modulus E is given by the semiempirical relations

$$\sigma_a = \frac{\left[1 - \dfrac{(KL/r)^2}{2C_c^2}\right]\sigma_{yp}}{\left[\dfrac{5}{3} + \dfrac{3(KL/r)}{8C_c} - \dfrac{(KL/r)^3}{8C_c^3}\right]} \qquad \text{for } \frac{KL}{r} < C_c \tag{1}$$

$$\sigma_a = \frac{\pi^2 E}{(\frac{23}{12})(KL/r)^2} \qquad \text{for } \frac{KL}{r} > C_c \tag{2}$$

where

$$C_c = \sqrt{\frac{2\pi^2 E}{\sigma_{yp}}} \tag{3}$$

Here K is the end fixity coefficient introduced in Problem 14.1. These equations may be used with either the SI or USCS systems of units. In Eqs. (1) and (2) the denominators represent safety factors which clearly increase with increasing values of the slenderness ratio L/r.

The second approach, which is perhaps in best agreement with experimental evidence, is due to R. Bjorhovde* who, in 1971, analyzed the behavior of a large number of full-scale test columns all having measured initial imperfections from perfect straightness as well as residual (fabrication) stresses. These columns were relatively light- or medium-weight hot-rolled wide-flange W sections having flange thicknesses less than 2 in (50.8 mm) and material yield points less than approximately 49,000 lb/in (335 MPa). He found that the mean (over the cross section) axial compressive stress σ_u just prior to collapse is given by the expressions

$$\begin{aligned}
\sigma_u &= \sigma_{yp} && \text{for } 0 < \lambda < 0.15 \\
\sigma_u &= \sigma_{yp}(1.035 - 0.202\lambda - 0.222\lambda^2) && \text{for } 0.15 \leq \lambda \leq 1.0 \\
\sigma_u &= \sigma_{yp}(-0.111 + 0.636\lambda^{-1} + 0.087\lambda^{-2}) && \text{for } 1.0 \leq \lambda \leq 2.0 \\
\sigma_u &= \sigma_{yp}(0.009 + 0.877\lambda^{-2}) && \text{for } 2.0 \leq \lambda < 3.6 \\
\sigma_u &= \sigma_{yp}\lambda^{-2}\ \text{(Euler's curve)} && \text{for } \lambda \geq 3.6
\end{aligned} \tag{4}$$

*R. Bjorhovde and L. Tall, "Minimum Column Strength and Multiple Column Curve Concept," Report 337.29, Lehigh University, Fritz Eng. Lab, Bethlehem, PA, 1971. R. Bjorhovde, "Deterministic and Probabilistic Approaches to the Strength of Steel Columns," Ph.D. dissertation, Lehigh University, Bethlehem, PA, 1972.

where
$$\lambda = \frac{L}{\pi r}\sqrt{\frac{\sigma_{yp}}{E}} \tag{5}$$

These results, known in graphical form as the Structural Stability Research Council Curve No. 2, represent prototype behavior of steel columns in region ⓑ of Fig. 14-8. The equations may be used with either the SI or USCS systems of units. Since the stress σ_u in Eqs. (*4*) is that existing just prior to collapse, no safety factor is present but instead must be introduced by the designer. Two comparable sets of equations were given by Bjorhovde for other types of steel sections.

14.12. Use the AISC design recommendation discussed in Problem 14.11 to determine the allowable axial load on a W8 × 19 section 10 ft long. The ends are pinned, the yield point is 36,000 lb/in^2, and $E = 30 \times 10^2$ lb/in.

From Table 8-1 Chap. 8 we have the properties of the cross section as

$$I_{\min} = 7.9\ \text{in}^4 \qquad A = 5.59\ \text{in}^2$$

The radius of gyration is found by the method of Chap. 7 to be

$$r = \sqrt{\frac{7.9\ \text{in}^4}{5.59\ \text{in}^2}} = 1.189\ \text{in}$$

Thus,
$$\frac{L}{r} = \frac{(10)(12)}{1.189} = 100.9$$

From Problem 14.11 we have from Eq. (*3*)

$$C_c = \sqrt{\frac{2\pi^2 E}{\sigma_{yp}}} = \sqrt{\frac{2\pi^2(30\times 10^6\ \text{lb/in}^2)}{36{,}000\ \text{lb/in}^2}} = 128.26$$

For both ends pinned, $K = 1$ and thus $K(L/r) < C_c$ so that the allowable axial stress is given by Eq. (*1*) of Problem 14.11 to be

$$\sigma_a = \frac{\left[1 - \frac{(KL/r)^2}{2C_c^2}\right]\sigma_{yp}}{\frac{5}{3} + \frac{3(KL/r)}{8C_c} - \frac{(KL/r)^3}{8C_c^3}} = \frac{\left[1 - \frac{(100.9)^2}{2(128.26)^2}\right](36{,}000)}{\frac{5}{3} + \frac{3(100.9)}{8(128.26)} - \frac{(100.0)^3}{8(128.26)^3}}$$
$$= 13{,}100\ \text{lb/in}^2$$

The allowable axial load is

$$P_a = (5.59\ \text{in}^2)(13{,}100\ \text{lb/in}^2) = 73{,}100\ \text{lb}$$

14.13. Reconsider the column of Problem 14.12 but now with a length of 15 ft. Use the AISC design recommendation to determine the allowable axial load. Both ends are pinned.

Now we have $L/r = (15)(12)/1.189 = 151.4$. Thus the increased length (in comparison to that of Problem 14.12) leads to

$$K\frac{L}{r}(= 151.4) > C_c(= 128.26)$$

so that we must compute the allowable axial stress from Eq. (*2*) of Problem 14.11:

$$\sigma_a = \frac{12\pi^2 E}{23(KL/r)^2}$$
$$= \frac{12\pi^2(30\times 10^6\ \text{lb/in}^2)}{23(151.4)^2} = 6740\ \text{lb/in}^2$$

The allowable axial load is thus

$$P_a = (5.59\ \text{in}^2)(6740\ \text{lb/in}^2) = 37{,}670\ \text{lb}$$

14.14. Use the AISC recommendation to determine the allowable axial load on a W203 × 28 section 3 m long. The ends are pinned. The material yield point is 250 MPa and $E = 200$ GPa.

From Table 8-2 of Chap. 8 we have the sectional properties as

$$I_{\min} = 3.28 \times 10^6\ \text{mm}^4 \qquad A = 3600\ \text{mm}^2$$

The radius of gyration is found to be

$$r = \sqrt{\frac{3.28 \times 10^6\ \text{mm}^4}{3600\ \text{mm}^2}} = 30.18\ \text{mm}$$

Thus

$$\frac{L}{r} = \frac{3000\ \text{mm}}{30.18\ \text{mm}} = 99.4$$

From Problem 14.11, Eq. (*3*), we have

$$C_c = \sqrt{\frac{2\pi^2 E}{\sigma_{yp}}} = \sqrt{\frac{2\pi^2(200 \times 10^9\ \text{N/m}^2)}{250 \times 10^6\ \text{N/m}^2}} = 125.7$$

For both ends pinned, $K = 1$ and thus $K(L/r) < C_c$ so that the allowable axial stress is given by Eq. (*1*) of Problem 14.11 to be

$$\sigma_a = \frac{\left[1 - \dfrac{(KL/r)^2}{2C_c^2}\right]\sigma_{yp}}{\dfrac{5}{3} + \dfrac{3(KL/r)}{8C_c} - \dfrac{(KL/r)^3}{8C_c^3}} = \frac{\left[1 - \dfrac{(99.4)^2}{2(125.7)^2}\right]250 \times 10^6\ \text{N/m}^2}{\dfrac{5}{3} + \dfrac{3(99.4)}{8(125.7)} - \dfrac{(99.4)^3}{8(125.7)^3}}$$

$$= 90.35\ \text{MPa}$$

The allowable axial load is

$$P = (3600\ \text{mm}^2)\left(\frac{\text{m}}{10^3\ \text{mm}}\right)^2(90.35 \times 10^6\ \text{N/m}^2)$$

$$= 325{,}000\ \text{N} \quad \text{or} \quad 325\ \text{kN}$$

14.15. Reconsider the column of Problem 14.12 but now use the SSRC recommendation discussed in Problem 14.11 to estimate the maximum load-carrying capacity of the column.

As discussed in Problem 14.11, we must first compute the parameter

$$\lambda = \frac{KL}{r} \cdot \frac{1}{\pi}\sqrt{\frac{\sigma_{yp}}{E}}$$

Here,

$$\lambda = \frac{(1)(10\ \text{ft})(12\ \text{in/ft})}{(1.189)} \cdot \frac{1}{\pi}\sqrt{\frac{36{,}000\ \text{lb/in}^2}{30 \times 10^6\ \text{lb/in}^2}} = 1.113$$

From Problem 14.11, for this value of λ we must determine the ultimate (peak) axial stress in the column from the semiempirical relation

$$\sigma_u = \sigma_{yp}\left[-0.111 + \frac{0.636}{\lambda} + \frac{0.087}{\lambda^2}\right]$$

$$= (36{,}000\ \text{lb/in}^2)\left[-0.111 + \frac{0.636}{1.113} + \frac{0.087}{(1.113)^2}\right] = 19{,}000\ \text{lb/in}^2$$

The axial load corresponding to this stress is

$$P_{max} = (5.59\ \text{in}^2)(19{,}000\ \text{lb/in}^2) = 106{,}200\ \text{lb}$$

This load represents the average of actual test values of peak loads that columns of this type were found to carry. It is to be noted that no safety factor is incorporated into these computations, so that the design load for this member is less than the 106,600 lb.

14.16. Select a wide-flange section from Table 8-2 of Chap. 8 to carry an axial compressive load of 750 kN. The column is 3.5 m long with a yield point of 250 MPa and a modulus of 200 GPa. Use the AISC specifications. The bar is pinned at each end.

To get a first approximation, let us merely use $P = A\sigma$, from which we have

$$A = \frac{750{,}000\ \text{N}}{250 \times 10^6\ \text{N/m}^2} = 0.0030\ \text{m}^2 \qquad \text{or} \qquad 3000\ \text{mm}^2$$

This tells us that any wide-flange section having an area smaller than 3000 mm is unacceptable.

Next, let us try the W203 × 28 section. From Table 8-2 we find area = 3600 mm^2 and $I_{min} = 3.28 \times 10^6\ \text{mm}^4$. The minimum radius of gyration is thus

$$r = \sqrt{\frac{3.28 \times 10^6\ \text{mm}^4}{3600\ \text{mm}^2}} = 30.2\ \text{mm}$$

from which the slenderness ratio is $L/r = 3500/30.2 = 116$.

From Problem 14.11, (Eq. (*3*)), we have

$$C_c = \sqrt{\frac{2\pi^2(200 \times 10^9\ \text{N/m}^2)}{250 \times 10^6\ \text{N/m}^2}} = 125.6$$

Thus, since $K = 1$ for both ends pinned,

$$K\frac{L}{r}(= 116) < C_c(= 125.6)$$

So, we must employ Eq. (*1*) of Problem 14.11. This leads to

$$\sigma_a = \frac{\left[1 - \dfrac{(116)^2}{2(125.6)^2}\right]250}{\left[\dfrac{5}{3} + \dfrac{3(116)}{8(125.6)} - \dfrac{(116)^3}{8(125.6)^3}\right]} = 74.95\ \text{MPa}$$

from which $$P_a = (3600\ \text{mm}^2)\left(\frac{\text{m}}{10^3\ \text{mm}}\right)^2(74.95 \times 10^6\ \text{N/m}^2) = 270{,}000\ \text{N} \qquad \text{or} \qquad 270\ \text{kN}$$

which indicates that this is far too light a section.

Next, let us try the section W254 × 72 having an area of 9280 mm^2 and $I_{min} = 38.6 \times 10^6\ \text{mm}^4$. The minimum radius of gyration is found to be

$$r = \sqrt{\frac{38.6 \times 10^6\ \text{mm}^4}{9280\ \text{mm}^2}} = 64.5\ \text{mm}$$

from which the slenderness ratio is $3500/64.5 = 54.26$. Again we have

$$K\frac{L}{R}(= 54.26) < C_c(= 125.6)$$

so that we must again use Eq. (*1*) of Problem 14.11 to find the allowable stress which is

$$\sigma_a = \frac{\left[1 - \frac{(54.26)^2}{2(125.6)^2}\right]250}{\left[\frac{5}{3} + \frac{3(54.26)}{8(125.6)} - \frac{(54.26)^3}{8(125.6)^3}\right]} = 124.6\text{ MPa}$$

for which $$P_a = (9280\text{ mm}^2)\left(\frac{1}{10^3\text{ mm}}\right)^2(124.6 \times 10^6\text{ N/m}^2) = 1.15 \times 10^6\text{ N} \quad \text{or} \quad 1150\text{ kN}$$

This section is rather heavy, so let us investigate the W254 × 54. Here, the area is 7010 mm^2 and $I_{min} = 17.5 \times 10^6$ mm^4. So, the minimum radius of gyration is found to be 50.0 mm and the slenderness ratio is 3500/50 = 70. Again using Eq. (*1*) of Problem 14.11 we find $\sigma_a = 114$ MPa, from which the allowable load is $P_a = 799$ kN.

Investigation of the next lighter section, W254 × 43, by the above method indicates that it can carry only 478 kN.

Thus, the desired section is the W254 × 54, which can carry an axial load of 799 kN, which is in excess of the 750 kN required. A more complete table of structural shapes might well indicate a slightly lighter section than the W254 × 54.

14.17. Develop a FORTRAN program to represent the AISC value of allowable axial load on a steel column as discussed in Problem 14.11.

The symbols are defined in Problem 14.11 and Eqs. (*1*) and (*2*) of that problem indicate allowable axial compressive stress for values of KL/r less than or greater than the dimensionless parameter C_c. The program listing is

```
00010***********************************************************************
00020          PROGRAM STEELCL (INPUT,OUTPUT)
00030*                 (AMERICAN INSTITUTE OF STEEL CONSTRUCTION)
00040***********************************************************************
00050*
00060*         AUTHOR: KATHLEEN DERWIN
00070*         DATE  : JANUARY 24, 1989
00080*
00090*   BRIEF DESCRIPTION:
00100*       ONE APPROACH TO CONSIDERING DESIGN CRITERIA FOR STRUCTURAL
00110*   STEEL COLUMNS IS GIVEN BY THE A.I.S.C. (AMERICAN INSTITUTE OF
00120*   STEEL CONSTRUCTION). THIS PROGRAM DETERMINES THE ALLOWABLE AXIAL
00130*   COMPRESSIVE STRESS AND LOADING OF A STEEL COLUMN USING THE RELATIO
00140*   DEVELOPED AND ACCEPTED BY THE A.I.S.C.
00150*
00160*   INPUT:
00170*       THE USER IS FIRST ASKED IF USCS OR SI UNITS WILL BE USED. THEN,
00180*   THE COLUMN LENGTH, THE MINIMUM MOMENT OF INERTIA AND AREA OF THE
00190*   COLUMN CROSS SECTION, THE MATERIAL YIELD POINT, AND YOUNG'S MODULUS
00200*   ARE INPUTTED. ALSO, THE END FIXITY COEFFICIENT IS ENTERED.
00210*
00220*   OUTPUT:
00230*       THE ALLOWABLE AXIAL COMPRESSIVE STRESS AND LOADING OF THE COLUMN
00240*   IS DETERMINED.
00250*
00260*   VARIABLES:
00270*       ANS     ---     DENOTES IF USCS OR SI UNITS ARE DESIRED
00280*     L,I,A     ---     LENGTH, MIN.MOMENT OF INERTIA, AREA OF COLUMN X-SECT
00290*    SIGYP,E    ---     YIELD POINT, YOUNG'S MODULUS OF THE MATERIAL
00300*       R       ---     MIN. RADIUS OF GYRATION AS CALCULATED FROM THE
00310*                       CROSS-SECTIONAL AREA AND MOMENT OF INERTIA
00320*       CC      ---     CRITICAL CONSTANT OF THE COLUMN...A FUNCTION OF ITS
00330*                       PHYSICAL AND MATERIAL PROPERTIES
00340*     CHECK     ---     THE COLUMN CONSTANT AS CALCULATED FOR THE SPECIFIC
```

```
00350*                        CASE CONSIDERED. THIS IS COMPARED TO THE CRITICAL
00360*                        CONSTANT TO DETERMINE WHICH OF TWO RELATIONS TO USE
00370*      K       ---       END FIXITY COEFFICIENT OF THE COLUMN
00380* HOLD1,HOLD2---         PARTIAL CALCULATIONS OF THE MORE COMPLICATED FUNCTIO
00390*                        (USED FOR EASE IN PROGRAMMING)
00400*     SIGA     ---       ALLOWABLE AXIAL COMPRESSIVE STRESS
00410*    LOADA     ---       ALLOWABLE AXIAL LOAD
00420*      PI      ---       3.14159
00430*
00440*********************************************************************
00450******                    MAIN PROGRAM                          *****
00460*********************************************************************
00470*
00480*         VARIABLE DECLARATIONS
00490*
00500      REAL L,I,A,SIGYP,E,R,CHECK,CC,K,SIGA,LOADA,PI,HOLD1,HOLD2
00510      INTEGER ANS
00520*
00530      PI = 3.14159
00540*
00550*        USER INPUT
00560*
00570      PRINT*,'PLEASE INDICATE YOUR CHOICE OF UNITS:'
00580      PRINT*,'1 - USCS'
00590      PRINT*,'2 - SI'
00600      PRINT*,' '
00610      PRINT*,'ENTER 1,2'
00620      READ*,ANS
00630      IF (ANS.EQ.1) THEN
00640         PRINT*,'PLEASE INPUT ALL DATA IN UNITS OF POUND AND/OR INCH...'
00650      ELSE
00660         PRINT*,'PLEASE INPUT ALL DATA IN UNITS OF NEWTON AND/OR METER..
00670      ENDIF
00680      PRINT*,' '
00690      PRINT*,'ENTER COLUMN LENGTH:'
00700      READ*,L
00710      PRINT*,'ENTER THE CROSS-SECTIONAL PROPERTIES...'
00720      PRINT*,'MOMENT OF INERTIA, I:'
00730      READ*,I
00740      PRINT*,'AREA:'
00750      READ*,A
00760      PRINT*,'ENTER THE MATERIAL YIELD POINT:'
00770      READ*,SIGYP
00780      PRINT*,'ENTER THE VALUE FOR YOUNG'S MODULUS:'
00790      READ*,E
00800      PRINT*,'FINALLY, ENTER THE END FIXITY COEFFICIENT, K:'
00810      READ*,K
00820*
00830*           END USER INPUT
00840*
00850*
00860******          CALCULATIONS          ******
00870*
00880*       MINIMUM RADIUS OF GYRATION
00890*
00900     R = (I/A)**0.5
00910*
00920*         CRITICAL CONSTANT FOR THIS COLUMN SPECIFICATION
00930*
00940     CHECK = (L/R)*K
00950*
00960*         THE CRITICAL CONSTANT FOR ALL COLUMNS OF THIS MATERIAL
00970*
00980     CC = ((2 * (PI**2) * E)/SIGYP)**0.5
00990*
01000*         COMPARE CC AND CHECK TO DETERMINE WHICH RELATION TO USE
01010*
```

```
01020       IF (CHECK.LT.CC) THEN
01030          HOLD1 = (1 - ((CHECK**2)/(2*(CC**2))))*SIGYP
01040          HOLD2 = ((5./3)+((3*CHECK)/(8*CC)) - ((CHECK**3)/(8*(CC**3))))
01050       ELSE
01060          HOLD1 = (PI**2)*E
01070          HOLD2 = (23./12)*(CHECK**2)
01080       ENDIF
01090*
01100*          THE ALLOWABLE AXIAL STRESS AND LOADING
01110*
01120       SIGA = HOLD1/HOLD2
01130       LOADA = SIGA*A
01140*
01150******          PRINTING OUTPUT         ******
01160*
01170       PRINT*,' '
01180       PRINT*,' '
01190       PRINT*,'AMERICAN INSTITUTE OF STEEL CONSTRUCTION (AISC) STANDARDS:'
01200       PRINT*,' '
01210       IF (ANS.EQ.1) THEN
01220          PRINT 10,SIGA,'PSI.'
01230          PRINT 20,LOADA,'LB.'
01240       ELSE
01250          SIGA=SIGA/1000000.0
01260          PRINT 10,SIGA,'MPA.'
01270          PRINT 20,LOADA,'NEWTONS.'
01280       ENDIF
01290*
01300*              FORMAT STATEMENTS
01310*
01320 10    FORMAT(2X,'THE ALLOWABLE AXIAL COMPRESSIVE STRESS  IS',F10.1,
01330+              1X,A4)
01340 20    FORMAT(2X,'THE ALLOWABLE AXIAL LOAD  IS',F10.1,1X,A)
01350*
01360       STOP
01370       END
```

14.18. A pinned end W8 × 19 steel column has a yield point of 33,000 lb/in^2 and a modulus of 30×10^6 lb/in^2. The length of the column is 15 ft. Use the FORTRAN program of Problem 14.17 to determine the allowable axial stress and also the load based on AISC specifications.

From Table 8-1 of Chap. 8 we find $I_{min} = 7.9$ in^4 and $A = 5.59$ in^2. The self-prompting program and computer run is

```
run
 PLEASE INDICATE YOUR CHOICE OF UNITS:
 1 - USCS
 2 - SI

 ENTER 1,2
? 1
 PLEASE INPUT ALL DATA IN UNITS OF POUND AND/OR INCH...

 ENTER COLUMN LENGTH:
? 180
 ENTER THE CROSS-SECTIONAL PROPERTIES...
 MOMENT OF INERTIA, I:
? 7.9
 AREA:
? 5.59
 ENTER THE MATERIAL YIELD POINT:
? 33000
```

```
 ENTER THE VALUE FOR YOUNG'S MODULUS:
? 30E+6
 FINALLY, ENTER THE END FIXITY COEFFICIENT, K:
? 1


 AMERICAN INSTITUTE OF STEEL CONSTRUCTION (AISC) STANDARDS:

  THE ALLOWABLE AXIAL COMPRESSIVE STRESS  IS      6738.2 PSI.
  THE ALLOWABLE AXIAL LOAD  IS   37666.5 LB.

SRU       0.780 UNTS.
```

14.19. Consider a pin-ended W305 × 37 column made of steel having a yield point of 270 MPa and a modulus of 200 GPa. The length of the column is 10 m. From Table 8-2 of Chap. 8 we find $I_{min} = 6.02 \times 10^{-6}\ m^4$ and $A = 4760 \times 10^{-6}\ m^2$. Use the FORTRAN program of Problem 14.17 to determine the allowable axial stress and load based on AISC specifications.

Using these input data, the computer run is

```
run
 PLEASE INDICATE YOUR CHOICE OF UNITS:
 1 - USCS
 2 - SI

 ENTER 1,2
? 2
 PLEASE INPUT ALL DATA IN UNITS OF NEWTON AND/OR METER...

 ENTER COLUMN LENGTH:
? 10
 ENTER THE CROSS-SECTIONAL PROPERTIES...
 MOMENT OF INERTIA, I:
? 6.02E-6
 AREA:
? 4760E-6
 ENTER THE MATERIAL YIELD POINT:
? 270E+6
 ENTER THE VALUE FOR YOUNG'S MODULUS:
? 200E+9
 FINALLY, ENTER THE END FIXITY COEFFICIENT, K:
? 1


 AMERICAN INSTITUTE OF STEEL CONSTRUCTION (AISC) STANDARDS:

  THE ALLOWABLE AXIAL COMPRESSIVE STRESS  IS      13.0 MPA.
  THE ALLOWABLE AXIAL LOAD  IS   61998.2 NEWTONS.

SRU       0.777 UNTS.
```

14.20. Develop a FORTRAN program to represent the SSRC values of mean axial compressive stress just prior to collapse as discussed in Problem 14.11.

The symbols are defined in Problem 14.11 and the Eqs. (*4*) of that problem indicate axial stress just prior to collapse for various values of λ given by Eq. (*5*). The program listing is

```
00010*****************************************************************
00020           PROGRAM STEELCL (INPUT,OUTPUT)
00030*                  (BJORHOVDE, STRUCTURAL STABILITY RESEARCH COUNCIL)
00040*****************************************************************
00050*
00060*          AUTHOR: KATHLEEN DERWIN
00070*          DATE  : JANUARY 24, 1989
00080*
00090*  BRIEF DESCRIPTION:
00100*      ONE APPROACH TO CONSIDERING DESIGN CRITERIA FOR STRUCTURAL
00110*  STEEL COLUMNS WAS DEVELOPED BY  R. BJORHOVDE, AND IS POSSIBLY
00120*  IN THE BEST AGREEMENT WITH EXPERIMENTAL EVIDENCE. THE MEAN AXIAL
00130*  COMPRESSIVE STRESS JUST PRIOR TO COLLAPSE CAN BE OBTAINED FOR THE
00140*  SPECIFIC COLUMN BY FIRST CALCULATING THE 'COLUMN CONSTANT' AND THEN
00150*  DETERMINING THE MEAN STRESS AT FAILURE FROM THE APPROPRIATE RELATION.
00160*
00170*  INPUT:
00180*      THE USER IS FIRST ASKED IF USCS OR SI UNITS WILL BE USED. THEN,
00190*  THE COLUMN LENGTH, THE MINIMUM MOMENT OF INERTIA AND AREA OF THE
00200*  COLUMN CROSS SECTION, THE MATERIAL YIELD POINT, AND YOUNG'S MODULUS
00210*  ARE INPUTTED. ALSO, THE END FIXITY COEFFICIENT IS ENTERED.
00220*
00230*  OUTPUT:
00240*      THE MEAN (OVER THE CROSS SECTION) AXIAL COMPRESSIVE STRESS AND
00250*  THE MEAN PEAK LOADING CONDITIONS ARE DETERMINED.
00260*
00270*  VARIABLES:
00280*       ANS    ---    DENOTES IF USCS OR SI UNITS ARE DESIRED
00290*     L,I,A    ---    LENGTH, MIN.MOMENT OF INERTIA, AREA OF COLUMN X-SECT
00300*    SIGYP,E   ---    YIELD POINT, YOUNG'S MODULUS OF THE MATERIAL
00310*       R      ---    MIN. RADIUS OF GYRATION AS CALCULATED FROM THE
00320*                     X-SECTIONAL AREA AND MOMENT OF INERTIA
00330*     LAMDA    ---    CRITICAL CONSTANT OF THE COLUMN...A FUNCTION OF ITS
00340*                     PHYSICAL AND MATERIAL PROPERTIES
00350*       K      ---    END FIXITY COEFFICIENT OF THE COLUMN
00360*      SIGU    ---    MEAN AXIAL COMPRESSIVE STRESS AT FAILURE
00370*     LOADU    ---    MEAN AXIAL LOAD AT FAILURE
00380*       PI     ---    3.14159
00390*
00400*********************************************************************
00410******                     MAIN PROGRAM                        *****
00420*********************************************************************
00430*
00440*           VARIABLE DECLARATIONS
00450*
00460       REAL L,I,A,SIGYP,E,R,LAMDA,K,SIGU,LOADU,PI
00470       INTEGER ANS
00480*
00490       PI = 3.14159
00500*
00510*          USER INPUT
00520*
00530       PRINT*,'PLEASE INDICATE YOUR CHOICE OF UNITS:'
00540       PRINT*,'1 - USCS'
00550       PRINT*,'2 - SI'
00560       PRINT*,' '
00570       PRINT*,'ENTER 1,2'
00580       READ*,ANS
00590       IF (ANS.EQ.1) THEN
00600          PRINT*,'PLEASE INPUT ALL DATA IN UNITS OF POUND AND/OR INCH...'
00610       ELSE
00620          PRINT*,'PLEASE INPUT ALL DATA IN UNITS OF NEWTON AND/OR METER..
00630       ENDIF
00640       PRINT*,' '
00650       PRINT*,'ENTER COLUMN LENGTH:'
00660       READ*,L
```

```
00670       PRINT*,'ENTER THE CROSS-SECTIONAL PROPERTIES...'
00680       PRINT*,'MOMENT OF INERTIA, I:'
00690       READ*,I
00700       PRINT*,'AREA:'
00710       READ*,A
00720       PRINT*,'ENTER THE MATERIAL YIELD POINT:'
00730       READ*,SIGYP
00740       PRINT*,'ENTER THE VALUE FOR YOUNG'S MODULUS:'
00750       READ*,E
00760       PRINT*,'FINALLY, ENTER THE END FIXITY COEFFICIENT, K:'
00770       READ*,K
00780*
00790*              END USER INPUT
00800*
00810*
00820******          CALCULATIONS         ******
00830*
00840*         MINIMUM RADIUS OF GYRATION
00850*
00860      R = (I/A)**0.5
00870*
00880*      CRITICAL CONSTANT FOR THIS COLUMN SPECIFICATION
00890*
00900      LAMDA = ((K*L)/(R*PI))*((SIGYP/E)**0.5)
00910*
00920*      MEAN AXIAL COMPRESSIVE STRESS AND LOADING
00930*
00940      IF (LAMDA.LT.0.15) THEN
00950         SIGU = SIGYP
00960      ELSEIF (LAMDA.GE.0.15 .AND. LAMDA.LT.1.0) THEN
00970         SIGU = SIGYP*(1.035 - 0.202*LAMDA - 0.222*(LAMDA**2))
00980      ELSEIF (LAMDA.GE.1.0 .AND. LAMDA.LT.2.0) THEN
00990         SIGU = SIGYP*(-0.111 + 0.636/LAMDA + 0.0872/(LAMDA**2))
01000      ELSEIF (LAMDA.GE.2.0 .AND. LAMDA.LT.3.6) THEN
01010         SIGU = SIGYP*(0.009 + 0.877/(LAMDA**2))
01020      ELSEIF (LAMDA.GE.3.6) THEN
01030         SIGU = SIGYP/(LAMDA**2)
01040      ENDIF
01050*
01060      LOADU = SIGU*A
01070*
01080******         PRINTING OUTPUT         ******
01090*
01100      PRINT*,' '
01110      PRINT*,' '
01120      PRINT*,'STRUCTURAL STABILITY RESEARCH COUNCIL (BJORHOVDE) STANDARDS
01130      PRINT*,' '
01140      IF (ANS.EQ.1) THEN
01150         PRINT 10,SIGU,'PSI'
01160         PRINT 20,LOADU,'LB'
01170      ELSE
01180         SIGU=SIGU/1000000.0
01190         PRINT 10,SIGU,'MPA'
01200         PRINT 20,LOADU,'NEWTONS'
01210      ENDIF
01220*
01230*              FORMAT STATEMENTS
01240*
01250 10   FORMAT(2X,'THE MEAN AXIAL COMPRESSIVE STRESS AT FAILURE IS',F10.1,
01260+            1X,A3)
01270 20   FORMAT(2X,'THE MEAN AXIAL LOAD AT FAILURE IS',F10.1,1X,A)
01280*
01290      STOP
01300      END
```

14.21. Consider a 3.5-m-long pinned end steel column of wide-flange type W254 × 79. The material has a yield point of 250 MPa and a modulus of 200 GPa. Use the FORTRAN program of Problem 14.20 to determine the mean axial compressive stress just prior to collapse as indicated by the SSRC relations.

The constants of this cross section are found from Table 8-2 of Chap. 8 to be $I = 43.1 \times 10^{-6}\ \text{m}^4$ and $A = 10{,}200 \times 10^{-6}\ \text{m}^2$. Using these values, together with the designated length, yield point, and modulus, the self-prompting program prints as follows:

```
run
 PLEASE INDICATE YOUR CHOICE OF UNITS:
 1 - USCS
 2 - SI

 ENTER 1,2
? 2
 PLEASE INPUT ALL DATA IN UNITS OF NEWTON AND/OR METER...

 ENTER COLUMN LENGTH:
? 3.5
 ENTER THE CROSS-SECTIONAL PROPERTIES...
 MOMENT OF INERTIA, I:
? 43.1E-6
 AREA:
? 10200E-6
 ENTER THE MATERIAL YIELD POINT:
? 250E+6
 ENTER THE VALUE FOR YOUNG'S MODULUS:
? 200E+9
 FINALLY, ENTER THE END FIXITY COEFFICIENT, K:
? 1

 STRUCTURAL STABILITY RESEARCH COUNCIL (BJORHOVDE) STANDARDS:

  THE MEAN AXIAL COMPRESSIVE STRESS AT FAILURE IS     207.8 MPA
  THE MEAN AXIAL LOAD AT FAILURE IS 2119270.2 NEWTONS

SRU       0.786 UNTS.
```

14.22. Consider an initially straight, pin-ended column subject to an axial compressive force applied with known eccentricity e (see Fig. 14-11). Determine the maximum compressive stress in the column.

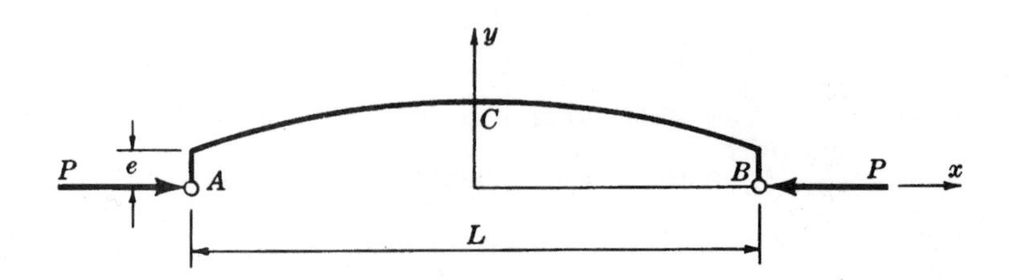

Fig. 14-11

The differential equation of the bar in its deflected configuration is

$$EI\frac{d^2y}{dx^2} = -Py$$

which has the standard solution

$$y = C_1 \sin\left(\sqrt{\frac{P}{EI}}x\right) + C_2 \cos\left(\sqrt{\frac{P}{EI}}x\right)$$

Since $y = e$ at each of the ends $x = -L/2$ and $x = L/2$, the values of the two constants of integration are readily found to be

$$C_1 = 0 \qquad C_2 = \frac{e}{\cos\left(\sqrt{\frac{P}{EI}}\frac{L}{2}\right)}$$

Thus, the deflection curve of the bent bar is

$$y = \frac{e}{\cos\left(\sqrt{\frac{P}{EI}}\frac{L}{2}\right)}\cos\left(\sqrt{\frac{P}{EI}}x\right)$$

The maximum value of deflection occurs at $x = 0$, by symmetry, and is

$$y_{\max} = e\sec\left(\sqrt{\frac{P}{EI}}\frac{L}{2}\right)$$

Introducing the value of the critical load P_{cr} as given by (9) of Problem 14.1, this becomes

$$y_{\max} = e\sec\left(\frac{\pi}{2}\sqrt{\frac{P}{P_{cr}}}\right)$$

Evidently the maximum deflection, which occurs at the center of the bar, becomes very great as the load P approaches the critical value. The phenomenon is one of gradually increasing lateral deflections, not buckling. The maximum compressive stress occurs on the concave side of the bar at C and is given by

$$\sigma_{\max} = \frac{P}{A} + \frac{M_{\max}c}{I} = \frac{P}{A} + \frac{Pec}{I}\sec\left(\frac{\pi}{2}\sqrt{\frac{P}{P_{cr}}}\right)$$

where c denotes the distance from the neutral axis to the outer fibers of the bar. If we now introduce the radius of gyration r of the cross section, this becomes

$$\sigma_{\max} = \frac{P}{A}\left[1 + \frac{ec}{r^2}\sec\left(\frac{L}{2r}\sqrt{\frac{P}{AE}}\right)\right]$$

This is the *secant formula* for an eccentrically loaded long column. In it, P/A is the average compressive stress. If the maximum stress is specified to be the yield point of the material, then the corresponding average compressive stress which will first produce yielding may be found from the equation

$$\frac{P_{yp}}{A} = \frac{\sigma_{yp}}{1 + \frac{ec}{r^2}\sec\left(\frac{L}{2r}\sqrt{\frac{P_{yp}}{AE}}\right)}$$

For any designated value of the ratio ec/r^2, this equation may be solved by trial and error and a curve of P/A versus L/r plotted to indicate the value of P/A at which yielding first begins in the extreme fibers.

14.23. Obtain the load-deflection relation for a pin-ended column subject to axial compression and undergoing finite lateral displacements.

The treatment presented in Problem 14.1 is restricted to extremely small lateral deflections because

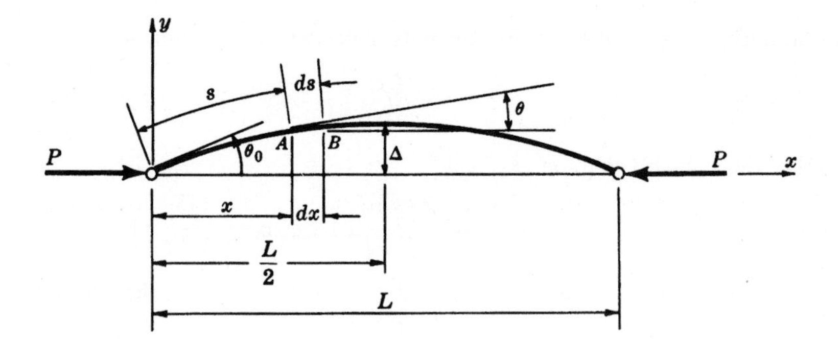

Fig. 14-12

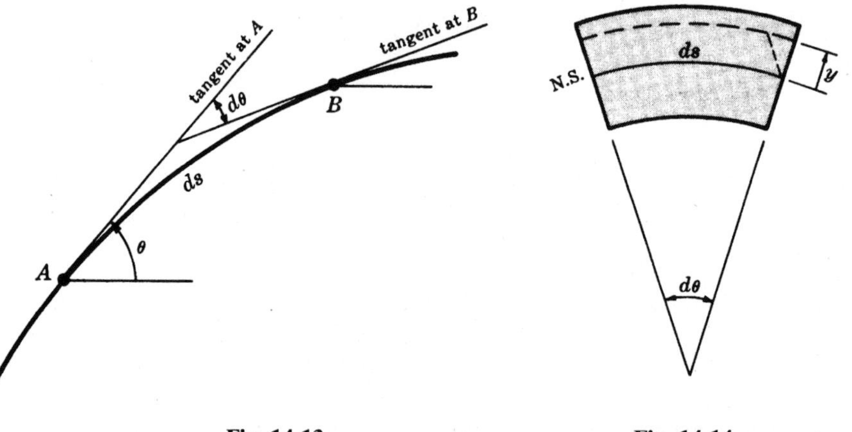

Fig. 14-13

Fig. 14-14

this was the assumption made in deriving Eq. (*1*), the Euler-Bernoulli equation. To obtain a more general representation let us introduce the angular coordinate θ and arc length s, in addition to the x- and y-coordinates (see Fig. 14-12).

An enlarged view of the deformed bar illustrates the angular coordinates more clearly (Fig. 14-13). Note that $d\theta$ is negative. Let us now examine an element of arc length ds bounded by two adjacent cross sections of the bar. Prior to loading these cross sections are parallel to each other but after the bar has deflected laterally they have the appearance shown in Fig. 14-14 in which they subtend a central angle $d\theta$. In a manner similar to that used in Problem 8.1, we may determine the normal strain of a fiber a distance y from the neutral surface to be

$$\epsilon = \frac{y\,d\theta}{ds} = \frac{\sigma}{E}$$

where σ is the longitudinal stress acting on this fiber. But from Problem 8.1 we have $\sigma = My/I$. Thus

$$\frac{y\,d\theta}{ds} = \frac{My}{EI}$$

or, since $M = -Py$ for the bar,

$$\frac{d\theta}{ds} = -\frac{Py}{EI} \tag{1}$$

If we let $a^2 = P/EI$ then

$$\frac{d\theta}{ds} = -\alpha^2 y \tag{2}$$

from which

$$\frac{d^2\theta}{ds^2} = -\alpha^2 \frac{dy}{ds} = -\alpha^2 \sin\theta \tag{3}$$

This equation is valid for large, finite lateral deflections of the bar in contrast to (5) of Problem 9.1 which is limited to very small values of deflection. To solve (3), let us multiply through by the integrating factor $2(d\theta/ds)$:

$$2\frac{d\theta}{ds}\frac{d^2\theta}{ds^2} = -2\alpha^2(\sin\theta)\frac{d\theta}{ds} \tag{4}$$

Integrating,

$$\left(\frac{d\theta}{ds}\right)^2 = 2\alpha^2\cos\theta + C_1 \tag{5}$$

When $x = 0$, $\theta = \theta_0$ (the initial slope) and at this same point $y = 0$; hence $d\theta/ds = 0$ from (2). Thus, from (2),

$$0 = 2\alpha^2\cos\theta + C_1$$

so that

$$\frac{d\theta}{ds} = -\sqrt{2}\,\alpha\sqrt{\cos\theta - \cos\theta_0} \tag{6}$$

where the negative square root is taken because $d\theta$ is always negative. This may be transformed to

$$\frac{d\theta}{ds} = -2\alpha\sqrt{\sin^2\frac{\theta_0}{2} - \sin^2\frac{\theta}{2}} \tag{7}$$

We next introduce the change of variables

$$\sin\frac{\theta}{2} = k\sin\phi \tag{8}$$

where ϕ is a parameter assuming the value $\pi/2$ when $x = 0$ and the value 0 when $x = L/2$, from which

$$k = \sin\frac{\theta_0}{2} \tag{9}$$

Then

$$\theta = 2\arcsin(k\sin\phi)$$

and

$$d\theta = \frac{2k\cos\phi\,d\phi}{\sqrt{1 - k^2\sin^2\phi}} \tag{10}$$

From (7), (8), (9), and (10) we have

$$\frac{d\phi}{\sqrt{1 - k^2\sin^2\phi}} = -\alpha\,ds \tag{11}$$

Integrating the last equation and remembering the definition of ϕ at its endpoint values,

$$\alpha\int_0^{L/2} ds = -\int_{\pi/2}^{0} \frac{d\phi}{\sqrt{1 - k^2\sin^2\phi}}$$

or

$$\alpha\frac{L}{2} = \int_0^{\pi 2} \frac{d\phi}{\sqrt{1 - k^2\sin^2\phi}} \tag{12}$$

The right-hand side of (12) is termed the *complete elliptic integral of the first kind* with modulus k and argument ϕ. Tabulated values of the integral for any specified value of k are readily available; see for example B. O. Peirce, *A Short Table of Integrals*, 4th ed., Ginn, 1957. To employ these tables we must select a value of θ_0 thus fixing k from Eq. (*9*). Then (*12*) may be rewritten in the form

$$P = \frac{4EI}{L^2}\left[\int_0^{\pi/2} \frac{d\phi}{\sqrt{1-k^2\sin^2\phi}}\right]^2 \tag{13}$$

to determine the axial load P corresponding to this assumed value of θ_0. To find the maximum deflection occurring at $x = L/2$, we have from geometry

$$\frac{dy}{ds} = \sin\theta = 2\sin\frac{\theta}{2}\cos\frac{\theta}{2} \tag{14}$$

From (*11*) this becomes

$$\frac{dy}{ds} = -\frac{\alpha\, dy\sqrt{1-k^2\sin^2\phi}}{d\phi} \tag{15}$$

Equating the right sides of (*14*) and (*15*),

$$\frac{-\alpha\, dy\sqrt{1-k^2\sin^2\phi}}{d\phi} = 2k(\sin\phi)\sqrt{1-k^2\sin^2\phi}$$

or

$$\alpha\, dy = -2k\sin\phi\, d\phi \tag{16}$$

Integrating,

$$\alpha y = 2k\cos\phi + C_2$$

When $y = 0$, $\phi = \pi/2$ from which $C_2 = 0$. When $x = L/2$, $\phi = 0$ and $y = y_{max} = \Delta$. Thus $\alpha\Delta = 2k$ or

$$\Delta = \frac{2k}{\alpha} = \frac{2k}{\sqrt{\dfrac{P}{EI}}} = \frac{kL}{\displaystyle\int_0^{\pi/2} \frac{d\phi}{\sqrt{1-k^2\sin^2\phi}}} \tag{17}$$

The procedure is as follows:

1. Select a value of θ_0 and determine k from Eq. (*9*).
2. Ascertain the value of $\displaystyle\int_0^{\pi/2} \frac{d\phi}{\sqrt{1-k^2\sin^2\phi}}$ from tabulated values in, for example, B. O. Peirce, and then calculate the axial force P corresponding to this value of θ_0 from Eq. (*13*).
3. Calculate the central deflection Δ from Eq. (*17*).

Results of this computation for selected values appear in Table 14-1 in which the starred value 9.87 ($= \pi^2$) indicates that the simple theory of Problem 14.1 actually gives an exact result if it is assumed that the end slopes are zero.

Table 14-1

θ_0, degrees	k	$\int_0^{\pi/2} \frac{d\phi}{\sqrt{1-k^2\sin^2\phi}}$	$\frac{PL^2}{EI}$	$\frac{\Delta}{L}$
0	0	$\pi/2$	9.87 ($= \pi^2$)*	0
40	0.342	1.6200	10.50	0.211
80	0.643	1.7868	12.75	0.360
120	0.866	2.1565	18.56	0.403
160	0.985	3.1534	39.76	0.313

From the above the progressive states of deformation of the bar are as shown in Fig. 14-15.

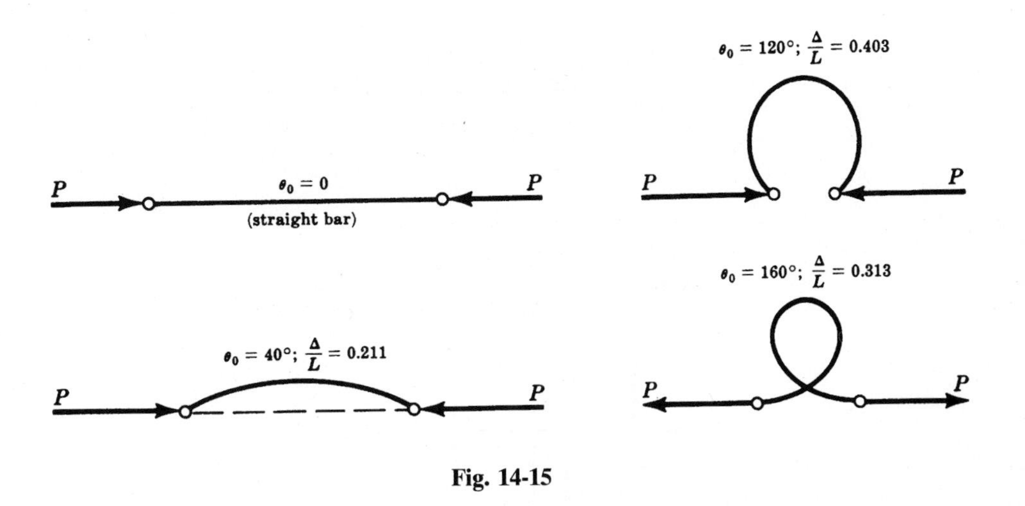

Fig. 14-15

This problem was first investigated by L. Euler in 1744 and the shape of the elastic curve is termed the *elastica*. It is only through use of this more exact finite-deflection theory that the amplitude of the lateral deflection may be determined. The approximate small-deflection treatment of Problem 14.1 does not permit determination of this quantity.

14.24. A problem that arises in insertion of a fiber-optic cable in a surrounding rigid conduit is that the cable buckles under certain axial "pushing" forces. This situation is represented in Fig. 14-16(*a*) by a long slender bar (the cable) having simply supported ends (and represented by a line element) with a clearance Δ between the bar and the inside of the surrounding rectangular conduit. Assume that the behavior of the cable in this conduit is two-dimensional and determine the behavior of the cable under increasing axial compressive forces P.*

From Problem 14.1 when the axial force $P = EI/L^2$, the bar buckles and touches the conduit walls in the central region which is of unknown length L_2. For equilibrium of the left region of length L_1, there is a concentrated force R acting at the pin, as well as another at $x = L_1$, as indicated in Fig. 14-16(*b*). The differential equation of the deformed bar in the region $x \leqslant L_1$ is

$$EI\frac{d^2y}{dx^2} + Py = Rx$$

where the transverse force R must be considered to be exerted on the bar by the pin at A. The solution of this equation is found as the sum of the general solution to the homogeneous equation plus a particular solution to the nonhomogeneous equation, as in Problem 14.3. Thus we have

$$y = A\sin\alpha x + B\cos\alpha x + \frac{Rx}{P} \qquad (1)$$

where $\alpha^2 = P/EI$. The boundary conditions for the region of length L_1 are (*a*) when $x = 0$, $y = 0$; (*b*) when $x = L_1$, $y = \Delta$; and (*c*) when $x = L_1$, $dy/dx = 0$. From (*a*) we have $B = 0$. From (*b*) and (*c*) we get

$$A\sin\alpha L_1 + \frac{R}{P}L_1 = \Delta \qquad (2)$$

$$A\alpha\cos\alpha L_1 + \frac{R}{P} = 0 \qquad (3)$$

Since the region of the deformed bar between $x = L_1$ and $x < (L_1 + L_2)$ is in contact with the rigid conduit,

*The author is indebted to Professor V. I. Feodosyev of the Moscow Higher Technical School for suggesting this problem and for his discussions concerning it.

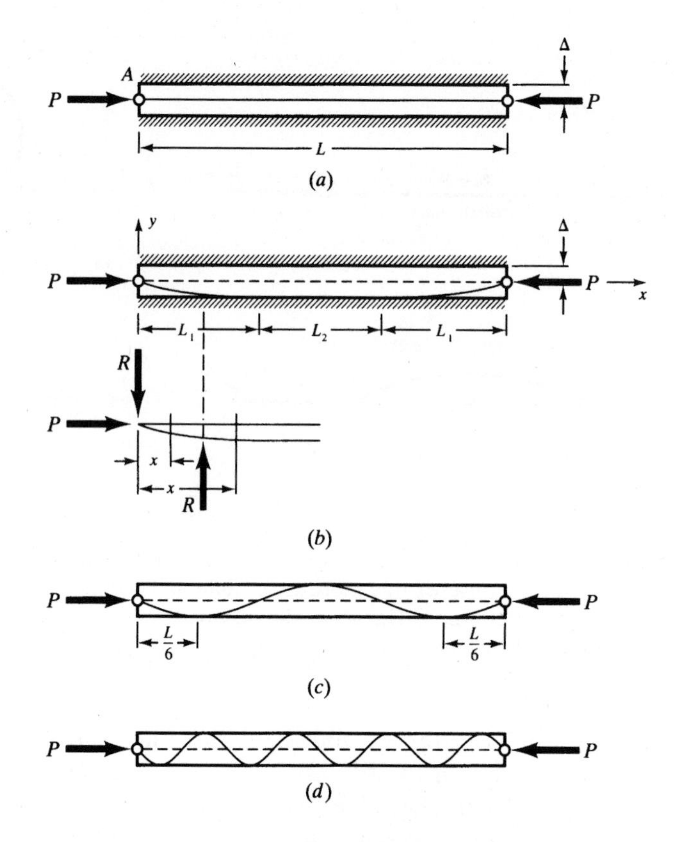

Fig. 14-16

the cable is straight in this region, and from Eq. (*5*) of Problem 9.1 the bending moment in that region is zero. Thus the Euler-Bernoulli equation of the beam in this central region of length L_2 becomes

$$P\Delta + Rx + R(x - L_1) = 0 \tag{4}$$

from which we have

$$R = P\left(\frac{\Delta}{L_1}\right) \tag{5}$$

From Eq. (*2*) we now have

$$A \sin \alpha L_1 + \Delta = \Delta \tag{6}$$

and thus $\alpha L_1 = \pi$ from which $\alpha = \pi/4$ or $L_1 = \pi/2$. Substituting (*5*) and (*6*) in (*2*), we obtain

$$A = \frac{\Delta}{\pi} \tag{7}$$

Since from (*1*) we have $\alpha^2 = P/EI$, we have from (*6*)

$$\left(\frac{\pi}{4}\right)^2 = \frac{P}{EI}$$

or

$$P = \frac{\pi^2 EI}{L_1} \tag{8}$$

When only the midpoint of the bar of length L is in contact with the interior wall of the conduit, i.e., when $L_1 = L/2$, Eq. (*8*) becomes

$$P = \frac{4\pi^2 EI}{L^2} \tag{9}$$

This indicates that for values of axial force lying between

$$\frac{\pi^2 EI}{L^2} < P < \frac{4\pi^2 EI}{L^2} \tag{10}$$

the flexible fiber-optic cable touches the rigid wall only at the midpoint of the length of the conduit, i.e., at $x = L/2$. Only for values of $P > 4\pi^2 EI/L^2$ is the flexible cable in contact with the conduit interior for a finite length. This is indicated in Fig. 14-16(*b*).

Next, the central region of length L_2 may buckle for sufficiently large values of compressive force P. The central portion obviously behaves as a clamped end column as shown in Problem 14.3, and it buckles at the load

$$P = \frac{4\pi^2 EI}{L_2^2} \tag{11}$$

into the configuration shown in Fig. 14-16(*c*). But from Fig. 14-16(*b*), we have

$$2L_1 + L_2 = L$$

so from Eq. (*5*) we have

$$L_2 = L - 2\left(\frac{\pi}{\alpha}\right) \tag{12}$$

If we now equate from the values of P from (*8*) and (*12*), we find $L_1 = L/4$, and from (*8*) we find for this value of L_1

$$P = \frac{16\pi^2 EI}{L^2} \tag{13}$$

By analyses such as the above, it can be shown that increases in axial force P over that given by Eq. (*13*) lead to the value $L_1 = L/6$, and that configuration is retained until the axial load is

$$P = \frac{36\pi^2 EI}{L^2} \tag{14}$$

Still greater values of axial load will lead to the configuration indicated in Fig. 14-16(*d*). Thus, simple buckling theory has led to the plausible configurations indicated in Fig. 14-16.

14.25. Determine the deflection curve of a pin-ended bar subject to combined axial compression P together with a uniform normal loading as shown in Fig. 14-17.

One convenient coordinate system to designate points on the deflected bar is shown in Fig. 14-17. There, the origin is situated at the point of maximum deflection. The bending moment at an arbitrary point (x, y) on the deflected bar is written most easily as the sum of the moment of all forces to the *right* of (x, y)

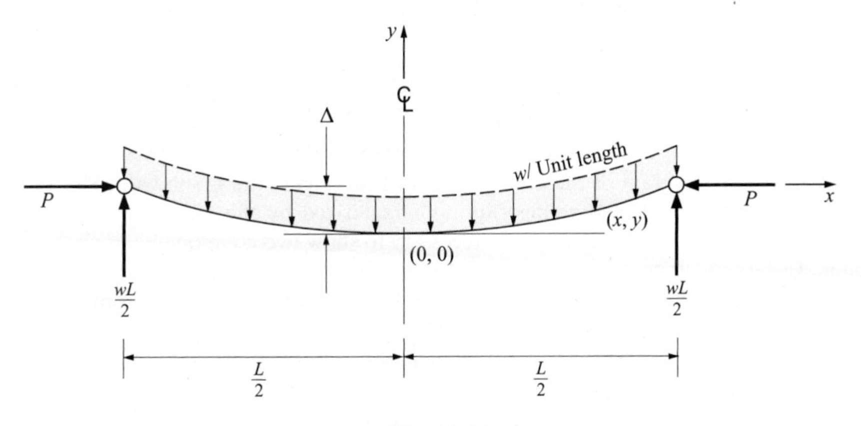

Fig. 14-17

and with algebraic signs consistent with the definitions of positive and negative bending introduced in Chap. 6.

The bending moment is thus

$$M = P(\Delta - y) + \frac{wL}{2}\left(\frac{L}{2} - x\right) - w\left(\frac{L}{2} - x\right)\left(\frac{\frac{L}{2} - x}{2}\right) \tag{1}$$

so that the differential equation of the deflected bar is

$$EI\frac{d^2y}{dx^2} = P\Delta - Py + \frac{w}{2}\left(\frac{L^2}{4} - x^2\right) \tag{2}$$

If we introduce the notation

$$n = \sqrt{\frac{P}{EI}}$$

we have the nonhomogeneous differential equation of the bar

$$\frac{d^2y}{dx^2} + n^2y = \frac{w}{2EI}\left(\frac{L^2}{4} - x^2\right) + n^2\Delta^2$$

The solution is given by the usual methods of differential equations as the sum of (*a*) the solution of the corresponding homogeneous equation, and (*b*) any particular solution of the entire nonhomogeneous equation. Thus we may write the solution as

$$y = A\cos nx + B\sin nx - \frac{w}{2P}\left(\frac{L^2}{4} - x^2\right) + \frac{2w}{2n^2P} + \Delta$$

where A and B are constants of integration. These are easily found by realizing that, because of symmetry of the bent bar, the deflection is Δ at $x = L/2$ and also the bar has a horizontal tangent at $x = 0$. This leads to

$$y = \Delta + \frac{w}{n^2P}\left[\left(\sec\frac{nL}{2}\cos nx - 1\right) - n^2\left(\frac{L^2}{8} - \frac{x^2}{2}\right)\right]$$

as the solution of the nonhomogeneous equation. The peak deflection occurs at the midpoint of the bar (the origin of our coordinate system) and is given by

$$\Delta = \frac{w}{n^2P}\left[\left(\sec\frac{nL}{2} - 1\right) - \frac{n^2L^2}{8}\right]$$

14.26. Two identical rigid bars AB and BC are pinned at B and C and supported at A by a pin in a frictionless roller that can only displace vertically. A spring of constant k is attached to bar BC, as shown in Fig. 14-18(*a*). Determine the critical load of the system.

A free-body diagram of the entire system of two rigid bars is shown in Fig. 14-18(*b*). The system is shown in a slightly deflected configuration characterized by the angle $\Delta\theta$ corresponding to its buckled shape. Ends A and C are pinned so it is necessary to show two components of pin reaction at each of these points. The spring elongates an amount $a(\Delta\theta)$ and consequently exerts a force $ka(\Delta\theta)$ on bar BC. The pin at B is internal to this free body; hence no pin forces should be shown. From statics,

$$+\uparrow\Sigma M_A = C_x(4a) - ka(\Delta\theta)(3a) = 0$$

$$C_x = \frac{3ka(\Delta\theta)}{4}$$

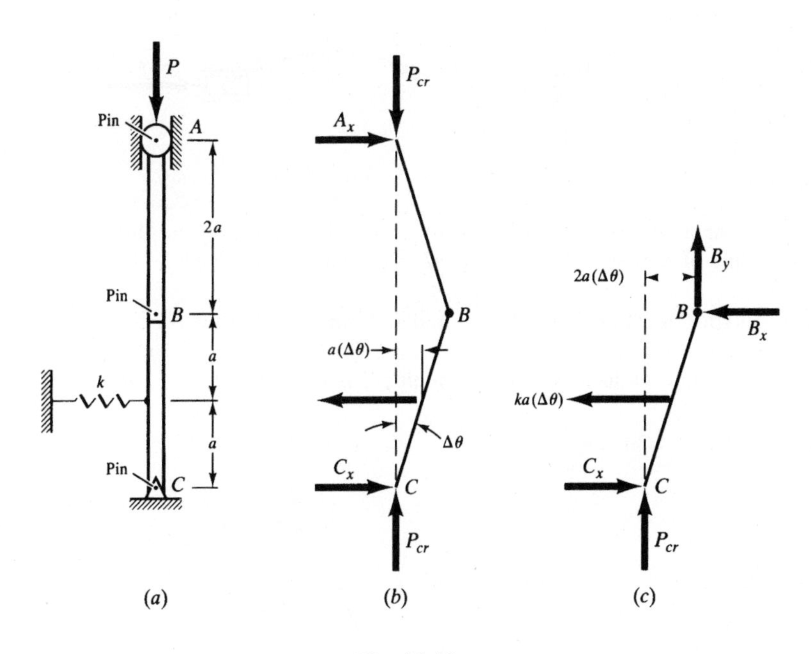

Fig. 14-18

Next, consider the free-body diagram of the lower bar BC shown in Fig. 14-18(c). Now the pin forces at B become external to this free body, and from statics we have

$$+\circlearrowright \Sigma M_B = \frac{3ka(\Delta\theta)}{4}(2a) - P_{cr}a(2a\,\Delta\theta) - [ka(\Delta\theta)]a = 0$$

$$P_{cr} = \frac{ka}{4}$$

It is impossible to determine $(\Delta\theta)$ by this approach.

Supplementary Problems

14.27. A steel bar of solid circular cross section is 50 mm in diameter. The bar is pinned at each end and subject to axial compression. If the proportional limit of the material is 210 MPa and $E = 200$ GPa, determine the minimum length for which Euler's formula is valid. Also, determine the value of the Euler buckling load if the column has this minimum length. *Ans.* 1.21 m, 412 kN

14.28. The column shown in Fig. 14-19 is pinned at both ends and is free to expand into the opening at the upper end. The bar is steel, is 25 mm in diameter, and occupies the position shown at 16 °C. Determine the temperature to which the column may be heated before it will buckle. Take $\alpha = 12 \times 10^{-6}$/°C and $E = 200$ GPa. Neglect the weight of the column. *Ans.* 29.3 °C

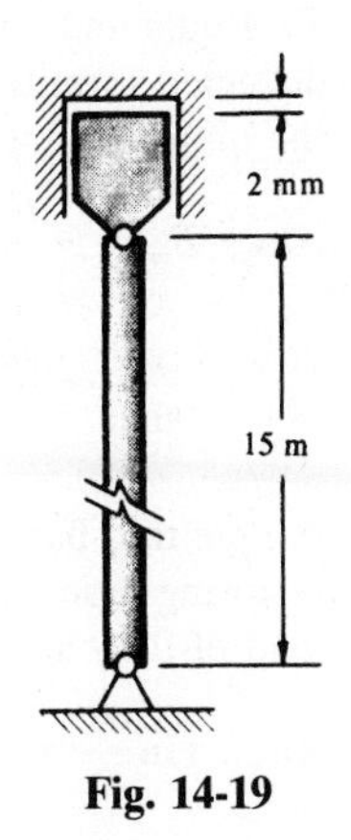

Fig. 14-19

14.29. A long slender bar AB is clamped at A and supported at B in such a way that transverse displacement is impossible as in Fig. 14-20, but the

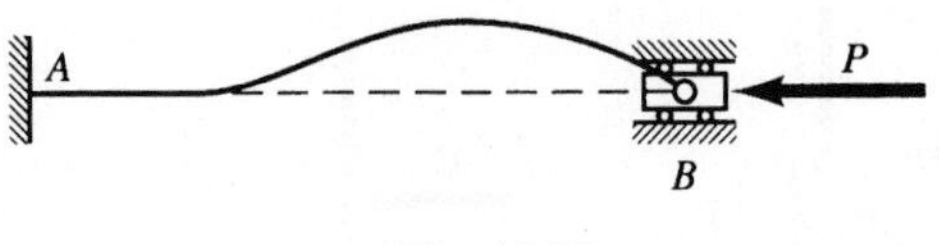

Fig. 14-20

end of the bar at B is capable of rotating about B. Determine the differential equation governing the buckled shape of the bar. *Ans.* $\tan nL = nL$ where $n^2 = P/EI$

14.30. A bar of length L is clamped at its lower end and subject to both vertical and horizontal forces at the upper end, as shown in Fig. 14-21. The vertical force P is equal to one-fourth of the Euler load for this bar. Determine the lateral displacement of the upper end of the bar. *Ans.* $16(4 - \pi)RL^3/\pi^3 EI$

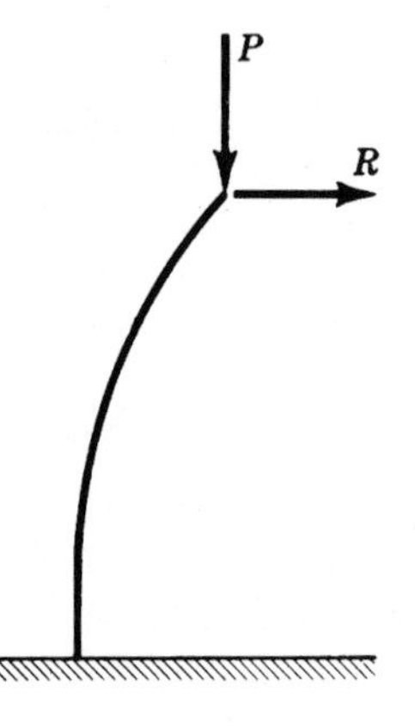

Fig. 14-21

14.31. A bar of length L and flexural rigidity EI has pinned ends. An axial compressive force

$$P = \frac{\pi^2 EI}{4L^2}$$

is applied to the beam and a bending moment M is applied at one end. Determine the rotational stiffness, i.e., applied moment per radian of rotation at that end of the bar. Rework the problem for the case of an axial tensile force of the same numerical value.

Ans. $\dfrac{2.47EI}{L}, \dfrac{3.47EI}{L}$

14.32. An initially straight bar AC is pinned at each end and supported at the midpoint B by a spring which resists any lateral movement δ of B with a lateral force $(kEI/L^3)\delta$. The bar is of length $2L$ and least flexural rigidity EI. Equal and opposite thrusts P are applied at the end C as well as at the centroid of the bar at B. In any deflected form the line of action of the thrust applied at B remains parallel to the chord AC. Determine the minimum buckling load of the system.

Ans. $P_{cr} = \beta^2 \dfrac{EI}{L^2}$ where β is the smallest positive root of the equation

$$\frac{\beta}{\tan\beta} = \frac{3k + (9 + k)\beta^2 - \beta^4}{3(k - \beta^2)}$$

14.33. A long thin bar of length L and rigidity EI is supported at each end in an elastic medium which exerts a restoring moment of magnitude λ per radian of angular rotation at the end. Determine the first buckling load of the bar.

Ans. $\tan\dfrac{\alpha L}{2} = -\dfrac{P}{\alpha\lambda}$ where $\alpha^2 = \dfrac{P}{EI}$

14.34. A long thin bar is pinned at each end and is embedded in an elastic packing which exerts a transverse force on the bar when it deflects laterally. When the transverse deflection at any point is given by y, the packing exerts a transverse force per unit length of the bar equal to ky. Determine the axial force required to buckle the bar.

Ans. $P_{cr} = \dfrac{\pi^2 EI}{L^2}\left(n^2 + \dfrac{kL^4}{n^2 \pi^4 EI}\right)$ where n is the integer for which P_{cr} is minimum

14.35. Use the AISC formula to determine the allowable axial load on a W10 × 54 column that is 22 ft long. The yield point of the material is 34,000 lb/in^2 and the modulus is 30×10^6 lb/in^2. *Ans.* 197,250 lb

14.36. Use the AISC formula to determine the allowable axial load on a W254 × 79 column that is 14 m long. The yield point of the material is 250 MPa and the modulus is 200 GPa. *Ans.* 226,500 N

14.37. A W12 × 25 pin-ended column made of steel having a yield point of 36,000 lb/in^2 and a modulus of 30×10^6 lb/in^2 is 30 ft long. Use the FORTRAN program of Problem 14.17 to determine the allowable axial stress and load based on AISC specifications. *Ans.* 2340 lb/in^2, 17,280 lb

14.38. A W254 × 79 pin-ended column made of steel having a yield point of 250 MPa and a modulus of 200 GPa is 14 m long. Use the FORTRAN program of Problem 14.17 to determine the allowable axial stress and load based on AISC specifications. *Ans.* 22.2 MPa, 226 kN

14.39. Consider a pinned end column 9 m long of wide flange designation W203 × 28. The yield point of the material is 250 MPa and the modulus is 200 GPa. Use the FORTRAN program of Problem 14.20 to determine the mean axial compressive stress as well as axial load just prior to collapse as indicated by the SSRC equations. *Ans.* 21.7 MPa, 78.2 kN

14.40. Consider a pinned end column 22 ft long of wide flange designation W10 × 54. The yield point of the steel is 34,000 lb/in^2 and the modulus is 30×10^6 lb/in^2. Use the FORTRAN program of Problem 14.20 to determine the mean axial compressive stress as well as load just prior to collapse as indicated by the SSRC equations. *Ans.* 18,200 lb/in^2, 289,000 lb

14.41. Determine the deflection curve of a pin-ended bar subject to axial compression together with a central transverse force as shown in Fig. 14-22.

Ans. $y = \dfrac{Q \sin nx}{2Pn \cos \dfrac{nL}{2}} - \dfrac{Q}{2P}x$ where $n = \sqrt{\dfrac{P}{EI}}$

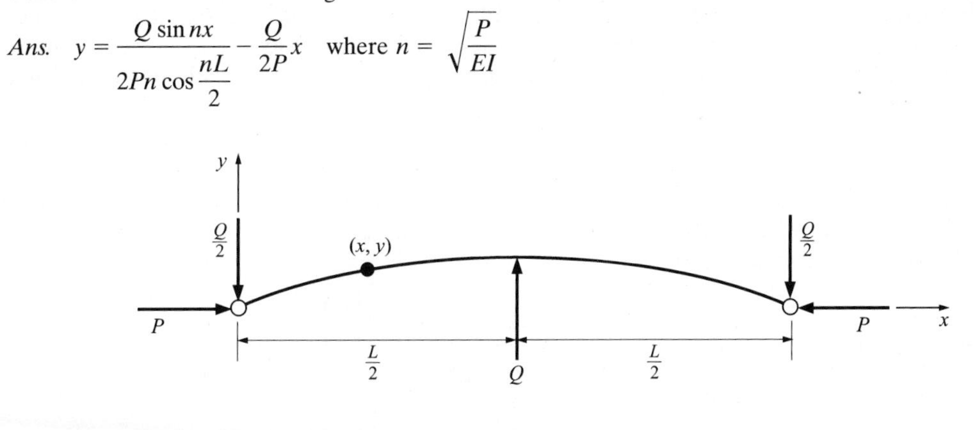

Fig. 14-22

14.42. A pin-ended bar of flexural rigidity EI is subject to the two transverse loads indicated in Fig. 14-23, each being one quarter of the Euler axial buckling load of the bar and simultaneously the axial loads each being half the Euler buckling load of the bar. Determine the peak transverse deflection of the bar.
Ans. $0.008L$

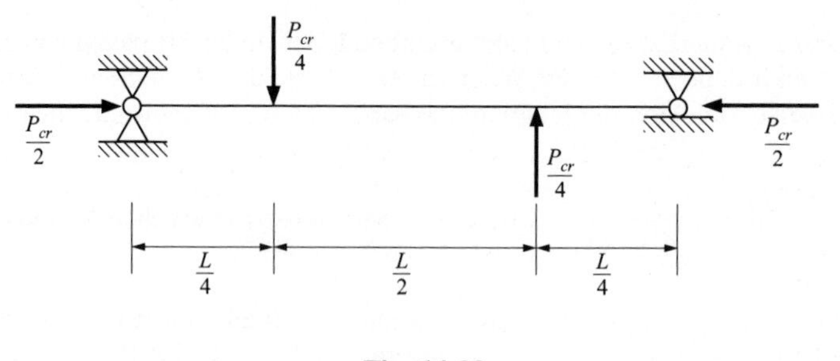

Fig. 14-23

14.43. The system of two rigid vertical bars AB and BC shown in Fig. 14-24 is pinned at the base C and restrained against lateral motion at the top A, but is free to rotate there. The bars are also pinned at B. The midpoint B is partially restrained against lateral displacement by the two linear springs, each offering k lb of resistance per inch of lateral movement. The springs are load free prior to application of P. Determine the buckling load P_{cr}. *Ans.* $P_{cr} = 12k$

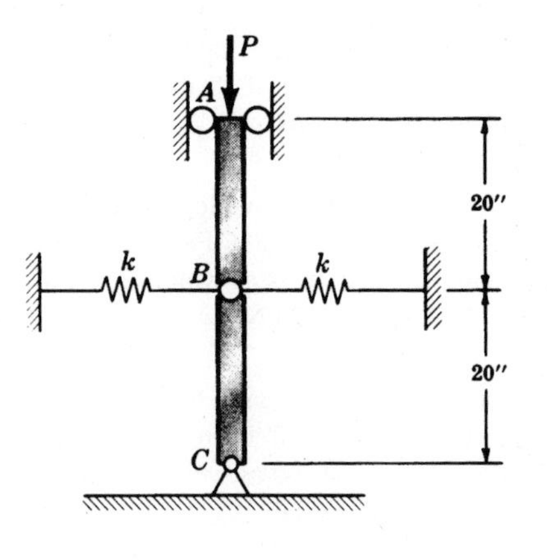

Fig. 14-24

14.44. The rigid bar OA in Fig. 14-25 is pinned at O and supports a vertical force P at the upper end A. Point A is tied back to the ground by a spring of constant k. The spring is load free when the rod OA is vertical. Weights of all members are to be neglected. Determine the load P at which the system becomes unstable. *Ans.* $kL/2$

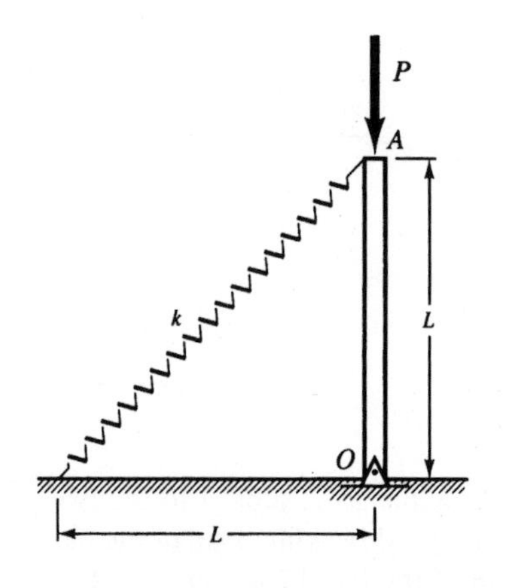

Fig. 14-25

14.45. The guyed steel mast AB in Fig. 14-26 is pinned at A and braced by a planar system of two thin wires BC and BD, as shown in Fig. 14-26. The moment of inertia of the mast is $3.00\ \text{in}^4$ and its height is 50 in. Its modulus of elasticity is $30 \times 10^6\ \text{lb/in}^2$. The wires are each of aluminum having modulus of $10 \times 10^6\ \text{lb/in}^2$ and cross-sectional area $0.10\ \text{in}^2$. The mast is subject to a vertical force P applied at B. Determine the magnitude of the buckling load. (*Hint:* It is necessary to consider rigid-body rotation of the mast about A to the configuration AB' as well as independently computing the Euler-type buckling load of the mast into one loop of a sine curve.) *Ans.* $P = 350{,}000\ \text{lb}$

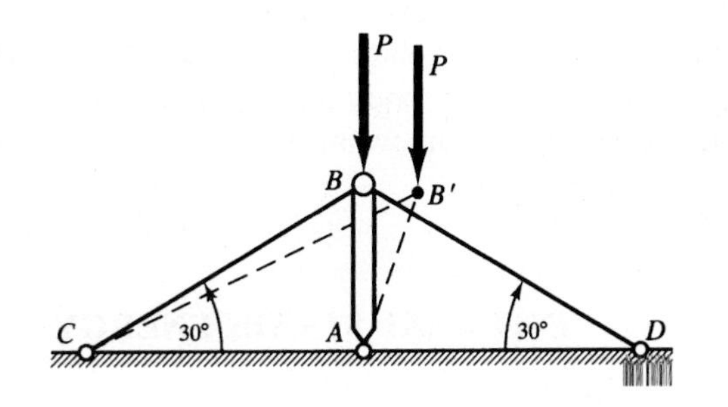

Fig. 14-26

Chapter 15

Strain Energy Methods

Thus far in this book various techniques have been discussed for finding deformations and determining values of indeterminate reactions. These techniques have essentially been based upon geometric considerations. There are, however, many types of problems that can be solved more efficiently through techniques based upon relations between the work done by the external forces and the internal strain energy stored within the body during the deformation process. The present chapter will discuss these techniques, which are somewhat more general and more powerful than the various geometric approaches.

INTERNAL STRAIN ENERGY

When an external force acts upon an elastic body and deforms it, the work done by the force is stored within the body in the form of strain energy. The strain energy is always a scalar quantity. For a straight bar subject to a tensile force P, the internal strain energy U is given by

$$U = \frac{P^2 L}{2AE}$$

where L represents the length of the bar, A is its cross-sectional area, and E is Young's modulus. This expression is derived in Problem 15.1.

For a circular bar of length L subject to a torque T, the internal strain energy U is given by

$$U = \frac{T^2 L}{2GJ}$$

where G is the modulus of elasticity in shear and J is the polar moment of inertia of the cross-sectional area. This expression is derived in Problem 15.2.

For a bar of length L subject to a bending moment M, the internal strain energy U is given by

$$U = \frac{M^2 L}{2EI}$$

where I is the moment of inertia of the cross-sectional area about the neutral axis. This is derived in Problem 15.3.

Note that in each of these expressions the external load always occurs in the form of a squared magnitude, hence each of these energy expressions is always a positive scalar quantity.

SIGN CONVENTIONS

Strain energy methods are particularly well suited to problems involving several structural members at various angles to one another. The fact that the members may be curved in their planes presents no additional difficulties. One of the great advantages of strain energy methods is that independent coordinate systems may be established for each member without regard for consistency of positive directions of the various coordinate systems. This advantage is essentially due to the fact that the strain energy is always a positive scalar quantity, and hence algebraic signs of external forces need be consistent only within each structural member.

CASTIGLIANO'S THEOREM

This theorem is extremely useful for finding displacements of elastic bodies subject to axial loads, torsion, bending, or any combination of these loadings. The theorem states that the partial derivative of the total internal strain energy with respect to any external applied force yields the displacement under the point of application of that force in the direction of that force. Here, the terms force and displacement are used in their generalized sense and could either indicate a usual force and its linear displacement, or a couple and the corresponding angular displacement. In equation form the displacement under the point of application of the force P_n is given according to this theorem by

$$\delta_n = \frac{\partial U}{\partial P_n}$$

This theorem is derived in Problem 15.8.

APPLICATION TO STATICALLY DETERMINATE PROBLEMS

In such problems all external reactions can be found by application of the equations of statics. After this has been done, the deflection under the point of application of any external applied force can be found directly by use of Castigliano's theorem. This is illustrated in Problems 15.9 and 15.10. If the deflection is desired at some point where there is no applied force, then it is necessary to introduce an auxiliary (i.e., fictitious) force at that point and, treating that force just as one of the real ones, use Castigliano's theorem to determine the deflection at that point. At the end of the problem the auxiliary force is set equal to zero. This is illustrated in Problems 15.9, 15.12, 15.13, and 15.19.

APPLICATION TO STATICALLY INDETERMINATE PROBLEMS

Castigliano's theorem is extremely useful for determining the indeterminate reactions in such problems. This is because the theorem can be applied to each reaction, and the displacement corresponding to each reaction is known beforehand and is usually zero. In this manner it is possible to establish as many equations as there are redundant reactions, and these equations together with those found from statics yield the solution for all reactions. After the values of all reactions have been found, the deflection at any desired point can be found by direct use of Castigliano's theorem. This is illustrated in Problems 15.16 through 15.18.

ASSUMPTIONS AND LIMITATIONS

Throughout this chapter it is assumed that the material is a linear elastic one obeying Hooke's law. Further, it is necessary that the entire system obey the law of superposition. This implies that certain unusual systems, such as that discussed in Problem 1.17, cannot be treated by the techniques discussed here.

Solved Problems

15.1. Determine the internal strain energy stored within an elastic bar subject to an axial tensile force P.

For such a bar the elongation Δ has been found in Problem 1.1 to be $\Delta = PL/AE$, where A represents the cross-sectional area, L is the length, and E is Young's modulus. The force-elongation diagram will

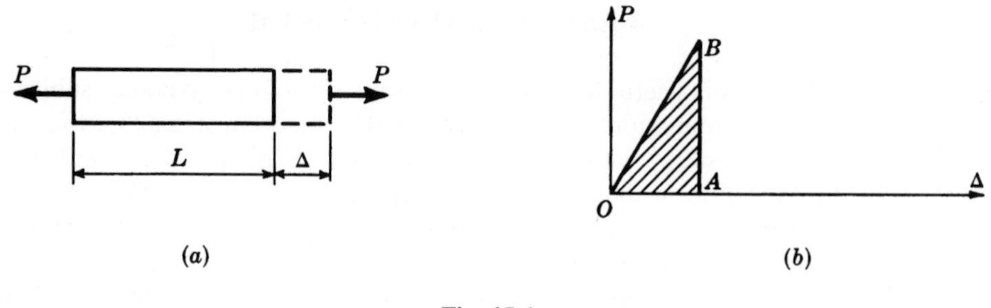

Fig. 15-1

consequently be linear, as shown in Fig. 15-1(b). For any specific value of the force P, such as that corresponding to point B in the force-elongation diagram, the force will have done positive work indicated by the shaded area OBA. This triangular area is given by $\frac{1}{2}P\Delta$. Replacing Δ by the value given above, this becomes $P^2L/2AE$. This is the work done by the external force and the work is stored within the bar in the form of internal strain energy, denoted by U. Hence

$$U = \frac{P^2 L}{2AE}$$

Essentially, the elastic bar is acting as a spring to store this energy. The same expression for internal strain energy applies if the load is compressive, since the axial force appears as a squared quantity and hence the final result is the same for either a positive or negative force.

If the axial force P varies along the length of the bar, then in an elemental length dx of the bar the strain energy is

$$dU = \frac{P^2\,dx}{2AE}$$

and the energy in the entire bar is found by integrating over the length:

$$U = \int_0^L \frac{P^2\,dx}{2AE}$$

15.2. Determine the internal strain energy stored within an elastic bar subject to a torque T as shown in Fig. 15-2(a).

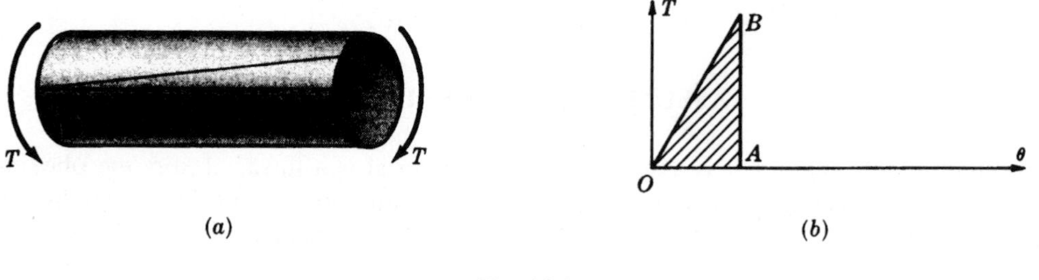

Fig. 15-2

In Problem 5.3, the angle of twist θ has been found to be $\theta = TL/GJ$, where G is the modulus of elasticity in shear, L is the length, and J is the polar moment of inertia of the cross-sectional area. According to this expression, the relation between torque and angle of twist is a linear one, as shown in Fig. 15-2(b). When the torque has reached a specific value such as that indicated by point B, it will have done positive work indicated by the shaded area OBA. This triangular area is given by $\frac{1}{2}T\theta$, or $T^2L/2GJ$. This work done by the external torque is stored within the bar as internal strain energy, denoted by U. Hence

$$U = \frac{T^2 L}{2GJ}$$

If the torque T varies along the length of the bar, then in an elemental length dx the strain energy is

$$dU = \frac{T^2\,dx}{2GJ}$$

and in the entire bar it is

$$U = \int_0^L \frac{T^2\,dx}{2GJ}$$

15.3. Determine the internal strain energy stored within an elastic bar subject to a pure bending moment M.

In Problem 8.1 is shown an initially straight bar subject to the pure bending moment M which deforms it into a circular arc of radius of curvature ρ. In Eq. *(7)* of that problem it was shown that $M = EI/\rho$, where I denotes the moment of inertia of the cross-sectional area about the neutral axis. But the length of the bar, L, is equal to the product of the central angle θ subtended by the circular arc and the radius ρ. Thus

$$\frac{M}{EI} = \frac{1}{\rho} = \frac{\theta}{L} \qquad \text{or} \qquad \theta = \frac{ML}{EI}$$

According to this the relation between moment and angle subtended is a linear one, and this is illustrated in Fig. 15-3. When the moment has reached a specific value M, such as that indicated by point B, it will have done work indicated by the shaded area OAB. This area is given by $\frac{1}{2}M\theta$, or $M^2L/2EI$. This work done by the external moment is stored within the bar as internal strain energy, denoted by U. Hence

$$U = \frac{M^2L}{2EI}$$

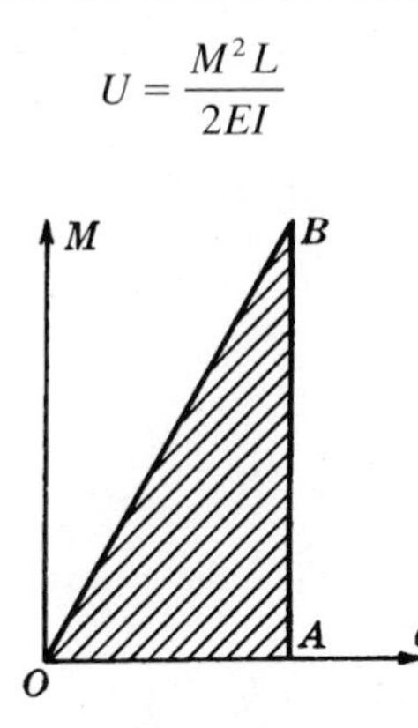

Fig. 15-3

If the bending moment M varies along the length of the bar, then in an elemental length dx the strain energy is

$$dU = \frac{M^2\,dx}{2EI}$$

and in the entire bar it is

$$U = \int_0^L \frac{M^2\,dx}{2EI}$$

15.4. Consider the two simply supported beams shown in Fig. 15-4. Both are of rectangular cross section and of equal width. The materials are identical. The first beam has constant height along the length, the second has a small groove in the center which reduces the height by one-fifth. The length of the groove along the axis is negligible. The maximum stress in each bar due to

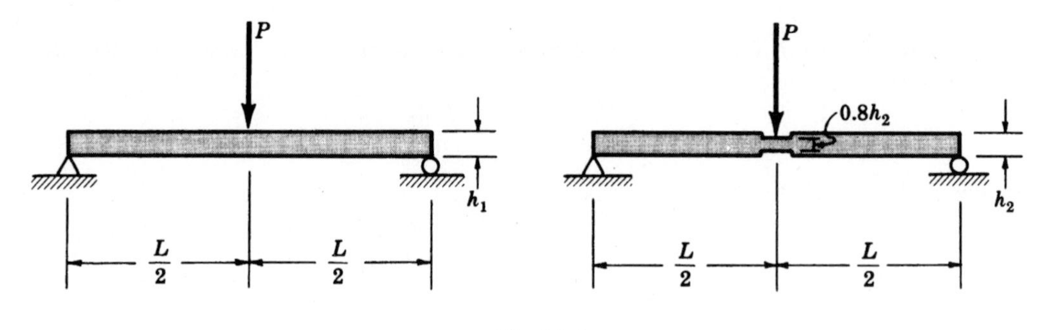

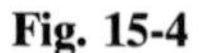

Fig. 15-4

the action of the central force P is the elastic limit of the material. Neglecting the effect of stress concentrations, determine the ratio of internal strain energies in the two bars.

For the first bar, the section modulus is

$$Z = \frac{I}{c} = \frac{\frac{1}{12}bh_1^3}{0.5h_1} = 0.167h_1^2 b$$

For the second bar, in the grooved region the section modulus is

$$Z = \frac{I}{c} = \frac{\frac{1}{12}b(0.8h_2)^3}{0.4h_2} = 0.107h_2^2 b$$

and in the thicker region of depth h_2 the section modulus is

$$Z = \frac{I}{c} = \frac{\frac{1}{12}bh_2^3}{0.5h_2} = 0.167h_2^2 b$$

In general, for bending we have the bending stress at the outer fibers of a bar given by the relation $\sigma = M/Z$. Since the maximum stresses in each bar are equal, we have

$$0.167h_1^2 b = 0.107h_2^2 b \qquad \text{or} \qquad h_2 = 1.25h_1$$

The strain energy in the first bar is

$$U_1 = \frac{M^2 L}{2EI} = \frac{M^2 L}{2E(\frac{1}{12}bh_1^3)}$$

The strain energy in the second bar, since the groove is of negligible length, is

$$U_2 = \frac{M^2 L}{2E[\frac{1}{12}b(1.25h_1)^3]}$$

The loadings and lengths are identical, hence we need not calculate $M^2 L$ to obtain the desired ratio, which is

$$U_2 : U_1 = 0.512$$

This indicates that a grooved bar is very ineffective in storing internal strain energy. This is an important consideration in the design of bars to withstand dynamic loadings.

15.5. Consider a vertical bar of uniform cross section with a flange at the lower end (Fig. 15-5). A weight W is released from the top of the bar and falls freely along the bar until it strikes the flange. Determine the maximum elongation of the bar and also the maximum stress.

To solve this problem we shall introduce several simplifying assumptions: (*a*) the weight of the vertical bar is very small compared to W, (*b*) there are no losses of energy due to friction or local distortion, and (*c*) the stress-strain diagram of the material of the bar is the same for dynamic loading as for static. Actually, a more sophisticated treatment would take strain wave propagation in the bar into account, but that is beyond the scope of the present study.

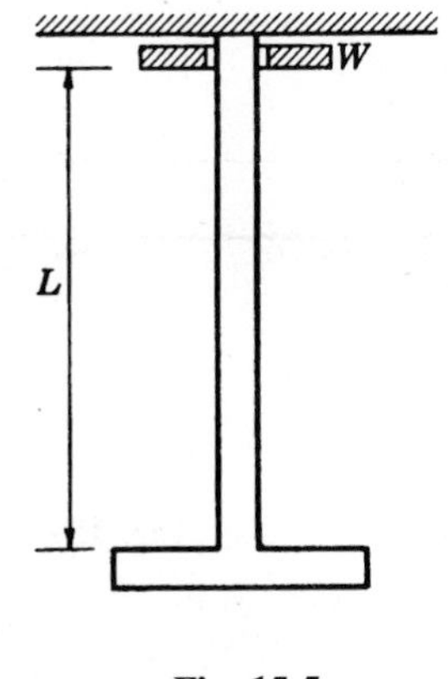

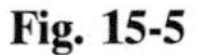

Fig. 15-5

The weight W falls through the distance L and after striking the flange extends the bar an unknown amount Δ. At this maximum extension the tension in the bar is maximum and the equation relating work done by W and the internal strain energy of extension at this instant of maximum deformation is

$$W(L+\Delta) = \frac{P^2 L}{2AE} \tag{1}$$

But $\Delta = PL/AE$, and substituting for P in (*1*), we get

$$W(L+\Delta) = \frac{AE\Delta^2}{2L} \tag{2}$$

The static extension of the bar due to the weight W would be $\Delta_{st} = WL/AE$. If the value of W from this expression is introduced in the above equation and the resulting quadratic equation solved for the unknown extension Δ, we get

$$\Delta = \Delta_{st} + \sqrt{\Delta_{st}^2 + \frac{\Delta_{st}}{g} v^2} \tag{3}$$

where g is the acceleration due to gravity and $v = \sqrt{2gL}$ is the velocity with which W strikes the flange. If the length of the bar, L, is very large compared to Δ_{st}, then the above expression becomes approximately

$$\Delta = \sqrt{\frac{\Delta_{st}}{g} v^2} \tag{4}$$

In this case the axial stress is given by

$$\sigma = \frac{P}{A} = \frac{\Delta E}{L} = \frac{E}{L}\sqrt{\frac{\Delta_{st}}{g} v^2} = \sqrt{\frac{Wv^2}{2g}\frac{2E}{AL}} \tag{5}$$

It is of interest to note that in the dynamic case the stress depends upon the length L as well as the Young's modulus E. The corresponding static stress does not involve either of these factors.

For the special case of a suddenly applied load W acting on the flange, the length L through which the weight falls may be set equal to zero in (*3*) to obtain

$$\Delta = 2\Delta_{st} \tag{6}$$

Thus, for this particular problem, a suddenly applied load produces a deflection twice as great as would be produced by a gradually applied load.

15.6. A cantilever beam is struck at its tip by a body of weight W falling freely through a height h above the beam, as shown in Fig. 15-6. Neglecting the weight of the beam, determine the total deflection at the tip.

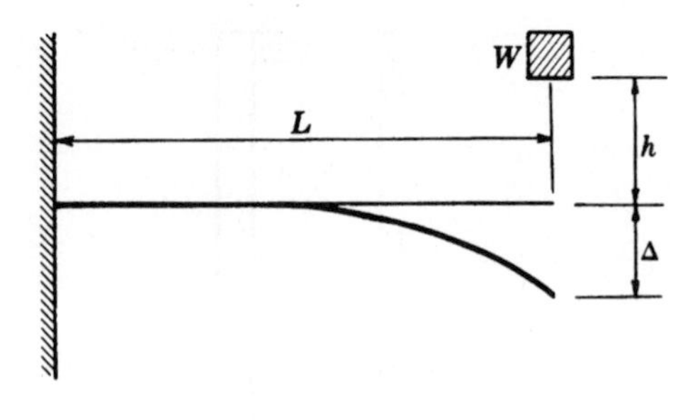

Fig. 15-6

By the time the weight has deflected the tip of the beam to its maximum value, the weight will have done an amount of work given by

$$W(h + \Delta) \tag{1}$$

If we let P denote the force exerted by the weight on the beam at the time of peak deflection, then at this moment the strain energy in the beam is given by $P\Delta/2$. Thus, once the work done by the external force is stored within the beam as internal strain energy we have

$$W(h + \Delta) = \frac{P\Delta}{2} \tag{2}$$

or

$$P = \frac{2W}{\Delta}(h + \Delta) \tag{3}$$

But from Problem 9.2 we know that if this force P acts at the tip of a cantilever beam the deflection at that point is

$$\Delta = \left[\frac{2W}{\Delta}(h + \Delta)\right]\frac{L^3}{3EI} \tag{4}$$

where I is the moment of inertia of the cross section about the neutral axis through the centroid. However, the deflection due to the weight W, if it were statically applied, is

$$\Delta_{st} = \frac{WL^3}{3EI} \tag{5}$$

and hence (*4*) becomes

$$\Delta^2 - 2\Delta_{st}\Delta - 2h\Delta_{st} = 0 \tag{6}$$

Solving,

$$\Delta = \Delta_{st} + \sqrt{\Delta_{st}^2 + 2h\Delta_{st}} \tag{7}$$

where the positive square root is taken so as to obtain the maximum deflection. For the special case of a suddenly applied load at the tip, $h = 0$, and (*7*) yields $\Delta = 2\Delta_{st}$. Just as in Problem 15.5, a load suddenly applied produces twice the deflection it would if it were applied gradually.

15.7. A simply supported beam is struck at its midpoint by a weight $W = 1$ kN falling freely from a height of $h = 100$ mm above the top of the beam. The beam is 5 m long and of circular cross section 100 mm in diameter. Take $E = 200$ GPa. Determine the maximum deflection of the beam.

The work done by the falling weight in producing the maximum central deflection Δ is

$$W(h + \Delta) \tag{1}$$

If P denotes the force exerted by the weight on the beam during the moment of maximum deflection, then the strain energy in the beam is $P\Delta/2$. Thus

$$\frac{P\Delta}{2} = W(h + \Delta) \tag{2}$$

or

$$P = \frac{2W(h + \Delta)}{\Delta} \tag{3}$$

But the central deflection of a centrally loaded, simply supported beam is given in Problem 9.12 as

$$\Delta = \frac{PL^3}{48EI} \tag{4}$$

Substituting the above value of P, this becomes

$$\Delta = \frac{2W(h + \Delta)}{\Delta} \frac{L^3}{48EI} \tag{5}$$

But the static deflection corresponding to W is $\Delta_{st} = WL^3/48EI$, and hence (5) can be written in the form

$$\Delta^2 - 2\Delta_{st}\Delta - 2h\Delta_{st} = 0 \tag{6}$$

Solving,

$$\Delta = \Delta_{st} + \sqrt{\Delta_{st}^2 + 2h\Delta_{st}} \tag{7}$$

For the beam under consideration,

$$I = \frac{\pi D^4}{64} = 4.9 \times 10^6 \text{ mm}^4$$

The maximum deflection is found from (7) as

$$\Delta_{st} = \frac{(1000)(5)(10^3)^3}{48(200 \times 10^9 \times 10^{-6})(4.9 \times 10^6)} = 2.66 \text{ mm}$$

Thus,

$$\Delta = 2.66 + \sqrt{(2.66)^2 + 2(100)(2.66)} = 25.9 \text{ mm}$$

15.8. Derive Castigliano's theorem.

Let us consider a general three-dimensional elastic body loaded by the forces P_1, P_2, etc. (Fig. 15-7). These would include forces exerted on the body by the various supports. We shall denote the displacement under P_1 *in the direction* of P_1 by Δ_1, that under P_2 in the direction of P_2 by Δ_2, etc. If we assume that all

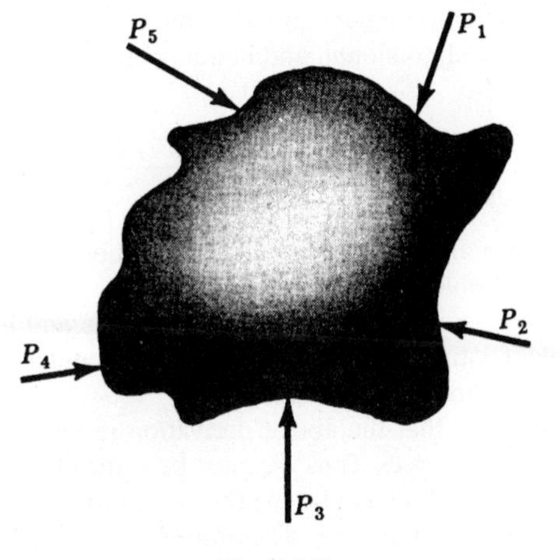

Fig. 15-7

forces are applied simultaneously and gradually increased from zero to their final values given by P_1, P_2, etc., then the work done by the totality of forces will be

$$U = \frac{P_1}{2}\Delta_1 + \frac{P_2}{2}\Delta_2 + \frac{P_3}{2}\Delta_3 + \cdots \tag{1}$$

This work is stored within the body as elastic strain energy.

Let us now increase the nth force by an amount dP_n. This changes both the state of deformation and also the internal strain energy slightly. The increase in the latter is given by

$$\frac{\partial U}{\partial P_n} dP_n \tag{2}$$

Thus, the total strain energy after the increase in the nth force is

$$U + \frac{\partial U}{\partial P_n} dP_n \tag{3}$$

Let us reconsider this problem by first applying a very small force dP_n alone to the elastic body. Then, we apply the same forces as before, namely, P_1, P_2, P_3, etc. Due to the application of dP_n there is a displacement in the direction of dP_n which is infinitesimal and may be denoted by $d\Delta_n$. Now, when P_1, P_2, P_3, etc., are applied, their effect on the body will not be changed by the presence of dP_n and the internal strain energy arising from application of P_1, P_2, P_3, etc., will be that indicated in (*1*). But as these forces are being applied the small force dP_n goes through the additional displacement Δ_n caused by the forces P_1, P_2, P_3, etc. Thus, it gives rise to additional work $(dP_n)\Delta_n$ which is stored as internal strain energy and hence the total strain energy in this case is

$$U + (dP_n)\Delta_n \tag{4}$$

Since the final strain energy must be independent of the order in which the forces are applied, we may equate (*3*) and (*4*):

$$U + \frac{\partial U}{\partial P_n} dP_n = U + (dP_n)\Delta_n$$

or

$$\Delta_n = \frac{\partial U}{\partial P_n} \tag{5}$$

This is *Castigliano's theorem*; i.e., the displacement of an elastic body under the point of application of any force, in the direction of that force, is given by the partial derivative of the total internal strain energy with respect to that force. Equations for U are given in Problems 15.1, 15.2, and 15.3 for axial, torsional, and bending loadings, respectively. However, instead of using the integral forms of the equations in those problems, it is usually more convenient to differentiate through the integral signs, and thus for a body subject to combined axial, torsional, and bending effects, we have for the displacement Δ_n under the force P_n

$$\Delta_n = \int \frac{P(\partial P/\partial P_n)\,ds}{AE} + \int \frac{T(\partial T/\partial P_n)\,ds}{GJ} + \int \frac{M(\partial M/\partial P_n)\,ds}{EI}$$

For a body composed of a finite number of elastic subbodies, these integrals are replaced by finite summations, as shown in Problem 15.9.

The term "force" here is used in its most general sense and implies either a true force or a couple. For the case of a couple, Castigliano's theorem gives the angular rotation under the point of application of the couple in the sense of rotation of the couple.

It is important to observe that the above derivation required that we be able to vary the nth force, P_n, independently of the other forces. Thus, P_n must be statically independent of the other external forces, implying that the energy U must always be expressed in terms of the statically independent forces of the system. Obviously, reactions that can be determined by statics cannot be considered as independent forces.

15.9. The bars AB and CB of Fig. 15-8 are pinned at A, C, and B and subject to the horizontal applied load P acting at B. Use Castigliano's theorem to determine the horizontal and vertical components of displacement of pin B.

In order to use Castigliano's theorem, we must have a force at B acting in each of the directions in which we seek the displacement. Since the real force P acts horizontally, we must consider that force as well as an auxiliary force Q that we introduce in the vertical direction at B. Thus, the free-body diagram of the pin at B appears as shown in Fig. 15-9.

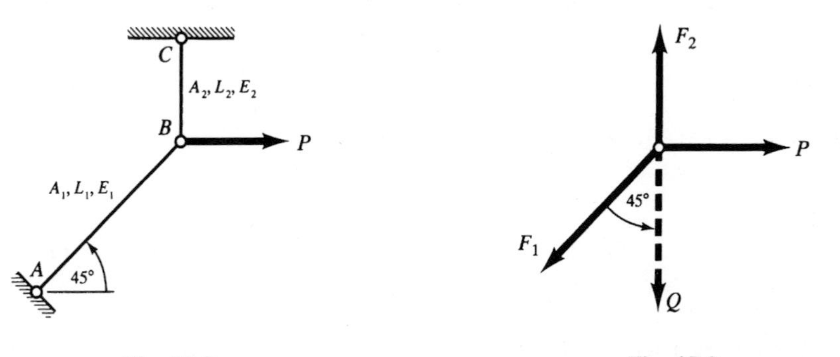

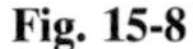
Fig. 15-8

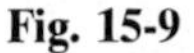
Fig. 15-9

For equilibrium we have

$$\Sigma F_x = P - F_1 \sin 45^\circ = 0 \qquad \text{and therefore} \qquad F_1 = P\sqrt{2} \tag{1}$$

$$\Sigma F_y = F_2 - Q - F_1 \cos 45^\circ = 0 \qquad \text{and therefore} \qquad F_2 = P + Q \tag{2}$$

Castigliano's theorem applied to a bar system states that

$$\Delta_x = \sum_{i=1,2} \frac{F_i(\partial F_i/\partial P)L_i}{A_i E_i} \qquad \Delta_y = \sum_{i=1,2} \frac{F_i(\partial F_i/\partial Q)L_i}{A_i E_i} \tag{3}$$

For our bar forces we have

$$F_1 = P\sqrt{2} \qquad \frac{\partial F_1}{\partial P} = \sqrt{2} \qquad \frac{\partial F_1}{\partial Q} = 0$$

$$F_2 = P + Q \qquad \frac{\partial F_2}{\partial P} = 1 \qquad \frac{\partial F_2}{\partial Q} = 1$$

Now that we have taken the partial derivatives with respect to P and Q, we may set $Q = 0$.

Substituting in (3),

$$\Delta_x = \frac{(P\sqrt{2})(\sqrt{2})L_1}{A_1 E_1} + \frac{(P)(1)L_2}{A_2 E_2} = \frac{2PL_1}{A_1 E_1} + \frac{PL_2}{A_2 E_2}$$

$$\Delta_y = \frac{(P\sqrt{2})(0)L_1}{A_1 E_1} + \frac{P(1)L_2}{A_2 E_2} = \frac{PL_2}{A_2 E_2}$$

which agree with the results found using a geometric approach in Problem 1.12.

15.10. The system shown in Fig. 15-10 consists of a horizontal bar CDF of bending rigidity EI_1 and torsional rigidity GJ_1 which is rigidly welded at D to bar DB of bending rigidity EI_2. At point B the horizontal bar DB is attached to the vertical bar AB of cross-sectional area A and Young's modulus E. The support at C permits only rotation in the x-y plane about the z-axis and the end F is restrained against angular rotation about the x-axis and can deflect only vertically. Determine the vertical deflection at F due to the application of the load P acting parallel to the y-axis.

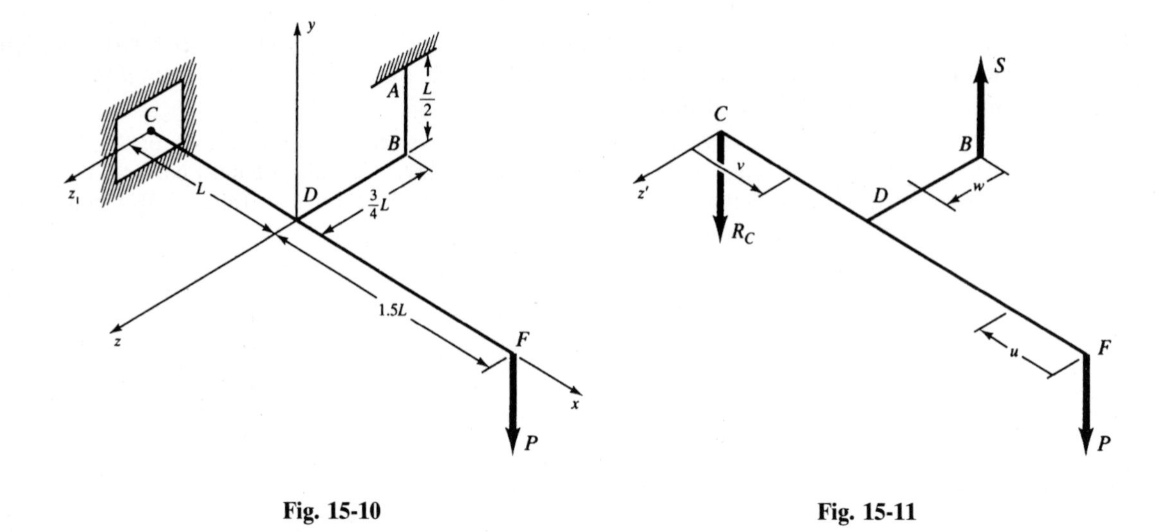

Fig. 15-10 Fig. 15-11

A free-body diagram of the bars CDF and DB is shown in Fig. 15-11, where R_C is the vertical reaction at C and S is the axial force in bar AB. For equilibrium about the z_1-axis, we have

$$\Sigma M_{z_1} = S(L) - (2.5L)P = 0$$

$$S = 2.5P$$

and for equilibrium in the y-direction

$$-R_C - P + S = 0$$

$$R_C = 1.5P$$

Let us introduce the variables u, v, and w as shown in Fig. 15-11 to denote positions of points in regions FD, CD, and BD of the system. The bending and twisting moments are then given by

In FD: $$M = Pu \qquad \frac{\partial M}{\partial P} = u$$

In CD: $$M = R_C v = 1.5Pv \qquad \frac{\partial M}{\partial P} = 1.5v$$

In CDF: $$T = S\left(\frac{3}{4}L\right) = (2.5P)\left(\frac{3}{4}L\right) = \frac{15}{8}PL \qquad \frac{\partial T}{\partial P} = \frac{15}{8}L$$

In BD: $$M = Sw = (2.5P)w \qquad \frac{\partial M}{\partial P} = 2.5w$$

Castigliano's theorem gives the deflection at F as the partial derivative of the total internal strain energy with respect to F. As indicated by the bending moments in FD, CD, and BD, as well as the twisting moment in CDF, and the axial force in AB, this becomes

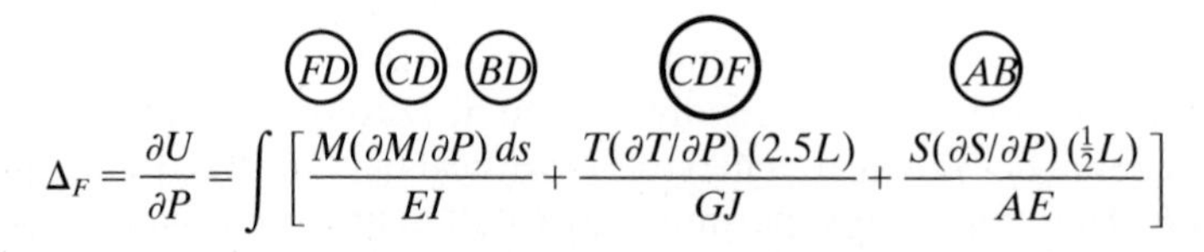

$$\Delta_F = \frac{\partial U}{\partial P} = \int \left[\frac{M(\partial M/\partial P)\,ds}{EI} + \frac{T(\partial T/\partial P)\,(2.5L)}{GJ} + \frac{S(\partial S/\partial P)\,(\frac{1}{2}L)}{AE} \right]$$

where s is a coordinate of length used as a variable of integration over the appropriate variable in each of the bars indicated by the circled bar designators above the integrals. The twisting moment is constant in CDF and the axial force is constant in AB, so there is no need to integrate to obtain the strain energy corresponding to these loads. Substituting the above values of bending and twisting moments and axial force, we find

$$\Delta_F = \overset{(FD)}{\int_0^{1.5L} \frac{(Pu)(u)\,du}{EI_1}} + \overset{(CD)}{\int_0^{L} \frac{(1.5Pv)(1.5v)\,dv}{EI_1}} + \overset{(BD)}{\int_0^{(3/4)L} \frac{(2.5Pw)(2.5w)\,dw}{EI_2}}$$

$$+ \overset{(CDF)}{\frac{(\frac{15}{8}PL)(\frac{15}{8}L)(2.5L)}{GJ_1}} + \overset{(AB)}{\frac{(2.5P)(2.5)(\frac{1}{2}L)}{AE}}$$

$$= 1.875\frac{PL^3}{EI_1} + 0.879\frac{PL^3}{EI_2} + 8.79\frac{PL^3}{GJ_1} + 3.13\frac{PL}{AE}$$

15.11. The pin-connected framework shown in Fig. 15-12 consists of two identical upper rods AB and AC, two shorter, lower rods BD and DC, together with a rigid horizontal brace BC. All bars have cross-sectional area A and modulus of elasticity E. Determine the vertical displacement of point D due to the action of the vertical load applied there.

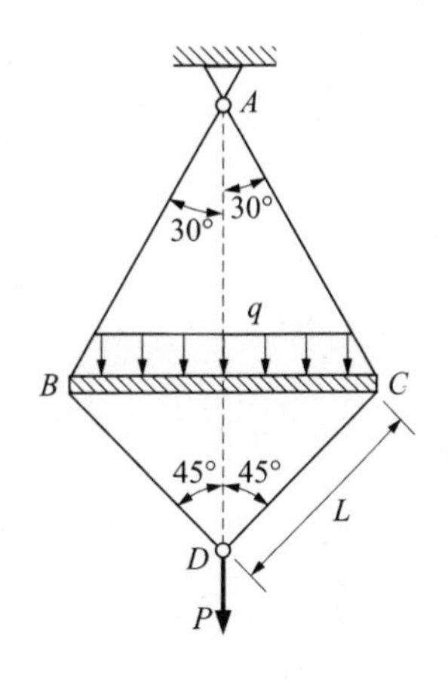

Fig. 15-12

This problem was considered by an approach involving the geometry of displacement in Problem 1.11. Let us consider it now using Castigliano's approach. We have already used statics to find bar forces in Problem 1.11, and these are

$$F_{DB} = F_{DC} = \frac{P\sqrt{2}}{2}$$

$$F_{AB} = F_{AC} = \frac{P + \left(\dfrac{2}{\sqrt{2}}\right)qL}{\sqrt{3}}$$

The deflection of D in the direction of P is given by Castigliano's theorem as

$$\Delta_D = \sum \frac{S\left(\dfrac{\partial S}{\partial P}\right)L}{AE}$$

Substituting, we have

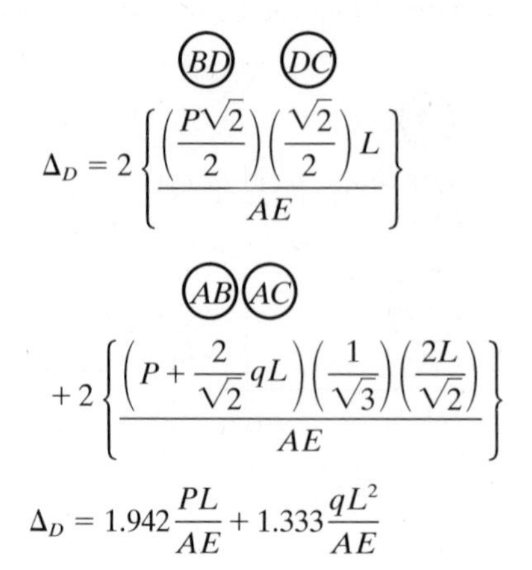

$$\Delta_D = 2\left\{\frac{\left(\frac{P\sqrt{2}}{2}\right)\left(\frac{\sqrt{2}}{2}\right)L}{AE}\right\}$$

$$+2\left\{\frac{\left(P+\frac{2}{\sqrt{2}}qL\right)\left(\frac{1}{\sqrt{3}}\right)\left(\frac{2L}{\sqrt{2}}\right)}{AE}\right\}$$

$$\Delta_D = 1.942\frac{PL}{AE} + 1.333\frac{qL^2}{AE}$$

which agrees with the result found by the geometric approach in Problem 1.11.

15.12. A structure is in the form of one quadrant of a thin circular ring of radius R. One end is clamped and the other end is loaded by a vertical force P (see Fig. 15-13). Determine the vertical displacement under the point of application of the force P. Consider only strain energy of bending.

From statics, the reactions at the clamped end consist of a vertical force P and a couple PR. The bending moment at the section in the ring located by the angle θ is given by

$$M = PR - P(R - R\cos\theta) = PR\cos\theta \qquad \text{from which} \qquad \frac{\partial M}{\partial P} = R\cos\theta$$

Castigliano's theorem states that the vertical deflection at A is given by

$$\Delta_v = \frac{\partial U}{\partial P} = \int_0^{\pi/2}\frac{M(\partial M/\partial P)R\,d\theta}{EI} = \int_0^{\pi/2}\frac{(PR\cos\theta)(R\cos\theta)R\,d\theta}{EI} = \frac{P\pi R^3}{4EI}$$

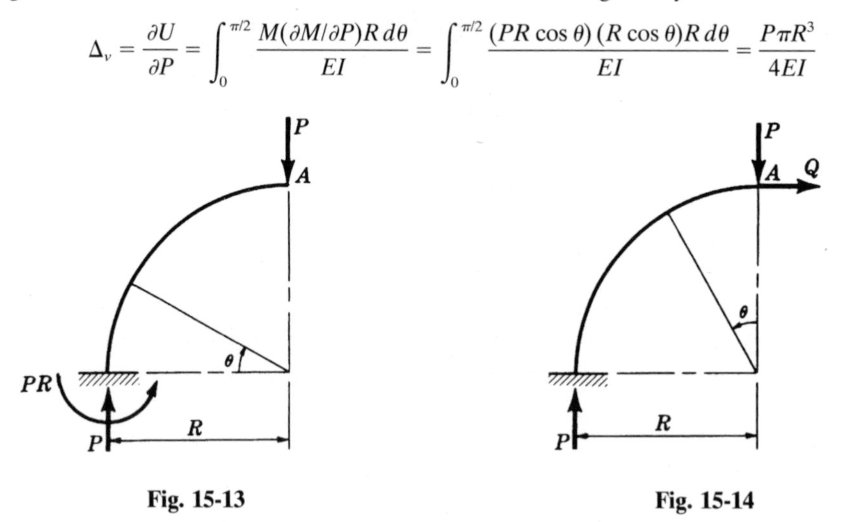

Fig. 15-13

Fig. 15-14

15.13. Determine the horizontal displacement of point A in Problem 15.12.

Since there is no horizontal force applied at A, we must temporarily introduce an auxiliary force Q shown in Fig. 15-14 in order to be able to use Castigliano's theorem. This time, let us measure θ from the vertical, making it unnecessary to determine reactions at B. Thus, at the section denoted by θ the bending moment is

$$M = PR\sin\theta + Q(R - R\cos\theta) \qquad \text{from which} \qquad \frac{\partial M}{\partial Q} = R - R\cos\theta$$

The horizontal displacement at A is given by

$$\Delta_h = \frac{\partial U}{\partial Q} = \int_0^{\pi/2} \frac{M(\partial M/\partial Q)R\,d\theta}{EI}$$

Now that the partial derivative has been taken, Q may be set equal to zero, yielding

$$\Delta_h = \int_0^{\pi/2} \frac{(PR\sin\theta)(R - R\cos\theta)R\,d\theta}{EI} = \frac{PR^3}{2EI}$$

15.14. A thin circular ring in the form of one quadrant OA of a circle lies in the x-z plane and has rigidly attached to it at the point A a straight bar AB also in the x-z plane. Both the ring and the bar have bending rigidity EI and torsional rigidity GJ. The unsupported end B is loaded by a twisting moment represented by the vector T_B directed parallel to the x-axis as shown in Fig. 15-15. Determine the y-component of displacement of point B.

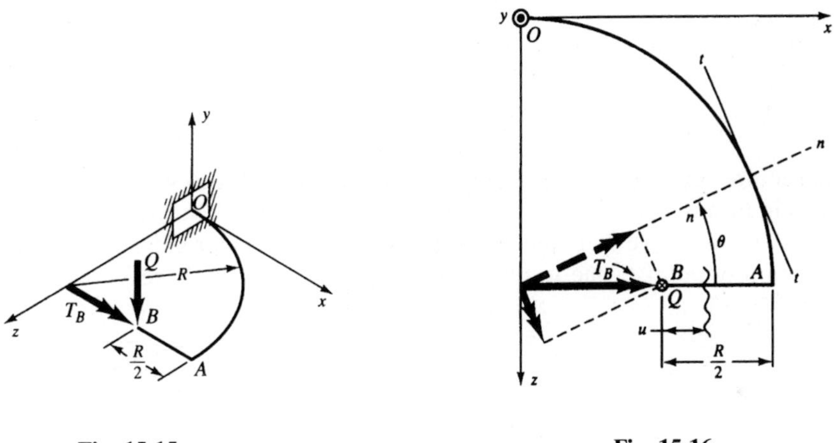

Fig. 15-15

Fig. 15-16

To utilize Castigliano's theorem, we must introduce an auxiliary force Q in the direction of the desired displacement; that is, Q must be directed downward and parallel to the y-axis. The view of the system looking from the positive end of the y-axis toward the x-z plane appears as in Fig. 15-16, where n-n and t-t denote axes normal and tangential, respectively, to the ring at an arbitrary location denoted by the angle θ. In that figure the applied twisting moment T_B is shown, along with its components oriented in the n-n and t-t directions. The auxiliary force Q is represented by the tail of its vector representation at B to denote its downward direction.

From Fig. 15-16 we have in the straight bar BA

$$M = Qu \qquad \frac{\partial M}{\partial Q} = u$$

$$T = T_B \qquad \frac{\partial T}{\partial Q} = 0$$

In the quadrant AO from the geometry of the figure

$$M = T_B\cos\theta + Q\left(\frac{R}{2}\sin\theta\right)$$

$$\frac{\partial M}{\partial Q} = \frac{R}{2}\sin\theta$$

and
$$T = T_B \sin\theta + Q\left(R - \frac{R}{2}\cos\theta\right)$$

$$\frac{\partial T}{\partial Q} = R - \frac{R}{2}\cos\theta$$

Using Castigliano's theorem, we find when we set $Q = 0$ after taking the partial derivatives

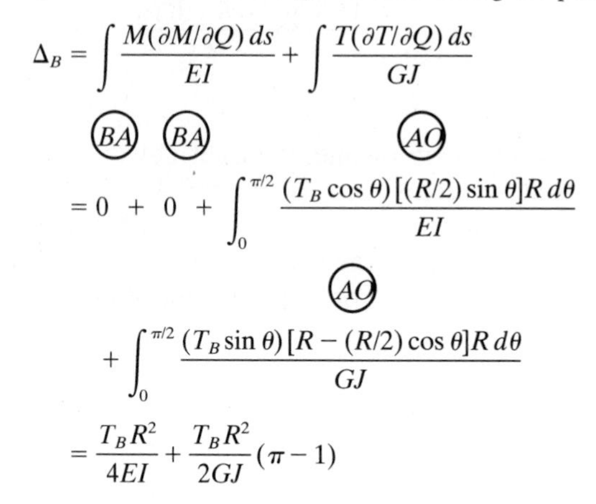

$$\Delta_B = \int \frac{M(\partial M/\partial Q)\,ds}{EI} + \int \frac{T(\partial T/\partial Q)\,ds}{GJ}$$

$$= 0 + 0 + \int_0^{\pi/2} \frac{(T_B\cos\theta)[(R/2)\sin\theta]R\,d\theta}{EI}$$

$$+ \int_0^{\pi/2} \frac{(T_B\sin\theta)[R - (R/2)\cos\theta]R\,d\theta}{GJ}$$

$$= \frac{T_B R^2}{4EI} + \frac{T_B R^2}{2GJ}(\pi - 1)$$

15.15. A structure consists of a quadrant of a circular ring OA, to which is rigidly attached a bar BA which in turn is welded to bar CB. These bars all lie in the horizontal plane x-z, as shown in Fig. 15-17, and all have bending rigidity EI and torsional rigidity GJ. Determine the vertical deflection of point C due to the load P applied vertically there.

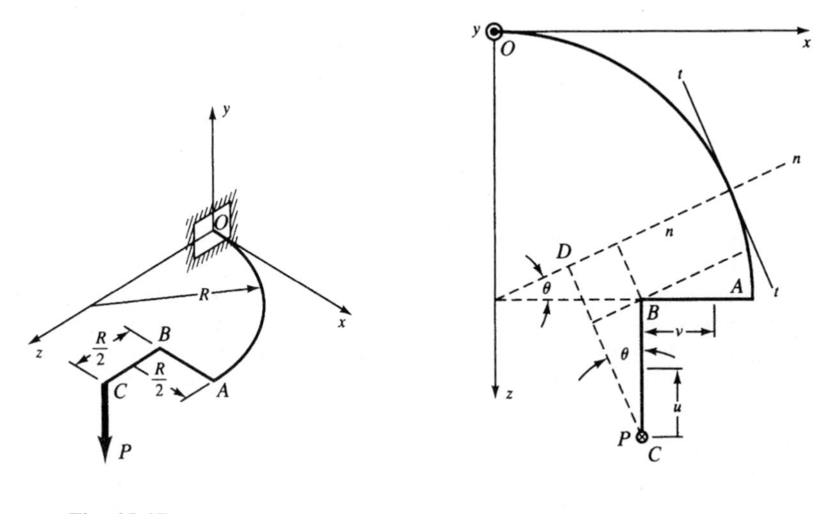

Fig. 15-17 **Fig. 15-18**

It is first necessary to determine the bending and twisting moments at an arbitrary point in the quadrant OA. Let us introduce the coordinate system shown in Fig. 15-18, where θ denotes the angular coordinate of this arbitrary point. The axes n-n and t-t represent normal and tangential directions to the circular ring at the point represented by the angle θ. From the geometry of Fig. 15-18, we have the bending moment about n-n to be

$$M = M_{u\text{-}u} = P(\overline{CD}) = P\left(\frac{R}{2}\cos\theta + \frac{R}{2}\sin\theta\right)$$

and the twisting moment about t-t to be

$$T = T_{t\text{-}t} = P\left[\left(R - \frac{R}{2}\cos\theta\right) + \frac{R}{2}\sin\theta\right]$$

From these equations we thus have in the ring OA

$$\frac{\partial M}{\partial P} = \frac{R}{2}(\sin\theta + \cos\theta)$$

$$\frac{\partial T}{\partial P} = \frac{R}{2}(1 + \sin\theta - \cos\theta)$$

Next, in bar CB from Fig. 15-18 we have the bending moment at an arbitrary point represented by u to be

$$M = Pu \qquad \text{so} \qquad \frac{\partial M}{\partial P} = u$$

and the twisting moment T in this bar is zero.

In bar BA the bending moment from Fig. 15-18 is $M = Pv$ and the twisting moment is $T = PR/2$. Thus, for BA

$$\frac{\partial M}{\partial P} = v \qquad \frac{\partial T}{\partial P} = \frac{R}{2}$$

By Castigliano's theorem, the deflection of point C due to the force P is

$$\Delta_C = \int \frac{M(\partial M/\partial P)\,ds}{EI} + \int \frac{T(\partial T/\partial P)\,ds}{GJ}$$

$$= \underbrace{\frac{1}{EI}\int_0^{\pi/2} \frac{PR^2}{4}(\sin\theta + \cos\theta)^2 R\,d\theta}_{OA}$$

$$+ \underbrace{\frac{1}{GJ}\int_0^{\pi/4} \frac{PR^2}{4}(1 + \sin\theta - \cos\theta)^2 R\,d\theta}_{OA} + \underbrace{\int_0^{R/2} \frac{(Pu)u\,du}{EI}}_{CB}$$

$$+ \underbrace{\int_0^{R/2} \frac{(Pv)v\,dv}{EI}}_{BA} + \underbrace{\frac{(PR/2)(R/2)(R/2)}{GJ}}_{BA}$$

Therefore

$$\Delta_C = \frac{PR^3}{4EI}\left(\frac{\pi}{2} + \frac{4}{3}\right) + \frac{PR^3}{4GJ}\left(\pi - \frac{1}{2}\right)$$

15.16. A thin semicircular ring is hinged at each end and loaded by a central concentrated force P, as shown in Fig. 15-19. Determine the horizontal reaction at each hinge.

A free-body diagram of this ring, Fig. 15-20, indicates that the desired reaction H is statically indeterminate. We may formulate the bending moment in the right half of the ring as follows:

$$M = \frac{P}{2}(R - R\cos\theta) - HR\sin\theta \qquad \text{and} \qquad \frac{\partial M}{\partial H} = -R\sin\theta \qquad \text{for} \qquad 0 < \theta < \frac{\pi}{2}$$

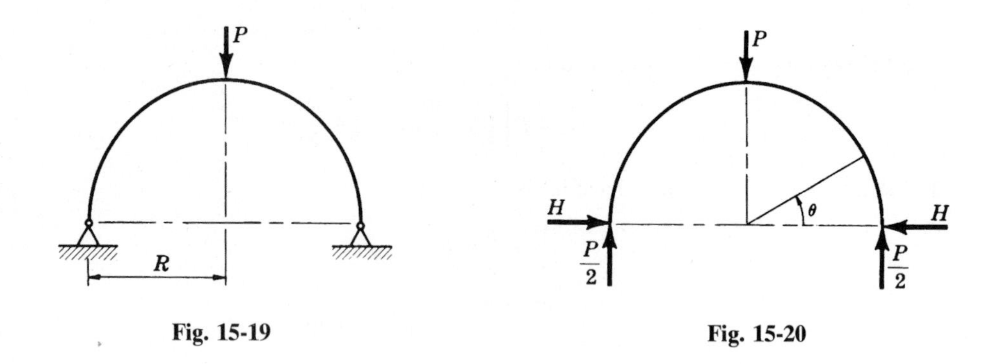

Fig. 15-19

Fig. 15-20

According to Castigliano's theorem, the horizontal displacement at the pin is given by

$$\Delta_H = \frac{\partial U}{\partial H}$$

But we know that this displacement is zero. Taking advantage of the symmetry about the centerline, we may now write

$$0 = \Delta_H = \frac{\partial U}{\partial H} = 2\int_0^{\pi/2} \frac{M(\partial M/\partial H)R\,d\theta}{EI} = \int_0^{\pi/2} \frac{[(P/2)(R - R\cos\theta) - HR\sin\theta](-R\sin\theta)R\,d\theta}{EI}$$

Solving for the unknown H: $H = P/\pi$.

15.17. In Problem 15.16, determine the vertical displacement under the point of application of the central force P.

In almost all statically indeterminate problems it is necessary first to determine the redundant reactions before any displacements can be found. For the present ring this has already been done in Problem 15.16.

In the right half of the ring, the bending moment is

$$M = \frac{P}{2}(R - R\cos\theta) - \frac{P}{\pi}R\sin\theta \qquad \text{for} \qquad 0 < \theta < \frac{\pi}{2}$$

and

$$\frac{\partial M}{\partial P} = \frac{1}{2}(R - R\cos\theta) - \frac{R}{\pi}\sin\theta$$

By Castigliano's theorem, the vertical displacement under the point of application of P is

$$\Delta = \frac{\partial U}{\partial P} = 2\int_0^{\pi/2} \frac{M(\partial M/\partial P)R\,d\theta}{EI}$$

where we have taken advantage of symmetry. Thus

$$\Delta = 2\int_0^{\pi/2} \frac{[(P/2)(R - R\cos\theta) - (PR/\pi)\sin\theta]\,[\tfrac{1}{2}(R - R\cos\theta) - (R/\pi)\sin\theta]R\,d\theta}{EI} = \frac{PR^3}{EI}\left(\frac{3\pi}{8} + \frac{3}{2\pi} - 1\right)$$

15.18. A structure in the form of a thin semicircular ring lies in a horizontal plane, has both ends clamped, and is subjected to a central vertical force P, as shown in Fig. 15-21. Determine the various reactions.

The vertical force reactions at A and C are each $P/2$ and the bending moment exerted by the support on the ring at each of these points is found from statics to be $PR/2$. There is also another component of reaction exerted by the support on the ring, i.e., a twisting moment T_0 acting at each of the points A and C. These two types of moment reaction are best illustrated by the vector representation of moment in Fig. 15-22, where a double-headed arrow indicates a moment in the usual sense of the right-hand rule for vector

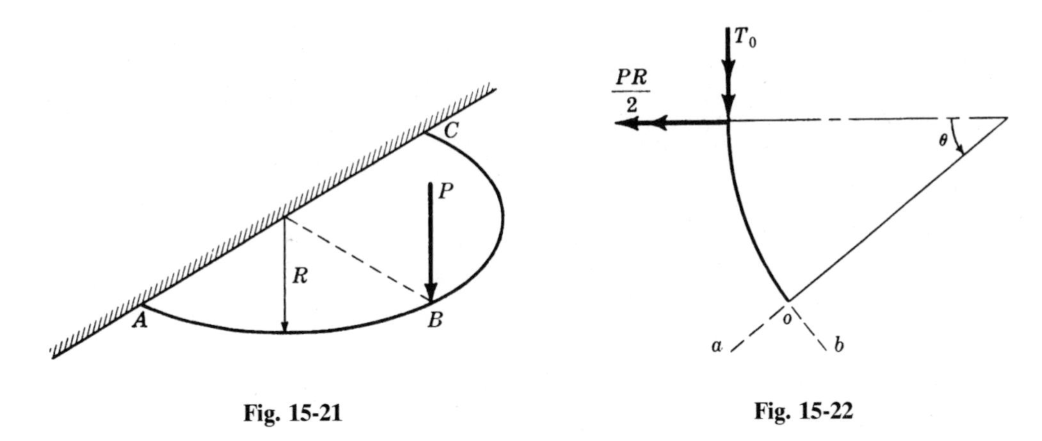

Fig. 15-21

Fig. 15-22

representation of moment. A segment of the ring to the arbitrary point represented by θ $(0<\theta<\pi/2)$ is shown and at this cross section given by θ there is a bending moment about the oa-axis given by

$$M=\frac{P}{2}R\sin\theta-\frac{PR}{2}\cos\theta-T_0\sin\theta$$

There is a twisting moment about the ob-axis given by

$$T=\frac{P}{2}(R-R\cos\theta)-\frac{PR}{2}\sin\theta+T_0\cos\theta$$

From these,

$$\frac{\partial M}{\partial T_0}=-\sin\theta \qquad \frac{\partial T}{\partial T_0}=\cos\theta$$

Since the ring is completely restrained at points A and C, we may write (taking advantage of symmetry)

$$0=\phi_A=\phi_C=2\int_0^{\pi/2}\frac{M(\partial M/\partial T_0)R\,d\theta}{EI}+2\int_0^{\pi/2}\frac{T(\partial T/\partial T_0)R\,d\theta}{GJ}$$

where ϕ is used to denote angular rotation of an arbitrary point of the bar, and ϕ_A and ϕ_C are the zero values of this quantity at the points A and C. Substituting,

$$0=\int_0^{\pi/2}\frac{\left(\frac{PR}{2}\sin\theta-\frac{PR}{2}\cos\theta-T_0\sin\theta\right)(-\sin\theta)R\,d\theta}{EI}$$
$$+\int_0^{\pi/2}\frac{\left[\frac{P}{2}(R-R\cos\theta)-\frac{PR}{2}\sin\theta+T_0\cos\theta\right](\cos\theta)R\,d\theta}{GJ}$$

Solving,

$$T_0=\frac{\frac{PR}{2}\left(\frac{2-\pi}{EI}+\frac{2-\pi}{GJ}\right)}{\left(\frac{\pi}{EI}-\frac{\pi}{GJ}\right)}$$

15.19. The thin rod shown in Fig. 15-23 consists of the straight bar GFD attached to semicircular end bars BCD and GHJ, together with two more straight bars JK and AB as indicated. There exists a very small gap $2\Delta_A$ between points A and K. Determine the magnitude of this gap when the forces Q are applied.

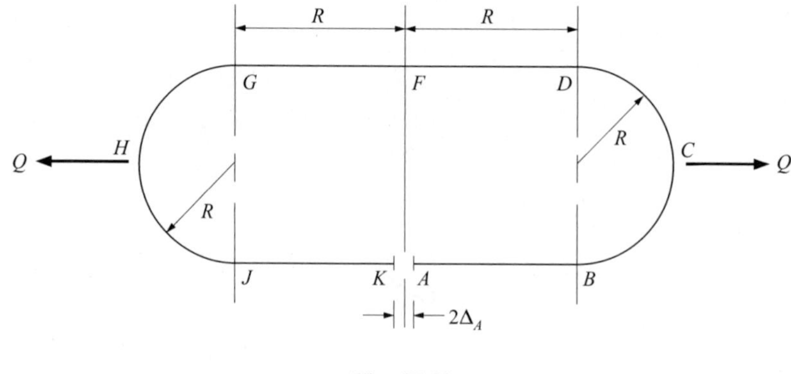

Fig. 15-23

Because of the symmetry of both structure and loading about a centerline extending through F, it is possible to examine the structural behavior of only one half of the system, say the right-hand half as shown in Fig. 15-23. Because of the symmetry, point F in Fig. 15-23 behaves as if it were clamped. The real load on this half is Q, and to determine displacement at the gap we introduce an auxiliary force P as shown in Fig. 15-24.

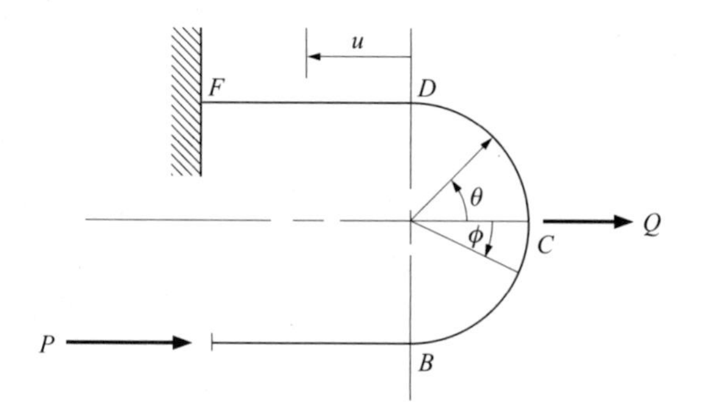

Fig. 15-24

Considering only bending action, in the entire system we have the bending moment in the various regions given by

In BC:

$$M = P(R - R\sin\phi); \qquad \frac{\partial M}{\partial P} = R(1 - \sin\phi)$$

In CD:

$$M = P(R + R\sin\theta) + QR\sin\theta$$

$$\frac{\partial M}{\partial P} = R + R\sin\theta$$

In DF:

$$M = 2PR + QR; \qquad \frac{\partial M}{\partial P} = 2R$$

The deflection at A in the direction of P is given by Castigliano's theorem as

$$\Delta_A = \frac{\partial V}{\partial P} = \int \frac{M\left(\dfrac{\partial M}{\partial P}\right) ds}{EI}$$

Substituting, we have

$$\Delta_A = \overset{\textcircled{BC}}{\int_0^{\pi/2} \frac{P(R - R\sin\phi)R(1 - \sin\phi)R\,d\phi}{EI}}$$

$$+ \overset{\textcircled{CD}}{\int_0^{\pi/2} \frac{[P(R + R\sin\theta) + QR\sin\theta](R + R\sin\theta)R\,d\theta}{EI}}$$

$$+ \overset{\textcircled{DF}}{\int_0^{R} \frac{(2PR + QR)(du)}{EI}}$$

This integrates to

$$\Delta_A = \frac{PR^3}{EI}\left[\frac{\pi}{2} - 2 + \frac{\pi}{4}\right] + \frac{R^3}{EI}\left[P\left(\frac{\pi}{2}\right) + P + Q + P + P\left(\frac{\pi}{4}\right) + Q\frac{\pi}{4}\right] + \frac{4PR^3}{EI} + \frac{2QR^3}{EI}$$

Now that the integration has been carried out, we may set $P = 0$ to find

$$\Delta_A = \frac{QR^3}{EI} + \frac{Q\pi R^3}{4EI} + \frac{2QR^3}{EI}$$

The gap at A is twice this because of the deformation of the left half of the system, so that the gap is

$$\frac{QR^3}{2EI}(12 + \pi)$$

15.20. The elastic beam $FDCG$ of bending rigidity EI shown in Fig. 15-25 is supported by pinned elastic bars AB, BC, and BD, each of extensional rigidity AE. These bars are incapable of resisting bending effects. The load on the system consists of a single concentrated force P applied at the free end F. Determine the vertical displacement of F.

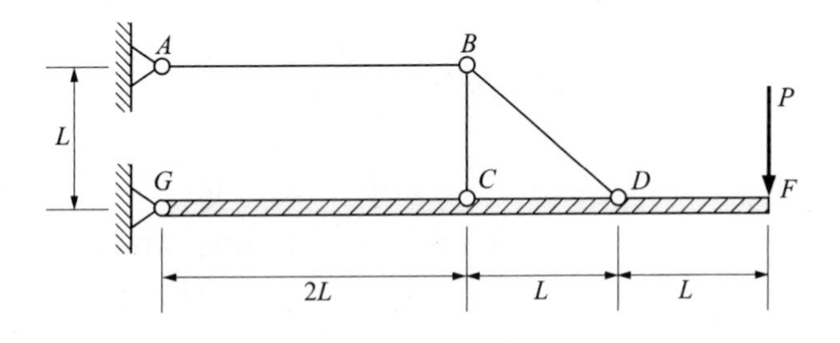

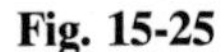
Fig. 15-25

This solution is best carried out by Castigliano's method since both bending as well as extensional energies are involved. We must first determine external reactions. A free-body diagram of the system is shown in Fig. 15-26. There is no vertical reaction at A since bar AB is not able to resist transverse (bending) loads. From statics,

$$\Sigma M_A = G_x(L) - P(4L) = 0 \qquad \therefore G_x = 4P$$

$$\Sigma F_H = -A_x + 4P = 0 \qquad \therefore A_x = 4P$$

$$\Sigma F_V = G_y - P = 0 \qquad \therefore G_y = P$$

Thus, bar AB carries a tensile force $4P$.

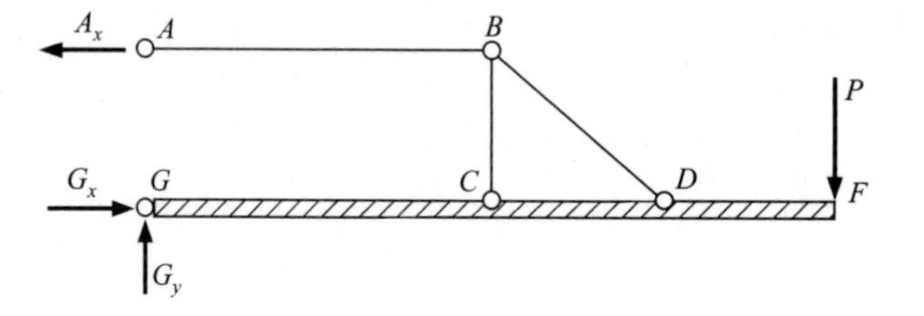

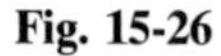

Fig. 15-26

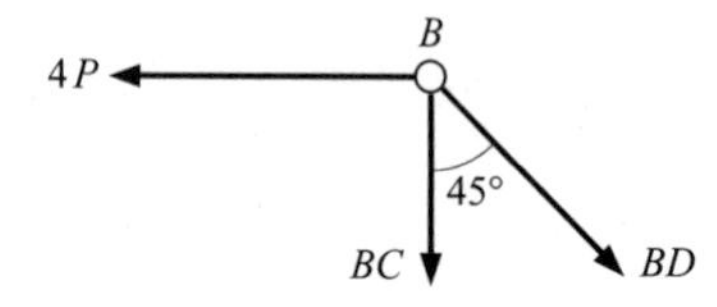

Fig. 15-27

Next, we show in Fig. 15-27 a free-body diagram of the system where a section has been passed through the three bars and axial forces are represented by BD and BC in those bars. From statics,

$$\Sigma F_H = -4P + BD \sin 45^\circ = 0$$

$$\Sigma F_V = -BC - BD \cos 45^\circ = 0$$

Consequently, the axial forces in the three bars are

$$AB = 4P$$

$$BC = -4P$$

$$BD = 4\sqrt{2}P$$

From Problem 15.8 we may determine the deflection at F due to axial loading (only) in these three bars to be

$$\Delta_1 = \sum \frac{S\left(\dfrac{\partial S}{\partial P}\right)L}{AE}$$

$$= \underbrace{\frac{(4P)(4)(2L)}{AE}}_{\textcircled{AB}} + \underbrace{\frac{(-4P)(-4)(L)}{AE}}_{\textcircled{BC}}$$

$$+ \underbrace{\frac{(4\sqrt{2}P)(4\sqrt{2})(L\sqrt{2})}{AE}}_{\textcircled{BD}} \tag{1}$$

$$= \frac{PL}{AE}[24 + 32\sqrt{2}] = 69.2\frac{PL}{AE} \tag{2}$$

Finally, we determine the deflection at F due only to the bending effects in beam $FDCG$. This was shown in Problem 15.8 to be

$$\Delta_z = \int \frac{M\dfrac{\partial M}{\partial P}ds}{EI} \tag{3}$$

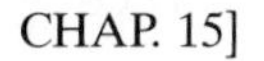

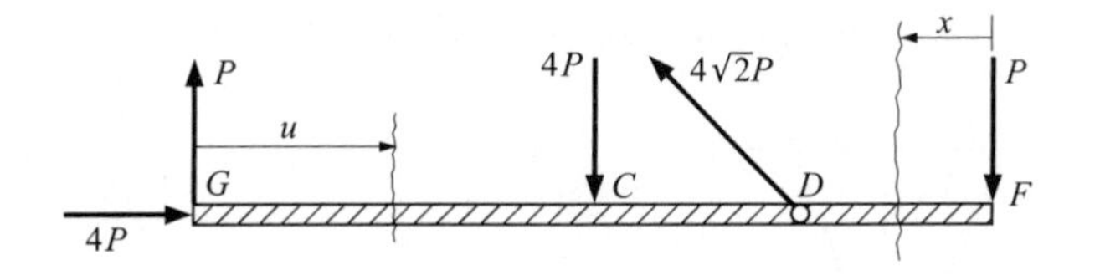

Fig. 15-28

Figure 15-28 shows a free-body diagram of the beam with all forces acting upon it. Coordinates u and x are introduced to permit evaluation of the integral in (3). The bending moments are given by

$$FD: \qquad M = Px; \qquad \frac{\partial M}{\partial P} = x$$

$$DC: \qquad M = Px - (4\sqrt{2}P)\left(\frac{1}{\sqrt{2}}\right)(x - L) = -3P_x + 4PL; \qquad \frac{\partial M}{\partial P} = -3x + 4L$$

$$GC: \qquad M = Pu; \qquad \frac{\partial M}{\partial P} = u$$

Thus, for bending effects only, (3) becomes

$$\Delta_2 = \overset{\textcircled{FD}}{\int_{x=0}^{L} \frac{(Px)(x)\,dx}{EI}} + \overset{\textcircled{CD}}{\int_{x=L}^{2L} \frac{(-3Px + 4PL)(-3x + 4L)\,dx}{EI}}$$

$$+ \overset{\textcircled{CG}}{\int_{u=0}^{u=2L} \frac{(Pu)(u)\,du}{EI}}$$

$$= 3.67\frac{PL^3}{EI}$$

The true deflection at F is the sum of Δ_1 and Δ_2:

$$\Delta_F = 69.2\frac{PL}{AE} + 3.67\frac{PL^3}{EI}$$

Supplementary Problems

15.21. A solid conical bar of circular cross section (Fig. 15-29) hangs vertically, subjected only to its own weight, which is γ per unit volume. Determine the strain energy stored within the bar.
Ans. $U = \pi D^2 L^3 \gamma^2 / 360E$

15.22. The two bars AB and CB of Fig. 15-30 are pinned at each end and subject to a single vertical force P. The geometric and elastic constants of each bar are as indicated. Use Castigliano's theorem to determine the horizontal and vertical components of displacement of pin B.

Ans. $\Delta_x = -\dfrac{PL_1}{\sqrt{3}A_1E_1} + \dfrac{PL_2}{\sqrt{3}A_2E_2}, \ \Delta_y = \dfrac{PL_1}{3A_1E_1} + \dfrac{PL_2}{3A_2E_2}$

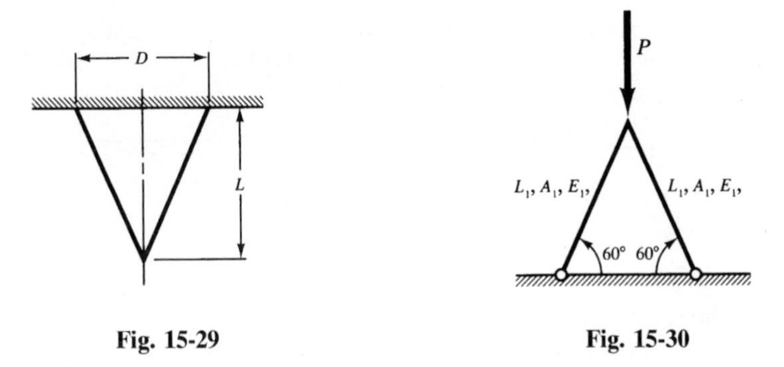

Fig. 15-29 Fig. 15-30

15.23. The pin-connected truss shown in Fig. 15-31 is composed of five bars, each of area A and modulus of elasticity E. Determine the vertical displacement of point B due to the load Q by equating the work done by Q to the internal strain energy. *Ans.* $\Delta = 2.914QL/AE$

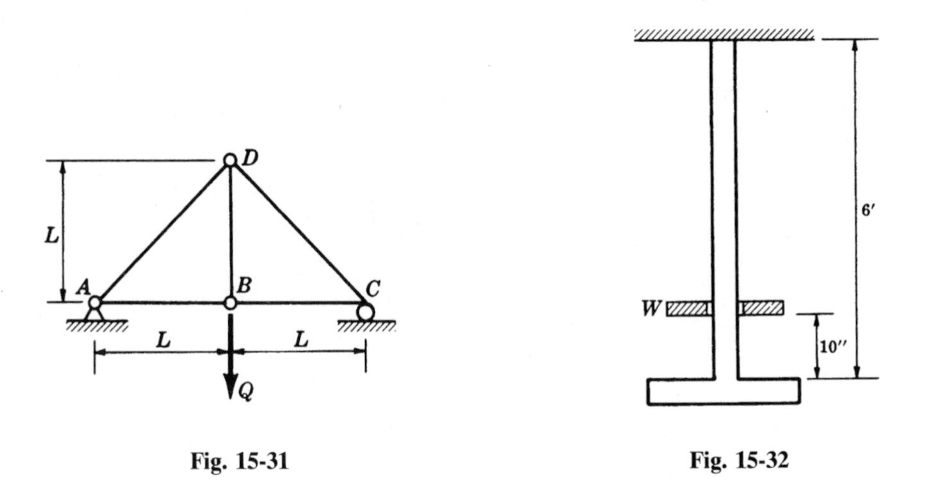

Fig. 15-31 Fig. 15-32

15.24. Determine the maximum weight W that can be dropped 10 in onto the flange at the end of the steel bar shown in Fig. 15-32. The bar is 1 in × 2 in in cross section and 6 ft in length. The axial stress is not to exceed 20,000 lb/in². Take $E = 30 \times 10^6$ lb/in². *Ans.* $W = 96$ lb

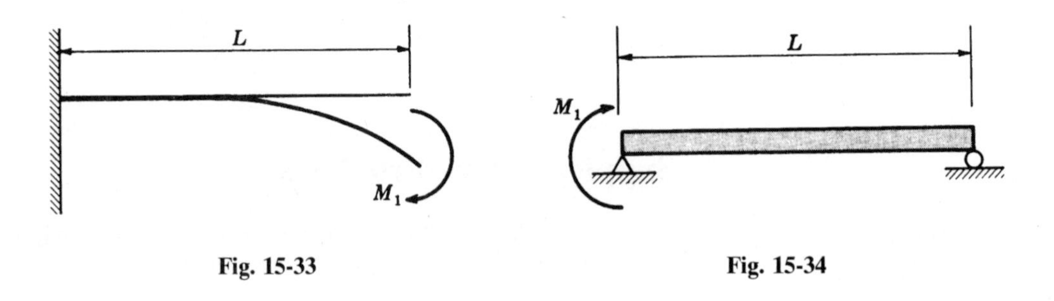

Fig. 15-33 Fig. 15-34

15.25. A cantilever beam is loaded by a moment M_1 applied at the tip (Fig. 15-33). Determine by Castigliano's theorem the deflection of the tip. *Ans.* $M_1L^2/2EI$

15.26. A simply supported beam is loaded by a moment M_1 at the left end, as shown in Fig. 15-34. Use Castigliano's theorem to determine the deflection at the midpoint of the bar. *Ans.* $M_1L^2/16EI$

15.27. A W203 × 28 steel wide-flange section is used as a cantilever beam of length 4 m. A weight W of 1 kN falls freely through a distance of 150 mm before striking the tip of the beam. Find the maximum deflection of the beam. Take $E = 200$ GPa. Use beam parameters given in Table 8-2 of Chap. 8. *Ans.* 38.8 mm

15.28. A structure lies in a vertical plane and is in the form of three quadrants of a thin ring (see Fig. 15-35). One end is clamped, the other is loaded by a vertical force P. Determine the horizontal displacement of point A. Consider only bending energy. *Ans.* $PR^3/2EI$

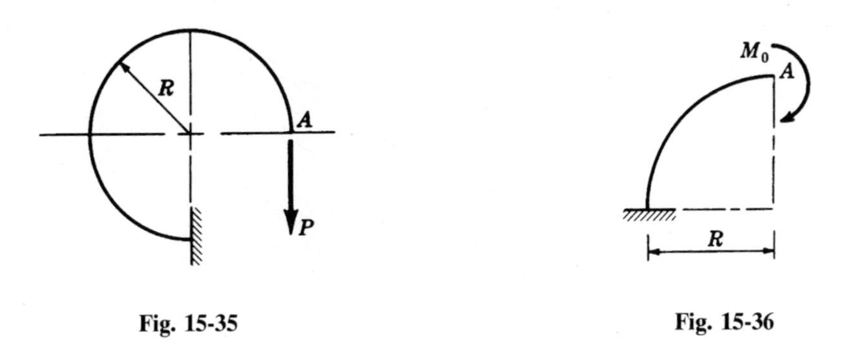

Fig. 15-35 **Fig. 15-36**

15.29. A structure is in the form of one quadrant of a thin circular ring of radius R. One end is clamped and the other is subject to a couple M_0, as shown in Fig. 15-36. Determine the angular rotation, as well as the vertical and horizontal components of displacement of point A.

Ans. $\dfrac{M_0 \pi R}{2EI}$; $\dfrac{M_0 R^2}{EI}$; $0.571\dfrac{M_0 R^2}{EI}$

15.30. The two-sided framework shown in Fig. 15-37 is loaded by a uniformly distributed load q per unit length in region AB together with a couple M_0 at the midpoint of BC. Determine the vertical displacement of point A.

Ans. $\dfrac{qL^4}{8EI} + \dfrac{2qL^3 H}{3EI} - \dfrac{M_0 LH}{2EI}$

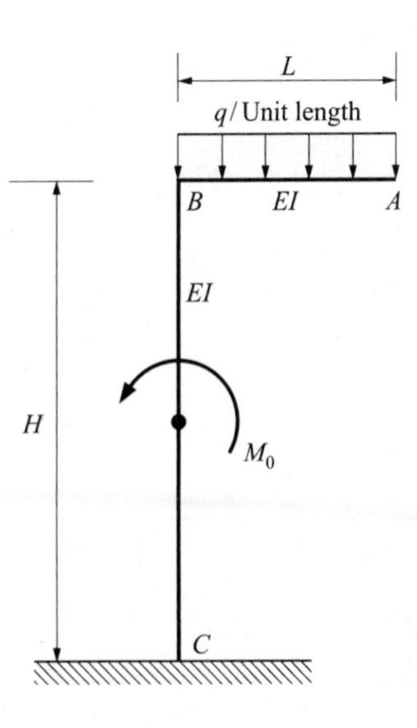

Fig. 15-37

15.31. The straight bar AC of Fig. 15-38 is rigidly attached at its midpoint B to another rod BD which has end D unsupported but subject to a vertical force P. The flexural rigidity of each bar is EI and the torsional rigidity is GJ. Bar AC is rigidly clamped at ends A and C, and AC and BD lie in a horizontal plane. Determine the deflection under the load P.

Ans. $\dfrac{3PL^3}{8EI} + \dfrac{PL^3}{4GJ}$

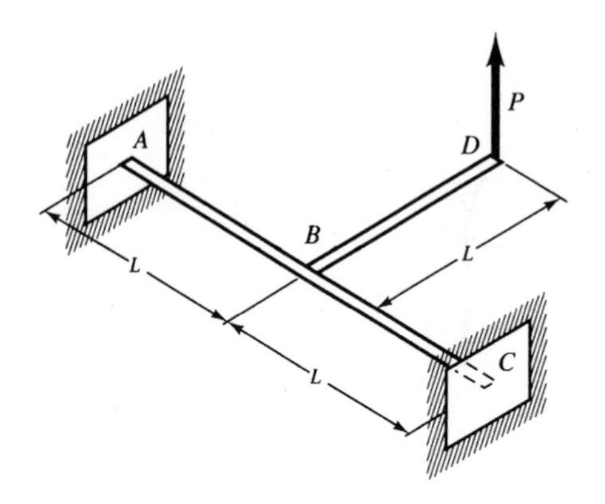

Fig. 15-38

15.32. Figure 15-39 shows a thin ring in the form of one quadrant of a circle. One end is fixed, the other is free, and the system is loaded by a moment at the midpoint. Determine the vertical component of displacement of point A.

Ans. $\dfrac{M_0 R^2}{\sqrt{2}EI}$

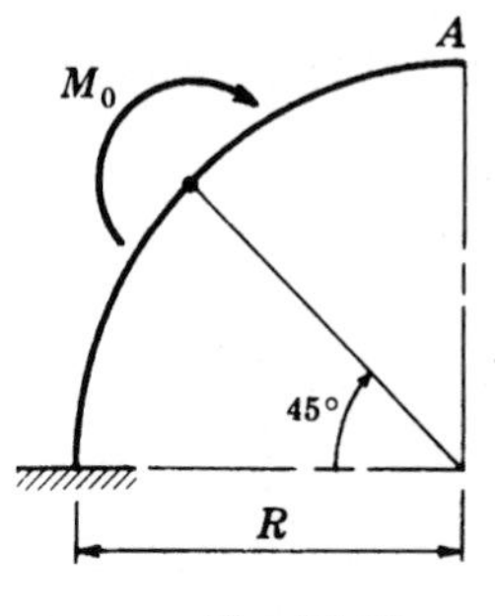

Fig. 15-39

15.33. The beam of Fig. 15-40 is supported at the left end, clamped at the right end and subject to a concentrated load. Determine the reaction at the left support by Castigliano's theorem. *Ans.* $Pb^2(2L + a)/2L^3$

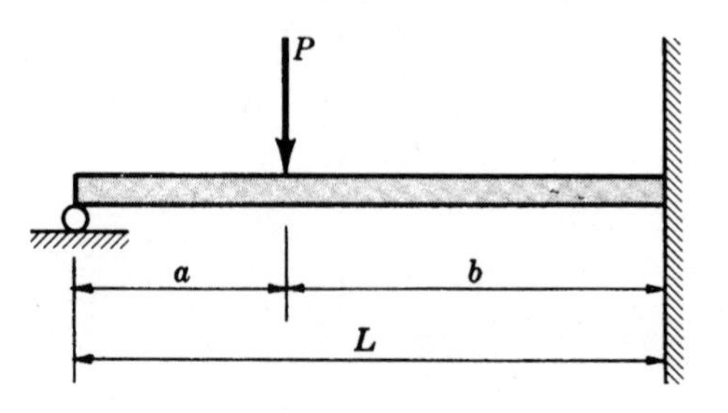

Fig. 15-40

15.34. A thin ring forms one quadrant of a circle and is loaded as shown in Fig. 15-41. One end is fixed and the other is pinned so as to prevent horizontal and vertical displacements. Find the components of reaction at the pin. *Ans.* $B_v = 0.19M_0/R$, $B_h = 1.12M_0/R$

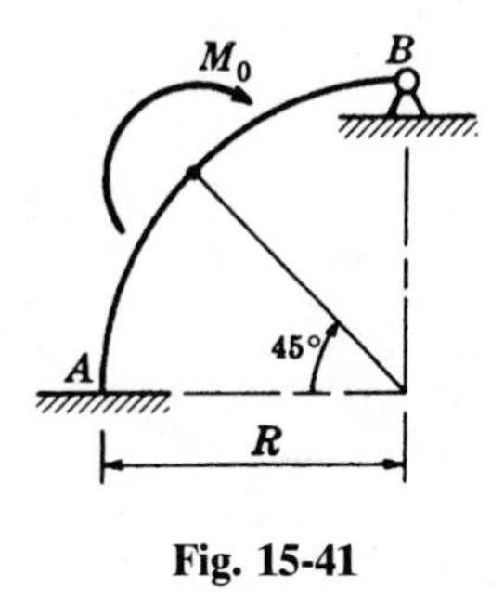

Fig. 15-41

15.35. A thin ring in the form of one quadrant of a circle lies in a vertical plane and is subject to uniform radial loading, as shown in Fig. 15-42. One end, A, is rigidly clamped and the other end, C, is unsupported. Determine the horizontal and vertical components of displacement of point C.

Ans. $\Delta_x = 0.500\dfrac{qR^4}{EI}$, $\Delta_y = 0.36\dfrac{qR^4}{EI}$

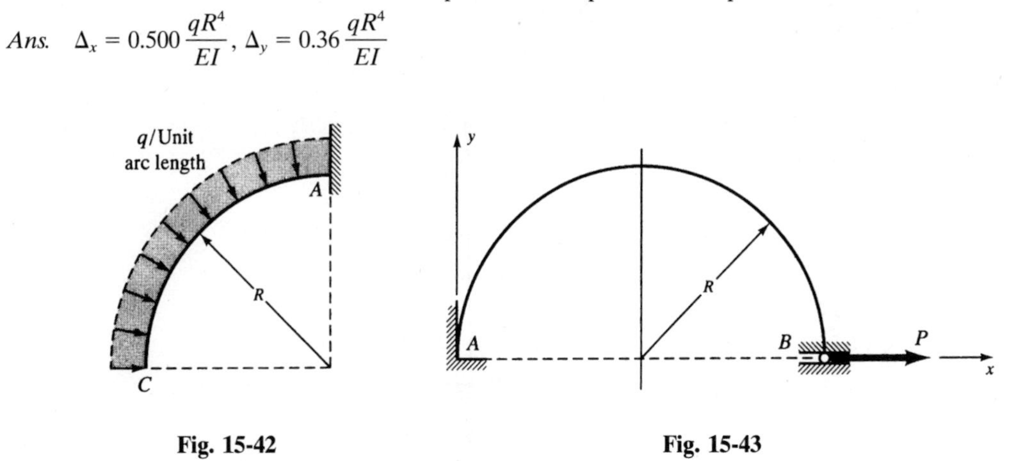

Fig. 15-42 **Fig. 15-43**

15.36. A thin semicircular ring (see Fig. 15-43) of bending rigidity EI lies in a vertical plane, is clamped at end A, and may move in a horizontal, frictionless guide at end B. The load is P, applied horizontally at end B. Determine the horizontal displacement of end B of the ring. Also, determine the vertical displacement due to the same load at B if the guide is removed.

Ans. $\Delta_{B_x} = 0.14\dfrac{PR^3}{EI}$, $\Delta_{B_y} = \dfrac{2PR^3}{EI}$

15.37. The structure of Fig. 15-44 is in the form of one quadrant of a thin circular ring AB together with a straight bar BC rigidly joined at B so that AB is tangent to the ring. A load P acts parallel to the y-axis at B. The end C is unsupported. Determine the y-component of displacement of point C. The bending rigidity of both regions is EI and the torsional rigidity is GJ.

$$Ans.\quad \Delta_C = \frac{1}{EI}\left[\frac{\pi}{4}PR^3 + \frac{1}{2}PR^2L\right] + \frac{1}{GJ}\left[\left(\frac{3\pi}{4} - 2\right)PR^3 - \frac{PR^2L}{2}\right]$$

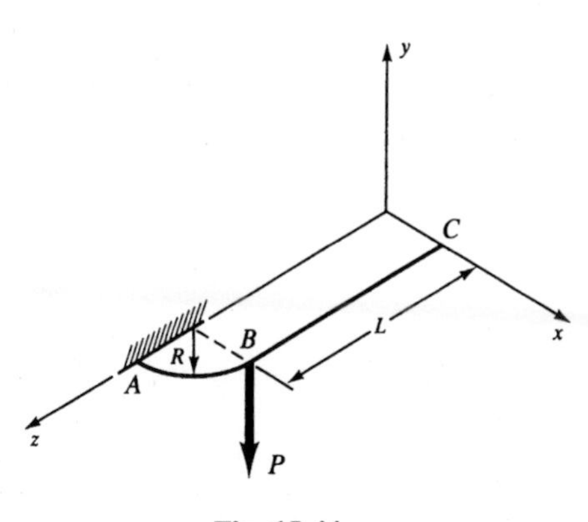

Fig. 15-44

15.38. The balcony-like structure of Fig. 15-45 is in the form of a semicircular ring, lies in a horizontal plane, and is subject to a twisting moment T_B at its midpoint. Determine the reactive twisting moment at each end A and C. *Ans.* $T_B/9\pi$

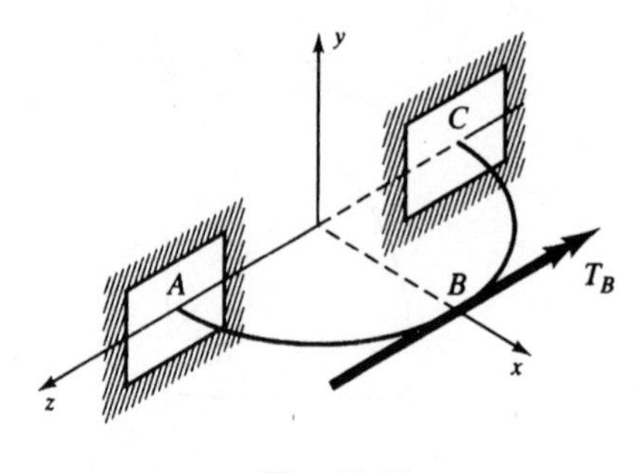

Fig. 15-45

15.39. A thin ring is subjected to the equal and opposite diametral forces indicated in Fig. 15-46. Determine the bending moment at A and also the increase in diameter of the ring along the diameter CD.

Ans. $M_A = \dfrac{PR}{2}\left(\dfrac{\pi - 2}{2}\right)$, $\Delta = 0.149\dfrac{PR^3}{EI}$

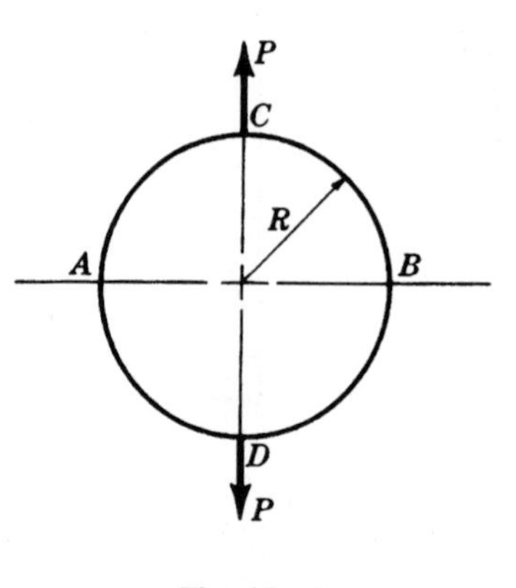

Fig. 15-46

15.40. A thin ring is loaded by forces which are uniformly distributed along the horizontal projection of the ring (see Fig. 15-47). Determine the decrease in the vertical diameter. *Ans.* $wR^4/6EI$

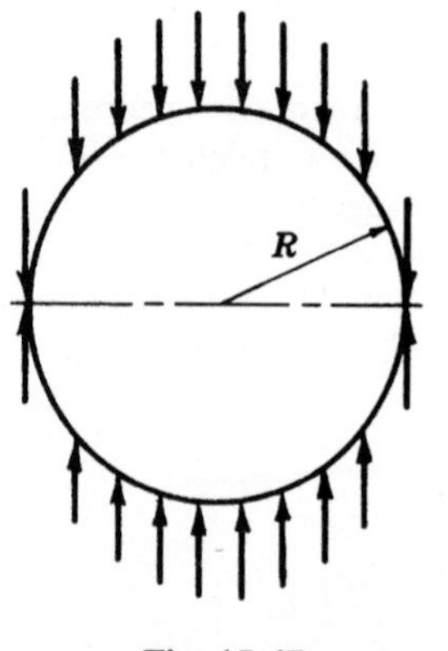

Fig. 15-47

15.41. A thin semicircular ring shown in Fig. 15-48 of bending rigidity EI lies in a vertical plane; it is clamped at end A and unsupported at B. It is loaded by a horizontal force P at end B. Determine horizontal and vertical components of displacement of end B of the ring.

Ans. $\Delta_{B_x} = \dfrac{PR^3\pi}{2EI}, \ \Delta_{B_y} = \dfrac{2PR^3}{EI}$

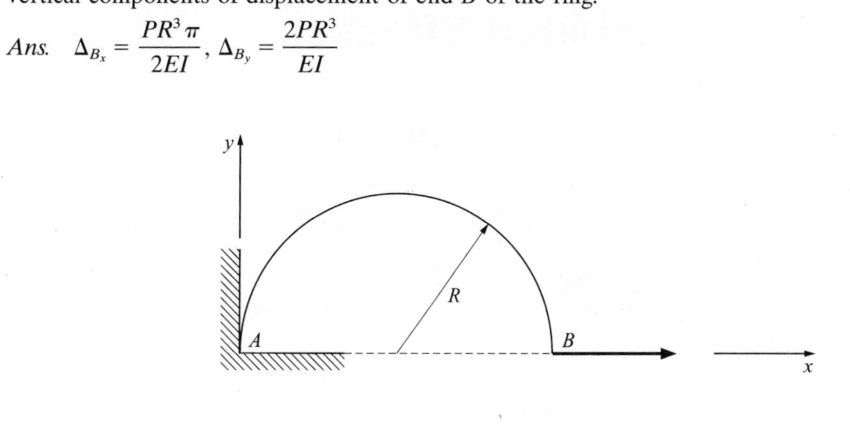

Fig. 15-48

15.42. A structure in the form of a thin three-sided rectangular frame lies in a horizontal plane, has both ends clamped, and is subject to a central vertical force P, as shown in Fig. 15-49. Determine the reactive torque at each support. The frame is of constant cross section throughout.

Ans. $\dfrac{Pb^2/EI}{\dfrac{b}{EI} + \dfrac{a}{GJ}}$

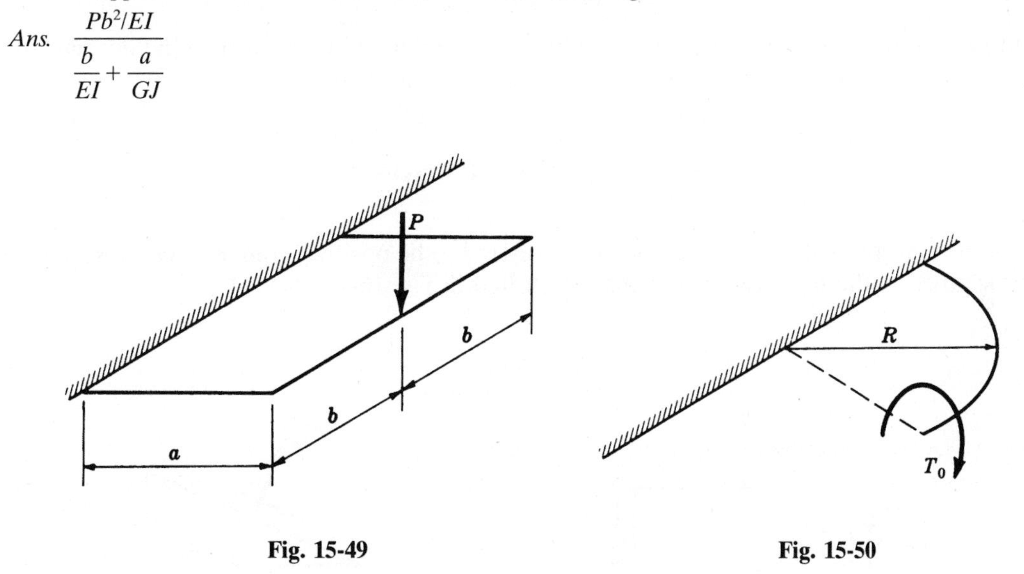

Fig. 15-49 Fig. 15-50

15.43. A thin structure in the form of one quadrant of a circle (Fig. 15-50) lies in a horizontal plane and is subject to a torque T_0 at the free end. The other end is clamped. Determine the vertical displacement of the free end.

Ans. $T_0R^2\left(\dfrac{\pi}{4EI} + \dfrac{\pi}{4GJ} - \dfrac{1}{GJ}\right)$

Chapter 16

Combined Stresses

INTRODUCTION

Previously in this book we have considered stresses arising in bars subject to axial loading, shafts subject to torsion, and beams subject to bending, as well as several cases involving thin-walled pressure vessels. It is to be noted that we have considered a bar, for example, to be subject to only *one* loading at a time, such as bending. But frequently such bars are simultaneously subject to several of the previously mentioned loadings, and it is required to determine the state of stress under these conditions. Since normal and shearing stress are vector quantities, considerable care must be exercised in combining the stresses given by the expressions for single loadings as derived in previous chapters. It is the purpose of this chapter to investigate the state of stress on an arbitrary plane through an element in a body subject to several simultaneous loadings.

GENERAL CASE OF TWO-DIMENSIONAL STRESS

In general if a plane element is removed from a body it will be subject to the normal stresses σ_x and σ_y together with the shearing stress τ_{xy}, as shown in Fig. 16-1.

SIGN CONVENTION

For normal stresses, tensile stresses are considered to be positive, compressive stress negative. For shearing stresses, the positive sense is that illustrated in Fig. 16-1.

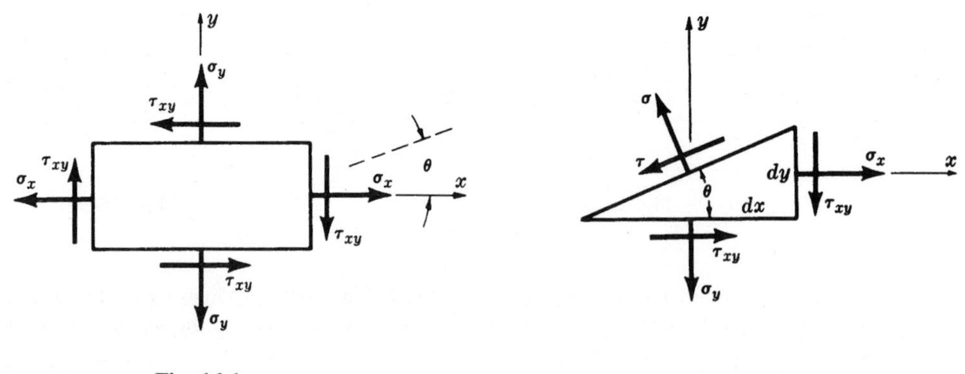

Fig. 16-1

Fig. 16-2

STRESSES ON AN INCLINED PLANE

We shall assume that the stresses σ_x, σ_y, and τ_{xy} are known. (Their determination will be discussed in Chap. 17.) Frequently it is desirable to investigate the state of stress on a plane inclined at an angle

θ to the x-axis, as shown in Fig. 16-1. The normal and shearing stresses on such a plane are denoted by σ and τ and appear as in Fig. 16-2. In Problem 16.13 it is shown that

$$\sigma = \frac{\sigma_x + \sigma_y}{2} - \frac{\sigma_x - \sigma_y}{2}\cos 2\theta + \tau_{xy}\sin 2\theta \qquad (16.1)$$

$$\tau = \frac{\sigma_x - \sigma_y}{2}\sin 2\theta + \tau_{xy}\cos 2\theta \qquad (16.2)$$

Thus, for any value of θ, σ and τ may be obtained from these expressions. For applications see Problems 16.15, 16.17, and 16.18.

PRINCIPAL STRESSES

There are certain values of the angle θ that lead to maximum and minimum values of σ for a given set of stresses σ_x, σ_y, and τ_{xy}. These maximum and minimum values that σ may assume are termed *principal stresses* and are given by

$$\sigma_{\max} = \frac{\sigma_x + \sigma_y}{2} + \sqrt{\left(\frac{\sigma_x - \sigma_y}{2}\right)^2 + (\tau_{xy})^2} \qquad (16.3)$$

$$\sigma_{\min} = \frac{\sigma_x + \sigma_y}{2} - \sqrt{\left(\frac{\sigma_x - \sigma_y}{2}\right)^2 + (\tau_{xy})^2} \qquad (16.4)$$

These expressions are derived in Problem 16.13. For applications see Problems 16.15 and 16.18.

DIRECTIONS OF PRINCIPAL STRESSES; PRINCIPAL PLANES

The angles designated as θ_p between the x-axis and the planes on which the principal stresses occur are given by the equation

$$\tan 2\theta_p = \frac{-\tau_{xy}}{\left(\dfrac{\sigma_x - \sigma_y}{2}\right)} \qquad (16.5)$$

This expression also is derived in Problem 16.13. For applications see Problems 16.15 and 16.18. As shown there, we always have two values of θ_p satisfying this equation. The stress $\sigma_{\max}$ occurs on one of these planes, and the stress $\sigma_{\min}$ occurs on the other. The planes defined by the angles θ_p are known as *principal planes*.

COMPUTER IMPLEMENTATION

For this two-dimensional situation, a simple FORTRAN program may be written to indicate the values of the principal stresses indicated by Eqs. (*16.3*) and (*16.4*) as well as directions of these stresses as given by Eq. (*16.5*). Such a program is developed in Problem 16.20 and an application is found in Problem 16.21.

SHEARING STRESSES ON PRINCIPAL PLANES

In Problem 16.13 it is demonstrated that the shearing stresses on the planes on which $\sigma_{\max}$ and $\sigma_{\min}$ occur are always zero, regardless of the values of σ_x, σ_y, and τ_{xy}. Thus, an element oriented along the principal planes and subject to the principal stresses appears as in Fig. 16-3.

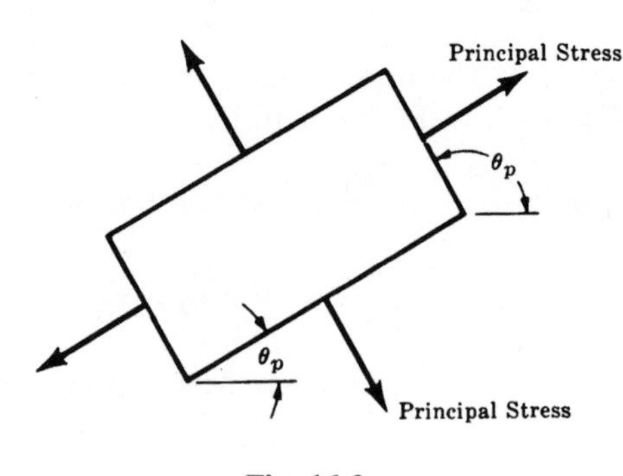

Fig. 16-3

MAXIMUM SHEARING STRESSES

There are certain values of the angle θ that lead to a maximum value of τ for a given set of stresses σ_x, σ_y, and τ_{xy}. The maximum and minimum values of the shearing stress are given by

$$\tau_{\substack{\max \\ \min}} = \pm\sqrt{\left(\frac{\sigma_x - \sigma_y}{2}\right)^2 + (\tau_{xy})^2} \tag{16.6}$$

This expression is derived in Problem 16.13. For applications see Problems 16.3, 16.10, 16.18, and 16.19.

DIRECTIONS OF MAXIMUM SHEARING STRESS

The angles θ_s between the x-axis and the planes on which the maximum shearing stresses occur are given by the equation

$$\tan 2\theta_s = \frac{\left(\dfrac{\sigma_x - \sigma_y}{2}\right)}{\tau_{xy}} \tag{16.7}$$

This expression also is derived in Problem 16.13. For applications see Problems 16.3, 16.10, 16.18, and 16.19. There are always two values of θ_s satisfying this equation. The shearing stress corresponding to the positive square root given above occurs on one of the planes designated by θ_s, while the shearing stress corresponding to the negative square root occurs on the other plane.

NORMAL STRESSES ON PLANES OF MAXIMUM SHEARING STRESS

In Problem 16.13, it is demonstrated that the normal stress on each of the planes of maximum shearing stress (which are of course 90° apart) is given by

$$\tau' = \frac{\sigma_x + \sigma_y}{2}$$

Thus an element oriented along the planes of maximum shearing stress appears as in Fig. 16-4. This is illustrated in Problems 16.7, 16.9, and 16.15.

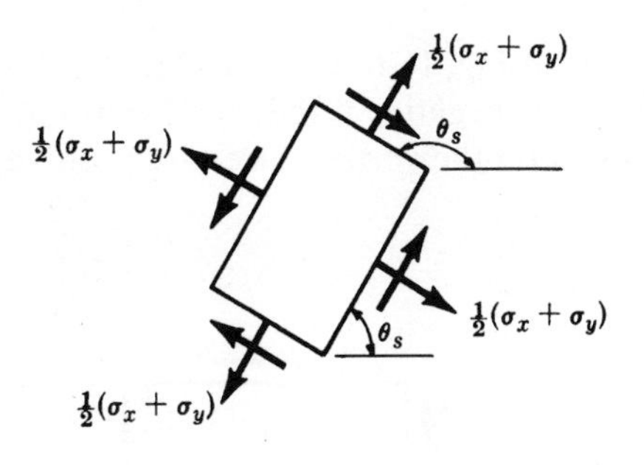

Fig. 16-4

MOHR'S CIRCLE

All the information contained in the above equations may be presented in a convenient graphical form known as *Mohr's circle*. In this representation normal stresses are plotted along the horizontal axis and shearing stresses along the vertical axis. The stresses σ_x, σ_y, and τ_{xy} are plotted to scale and a circle is drawn through these points having its center on the horizontal axis. Figure 16-5 shows Mohr's circle for an element subject to the general case of plane stress. For applications see Problems 16.4, 16.5, 16.8, 16.10, 16.12, 16.14, 16.16, 16.17, and 16.19.

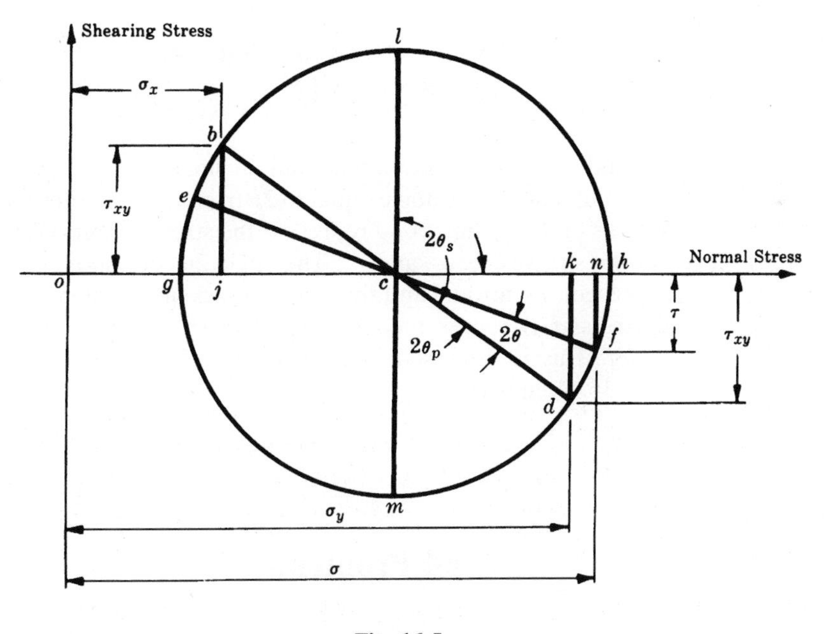

Fig. 16-5

SIGN CONVENTIONS USED WITH MOHR'S CIRCLE

Tensile stresses are considered to be positive and compressive stresses negative. Thus tensile stresses are plotted to the right of the origin in Fig. 16-5 and compressive stresses to the left. With regard to shearing stresses it is to be carefully noted that a different sign convention exists than is used in connection with the above-mentioned equations. We shall refer to a plane element subject to shearing stresses and appearing as in Fig. 16-6. We shall say that shearing stresses are positive if they tend to rotate the element clockwise, negative if they tend to rotate it counterclockwise. Thus for the above element the shearing stresses on the vertical faces are positive, those on the horizontal faces are negative.

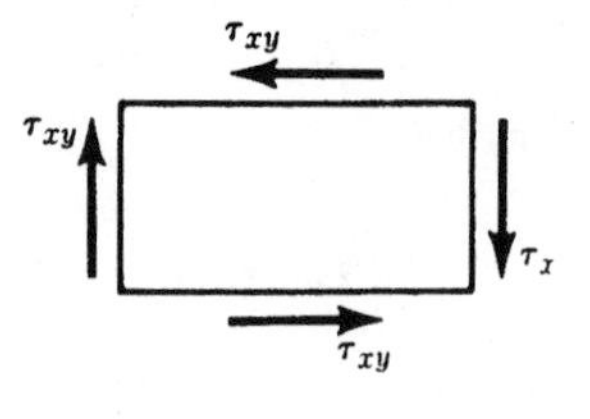

Fig. 16-6

DETERMINATION OF PRINCIPAL STRESSES BY MEANS OF MOHR'S CIRCLE

When Mohr's circle has been drawn as in Fig. 16-5, the principal stresses are represented by the line segments *og* and *oh*. These may either be scaled from the diagram or determined from the geometry of the figure. This is explained in detail in Problem 16.14. For application see Problems 16.4, 16.5, 16.8, 16.10, 16.12, 16.14, 16.16, 16.17, and 16.19.

DETERMINATION OF STRESSES ON AN ARBITRARY PLANE BY MEANS OF MOHR'S CIRCLE

To determine the normal and shearing stresses on a plane inclined at a counterclockwise angle θ with the x-axis, we measure a counterclockwise angle equal to 2θ from the diameter *bd* of Mohr's circle shown in Fig. 16-5. The endpoints of this diameter *bd* represent the stress conditions in the original x-y directions; i.e., they represent the stresses σ_x, σ_y, and τ_{xy}. The angle 2θ corresponds to the diameter *ef*. The coordinates of point *f* represent the normal and shearing stresses on the plane at an angle θ to the x-axis. That is, the normal stress σ is represented by the abscissa *on* and the shearing stress is represented by the ordinate *nf*. This is discussed in detail in Problem 16.14. For applications see Problems 16.4, 16.5, 16.6, 16.8, 16.14, and 16.17.

Solved Problems

16.1. Let us consider a straight bar of uniform cross section loaded in axial tension. Determine the normal and shearing stress intensities on a plane inclined at an angle θ to the axis of the bar. Also, determine the magnitude and direction of the maximum shearing stress in the bar.

This is the same elastic body that was considered in Chap. 1, but there the stresses studied were normal stresses in the direction of the axial force acting on the bar. In Fig. 16-7(a), P denotes the axial force acting on the bar, A the area of the cross section perpendicular to the axis of the bar, and from Chap. 1 the normal stress σ_x is given by $\sigma_x = P/A$.

Suppose now that instead of using a cutting plane which is perpendicular to the axis of the bar, we pass a plane through the bar at an angle θ with the axis of the bar. Such a plane *mn* is shown in Fig. 16-7(b).

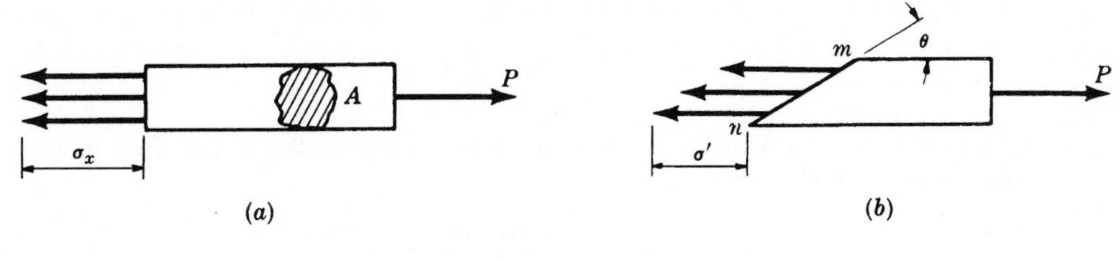

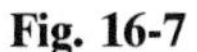
Fig. 16-7

Since we must still have equilibrium of the bar in the horizontal direction, there must evidently be distributed horizontal stresses acting over this inclined plane as shown. Let us designate the magnitude of these stresses by σ'. Evidently the area of the inclined cross section is $A/\sin\theta$ and for equilibrium of forces in the horizontal direction we have

$$\sigma'\left(\frac{A}{\sin\theta}\right) = P \qquad \text{or} \qquad \sigma' = \frac{P\sin\theta}{A}$$

In Fig. 16-8, we consider only a single stress vector σ' and resolve it into two components, one normal to the inclined plane mn and one tangential to this plane. We shall label the first of these components σ to denote a normal stress, and the second τ to represent a shearing stress.

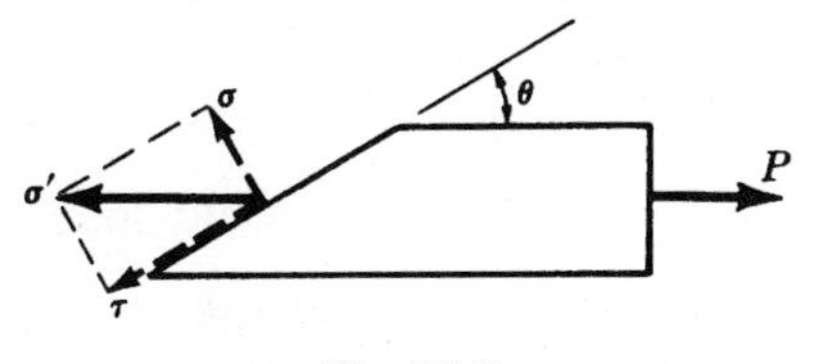

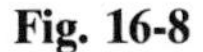
Fig. 16-8

Since the angle between σ' and τ is θ, we immediately have the relations

$$\tau = \sigma'\cos\theta \qquad \text{and} \qquad \sigma = \sigma'\sin\theta$$

But $\sigma' = (P\sin\theta)/A$. Substituting this value in the above equations, we obtain

$$\tau = \frac{P\sin\theta\cos\theta}{A} \qquad \text{and} \qquad \sigma = \frac{P\sin^2\theta}{A}$$

But $\sigma_x = P/A$. Hence we may write these in the form

$$\tau = \sigma_x\sin\theta\cos\theta \qquad \text{and} \qquad \sigma = \sigma_x\sin^2\theta$$

Now, employing the trigonometric identities

$$\sin 2\theta = 2\sin\theta\cos\theta \qquad \text{and} \qquad \sin^2\theta = \frac{1-\cos 2\theta}{2}$$

we may write

$$\tau = \tfrac{1}{2}\sigma_x\sin 2\theta \qquad (1)$$

$$\sigma = \tfrac{1}{2}\sigma_x(1-\cos 2\theta) \qquad (2)$$

These expressions give the normal and shearing stresses on a plane inclined at an angle θ to the axis of the bar.

16.2. A bar of cross section 850 mm^2 is acted upon by axial tensile forces of 60 kN applied at each end of the bar. Determine the normal and shearing stresses on a plane inclined at 30° to the direction of loading.

From Problem 16.1 the normal stress on a cross section perpendicular to the axis of the bar is

$$\sigma_x = \frac{P}{A} = \frac{60 \times 10^3}{850} = 70.6 \text{ MPa}$$

The normal stress on a plane at an angle θ with the direction of loading was found in Problem 16.1 to be $\sigma = \frac{1}{2}\sigma_x(1 - \cos 2\theta)$. For $\theta = 30°$ this becomes

$$\sigma = \tfrac{1}{2}(70.6)(1 - \cos 60°) = 17.65 \text{ MPa}$$

The shearing stress on a plane at an angle θ with the direction of loading was found in Problem 16.1 to be $\tau = \frac{1}{2}\sigma_x \sin 2\theta$. For $\theta = 30°$ this becomes

$$\tau = \tfrac{1}{2}(70.6)(\sin 60°) = 30.6 \text{ MPa}$$

These stresses together with the axial load of 60 kN are represented in Fig. 16-9.

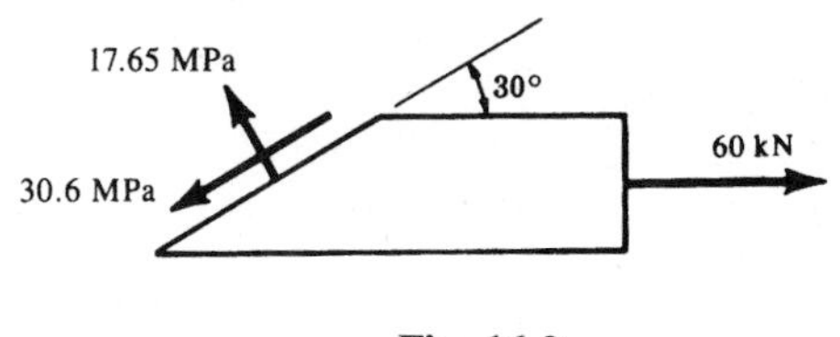

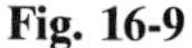

Fig. 16-9

16.3. Determine the maximum shearing stress in the axially loaded bar described in Problem 16.2.

The shearing stress on a plane at an angle θ with the direction of the load was shown in Problem 16.1 to be $\tau = \frac{1}{2}\sigma_x \sin 2\theta$. This is maximum when $2\theta = 90°$, that is, when $\theta = 45°$. For this loading we have $\sigma_x = 70.6$ MPa and when $\theta = 45°$ the shear stress is

$$\tau = \tfrac{1}{2}(70.6) \sin 90° = 35.3 \text{ MPa}$$

That is, the maximum shearing stress is equal to one-half of the maximum normal stress.

The normal stress on this 45° plane may be found from the expression

$$\sigma = \tfrac{1}{2}\sigma_x(1 - \cos 2\theta) = \tfrac{1}{2}(70.6)(1 - \cos 90°) = 35.3 \text{ MPa}$$

16.4. Discuss a graphical representation of Eqs. (*1*) and (*2*) of Problem 16.1.

According to these equations the normal and shearing stresses on a plane inclined at an angle θ to the direction of loading are given by

$$\sigma = \tfrac{1}{2}\sigma_x(1 - \cos 2\theta) \qquad \text{and} \qquad \tau = \tfrac{1}{2}\sigma_x \sin 2\theta$$

To represent these relations graphically it is customary to introduce a rectangular cartesian coordinate system, plotting normal stresses as abscissas and shearing stresses as ordinates.

Let us proceed by first laying off to some convenient scale the normal stress σ_x (taken to be tensile) along the positive horizontal axis. The midpoint of this line segment, point c in Fig. 16-10, serves as the center of a circle whose diameter is σ_x. The radius of this circle, denoted by $\overline{oc}$, $\overline{ch}$, and $\overline{cd}$, is $\frac{1}{2}\sigma_x$. The angle 2θ is measured positive in a counterclockwise direction from the radial line $\overline{oc}$. From the figure we immediately have the relations

$$\overline{kd} = \tau = \tfrac{1}{2}\sigma_x \sin 2\theta \qquad \overline{ok} = \overline{oc} - \overline{kc} = \tfrac{1}{2}\sigma_x - \tfrac{1}{2}\sigma_x \cos 2\theta = \sigma = \tfrac{1}{2}\sigma_x(1 - \cos 2\theta)$$

It is to be noted that the scales used in the horizontal and vertical directions are equal.

Thus the abscissa and ordinate of point d represent, respectively, the normal stress and the shearing stress acting on a plane at an angle θ with the axis of the bar subject to tension. In plotting this diagram tensile stresses are regarded as positive in algebraic sign and compressive stresses are taken to be negative.

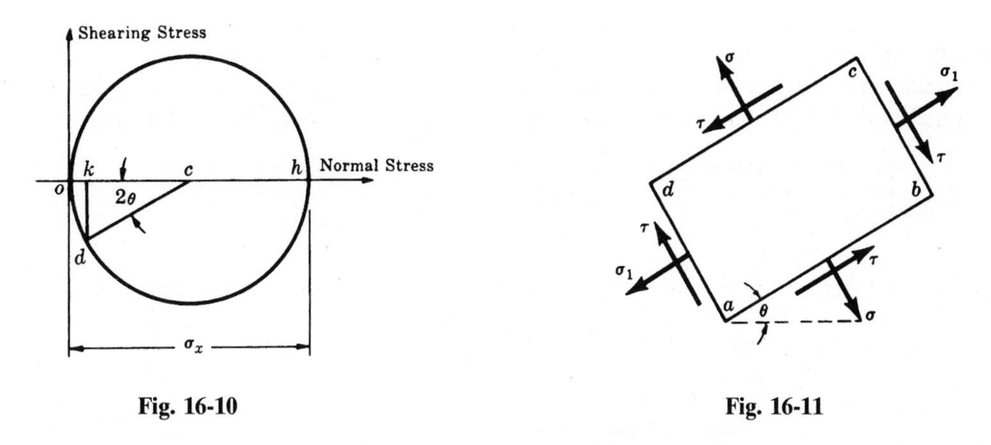

Fig. 16-10 **Fig. 16-11**

Let us return to Problem 16.1 and examine a free-body diagram (Fig. 16-11) of an element taken from the surface of the inclined section on which the stresses σ and τ act. We shall consider shearing stresses to be positive if they tend to rotate the element clockwise, negative if they tend to rotate the element counterclockwise. This sign convention is used only in this graphical representation, not in the analytical treatment of Problem 16.1. Since the shearing stresses found in Problem 16.1 were actually those acting on face *dc* of the above element, they should be regarded as negative. Hence in the circular diagram representing normal and shearing stresses in Fig. 16-10, the shearing stress on plane *dc* appears as an ordinate $\overline{kd}$ plotted in the negative sense.

This diagram, termed *Mohr's circle* as noted earlier, was first presented by O. Mohr in 1882. It represents the variation of normal and shearing stresses on all inclined planes passing through a given point in the body. It is a convenient graphical representation of Eqs. (*1*) and (*2*) of Problem 16.1.

16.5. Consider again the axially loaded bar discussed in Problem 16.2. Use Mohr's circle to determine the normal and shearing stresses on the 30° plane.

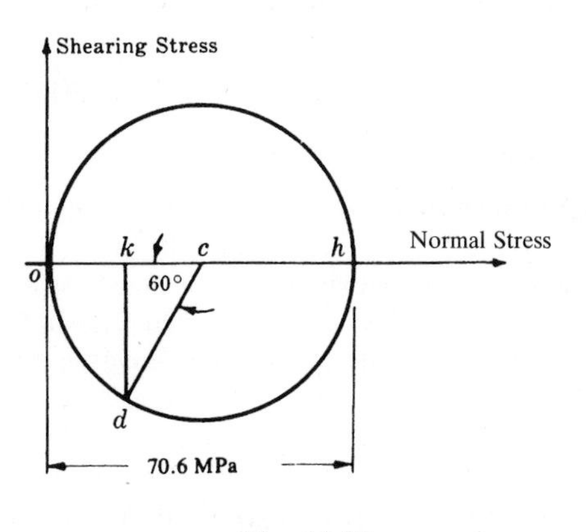

Fig. 16-12

In Fig. 16-12, the normal stress of 70.6 MPa is laid off along the horizontal axis to some convenient scale and a circle is drawn with this line as a diameter. The angle $2\theta = 2(30°) = 60°$ is measured counterclockwise from $\overline{oc}$. The coordinates of the point *d* are

$$\overline{kd} = \tau = -\tfrac{1}{2}(70.6)\sin 60° = -30.6 \text{ MPa}$$

$$\overline{ok} = \sigma = \overline{oc} - \overline{kc} = \tfrac{1}{2}(70.6) - \tfrac{1}{2}(70.6)\cos 60° = 17.65 \text{ MPa}$$

The negative sign accompanying the value of the shearing stress indicates that the shearing stress on this 30° plane tends to rotate an element bounded by this plane in a counterclockwise direction. This is in agreement with the direction of the shearing stress illustrated in Fig. 16-9.

16.6. A bar of cross section 1.3 in^2 is acted upon by axial compressive forces of 15,000 lb applied to each end of the bar. Using Mohr's circle, find the normal and shearing stresses on a plane inclined at 30° to the direction of loading. Neglect the possibility of buckling of the bar.

The normal stress on a cross-section perpendicular to the axis of the bar is

$$\sigma_x = \frac{P}{A} = \frac{-15{,}000}{1.3} = -11{,}500\ \text{lb/in}^2$$

We shall first lay off this compressive normal stess to some convenient scale along the negative end of the horizontal axis. The midpoint of the line segment, point c in Fig. 16-13, serves as the center of a circle whose diameter is 11,500 lb/in^2 to the scale chosen.

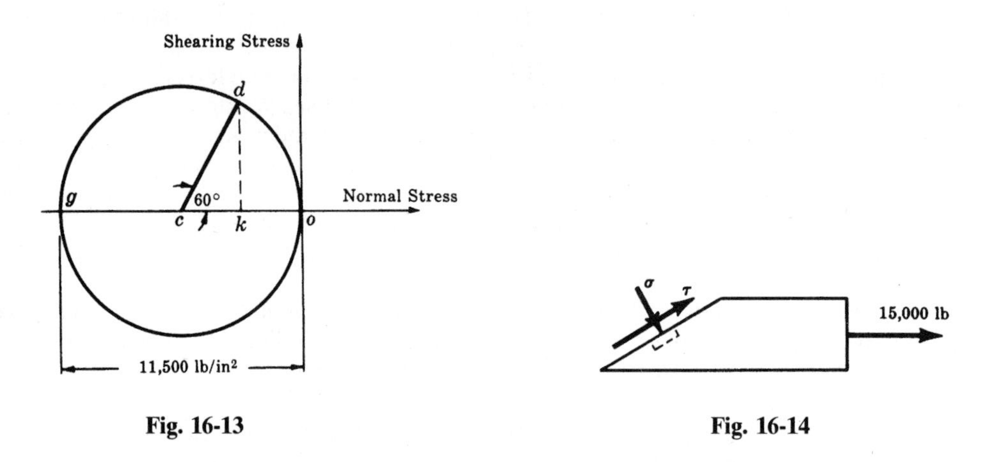

Fig. 16-13

Fig. 16-14

The angle $2\theta = 2(30°) = 60°$ with the vertex at c is measured counterclockwise from $\overline{co}$ as shown. The abscissa of point d represents the normal stress and the ordinate the shearing stress on the desired 30° plane. The coordinates of point d are

$$\overline{kd} = \tau = \tfrac{1}{2}(11{,}500)\sin 60° = 4940\ \text{lb/in}^2$$

$$\overline{ok} = \sigma = \overline{oc} - \overline{ck} = \tfrac{1}{2}(11{,}500) - \tfrac{1}{2}(11{,}500)\cos 60° = 2870\ \text{lb/in}^2$$

It is to be noted that line segment $\overline{ok}$ lies to the left of the origin of coordinates; hence this normal stress is compressive.

The positive algebraic sign accompanying the shearing stress indicates that the shearing stress on the 30° plane tends to rotate an element (denoted by dashed lines in Fig. 16-14) bounded by this plane in a clockwise direction. The directions of the normal and shearing stresses together with the axial load of 15,000 lb are shown in the figure.

16.7. Consider a plane element removed from a stressed elastic body and subject to the normal and shearing stresses σ_x and τ_{xy}, respectively, as shown in Fig. 16-15. (*a*) Determine the normal and shearing stress intensities on a plane inclined at an angle θ to the normal stress σ_x. (*b*) Determine the maximum and minimum values of the normal stress that may exist on inclined planes and find the directions of these stresses. (*c*) Determine the magnitude and direction of the maximum shearing stress that may exist on an inclined plane.

(*a*) The desired normal and shearing stresses acting on an inclined plane are internal quantities with respect to the element shown in Fig. 16-15. We shall follow the customary procedure of cutting this element with a plane in such a manner as to render the desired stresses external to the new body; that is, we will cut the originally rectangular element along the plane inclined at an angle θ with the x-axis and thus obtain a triangular element as shown in Fig. 16-16. The normal and shearing stresses, designated as σ and τ, respectively, represent the effect of the remaining portion of the

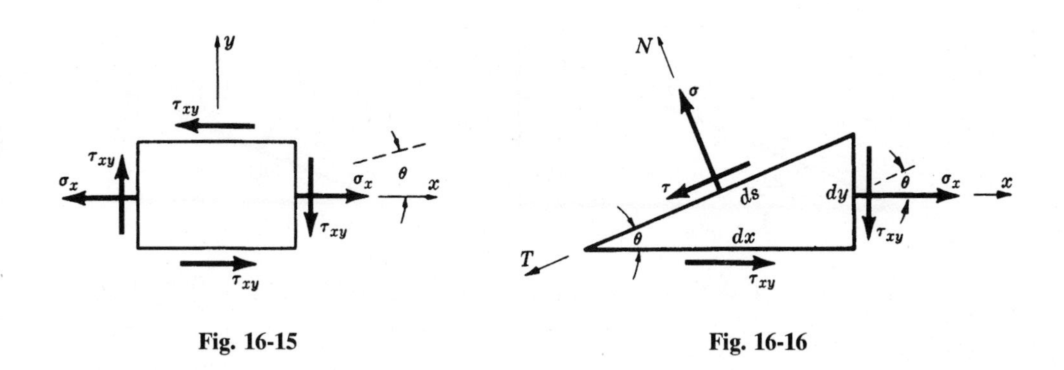

Fig. 16-15 **Fig. 16-16**

originally rectangular block that has been removed. Consequently, the problem reduces to finding the unknown stresses σ and τ in terms of the known stresses σ_x and τ_{xy}. It is to be observed that in the free-body diagram of the triangular element, the vectors indicate stresses acting on the various faces of the element and not forces. Each of these stresses is assumed to be uniformly distributed over the area upon which it acts. The thickness of the element perpendicular to the plane of the paper is denoted by t.

Let us introduce N- and T-axes normal and tangent to the inclined plane, as shown in Fig. 16-16. First, we shall sum forces in the N-direction. For equilibrium we have

$$\Sigma F_N = \sigma t\, ds - \sigma_x t\, dy \sin\theta - \tau_{xy} t\, dy \cos\theta - \tau_{xy} t\, dx \sin\theta = 0$$

But $dy = ds \sin\theta$, $dx = ds\cos\theta$. Substituting these relations in the equilibrium equation above, we find

$$\sigma(ds) = \sigma_x(ds)\sin^2\theta + 2\tau_{xy}(ds)\sin\theta\cos\theta$$

Next, employing the identities $\sin^2\theta = \frac{1}{2}(1-\cos 2\theta)$ and $\sin 2\theta = 2\sin\theta\cos\theta$, we obtain

$$\sigma = \tfrac{1}{2}\sigma_x(1-\cos 2\theta) + \tau_{xy}\sin 2\theta = \tfrac{1}{2}\sigma_x - \tfrac{1}{2}\sigma_x\cos 2\theta + \tau_{xy}\sin 2\theta \tag{1}$$

Thus the normal stress σ on any plane inclined at an angle θ with the x-axis is known as a function of σ_x, τ_{xy}, and θ.

Next we shall consider the equilibrium of the forces acting on the triangular element in the T-direction. This leads to the equation

$$\Sigma F_T = \tau t\, ds - \sigma_x t\, dy\cos\theta + \tau_{xy} t\, dy \sin\theta - \tau_{xy} t\, dx\cos\theta = 0$$

Substituting $dy = ds\sin\theta$ and $dx = ds\cos\theta$, we obtain

$$\tau(ds) = +\sigma_x(ds)\sin\theta\cos\theta - \tau_{xy}(ds)\sin^2\theta + \tau_{xy}(ds)\cos^2\theta$$

Employing the identities $\cos 2\theta = \cos^2\theta - \sin^2\theta$ and $\sin 2\theta = 2\sin\theta\cos\theta$, this becomes

$$\tau = \tfrac{1}{2}\sigma_x \sin 2\theta + \tau_{xy}\cos 2\theta \tag{2}$$

Thus the shearing stress τ on any plane inclined at an angle θ with the x-axis is known as a function of σ_x, τ_{xy}, and θ.

(*b*) To determine the maximum value that the normal stress σ may assume as the angle θ varies, we shall differentiate Eq. (*1*) with respect to θ and set this derivative equal to zero. Thus

$$\frac{d\sigma}{d\theta} = +\sigma_x\sin 2\theta + 2\tau_{xy}\cos 2\theta = 0$$

The values of θ leading to maximum and minimum values of the normal stress are consequently

$$\tan 2\theta_p = \frac{-\tau_{xy}}{\frac{1}{2}\sigma_x} \tag{3}$$

The planes defined by the angles θ_p are called *principal planes*. The normal stresses that exist on these planes are designated as *principal stresses*. They are the maximum and minimum values that the normal stress may assume in the element under consideration. The values of the principal stresses

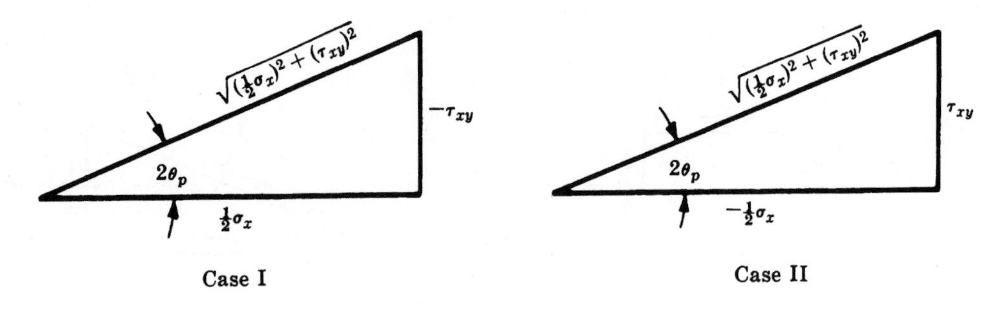

Fig. 16-17

may easily be found by interpreting Eq. (*3*) graphically, as in Fig. 16-17. Evidently the tangent of either of the angles designated as $2\theta_p$ has the value given in (*3*). Thus there are two solutions to (*3*), and consequently two values of $2\theta_p$ (differing by 180°) and also two values of θ_p. These values of θ_p differ by 90°. It is to be noted that the triangles of Fig. 16-17 bear no direct relationship to the triangular element whose free-body diagram was considered earlier.

The values of $\sin 2\theta_p$ and $\cos 2\theta_p$ as found from Fig. 16-17 may now be substituted in (*1*) to yield the maximum and minimum values of the normal stresses. Observing that

$$\sin 2\theta_p = \frac{\mp\tau_{xy}}{\sqrt{(\frac{1}{2}\sigma_x)^2 + (\tau_{xy})^2}} \qquad \cos 2\theta_p = \frac{\pm\frac{1}{2}\sigma_x}{\sqrt{(\frac{1}{2}\sigma_x)^2 + (\tau_{xy})^2}}$$

where the upper signs pertain to Case I and the lower signs to Case II, we obtain from (*1*)

$$\sigma = \tfrac{1}{2}\sigma_x \mp \tfrac{1}{2}\sigma_x \frac{\frac{1}{2}\sigma_x}{\sqrt{(\frac{1}{2}\sigma_x)^2 + (\tau_{xy})^2}} \mp \frac{(\tau_{xy})^2}{\sqrt{(\frac{1}{2}\sigma_x)^2 + (\tau_{xy})^2}} = \tfrac{1}{2}\sigma_x \pm \sqrt{(\tfrac{1}{2}\sigma_x)^2 + (\tau_{xy})^2} \qquad (4)$$

The maximum normal stress is

$$\sigma_{\max} = \tfrac{1}{2}\sigma_x + \sqrt{(\tfrac{1}{2}\sigma_x)^2 + (\tau_{xy})^2} \qquad (5)$$

The minimum normal stress is

$$\sigma_{\min} = \tfrac{1}{2}\sigma - \sqrt{(\tfrac{1}{2}\sigma_x)^2 + (\tau_{xy})^2} \qquad (6)$$

The stresses given by (*5*) and (*6*) are the principal stresses and they occur on the principal planes defined by (*3*). By substituting one of the values of θ_p from (*3*) into (*1*), one may readily determine which of the two principal stresses is acting on that plane. The other principal stress naturally acts on the other principal plane.

By substituting the values of the angles $2\theta_p$ as given by (*3*) and Fig. 16-17 into (*2*), it is readily seen that the shearing stresses τ on the principal planes are zero.

(*c*) To determine the maximum value the shearing stress τ may assume as the angle θ varies, we shall differentiate Eq. (*2*) with respect to θ and set this derivative equal to zero. Thus

$$\frac{d\tau}{d\theta} = \sigma_x \cos 2\theta - 2\tau_{xy} \sin 2\theta = 0$$

The values of θ leading to maximum values of the shearing stress are consequently

$$\tan 2\theta_s = \frac{\frac{1}{2}\sigma_x}{\tau_{xy}} \qquad (7)$$

The planes defined by the two solutions to this equation are the planes of maximum shearing stress.

Again, a graphical interpretation of (*7*) is convenient. The two values of the angle $2\theta_s$ satisfying this equation may be represented as in Fig. 16-18. We see that

$$\sin 2\theta_s = \frac{\pm\frac{1}{2}\sigma_x}{\sqrt{(\frac{1}{2}\sigma_x)^2 + (\tau_{xy})^2}} \qquad \cos 2\theta_s = \frac{\pm\tau_{xy}}{\sqrt{(\frac{1}{2}\sigma_x)^2 + (\tau_{xy})^2}}$$

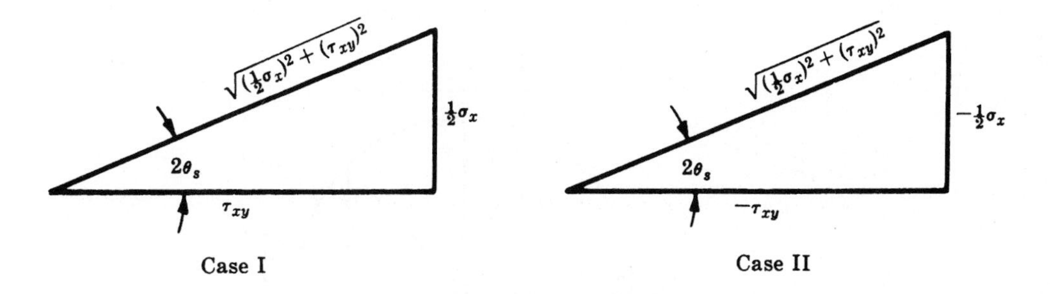

Fig. 16-18

where the upper (positive) signs pertain to Case I and the lower (negative) signs apply to Case II. Substituting these values in (2), we obtain

$$\tau_{\substack{\max\\\min}} = \tfrac{1}{2}\sigma_x \frac{\pm\frac{1}{2}\sigma_x}{\sqrt{(\frac{1}{2}\sigma_x)^2 + (\tau_{xy})^2}} + (\tau_{xy}) \frac{\pm\tau_{xy}}{\sqrt{(\frac{1}{2}\sigma_x)^2 + (\tau_{xy})^2}} = \pm\sqrt{(\tfrac{1}{2}\sigma_x)^2 + (\tau_{xy})^2} \qquad (8)$$

Here the positive sign represents the maximum shearing stress, the negative sign the minimum shearing stress.

If we compare (3) and (7), it is evident that the angles $2\theta_p$ and $2\theta_s$ differ by 90°, since the tangents of these angles are the negative reciprocals of one another. Hence the planes defined by the angles θ_p and θ_s differ from one another by 45°; that is, the planes of maximum shearing stress are oriented 45° from the planes of maximum normal stress.

It is also of interest to determine the normal stresses on the planes of maximum shearing stress. These planes are defined by (7). If we now substitute these values of $\sin 2\theta_s$ and $\cos 2\theta_s$ in (1) for the normal stress, we find

$$\sigma = \tfrac{1}{2}\sigma_x - \tfrac{1}{2}\sigma_x \frac{\pm\tau_{xy}}{\sqrt{(\frac{1}{2}\sigma_x)^2 + (\tau_{xy})^2}} + (\tau_{xy}) \frac{\pm\frac{1}{2}\sigma_x}{\sqrt{(\frac{1}{2}\sigma_x)^2 + (\tau_{xy})^2}} = \tfrac{1}{2}\sigma_x \qquad (9)$$

Thus on each plane of maximum shearing stress we have a normal stress of magnitude $\frac{1}{2}\sigma_x$.

16.8. Discuss a graphical representation of the analysis presented in Problem 16.7.

For given values of σ_x and τ_{xy} proceed as follows:

1. Introduce a rectangular coordinate system in which normal stresses are represented along the horizontal axis and shearing stresses along the vertical axis. The scales used on these two axes must be equal.
2. With reference to the original rectangular element considered in Problem 16.7 and reproduced in Fig. 16-19, we shall introduce the sign convention that shearing stresses are positive if they tend to rotate the element clockwise, negative if they tend to rotate it counterclockwise. Here the shearing stresses on the vertical faces are positive, those on the horizontal faces are negative. Also, tensile stresses are considered to be positive and compressive stresses negative.
3. We first locate point b by laying out σ_x and τ_{xy} to their given values. The shear stress τ_{xy} on the vertical faces on which σ_x acts is positive; hence this value is plotted as positive in Fig. 16-20. This is drawn on

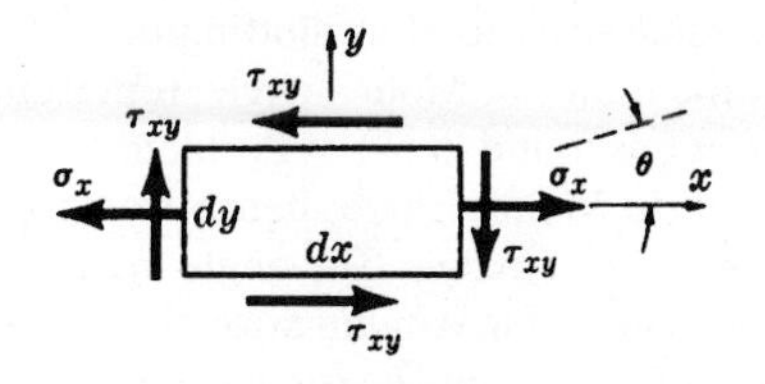

Fig. 16-19

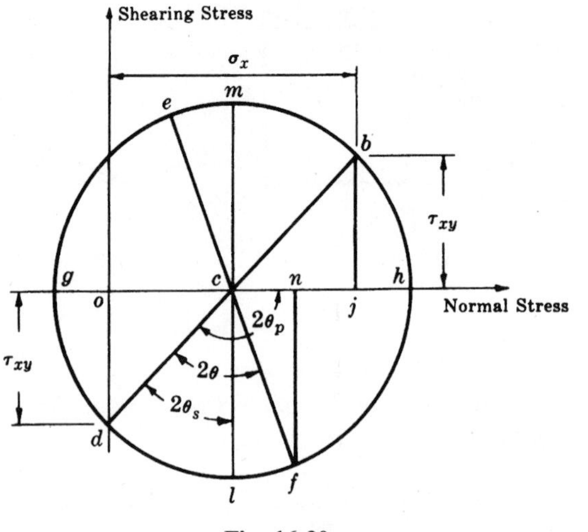

Fig. 16-20

the assumption that σ_x is a tensile stress, although the treatment presented here is valid if σ_x is compressive.

4. We next locate point d in a similar manner by laying off τ_{xy} on the negative side of the vertical axis. Actually, this point d corresponds to the negative shearing stresses τ_{xy} existing on the horizontal faces of the element together with a zero normal stress acting on those same faces.
5. Next, we draw line $\overline{bd}$, locate the midpoint c, and draw a circle having its center at c and radius equal to $\overline{cb}$. This is known as Mohr's circle.

We shall first show that the points g and h along the horizontal diameter of the circle represent the principal stresses. To do this we note that the point c lies at a distance $\frac{1}{2}\sigma_x$ from the origin of the coordinate system. From the right-triangle relationship we have

$$(\overline{cd})^2 = (\overline{oc})^2 + (\overline{od})^2 \qquad \text{or} \qquad \overline{cd} = \sqrt{(\tfrac{1}{2}\sigma_x)^2 + (\tau_{xy})^2}$$

Also, we have $\overline{cd} = \overline{ch} = \overline{cg}$. Hence, the x-coordinate of point h is $\overline{oc} + \overline{ch}$ or

$$\tfrac{1}{2}\sigma_x + \sqrt{(\tfrac{1}{2}\sigma_x)^2 + (\tau_{xy})^2}$$

But this expression is exactly the maximum principal stress, as found in (5) of Problem 16.7. Likewise the x-coordinate of point g is $\overline{oc}$-$\overline{cg}$. But this quantity is negative; hence $\overline{og}$ lies to the left of the origin, and point g symbolizes a compressive stress. This stress becomes

$$\tfrac{1}{2}\sigma_x - \sqrt{(\tfrac{1}{2}\sigma_x)^2 + (\tau_{xy})^2}$$

But this expression is exactly the minimum principal stress, as found in (6) of Problem 16.7. Consequently the points g and h represent the principal stresses existing in the original element. We see that the tangent of $\angle ocd$ is $\tau_{xy}/(\frac{1}{2}\sigma_x)$. But from (3) of Problem 16.7, $\tan 2\theta_p = -\tau_{xy}/\frac{1}{2}\sigma_x$; and by comparison of these two relations we see that $\angle hcd = 2\theta_p$, since $\tan(180° - \theta) = -\tan\theta$. Thus a counterclockwise rotation from the diameter $\overline{bd}$ (corresponding to the stresses in the x- and y-directions) leads us to the diameter $\overline{gh}$, representing the principal planes, on which the principal stresses occur. The principal planes lie at an angle θ_p from the x-direction.

Thus Mohr's circle is a convenient device for finding the principal stresses, since one can merely establish the circle for a given set of stresses σ_x and τ_{xy} then measure $\overline{og}$ and $\overline{oh}$. These abscissas represent the principal stresses to the same scale used in plotting σ_x and τ_{xy}.

It is now apparent that the radius of Mohr's circle, represented by $\overline{cd} = \sqrt{(\frac{1}{2}\sigma_x)^2 + (\tau_{xy})^2}$, corresponds to the maximum shearing stress, as found in (8) of Problem 16.7. Actually, the shearing stress on any plane is represented by the ordinate to Mohr's circle; hence we should consider the radial lines $\overline{cl}$ and $\overline{cm}$ as representing the maximum shearing stresses. The angle dcl is evidently $2\theta_s$ and hence it is apparent that the double angle between the planes of maximum normal stress and the planes of maximum shearing stress ($\angle lch$) is 90°; thus the planes of maximum shearing stress are oriented 45° from the planes of maximum normal stress.

Evidently the endpoints of the diameter $\overline{bd}$ represent the stresses acting in the original x- and y-directions. We shall now demonstrate that the endpoints of any other diameter, such as $\overline{ef}$ (at any angle 2θ with $\overline{bd}$), represent the stresses on a plane inclined at an angle θ to the x-axis. To do this we note that the abscissa of point f is given by

$$\begin{aligned}\sigma = \overline{oc} + \overline{cn} &= \tfrac{1}{2}\sigma_x + \overline{cf}\cos(2\theta_p - 2\theta)\\ &= \tfrac{1}{2}\sigma_x + \overline{cf}(\cos 2\theta_p \cos 2\theta + \sin 2\theta_p \sin 2\theta)\\ &= \tfrac{1}{2}\sigma_x + \sqrt{(\tfrac{1}{2}\sigma_x)^2 + (\tau_{xy})^2}\,(\cos 2\theta_p \cos 2\theta + \sin 2\theta_p \sin 2\theta)\end{aligned}$$

But from inspection of triangle cod appearing in Mohr's circle it is evident that

$$\sin 2\theta_p = \frac{\tau_{xy}}{\sqrt{(\frac{1}{2}\sigma_x)^2 + (\tau_{xy})^2}} \qquad \text{and} \qquad \cos 2\theta_p = \frac{-\frac{1}{2}\sigma_x}{\sqrt{(\frac{1}{2}\sigma_x)^2 + (\tau_{xy})^2}} \tag{1}$$

Substituting the values of τ_{xy} and $\frac{1}{2}\sigma_x$ from these two equations into the previous equation, we find

$$\sigma = \tfrac{1}{2}\sigma_x - \tfrac{1}{2}\sigma_x \cos 2\theta + \tau_{xy} \sin 2\theta$$

But this is exactly the normal stress on a plane inclined at an angle θ to the x-axis as derived in (*1*) of Problem 16.7.

Next we observe that the ordinate of point f is given by

$$\begin{aligned}\tau = \overline{nf} &= \overline{cf}\sin(2\theta_p - 2\theta)\\ &= \sqrt{(\tfrac{1}{2}\sigma_x)^2 + (\tau_{xy})^2}\,(\sin 2\theta_p \cos 2\theta - \cos 2\theta_p \sin 2\theta)\end{aligned}$$

Again, substituting the values of τ_{xy} and $\frac{1}{2}\sigma_x$ from Eqs. (*1*) into this equation, we find

$$\tau = \tfrac{1}{2}\sigma_x \sin 2\theta + \tau_{xy} \cos 2\theta$$

But this is exactly the shearing stress on a plane inclined at an angle θ to the x-axis as derived in (*2*) of Problem 16.7.

Hence the coordinates of point f on Mohr's circle represent the normal and shearing stresses on a plane inclined at an angle θ to the x-axis.

16.9. A plane element in a body is subjected to a normal stress in the x-direction of 12,000 lb/in^2, as well as a shearing stress of 4000 lb/in^2, as shown in Fig. 16-21. (*a*) Determine the normal and shearing stress intensities on a plane inclined at an angle of 30° to the normal stress. (*b*) Determine the maximum and minimum values of the normal stress that may exist on inclined planes and the directions of these stresses. (*c*) Determine the magnitude and direction of the maximum shearing stress that may exist on an inclined plane.

(*a*) In accordance with the notation of Problem 16.7, we have $\sigma_x = 12{,}000$ lb/in^2 and $\tau_{xy} = 4000$ lb/in^2. From (*1*) of Problem 16.7, the normal stress on a plane inclined at an angle θ to the x-axis is

$$\sigma = \tfrac{1}{2}\sigma_x - \tfrac{1}{2}\sigma_x \cos 2\theta + \tau_{xy} \sin 2\theta$$

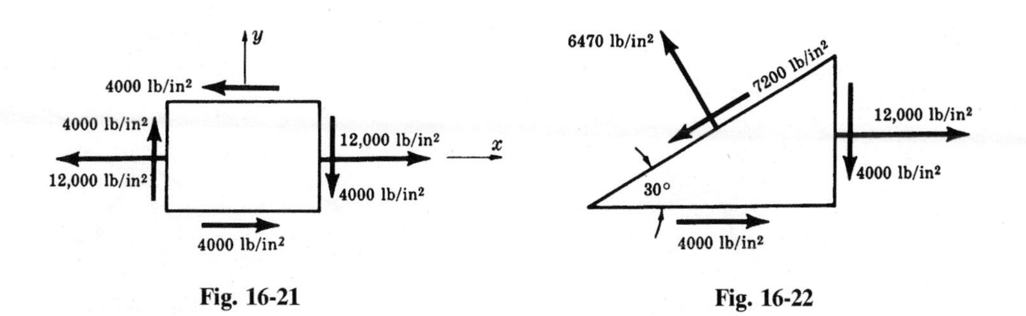

Fig. 16-21 Fig. 16-22

Substituting the above values of σ_x and τ_{xy}, when $\theta = 30°$ this becomes

$$\sigma = \tfrac{1}{2}(12{,}000) - \tfrac{1}{2}(12{,}000)\cos 60° + 4000 \sin 60° = 6470 \text{ lb/in}^2$$

From (2) of Problem 16.7, the shearing stress on any plane inclined at an angle θ to the x-axis is

$$\tau = \tfrac{1}{2}\sigma_x \sin 2\theta + \tau_{xy} \cos 2\theta$$

Substituting the above values of σ_x and τ_{xy}, when $\theta = 30°$ this becomes

$$\tau = \tfrac{1}{2}(12{,}000)\sin 60° + 4000 \cos 60° = 5200 + 2000 = 7200 \text{ lb/in}^2$$

The positive directions of the normal and shearing stresses on an inclined plane were illustrated in Fig. 16-16. In accordance with this sign convention the stresses on the 30° plane appear as in Fig. 16-22.

(*b*) The values of the principal stresses, that is, the maximum and minimum values of the normal stresses existing in this element, were given by (5) and (6) of Problem 16.7. From (5) for the maximum normal stress, we have

$$\sigma_{max} = \tfrac{1}{2}\sigma_x + \sqrt{(\tfrac{1}{2}\sigma_x)^2 + \tau_{xy})^2} = 6000 + \sqrt{(6000)^2 + (4000)^2} = 13{,}220 \text{ lb/in}^2$$

From (6) for the minimum normal stress, we have

$$\sigma_{min} = \tfrac{1}{2}\sigma_x - \sqrt{(\tfrac{1}{2}\sigma_x)^2 + (\tau_{xy})^2} = 6000 - \sqrt{(6000)^2 + (4000)^2} = -1220 \text{ lb/in}^2$$

The directions of the planes on which these principal stresses occur were found in (3) of Problem 16.7 to be

$$\tan 2\theta_p = -\frac{\tau_{xy}}{\tfrac{1}{2}\sigma_x} = -\frac{4000}{6000} = -\frac{2}{3}$$

Since the tangent of the angle $2\theta_p$ is negative, the two values of $2\theta_p$ lie in the second and fourth quadrants. In the second quadrant, $2\theta_p = 146°20'$; in the fourth quadrant, $2\theta_p' = 326°20'$. Consequently we have the principal planes defined by $\theta_p = 73°10'$ and $\theta_p' = 163°10'$. If $\theta_p = 73°10'$, together with the given values of σ_x and τ_{xy}, is now substituted in (1) of Problem 16.7, we find

$$\sigma = \tfrac{1}{2}\sigma_x - \tfrac{1}{2}\sigma_x \cos 2\theta + \tau_{xy} \sin 2\theta = 6000 - 6000 \cos 146°20' + 4000 \sin 146°20'$$

$$= 6000 - 6000(-0.833) + 4000(0.554) = 13{,}220 \text{ lb/in}^2$$

Thus the principal stress of 13,220 lb/in² occurs on the principal plane oriented at 73°10′ to the x-axis. The principal stresses thus appear as in Fig. 16-23. As stated in Problem 16.7, the shearing stresses on these principal planes are zero.

(*c*) The values of the maximum and minimum shearing stresses were found in (8) of Problem 16.7 to be

$$\tau_{\substack{max \\ min}} = \pm\sqrt{(\tfrac{1}{2}\sigma_x)^2 + (\tau_{xy})^2} = \pm\sqrt{(6000)^2 + (4000)^2} = \pm 7220 \text{ lb/in}^2$$

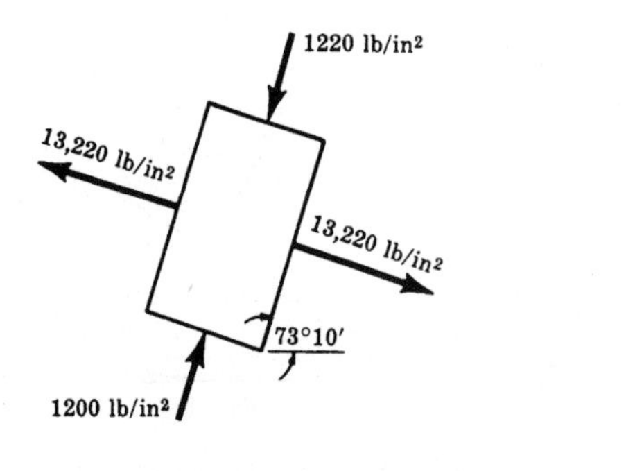

Fig. 16-23

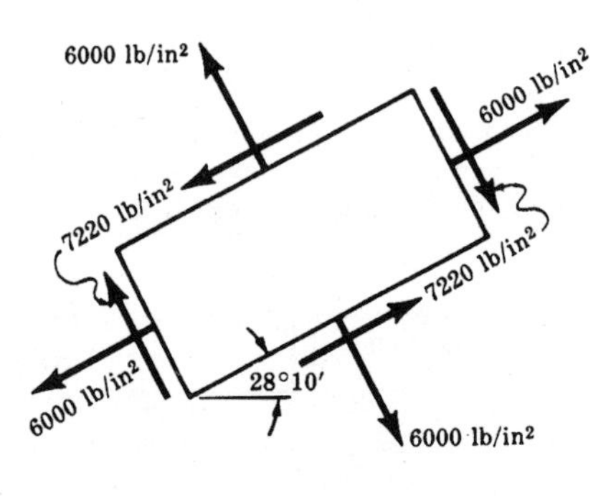

Fig. 16-24

The directions of the planes on which these maximum shearing stresses occur were found in (7) of Problem 16.7 to be given by

$$\tan 2\theta_s = \frac{\frac{1}{2}\sigma_x}{\tau_{xy}} = \frac{6000}{4000} = \frac{3}{2}$$

The angles $2\theta_s$ are consequently in the first and third quadrants, since the tangent is positive. Thus we have $2\theta_s = 56°20'$ and $2\theta_s' = 236°20'$, or $\theta_s = 28°10'$ and $\theta_s' = 118°10'$. The shearing stress on any plane inclined at an angle θ with the x-axis was found in (2) of Problem 16.7 to be

$$\tau = \tfrac{1}{2}\sigma_x \sin 2\theta + \tau_{xy} \cos 2\theta$$

Substituting $\sigma_x = 12{,}000 \text{ lb/in}^2$, $\tau_{xy} = 4000 \text{ lb/in}^2$, and $\theta = 28°10'$, we find

$$\tau = \tfrac{1}{2}(12{,}000) \sin 56°20' + 4000 \cos 56°20' = +7220 \text{ lb/in}^2$$

Thus the shearing stress on the 28°10′ plane is positive. The positive sense of shearing stress was shown in Fig. 16-6.

The normal stresses on the planes of maximum shearing stress are found from (9) of Problem 16.7 to be

$$\sigma = \tfrac{1}{2}\sigma_x = \tfrac{1}{2}(12{,}000) = 6000 \text{ lb/in}^2$$

This normal stress acts on each of the planes of maximum shearing stress, as shown in Fig. 16-24.

16.10. A plane element is subject to the stresses shown in Fig. 16-25. Using Mohr's circle, determine (*a*) the principal stresses and their directions and (*b*) the maximum shearing stresses and the directions of the planes on which they occur.

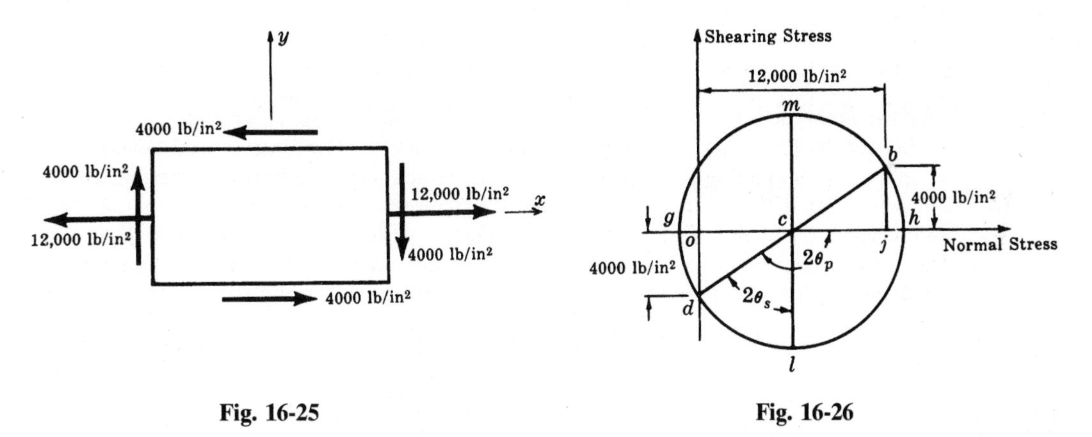

Fig. 16-25 **Fig. 16-26**

Following the procedure for the construction of Mohr's circle outlined in Problem 16.8, we realize that the shearing stress on the vertical faces of the given element are positive, whereas those on the horizontal faces are negative. Thus the stress condition of $\sigma_x = 12{,}000 \text{ lb/in}^2$, $\tau_{xy} = 4000 \text{ lb/in}^2$ existing on the vertical faces of the element plots as point b in Fig. 16-26. The stress condition of $\tau_{xy} = -4000 \text{ lb/in}^2$ together with a zero normal stress on the horizontal faces plots as point d. Line $\overline{bd}$ is drawn, its midpoint c is located, and a circle of radius $\overline{cb} = \overline{cd}$ is drawn with c as a center. This is Mohr's circle. The endpoints of the diameter $\overline{bd}$ represent the stress conditions existing in the element if it has the original orientation shown above.

(*a*) The principal stresses are represented by points g and h, as shown in Problem 16.8. The principal stresses may be determined either by direct measurement from Fig. 16-26 or by realizing that the coordinate of c is 6000, and that $\overline{cd} = \sqrt{(6000)^2 + (4000)^2} = 7220$. Therefore the minimum principal stress is

$$\sigma_{\min} = \overline{og} = \overline{oc} - \overline{cg} = 6000 - 7200 = -1220 \text{ lb/in}^2$$

Also, the maximum principal stress is

$$\sigma_{\max} = \overline{oh} = \overline{oc} + \overline{ch} = 6000 + 7220 = 13{,}220 \text{ lb/in}^2$$

The angle $2\theta_p$ designated above is given by

$$\tan 2\theta_p = -\frac{4000}{6000} = -\frac{2}{3} \qquad \text{or} \qquad \theta_p = 73°10'$$

This value could also be obtained by measurement of $\angle dch$ in Mohr's circle. From this it is readily seen that the principal stress represented by point h acts on a plane oriented 73°10′ from the original x-axis. The principal stresses thus appear as in Fig. 16-27(a). It is evident from Mohr's circle that the shearing stresses on these planes are zero, since points g and h lie on the horizontal axis of Mohr's circle.

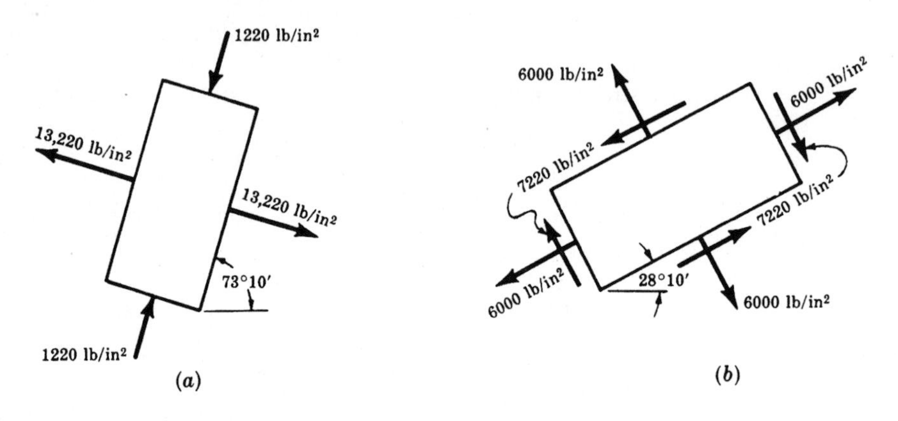

Fig. 16-27

(b) The maximum shearing stress is represented by $\overline{cl}$ in Mohr's circle. This radius has already been found to be equal to 7220 lb/in². The angle $2\theta_s$ may be found either by direct measurement from Fig. 16-26 or simply by subtracting 90° from the angle $2\theta_p$, which has already been determined. This leads to $2\theta_s = 56°20'$ and $\theta_s = 28°10'$. The shearing stress represented by point l is negative; hence on this 28°10′ plane the shearing stress tends to rotate the element in a counterclockwise direction. Also, from Mohr's circle the abscissa of point l is 6000 lb/in² and this represents the normal stress occurring on the planes of maximum shearing stress. The maximum shearing stresses thus appear as in Fig. 16-27(b).

16.11. A plane element in a body is subject to a normal compressive stress in the x-direction of 12,000 lb/in² as well as a shearing stress of 4000 lb/in², as shown in Fig. 16-28. (a) Determine the normal and shearing stress intensities on a plane inclined at an angle of 30° to the normal stress. (b) Determine the maximum and minimum values of the normal stress that may exist on

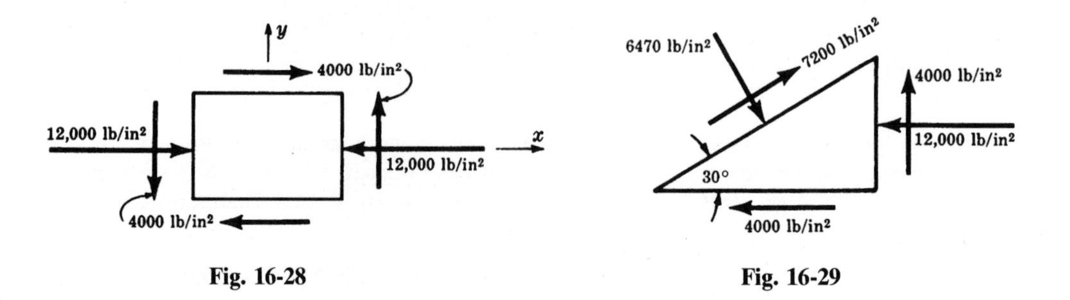

Fig. 16-28 Fig. 16-29

inclined planes and the direction of these stresses. (*c*) Find the magnitude and direction of the maximum shearing stress that may exist on an inclined plane.

(*a*) By the sign convention for normal and shearing stresses adopted in Problem 16.7, we have here $\sigma_x = -12{,}000\ \text{lb/in}^2$, $\tau_{xy} = -4000\ \text{lb/in}^2$. From (*1*) of Problem 16.7, the normal stress on the 30° plane is

$$\sigma = -12{,}000/2 - (-12{,}000/2)\cos 60° - 4000 \sin 60° = -6470\ \text{lb/in}^2$$

From (*2*) of Problem 16.7, the shearing stress on the 30° plane is

$$\tau = \tfrac{1}{2}(-12{,}000)\sin 60° - 4000\cos 60° = -7200\ \text{lb/in}^2$$

The positive directions of the normal and shearing stresses on an inclined plane were illustrated in Fig. 16-16. By this sign convention the stresses on the 30° plane appear as in Fig. 16-29.

(*b*) The values of the principal stresses were given by (*5*) and (*6*) of Problem 16.7. From (*5*),

$$\sigma_{\max} = -12{,}000/2 + \sqrt{(-12{,}000/2)^2 + (-4000)^2} = 1220\ \text{lb/in}^2$$

From (*6*),

$$\sigma_{\min} = -12{,}000/2 - \sqrt{(-12{,}000/2)^2 + (-4000)^2} = -13{,}220\ \text{lb/in}^2$$

The tensile principal stress is usually referred to as the maximum, even though its absolute value is smaller than that of the compressive stress.

The directions of the planes on which these principal stresses occur are given by (*3*) of Problem 16.7 to be

$$\tan 2\theta_p = -\frac{\tau_{xy}}{\frac{1}{2}\sigma_x} = -\frac{-4000}{-12{,}000/2} = -2/3$$

The angles defined by $2\theta_p$ lie in the second and fourth quadrants since the tangent is negative. Hence $2\theta_p = 146°20'$ and $2\theta'_p = 326°20'$. Thus the principal planes are defined by $\theta_p = 73°10'$ and $\theta'_p = 163°10'$. If $\theta_p = 73°10'$, together with the given values of σ_x and τ_{xy}, is now substituted in (*1*) of Problem 16.7, we find

$$\begin{aligned}\sigma &= \tfrac{1}{2}\sigma_x - \tfrac{1}{2}\sigma_x \cos 2\theta + \tau_{xy}\sin 2\theta \\ &= -12{,}000/2 - (-12{,}000/2)\cos 146°20' - 4000 \sin 146°20' = -13{,}220\ \text{lb/in}^2\end{aligned}$$

Thus the principal stress of $-13{,}220\ \text{lb/in}^2$ occurs on the principal plane oriented at 73°10′ to the *x*-axis. The principal stresses are shown in Fig. 16-30. The shearing stresses on these principal planes are zero.

(*c*) The value of the maximum shearing stress is found from (*8*) of Problem 16.7 to be

$$\tau_{\substack{\max\\\min}} = \pm\sqrt{(\tfrac{1}{2}\sigma_x)^2 + (\tau_{xy})^2} = \pm\sqrt{(-12{,}000/2)^2 + (-4000)^2} = \pm 7220\ \text{lb/in}^2$$

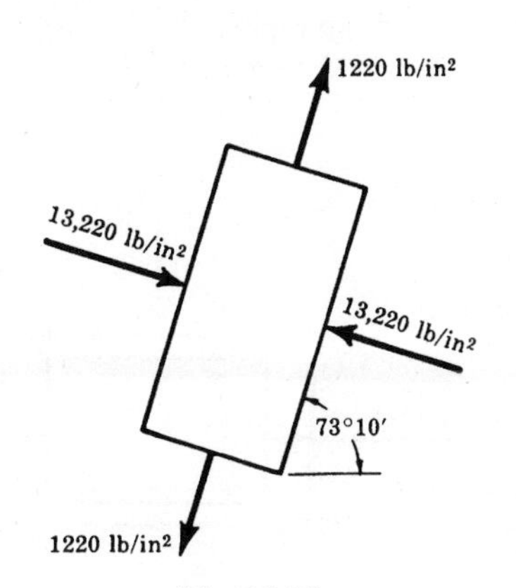

Fig. 16-30

The directions of the planes on which these shearing stresses occur was found in (7) of Problem 16.7 to be

$$\tan 2\theta_s = \frac{\frac{1}{2}\sigma_x}{\tau_{xy}} = \frac{-12{,}000/2}{-4000} = \frac{3}{2}$$

Thus $2\theta_s = 56°20'$ and $2\theta_s' = 236°20'$; or $\theta_s = 28°10'$ and $\theta_s' = 118°10'$. From (2) of Problem 16.7, the shearing stress on any plane inclined at an angle θ with the x-axis is

$$\tau = \tfrac{1}{2}\sigma_x \sin 2\theta + \tau_{xy}\cos 2\theta = \tfrac{1}{2}(-12{,}000)\sin 56°20' - 4000\cos 56°20' = -7220 \text{ lb/in}^2$$

Thus the shearing stress on the 28°10′ plane is negative. The positive sense of shearing stress was shown in Fig. 16-16.

The normal stresses on the planes of maximum shearing stress were found in (9) of Problem 16.7 to be

$$\sigma = \tfrac{1}{2}\sigma_x = -12{,}000/2 = -6000 \text{ lb/in}^2$$

This normal stress acts on each of the planes of maximum shearing stress, as shown in Fig. 16-31.

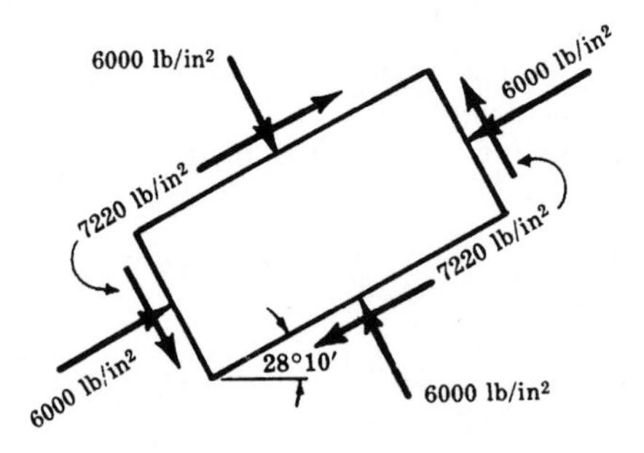

Fig. 16-31

16.12. A plane element is subject to the stresses shown in Fig. 16-32. Using Mohr's circle, determine (a) the principal stresses and their directions and (b) the maximum shearing stresses and the directions of the planes on which they occur.

The procedure for the construction of Mohr's circle was outlined in Problem 16.8. Following the instructions there, the shearing stresses on the vertical faces of the above element are negative, those on the horizontal faces are positive. Thus the stress condition of $\sigma_x = -12{,}000$ lb/in², $\tau_{xy} = -4000$ lb/in² existing on the vertical faces of the element plots as point b in Fig. 16-33. The stress condition of $\tau_{xy} = 4000$ lb/in², together with a zero normal stress on the horizontal faces, plots as point d. Line $\overline{bd}$ is drawn, its midpoint c is located, and a circle of radius $\overline{cb} = \overline{cd}$ is drawn with c as a center. This is Mohr's circle. The endpoints of the diameter $\overline{bd}$ represent the stress conditions existing in the element if it has the original orientation shown in Fig. 16-32.

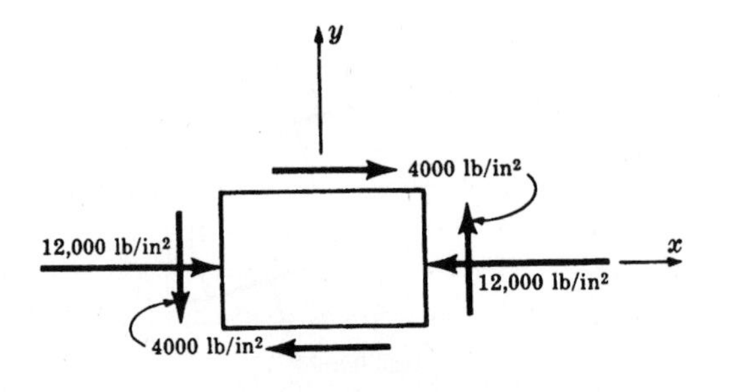

Fig. 16-32

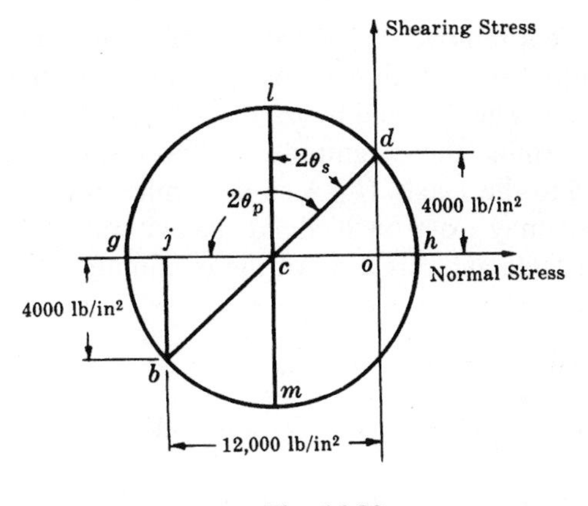

Fig. 16-33

(*a*) The principal stresses are represented by points g and h (Fig. 16-33), as demonstrated in Problem 16.8. They may be determined either by direct measurement from the above diagram or by realizing that the coordinate of c is -6000, and that $\overline{cd} = \sqrt{(6000)^2 + (4000)^2} = 7220$. Thus the minimum principal stress is

$$\sigma_{\min} = \overline{og} = +(\overline{oc} + \overline{cg}) = -6000 - 7220 = -13{,}220 \text{ lb/in}^2$$

The maximum principal stress is

$$\sigma_{\max} = \overline{oh} = \overline{ch} - \overline{co} = 7220 - 6000 = 1220 \text{ lb/in}^2$$

The angle $2\theta_p$ designated above is given by $\tan 2\theta_p = -4000/6000 = -2/3$ since $\tan(180° - \theta) = -\tan\theta$. Hence $2\theta_p = 146°20'$, and $\theta_p = 73°10'$. This value could of course have been obtained by direct measurement of angle dcg in Mohr's circle. Thus the principal stress of $-13{,}220$ lb/in² represented by point g acts on a plane oriented 73°10′ from the original x-axis. The principal stresses thus appear as in Fig. 16-34. It is evident from Mohr's circle that the shearing stresses on these planes are zero, since points g and h lie on the horizontal axis of Mohr's circle.

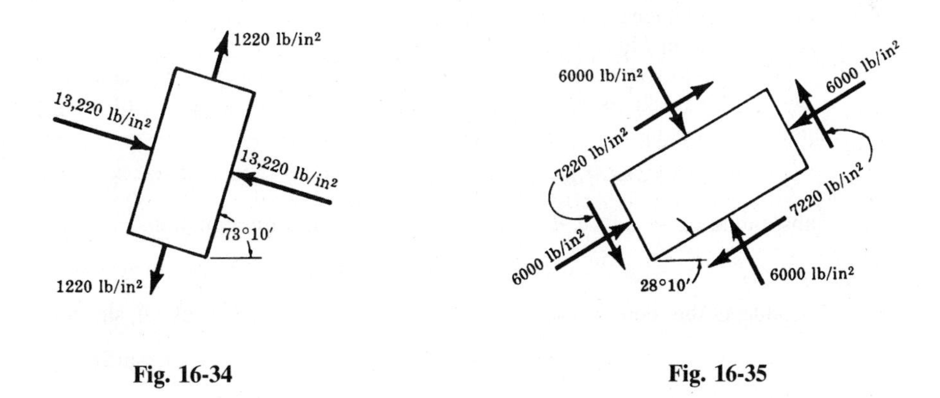

Fig. 16-34 **Fig. 16-35**

(*b*) The maximum shearing stress is represented by $\overline{cl}$ in Mohr's circle. This radius has already been found to be equal to 7220 lb/in². The angle $2\theta_s$ may be found either by direct measurement from Mohr's circle or simply by subtracting 90° from the above value of $2\theta_p$. This leads to $\theta_s = 28°10'$. The shearing stress represented by point l is positive; hence on this 28°10′ plane the shearing stress tends to rotate the element in a clockwise direction. Also, from Mohr's circle the abscissa of point l is -6000 lb/in² and this represents the normal stress occurring on the planes of maximum shearing stresses, as shown in Fig. 16-35.

16.13. Consider a plane element removed from a stressed elastic member. In general such an element will be subject to normal stresses in each of two perpendicular directions, as well as shearing stresses. Let these stresses be denoted by σ_x, σ_y, and τ_{xy} and have the positive directions shown in Fig. 16-36. (*a*) Determine the magnitudes of the normal and shearing stresses on a plane inclined at an angle θ to the x-axis. (*b*) Also determine the maximum and minimum values of the normal stress that may exist on inclined planes and the directions of these stresses. (*c*) Finally, find the magnitude and direction of the maximum shearing stress that may exist on an inclined plane.

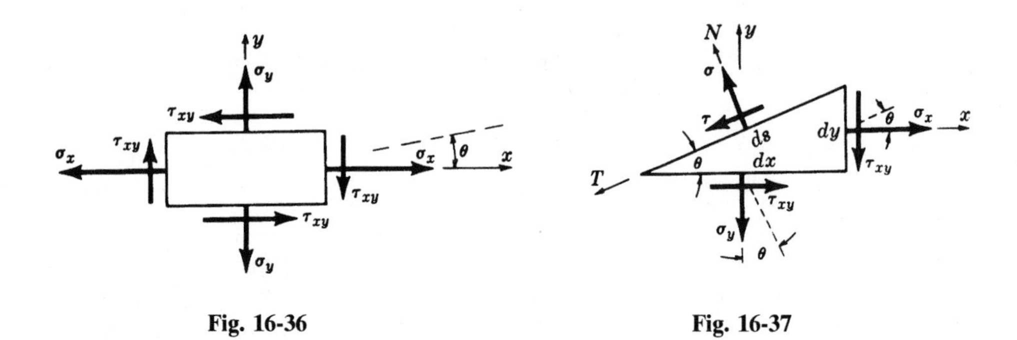

Fig. 16-36 **Fig. 16-37**

(*a*) Evidently the desired stresses acting on the inclined planes are internal quantities with respect to the element shown in Fig. 16-36. Following the usual procedure of introducing a cutting plane so as to render the desired quantities external to the new section, we cut the originally rectangular element along the plane inclined at the angle θ to the x-axis and thus obtain the triangular element shown in Fig. 16-37. Since we have removed half of the material in the rectangular element, we must replace it by the effect that it exerted upon the remaining lower triangle shown and this effect in general consists of both normal and shearing forces acting along the inclined plane. We shall designate the magnitudes of the normal and shearing stresses corresponding to these forces by σ and τ, respectively. Thus our problem reduces to finding the unknown stresses σ and τ in terms of the known stresses σ_x, σ_y, and τ_{xy}. Chapter 17 illustrates the manner of determination of the stresses σ_x, σ_y, and τ_{xy}. It is to be carefully noted that the free-body diagram, Fig. 16-37, indicates stresses acting on the various faces of the element, and not forces. Each of these stresses is assumed to be uniformly distributed over the area on which it acts.

We shall introduce the N- and T-axes normal and tangential to the inclined plane as shown. Let t denote the thickness of the element perpendicular to the plane of the page. Let us begin, by summing forces in the N-direction. For equilibrium we have

$$\Sigma F_N = \sigma t\,ds - \sigma_x t\,dy \sin\theta - \tau_{xy} t\,dy \cos\theta - \sigma_y t\,dx \cos\theta - \tau_{xy} t\,dx \sin\theta = 0$$

Substituting $dy = ds \sin\theta$, $dx = ds \cos\theta$ in the equilibrium equation,

$$\sigma\,ds = \sigma_x\,ds \sin^2\theta + \sigma_y\,ds \cos^2\theta + 2\tau_{xy}\,ds \sin\theta\cos\theta$$

Introducing the identities $\sin^2\theta = \frac{1}{2}(1 - \cos 2\theta)$, $\cos^2\theta = \frac{1}{2}(1 + \cos 2\theta)$, $\sin 2\theta = 2\sin\theta\cos\theta$, we find

$$\sigma = \tfrac{1}{2}\sigma_x(1 - \cos 2\theta) + \tfrac{1}{2}\sigma_y(1 + \cos 2\theta) + \tau_{xy} \sin 2\theta$$

or

$$\sigma = \tfrac{1}{2}(\sigma_x + \sigma_y) - \tfrac{1}{2}(\sigma_x - \sigma_y)\cos 2\theta + \tau_{xy} \sin 2\theta \qquad (1)$$

Thus the normal stress σ on any plane inclined at an angle θ with the x-axis is known as a function of σ_x, σ_y, τ_{xy}, and θ.

Next, summing forces acting on the element in the T-direction, we find

$$\Sigma F_T = \tau t\,ds - \sigma_x t\,dy \cos\theta + \tau_{xy} t\,dy \sin\theta - \tau_{xy} t\,dx \cos\theta + \sigma_y t\,dx \sin\theta = 0$$

Substituting for dx and dy as before, we get

$$\tau\,ds = \sigma_x\,ds \sin\theta\cos\theta - \tau_{xy}\,ds \sin^2\theta + \tau_{xy}\,ds \cos^2\theta - \sigma_y\,ds \sin\theta\cos\theta$$

Introducing the previous identities and the relation $\cos 2\theta = \cos^2\theta - \sin^2\theta$, this last equation becomes

$$\tau = \tfrac{1}{2}(\sigma_x - \sigma_y)\sin 2\theta + \tau_{xy}\cos 2\theta \qquad (2)$$

Thus the shearing stress τ on any plane inclined at an angle θ with the x-axis is known as a function of σ_x, σ_y, τ_{xy}, and θ.

(*b*) To determine the maximum value that the normal stress σ may assume as the angle θ varies, we shall differentiate Eq. (*1*) with respect to θ and set this derivative equal to zero. Thus

$$\frac{d\sigma}{d\theta} = (\sigma_x - \sigma_y)\sin 2\theta + 2\tau_{xy}\cos 2\theta = 0$$

Hence the values of θ leading to maximum and minimum values of the normal stress are given by

$$\tan 2\theta_p = -\frac{\tau_{xy}}{\tfrac{1}{2}(\sigma_x - \sigma_y)} \qquad (3)$$

The planes defined by the angles θ_p are called *principal planes*. The normal stresses that exist on these planes are designated as *principal stresses*. They are the maximum and minimum values that the normal stress may assume in the element under consideration. The values of the principal stresses may easily be found by considering the graphical interpretation of (*3*) given in Fig. 16-38. Evidently the tangent of either of the angles designated as $2\theta_p$ has the value given in (*3*). Thus there are two solutions of (*3*), and consequently two values of $2\theta_p$ (differing by 180°) and also two values of θ_p (differing by 90°). It is to be noted that Fig. 16-38 bears no direct relationship to the triangular element whose free-body diagram was given in Fig. 16-37.

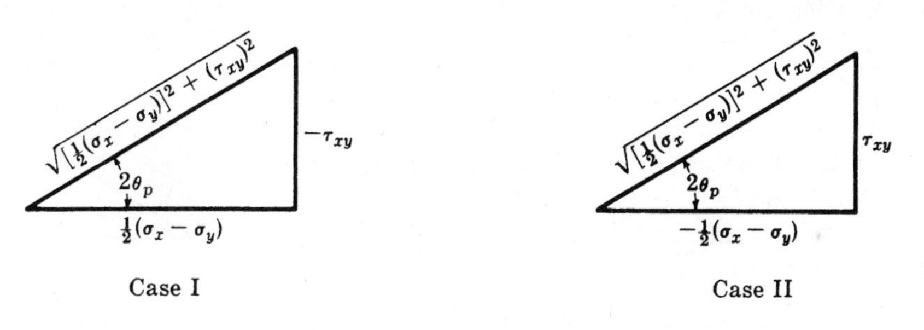

Fig. 16-38

The values of $\sin 2\theta_p$ and $\cos 2\theta_p$ as found from the above two diagrams may now be substituted in (*1*) to yield the maximum and minimum values of the normal stresses. Observing that

$$\sin 2\theta_p = \frac{\mp\tau_{xy}}{\sqrt{[\tfrac{1}{2}(\sigma_x - \sigma_y)]^2 + (\tau_{xy})^2}} \qquad \cos 2\theta_p = \frac{\pm\tfrac{1}{2}(\sigma_x - \sigma_y)}{\sqrt{[\tfrac{1}{2}(\sigma_x - \sigma_y)]^2 + (\tau_{xy})^2}}$$

where the upper signs pertain to Case I and the lower to Case II, we obtain from (*1*)

$$\sigma = \tfrac{1}{2}(\sigma_x + \sigma_y) \pm \sqrt{[\tfrac{1}{2}(\sigma_x - \sigma_y)]^2 + (\tau_{xy})^2} \qquad (4)$$

The maximum normal stress is

$$\sigma_{max} = \tfrac{1}{2}(\sigma_x + \sigma_y) + \sqrt{[\tfrac{1}{2}(\sigma_x - \sigma_y)]^2 + (\tau_{xy})^2} \qquad (5)$$

The minimum normal stress is

$$\sigma_{min} = \tfrac{1}{2}(\sigma_x + \sigma_y) - \sqrt{[\tfrac{1}{2}(\sigma_x - \sigma_y)]^2 + (\tau_{xy})^2} \qquad (6)$$

The stresses given by (*5*) and (*6*) are the principal stresses and they occur on the principal planes defined by (*3*). By substituting one of the values of θ_p from (*3*) into Eq. (*1*), one may readily determine which of the two principal stresses is acting on that plane. The other principal stress naturally acts on the other principal plane.

By substituting the values of the angle $2\theta_p$ as given by (*3*) or by Fig. 16-38 into (*2*), it is readily seen that the shearing stresses τ on the principal planes are zero.

(*c*) To determine the maximum value that the shearing stress τ may assume as the angle θ varies, we shall differentiate Eq. (*2*) with respect to θ and set this derivative equal to zero. Thus

$$\frac{d\tau}{d\theta} = (\sigma_x - \sigma_y)\cos 2\theta - 2\tau_{xy}\sin 2\theta = 0$$

The values of θ leading to the maximum values of the shearing stress are thus

$$\tan 2\theta_s = \frac{\frac{1}{2}(\sigma_x - \sigma_y)}{\tau_{xy}} \tag{7}$$

The planes defined by the two solutions to this equation are the planes of maximum shearing stress.

Again, a graphical interpretation of (*7*) is convenient. The two values of the angle $2\theta_s$ satisfying this equation may be represented as in Fig. 16-39. From these diagrams we have

$$\sin 2\theta_s = \frac{\pm\frac{1}{2}(\sigma_x - \sigma_y)}{\sqrt{[\frac{1}{2}(\sigma_x - \sigma_y)]^2 + (\tau_{xy})^2}} \qquad \cos 2\theta_s = \frac{\pm\tau_{xy}}{\sqrt{[\frac{1}{2}(\sigma_x - \sigma_y)]^2 + (\tau_{xy})^2}}$$

where the upper (positive) sign refers to Case I and the lower (negative) sign applies to Case II. Substituting these values in (*2*) we find

$$\tau_{\substack{\max\\ \min}} = \pm\sqrt{[\tfrac{1}{2}(\sigma_x - \sigma_y)]^2 + (\tau_{xy})^2} \tag{8}$$

Here the positive sign represents the maximum shearing stress, the negative sign the minimum shearing stress.

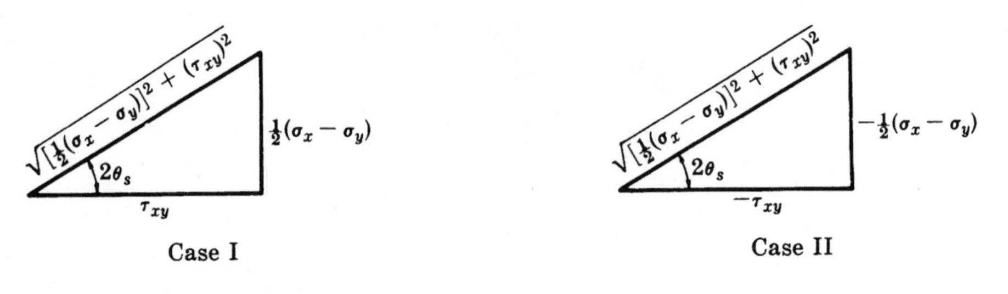

Fig. 16-39

If we compare (*3*) and (*7*), it is evident that the angles $2\theta_p$ and $2\theta_s$ differ by 90°, since the tangents of these angles are the negative reciprocals of one another. Hence the planes defined by the angles θ_p and θ_s differ by 45°; that is, the planes of maximum shearing stress are oriented 45° from the planes of maximum normal stress.

It is also of interest to determine the normal stresses on the planes of maximum shearing stress. These planes are defined by (*7*). If we now substitute the values of $\sin 2\theta_s$ and $\cos 2\theta_s$ in Eq. (*1*) for normal stress, we find

$$\sigma = \tfrac{1}{2}(\sigma_x + \sigma_y) \tag{9}$$

Thus on each of the planes of maximum shearing stress is a normal stress of magnitude $\frac{1}{2}(\sigma_x + \sigma_y)$.

16.14. Discuss a graphical representation of the analysis presented in Problem 16.13.

For given values of σ_x, σ_y, and τ_{xy} we proceed this way:

1. Introduce a rectangular coordinate system in which normal stresses are represented along the horizontal axis and shearing stresses along the vertical axis. The scales used on these two axes must be equal.

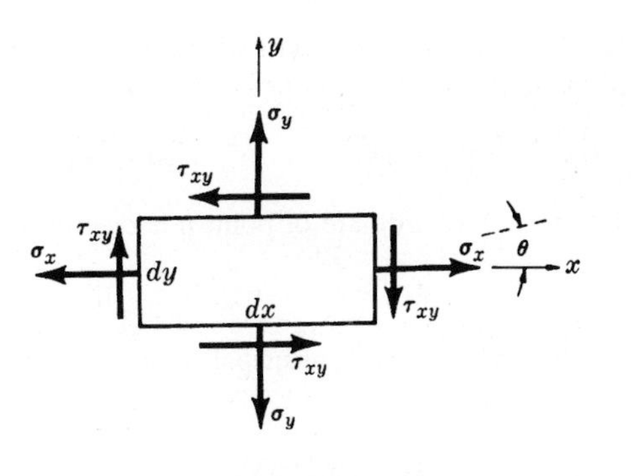

Fig. 16-40

2. With reference to the original rectangular element considered in Problem 16.13 and reproduced in Fig. 16-40, we shall introduce the sign convention that shearing stresses are positive if they tend to rotate the element clockwise, and negative if they tend to rotate it counterclockwise. Here the shearing stresses on the vertical faces are positive, those on the horizontal faces are negative. Also, tensile normal stresses are considered to be positive, compressive stresses negative.
3. We first locate point b by laying out σ_x and τ_{xy} to their given values. The shear stress τ_{xy} on the vertical faces on which σ_x acts is positive; hence this value is plotted as positive in Fig. 16-41.

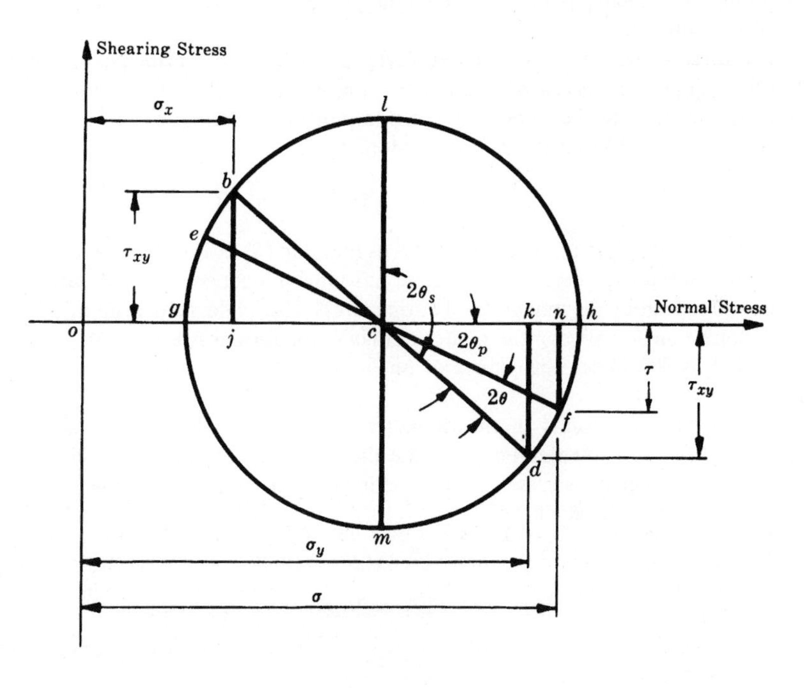

Fig. 16-41

4. We next locate point d in a similar manner by laying off σ_y and τ_{xy} to their given values. Figure 16-41 is drawn on the assumption that $\sigma_y > \sigma_x$ although the treatment presented here holds if $\sigma_y < \sigma_x$. The shear stress τ_{xy} on the horizontal faces on which σ_y acts is negative; hence this value is plotted below the reference axis.
5. Next, we draw line $\overline{bd}$, locate midpoint c, and draw a circle having its center at c and radius equal to $\overline{cb}$. This is known as Mohr's circle.

We shall first show that the points g and h along the horizontal diameter of the circle represent the principal stresses. To do this we note that the point c lies at a distance $\frac{1}{2}(\sigma_x + \sigma_y)$ from the origin of the coordinate system. Also, the line segment $\overline{jk}$ is of length $\sigma_y - \sigma_x$; hence $\overline{ck}$ is of length $\frac{1}{2}(\sigma_y - \sigma_x)$. From the right triangle relationship we have

$$(\overline{cd})^2 = (\overline{ck})^2 + (\overline{kd})^2 \qquad \text{or} \qquad \overline{cd} = \sqrt{[\tfrac{1}{2}(\sigma_x - \sigma_y)]^2 + (\tau_{xy})^2}$$

Also, $\overline{cg} = \overline{ch} = \overline{cd}$. Hence the x-coordinate of point h is $\overline{oc} + \overline{ch}$ or

$$\tfrac{1}{2}(\sigma_x + \sigma_y) + \sqrt{[\tfrac{1}{2}(\sigma_x - \sigma_y)]^2 + (\tau_{xy})^2}$$

But this expression is exactly the maximum principal stress, as found in (5) of Problem 16.13. Likewise the x-coordinate of point g is $\overline{oc} - \overline{gc}$ or

$$\tfrac{1}{2}(\sigma_x + \sigma_y) - \sqrt{[\tfrac{1}{2}(\sigma_x - \sigma_y)]^2 + (\tau_{xy})^2}$$

and this expression is exactly the minimum principal stress, as found in (6) of Problem 16.13. Consequently the points g and h represent the principal stresses existing in the original element. We see that the tangent of $\angle kcd = \overline{dk}/\overline{ck} = \tau_{xy}/\frac{1}{2}(\sigma_y - \sigma_x)$. But from (3) of Problem 16.13 we had

$$\tan 2\theta_p = -\frac{\tau_{xy}}{\frac{1}{2}(\sigma_x - \sigma_y)}$$

and by comparison of these two relations we see that $\angle kcd = 2\theta_p$; that is, a counterclockwise rotation from the diameter $\overline{bd}$ (corresponding to the stresses in the x- and y-directions) leads us to the diameter $\overline{gh}$, representing the principal planes, on which the principal stresses occur. The principal planes lie at an angle θ_p from the x-direction.

Thus Mohr's circle is a convenient device for finding the principal stresses, since one can merely establish the circle for a given set of stresses σ_x, σ_y, τ_{xy}, then measure $\overline{og}$ and $\overline{oh}$. These abscissas represent the principal stresses to the same scale used in plotting σ_x, σ_y, τ_{xy}.

It is now apparent that the radius of Mohr's circle,

$$\overline{cd} = \sqrt{[\tfrac{1}{2}(\sigma_x - \sigma_y)]^2 + (\tau_{xy})^2}$$

corresponds to the maximum shearing stress as found in (8) of Problem 16.13. Actually, the shearing stress on any plane is represented by the ordinate to Mohr's circle; hence we should consider the radial lines $\overline{cl}$ and $\overline{cm}$ as representing the maximum shearing stress. The angle dcl is evidently $2\theta_s$ and hence it is apparent that the double angle between the planes of maximum normal stress and the planes of maximum shearing stress ($\angle kcl$) is 90°; hence the planes of maximum shearing stress are oriented 45° from the planes of maximum normal stress.

Evidently the endpoints of the diameter $\overline{bd}$ represent the stresses acting in the original x- and y-directions. We shall now demonstrate that the endpoints of any other diameter such as $\overline{ef}$ (at an angle 2θ with $\overline{bd}$) represent the stresses on a plane inclined at an angle θ to the x-axis. To do this we note that the abscissa of point f is given by

$$\begin{aligned}\sigma = \overline{oc} + \overline{cn} &= \tfrac{1}{2}(\sigma_x + \sigma_y) + \overline{cf}\cos(2\theta_p - 2\theta)\\ &= \tfrac{1}{2}(\sigma_x + \sigma_y) + \overline{cf}(\cos 2\theta_p \cos 2\theta + \sin 2\theta_p \sin 2\theta)\\ &= \tfrac{1}{2}(\sigma_x + \sigma_y) + \sqrt{[\tfrac{1}{2}(\sigma_x - \sigma_y)]^2 + (\tau_{xy})^2}\,(\cos 2\theta_p \cos 2\theta + \sin 2\theta_p \sin 2\theta)\end{aligned}$$

But from an inspection of triangle ckd in Mohr's circle it is evident that

$$\sin 2\theta_p = \frac{\tau_{xy}}{\sqrt{[\frac{1}{2}(\sigma_x - \sigma_y)]^2 + (\tau_{xy})^2}} \qquad \cos 2\theta_p = \frac{\frac{1}{2}(\sigma_y - \sigma_x)}{\sqrt{[\frac{1}{2}(\sigma_x - \sigma_y)]^2 + (\tau_{xy})^2}} \tag{1}$$

Substituting the values of τ_{xy} and $\frac{1}{2}(\sigma_y - \sigma_x)$ from these last two equations into the previous equation, we find

$$\sigma = \tfrac{1}{2}(\sigma_x + \sigma_y) - \tfrac{1}{2}(\sigma_x - \sigma_y)\cos 2\theta + \tau_{xy}\sin 2\theta$$

But this is exactly the normal stress on a plane inclined at an angle θ to the x-axis as derived in (1) of Problem 16.13.

Next we observe that the ordinate of point f is given by

$$\tau = \overline{nf} = \overline{cf}\sin(2\theta_p - 2\theta) = \overline{cf}(\sin 2\theta_p \cos 2\theta - \cos 2\theta_p \sin 2\theta)$$
$$= \sqrt{[\tfrac{1}{2}(\sigma_x - \sigma_y)]^2 + (\tau_{xy})^2}\,(\sin 2\theta_p \cos 2\theta - \cos 2\theta_p \sin 2\theta)$$

Again, substituting the values of τ_{xy} and $\frac{1}{2}(\sigma_y - \sigma_x)$ from (*1*) into this equation, we find

$$\tau = \tau_{xy}\cos 2\theta + \tfrac{1}{2}(\sigma_x - \sigma_y)\sin 2\theta$$

But this is exactly the shearing stress on a plane inclined at an angle θ to the x-axis as derived in (*2*) of Problem 16.13.

Hence the coordinates of point f on Mohr's circle represent the normal and shearing stresses on a plane inclined at an angle θ to the x-axis.

16.15. A plane element is subject to the stresses shown in Fig. 16-42. Determine (*a*) the principal stresses and their directions, (*b*) the maximum shearing stresses and the directions of the planes on which they occur.

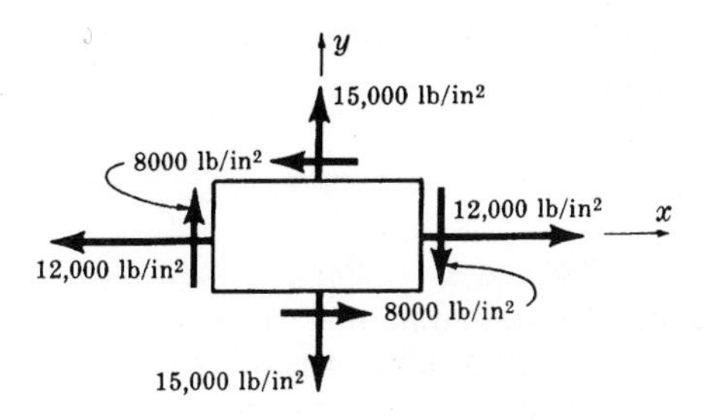

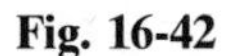
Fig. 16-42

(*a*) In accordance with the notation of Problem 16.13, we have $\sigma_x = 12{,}000$ lb/in^2, $\sigma_y = 15{,}000$ lb/in^2, and $\tau_{xy} = 8000$ lb/in^2. The maximum normal stress is, by (*5*) of Problem 16.13,

$$\sigma_{\max} = \tfrac{1}{2}(\sigma_x + \sigma_y) + \sqrt{[\tfrac{1}{2}(\sigma_x - \sigma_y)]^2 + (\tau_{xy})^2}$$
$$= \tfrac{1}{2}(12{,}000 + 15{,}000) + \sqrt{[\tfrac{1}{2}(12{,}000 - 15{,}000)]^2 + (8000)^2}$$
$$= 13{,}500 + 8150 = 21{,}650 \text{ lb/in}^2$$

The minimum normal stress is given by (*6*) of Problem 16.13 to be

$$\sigma_{\min} = \tfrac{1}{2}(\sigma_x + \sigma_y) - \sqrt{[\tfrac{1}{2}(\sigma_x - \sigma_y)]^2 + (\tau_{xy})^2} = 13{,}500 - 8150 = 5350 \text{ lb/in}^2$$

From (*3*) of Problem 16.13 the directions of the principal planes on which these stresses of 21,650 lb/in^2 and 5350 lb/in^2 occur are given by

$$\tan 2\theta_p = -\frac{\tau_{xy}}{\frac{1}{2}(\sigma_x - \sigma_y)} = -\frac{8000}{\frac{1}{2}(12{,}000 - 15{,}000)} = 5.33$$

Then $2\theta_p = 79°24', 259°24'$ and $\theta_p = 39°42', 129°42'$.

To determine which of the above principal stresses occurs on each of these planes, we return to (*1*) of Problem 16.13, namely,

$$\sigma = \tfrac{1}{2}(\sigma_x + \sigma_y) - \tfrac{1}{2}(\sigma_x - \sigma_y)\cos 2\theta + \tau_{xy}\sin 2\theta$$

and substitute $\theta = 39°42'$ together with the given values of σ_x, σ_y, and τ_{xy} to obtain

$$\sigma = \tfrac{1}{2}(12{,}000 + 15{,}000) - \tfrac{1}{2}(12{,}000 - 15{,}000)\cos 79°24' + 8000 \sin 79°24' = 21{,}650 \text{ lb/in}^2$$

Thus an element oriented along the principal planes and subject to the above principal stresses appears as in Fig. 16-43. The shearing stresses on these planes are zero.

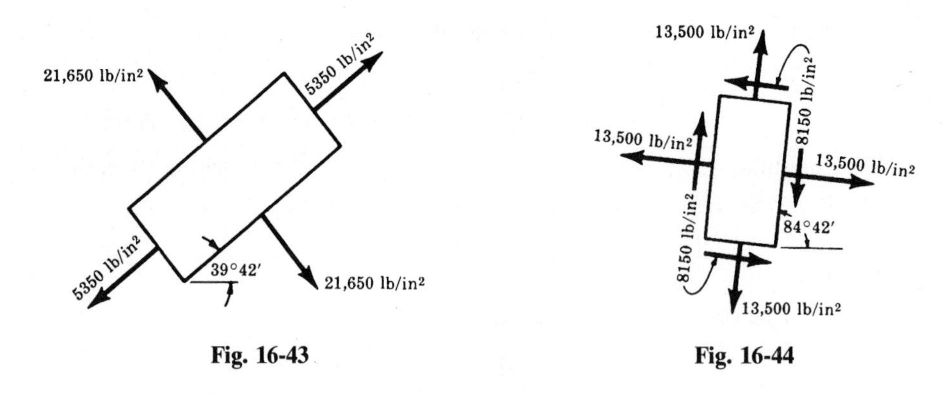

Fig. 16-43

Fig. 16-44

(*b*) The maximum and minimum shearing stresses were found in (*8*) of Problem 16.13 to be

$$\tau_{\substack{\max\\ \min}} = \pm\sqrt{[\tfrac{1}{2}(\sigma_x - \sigma_y)]^2 + (\tau_{xy})^2}$$

$$= \pm\sqrt{[\tfrac{1}{2}(12{,}000 - 15{,}000)]^2 + (8000)^2} = \pm 8150\ \text{lb/in}^2$$

From (*7*) of Problem 16.13 the planes on which these maximum shearing stresses occur are defined by the equation

$$\tan 2\theta_s = \frac{\tfrac{1}{2}(\sigma_x - \sigma_y)}{\tau_{xy}} = -0.188$$

Then $2\theta_s = 169°24', 349°24'$ and $\theta_s = 84°42', 174°42'$. Evidently these planes are located 45° from the planes of maximum and minimum normal stress.

To determine whether the shearing stress is positive or negative on the 84°42′ plane, we return to (*2*) of Problem 16.13, namely,

$$\tau = \tfrac{1}{2}(\sigma_x - \sigma_y)\sin 2\theta + \tau_{xy}\cos 2\theta$$

and substitute $\theta = 84°42'$ together with the given values of σ_x, σ_y, and τ_{xy} to obtain

$$\tau = \tfrac{1}{2}(12{,}000 - 15{,}000)\sin 169°24' + 8000\cos 169°24' = -8150\ \text{lb/in}^2$$

The negative sign indicates that the shearing stress is directed oppositely to the assumed positive direction shown in Fig. 16-36. Finally, the normal stresses on these planes of maximum shearing stress are found from (*9*) of Problem 16.13 to be

$$\sigma = \tfrac{1}{2}(\sigma_x + \sigma_y) = \tfrac{1}{2}(12{,}000 + 15{,}000) = 13{,}500\ \text{lb/in}^2$$

The orientation of the element for which the shearing stresses are maximum is as in Fig. 16-44.

16.16. A plane element is subject to the stresses shown in Fig. 16-45. Using Mohr's circle, determine (*a*) the principal stresses and their directions and (*b*) the maximum shearing stresses and the directions of the planes on which they occur.

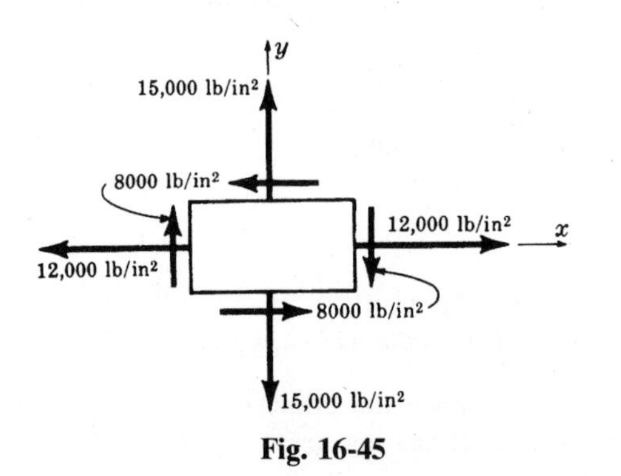

Fig. 16-45

The procedure for the construction of Mohr's circle was outlined in Problem 16.14. Following the instructions there, we realized that the shearing stresses on the vertical faces of the given element are positive, whereas those on the horizontal faces are negative. Thus the stress condition of $\sigma_x = 12{,}000\ \text{lb/in}^2$, $\tau_{xy} = 8000\ \text{lb/in}^2$ existing on the vertical faces of the element plots as point b in Fig. 16-46. The stress condition of $\sigma_y = 15{,}000\ \text{lb/in}^2$, $\tau_{xy} = -8000\ \text{lb/in}^2$ existing on the horizontal faces plots as point d. Line $\overline{bd}$ is drawn, its midpoint c is located, and a circle of radius $\overline{cb} = \overline{cd}$ is drawn with c as a center. This is Mohr's circle. The endpoints of the diameter $\overline{bd}$ represent the stress conditions existing in the element if it has the original orientation of Fig. 16-45.

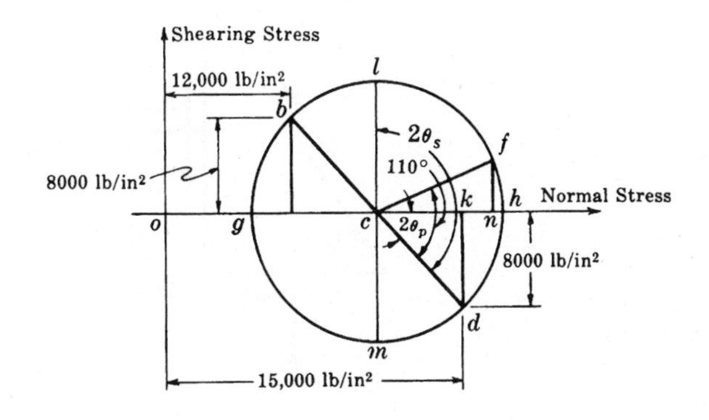

Fig. 16-46

(*a*) The principal stresses are represented by points g and h, as demonstrated in Problem 16.14. The principal stress may be determined either by direct measurement from Fig. 16-46 or by realizing that the coordinate of c is 13,500, that $\overline{ck} = 1500$, and that $\overline{cd} = \sqrt{(1500)^2 + (8000)^2} = 8150$. Thus the minimum principal stress is

$$\sigma_{\min} = \overline{og} = \overline{oc} - \overline{cg} = 13{,}500 - 8150 = 5350\ \text{lb/in}^2$$

Also, the maximum principal stress is

$$\sigma_{\max} = \overline{oh} = \overline{oc} + \overline{ch} = 13{,}500 + 8150 = 21{,}650\ \text{lb/in}^2$$

The angle $2\theta_p$ is given by $\tan 2\theta_p = 8000/1500 = 5.33$ from which $\theta_p = 39°42'$. This value could also be obtained by measurement of $\angle dck$ in Mohr's circle. From this it is readily seen that the principal stress represented by point h acts on a plane oriented $39°42'$ from the original x-axis. The principal stresses thus appear as in Fig. 16-47. It is evident that the shearing stresses on these planes are zero, since points g and h lie on the horizontal axis of Mohr's circle.

(*b*) The maximum shearing stress is represented by $\overline{cl}$ in Mohr's circle. This radius has already been found to represent $8150\ \text{lb/in}^2$. The angle $2\theta_s$ may be found either by direct measurement from the above plot or simply by adding 90° to the angle $2\theta_p$, which has already been determined. This leads to $2\theta_s = 169°24'$ and $\theta_s = 84°42'$. The shearing stress represented by point l is positive; hence on this $84°42'$ plane the shearing stress tends to rotate the element in a clockwise direction.

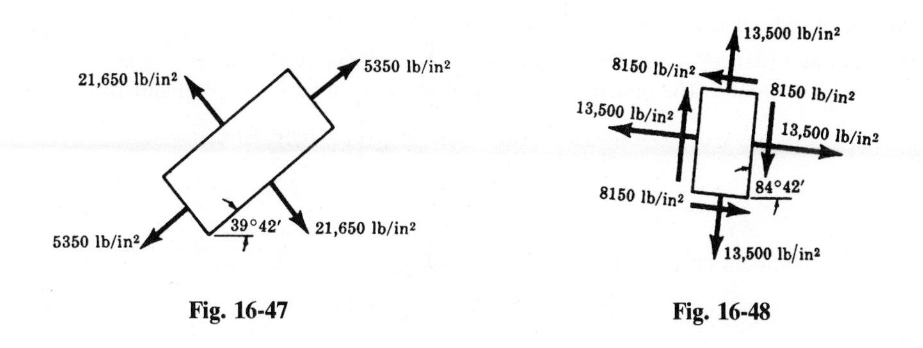

Fig. 16-47 Fig. 16-48

Also, from Mohr's circle the abscissa of point l is 13,500 lb/in^2 and this represents the normal stress occurring on the planes of maximum shearing stress. The maximum shearing stresses thus appear as in Fig. 16-48.

16.17. For the element discussed in Problem 16.16, determine the normal and shearing stresses on a plane making an angle of 55° measured counterclockwise from the positive end of the x-axis.

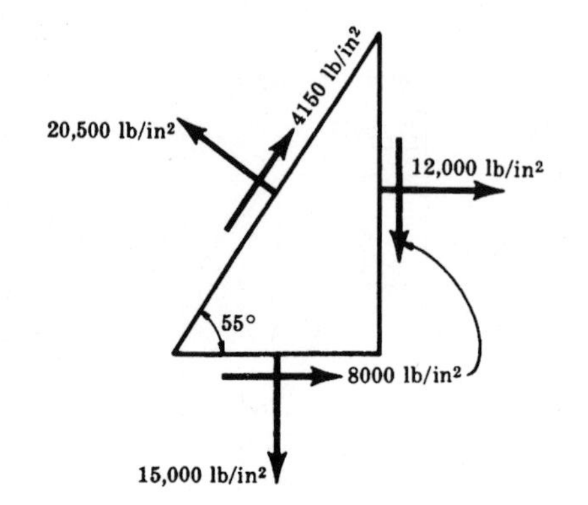

Fig. 16-49

According to the properties of Mohr's circle discussed in Problem 16.14, we realize that the endpoints of the diameter $\overline{bd}$ represent the stress conditions occurring on the original x-y plane. On any plane inclined at an angle θ to the x-axis the stress conditions are represented by the coordinates of a point f, where the radius $\overline{cf}$ makes an angle of 2θ with the original diameter $\overline{bd}$. This angle 2θ appearing in Mohr's circle is measured in the same direction as the angle representing the inclined plane, namely, counterclockwise.

Hence in the Mohr's circle appearing in Problem 16.16, we merely measure a counterclockwise angle of $2(55°) = 110°$ from line $\overline{cd}$. This locates point f. The abscissa of point f represents the normal stress on the desired 55° plane and may be found either by direct measurement or by realizing that

$$\overline{on} = \overline{oc} + \overline{cn} = 13{,}500 + 8150\cos(110° - 79°24') = 20{,}500\ \text{lb/in}^2$$

The ordinate of point f represents the shearing stress on the desired 55° plane and may be found from the relation

$$\overline{fn} = 8150\sin(110° - 79°24') = 4150\ \text{lb/in}^2$$

The stresses acting on the 55° plane may thus be represented as in Fig. 16-49.

16.18. A plane element is subject to the stresses shown in Fig. 16-50. Determine (a) the principal stresses and their directions and (b) the maximum shearing stresses and the directions of the planes on which they occur.

(a) In accordance with the notation of Problem 16.13, $\sigma_x = -75$ MPa, $\sigma_y = 100$ MPa, and $\tau_{xy} = -50$ MPa. The maximum normal stress is given by (5) of Problem 16.13 to be

$$\begin{aligned}\sigma_{\max} &= \tfrac{1}{2}(\sigma_x + \sigma_y) + \sqrt{[\tfrac{1}{2}(\sigma_x - \sigma_y)]^2 + (\tau_{xy})^2} \\ &= \tfrac{1}{2}(-75 + 100) + \sqrt{[\tfrac{1}{2}(-75 - 100)]^2 + (-50)^2} \\ &= 12.5 + 100.8 = 113.3\ \text{MPa}\end{aligned}$$

The minimum normal stress is given by (6) of Problem 16.13 to be

$$\sigma_{\min} = \tfrac{1}{2}(\sigma_x + \sigma_y) - \sqrt{[\tfrac{1}{2}(\sigma_x - \sigma_y)]^2 + (\tau_{xy})^2} = 12.5 - 100.8 = -88.3\ \text{MPa}$$

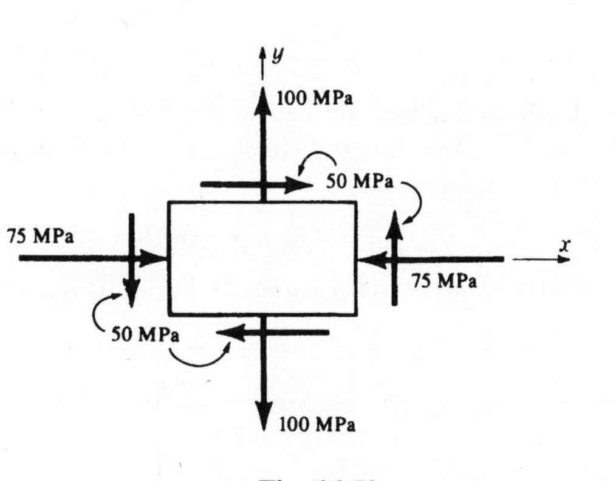

Fig. 16-50

From (3) of Problem 16.13 the directions of the principal planes on which these stresses of 113.3 MPa and −88.3 MPa occur are given by

$$\tan 2\theta_p = -\frac{\tau_{xy}}{\frac{1}{2}(\sigma_x - \sigma_y)} = -\frac{-50}{\frac{1}{2}(-75-100)} = -0.571$$

Then $2\theta_p = 150°15', 330°15'$ and $\theta_p = 75°8', 165°8'$.

To determine which of the above principal stresses occurs on each of these planes, we return to (1) of Problem 16.13, namely,

$$\sigma = \tfrac{1}{2}(\sigma_x + \sigma_y) - \tfrac{1}{2}(\sigma_x - \sigma_y)\cos 2\theta + \tau_{xy}\sin 2\theta$$

and substitute $\theta = 75°8'$ together with the given values of σ_x, σ_y, and τ_{xy} to obtain

$$\sigma = \tfrac{1}{2}(-75+100) - \tfrac{1}{2}(-75-100)\cos 150°15' - 50\sin 150°15' = 88.3 \text{ MPa}$$

Consequently an element oriented along the principal planes and subject to the above principal stresses appears as in Fig. 16-51. The shearing stresses on these planes are zero.

(b) The maximum and minimum shearing stresses were found in (8) of Problem 16.13 to be

$$\tau_{\substack{\max\\ \min}} = \pm\sqrt{[\tfrac{1}{2}(\sigma_x - \sigma_y)]^2 + (\tau_{xy})^2} = \pm\sqrt{[\tfrac{1}{2}(-75-100)]^2 + (-50)^2} = \pm 100.8 \text{ MPa}$$

From (7) of Problem 16.13, the planes on which these maximum shearing stresses occur are defined by

$$\tan 2\theta_s = \frac{\frac{1}{2}(\sigma_x - \sigma_y)}{\tau_{xy}} = 1.75$$

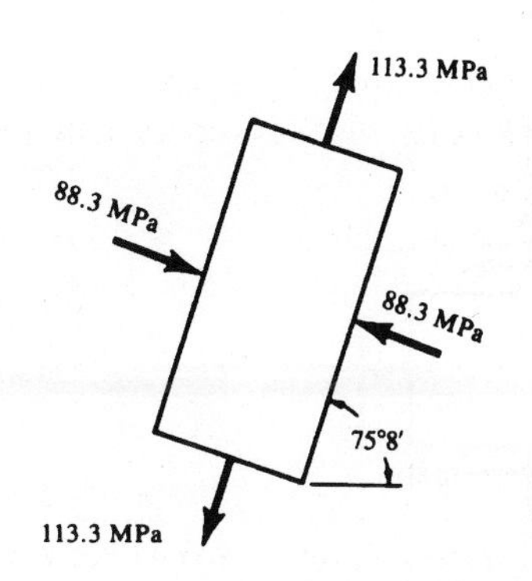

Fig. 16-51

Then $2\sigma_s = 60°15', 240°15'$ and $\theta_s = 30°8', 120°8'$. It is apparent that these planes are located 45° from the planes of maximum and minimum normal stress.

To determine whether the shearing stress is positive or negative on the 30°8′ plane, we return to (2) of Problem 16.13, namely,

$$\tau = \tfrac{1}{2}(\sigma_x - \sigma_y)\sin 2\theta + \tau_{xy}\cos 2\theta$$

and substitute $\theta = 30°8'$ together with the given values of σ_x, σ_y, and τ_{xy} to obtain

$$\tau = \tfrac{1}{2}(-75 - 100)\sin 60°15' - 50\cos 60°15' = -100.8\ \text{MPa}$$

The negative sign indicates that the shearing stress on the 30°8′ plane is directed oppositely to the assumed positive direction shown in Fig. 16-36. The normal stresses on these planes of maximum shearing stress were found in (9) of Problem 16.13 to be

$$\sigma = \tfrac{1}{2}(\sigma_x + \sigma_y)$$
$$= \tfrac{1}{2}(-75 + 100) = 12.5\ \text{MPa}$$

Consequently, the orientation of the element for which the shearing stresses are a maximum appears as in Fig. 16-52.

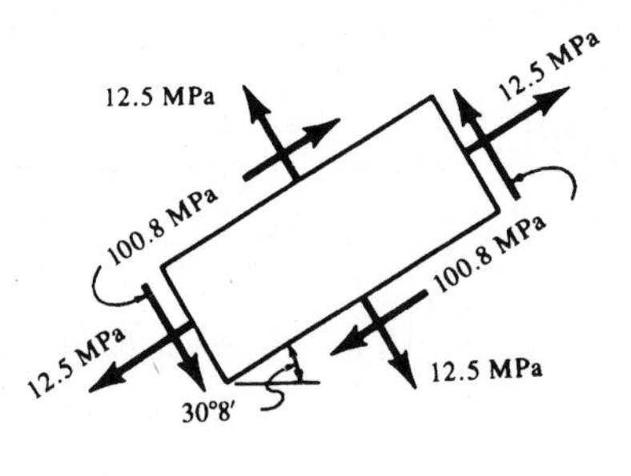

Fig. 16-52

16.19. A plane element is subject to the stresses shown in Fig. 16-53. Using Mohr's circle, determine (*a*) the principal stresses and their directions and (*b*) the maximum shearing stresses and the directions of the planes on which they occur.

Again we refer to Problem 16.14 for the procedure for constructing Mohr's circle. In accordance with the sign convention outlined there, the shearing stresses on the vertical faces of the element are negative, those on the horizontal faces positive. Thus the stress condition of $\sigma_x = -75$ MPa, $\tau_{xy} = -50$ MPa existing on the vertical faces of the element plots as point b in Fig. 16-54. The stress condition of $\sigma_y = 100$ MPa, $\tau_{xy} = 50$ MPa existing on the horizontal faces plots as point d. Line $\overline{bd}$ is drawn, its midpoint c is located, and a circle of radius $\overline{cb} = \overline{cd}$ is drawn with c as a center. This is Mohr's circle. The endpoints of the

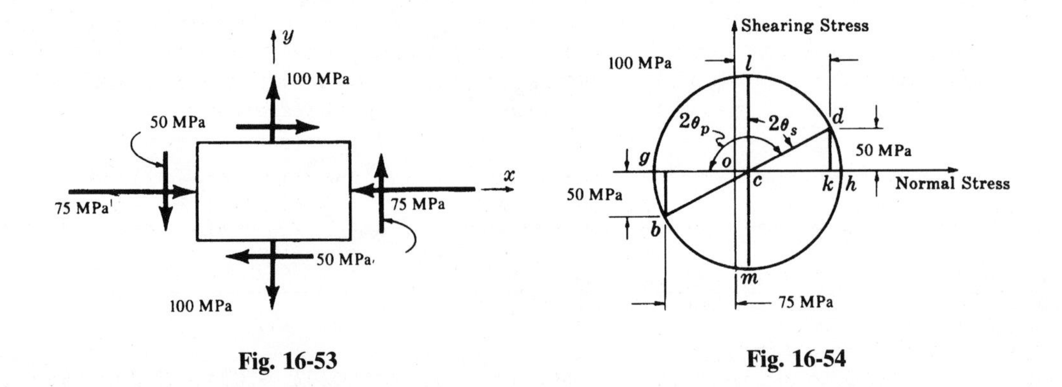

Fig. 16-53 **Fig. 16-54**

diameter $\overline{bd}$ represent the stress conditions existing in the element if it has the original orientation shown above.

(*a*) The principal stresses are represented by points g and h, as shown in Problem 16.14. They may be found either by direct measurement from the above diagram or by realizing that the coordinate of c is 12.5, that $ck = 87.5$, and that $\overline{cd} = \sqrt{(87.5)^2 + (50)^2} = 100.8$ MPa. Thus the minimum principal stress is

$$\sigma_{\min} = \overline{og} = \overline{oc} - \overline{cg} = 12.5 - 100.8 = 88.3 \text{ MPa}$$

Also, the maximum principal stress is

$$\sigma_{\max} = \overline{oh} = \overline{oc} + \overline{ch} = 12.5 + 100.8 = 113.3 \text{ MPa}$$

The angle $2\theta_p$ is given by $\tan 2\theta_p = -50/87.5 = -0.571$ from which $\theta_p = 75°8'$. This value could also be obtained by measurement of $\angle dcg$ in Mohr's circle. From this it is readily seen that the principal stress represented by point g acts on a plane oriented 75°8′ from the original x-axis. The principal stresses thus appear as in Fig. 16-55. Since the ordinates of points g and h are each zero, the shearing stresses on these planes are zero.

(*b*) The maximum shearing stress is represented by $\overline{cl}$ in Mohr's circle. This radius has already been found to represent 100.8 MPa. The angle $2\theta_s$ may be found either by direct measurement from the above plot or simply by subtracting 90° from the angle $2\theta_p$ which has already been determined. This leads to $2\theta_s = 60°15'$ and $\theta_s = 30°8'$. The shearing stress represented by point l is positive, hence on this 30°8′ plane the shearing stress tends to rotate the element in a clockwise direction.

Also, from Mohr's circle the abscissa of point l is 12.5 MPa and this represents the normal stress occurring on the planes of maximum shearing stress. The maximum shearing stresses thus appear as in Fig. 16-56.

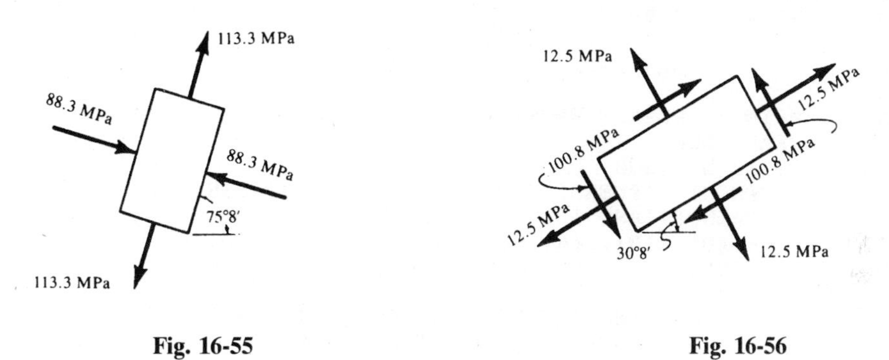

Fig. 16-55

Fig. 16-56

16.20. Develop a FORTRAN program to indicate the principal stresses as well as their directions for an element subject to the stresses shown in Fig. 16-36.

The input to the program consists of the two normal stresses and one shearing stress, as indicated in Fig. 16-36. The normal stresses, for purposes of developing a program, are, as before, taken to be positive if tensile. The simplest sign convention for shearing stresses is to regard the horizontally directed shears as positive if they tend to produce clockwise rotation of the element, i.e., opposite to the convention associated with Problem 16.13. In Problem 16.13 we found the principal stresses to be given by Eqs. (*5*) and (*6*) and their directions by Eq. (*3*). The desired program is listed below.

```
00010*********************************************************************
00020                         PROGRAM STRES2D (INPUT,OUTPUT)
00030*********************************************************************
00040*
00050*          AUTHOR: KATHLEEN DERWIN
00060*          DATE  : JANUARY 26,1989
00070*
00080*   BRIEF DESCRIPTION:
00090*      THIS FORTRAN PROGRAM  MAY BE USED TO SOLVE A SIMPLE 2-D STRESS
```

```
00100*   PROBLEM WHERE THE USER IS PROMPTED FOR THE STRESS CONDITIONS FOR A
00110*   SINGLE OR SET OF POINTS, AND THE PRINCIPAL STRESS  AND ROTATING ANGLE
00120*   ARE CALCULATED.
00130*
00140*   INPUT:
00150*      THE USER WILL BE ASKED TO INPUT THE NUMBER OF STRESS SETS AND THE
00160*   NORMAL AND SHEAR STRESSES AT EACH POINT.
00170*
00180*   OUTPUT:
00185*      THE PRINCIPAL STRESSES AND ROTATING ANGLE FOR EACH SET OF PTS. WIL
00190*   BE PRINTED.
00200*
00210*   VARIABLES:
00220*      X(100),Y(100),S(100)   --- NORMAL AND SHEAR STRESS ARRAYS
00230*              NUM            --- THE NUMBER OF STRESS SETS
00240*              PI             --- 3.14159
00250*
00260*   SUBROUTINES CALLED:
00270*      PRINCIP --- CALCULATES THE PRINCIPAL STRESSES AND THE ROTATING
00280*                  ANGLE FOR A SINGLE OR SET OF POINTS.
00290*
00300***********************************************************************
00310**********                 MAIN PROGRAM                    ************
00320***********************************************************************
00330*
00340*          VARIABLE DECLARATIONS
00350*
00360       REAL X(100),Y(100),S(100),PI
00370       INTEGER NUM
00380*
00390      PI = 3.14159
00400*
00410*           USER INPUT
00420*
00430      PRINT*,'PLEASE ENTER THE NUMBER OF STRESS SETS:'
00440      READ*,NUM
00450      DO 10 N=1,NUM
00460         PRINT*,'PLEASE ENTER THE NORMAL STRESSES IN THE X,Y DIRECTIONS'
00470         PRINT*,'AND THE SHEAR STRESS:'
00480         READ*,X(N),Y(N),S(N)
00490 10   CONTINUE
00500*
00510*           CALLING SUBROUTINE PRINCIP TO CALCULATE THE PRINCIPAL
00520*           STRESSES AND THE ROTATING ANGLE
00530*
00540      CALL PRINCIP(X,Y,S,NUM)
00550*
00560      STOP
00570      END
00580***********************************************************************
00590         SUBROUTINE PRINCIP(XX,YY,SS,NUM)
00600*
00610*      THIS SUBROUTINE WILL EVALUATE THE PRINCIPAL STRESSES AND ROTATING
00620*      ANGLE FOR A SINGLE OR SET OF POINTS.
00630*
00640*          VARIABLE DECLARATIONS
00650*
00660      REAL PI,XX(100),YY(100),SS(100),P1(100),P2(100),T(100)
00670      INTEGER NUM
00680*
00690*          CALCULATIONS
00700*
00710      PI = 3.14159
00720      DO 15 N=1,NUM
00730         A=((XX(N)-YY(N))/2.0)**2
00740         B=SQRT(A+(SS(N)**2))
```

```
00750          C=(XX(N)+YY(N))/2.0
00760          P1(N)=C+B
00770          P2(N)=C-B
00780          A1=2*SS(N)/(XX(N)-YY(N))
00790          T(N)=90*ATAN(A1)/PI
00800          IF (XX(N).EQ.YY(N)) THEN
00810             T(N) = 45.0
00820          ENDIF
00830 15    CONTINUE
00840*
00850*            PRINTING OUTPUT
00860*
00870       PRINT 30
00880       DO 20 N=1,NUM
00890          PRINT 40,N,XX(N),YY(N),SS(N),P1(N),P2(N),T(N)
00900 20    CONTINUE
00910*
00920*            FORMAT STATEMENTS
00930*
00940 30    FORMAT(/,2X,'NO.',5X,'SIGXX',7X,'SIGYY',7X,'SIGXY',7X,'SIG(1)',
00950+            7X,'SIG(2)',7X,'THETA',/)
00960 40    FORMAT(2X,I2,3X,5(F9.2,3X),F9.2)
00970*
00980*            END SUBROUTINE PRINCIP
00990*
01000       RETURN
01010       END
```

16.21. Use the FORTRAN program of Problem 16.20 to determine principal stresses and their directions for an element subject to the stresses indicated in Fig. 16-57.

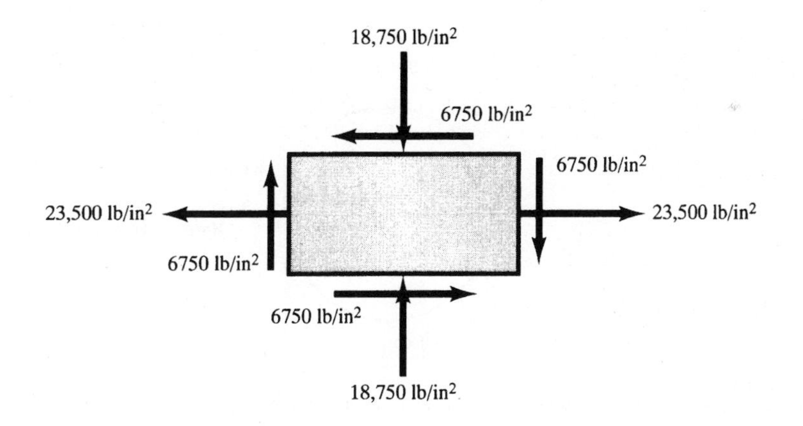

Fig. 16-57

If we use the notation of Problem 16.20 together with the directions of stresses shown in Fig. 16-57, we have $\sigma_x = 23{,}500\ \text{lb/in}^2$, $\sigma_y = -18{,}750\ \text{lb/in}^2$, and $\tau_{xy} = -6750\ \text{lb/in}^2$. Substituting these values into the self-prompting program of Problem 16.20, we get the following computer run.

```
 READY.
run
 PLEASE ENTER THE NUMBER OF STRESS SETS:
? 1
 PLEASE ENTER THE NORMAL STRESSES IN THE X,Y DIRECTIONS
 AND THE SHEAR STRESS:
? 23500,-18750,-6750
```

NO.	SIGXX	SIGYY	SIGXY	SIG(1)	SIG(2)	THETA
1	23500.00	-18750.00	-6750.00	24552.20	-19802.20	-8.86

```
SRU       0.734 UNTS.

RUN COMPLETE.
```

Supplementary Problems

16.22. A bar of uniform cross section 50 mm × 75 mm is subject to an axial tensile force of 500 kN applied at each end of the bar. Determine the maximum shearing stress existing in the bar. *Ans.* 66.7 MPa

16.23. In Problem 16.22 determine the normal and shearing stresses acting on a plane inclined at 11° to the line of action of the axial loads. *Ans.* 4.87 MPa, 24.97 MPa

16.24. A square steel bar 1 in on a side is subject to an axial compressive load of 8000 lb. Determine the normal and shearing stresses acting on a plane inclined at 30° to the line of action of the axial loads. The bar is so short that the possibility of buckling as a column may be neglected.
Ans. $\sigma = -2000\ \text{lb/in}^2$, $\tau = -3460\ \text{lb/in}^2$

16.25. Rework Problem 16.24 by use of Mohr's circle.
Ans. See Fig. 16.58. $\sigma = \overline{ko} = -2000\ \text{lb/in}^2$, $\tau = \overline{dk} = 3460\ \text{lb/in}^2$

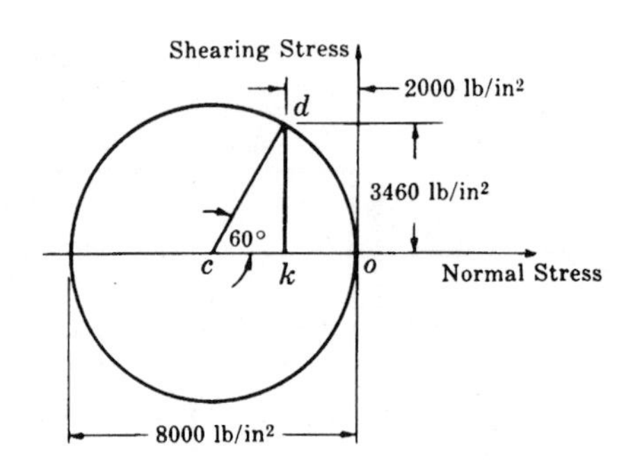

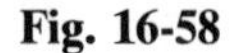
Fig. 16-58

16.26. A plane element in a body is subject to the stresses $\sigma_x = 20$ MPa, $\sigma_y = 0$, and $\tau_{xy} = 30$ MPa. Determine analytically the normal and shearing stresses existing on a plane inclined at 45° to the x-axis.
Ans. $\sigma = 40$ MPa, $\tau = 10$ MPa

16.27. A plane element is subject to the stresses $\sigma_x = 50$ MPa and $\sigma_y = 50$ MPa. Determine analytically the maximum shearing stress existing in the element. *Ans.* 0

16.28. A plane element is subject to the stresses $\sigma_x = 12{,}000\ \text{lb/in}^2$ and $\sigma_y = -12{,}000\ \text{lb/in}^2$. Determine analytically the maximum shearing stress existing in the element. What is the direction of the planes on which the maximum shearing stresses occur? *Ans.* $12{,}000\ \text{lb/in}^2$ at 45°

16.29. For the element described in Problem 16.28 determine analytically the normal and shearing stresses acting on a plane inclined at 30° to the x-axis. *Ans.* $\sigma = -6000\ \text{lb/in}^2$, $\tau = 10{,}400\ \text{lb/in}^2$

16.30. Draw Mohr's circle for a plane element subject to the stresses $\sigma_x = 8000\ \text{lb/in}^2$ and $\sigma_y = -8000\ \text{lb/in}^2$. From Mohr's circle determine the stresses acting on a plane inclined at 20° to the x-axis.
Ans. See Fig. 16-59. $\sigma = \overline{on} = -6130\ \text{lb/in}^2$, $\tau = \overline{nf} = -5130\ \text{lb/in}^2$

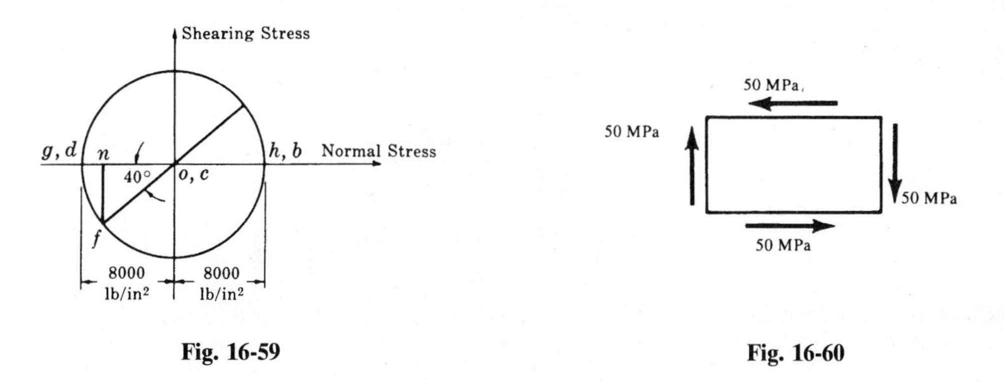

Fig. 16-59 **Fig. 16-60**

16.31. A plane element removed from a thin-walled cylindrical shell loaded in torsion is subject to the shearing stresses shown in Fig. 16-60. Determine the principal stresses existing in this element and the directions of the planes on which they occur. *Ans.* 50 MPa at 45°

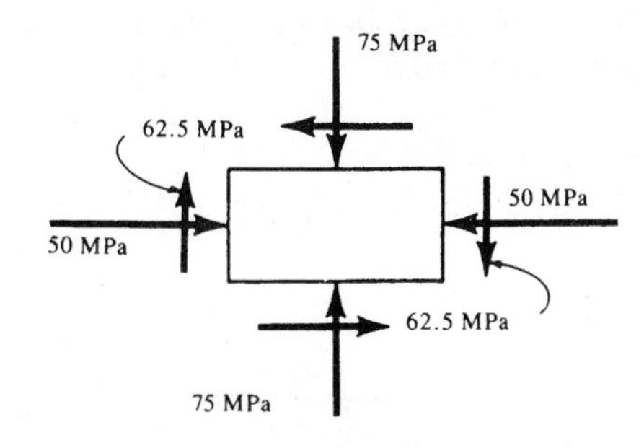

Fig. 16-61

16.32. A plane element is subject to the stresses shown in Fig. 16-61. Determine analytically (a) the principal stresses and their directions and (b) the maximum shearing stresses and the directions of the planes on which they act.
Ans. (a) $\sigma_{max} = 1.2$ MPa at 50°40′, $\sigma_{min} = -126.2$ MPa at 140°40′; (b) $\tau_{max} = 63.7$ MPa at 5°40′

16.33. Rework Problem 16.32 by the use of Mohr's circle. *Ans.* See Fig. 16-62.

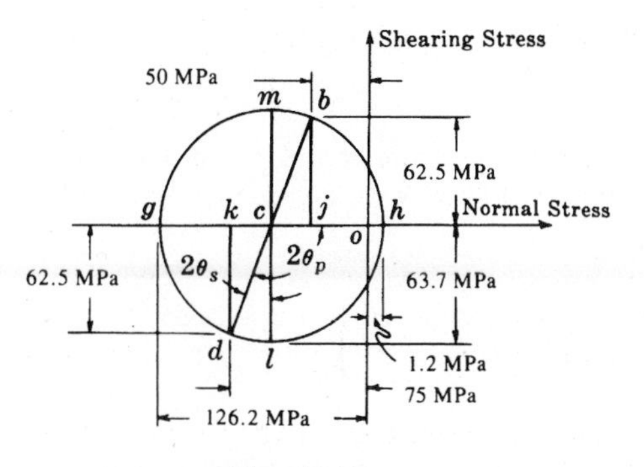

Fig. 16-62

16.34. A plane element is subject to the stresses indicated in Fig. 16-63. Use the FORTRAN program of Problem 16.20 to determine principal stresses together with their orientation.
Ans. SIG(1): 198.12; SIG(2): 66.88; THETA: 24.82

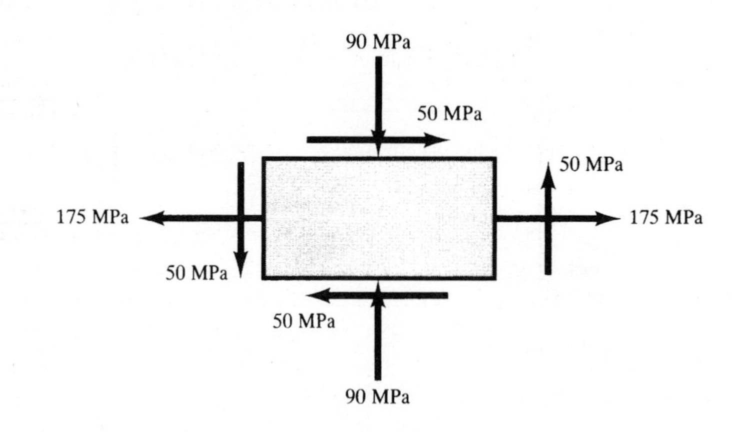

Fig. 16-63

16.35. A plane element is subject to the stresses indicated in Fig. 16-64. Use the FORTRAN program of Problem 16.20 to determine principal stresses together with their orientation.
Ans. SIG(1): 20,388.68; SIG(2): −31,738.68; THETA: 14.20

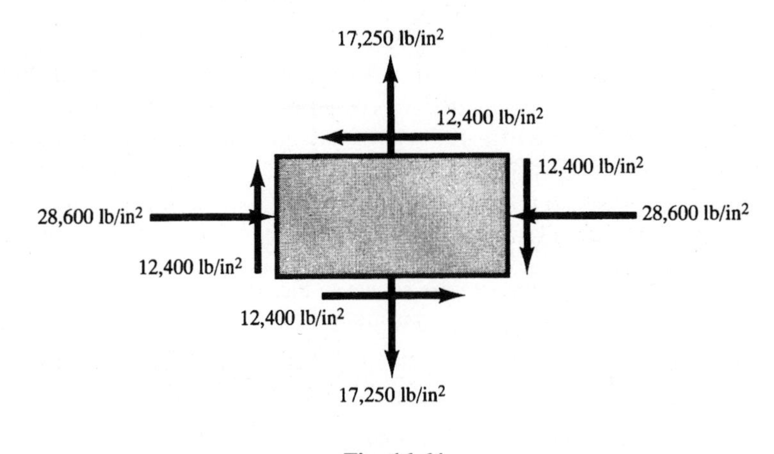

Fig. 16-64

Chapter 17

Members Subject to Combined Loadings; Theories of Failure

AXIALLY LOADED MEMBERS SUBJECT TO ECCENTRIC LOADS

In Chaps. 1 and 2, where we considered straight bars subject to either tensile or compressive loads, it was always required that the action line of the applied force pass through the centroid of the cross section of the member. In the present chapter we shall consider those cases where the action line of the applied force acting on a bar in either tension or compression does *not* pass through the centroid of the cross section. A typical example of such an eccentric loading is shown in Fig. 17-1. For those cross sections of the bar that are perpendicular to the direction of the load, the resultant stress at any point is the sum of the direct stress due to a concentric load of equal magnitude P plus a bending stress due to a couple of moment Pe. This first stress is found from the expression derived in Chap. 1, namely, $\sigma = P/A$. The second stress is found from the formula for bending stress presented in Chap. 8, namely, $\sigma = My/I$. An application may be found in Problem 17.1.

CYLINDRICAL SHELLS SUBJECT TO COMBINED INTERNAL PRESSURE AND AXIAL TENSION

In Chap. 3 we considered the stresses arising in a thin-walled cylindrical shell subject to uniform internal pressure. There it was shown that a longitudinal stress given by $\sigma = pr/2t$, as well as a circumferential stress given by $\sigma = pr/t$, exists because of the internal pressure p. If in addition an axial tension P is acting simultaneously with the internal pressure, then there arises an additional longitudinal stress given by $\sigma = P/A$ where A denotes the cross-sectional area of the shell. The resultant stress in the longitudinal direction is thus the algebraic sum of these two longitudinal stresses, and the resultant stress in the circumferential direction is equal to that due to the internal pressure.

CYLINDRICAL SHELLS SUBJECT TO COMBINED TORSION AND AXIAL TENSION/COMPRESSION

In Chap. 5 we considered the stresses arising in a thin-walled cylindrical shell subject to torsion. There it was shown that a shearing stress given by $\tau_{xy} = T\rho/J$ exists on cross sections perpendicular to the axis of the cylinder. If in addition an axial tension P is acting simultaneously with the torque, then there arises a longitudinal stress given by $\sigma = P/A$. This loading is illustrated in Fig. 17-2. In this case the stresses due to these two loadings are acting in different directions and use must be made of the results obtained in Chap. 16. In this manner it will be possible to obtain the principal stresses due to these two loads acting simultaneously. For an application see Problem 17.2.

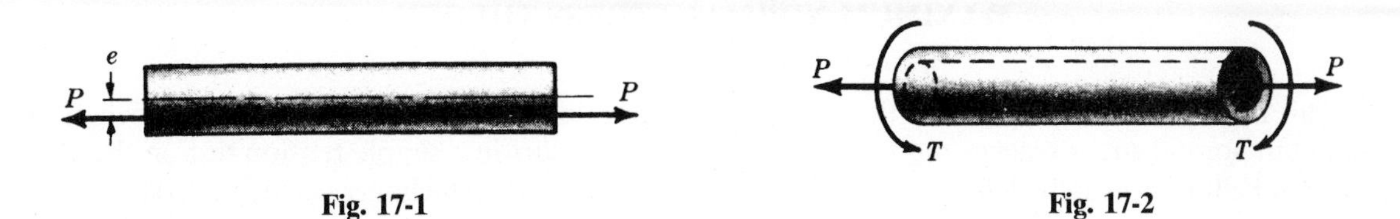

Fig. 17-1

Fig. 17-2

CIRCULAR SHAFT SUBJECT TO COMBINED AXIAL TENSION AND TORSION

This loading is illustrated in Fig. 17-3. Due to the axial tensile force P, there exists a uniform longitudinal tensile stress given by $\sigma = P/A$, where A denotes the cross-sectional area of the bar. From Chap. 5 we know that there exists a torsional shearing stress over any cross section perpendicular to the axis given by $\tau_{xy} = T\rho/J$. Again, the stresses due to these two loadings are acting in different directions and the results of Chap. 16 must be employed to obtain the values of the principal stresses at any point or to obtain the state of stress on any plane inclined at some angle to a generator of the shaft.

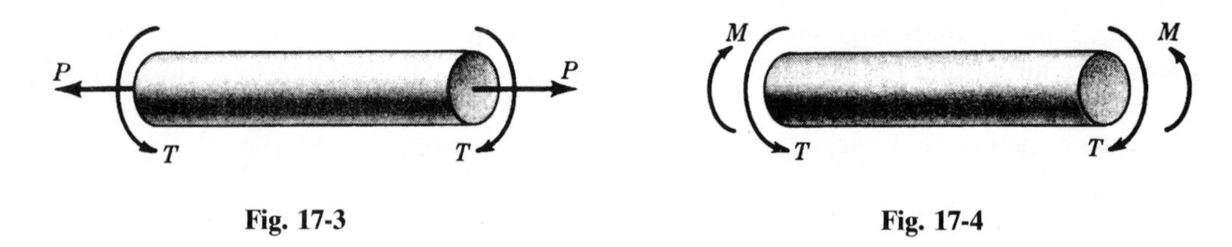

Fig. 17-3

Fig. 17-4

CIRCULAR SHAFT SUBJECT TO COMBINED BENDING AND TORSION

This loading is illustrated in Fig. 17-4. Again from Chap. 5 we know that there exists a torsional shearing stress over any cross section perpendicular to the axis given by $\tau_{xy} = T\rho/J$. From Chap. 8 we know that there also exists a bending stress perpendicular to this cross section, i.e., in the direction of the axis of the shaft, given by $\sigma = My/I$. Since these stresses are acting in different directions the results of Chap. 16 must be employed to obtain the values of the principal stresses at any point in the shaft or to obtain the state of stress on any plane inclined to a generator of the shaft. For applications see Problem 17.3.

DESIGN OF MEMBERS SUBJECT TO COMBINED LOADINGS

So far we have discussed only *analysis*, i.e., determination of principal stresses in a member subject to combined loadings. The inverse problem, i.e., *design* of a member to withstand combined loads, is somewhat more complex and must necessarily be related to experimentally determined mechanical properties of the materials. Because such properties cannot be determined for all possible combinations of loadings, the mechanical characteristics are usually determined in very simple tensile, compressive, or shear tests. The problem then arises as to how to relate the strength of an elastic body subject to combined loadings to these known strength characteristics under the simpler loading conditions. Relations between strength under various combined loads and simple mechanical properties of the material are termed *theories of failure*. Many such theories are available but we shall discuss only the three most commonly used, one applicable to brittle materials and two suitable for use in design of ductile members.

MAXIMUM NORMAL STRESS THEORY

This theory states that failure of the material subject to biaxial or triaxial stresses occurs when the maximum normal stress reaches the value at which failure occurs in a simple tension test on the same material. Failure is usually defined as either yielding or fracture — whichever occurs first. This theory is in good agreement with experimental evidence on brittle materials. For applications, see Problems 17.9 and 17.10.

MAXIMUM SHEARING STRESS THEORY

This theory states that failure of the material subject to biaxial or triaxial stresses occurs when the maximum shearing stress reaches the value of the shearing stress at failure in a simple tension or compression test on the same material. The theory is widely used for design of ductile materials. For applications see Problem 17.11.

HUBER–VON MISES–HENCKY (MAXIMUM ENERGY OF DISTORTION) THEORY

For an element subject to the principal stresses σ_1, σ_2, σ_3 this theory states that yielding begins when

$$(\sigma_1 - \sigma_2)^2 + (\sigma_2 - \sigma_3)^2 + (\sigma_1 - \sigma_3)^2 = 2(\sigma_{yp})^2$$

where σ_{yp} is the yield point of the material. This theory is in excellent agreement with experiments on ductile materials. For applications see Problem 17.12.

Solved Problems

17.1. The rectangular block shown in Fig. 17-5 has its axis of symmetry oriented vertically, is clamped at its lower base, and is subject to a concentric compressive force of 220 kN together with a couple M at point C, the midpoint of the top cross section. If the peak allowable compressive stress is 180 MPa, determine the allowable magnitude of the couple.

The compressive force gives rise to a compressive stress that is uniform over any horizontal cross section. From Chap. 1 this vertically directed stress is

$$\sigma_1 = \frac{P}{A} = \frac{220{,}000\ \text{N}}{(0.07\ \text{m})(0.05\ \text{m})} = 62.86\ \text{MPa}$$

The couple (located in the x-y plane) gives rise to bending about the z-axis (as a neutral axis) and from Chap. 8 creates a compressive stress everywhere to the right of the z-axis. At point A this is given by

$$\sigma_2 = \frac{Mc}{I} = \frac{M(0.035\ \text{m})}{\frac{1}{12}(0.05\ \text{m})(0.07\ \text{m})^3}$$

The resultant compressive stress at A is $(\sigma_1 + \sigma_2)$ and since this must not exceed 180 MPa, we have at A

$$180 \times 10^6\ \text{N/m}^2 = 62.86 \times 10^6\ \text{N/m}^2 + \frac{M(0.035\ \text{m})}{\frac{1}{12}(0.05\ \text{m})(0.07\ \text{m})^3}$$

Solving,

$$M = 4.70\ \text{kN} \cdot \text{m}$$

17.2. Consider a hollow cylindrical shell of outer radius $R_o = 140$ mm and inner radius $R_i = 125$ mm. It is subject to an axial compressive force of 68 kN together with a torque of 35 kN · m, as shown in Fig. 17-6. Determine the principal stresses as well as the peak shearing stress in the shell.

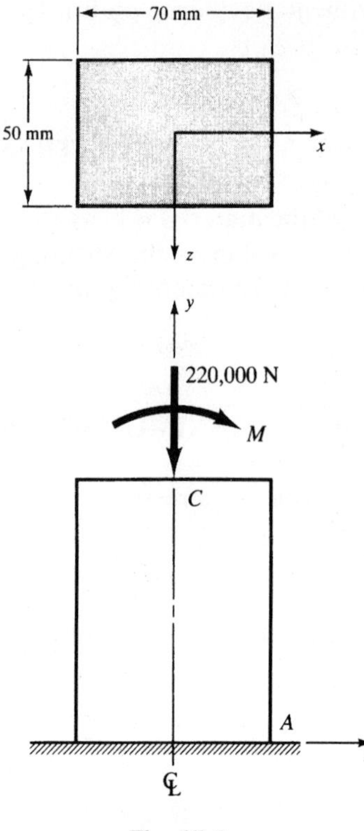

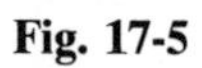
Fig. 17-5

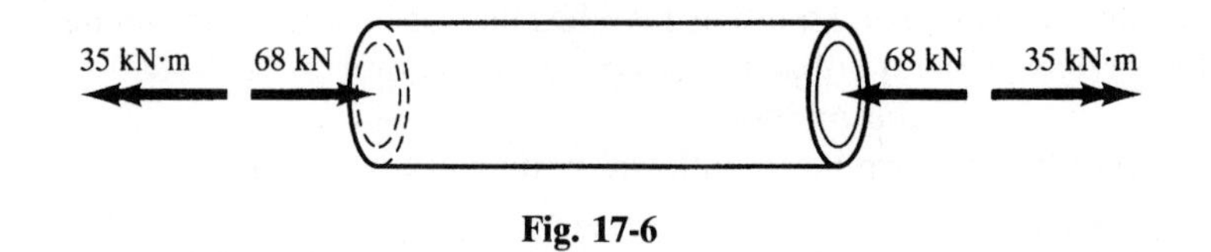

Fig. 17-6

The 68-kN force produces a uniformly distributed compressive stress given by

$$\sigma_1 = \frac{-68{,}000\ \text{N}}{\pi[(0.140\ \text{m})^2 - (0.125\ \text{m})^2]} = -5.44\ \text{MPa}$$

as shown in Fig. 17-7. The torsional shearing stresses due to the 35-kN · m torque were found in Problem 5.2 to be $\tau = T\rho/J$. Here, the polar moment of inertia is

$$J = \frac{\pi}{2}[(0.140\ \text{m})^4 - (0.125)^4] = 0.0002199\ \text{m}^4$$

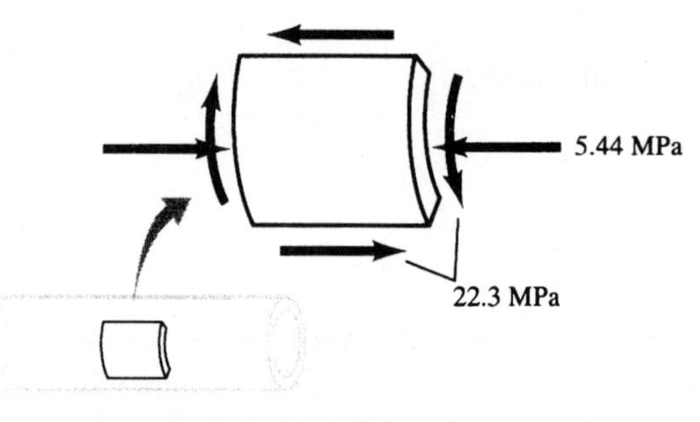

Fig. 17-7

If the approximate expression of Problem 5.6 is used, we find 0.0002191 m^4. Thus, the shearing stresses at the outer fibers of the shell are given by

$$\tau = \frac{T\rho}{J} = \frac{(35{,}000\ \mathrm{N\cdot m})(0.140\ \mathrm{m})}{0.0002199} = 22.3\ \mathrm{MPa}$$

and these are shown in Fig. 17-7.

From Problem 16.13 the principal stresses are found to be

$$\sigma = \frac{-5.44+0}{2} \pm \sqrt{\left(-\frac{5.44-0}{2}\right)^2 + (22.3)^2}$$

$$\sigma_{\max} = 19.75\ \mathrm{MPa}$$

$$\sigma_{\min} = -25.19\ \mathrm{MPa}$$

and the peak shearing stress is 22.47 MPa.

17.3. Consider a hollow circular shaft whose outside diameter is 3 in and whose inside diameter is equal to one-half the outside diameter. The shaft is subject to a twisting moment of 20,000 lb · in as well as a bending moment of 30,000 lb · in. Determine the principal stresses in the body. Also, determine the maximum shearing stress.

The twisting moment gives rise to shearing stresses that attain their peak values in the outer fibers of the shaft. From Problem 5.2 these shearing stresses are given by $\tau_{xy} = T\rho/J$. From Problem 5.1 it is seen that for the hollow circular area

$$J = \frac{\pi}{32}(D_o^4 - D_i^4) = \frac{\pi}{32}[3^4 - (1.5)^4] = 7.46\ \mathrm{in}^4$$

where D_o denotes the outer diameter of the section and D_i represents the inner diameter. At the outer fibers the torsional shearing stresses are thus

$$\tau_{xy} = \frac{T\rho}{J} = \frac{20{,}000(1.5)}{7.46} = 4000\ \mathrm{lb/in^2}$$

Let the bending moments lie in a vertical plane. Then the upper and lower fibers of the beam are subject to the peak bending stresses. These are found from the expression $\sigma_x = My/I$. The moment of inertia I for the hollow circular cross section may be seen from Problem 7.9 to be

$$I = \frac{\pi}{64}(D_o^4 - D_i^4) = \frac{\pi}{64}[3^4 - (1.5)^4] = 3.73\ \mathrm{in}^4$$

Substituting,

$$\sigma_x = \frac{My}{I} = \frac{30{,}000(1.5)}{3.73} = 12{,}000\ \mathrm{lb/in^2}$$

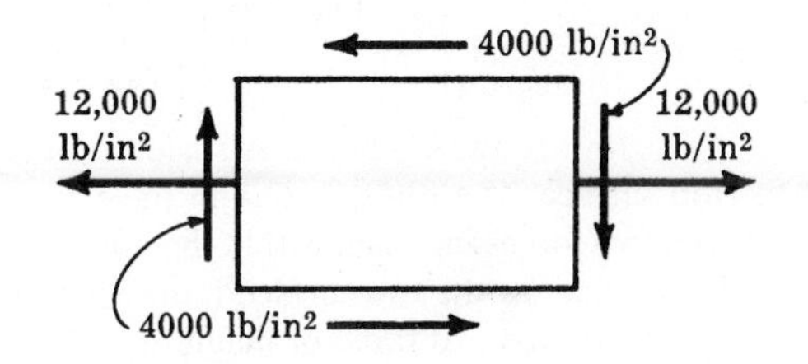

Fig. 17-8

Thus an element located at the lower extremity of the shaft is subject to the stresses shown in Fig. 17-8. From Problem 16.7 the principal stresses for this element are

$$\sigma_{\max} = \tfrac{1}{2}\sigma_x + \sqrt{(\tfrac{1}{2}\sigma_x)^2 + (\tau_{xy})^2} = 12{,}000/2 + \sqrt{(12{,}000/2)^2 + (4000)^2} = 13{,}200 \text{ lb/in}^2$$

$$\sigma_{\min} = \tfrac{1}{2}\sigma_x + \sqrt{(\tfrac{1}{2}\sigma_x)^2 + (\tau_{xy})^2} = 12{,}000/2 - \sqrt{(12{,}000/2)^2 + (4000)^2} = -1200 \text{ lb/in}^2$$

These stresses occur on planes defined by (*3*) of Problem 16.7:

$$\tan 2\theta_p = -\frac{\tau_{xy}}{\frac{1}{2}\sigma_x} = -\frac{4000}{12{,}000/2} = -\frac{2}{3} \qquad \text{or} \qquad \theta_p = 73°10', 163°10'$$

Substituting in (*1*) of Problem 16.7 and letting $\theta = 73°10'$, we have

$$\sigma = 12{,}000/2 - (12{,}000/2)\cos 146°20' + 4000 \sin 146°20' = 13{,}200 \text{ lb/in}^2$$

Thus the maximum tensile stress is 13,200 lb/in^2, occurring on a plane oriented 73°10′ to the geometric axis of the shaft. The other principal stress, $\sigma_{\min} = -1200$ lb/in^2, occurs on a plane oriented 163°10′ to the axis.

The maximum shearing stress is given by (*8*) of Problem 16.7. It is

$$\tau = \pm\sqrt{(\tfrac{1}{2}\sigma_x)^2 + (\tau_{xy})^2} = \pm\sqrt{(12{,}000/2)^2 + (4000)^2} = \pm 7200 \text{ lb/in}^2$$

and occurs on planes oriented at 45° to the planes found above on which the principal stresses act.

17.4. The thick-walled cylindrical shell shown in Fig. 17-9 has its axis of symmetry oriented vertically. It is clamped at its lower extremity and subject to the three concentrated forces indicated. Determine the normal stresses at points *A*, *B*, *C*, and *D*.

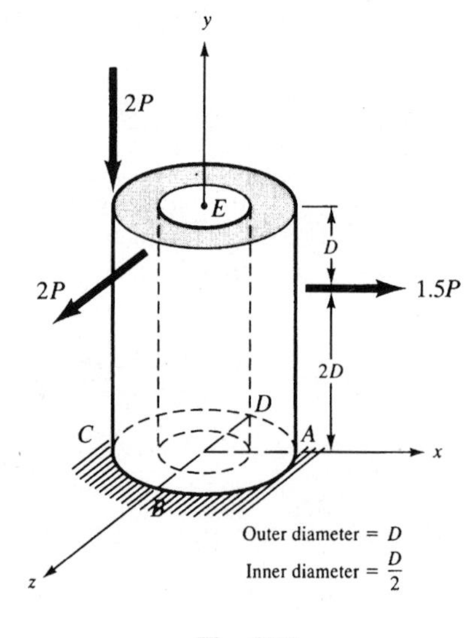

Fig. 17-9

Let us look down the *z*-axis toward the *x*-*y* plane. Also, let us introduce two forces, each of magnitude $2P$, at the center *E* of the top surface. The force system in the *x*-*y* plane for this set of three forces thus appears as in Fig. 17-10(*a*). The two forces included within the dotted lines constitute a couple of magnitude $(2P)(D/2) = PD$, so that the loading on the top surface (corresponding to the original force $2P$) may be considered to consist of a central downward force of magnitude $2P$ together with a couple of magnitude PD, as shown in Fig. 17-10(*b*). The total loading on the shell thus consists of the concentric force $2P$, the couple PD, and the two concentrated forces of magnitudes $1.5P$ and $2P$.

The effects of these four forces are:

(*a*) The central downward force $2P$ gives rise to uniform compressive stresses over any horizontal cross section.

(*b*) The couple PD shown in Fig. 17-10(*b*) gives rise to bending about an axis parallel to the z-axis as a neutral axis.

(*c*) The force $1.5P$ gives rise to bending about an axis parallel to the z-axis as a neutral axis.

(*d*) The force $2P$ gives rise to bending about an axis parallel to the x-axis as a neutral axis.

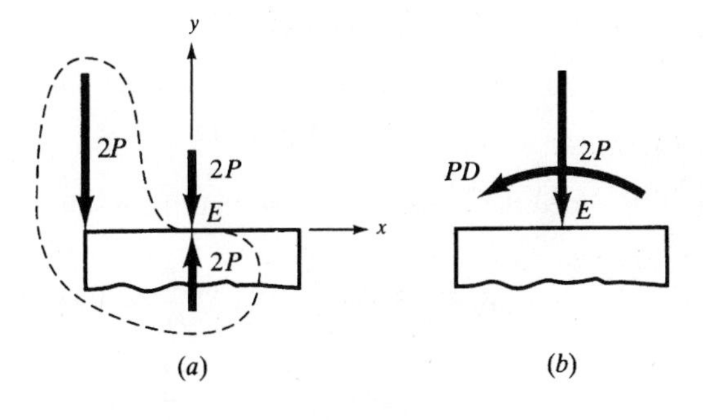

Fig. 17-10

From the geometry of the cross section, we find $A = 0.589D^2$ in and $I_x = I_z = 0.0460D^4$ in^4. From effect (*a*), we have

$$\sigma_1 = \frac{P}{A} = -\frac{2P}{0.589D^2} = -3.396\frac{P}{D^2}$$

From (*b*), the bending stresses are

$$\sigma'_A = \frac{Mc}{I} = \frac{(PD)(D/2)}{0.0460D^4} = 10.87\frac{P}{D^2}$$

$$\sigma'_C = \frac{Mc}{I} = -\frac{(PD)(D/2)}{0.0460D^4} = -10.87\frac{P}{D^2}$$

From (*c*), the bending stresses are

$$\sigma''_A = \frac{Mc}{I} = -\frac{(1.5P)(2D)(D/2)}{0.0460D^4} = -32.61\frac{P}{D^2}$$

$$\sigma''_C = \frac{Mc}{I} = \frac{(1.5P)(2D)(D/2)}{0.0460D^4} = 32.61\frac{P}{D^2}$$

These stresses appear at A and C as shown in Fig. 17-11, for which

$$\sigma_A = -3.396\frac{P}{D^2} + 10.87\frac{P}{D^2} - 32.61\frac{P}{D^2} = -25.14\frac{P}{D^2}$$

$$\sigma_C = -3.396\frac{P}{D^2} - 10.87\frac{P}{D^2} + 32.61\frac{P}{D^2} = -18.34\frac{P}{D^2}$$

From effect (*d*), we have the bending as

$$\sigma'''_B = \frac{Mc}{I} = \frac{-(2P)(3D)(D/2)}{0.046D^4} = -65.22\frac{P}{D^2}$$

$$\sigma'''_D = \frac{Mc}{I} = \frac{(2P)(3D)(D/2)}{0.046D^4} = 65.22\frac{P}{D^2}$$

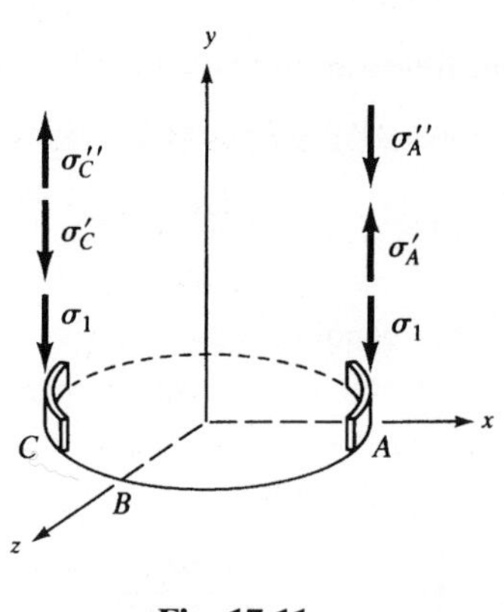

Fig. 17-11

To these values must be added the direct stresses so that the resultant vertical normal stresses at B and D are

$$\sigma_B = -3.396\frac{P}{D^2} - 65.22\frac{P}{D^2} = -68.62\frac{P}{D^2}$$

$$\sigma_D = -3.396\frac{P}{D^2} + 65.22\frac{P}{D^2} = 61.82\frac{P}{D^2}$$

17.5. The shaft shown in Fig. 17-12(a) rotates with constant angular velocity. The belt pulls create a state of combined bending and torsion. Neglect the weights of the shaft and pulleys and assume that the bearings can exert only concentrated force reactions. The diameter of the shaft is 1.25 in. Determine the principal stresses in the shaft.

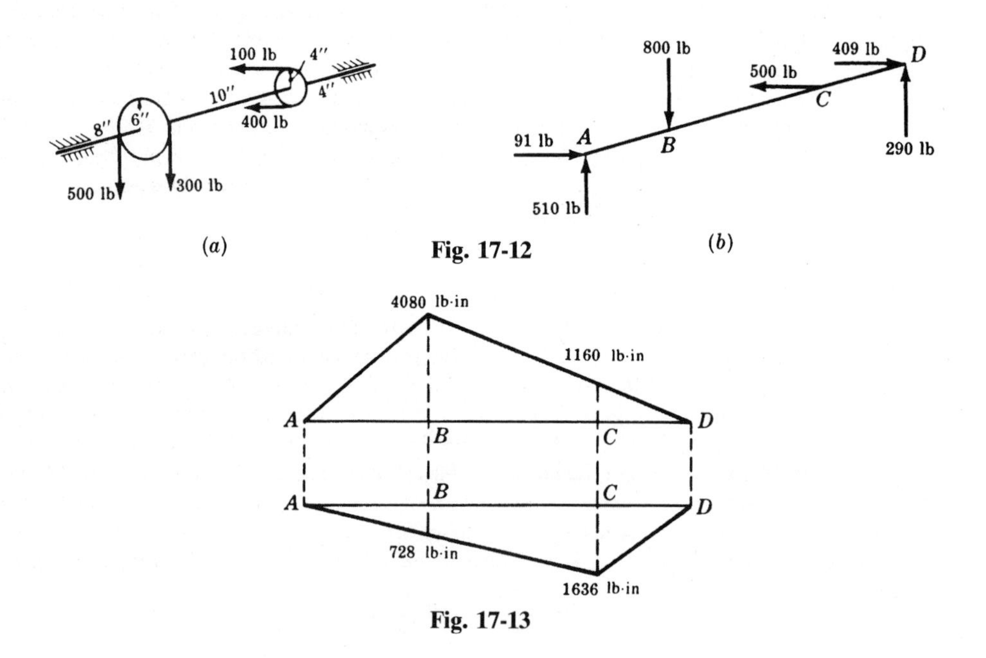

Fig. 17-12

Fig. 17-13

The transverse forces acting on the shaft are not parallel and the bending moments caused by them must be added vectorially to obtain the resultant bending moment. This vector addition need be carried out at only a few apparently critical points along the length of the shaft. The loads causing bending, together with the reactions they produce, are shown above in Fig. 17-12(b). They are considered as passing through the axis of the shaft. The upper and lower shaded portions of Fig. 17-13, respectively, represent the bending moment diagrams for a vertical and for a horizontal plane.

The resultant bending moments at B and C are

$$M_B = \sqrt{(4080)^2 + (728)^2} = 4140 \text{ lb} \cdot \text{in}$$

$$M_C = \sqrt{(1160)^2 + (1636)^2} = 2000 \text{ lb} \cdot \text{in}$$

The twisting moment between the two pulleys is constant and equal to

$$T = (400 - 100)(4) = 1200 \text{ lb} \cdot \text{in}$$

Since the torque is the same at B and C, the critical element lies at the outer fibers of the shaft at point B. The maximum bending stress is given by

$$\sigma_x = \frac{My}{I} = \frac{(4140)(1.25/2)}{\pi(1.25)^4/64} = 21{,}500 \text{ lb/in}^2$$

The maximum shearing stress, occurring at the outer fibers of the shaft, is given by

$$\tau_{xy} = \frac{T_\rho}{J} = \frac{1200(1.25/2)}{\pi(1.25)^4/32} = 3100 \text{ lb/in}^2$$

The principal stresses were found in Problem 16.13 to be

$$\sigma_{\text{max}} = \tfrac{1}{2}\sigma_x + \sqrt{(\tfrac{1}{2}\sigma_x)^2 + (\tau_{xy})^2} = 21{,}500/2 + \sqrt{(21{,}500/2)^2 + (3100)^2} = 22{,}000 \text{ lb/in}^2$$

$$\sigma_{\text{min}} = \tfrac{1}{2}\sigma_x - \sqrt{(\tfrac{1}{2}\sigma_x)^2 + (\tau_{xy})^2} = 21{,}500/2 - \sqrt{(21{,}500/2)^2 + (3100)^2} = -400 \text{ lb/in}^2$$

17.6. Discuss a failure criterion for *brittle* materials.

The criterion which is in best agreement with experimental evidence was advanced by the English engineer W. J. M. Rankine and is termed the *maximum normal stress theory*. It states that failure of the material (i.e., either yielding or fracture — whichever occurs first) occurs when the maximum normal stress reaches the value at which failure occurs in a simple tension test on the same material. Alternatively, if the loading is compressive, failure occurs when the minimum normal stress reaches the value at which failure occurs in a simple compression test. Evidently this criterion considers only the greatest (or smallest) of the principal stresses and disregards the influence of the other principal stresses.

17.7. Discuss the *maximum shearing stress* failure criterion for *ductile* materials.

This criterion is in good agreement with experimental evidence, provided the yield point of the material in tension is equal to that in compression. It was advanced first by C. A. Coulomb in 1773 and later by H. Tresca in 1864; in fact, it is often called the *Tresca criterion*. The criterion states that failure of the material subject to biaxial or triaxial stress occurs when the maximum shearing stress at any point reaches the value of the shearing stress at failure in a simple tension or compression test on the same material. In Problem 16.13 it was shown that the maximum shear stress is one-half the difference between the maximum and minimum principal stresses and always occurs on a plane inclined at 45° to the principal planes. Thus, if σ_{yp} denotes the yield point of the material in simple tension or compression, then the corresponding maximum shear stress is $\sigma_{yp}/2$. Accordingly, the maximum shearing stress criterion may be formulated as

$$\frac{\sigma_{\text{max}} - \sigma_{\text{min}}}{2} = \frac{\sigma_{yp}}{2}$$

or

$$\sigma_{\text{max}} - \sigma_{\text{min}} = \sigma_{yp} \tag{1}$$

where σ_{max} and σ_{min} are maximum and minimum principal stresses, respectively. It is to be observed that judgment must be used in analysis of three-dimensional situations to determine which of the three principal stresses lead to the greatest difference on the left-hand side of (*1*).

17.8. Discuss the *Huber–von Mises–Hencky* failure criterion for *ductile* materials.

This theory was advanced by M. T. Huber in Poland in 1904 and independently by R. von Mises in Germany in 1913 and H. Hencky in 1925. It is in even better agreement with experimental evidence concerning failure of ductile materials subject to biaxial or triaxial stresses than the maximum shearing stress theory discussed in Problem 17.7.

Development of this widely accepted criterion first necessitates determination of the strain energy per unit volume in a simple tension specimen. If the axial tensile stress arising in this test is σ_1 and the corresponding axial strain is ϵ_1, then the work done on a unit volume of the test specimen is the product of the mean value of force per unit area, that is, $\sigma_1/2$, times the displacement in the direction of the force, or ϵ_1. The work is thus $U = \sigma_1 \epsilon_1/2$ and this work is stored as internal strain energy.

The strain energy per unit volume in an element subject to triaxial *principal stresses* σ_1, σ_2, σ_3 is readily found by superposition (since energy is a scalar quantity) to be

$$U = \tfrac{1}{2}\sigma_1 \epsilon_1 + \tfrac{1}{2}\sigma_2 \epsilon_2 + \tfrac{1}{2}\sigma_3 \epsilon_3 \qquad (a)$$

where ϵ_1, ϵ_2, ϵ_3 are the normal strains in the directions of the principal stresses, respectively. If the strains are expressed in terms of the stresses according to the relations given in Problem 1.23, Eq. (*a*) becomes

$$U = \frac{1}{2E}[(\sigma_1^2 + \sigma_2^2 + \sigma_3^2) - 2\mu(\sigma_1\sigma_2 + \sigma_1\sigma_3 + \sigma_2\sigma_3)] \qquad (b)$$

The triaxial principal stresses may be represented as in Fig. 17-14(*a*). Alternatively, this general state of stress may be represented as the sum of the two triaxial states shown in Figs. 17-14(*b*) and 17-14(*c*).

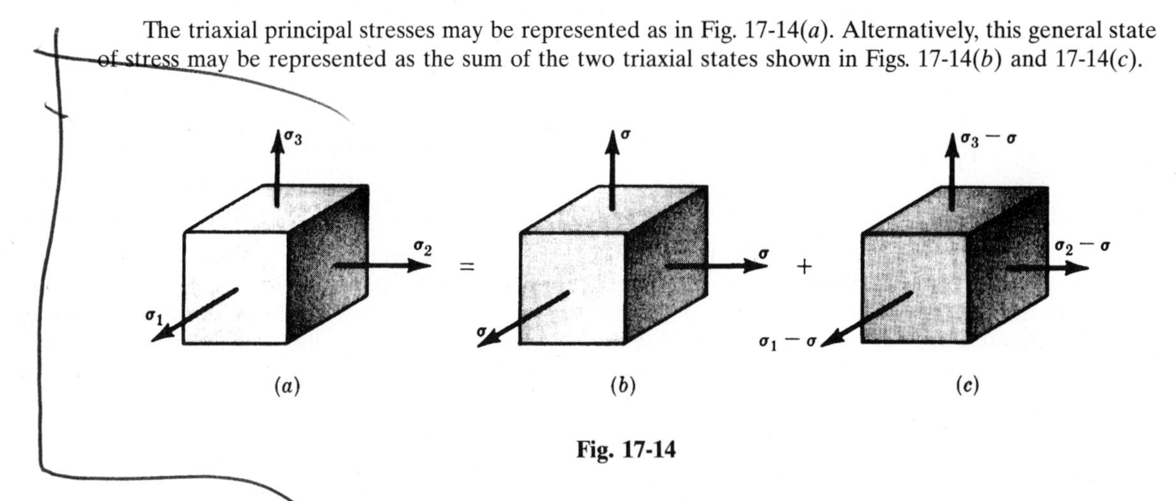

Fig. 17-14

The strain energy U given by Eq. (*b*) may be resolved into two components, one portion U_v corresponding to a change of volume with no distortion of the element, the other, U_d, corresponding to distortion of the element with no change of volume. The stresses indicated in Fig. 17-14(*c*) represent *distortion only* with no change of volume, provided the expression for *dilatation* given in Problem 1.23 is set equal to zero. Thus

$$\epsilon_1 + \epsilon_2 + \epsilon_3 = \frac{1}{E}[(\sigma_1 - \sigma) - \mu(\sigma_2 + \sigma_3 - 2\sigma) + (\sigma_2 - \sigma) - \mu(\sigma_1 + \sigma_3 - 2\sigma)$$
$$+ (\sigma_3 - \sigma) - \mu(\sigma_1 + \sigma_2 - 2\sigma)] = 0 \qquad (c)$$

Solving (*c*), we find

$$\sigma = \frac{\sigma_1 + \sigma_2 + \sigma_3}{3} \qquad (d)$$

for the uniform stresses in Fig. 17-14(*b*) which correspond to change of volume with no distortion. The normal strains corresponding to the stresses given in (*d*) are readily found from the three-dimensional form of Hooke's law given in Problem 1.23 to be

$$\epsilon = \frac{(1 - 2\mu)\sigma}{E} \qquad (e)$$

Thus, the internal strain energy corresponding to the unit volume indicated in Fig. 17-14(*b*) is found by substituting the expressions (*d*) and (*e*) in (*a*), with $\sigma_1 = \sigma_2 = \sigma_3 = \sigma$ and $\epsilon_1 = \epsilon_2 = \epsilon_3 = \epsilon$, to obtain

$$U_v = 3\left(\frac{\sigma\epsilon}{2}\right) = \frac{1-2\mu}{6E}(\sigma_1 + \sigma_2 + \sigma_3)^2 \qquad (f)$$

The strain energy corresponding to *distortion only*, with no change of volume, is now found to be

$$U_d = U - U_v = \frac{1+\mu}{6E}[(\sigma_1 - \sigma_2)^2 + (\sigma_2 - \sigma_3)^2 + (\sigma_1 - \sigma_3)^2] \qquad (g)$$

The Huber–von Mises–Hencky theory assumes that failure takes place when the internal strain energy of distortion given by (*g*) is equal to that at which failure occurs in a simple tension test. In such a test $\sigma_2 = \sigma_3 = 0$, $\sigma_1 = \sigma_{yp}$ and the right side of (*g*) becomes

$$\frac{1+\mu}{6E}[2\sigma_{yp}^2] \qquad (h)$$

Equating the right side of (*g*) to (*h*), we find

$$(\sigma_1 - \sigma_2)^2 + (\sigma_2 - \sigma_3)^2 + (\sigma_1 - \sigma_3)^2 = 2\sigma_{yp}^2 \qquad (i)$$

as the criterion for failure. This is sometimes called the *maximum energy of distortion theory*. It assumes that U_v is ineffective in causing failure.

17.9. A thin-walled cylindrical pressure vessel is subject to an internal pressure of 5 MPa. The mean radius of the cylinder is 400 mm. If the material has a yield point of 300 MPa and a safety factor of 3 is employed, determine the required wall thickness using (*a*) the maximum normal stress theory, and (*b*) the Huber–von Mises–Hencky theory.

The stresses determined in Problem 3.1 are principal stresses. Thus we have

$$\sigma_1 = \sigma_c = \frac{pr}{h} = \frac{5(400)}{h} = \frac{2000}{h}$$

$$\sigma_2 = \sigma_l = \frac{pr}{2h} = \frac{5(400)}{2h} = \frac{1000}{h}$$

The third principal stress varies from zero at the outside of the shell to the value $-p$ at the inside. It is customary to neglect this third component in thin-shell design, so we shall assume that $\sigma_3 = 0$.

(*a*) Using the maximum normal stress theory we have

$$\frac{2000}{h} = \frac{300}{3} \qquad \text{from which} \qquad h = 20\text{ mm}$$

(*b*) Using the Huber-von Mises-Hencky theory we have, from (*i*) of Problem 17.8,

$$\left(\frac{2000}{h} - \frac{1000}{h}\right)^2 + \left(\frac{1000}{h} - 0\right)^2 + \left(\frac{2000}{h} - 0\right)^2 = 2\left(\frac{300}{3}\right)^2$$

whence $h = 17.3$ mm.

17.10. The solid circular shaft in Fig. 17-15(*a*) is subject to belt pulls at each end and is simply supported at the two bearings. The material has a yield point of 250 MPa. Determine the required diameter of the shaft using the maximum normal stress theory together with a safety factor of 3.

The bearing reactions, which are in a vertical plane, are denoted by R_B and R_C in the free-body diagram, Fig. 17-15(*b*). From statics it is found that $R_B = 2.83$ kN and $R_C = 3.67$ kN. The variation of

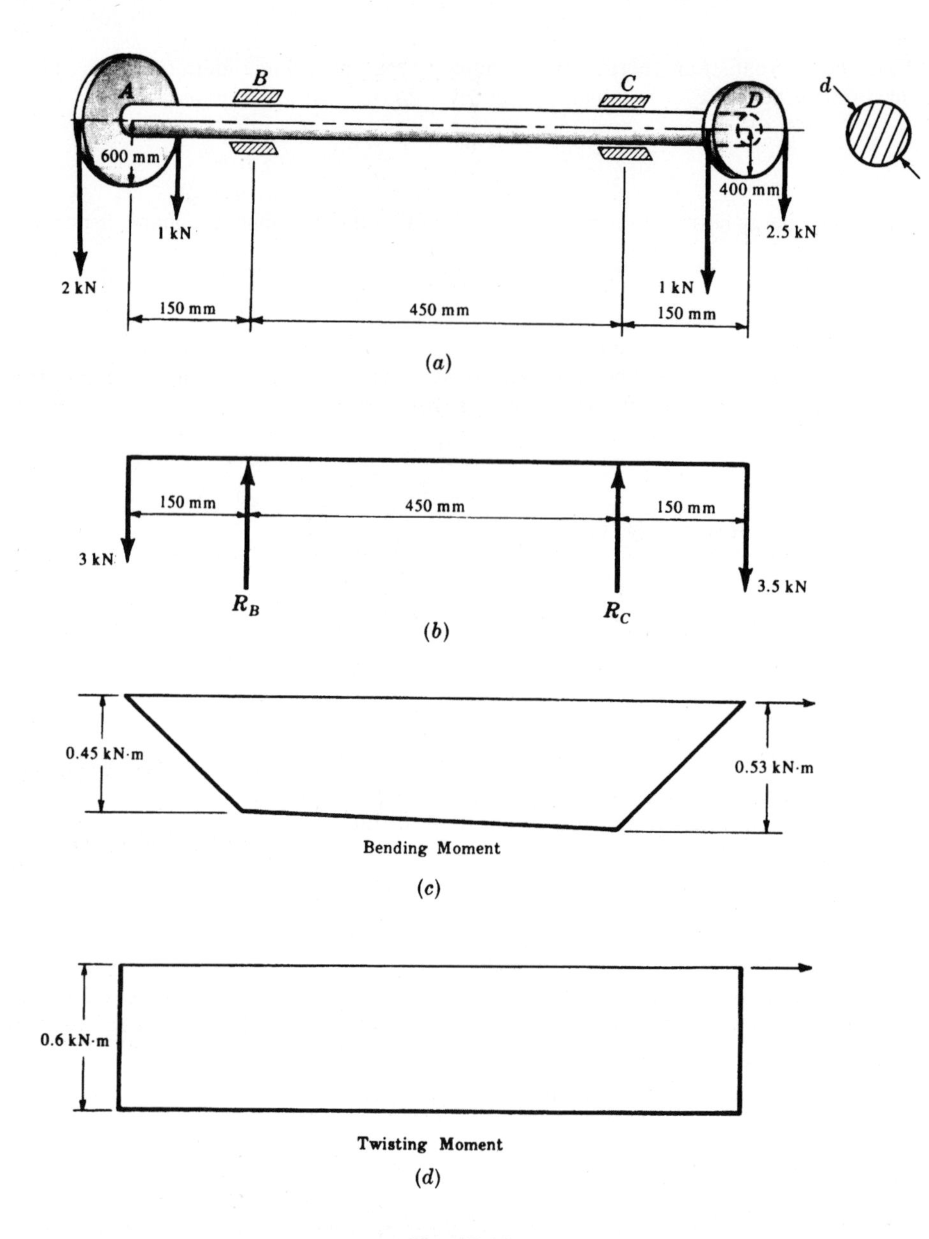

Fig. 17-15

bending moment along the length of the shaft is shown in Fig. 17-15(c). Similarly, the twisting moment along the length of the shaft may be depicted as a constant, as in Fig. 17-15(d).

Evidently the shaft is most critically stressed at its outer fibers at point C, where a top view of the uppermost element indicates the stresses σ_x and τ_{xy} shown in Fig. 17-16. The normal stress σ_x arises because of bending action, and is found from Problem 8.1 to be

$$\sigma_x = \frac{Mc}{I} = \frac{(0.53 \times 10^3)(10^3)(d/2)}{\pi d^4/64} = \frac{5.4 \times 10^6}{d^3}\text{MPa} \tag{a}$$

The other normal stresses, σ_y and σ_z, are zero. The shearing stresses τ_{xy} arise from the torsion due to the unequal belt pulls, and are found from Problem 5.2 to be

$$\tau_{xy} = \frac{Tr}{J} = \frac{(0.6 \times 10^3)(10^3)(d/2)}{\pi d^4/32} = \frac{3.06 \times 10^6}{d^3}\text{MPa} \tag{b}$$

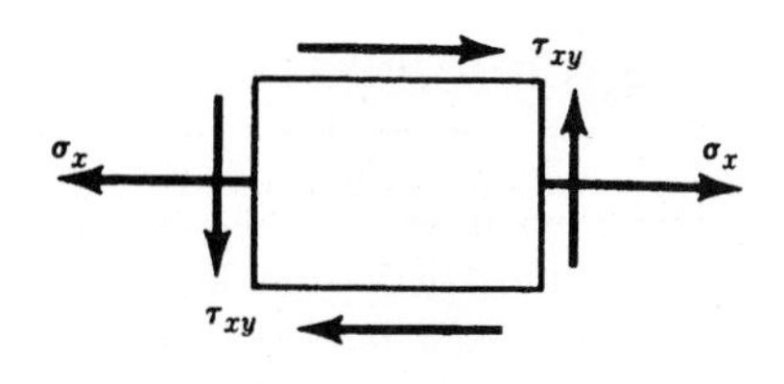

Fig. 17-16

According to the maximum normal stress theory, yielding of the shaft occurs when the maximum normal stress reaches the value at which yielding occurs in a simple tensile test. The maximum normal stress is found as the maximum principal stress of Problem 16.13 to be

$$\sigma_{max} = \frac{\sigma_x + \sigma_y}{2} + \sqrt{\left(\frac{\sigma_x - \sigma_y}{2}\right)^2 + (\tau_{xy})^2} \qquad (c)$$

Substituting the results of (a) and (b) into (c), and introducing the safety factor of 3, yields

$$\frac{250}{3} = \frac{5.4 \times 10^6 + 0}{2d^3} + \sqrt{\left(\frac{5.4 \times 10^6 - 0}{2d^3}\right)^2 + \left(\frac{3.06 \times 10^6}{d^3}\right)^2}$$

from which $d = 43$ mm.

17.11. For the shaft loaded as in Problem 17.10 determine the required diameter using the maximum shearing stress theory together with a safety factor of 3.

The maximum normal stress is given in (c) of Problem 16.13. The minimum normal stress is given by

$$\sigma_{min} = \frac{\sigma_x + \sigma_y}{2} - \sqrt{\left(\frac{\sigma_x - \sigma_y}{2}\right)^2 + (\tau_{xy})^2} \qquad (a)$$

It is to be carefully noted that the difference between the σ_{max} and σ_{min} indicated above leads to the *greatest* possible difference, since the third principal stress is zero and σ_{min} is evidently negative. Substituting in (1) of Problem 17.7, we have

$$2\sqrt{\left(\frac{5.4 \times 10^6 - 0}{2d^3}\right)^2 + \left(\frac{3.06 \times 10^6}{d^3}\right)^2} = \frac{250}{3} \qquad \text{or} \qquad d = 46 \text{ mm}$$

17.12. For the shaft loaded as in Problem 17.10 determine the required diameter using the Huber–von Mises–Hencky theory together with a safety factor of 3.

The criterion is expressed by (i) of Problem 17.8, where σ_1, σ_2, and σ_3 are principal stresses. We take these principal stresses to be

$$\sigma_1 = \sigma_{max} = \left(\frac{5.4 \times 10^6 + 0}{2d^3}\right) + \sqrt{\left(\frac{5.4 \times 10^6 - 0}{2d^3}\right)^2 + \left(\frac{3.06 \times 10^6}{d^3}\right)^2} = \frac{6.8 \times 10^6}{d^3}$$

$$\sigma_2 = 0$$

$$\sigma_3 = \sigma_{min} = \left(\frac{5.4 \times 10^6 + 0}{2d^3}\right) - \sqrt{\left(\frac{5.4 \times 10^6 - 0}{2d^3}\right)^2 + \left(\frac{3.06 \times 10^6}{d^3}\right)^2} = -\frac{1.4 \times 10^6}{d^3}$$

Substituting in (i) of Problem 17.8, we have

$$\left[\frac{6.8 \times 10^6}{d^3} - 0\right]^2 + \left[0 - \left(\frac{1.4 \times 10^6}{d^3}\right)\right]^2 + \left[\frac{6.8 \times 10^6}{d^3} - \left(\frac{1.4 \times 10^6}{d^3}\right)\right]^2 = 2\left(\frac{250}{3}\right)^2$$

Solving, $d = 45$ mm.

Supplementary Problems

17.13. A short block is loaded by a comprehensive force of 1.5 MN. The force is applied with an eccentricity of 60 mm, as shown in Fig. 17-17. The block is 300 mm in cross section. Determine the stresses at the outer fibers m and n. *Ans.* $\sigma_m = -36.7$ MPa, $\sigma_n = +3.3$ MPa

17.14. In Problem 17.13 how large an eccentricity must exist if the resultant stress at fiber m is to be zero? *Ans.* 50 mm

17.15. A short block is loaded by a compressive force of 500 kN acting 50 mm from one axis and 75 mm from another axis of a 200-mm × 200-mm cross section, as shown in Fig. 17-18. Determine the peak tensile and compressive stresses in the cross section. *Ans.* 34.75 MPa, −59.0 MPa

17.16. The hollow rectangular block shown in Fig. 17-19 has its vertical axis of symmetry parallel to the y-direction, is clamped at its lower extremity, and is subject to a single vertical concentrated load $P = 180$ kN as indicated. Determine the resultant vertical stress at point A lying at the remote corner of the lower extremity of the block. *Ans.* −111.9 MPa

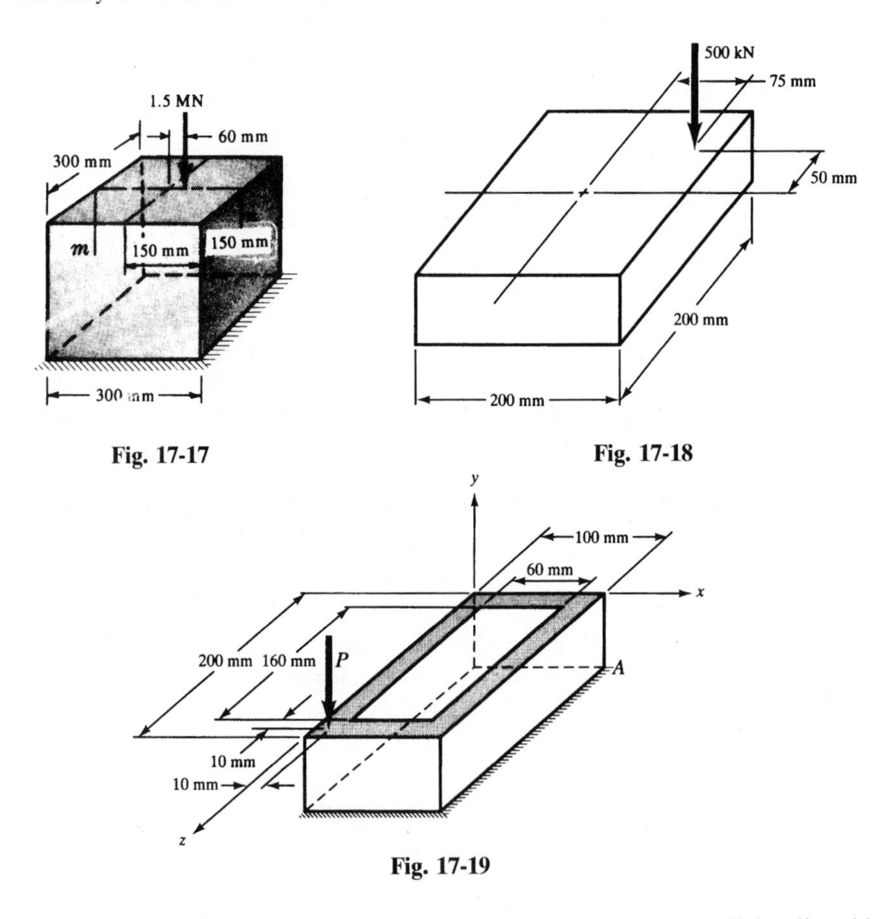

Fig. 17-17

Fig. 17-18

Fig. 17-19

17.17. In Problem 17.2, if the axial compressive force is 200 kN, find the allowable torque if the allowable shearing stress is 100 MPa. *Ans.* 1570 kN · m

17.18. A thin-walled cylinder is 10 in in diameter and of wall thickness 0.10 in. The cylinder is subject to a uniform internal pressure of 100 lb/in². What additional axial tension may act simultaneously without the maximum tensile stress exceeding 20,000 lb/in²? *Ans.* 55,000 lb

17.19. A thin-walled cylindrical shell is subject to an axial compression of 50,000 lb together with a torsional moment of 30,000 lb · in. The diameter of the cylinder is 12 in and the wall thickness 0.125 in. Determine the principal stresses in the shell. Also determine the maximum shearing stress. Neglect the possibility of buckling of the shell. *Ans.* $\sigma_{max} = 120\ \text{lb/in}^2$, $\sigma_{min} = -10{,}680\ \text{lb/in}^2$, $\tau = 5400\ \text{lb/in}^2$

17.20. A shaft 2.50 in in diameter is subject to an axial tension of 40,000 lb together with a twisting moment of 35,000 lb · in. Determine the principal stresses in the shaft. Also determine the maximum shearing stress. *Ans.* $\sigma_{max} = 16{,}180\ \text{lb/in}^2$, $\sigma_{min} = -8020\ \text{lb/in}^2$, $\tau = 12{,}100\ \text{lb/in}^2$

17.21. Consider a solid circular shaft subject to a twisting moment of 20,000 lb · in together with a bending moment of 30,000 lb · in. The diameter of the shaft is 3 in. Determine the principal stresses, as well as the maximum shearing stress in the shaft. *Ans.* $\sigma_{max} = 12{,}450\ \text{lb/in}^2$, $\sigma_{min} = -1150\ \text{lb/in}^2$, $\tau = 6800\ \text{lb/in}^2$

17.22. The shaft shown in Fig. 17-20 rotates with constant angular velocity and is subject to combined bending and torsion due to the indicated belt pulls. The weights of the shaft and pulleys may be neglected and the bearings can exert only concentrated force reactions. The diameter of the shaft is 1.75 in. Determine the principal stresses in the shaft. *Ans.* $\sigma_{max} = 16{,}600\ \text{lb/in}^2$, $\sigma_{min} = -750\ \text{lb/in}^2$

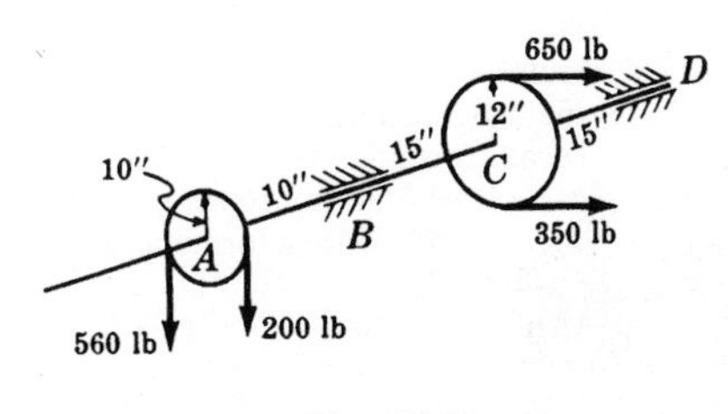

Fig. 17-20

17.23. Consider a thin-walled cylindrical pressure vessel with mean diameter 150 mm subject to a twisting moment of 1 kN · m together with an internal pressure of 3 MPa. If the allowable working stress in tension is 150 MPa, determine the wall thickness as required by the maximum normal stress theory.
Ans. 1.55 mm

17.24. For Problem 17.23 determine the wall thickness as required by the maximum shearing stress theory.
Ans. 1.55 mm

17.25. For Problem 17.23 determine the wall thickness as required by the Huber–von Mises–Hencky theory.
Ans. 1.34 mm

Index